Explorations

An Introduction to Astronomy

Eighth Edition

Thomas T. Arny & Stephen E. Schneider

The nine "Looking Up" figures on the following pages explore a variety of the amazing objects that can be spotted in the night sky. Brief descriptions of each also list the chapter where you can learn more about them.

Mc
Graw
Hill
Education

LOOKING UP #1
Northern Circumpolar Constellations

For observers over most of the northern hemisphere, there are five constellations that are circumpolar, remaining visible all night long: Ursa Major (the Big Bear), Ursa Minor (the Little Bear), Cepheus (the King), Cassiopeia (the Queen), and Draco (the Dragon). The brightest stars in Ursa Major and Ursa Minor form two well-known asterisms: the Big and Little Dippers.

Delta Cephei
A pulsating variable star (chapter 14) at a distance of 980 ly.

Cassiopeia

Cepheus

~12 ly

M52
This is an open star cluster (chapter 16). Its distance is uncertain— perhaps 3000 to 5000 ly.

Draco

Polaris — The North Star
This star lies about 430 ly away, almost directly above the Earth's North Pole, making it an important aid for navigation (chapter 1).

Little Dipper

•170,000 ly

M101
This spiral galaxy is ~27 million light-years away from us (chapter 17).

Thuban
This was the north star when the pyramids were built in ancient Egypt (chapter 6).

Cassiopeia in 3-D

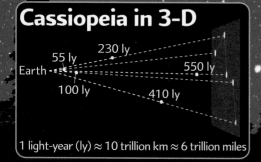

230 ly

55 ly

Earth

550 ly

100 ly

410 ly

1 light-year (ly) ≈ 10 trillion km ≈ 6 trillion miles

Big Dipper

M81 and M82
Gravitational interactions between these two galaxies have triggered star formation (chapter 17).

North

Circling in the northern sky is the Big Dipper, part of the well-known constellation Ursa Major, the Big Bear. The Big Dipper is technically not a constellation, but just an asterism—a star grouping. It is easy to see in the early evening looking north from mid-March through mid-September. The Big Dipper can help you find the North Star, and with a telescope on a dark, clear night, you can find several other intriguing objects as shown below.

Pointer stars

Big Dipper

Over the course of a night, stars appear to rotate counterclockwise around the star Polaris, which remains nearly stationary because it lies almost directly above Earth's North Pole. Polaris is not especially bright, but you can easily find Polaris by extending a line from the two stars at the end of the bowl of the Big Dipper, the pointer stars, as shown by the dashed yellow line (chapter 1).

Location of the Hubble Deep Field (chapter 17)

Polaris

Little Dipper

~1.6 ly

M97 — The Owl
This planetary nebula (chapter 14) is ~2500 ly away.

Mizar and Alcor
If you look closely at it, you may notice that the middle star in the "handle" is actually two stars—Mizar and Alcor. Despite appearing close together in the sky, they are probably not in orbit around each other. However, with a small telescope, you can see that Mizar (the brighter of the star pair) has a faint companion star. This companion does in fact orbit Mizar. Moreover, each of Mizar's stars is itself a binary star, making Mizar a quadruple system (chapter 13).

170,000 ly

M51
The Whirlpool Galaxy can be seen as a dim patch of light with a small telescope. M51 is about 37 million ly away from Earth (chapter 17).

Big Dipper in 3-D

80 ly
83 ly   123 ly
Earth   80 ly
83 ly
86 ly
104 ly

1 light-year (ly) ≈ 10 trillion km ≈ 6 trillion miles

LOOKING UP #3
M31 & Perseus

The galaxy M31 lies in the constellation Andromeda, near the constellations Perseus and Cassiopeia. It is about 2.5 million ly from us, the most distant object visible with the naked eye. Northern hemisphere viewers can see M31 in the evening sky from August through December.

Andromeda

M31 — Andromeda Galaxy (chapter 1

~150,000 ly

~200 ly

The Double Cluster

If you scan with binoculars from M31 toward the space between Perseus and Cassiopeia, you will see the Double Cluster—two groups of massive, luminous but very distant stars. The Double Cluster is best seen with binoculars. The two clusters are about 7000 ly away and a few hundred light-years apart (chapter 16).

Perseus

Algol

Algol, the "demon star," dims for about 10 hours every few days as its companion eclipses it (chapter 13).

California Nebula

An emission nebula (chapter 16) with a shape like the state of California.

Capella

The brightest star in the constellation Auriga, the Charioteer. A binary star (chapter 13).

Auriga

M45 — Pleiades

Perseus in 3-D

880 ly

34 ly   113 ly   240 ly   250 ly

Earth

90 ly

520 ly

520 ly

640 ly

750 ly

1 light-year (ly) ≈ 10 trillion km ≈ 6 trillion miles

The Summer Triangle consists of the three bright stars Deneb, Vega, and Altair, the brightest stars in the constellations Cygnus, Lyra, and Aquila, respectively. They rise in the east shortly after sunset in late June and are visible throughout the northern summer and into late October (when they set in the west in the early evening). Vega looks the brightest to us, but Deneb produces the most light, only looking dimmer because it is so much farther from us.

Vega

Lyra

Epsilon Lyra
A double, double star

Cygnus

M57 — Ring Nebula

This planetary nebula (chapter 14) is about 2300 ly distant. From its observed expansion rate it is estimated to be 7000 years old.

1 ly

Deneb

Deneb is a blue supergiant (chapter 13), one of the most luminous stars we can see, Deneb emits ~50,000 times more light than the Sun.

Altair

Albireo

Through a small telescope this star pair shows a strong color contrast between the orange red giant and blue main-sequence star (chapter 13). These stars may orbit each other every few hundred thousand years, but they are far enough apart that they may not be in orbit.

M27 — Dumbbell Nebula

Another planetary nebula (chapter 14), the Dumbbell is about 1200 ly distant and is about 2.5 ly in diameter.

~2.5 ly

The Summer Triangle in 3-D

Vega
25 ly

Albireo
430 ly

Earth

17 ly
Altair

1400 ly
Deneb

1 light-year (ly) ≈ 10 trillion km ≈ 6 trillion miles

LOOKING UP #5
Taurus

Taurus, the Bull, is one of the constellations of the zodiac and one of the creatures hunted by Orion in mythology. Taurus is visible in the evening sky from November through March. The brightest star in Taurus is Aldebaran, the eye of the bull. The nebula and two star clusters highlighted below have been critical in the history of astronomy for understanding the distances and fates of stars.

M45 — Pleiades

This open star cluster (chapter 16) is easy to see with the naked eye and looks like a tiny dipper. It is about 400 ly from Earth.

~8 ly

Aldebaran

Aldebaran is a red giant star (chapter 13). It is about 67 ly away from Earth and has a diameter about 45 times larger than the Sun's. Although it appears to be part of the Hyades, it is less than half as distant.

~1.0 ly

M1 — Crab Nebula

The Crab Nebula is the remnant of a star that blew up in the year A.D. 1054 as a supernova. At its center is a pulsar (chapter 15). It is about 6500 ly away from us.

T Tauri

T Tauri is an erratically varying pre-main-sequence star, prototype of a class of forming stars (chapter 14). It is about 600 ly distant.

Hyades

The "V" in Taurus is another nearby star cluster, measured to be 151 ly away by the *Hipparcos* satellite (chapter 13). It is easy to see its many stars with binoculars.

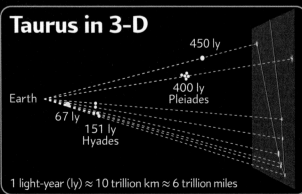

Taurus in 3-D

450 ly

Earth

400 ly
Pleiades

67 ly

151 ly
Hyades

1 light-year (ly) ≈ 10 trillion km ≈ 6 trillion miles

Orion is easy to identify because of the three bright stars of his "belt." You can see Orion in the evening sky from November to April, and before dawn from August through September.

Betelgeuse

Betelgeuse is a red supergiant star (chapter 13) that has swelled to a size that is larger than the orbit of Mars. Its red color indicates that it is relatively cool for a star, about 3500 kelvin.

Horsehead Nebula

The horsehead shape is produced by dust in an interstellar cloud blocking background light (chapter 16).

3 ly

M42 —
Orion Nebula

The Orion Nebula is an active star-forming region rich with dust and gas (chapter 14).

10 ly

Rigel

Rigel is a Blue Supergiant star (chapter 13). Its blue color indicates a surface temperature of about 10,000 kelvin.

Protoplanetary disk

This is the beginning of a star; our early Solar System may have looked like this (chapter 8).

Orion in 3-D

Earth

250 ly
640 ly
690 ly
740 ly
650 ly
860 ly
1300 ly
1340 ly

1 light-year (ly) ≈ 10 trillion km ≈ 6 trillion miles

Sagittarius marks the direction to the center of the Milky Way. It can be identified by its "teapot" shape, with the Milky Way seeming to rise like steam from the spout. From northern latitudes, the constellation is best seen July to September, when it is above the southern horizon in the evening. Many star-forming nebulae are visible in this region (chapter 16).

M16 — Eagle Nebula

This young star cluster and the hot gas around it lie about 7000 ly from Earth.

~1 ly

~70 ly

M22

M22 is one of many globular clusters (chapter 16) concentrated toward the center of our Galaxy. Easy to see with binoculars, it is just barely visible to the naked eye. It is about 11,000 ly away from us.

~100 ly

M17 — Swan Nebula

~50 ly

M20 — Trifid Nebula

The name Trifid was given because of the dark streaks that divide it into thirds. The distance of this nebula is uncertain, approximately 5000 ly away, making its size uncertain too.

The "teapot" of Sagittarius

M8 — Lagoon Nebula

Center of the Milky Way

(chapter 16)

Sagittarius in 3-D

Earth

78 ly
97 ly
230 ly 240 ly
122 ly
350 ly
88 ly 143 ly

These constellations are best observed from the southern hemisphere. Northern hemisphere viewers can see Centaurus low in the southern sky during evenings in May–July, but the Southern Cross rises above the horizon only for viewers south of latitude ~25° N (Key West, South Texas, and Hawaii in the U.S.).

Proxima Centauri
This dim star is the nearest star to the Sun, 4.22 ly distant (chapter 13).

Alpha Centauri

Centaurus

~50 ly

~200 ly

Omega Centauri
This is the largest globular cluster (chapter 16) in the Milky Way, ~16,000 ly distant and containing millions of stars.

The Jewel Box
NGC 4755, an open star cluster (chapter 16) ~500 ly from us.

The Coal Sack
An interstellar dust cloud (chapter 16)

Crux
The Southern Cross

50,000 ly

Centaurus A
This active galaxy (chapter 17), ~11 million ly distant, is one of the brightest radio sources in the sky.

Eta Carinae
At over 100 times the mass of the Sun, this is one of the highest-mass stars known and doomed to die young (chapter 14). It is about 8000 ly distant.

Southern Cross in 3-D

Earth

280 ly

89 ly

230 ly

320 ly

345 ly

1 light-year (ly) ≈ 10 trillion km ≈ 6 trillion miles

Most of the constellations in this part of the sky are dim, but observers in much of the southern hemisphere can see the Magellanic Clouds circling the south celestial pole throughout the night.

Crux
The Southern Cross

Musca

Hourglass Nebula

A planetary nebula (chapter 14) ~8000 ly distant

Apus

~0.51 ly

Octans

The constellation closest to the south celestial pole is named after a navigational instrument, the octant.

Chamaeleon

The South Celestial Pole

No bright stars lie near the south celestial pole (chapter 1), but the Southern Cross points toward it.

Thumbprint Nebula

A Bok globule (chapter 14) about 600 ly distant

Volans

Mensa

Hydrus

~1000 ly

Small Magellanic Cloud

A dwarf galaxy orbiting the Milky Way at a distance of ~200,000 ly (chapter 17).

Large Magellanic Cloud

A small galaxy orbiting the Milky Way at a distance of ~160,000 ly (chapter 17).

Tarantula Nebula

A star-formation region (chapter 16) in the Large Magellanic Cloud larger than any known in the Milky Way.

Explorations

An Introduction to Astronomy

Eighth Edition

Thomas T. Arny

Professor Emeritus
Department of Astronomy
University of Massachusetts, Amherst

Stephen E. Schneider

Professor of Astronomy
University of Massachusetts, Amherst

McGraw Hill Education

EXPLORATIONS: AN INTRODUCTION TO ASTRONOMY, EIGHTH EDITION

Published by McGraw-Hill Education, 2 Penn Plaza, New York, NY 10121. Copyright © 2017 by McGraw-Hill Education. All rights reserved. Printed in the United States of America. Previous editions © 2014, 2010, and 2008. No part of this publication may be reproduced or distributed in any form or by any means, or stored in a database or retrieval system, without the prior written consent of McGraw-Hill Education, including, but not limited to, in any network or other electronic storage or transmission, or broadcast for distance learning.

Some ancillaries, including electronic and print components, may not be available to customers outside the United States.

This book is printed on acid-free paper.

2 3 4 5 6 7 8 9 DOW 21 20 19 18 17 16

ISBN   978-0-07-351391-1
MHID   0-07-351391-1

Senior Vice President, Products & Markets: *Kurt L. Strand*
Vice President, General Manager, Products & Markets: *Marty Lange*
Vice President, Content Design & Delivery: *Kimberly Meriwether David*
Managing Director: *Thomas Timp*
Brand Manager: *Andrea M. Pellerito, Ph.D.*
Director, Product Development: *Rose Koos*
Product Developer: *Robin Reed*
Marketing Manager: *Danielle Dodds*
Director of Digital Content: *Justin Wyatt, Ph.D.*
Digital Product Analyst: *Patrick Diller*
Director, Content Design & Delivery: *Linda Avenarius*
Program Manager: *Lora Neyens*
Content Project Managers: *Laura Bies, Tammy Juran, & Sandra Schnee*
Buyer: *Jennifer Pickel*
Design: *David Hash*
Content Licensing Specialists: *Carrie Burger & Lorraine Buczek*
Cover Image: *Malibu Sea Cave © Jack Fusco*
Compositor: *MPS Limited*
Printer: *R. R. Donnelley*

All credits appearing on page or at the end of the book are considered to be an extension of the copyright page.

Library of Congress Cataloging-in-Publication Data

Arny, Thomas.
    Explorations : an introduction to astronomy / Thomas T. Arny, professor
emeritus, Department of Astronomy, University of Massachusetts, Amherst,
Stephen E. Schneider, professor of astronomy, University of Massachusetts,
Amherst.—Eighth edition.
        pages cm
    Includes index.
    ISBN 978-0-07-351391-1 (alk. paper)
    1. Astronomy—Textbooks. I. Schneider, Stephen E. (Stephen Ewing), 1957-
II. Title.
    QB45.2.A76 2017
    520—dc23

2015035340

The Internet addresses listed in the text were accurate at the time of publication. The inclusion of a website does not indicate an endorsement by the authors or McGraw-Hill Education, and McGraw-Hill Education does not guarantee the accuracy of the information presented at these sites.

mheducation.com/highered

Brief Contents

Contents

CHAPTER 15

Stellar Remnants: White Dwarfs, Neutron Stars, and Black Holes 386

CHAPTER 16

The Milky Way Galaxy 408

ESSAY 4

Life in the Universe 504

Answers to Test Yourself 516

Appendix

Glossary G-1

Credits C-1

Index I-1

Preface

Our motivations for writing *Explorations: An Introduction to Astronomy* are many, both personal and pedagogic. Perhaps foremost among these is a desire to share with students our own sense of wonder about the Universe.

That sense of wonder grows deeper when we begin to understand why things happen. Many astronomy books today seem to simply say, "This is how it is." We want instead to offer explanations that draw as much as possible on simple, everyday effects that students can see around them in the world. For example, why do some stars pulsate? A simple analogy of steam building up pressure under the lid of a pan offers a model of this phenomenon that is easy to understand and reasonably accurate. We have also tried to link complex physical processes to simple everyday experiences. Another example of this is that you can see the effects of differentiation in a previously-melted box of chocolate chip ice cream. When we can thus link physical principles to everyday observations, many of the more abstract and remote ideas become more familiar. Throughout the book we have made heavy use of analogies, along with carefully designed illustrations to make those analogies more concrete.

Knowing the facts about astronomical objects is important, but it is equally important to understand how astronomers deduce those facts. Thus, an additional aim throughout this text is to explain *how* astronomers have come to their understanding of our Universe. New observations can force astronomers to revise their ideas of how a given process occurs. As part of showing how scientists arrive at their ideas, we have set many of the modern discoveries in their historical context to illustrate that science is a dynamic process and subject to controversy—many ideas are not immediately accepted, even if they ultimately prove to be "correct." We hope that by seeing the arguments for and against various ideas, students will have a better understanding of how science works.

If we had attempted to make this textbook completely comprehensive, it would have been very long and overwhelming in detail. It is challenging to keep *Explorations* to a reasonable size because reviewers tend to suggest things that we should include, but rarely suggest things to omit. To solve this problem, we cover some topics, such as timekeeping and astrobiology, in essays that the instructor might choose to skip. We also cover some background topics in later chapters, in the astronomical context where they are most often encountered. This makes it possible to jump directly to some of the later chapters without having to work through the details of all the earlier chapters.

Some astronomy textbooks maintain brevity by omitting most of the mathematics, but we feel that math is essential for understanding many of the methods used by astronomers. We have therefore included the essential mathematics in a number of places. However, because math is so intimidating to so many readers, we begin these discussions by introducing the essence of the calculation in everyday language so that the basic idea can be understood without understanding the mathematics. For example, Wien's law relates the temperature of a hot object to its color by means of a mathematical law, but illustrations of the law can be seen in everyday life, as when we estimate how hot an electric stove burner is by the color of its glow. Where we do present the mathematics, we work through it step by step, explaining where terms must be cross-multiplied and so forth.

Because astronomical concepts often depend on a visual understanding of objects and phenomena, we pay very close attention to the figures. We have refined the illustrations to clarify the presentation, often making small changes to aid the viewer's ability to focus in on essential features while avoiding misconceptions. For example, we have converted all global maps of the planets to Mollweide projections. While no projection can perfectly represent a spherical surface, this one maintains equal areas and the consistent presentation helps the reader to compare features. We work very closely with the McGraw-Hill team throughout the design, layout, and composition process in an effort to make the book easier to read. For example, we often adjust figure labels and sizes to make sure they complement the text and fit very close to the spot where their content is discussed. This helps the reader to connect words, logic, images, and geometry.

NEW TO THE EIGHTH EDITION

In this eighth edition of *Explorations,* we have updated the art and text throughout the book in response to readers' comments and suggestions. Following are some of the highlights of these changes:

- Major update to Essay 1 ("Backyard Astronomy") with detailed advice on small telescopes and astrophotography.
- Major update to Chapter 8 ("Survey of Solar Systems") with latest results and analysis of exoplanets based on *Kepler* findings.
- Major update to Chapter 11 ("Small Bodies Orbiting the Sun") with latest images of Ceres, Pluto, and Comet Churyumov-Gerasimenko from the *Dawn, New Horizons,* and *Rosetta* spacecraft.
- The latest images and science results from planetary spacecraft and space telescopes, including *Hubble, Spitzer, Chandra, Messenger, Curiosity, Solar Dynamics Observatory, Fermi,* and others.
- New "Looking Up" icons in margins call attention to objects discussed in the text that are displayed in the Looking Up illustrations at the beginning of the book. Most of these objects can be seen in the night sky by eye, or with binoculars or a small telescope.

- New and updated images and art in every chapter not only add to the book's visual appeal but enhances student learning with clear, accurate representations that reflect the most current data in the field.

Detailed Revisions

- Chapter 1: Revised illustration of the zodiac to make clearer that it is part of the celestial sphere. Improved illustration of lunar phases with new images. Updated table of upcoming eclipses.
- Chapter 2: Rearranged section 1 to present early Greek astronomical findings in historical order. New figure and discussion of how Earth's curvature can be seen when looking across the surface of the ocean. Illustration of greatest elongations of Mercury and Venus moved here from essay 1 because of its importance to development of Copernican model. Added discussion and figure of orbital eccentricity. Added photos of Venus's phases. New Extending Our Reach box on astrology.
- Chapter 3: Added discussion and figure about Cavendish's experiment to measure value of the gravitational constant. Revised illustration of escape velocity to stress idea that it is based on an initial velocity (as opposed to a rocket that may apply thrust continuously).
- Chapter 4: New figure to illustrate relationship of frequency and wavelength in everyday experience. New infrared image of dog illustrating use of false colors to display "heat." New images of M31 to illustrate differences across wavebands. New figure of Sun with sunspots to illustrate Stefan-Boltzmann equation.
- Chapter 5: New images of M31 as it appears with different resolution and integration time. Added images of large radio telescopes.
- Chapter 6: Improved seismic wave illustration. New image of aurora from the International Space Station. Added graph of carbon dioxide and global temperatures since 1890.
- Chapter 7: New image of crater wall from the *Lunar Reconnaissance Orbiter*. Updated illustration and discussion of the formation of maria. New images of lunar rilles. Updated illustration of lunar interior based on recent reanalyses. New Astronomy by the Numbers box on the Moon's distance from the Earth in the past.
- Chapter 8: Reorganized chapter to introduce exoplanets and exoplanet systems after the Solar System, culminating with discussion of formation of planetary systems. Moved figure on the shape of small bodies here (from chapter 6). Exoplanet results are examined in much more detail. *Kepler* findings about multiple-planet systems, statistics of exoplanet sizes, and planet densities are explored. Added new images of protoplanetary disks, and expanded discussion of migrating planets and the possible early evolution of the Solar System.
- Chapter 9: New topographic map of Mercury based on *Messenger* data. Figure and discussion of radar evidence of ice at Mercury's poles. Expanded coverage of Mars *Curiosity* results. New images of Phobos and Deimos.
- Chapter 10: Expanded discussion of Jupiter's atmospheric circulation, and infrared images of the belts and zones. New

Hubble image of Jupiter's aurora. Added image of Galilean satellites seen through small telescope. New images and discussion of several interesting smaller satellites—Amalthea, Hyperion, Iapetus, and Enceladus. New images of major storm on Saturn and its polar vortex.
- Chapter 11: New *Dawn* image of Ceres, with comparison to the Moon and asteroids. First results on Pluto from *New Horizons*. New images and discussion of Comet Churyumov–Gerasimenko from early *Rosetta* results. New image of meteor from ISS, and of Chelyabinsk meteor and damage. Added image of Tunguska site.
- Chapter 12: New *Solar Dynamics Observatory* image of Sun. New diagram of *Voyager* findings about outer limits of the solar wind.
- Chapter 13: Section 2 is now split in half. The new section 2 now covers luminosity, inverse-square law, standard candles, and magnitudes. Section 3 focuses on determining stars' temperatures and radii. Added mention of new spectral type *Y*. Added discussion and figure about proper motion.
- Chapter 14: Added H-R diagram overviewing evolution of low- and high-mass stars. New *Hubble* image of Eagle nebula. Revised several illustrations of stellar interiors.
- Chapter 15: New X ray/optical images of type Ia supernova remnants.
- Chapter 16: New *Spitzer* image of Milky Way. Updated discussion of Galactic center, with new images and diagrams, including gamma-ray "bubbles" detected by *Fermi*. Revised discussion of future of Milky Way and added illustrations.
- Chapter 17: Expanded explanation of galaxy types. Added side-by-side comparison of optical and radio neutral hydrogen images of M81.
- Chapter 18: Added illustration from millennium simulation of growth of structure in the Universe. Revised presentation within section 1 to more strongly motivate need for modern model of expanding Universe. Latest *Planck* results for composition of the Universe. New table showing relationship of distances, times, and redshift for current cosmological parameters.
- Revised and updated the Moon and planet finder on the foldout chart.

FEATURES OF *EXPLORATIONS*

Explorations has been designed with a number of special features to help you better comprehend the many wide-ranging aspects of astronomy. Familiarize yourself with these features, then before you begin reading a chapter scan through to see what features and figures are present. This overview of the chapter will help deepen your understanding as you read.

Learning Objectives are presented at the start of each chapter. These identify the most important skills that the reader should gain upon completing the chapter. Use this as a checklist for successful completion of a chapter, as well as for identifying topics to reread or to seek further help about.

"What Is This?" questions are presented in each chapter to encourage deeper examination of photos and figures. At the beginning of each chapter, readers are presented with a mystery photo of an astronomical object and asked to guess what it is. After reading the chapter, they should have some idea of what is shown in the photo. In addition, there are questions in blue boxes about a number of other figures and images. The answers to these questions are provided at the end of each chapter under the heading "Figure Question Answers."

Q: WHAT IS THIS?

See end of chapter for the answer.

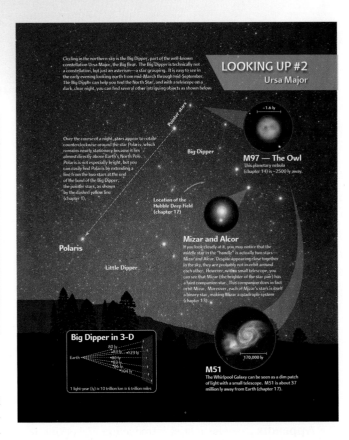

Concepts and Skills to Review are listed at the start of each chapter to provide quick pointers to earlier material that is critical for understanding the content of the chapter. If any look unfamiliar, you should review them before reading the chapter.

Astronomy by the Numbers boxes work through the details of some mathematical derivations and provide worked examples of typical calculations. Read these to gain a greater command of the mathematics behind the discussion in the text.

Extending Our Reach boxes present recent and advanced subjects that are not central to the main material in the text. These can be included for a deeper coverage of the topic.

Science at Work boxes discuss ideas, sometimes controversial, that illustrate how scientists examine new hypotheses.

Looking Up figures, each a full-page art piece, are located at the start of the book. These nine images of the night sky designed to show students how some of the astronomical objects discussed in the text connect with the real sky that they can see overhead at night. The figures cover nine especially interesting regions, ranging from the North Pole to the South Pole. In particular, they show where a variety of the frequently mentioned and important astronomical objects can be seen, many with binoculars or a small telescope. Each Looking Up figure presents a photograph of one or more constellations in which nebulas, star clusters, and other interesting objects are identified and illustrated, with references to the relevant chapter. These latter illustrations include scale factors to help students visualize how even immense objects many light-years across can appear as mere dots in the sky.

Along with the illustrated objects, most of the Looking Up features include a small insert to show how one of the constellation's stars are arranged in space.

When objects appearing in these figures are discussed in the text, Looking Up icons can be found in the margin. These point the reader to the appropriate Looking Up figure. We hope this connection to the night sky helps readers maintain or regain that sense of amazement when they view the sky.

Online Media are available on the *Explorations* website(www .mhhe.com/arny8e) to help students gain a better grasp of key concepts. Icons have been placed near figures and selections where students can gain additional understanding through Animations and Interactives. The Interactives are programmed in Flash, allowing users to manipulate parameters and gain a better

understanding of topics such as Blackbody Radiation, The Bohr Model, a Solar System Builder, Retrograde Motion, Cosmology, and the H-R Diagram by watching the effect of these manipulations.

Summary boxes at the end of each chapter give a brief review of the material covered. You also may want to read the summary *before* reading the chapter to get a general idea of the most important topics.

End-of-Chapter Questions are keyed to the relevant section numbers to help make connections between readings and problem solving. Use these cross references to delve back into the chapter if you are struggling with any of the questions.

When you finish a reading assignment, try to answer the "Questions for Review" for the sections you covered. They are short and are designed to help you see if you have assimilated the basic factual material in each section. Try to do this without looking back into the chapter, but if you can't remember, look it up rather than skip over the question. You might find it helpful to write out short answers to the questions.

Having worked your way through the material, go back and try to work through the other questions. "Thought Questions" challenge you to think more deeply about the readings. If you can't answer these on your own, talk them through with other students or your instructor. Then try some of the mathematical "Problems" and see if you can work through the material on your own. You may want to refer to the "Astronomy by the numbers" boxes in the chapter for ideas how to do these calculations. Finally, you can use the multiple-choice "Test Yourself" questions for a quick check of your understanding.

The **Appendix** contains a brief introduction to working with scientific notation and solving simple equations. It also contains 11 tables with important numbers and astronomical data, bringing together information about Solar System objects, other stars, and other galaxies so you can easily compare their properties.

The **Glossary** provides short definitions of all the key terms in the text. If you encounter words or terms as you read that you don't know, look them up in the glossary. If they are not included there, check the index or a dictionary or encyclopedia.

The **Foldout Star Chart** at the back of the book is useful for studying the sky and figuring out where the Moon and planets are located in any month. The chart is useful for projects such as plotting the changing location of the Moon and planets, or the paths of meteors.

Seeing a clear night sky spangled with stars is a wondrous experience. And yet the beauty and sense of wonder can be enriched even more by an appreciation of the complex processes that make the Universe work. We hope this book will similarly increase your appreciation of our Universe's wonders.

If you find mistakes or have suggestions about how to make this book better, please contact one of us. Write T. Arny at P.O. Box 545, Patagonia, AZ 85624, or by email at tarny@theriver.com; or S. Schneider at Department of Astronomy, University of Massachusetts, Lederle Tower, Amherst, MA 01003, or by email at schneider@astro.umass.edu.

ACKNOWLEDGMENTS

Many people have played an important role in bringing this book into being. We are also very grateful to many people at McGraw-Hill, but especially Robin Reed, Laura Bies, David Hash, Andrea Pellerito, and Carrie Burger, and to Mary Reeg for their help and patience.

The authors and McGraw-Hill would also like to extend a special thank-you to Patrick Koehn at *Eastern Michigan University,* who was instrumental in the development of this edition's Connect and SmartBook content as well as our instructor resources.

The following individuals also contributed their time and expertise to the preparation of SmartBook for *Explorations:*
Hugh H. *Crowl, Bennington College*
Gregory L. Dolise, *Harrisburg Area Community College*
Christopher C. Shope, *Harrisburg Area Community College*
Christopher L. Taylor, *California State University, Sacramento*

REVIEWERS OF THIS AND PREVIOUS EDITIONS

Special thanks and appreciation go out to reviewers of this and previous editions.

Eighth Edition Reviewers
Frank Bickford, Tompkins Cortland Community College
Jeffrey Butikofer, Upper Iowa University
Joseph DeRocher, Cuyahoga Community College
Sasa Dordevic, The University of Akron
Robert A. Egler, North Carolina State University
Gary Faraci, Western Iowa Tech Community College
Richard Gelderman, Western Kentucky University
Martin Hackworth, Idaho State University
Javier Hasbun, University of West Georgia
Earl Heath, Owens Community college and Lourdes University
David Hedin, Northern Illinois University
Dr. Terry L. Jenkins, Northeast Iowa Community College
James Mcateer, New Mexico State University
Stuart Mufson, Indiana University
Jeff Nelson, William & Mary
Keivan Stassun, Vanderbilt University
Aseem Talukdar, Madisonville Community College
Rico Tyler, Western Kentucky University
Steven S. Vogt, University of California at Santa Cruz
Anne G Young, Rochester Institute of Technology

Those who contributed to the seventh edition and earlier are too numerous to mention individually, but their contributions, constructive suggestions, new ideas, and invaluable advice played an important role in the development of this edition and its supplements.

Required=Results

McGraw-Hill Connect®
Learn Without Limits

Connect is a teaching and learning platform that is proven to deliver better results for students and instructors.

Connect empowers students by continually adapting to deliver precisely what they need, when they need it, and how they need it, so your class time is more engaging and effective.

Course outcomes improve with Connect.

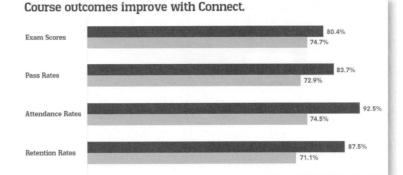

	With Connect	Without Connect
Exam Scores	80.4%	74.7%
Pass Rates	83.7%	72.9%
Attendance Rates	92.5%	74.5%
Retention Rates	87.5%	71.1%

88% of instructors who use **Connect** require it; instructor satisfaction **increases** by 38% when **Connect** is required.

Using **Connect** improves passing rates by **10.8%** and retention by **16.4%**.

Analytics

Connect Insight®

Connect Insight is Connect's new one-of-a-kind visual analytics dashboard—now available for both instructors and students—that provides at-a-glance information regarding student performance, which is immediately actionable. By presenting assignment, assessment, and topical performance results together with a time metric that is easily visible for aggregate or individual results, Connect Insight gives the user the ability to take a just-in-time approach to teaching and learning, which was never before available. Connect Insight presents data that empowers students and helps instructors improve class performance in a way that is efficient and effective.

Connect helps students achieve better grades

	A	B	C	D	F
With Connect	36%	29.5%	22%	4.3%	
Without Connect	27.2%	22.3%	25.6%	9.8%	

Based on McGraw-Hill Education Connect Effectiveness Study 2013

Students can view their results for any **Connect** course.

Mobile

Connect's new, intuitive mobile interface gives students and instructors flexible and convenient, anytime–anywhere access to all components of the Connect platform.

Adaptive

THE FIRST AND ONLY **ADAPTIVE READING EXPERIENCE** DESIGNED TO TRANSFORM THE WAY STUDENTS READ

> More students earn **A's** and **B's** when they use McGraw-Hill Education **Adaptive** products.

SmartBook®

Proven to help students improve grades and study more efficiently, SmartBook contains the same content within the print book, but actively tailors that content to the needs of the individual. SmartBook's adaptive technology provides precise, personalized instruction on what the student should do next, guiding the student to master and remember key concepts, targeting gaps in knowledge and offering customized feedback, and driving the student toward comprehension and retention of the subject matter. Available on smartphones and tablets, SmartBook puts learning at the student's fingertips—anywhere, anytime.

> Over **4 billion questions** have been answered, making McGraw-Hill Education products more intelligent, reliable, and precise.

STUDENTS WANT

Mc Graw Hill Education ISMARTBOOK®

95% of students reported **SmartBook** to be a more effective way of reading material

100% of students want to use the Practice Quiz feature available within **SmartBook** to help them study

100% of students reported having reliable access to off-campus wifi

90% of students say they would purchase **SmartBook** over print alone

95% reported that **SmartBook** would impact their study skills in a positive way

Mc Graw Hill Education

*Findings based on a 2015 focus group survey at Pellissippi State Community College administered by McGraw-Hill Education

The Cosmic Landscape

Astronomy is the study of the heavens, the realm extending from beyond the Earth's atmosphere to the most distant reaches of the Universe. Within this vast space we find an amazing diversity of planets, stars, and galaxies. It is amazing that creatures as tiny as ourselves not only can contemplate but also can understand such diversity and immensity. But even more amazing are the objects themselves: planets with dead volcanos whose summits dwarf Mount Everest, stars a hundred times the diameter of the Sun, and galaxies—slowly whirling clouds of stars—so vast that they make the Earth seem like a grain of sand in comparison. All this is the cosmic landscape in which we live, a landscape we will explore briefly here to familiarize ourselves with its features and to gain an appreciation for its vast scale.

THE EARTH, OUR HOME

We begin with the Earth, our home **planet** (fig. P.1). This spinning sphere of rock and iron circling the Sun is huge by human standards, but it is one of the smaller bodies in the cosmic landscape. Nevertheless, it is an appropriate place to start because, as the base from which we view the Universe, it influences what we can see. We cannot travel from object to object in our quest to understand the Universe. Instead, we are like children who know their neighborhood well but for whom the larger world is still a mystery, known only from books and television.

Just as children use knowledge of their neighborhood to build their image of the world, so astronomers use their knowledge of Earth as a guide to more exotic worlds. For example, we can deduce from the glowing lava of an erupting volcano and the boiling water shooting from a geyser that the interior of our planet is hot. That heat creates motion inside the Earth, much like the way heat makes soup in a pot bubble and churn. Although the motions inside Earth are far slower than those we see in bubbling soup, over millions of years they buckle the seemingly firm rock of our planet's crust to heave up mountains and volcanoes. Deeper inside Earth, similar motions generate magnetic forces that extend through the surface and into space. On Earth's surface these forces tug on the needle of a compass so that it points approximately north–south. High in our atmosphere, these same magnetic forces shape the northern lights.

Looking outward to our planetary neighbors, we find landscapes on Venus and Mars that bear evidence of many of the same processes that sculpt our planet and create its diversity. Likewise, when we look at the atmospheres of other planets, we see many of the same features that occur in our atmosphere. For example, winds in the thin envelope of gas that shelters us swirl around our planet much as similar winds sweep the alien landscapes of Venus and Mars.

FIGURE P.1
The planet Earth, our home, with blue oceans, white clouds, and multihued continents.

FIGURE P.2
The Moon as seen (A) with the unaided eye and (B) through a small telescope, and (C) *Apollo 17* astronauts on the surface.

THE MOON

The Moon is our nearest neighbor in space, a **satellite** that orbits the Earth some quarter million miles (384,000 km) away. Held in tow by the Earth's gravity, the Moon is much smaller than Earth—only about one-quarter our planet's diameter.

With the naked eye (fig. P.2A), and certainly with a pair of binoculars or small telescope (fig. P.2B), we can clearly see that the Moon's surface is totally unlike Earth's. Instead of white whirling clouds, green-covered hills, and blue oceans, we see an airless, pitted ball of rock that shows us the same face night after night.

Why are the Earth and Moon so different? Their differences arise in large part from the great disparity in their masses. The Moon's mass is only about 1/80th the Earth's, and it was therefore unable to retain an atmosphere. Without wind and rain, there has been relatively little erosion of the Moon's surface. Because of its smaller bulk, the Moon was also less able to retain heat. Without that strong internal heat, the crustal motions that are so important in shaping Earth are absent on the Moon. In fact, the Moon has changed so little for billions of years that its surface provides important clues to what Earth was like when it was young. In addition to this scientific importance, the Moon has symbolic significance for us—it is the farthest place from Earth that humans have traveled to (fig. P.2C).

THE PLANETS

Beyond the Moon, circling the Sun as the Earth does, are seven other planets, sister bodies of Earth. In the order of their average distance from the Sun, working outward, the eight planets are Mercury, Venus, Earth, Mars, Jupiter, Saturn, Uranus, and Neptune. These worlds have dramatically different sizes and landscapes. For example:

- Ancient craters blasted out by asteroid impacts scar the airless surface of Mercury.
- Dense clouds of sulfuric acid droplets completely shroud Venus.
- White clouds, blue oceans, green jungles, and red deserts tint Earth.
- Huge canyons and deserts spread across the ruddy face of Mars, but long ago there may have been lakes or even oceans.
- Immense atmospheric storms sweep across Jupiter—one storm almost as big as the whole Earth has lasted for centuries.
- Trillions of icy fragments orbit our second largest planet Saturn, forming its bright rings.
- Dark rings girdle Uranus, its spin tipped by some cosmic catastrophe in its distant past.
- Choking methane clouds whirl in the deep blue atmosphere of Neptune.

Figure P.3 shows pictures of these eight distinctive bodies and reveals something of their relative size and appearance. Mercury, Venus, Mars, Jupiter, and Saturn are visible to the naked eye

FIGURE P.3
The eight planets. *Top panel:* the four inner planets are shown to their correct relative size. *Lower panel:* the outer planets are shown to their correct relative size, with Earth for comparison.

Sun

FIGURE P.4
The Sun and the eight planets shown to the same scale. If separations were shown to the same scale, Earth would be about 30 feet away, and Neptune 1000 feet away. The image of the Sun was made through a filter that shows hot helium gas near its surface.

at night as bright points of light, much like stars. But whereas stars do not noticeably change their positions relative to one another, the planets, because of their orbital motion around the Sun, move slowly and regularly against the pattern of the background stars. This regular motion gave the planets a special significance to people in ancient times who named these moving "stars" after gods and goddesses—a significance that has been carried forward to today in the names of many of the days of the week. Saturday gets its name from Saturn, while in Spanish, miércoles (Wednesday) gets its name from Mercury.

Imagine how strange it must have seemed hundreds of years ago when astronomers first argued that the Earth was a "planet," one of those wandering stars seen in the night sky. Today with modern telescopes and spacecraft we can see that each planet is a unique, fascinating world. Some are airless while others have atmospheres so deep that they could swallow the Earth. As best we can tell, none other than Earth has given rise to life, but the characteristics of each planet offer us insights into our own planet's history and how we might maintain its unique environment.

Earth is a midsize planet. Jupiter is more than 300 times more massive, outweighing all of the other planets combined. However, all are dwarfed by the star they orbit: the Sun.

THE SUN

The Sun is a **star,** a huge ball of gas more than 100 times the diameter of the Earth and more than 300,000 times more massive: if the Sun were the size of a volleyball, the Earth would be about the size of a pinhead, and Jupiter roughly the size of a nickel (fig. P.4). The Sun contains about 1000 times more matter than all of the planets combined.

The Sun, of course, differs from the planets in more than just size: it generates energy in its core by nuclear reactions that convert hydrogen into helium. From the core, the energy flows to the Sun's surface, and from there it pours into space, illuminating and warming the planets.

The Sun's energy output cannot last forever. It has been warming the planets for more than 4 billion years—long enough for life to arise on Earth and for intelligent creatures to evolve who can marvel at such wonders. Studies of other stars teach us that the Sun will run out of fuel in another 5 or 6 billion years, then finally fade away like a cooling ember. Thus, astronomy helps us not only to examine unusual objects at huge distances, but to look deep into the past and far into the future.

THE SOLAR SYSTEM

The Sun and the eight planets orbiting it are the nine most massive bodies in the **Solar System.** Many less massive objects orbit the Sun as well. Among the most massive of these are the dwarf planets, and there are millions of smaller objects such as the asteroids and comets. There are also many satellites orbiting these bodies, some nearly as massive as Mercury.

Most asteroids orbit between Mars and Jupiter in the so-called asteroid belt (fig. P.5A), home to the dwarf planet Ceres. Ceres is similar to a planet in that its own gravity has forced it into a round shape and it orbits the Sun, but its orbit is strewn with thousands of other objects whose total mass actually exceeds the mass of Ceres. Unlike the major planets, Ceres has not "cleared its orbit" of material comparable to its own mass, so by a definition adopted in 2006, it is called a dwarf planet. The other objects orbiting in this belt are too small to have pulled themselves into a round shape and are called asteroids.

Over the last few decades, astronomers have discovered a vast number of objects orbiting beyond Neptune in what is known as the Kuiper belt (fig. P.5B). This realm is the home to uncounted icy bodies, large and small. There are probably dozens of dwarf planets, along with Pluto and the slightly more massive Eris, but it is very difficult to perform observations to confirm that gravity has given them a round shape. Millions of small comets also orbit in the outermost fringes of the Solar System, but we see them only when their orbits are disturbed, sending them to boil their ices away in the inner Solar System.

If the paths that the planets follow around the Sun were visible, we would see that the Solar System is like a huge set of nested, nearly circular rings, centered approximately on the Sun and extending about 3 billion miles outward to Neptune's orbit (fig. P.5B).

It is hard to imagine such immense distances measured in miles. In fact, using miles to measure the size of the Solar System is like using inches to measure the distance between New York and Tokyo. Whenever possible, astronomers try to use units appropriate to the scale of what they are measuring. For example, as we shall see in later chapters, the Earth's radius and mass are convenient units for measuring the sizes of other planets. Likewise, the Earth's distance from the Sun is a good unit for measuring the scale of the Solar System.

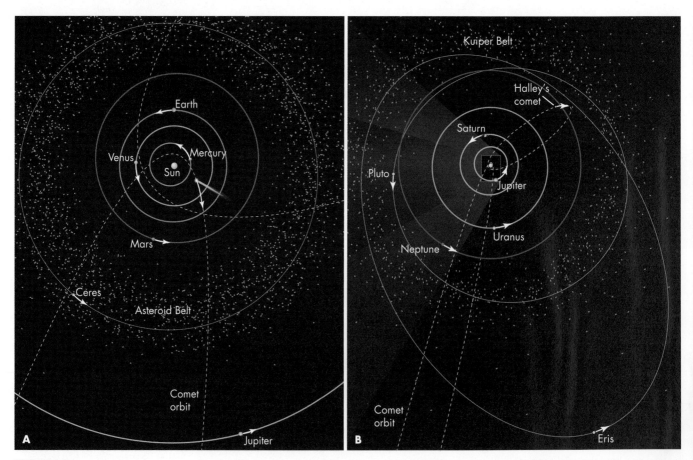

FIGURE P.5
Sketch of the positions and orbits of the planets and a variety of smaller bodies in our Solar System on March 20, 2011. The orbits of three of the largest "dwarf planets," Halley's comet, and another typical comet are also shown. The approximate location of small bodies in the asteroid belt and Kuiper belt are indicated. To show the orbits to scale, the (A) inner and (B) outer Solar System are shown separately.

FIGURE P.6

(A) This view of the Solar System is based on a series of real images made by the *Voyager 1* spacecraft. The craft was about 40 AU from the Sun and about 20 AU above Neptune's orbit. The images of the planets (mere dots because of their immense distance) and the Sun have been made bigger and brighter in this view to allow you to see them more clearly. Mercury is lost in the Sun's glare and Mars happened to lie nearly in front of the Sun at the time the image was made, so it too is invisible. (B) A sketch of the orbits of the planets, showing where each was located at the time the image was made in February 1990.

ASTRONOMICAL SIZES

The **astronomical unit,** abbreviated as **AU,** is the average distance from the Earth to the Sun.* This translates into about 93 million miles (150 million kilometers). If we use the AU to measure the scale of the Solar System, Mercury is 0.4 AU from the Sun, while Neptune is about 30 AU (fig. P.6). The Solar System extends far beyond the planets. Some comets drift along orbits that stretch up to about 100,000 AU away from the Sun.

Figure P.6 shows a picture of the Solar System made by the spacecraft *Voyager 1* after it passed Neptune. Notice how *empty* space is. The *Voyager* spacecraft is presently the fastest-moving and most distant probe we have yet launched. Even at this speed, it would take tens of thousands of years to reach a nearby star. Rather than spacecraft, we use telescopes to extend our view beyond the Solar System. And to describe the distances to stars, we need a far larger unit of measure—the light-year.

---

*Because the Earth's orbit is an ellipse, which we will discuss further when we consider planetary orbits, the AU is technically defined slightly differently.

Measuring a distance in terms of a time may at first sound peculiar, but we do it often. We may say, for example, that our town is a 2-hour drive from the city, or our dorm is a 5-minute walk from the library, but expressing a distance in this fashion implies that we have a standard speed.

Astronomers are fortunate to have a superb speed standard: the speed of light in empty space, which is a constant of nature and equal to 299,792,458 meters per second (about 186,000 miles per second). Moving at this constant and universal speed, light in 1 year travels a distance defined to be 1 **light-year,** abbreviated as ly. As we show in the Astronomy by the Numbers box below, this works out to be about 6 trillion miles (10 trillion kilometers).

Working with extremely large numbers is cumbersome, so astronomers use a more concise way to write them called **scientific notation** (also called powers-of-ten notation) in which we write numbers using ten to an exponent, or power. Thus we write $100 = 10 \times 10 = 10^2$ and 1 million (1,000,000) as $10 \times 10 \times 10 \times 10 \times 10 \times 10 = 10^6$. Instead of writing out all the zeros, therefore, we use the exponent to tell us the number of zeros. A number like the speed of light (186,000 miles

ASTRONOMY *by the numbers* — THE SIZE OF A LIGHT-YEAR

To find how far light travels in a year, we multiply its speed by the travel time. One year is approximately 31,600,000 (or 3.16×10^7) seconds. Multiplying this time by the speed of light gives the distance light travels in one year:

$$3.16 \times 10^7 \text{ seconds} \times 1.86 \times 10^5 \text{ miles/second}$$
$$= 3.16 \times 1.86 \times 10^{12} \text{ seconds} \times \text{miles/second}$$
$$= 5.88 \times 10^{12} \text{ miles,}$$

or about 6 trillion miles (about 10^{13} kilometers). In these units, the star nearest the Sun is 4.2 light-years away.

Although we achieve a major convenience in adopting such a huge distance for our scale unit when describing distances to stars, we should not lose sight of how truly immense such distances are. For example, if we were to count off the miles in a light-year, one every second, it would take us about 186,000 years!

per second) may also be written in scientific notation, becoming 1.86×10^5 miles per second. Likewise, the astronomical unit (150 million kilometers) can be written as 1.5×10^8 km.

One reason to use scientific notation is that multiplying and dividing becomes enormously easier. For example, to multiply two powers of ten we just add the exponents, and to divide we subtract them. Thus $10^2 \times 10^5 = 10^7$, and $10^8/10^3 = 10^5$. More details on using scientific notation are given in the appendix.

With the ability to describe these enormous interstellar distances, we are prepared to move beyond the Solar System. In this vastly larger realm, the Sun is but one of a vast swarm of stars orbiting the center of our galaxy, the Milky Way.

THE MILKY WAY

The **Milky Way Galaxy** is a cloud of several hundred billion stars with a flattened shape like the Solar System (fig. P.7), but about 100,000 ly across. The Sun orbits 27,000 ly from the center of the Milky Way at some 150 miles per second (240 kilometers per second), but so vast is our galaxy that it still takes the Sun about 210 million years to complete one trip around this immense disk. The Milky Way's myriad stars come in many varieties, some hundreds of times larger than the Sun, others hundreds of times smaller. Some stars are much hotter than the Sun and shine a dazzling blue-white, while others are cooler and glow a deep red.

In the Milky Way, as in other galaxies, stars intermingle with immense clouds of gas and dust. These clouds, enormously larger than the Solar System, are the sites of stellar birth and death. Deep within their cold, dark gas, gravity draws their matter into dense clumps that eventually turn into new stars, lighting the gas and dust around them. Some stars eventually burn themselves out and explode, spraying matter outward to mix with the surrounding clouds. This matter from exploded stars is ultimately recycled into new stars (fig. P.8).

In this huge swarm of stars and clouds, the Solar System is all but lost—like a single grain of sand on a vast beach—forcing us again to grapple with the problem of scale. Stars are almost unimaginably remote: the nearest one to the Sun is over 25 trillion miles away, or about 4.2 light-years. Such distances are so immense that analogy is often the only way to grasp them. For example, if we think of the Sun as a pinhead, the nearest star would be another pinhead about 35 miles away and the space between them would be nearly empty.

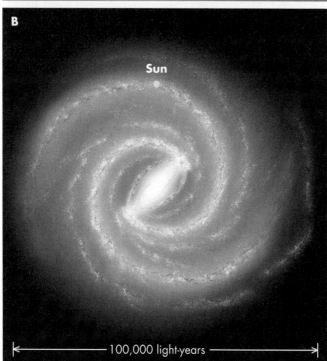

FIGURE P.7
The Milky Way Galaxy. (A) A side view made by plotting stars in the 2MASS star catalog. (B) The approximate structure of the Milky Way if it were seen from above, as mapped out by the Spitzer Space Telescope.

FIGURE P.8
An interstellar cloud in the Milky Way. Some stars are forming inside the dark cloud while other young stars heat the surrounding gas, making it glow. This Hubble Space Telescope image shows a region about 4 light-years across. At this scale the Solar System out to Neptune is about 100 times smaller than the period ending this sentence.

FIGURE P.9

(A) A sketch of the central region of the Local Group. (B) A sketch of the Virgo Supercluster. Only a few of the clusters of galaxies are labeled. The names of the galaxies M31, M33, M81, and M101 are from a list of galaxies and other astronomical objects that was compiled in the late 1700s by French astronomer Charles Messier ("*Mess-yay*").

GALAXY CLUSTERS AND THE UNIVERSE

Having gained some sense of scale for the Solar System and the Milky Way, we resume our exploration of the cosmic landscape, pushing out to the realm of other galaxies. Here we find that just as stars assemble into galaxies, so galaxies themselves assemble into **galaxy clusters.**

The cluster of galaxies to which the Milky Way belongs is called the **Local Group.** It is "local," of course, because it is the one we inhabit. It is termed a "group" because it is small as galaxy clusters go, containing just several dozen galaxies as members, but it is still a few million light-years in diameter. Despite such vast dimensions, the Local Group is itself part of a still larger assemblage of galaxies known as the **Virgo Supercluster.** Figure P.9 puts this in perspective.

Our supercluster consists of hundreds of galaxy groups and clusters, spread over some 100 million light-years, but it is perhaps itself part of an even larger structure known as the Great Attractor region, a cluster of superclusters, probably more than 300 million light-years across. Structures of such vast size are about the largest objects we can see before we take the final jump in scale to the **Universe** itself.

The visible Universe is the largest astronomical structure of which we have any knowledge. From the observations presently available to them, astronomers deduce that the Universe is about 13.8 billion years old. This limits the distance we can see, even in principle, to 13.8 billion light-years, a value we can use to describe the radius of the *visible* Universe. When we make an extremely deep photograph of the sky (fig. P.10), the light from the most distant visible galaxies takes nearly the age of the Universe to reach us, so we are seeing them when they first formed.

Although the visible Universe extends to 13.8 billion light-years from us, that does not mean the Universe ends there. Rather, it means we cannot yet see what lies beyond. But regardless of our uncertainty about the known Universe's size, we can observe that its structure is similar throughout the visible Universe. Small objects are clustered into larger systems, which are themselves clustered: planets around stars, stars in galaxies, galaxies in clusters, clusters in superclusters, and perhaps superclusters into even larger associations. Although astronomers do not yet understand completely how this orderly structure originated, they do know that gravity plays a crucial role.

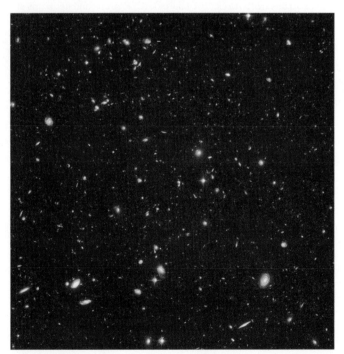

FIGURE P.10

A portion of the deepest image ever made with the Hubble Space Telescope. Virtually every one of the thousands of dots in the image is a galaxy—some near the edge of the visible Universe are seen when they were just beginning to form. A grain of sand held at arm's length would cover the tiny area imaged here.

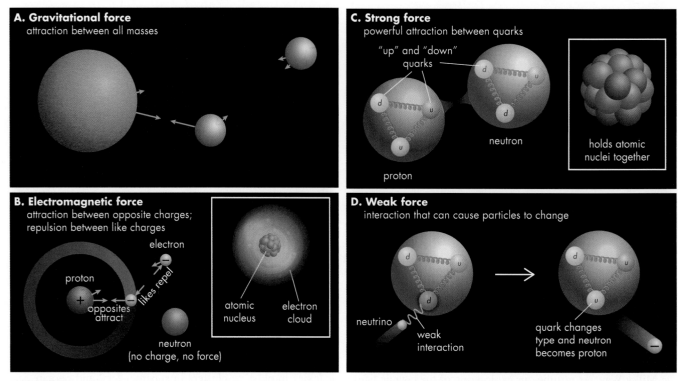

FIGURE P.11

The four fundamental forces. (A) The force of gravity is present between all objects with mass. The force, represented by green arrows in the figure, is always attractive, but grows weaker with distance. (B) The electromagnetic force arises between particles with an electric charge. It causes electrons (negative charge) to be attracted to protons (positive charge) to form atoms. The nucleus of an atom is made of protons and neutrons (with no electric charge). (C) Protons and neutrons are made of smaller particles called quarks, which are held together by the strong force. The strong attraction between quarks causes protons and neutrons to be attracted to each other, overcoming the electromagnetic repulsion between protons. (D) The weak force causes some particles to change into others as they interact. The weak force causes radioactive decay and plays a critical role in energy formation in stars.

FORCES AND MATTER

Gravity gives the Universe structure because it creates a force of attraction between *all* objects (fig. P.11A). You experience gravity's attraction in everyday life. For example, if you drop a book, the Earth's gravitational force makes the book fall. That same force spans the vast distance between the Earth and the Moon to hold our satellite in its orbit. Similarly, gravity holds our planet in its orbit around the Sun and the Sun in its orbit around the Milky Way.

Gravity may dominate the large-scale structure of the Universe, but other forces dominate on smaller scales. To understand these forces, we need to look at the small-scale structure of matter. Matter is composed of submicroscopic particles called **atoms.** Atoms are incredibly small. For example, a hydrogen atom is about one ten-billionth of a meter (10^{-10} m) in diameter. Ten million hydrogen atoms could be put in a line across the diameter of the period at the end of this sentence. But despite this tiny size, atoms themselves have structure. Every atom has a central core, called the **nucleus,** that is orbited by smaller particles called **electrons** (fig. P.11B). The nucleus is in turn composed of two other kinds of particles, called **protons** and **neutrons.**

Although the particles in an atom exert a gravitational attraction on one another, atoms are not held together by gravity. Instead, an electromagnetic force gives them their structure. That force arises because protons and electrons have a property called **electric charge.** A proton has a positive electric charge, and an electron has a negative electric charge. A neutron is "neutral" as its name suggests—it has no charge.

The **electromagnetic force** can either attract or repel, depending on the charges. Opposite charges attract, and like charges repel. Thus, two electrons (both negative) repel each other, while an electron and a proton (negative and positive) attract each other. That attraction is what holds the electrons in their orbits around the nucleus of an atom (fig. P.10B).

You can see the electric force at work in many ways. For example, the static electric charges generated when a clothes drier tumbles your laundry creates an attraction that may make clothes cling together. The crackling sound you hear as you pull fuzzy socks away from a shirt is the electric charges jumping and making tiny sparks.

The electric force is closely linked with the magnetic force that makes a compass work or holds the little magnets to the door of your refrigerator. In fact, the theory of relativity demonstrates that electric and magnetic forces are fundamentally the same, and scientists generally refer to them jointly as the electromagnetic force.

At yet a deeper level, protons and neutrons are made up of more basic particles called **quarks.** Quarks are attracted to each other by the **strong force,** which is so-named because its attraction can overcome the electromagnetic repulsion of like-charged particles. When protons and neutrons are very close to each other, the strong force between quarks can cause them to bind together, forming an atom's nucleus (fig. P.11C). Although the effects of the strong force cannot be seen directly in everyday life, without it the nuclei of atoms, and with them our familiar world, would disintegrate.

In addition, a fourth force, known as the **weak force***, operates on the subatomic scale and plays a role in radioactive decay (fig. P.10D). The weak force is so weak that interactions involving it are extremely rare. Their rareness is important in determining how long stars live. Stars would burn themselves out much more quickly, or would not shine at all, if the weak force were much stronger or weaker. Unlike the other forces, which produce attraction or repulsion between different kinds of matter, the weak force causes matter to change its form in fundamental ways. In fact, astronomers are beginning to suspect that the weak force plays a major role in shaping the kinds of matter that are present in the Universe.

The weak force earned its name because it is millions of times weaker than the electromagnetic and strong forces, but it is still trillions upon trillions of times stronger than gravity. Why then does gravity dominate the Universe? This genuinely weakest of the forces has the unique property that it always works in just one way, always pulling matter toward other matter. By contrast, the other forces sometimes push and sometimes pull, and the differently charged particles move about until the contrary forces cancel each other out. This leaves gravity as the only remaining force acting on the largest scales.

THE STILL-UNKNOWN UNIVERSE

Our quick trip from Earth outward has shown us a Universe of planets, stars, and galaxies. However, astronomers today have evidence that the bulk of the Universe must consist of something completely different. That evidence comes from many sources, the most convincing of which are the findings that stars within galaxies, and galaxies within clusters of galaxies, experience a far stronger gravitational force than can be explained by the directly observable matter. That is, both galaxies and galaxy clusters appear to contain huge amounts of what astronomers call **dark matter.**

Dark matter is so-named because it emits no as-yet-observed radiation. But from its gravitational effects, astronomers deduce that it outweighs luminous matter by a factor of about five to one. What is the dark matter? Astronomers do not know, but it may be made up of particles that interact only through the weak and gravitational forces. For example, there are billions of weakly interacting particles called **neutrinos** passing through your body each second. These were generated by the Universe in its early stages, by nuclear reactions in the Sun, and by other cosmic events. You do not sense these particles because normal matter is more transparent to them than a glass window is to light. Astronomers suspect that there may be particles much more massive than neutrinos that fill space, generating a much stronger gravitational pull than all of the stars in all of the galaxies that we can see.

On the largest scales, galaxies throughout the Universe are moving away from each other in a great cosmic expansion. This expansion began about 13.8 billion years ago in an unimaginable explosion called the **Big Bang** that created time and space and sent hot matter flying apart everywhere. During the last two decades, astronomers studying the expansion have discovered a great mystery—the rate of expansion is speeding up. Something is overcoming the gravitational attraction between galaxies, causing them to accelerate away from each other.

It is as if empty space contains a sort of energy that drives the expansion to grow faster. Because its nature is still unknown, astronomers have named it **dark energy.** If we compare the effective mass of dark energy and dark matter with the mass of the objects that we directly detect (such as stars, galaxies, and gas clouds), those luminous objects amount to a mere 1% of the Universe's total mass. What we see of the Universe is therefore much like the footprints of an invisible creature: a being who leaves tracks, but whose build and nature we do not yet know.

THE SCIENTIFIC METHOD

Our scientific understanding of the Universe has not come easily. It has grown out of the work of thousands of men and women over thousands of years. Their work is part of the broad field that we call science.

By "science" we mean the systematic study of things and the search for the underlying principles that govern them, be they living things, matter, or, in our case, the astronomical universe. An essential part of that study is the rigorous testing of ideas. We call the process of such testing the **scientific method.** In using the scientific method, a scientist typically proposes an idea—a hypothesis—about some property of the Universe and then tests that hypothesis by experiment. In fact, whether an idea is "scientific" depends to some extent on whether it can be verified by either a real or an imagined experiment. Ideally the experiment either confirms the hypothesis or refutes it. If refuted, the hypothesis is rejected. On the other hand, if the experiment confirms the hypothesis, the scientist may then go on to develop related hypotheses or perhaps to make predictions about some as-yet-undiscovered aspect of the subject.

Once a set of ideas has been thoroughly tested and verified, they may be incorporated into a theory or law. When

---

*The weak force is linked to the electromagnetic force, and their combination is known technically as the electroweak force.

we use the term *theory* here, we do *not* mean that the ideas are unproven or tentative. Rather, we mean that they have achieved wide acceptance by successful testing. For example, scientists have subjected the quantum theory of atomic structure and the theory of relativity to numerous tests, and these theories have passed all such tests with high precision.

The scientific method as usually described is an idealization of a much more complex process. In practice, scientists move back and forth between a variety of stages involving the gathering of data, the analysis of the data, and reformulating questions, all informed by interactions with the scientific community and society at large. A working description of the scientific method is shown in figure P.12.

Astronomers face a special difficulty in applying the scientific method because usually they cannot experiment with their subject matter directly: in virtually all cases, they can only passively observe. Nevertheless, they try—like all scientists—to use the scientific method. You will find some specific examples of this method in later chapters, where *Science at Work* boxes show how this process has led to new ideas and the revision of old ones.

Application of the scientific method is no guarantee that its results will be believed. For instance, we will see in chapter 2 that even before 300 B.C., the Greek philosopher Aristotle taught that the Earth is a sphere. Yet despite the proofs he offered to support that hypothesis, many people continued to believe the Earth to be flat. Today, too, some scientific hypotheses might be rejected despite their experimental verification, and others might be accepted though untrue. For example, one astronomer might find evidence supporting some hypothesis, but another astronomer might claim that the experiment was done incorrectly or the data were analyzed improperly. Therefore, throughout this book, whenever we discuss our knowledge of a given topic, we must keep in mind the fact that such knowledge is not always certain or even universally accepted. This is especially true of topics at the frontiers

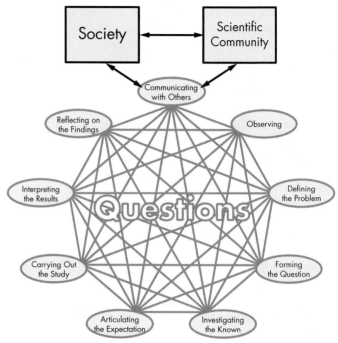

FIGURE P.12

A diagram illustrating the scientific method as described by scientists. The figure is based on interviews with scientists carried out by Reiff, Harwood, and Phillipson of Indiana University. Far from being a regular step-by-step procedure, the scientific method was described by scientists as a set of different activities and processes, which might be visited and revisited in a wide variety of orders as different questions arose. This is illustrated in the figure by lines connecting nine processes identified from the interviews.

of our understanding, such as the origin and structure of the Universe or the properties of black holes. Therefore, keep in mind that some of what we discuss in this book will be proved wrong in the future. That is not a failing of science, however. It is its strength.

SUMMARY

The Earth is one of eight planets orbiting the Sun, and the Sun is one of several hundred billion stars that make up the Milky Way Galaxy. The Milky Way, two other similar-size galaxies, and dozens of smaller galaxies compose the Local Group, which in turn is part of the Local Supercluster of galaxies. Superclusters seem to be grouped into even larger systems that fill the visible Universe. We can speak with some certainty about the size and properties of objects in our immediate neighborhood, but the farther we move from Earth, the less certain we become.

Astronomers use the astronomical unit (AU) and light-year (ly) to measure the immense sizes and distances of astronomical systems. The AU is defined by the average distance between the Earth and the Sun, and the light-year is

defined as the distance light travels in a year, which is about 10 trillion kilometers. Using these units, we can see the immense scale of the Universe in figure P.13 and table P.1. The former shows a series of images to help you visualize how enormous the Universe is.

Matter is made up of atoms in which charged particles called electrons orbit a nucleus. The nucleus is itself composed of smaller particles called protons and neutrons. Four forces give the Universe its structure: the electromagnetic, strong, and weak forces on the scale of atoms, and gravity on the cosmic scale. The whole Universe appears to be governed by these forces, yet there is also growing evidence that most of the Universe is made of types of matter and energy that we have not yet been able to detect.

FIGURE P.13

The Earth is but one of eight planets orbiting our star, the Sun. The Sun is but one of hundreds of billions of stars in our Galaxy, the Milky Way. The Milky Way is the second largest among many dozens of galaxies in our "Local Group." The Local Group is one of the smaller "clusters" of galaxies among hundreds of clusters that make up the "Virgo Supercluster." The Universe is filled with millions of other superclusters stretching to the limits of our vision.

Table P.1	The Scale of the Universe
Object	**Approximate Radius**
Earth	6400 km (~4000 miles)
Sun	700,000 km (~100 × radius of the Earth)
Earth's orbit	150 million km (~200 × radius of Sun) = 1 AU
Solar System to Neptune	30 AU (~6500 × radius of the Sun)
Milky Way Galaxy	50,000 ly (~10^8 × radius of Neptune's orbit)
Local Group	2.5 million ly (~50 × radius of the Milky Way)
Local Supercluster	50 million ly (~20 × radius of the Local Group)
Visible Universe	13.8 billion ly (~300 × radius of the Local Supercluster)

QUESTIONS FOR REVIEW

1. About how much bigger in radius is the Sun than the Earth?
2. How big is an astronomical unit?
3. Roughly how big across is the Milky Way Galaxy?
4. How is a light-year defined?
5. What force holds together the different astronomical systems described in this section? What other forces exist in nature?
6. What particles make up an atom?
7. What force holds the electrons to an atom's nucleus?
8. What was the Big Bang? What are dark matter and dark energy?
9. What is meant by the scientific method?
10. What is the difference between a hypothesis and a theory?

THOUGHT QUESTIONS

1. To what systems, in increasing order of size, does the Earth belong?
2. Propose a hypothesis about something you can experiment with in everyday life and try to verify or disprove the hypothesis. For example, what kind of surfaces will the little magnetic note holders people use on refrigerators stick to? Any smooth surface? Any metal surface?
3. If a new force were discovered, perhaps related somehow to dark energy or dark matter, how would this force and its effects need to "fit in" with the known four forces? Could it replace one of the existing forces as the explanation for some known phenomena? What kind of work would scientists need to do for this to happen? Apply this same logic to comment on what would be required to provide a scientific basis for ghosts or psychic powers.

PROBLEMS

1. The radius of the Sun is 7×10^5 kilometers, and that of the Earth is about 6.4×10^3 kilometers. Use scientific notation to show that the Sun's radius is about 100 times the Earth's radius.
2. Given that an astronomical unit is 1.5×10^8 kilometers and a light-year is about 10^{13} kilometers, how many AU are in a light-year?
3. What would be the circumference and diameter (circumference = $\pi \times$ diameter) of a ball that would represent the Moon if the Earth were a volleyball? What kind of ball or object matches this size?

4. Calculate approximately how long it takes light to travel from the Sun to the dwarf planet Eris.
5. If the Milky Way were the size of a nickel (about 2 cm), how big would the Local Group be? How big would the Local Supercluster be? How big would the visible Universe be? The data in table P.1 may help you here.
6. Suppose two galaxies move away from each other at 6000 km/sec and are 300 million (3×10^8) light-years apart. If their speed has remained constant, how long has it taken them to move that far apart? Express your answer in years.
7. A typical bacterium has a diameter of about 10^{-6} meters. A hydrogen atom has a diameter of about 10^{-10} meters. How many times smaller than a bacterium is a hydrogen atom?
8. Using scientific notation, numerically evaluate the expression $[10^5 \times (10^2)^3] / [100 \times 10^4 \times (10^8)^{1/2}]$.
9. Using scientific notation, numerically evaluate the expression $(8 \times 10^6)^2 / (2 \times 10^{-3})^3$.
10. Using scientific notation, numerically evaluate the expression $(3 \times 10^5)^2 / (4 \times 10^4)^{1/2}$.

TEST YOURSELF

1. Judging from the lower part of figure P.3, about how much larger is Jupiter's diameter than the Earth's?

   (a) 2 times   (c) 10 times   (e) 100 times
   (b) 5 times   (d) 25 times

2. Ancients believed the planets to be special compared to stars because

   (a) the surface of each planet is very different from Earth's.
   (b) planets repeat the same paths on the sky each week.
   (c) over time the planets appear to move against background stars.
   (d) they could see Jupiter's moons and Saturn's rings.

3. The light-year is a unit of

   (a) time.   (c) speed.   (e) weight.
   (b) distance. (d) age.

4. You write your home address in the order of street, town, state, and so on. Suppose you were writing your cosmic address in a similar manner. Which of the following is the correct order?

   (a) Earth, Milky Way, Solar System, Local Group
   (b) Earth, Solar System, Local Group, Milky Way
   (c) Earth, Solar System, Milky Way, Local Group
   (d) Solar System, Earth, Local Group, Milky Way
   (e) Solar System, Local Group, Milky Way, Earth

5. Which of the following astronomical systems is/are held together by gravity?

   (a) The Sun
   (b) The Solar System
   (c) The Milky Way
   (d) The Local Group
   (e) All of them are.

6. Which of the following statements can be tested for correctness using the scientific method? (There may be more than one correct answer.)

   (a) An astronaut cannot survive on the Moon without life-support systems.
   (b) The Moon is an uglier place than the Earth.
   (c) Electrons are charged particles.
   (d) The Sun's diameter is about 100 times larger than the Earth's diameter.
   (e) The sky is sometimes blue.

KEY TERMS

astronomical unit (AU), 5
atom, 8
Big Bang, 9
dark energy, 9
dark matter, 9
electric charge, 8
electromagnetic force, 8
electron, 8
galaxy cluster, 7
gravity, 8
light-year (ly), 5
Local Group, 7
Milky Way Galaxy, 6
neutrino, 9

neutron, 8
nucleus, 8
planet, 1
proton, 8
quark, 9
satellite, 2
scientific method, 9
scientific notation, 5
Solar System, 4
star, 3
strong force, 9
Universe, 7
Virgo Supercluster, 7
weak force, 9

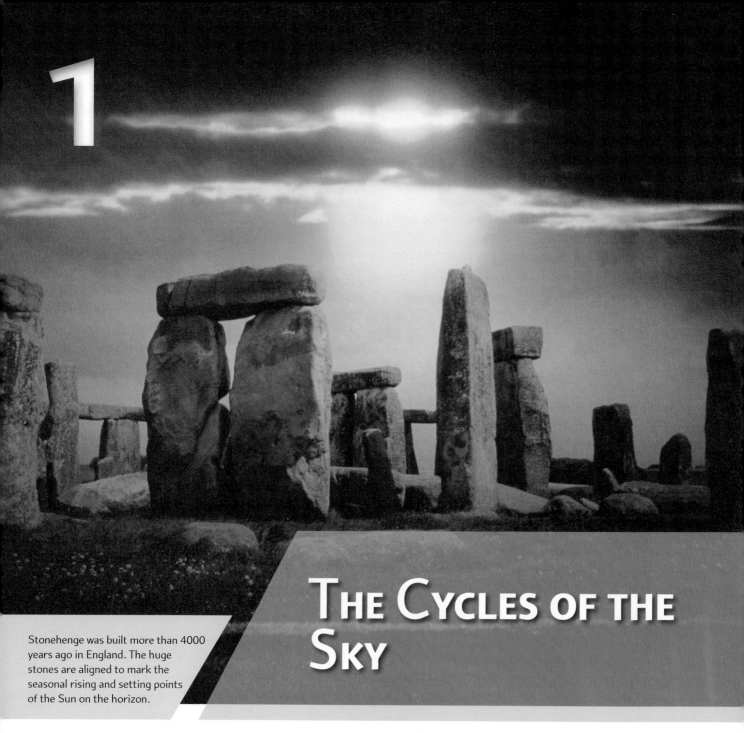

1

THE CYCLES OF THE SKY

Stonehenge was built more than 4000 years ago in England. The huge stones are aligned to mark the seasonal rising and setting points of the Sun on the horizon.

LEARNING OBJECTIVES

Upon completing this chapter you should be able to:

- Describe the motions of the Sun, Moon, and stars as they rise along the eastern horizon, move across the sky, and set along the western horizon.
- Recognize the kinds of fixed patterns of stars called constellations.
- Explain why different constellations are visible at different times of the year.
- Define the cycles of the Sun, Moon, and stars that are the basis for the day, month, and year.
- Describe how and why the shape of the lit portion of the

Moon seen from Earth changes during the month.
- Relate the tilt of the Earth's axis to the changes in the apparent daily path of the Sun during the course of the year.
- Explain why the tilt of the Earth's axis leads to seasonal changes of temperature on the Earth, and how the effects differ on different parts of the Earth.
- Describe where, and how frequently, lunar and solar eclipses occur, and describe the visual phenomena associated with each.
- Explain why eclipses are rare, and why their dates gradually shift.

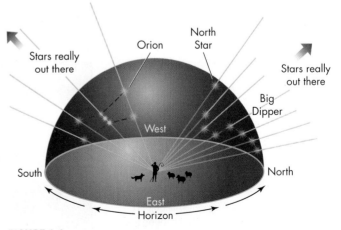

We do not know when people of antiquity first began studying the heavens, but it was certainly many thousands of years ago. Astronomical observations are part of virtually every culture and include events that anyone who watches the sky can see, such as the rising of the Sun in the eastern sky and its setting toward the west, the changing appearance of the Moon throughout the month, and the beautiful and awe-inspiring occurrences of eclipses.

For many prehistoric people, observations of the heavens had more than just curiosity value. Because so many astronomical phenomena are cyclic—that is, they repeat day after day and year after year—they can serve as timekeepers. For example, when is it safe to set out on a sea voyage? When is it time to harvest crops? When will an eclipse occur? Moreover, the cyclic behavior of the heavens implies that many events seen in the sky are predictable. The desire to foretell these changes in the sky and on Earth probably motivated early cultures to study the heavens, and it may have led them to build monumental stone structures such as Stonehenge (facing page).

Sadly, many of the astronomical phenomena well known to ancient people are not nearly so familiar to people living today, because the smog and bright lights of cities make it hard to see the sky and its rhythms. Perhaps more important, we no longer rely upon direct astronomical observations to tell us what season it is, when to plant, and so on. Therefore, if we are to appreciate the growth of astronomical ideas, we need to first understand what our distant ancestors knew and what we ourselves can learn by watching the sky over the course of a year.

In the following discussion, you might imagine yourself as a shepherd in the Middle East, a hunter-gatherer on the African plains, a trader sailing along the coast of the Mediterranean, or even a flight navigator in the early twentieth century. Whichever role you choose to assume, try to get out and actually look at the sky.

CONCEPTS AND SKILLS TO REVIEW

- The properties of Earth and Moon (Preview, pp. 1–2)
- The orbit of the Earth (Preview, pp. 4–5)

1.1   THE CELESTIAL SPHERE

One of nature's spectacles is the night sky seen from a clear, dark location with the stars scattered across the vault of the heavens (fig. 1.1*). Many of the patterns and motions of the stars have been all but forgotten in our hectic modern world, so our first goal is to familiarize ourselves with some general aspects of the sky at night.

Stars are at such huge distances that we cannot get any sense of their true three-dimensional arrangement in space when we view them. For purposes of naked-eye observations, we can therefore treat all stars as if they are at the same distance from the Earth, and imagine that they lie on the inside of a gigantic dome that stretches overhead. This dome seems to stretch to where the sky meets the ground along a horizontal circle that we call the **horizon.**

FIGURE 1.1
The stars appear to lie on a hemisphere over us that meets the ground at the horizon.

---

* In figure 1.1 and in many other figures throughout the book, distances and sizes of astronomical bodies are exaggerated for clarity.

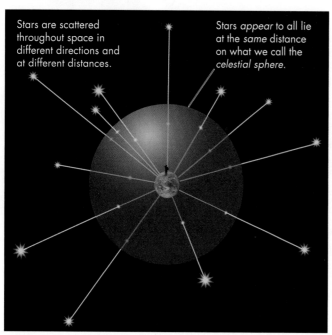

Stars are scattered throughout space in different directions and at different distances.

Stars *appear* to all lie at the *same* distance on what we call the *celestial sphere*.

FIGURE 1.2
The stars are scattered through space at very different distances, but they *appear* to lie at the same distance from us on what we call the celestial sphere.

Astronomers picture the dome of the night sky as half of the **celestial sphere,** which surrounds the Earth as depicted in figure 1.2. When we stand on the Earth, the ground blocks our view of approximately half the celestial sphere. If you were suspended in space far from Earth, you would see the entire celestial sphere surrounding you.

In reality, the thousands of stars visible on a clear night are at vastly different distances from us. The nearest is about 4 light-years away, so for the Earth at the size shown in figure 1.2, it would be about 6000 miles away. Other bright stars are thousands of times farther, millions of miles at the figure's scale!

Depicting the stars as though they lie on a celestial sphere is not physically realistic, but it serves as a useful **model** of the heavens—a way of simplifying the arrangement and motions of celestial bodies so they are easier to visualize. We use the term *model* to mean a representation of some aspect of the Universe. The celestial sphere represents a way of thinking about or picturing the location and motions of stars and planets for someone observing the sky from the Earth.

The celestial sphere is the first of many models we will encounter that humans have used to describe the Universe. In later chapters, we will use models to enhance our understanding whenever the size or other properties of what we study fall outside the range of everyday experience. We will speak of models of atoms, models of stars, and models of the Universe itself.

Constellations

As human beings, we seek order in what we see. When ancient people looked at the night sky, they noticed that the stars form fixed patterns on the celestial sphere, what we today call **constellations.** Some of these constellations resemble animals if we use a little imagination. For example, the pattern of stars in Leo looks a little like a lion, whereas that of Cygnus looks like a swan in flight, as depicted in figure 1.3. However, you will discover, as you learn to identify the constellations, that many have shapes that bear little resemblance to their namesakes.

FIGURE 1.3
The two constellations Leo (A) and Cygnus (B) with figures sketched in to help you visualize the animals they represent.

Some of the interesting celestial objects in and around Cygnus can be found in Looking Up #4 at the front of the book.

A

B

All stars move through space, but as seen from Earth, their positions change very slowly, taking tens of thousands of years to make any noticeable shift. Thus, we see today virtually the same pattern of stars that was seen by ancient peoples. A shepherd who lived 5000 years ago in the Middle East would have no trouble recognizing the star patterns of the night sky we see and might even call them by the same names.

We do not know how all the constellation names were chosen. Most date back thousands of years to prehistoric times. It seems likely that some names served as mnemonic devices for keeping track of the seasons and for navigating. For example, the beginning of the stormy winter months, when sailing was dangerous and ships were often wrecked, was foretold by the Sun's appearance in the constellations Pisces and Aquarius, the water constellations. Likewise, the harvest time was indicated by the Sun's appearance in Virgo, a constellation often depicted as the goddess Proserpine, holding a sheaf of grain.

Daily Motions of the Sun and Stars

Take a look at the night sky, and you will see stars rise along the eastern horizon, move across the sky, and set along the western horizon, just as the Sun does. You can verify this by watching the night sky for as little as 10 minutes. A star seen just above the eastern horizon will have risen noticeably higher, and stars near the western horizon will have sunk lower or disappeared (fig. 1.4A). Likewise, if you look at a constellation, you see its stars rise as a fixed pattern in the eastern sky, move across the sky, and set in the western sky.

In terms of our model of the heavens based on the celestial sphere, we can visualize the rising and setting of stars as rotation of the celestial sphere around us (fig. 1.4B). Ancient peoples would have found it far easier to believe in that rotation than to believe that the Earth moved. Thus, they attributed all celestial motion—that of the Sun, Moon, stars, and planets—to a vast sphere slowly turning overhead. Today we still say the Sun *rises* and *sets,* but of course we know that it is the Earth's rotation that makes the Sun, Moon, and stars rise and move westward across the sky each day. It is not the celestial sphere that spins but the Earth.

If you look at the celestial sphere turning overhead, two points on it do not move. These points are defined as the north and south **celestial poles.** The celestial poles lie

> *Q:* The stars appear to rotate counterclockwise around the north celestial pole. Which way does the Earth rotate as viewed from above the North Pole?

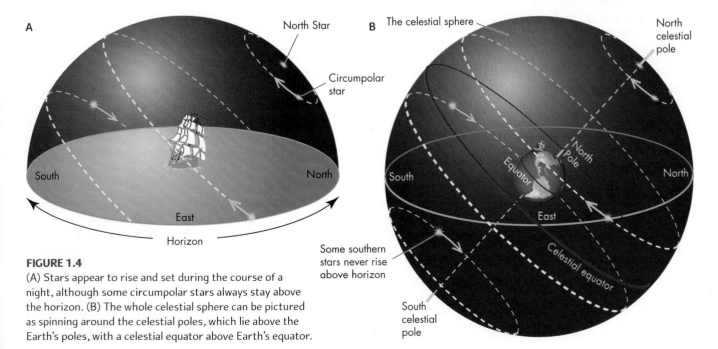

FIGURE 1.4

(A) Stars appear to rise and set during the course of a night, although some circumpolar stars always stay above the horizon. (B) The whole celestial sphere can be pictured as spinning around the celestial poles, which lie above the Earth's poles, with a celestial equator above Earth's equator.

LOOKING UP

The region of the north celestial pole is shown in Looking Up #1 at the front of the book. The region of the south celestial pole is shown in Looking Up #9.

ANIMATION

Star rise and set caused by Earth's rotation

FIGURE 1.5
The Sun appears to lie in Taurus on June 1, in Gemini on July 1, in Cancer on August 1, and so forth, making the constellations we see after sunset change with the seasons.

ANIMATION

Constellations by seasons

exactly above the North and South Poles of the Earth, and just as our planet turns about a line running from its North to South Poles, so the celestial sphere appears to rotate around the celestial poles, as illustrated in figure 1.4B. Over the course of a night, stars appear to circle the north celestial pole in a counterclockwise direction for observers in the Earth's northern hemisphere.

Because it lies directly above the Earth's North Pole, the north celestial pole always marks the direction of true north. Near the position of the north celestial pole, there happens to be a moderately bright star, Polaris, which is therefore known as the North Star. This is an important and widely used guide for travelers on land and sea, but it has not always been the same star throughout history. The direction of the Earth's axis gradually shifts or precesses over thousands of years, so different stars have served as the North Star in ancient times. No similarly bright star has happened to lie close to the south celestial pole for many thousands of years, so there is no equivalent "South Star." We examine the precession of the Earth's axis further in chapter 6.

Another important sky marker frequently used by astronomers is the **celestial equator.** The celestial equator lies directly above the Earth's equator, just as the celestial poles lie above the Earth's poles, as figure 1.4B shows. Only stars on the celestial equator rise due east and set due west. Stars north of the celestial equator rise in the northeast and set in the northwest, while stars south of the equator rise in the southeast and set in the southwest. For a northern observer some *circumpolar* stars near the north celestial pole never cross below the horizon, while stars close enough to the south celestial pole never rise above the horizon.

Annual Motion of the Sun

At the same time that the Earth's spin causes the apparent daily motion of the Sun and stars across the sky, the Earth's orbital motion around the Sun also causes changes in the parts of the sky we see on different nights of the year. If you compare the sky at the same time each evening for a few months, you will discover that different constellations are visible.

For example, in early June the Sun appears to lie in the direction of the constellation Taurus, so this constellation's stars are lost in the Sun's glare. After sunset, however, we can see the neighboring constellation, Gemini, just above the western horizon as illustrated in figure 1.5. By July, Gemini has disappeared behind the Sun, and instead Cancer is visible just above the horizon. And by August, Cancer has disappeared to be replaced by Leo. Around the rest of the sky we see a steady change of constellations throughout the course of the year. A year later, though, the same constellations will again be visible as they were originally.

The change of the constellations with the seasons is caused by the Earth's motion around the Sun. The Sun's glare blocks our view of the part of the celestial sphere that lies toward the Sun, making the stars that lie beyond the Sun invisible. If we picture the Earth orbiting the Sun within the celestial sphere, as illustrated in figure 1.6, month by month the Sun covers one constellation after another. It is like sitting around a campfire and not being able to see the faces of the people on the far side. But if we get up and walk around the fire, we can see faces that were previously hidden. Similarly the Earth's motion allows us to see stars previously hidden in the Sun's glare. Because these movements repeat in a yearly cycle, they are called *annual motions.*

Astronomers distinguish an object's spinning motion from its orbital motion with different terms. We say that the Earth rotates on its axis (spins) daily while it revolves around the Sun (moves along its orbit) annually. Because our planet orbits in the same direction as it spins, the Earth does not need to rotate quite as far each night to make a particular star visible as it does to face back toward the Sun. As a result, a star rises 3 minutes and 56 seconds earlier each night. That 3 minutes and 56 seconds, when added up each night over an entire year, amounts to 24 hours.

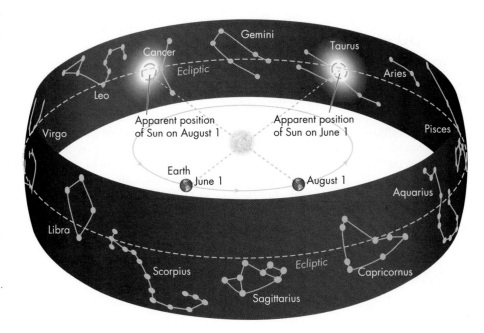

FIGURE 1.6
As the Earth orbits the Sun, the Sun appears to move around the celestial sphere through the background stars. The figure illustrates the portion of the celestial sphere on either side of the Sun's path, which is called the ecliptic. As the Earth orbits the Sun, the Sun appears to move through twelve constellations known as the zodiac that lie near the ecliptic. Note that the ecliptic is the extension of the Earth's orbital plane out to the celestial sphere.

This motion is slow and difficult to observe, but many ancient peoples developed techniques to keep track of these motions. This was extremely important to early people because it provided a way to measure the passage of time other than by carefully counting days. Moreover, the stars demonstrated that many celestial events are predictable and that they may be used to order our lives on Earth. For example, ancient Egyptians looked for the star Sirius near the Sun just before dawn as a way of predicting when the annual rising of the Nile would occur. Knowing the exact season can be crucial for such things as planting crops. A brief warm spell might have tricked an ancient farmer into sowing seeds too early, but by studying the sky for many years, she might have discovered that when the constellation Taurus is visible just before dawn, it is time to plant.

The Ecliptic and the Zodiac

If we could mark on the celestial sphere the path traced by the Sun as it moves through the constellations, we would see a line that runs around the celestial sphere, as illustrated in figure 1.6. Astronomers call the line that the Sun traces across the celestial sphere the **ecliptic.** The name *ecliptic* arises because only when the new or full moon crosses this line can an eclipse occur, as discussed in section 1.4. Examining figure 1.6, you can see that the ecliptic is the extension of the Earth's orbit onto the celestial sphere, just as the celestial equator is the extension of the Earth's equator onto the celestial sphere.

The belt-shaped region of the sky surrounding the ecliptic passes primarily through twelve constellations and is called the **zodiac.** The word *zodiac* is from the Greek *zoidion,* "little animal," and *kyklos,* "circle." That is, zodiac refers to a circle of animals, which the majority of its constellations represent. The names of these constellations are Aries (ram), Taurus (bull), Gemini (twins), Cancer (crab), Leo (lion), Virgo (maiden), Libra (scales), Scorpius (scorpion), Sagittarius (archer), Capricornus (sea-goat), Aquarius (water-bearer), and Pisces (fish).

The names of the constellations of the zodiac may look familiar from horoscope "signs," part of an ancient belief system of *astrology* that stars determined human destinies, much as they predicted the rising of the Nile. Astrology is today regarded as a pseudoscience, although horoscopes remain a popular entertainment (see Extending Our Reach: "Are You an Ophiuchan?").

EXTENDING
our reach

ARE YOU AN OPHIUCHAN?

The origin of horoscope signs dates back several thousand years. It is based on the notion that the location of the Sun along the zodiac at the time of people's birth (their "Sun sign") determines their basic personal traits. Astrologers often say things such as that a person born under the sign of Taurus is "strong and silent like a bull."

If you check where the Sun was actually located on the date of your birth, chances are that it was not in the constellation you would think based on your newspaper horoscope sign. This is because the dates of Sun signs were established thousands of years ago, but the precession of the Earth's axis (see chapter 6) has caused a shift in the dates of our calendar relative to the location of the Sun among the stars.

In fact, the Sun has shifted almost one full constellation, so if you think your sign is Aquarius, for example, the Sun was probably in Capricornus when you were born. In fact, the boundaries of the constellations are a little arbitrary, but the Sun actually moves through the constellation Ophiuchus, a snake charmer, during the first half of December. So many people who think they are Sagittarians are in fact "Ophiuchans"! Astronomers are not concerned about this, however, since there is no scientific evidence that astrology has any predictive power.

1.2   THE SEASONS

Many people mistakenly believe that we have seasons because the Earth's orbit is elliptical. They suppose that summer occurs when we are closest to the Sun and winter when we are farthest away. It turns out, however, that the Earth is nearest the Sun in early January, when the Northern Hemisphere is coldest. Clearly, then, seasons must have some other cause.

To see what *does* cause seasons, we need to look at how our planet is oriented in space. As the Earth orbits the Sun, our planet also spins. That spin is around an imaginary line—the **rotation axis**—that runs through the Earth from its North Pole to its South Pole. The Earth's rotation axis is *not* perpendicular to its orbit around the Sun. Rather, it is tipped by 23.5° from the vertical, as shown in figure 1.7A. As our planet moves along its orbit, its rotation axis maintains nearly exactly the same tilt and direction, as figure 1.7B shows. That is, the Earth behaves much like a giant gyroscope. The tendency of the Earth to preserve its tilt is shared by all spinning objects. For example,

The Earth's rotation axis

FIGURE 1.7
(A) The Earth's rotation axis is tilted 23.5° to the Earth's orbit around the Sun. (B) The Earth's rotation axis keeps the same tilt and direction as it moves around the Sun. (Sizes and distances are not to scale.)

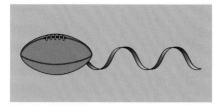

FIGURE 1.8
The tendency of a spinning object to keep its orientation is called "conservation of angular momentum," and it is the principle on which gyroscopes operate and the reason a quarterback puts "spin" on a football.

North Pole

North Pole

June
summer in Northern Hemisphere

December
summer in Southern Hemisphere

FIGURE 1.9
Because the Earth's rotation axis keeps the same tilt as we orbit the Sun, sunlight falls more directly on the Northern Hemisphere during part of the year and on the Southern Hemisphere during the other part of the year. (Sizes and distances are not to scale.)

it is what keeps a rolling coin upright, a Frisbee horizontal, and a thrown football pointed properly (fig. 1.8). You can easily feel this tendency of a spinning object to resist changes in its orientation by lifting a bicycle by the handlebars with the wheel spinning, then trying to twist it from side to side.

Because the Earth's tilt remains nearly constant as we move around the Sun, sunlight falls more directly on the Northern Hemisphere in June and surrounding months and more directly on the Southern Hemisphere around the month of December, as illustrated in figure 1.9. This causes a variation in the amount of heat each hemisphere receives from the Sun over the course of a year.

A surface facing directly toward a source of radiation is heated more than when the same surface is tilted. You take advantage of this effect instinctively when you warm your hands at a fire by holding your palms flat toward the fire, not edgewise. Figure 1.10 illustrates how this affects regions north and south of the equator. Equal areas of land do not receive the same amount of sunlight. When the North Pole is tilted toward the Sun in June, an area south of the equator receives an amount of radiation that is only a portion of the radiation intercepted by an equal area north of the equator. Therefore, over the course of a June day, the Northern Hemisphere is heated more than the Southern Hemisphere.

For the same reason, the Northern Hemisphere receives its greatest heating at the time of year when the Sun shines most directly on it, making it summer. Six months later, the Northern Hemisphere receives its sunlight least directly, and so it is colder and

Seasons

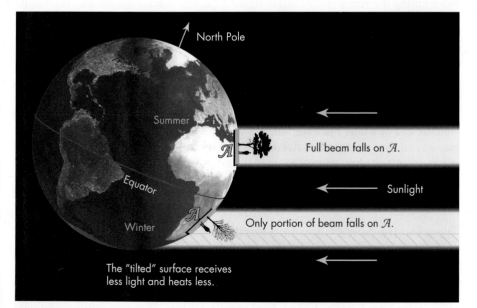

North Pole

Summer

$\mathcal{A}$ — Full beam falls on $\mathcal{A}$.

Equator

← Sunlight

Winter

$\mathcal{A}$ — Only portion of beam falls on $\mathcal{A}$.

The "tilted" surface receives less light and heats less.

FIGURE 1.10
A portion of the Earth's surface directly facing the Sun receives more concentrated light (and thus more heat) than other parts of the Earth's surface of equal area. The same size "beam" of sunlight (carrying the same amount of energy) spreads out over a larger area where the surface is "tilted."

Seasonal changes in daylight

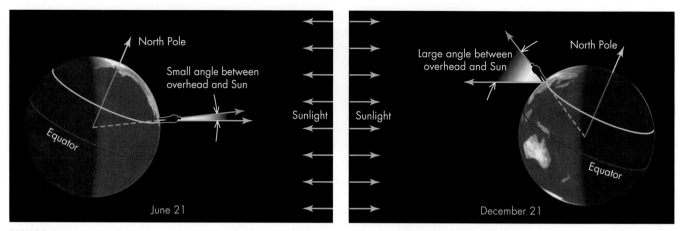

FIGURE 1.11
Between the extremes of the year six months apart, the angle at which sunshine strikes the ground at the same latitude can vary greatly.

therefore winter (fig. 1.11). This heating difference is enhanced because the Earth's tilt leads to many more hours of daylight in the summer than in the winter. As a result, not only do we receive the Sun's light more directly, we receive it for a longer time. Thus,

the seasons are caused by the tilt of the Earth's rotation axis.

From figure 1.11 it can be seen why the seasons are reversed between the Northern and Southern Hemispheres; when it is summer in one, it is winter in the other.

Solstices, Equinoxes, and the Ecliptic's Tilt

The tilt of the Earth's rotation axis not only causes seasons, it also is why the Sun's path across the celestial sphere—the ecliptic—is tilted with respect to the celestial equator. Because the Earth's axis remains oriented in a fixed direction, there is a point in its orbit when the North Pole is tipped most closely toward the Sun. This occurs on about June 21, as illustrated in figure 1.12. On this date the North Pole is tilted 23.5° toward the Sun, so the Sun lies 23.5° north of the celestial equator. (The date can vary from year to year, mostly because a year is about a quarter of a day longer than 365 days—which is also what causes us to insert leap years.) Half a year later, on about December 21, the Earth is on the other side of the Sun, and the Sun lies 23.5° south of the celestial equator.

As a result of this north–south motion, the Sun's path crosses the celestial equator twice during the year as illustrated in figure 1.12. The dates when the Sun reaches its extreme north and south positions are used to mark the beginning of summer and of winter, while the dates when the Sun crosses the celestial equator mark the beginning of spring and of autumn.

Astronomers give these dates special names. When the Sun is on the celestial equator, the days and nights are of equal length (approximately), so these dates are called the **equinoxes,** for "equal nights." The spring (or vernal) equinox occurs near March 20; the fall or autumnal equinox occurs near September 22. The beginning of summer and of winter mark the times of year when the Sun pauses in its north–south motion and changes direction. Accordingly, these times are called the **solstices,** meaning the Sun (sol) has stopped its motion north or south and is *static* and about to reverse direction. The dates of the solstices (summer and winter) also change slightly from one year to the next, but they are always close to June 21 and December 21.

Tracking the Sun's Changing Position

The motion of the Sun north and south in the sky over the course of the year causes the Sun to follow different paths through the sky each day as the Earth rotates. For a northern observer the Sun is high in the sky at noon on a summer day but low in the sky at noon on a winter day (fig. 1.13A). For example, on June 21 at a midnorthern latitude

The Sun's motion north and south in the sky as the seasons change

Although the seasons begin on the solstices and equinoxes, the hottest and coldest times of year occur roughly 6 weeks after the solstices. The delay, known as the lag of the seasons, results from the oceans and land being slow to warm up in summer and slow to cool down in winter.

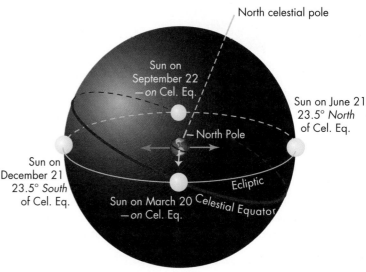

FIGURE 1.12

As the Earth orbits the Sun, the Sun's position with respect to the celestial equator changes. The Sun reaches 23.5° north of the celestial equator on June 21 but 23.5° south of the celestial equator on December 21. The Sun crosses the celestial equator on about March 20 and September 22 each year. The times when the Sun reaches its extremes are known as the solstices; the times when it crosses the celestial equator are the equinoxes. (The dates can vary because of the extra day inserted in leap years.)

of 40°, the noon Sun is about 73.5° above the horizon, or about 16.5° away from the **zenith**—the point in the sky straight overhead. On December 21 at this latitude, on the other hand, the highest point the Sun reaches is only about 26.5° above the horizon. See Astronomy by the Numbers: "The Angle of the Sun at Noon."

Because the Sun moves north and south of the celestial equator during the year, the Sun does *not* rise due east or set due west on most days. Rather, over a year, the direction to the rising and setting position of the Sun constantly changes (fig. 1.13B). On the vernal equinox the Sun is on the celestial equator, so it rises due east and sets due west. From this date up to the summer solstice, the Sun's rising and setting points shift northward each day. From the summer solstice to the winter solstice, the position shifts southward each day, rising and setting due east and due west again on the

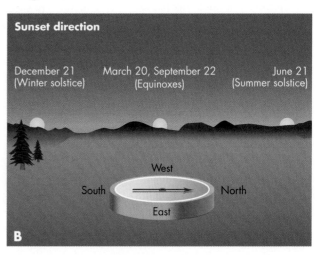

FIGURE 1.13

(A) The shifting location of the Sun north and south of the celestial equator causes it to reach different heights in the sky each day throughout the year. This diagram illustrates the Sun's path in the sky for an observer at about 40° northern latitude. (B) The motion of the Sun throughout the year results in the sunset (and sunrise) position shifting relative to features on the horizon each day.

ASTRONOMY
by the numbers

THE ANGLE OF THE SUN AT NOON

The angle of the Sun above the horizon at noon is almost never straight overhead, contrary to common belief. The only place the Sun ever passes straight overhead is in the tropics (between latitudes 23.5° South and 23.5° North), and this happens on only one or two days each year.

Because the celestial sphere's equator and poles lie directly above the Earth's equator and poles, an observer's zenith is as far north or south of the celestial equator as the observer's latitude is north or south of the Earth's equator. This tells you where the noon Sun will be on the equinoxes, when the Sun is on the celestial equator.

For example, consider Phoenix, Arizona, at latitude 33.5° North. At noon on the equinoxes, the Sun is 33.5° south of the zenith. Because the zenith is by definition 90° above the horizon, this means the Sun is 56.5° above the horizon. And the Sun is never straight overhead.

On the summer solstice in Phoenix, the Sun is 23.5° north of the celestial equator, so it is only 10° from the zenith, or 80° (= 90° − 10°) above the horizon. On the other hand, at the winter solstice, the Sun is 23.5° south of the celestial equator, so it is now 57° south of the zenith (33.5° + 23.5°), or only 33° above the horizon.

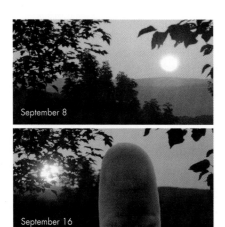

FIGURE 1.14
The sunset position shifted about 4° to the south between these two photos taken 8 days apart in September. The width of the outstretched thumb in the-lower picture indicates a scale of about 2°.

autumnal equinox. After the winter solstice, the Sun begins to move north again. The shift of the Sun's position is particularly obvious near the equinoxes, when the Sun's position on the horizon shifts by almost its own diameter each day (fig. 1.14).

The path the Sun follows each day can be quite different at different latitudes, as illustrated in figure 1.15. At the North Pole the Sun remains above the horizon for half the year, circling the sky above the horizon in each 24-hour period while gradually changing its height above the horizon. At the equator the Sun is up for 12 hours every day of the year, but it reaches its highest point in the sky on the equinoxes rather than one of the solstices. The Sun's path in equatorial regions is almost perpendicular to the horizon, so the Sun seems to set quickly and the period of twilight is short. At the edge of the tropics, the Sun reaches the zenith just on the day of one of the solstices.

Outside of the polar regions, the Sun's rising and setting positions on the horizon shift each day as the Sun travels northward and southward. And just as the changing position of the Sun against the constellations can be used as an indicator of the seasons, so too can the position of the rising and setting Sun. One well-known example is Stonehenge, the ancient stone circle in England (a photograph of which opens this chapter on page 14). Although we do not know for certain how this ancient monument was used, it was laid out so that such seasonal changes in the Sun's position could be observed by noting through which of the stone arches the Sun was visible when it rose or set. For example, on the summer solstice at sunrise, an observer standing at the center of this circle of immense standing stones would see the rising Sun framed by an

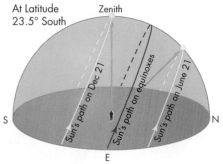

FIGURE 1.15
The path of the Sun in the sky differs depending on your latitude. At the North Pole, the Sun never sets for six months but gradually spirals up from the horizon from the vernal equinox to the summer solstice, then spirals back down to the horizon at the autumnal equinox before it disappears for six months. At the equator, the Sun rises straight upward from the horizon, but reaches the zenith only on the equinoxes. At 23.5° South, the Sun reaches the zenith at noon only on December 21, the start of summer in the Southern Hemisphere.

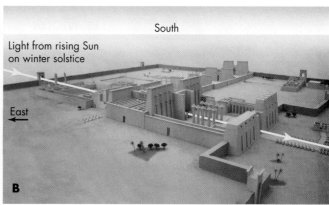

FIGURE 1.16
(A) Stonehenge, built more than 4000 years ago on the Salisbury plain in Britain. The enormous stones are arranged to frame various positions of the Sun on the horizon, helping to mark dates such as when the Sun reaches its point farthest north on the summer solstice. (B) The huge Karnak Temple complex in Luxor was built with its main axis aligned in the direction of the rising Sun on the winter solstice. It was begun almost 4000 years ago, and was expanded repeatedly.

arch, as illustrated in figure 1.16A. Similarly, some ancient Egyptian temples and pyramids have astronomical alignments, such as the Temple of Amun-Ra at Karnak, whose main axis points toward the position of sunrise at the winter solstice (fig. 1.16B).

Structures designed with astronomical alignments were built in many other places as well. For example, in Chankillo, Peru, a series of towers was built on a ridge about 2300 years ago. As viewed from an ancient observatory at the base of the ridge, the towers span the shift on the horizon of the rising Sun (fig. 1.17A). The Maya, native peoples of Central America, and their neighbors built pyramids from the summits of which they could get a clear view of the sky over the surrounding rain forest. The pyramid at Chichén Itzá was specially designed so that on the equinoxes, sunlight would create the image of a snake slithering down the steps (fig. 1.17B).

Many cultures also built monuments that appear to have been used to track another important celestial body: the Moon. Like the Sun, the Moon shifts relative to the stars, and its cyclic changes formed the basis for calendar systems around the world. Some archaeo-astronomers claim that sites such as Stonehenge were used to track the moonrises and moonsets and perhaps even used to predict eclipses.

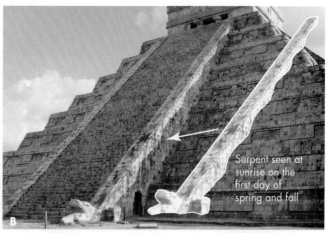

FIGURE 1.17
(A) The oldest known astronomical observatory in the Americas is found in Chankillo, Peru. This ancient observatory marked the shifting position of sunrise with a series of 13 towers built along a ridge about 2300 years ago. (B) At sunrise on the equinoxes, sunlight raking across the edge of the Mayan pyramid at Chichén Itzá creates a shape that resembles a serpent slithering down the steps. The head of the serpent is depicted in a sculpture at the base of the stairs.

1.3 THE MOON

INTERACTIVE

Lunar phases

Like all celestial objects, the Moon rises in the east and sets in the west. Also, like the Sun, the Moon shifts its position across the background stars from west to east. You can verify this motion by observing the Moon at the same time each evening and checking its position with respect to nearby stars. In fact, if the Moon happens to lie close to a bright star, its motion may be seen in a few minutes, because in 1 hour the Moon moves against the sky by approximately its own apparent diameter.

One of the most striking features of the Moon is that, unlike the Sun, its shape seems to change throughout the month in what is called the cycle of lunar **phases.** During a period of approximately 29.5 days, the Moon grows or *waxes* from invisibility (*new* phase), to a *crescent* shape, then *gibbous* when it is more than half lit, until it is a fully illuminated disk (*full*). Next it shrinks or *wanes* backward through this sequence until it is new again (fig. 1.18). This is the origin of the month as a time period and also the source of the name "month," which was derived from the word moon.

The cycle of the phases and the Moon's changing position against the stars are caused by the Moon's orbital motion around the Earth. Many people mistakenly believe that these changes in shape are caused by the Earth's shadow falling on the Moon. This clearly cannot be the explanation, because the crescent phases occur when the Moon and Sun lie approximately in the same direction in the sky and the Earth's shadow must therefore point *away* from the Moon. In fact, half of the Moon is always lit by the Sun, but as the Moon orbits around us, we see different amounts of its illuminated half. When the Moon lies approximately between us and the Sun, its fully lit side is turned nearly completely away from us, and therefore the side facing us is dark, as illustrated in figure 1.18. At the *first quarter* and *third quarter* points, the Moon is 90° from the Sun and appears half lit. When the Moon lies approximately opposite the Sun in the sky, the side of the Moon facing the Earth is fully lit. The alignment is rarely exact, so the Earth's shadow usually misses the Moon.

The Moon's motion around the Earth causes it to shift eastward through the stars. As a result, the Earth itself must rotate eastward a little extra each day to bring the Moon back above the horizon. This extra rotation takes about 50 minutes each day, on average. So if the Moon rises at 8 P.M. one evening, the next evening it will rise at

FIGURE 1.18

The cycle of the phases of the Moon, from new to full and back again. The phases are caused by our seeing different amounts of the half of the Moon's surface that is illuminated by the Sun. Images of the Moon's appearance in different phases are shown at right. Sizes and distances of objects are not to scale. In particular, the Moon is so small and far away that the Earth's shadow rarely falls upon it.

Appearance of the Moon from Earth

ASTRONOMY *by the numbers*

ESTIMATING WHEN THE MOON WILL RISE

If you know the Moon's phase, you can estimate the times when the Moon rises, sets, and is highest in the sky.

For example, when the Moon is at first quarter, it is one-quarter of the way around the sky, eastward of the Sun by about 90° (fig. 1.18). Therefore the Earth must turn about an additional 90° to bring the Moon to approximately the same position as the Sun. How long does it take the Earth to rotate those extra 90°? Since it takes the Earth 24 hours to rotate once (360°), to rotate 90° (= 360°/4) takes 6 hours (= 24 hours/4). Thus, the first-quarter Moon is highest in the sky at 6 hours after noon, or 6 P.M., rises about 6 hours earlier at about noon, and sets at about midnight. With similar reasoning, you can find when the Moon rises and sets in other phases.

As the Moon moves eastward from the Sun and its phase changes, it rises about 49 minutes later each night. This shift is simply the result of the Moon's orbital motion around the Earth, resulting in a complete cycle of phases over 29.5 days: 24 hours/29.5 days = 49 minutes/day.

Because the Moon orbits close to the plane of the ecliptic, it shifts north and south of the celestial equator during the month, just as the Sun does during the year. A consequence of this is that the full Moon's behavior is the opposite of the Sun's—the full Moon is relatively low in the sky in the summer and high in the sky in the winter. The Moon's position north or south of the celestial equator also affects the time between moonrise and moonset, just as the length of days depends on the Sun's position.

about 8:50 P.M., the following night at about 9:40 P.M., and so forth. See Astronomy by the Numbers: "Estimating When the Moon Will Rise."

The changing time of moonrise means that the Moon is visible at different times and places during the night or day depending on its phase. For example, shortly after the new phase you can see the Moon low in the western sky after sunset. A few hours later that same evening it will have set and become invisible. On the other hand, when the Moon is full, it rises at about sunset and doesn't set until dawn. Thus, the full moon is visible throughout the night. In most of its phases, you can see the Moon during some part of the day if you know where to look. The different times when the Moon is visible are explored further in Extending Our Reach: "Observing the Moon."

Because the Moon's orbit is close to the orbital plane of the Earth around the Sun, the Moon, like the Sun, moves through the constellations of the zodiac. While the Moon takes about 29.5 days to go through its cycle of phases, the combination of the Moon's and the Earth's orbits have the effect that the Moon requires only 27.3 days to complete its motion through the constellations of the zodiac. The reason for this is illustrated in figure 1.19, where you can see that after a month has passed the Earth has shifted its position in its orbit, so the Sun is in a different direction. After the Moon comes back into alignment with distant stars in 27.3 days, it must still travel farther around in its orbit two more days to come back into alignment with the Sun.

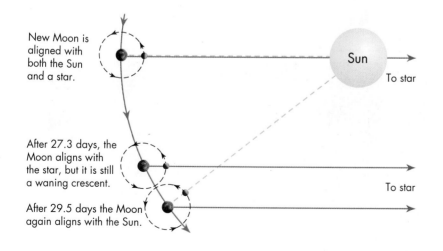

FIGURE 1.19

The sidereal month is the time the Moon takes to complete an orbit relative to the distant stars. This is about 27.3 days, less than the lunar month because as the Moon is orbiting the Earth, the Earth is orbiting the Sun. It takes about two additional days for the Moon to come back in alignment with the Sun.

EXTENDING
our reach

OBSERVING THE MOON

When the Moon is full, it lies approximately opposite to where the Sun lies, but when the Moon is a thin crescent, it lies in nearly the same direction as the Sun (see the middle of figure 1.20). These connections between the Moon's phase and its position with respect to the Sun are the key to understanding when the Moon is visible from Earth.

Because the full Moon is approximately opposite the Sun, it *rises* above the eastern horizon at about the same time that the Sun *sets* below the western horizon. Likewise, the full Moon *sets* at about the time the Sun *rises*. Therefore, the full Moon is visible all night and highest in the sky near midnight.

On the other hand, the crescent moon is not visible during most of the night. Because it lies in nearly the same direction as the Sun, once the Sun is well below the horizon, the crescent Moon must be below the horizon too. Moreover, the crescent Moon is hard to see during the day because it is only a sliver of light, so it is lost in the brightness of the daytime sky. Therefore, when the Moon is a few days past its new phase and is a thin crescent, you can see it low in the western sky at sunset. This crescent moon will set shortly after the Sun and not be visible again until after sunrise the next day.

FIGURE 1.20
Where do you look for the Moon and how does it appear at different times of day as it goes through its monthly cycle of phases? The central diagram shows a person standing on the Earth at five times of day: dawn, morning, afternoon, sunset, and midnight. The Moon's position in its orbit is shown on 7 days of the lunar cycle (days 4, 8, 11, 15, 19, 22, and 25). The five surrounding panels show what a person would see at each of those times of day as the Moon moves through its orbit.

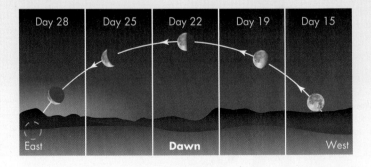

1.4   ECLIPSES

An eclipse occurs when the Earth lies directly between the Sun and the Moon, or when the Moon passes exactly between the Earth and the Sun so that all three bodies are on a straight line. Thus, there are two types of eclipse: lunar and solar.

A **lunar eclipse** occurs when the Earth passes between the Sun and the Moon and casts its shadow on the Moon, as shown in figure 1.21. A **solar eclipse** occurs whenever the Moon passes directly between the Sun and the Earth and blocks our view of the Sun, as depicted in figure 1.22.

INTERACTIVE

Eclipses

Appearance of Eclipses

Eclipses generally take a few hours from start to finish. Sometimes an eclipse is *partial,* with only a portion of the Moon or the Sun ever being covered over. These partial eclipses often pass unnoticed unless you know to look for them. However, *total* eclipses are beautiful and marvelous events.

As the Moon reaches the point along its orbit when it is full, it usually misses the Earth's shadow. If it happens to be crossing the ecliptic when it is full, however, the Moon will pass through the Earth's shadow, and a total lunar eclipse will occur. Total lunar eclipses are visible if you are anywhere on the night side of the Earth when the eclipse is occurring. As a total lunar eclipse begins, the Earth's shadow gradually spreads across the full Moon's face, cutting an ever deeper dark semicircle out of it. The shadow takes about an hour to completely cover the Moon and produce totality. At totality, the Moon generally appears a deep ruddy color, almost as if dipped in blood. Sometimes it becomes so dark that it may be hard to see at all. After totality, the Moon again becomes lit, bit by bit, reverting over the next hour to its unsullied, silvery light.

A little light falls on the Moon even at totality because the Earth's atmosphere bends some sunlight into the shadow. The light reaching the Moon is red because interactions with particles in the air remove the blue light as it passes through our

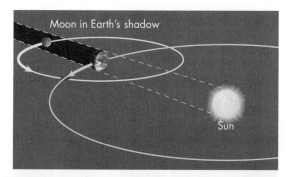

FIGURE 1.21
A lunar eclipse occurs when the Earth passes between the Sun and Moon, causing the Earth's shadow to fall on the Moon. Some sunlight leaks through the Earth's atmosphere, casting a deep reddish light on the Moon. The photo shows what the eclipse looks like from Earth.

FIGURE 1.22
A solar eclipse occurs when the Moon passes between the Sun and the Earth so that the Moon's shadow touches the Earth. The photo shows what the eclipse looks like from Earth.

Q: Sometimes you see clouds after sunset that are lit red. How is this like the red color you see on the totally eclipsed Moon?

FIGURE 1.23
As sunlight falls on the Earth, some passes through the Earth's atmosphere and is slightly bent so that it ends up in the Earth's shadow. In its passage through our atmosphere, most of the blue light is removed, leaving only the red. That red light then falls on the Moon, giving it its ruddy color at totality.

Be extremely careful when watching a partial solar eclipse. Looking at the Sun through improper filters will blind you. A safer way is to *not* look directly at the Sun but to use eyepiece projection to view the Sun. Hold a piece of paper about a foot from the eyepiece of a small telescope (or even binoculars), and a large image of the Sun will be visible on it. This method also allows many people to watch the eclipse simultaneously.

atmosphere, exactly as happens when we see the setting Sun, and the path of the light is bent by the atmosphere much as a prism bends the direction of light, as shown in figure 1.23. (The bending of light by the atmosphere is discussed further in chapter 5.)

It is far rarer to see a total solar eclipse because the Moon's shadow on the Earth is quite small. In fact, you are unlikely to ever see a total solar eclipse in your lifetime unless you travel to see it, because on average they occur in any location only once every several centuries. A total solar eclipse begins with a small black "bite" taken out of the Sun's edge as the Moon cuts across its disk (fig. 1.24A). Over the next hour or so, the Moon gradually covers over more and more of the Sun. While the Sun is only partially covered, you must be careful when viewing it, so you don't hurt your eyes. If you are fortunate enough to be at a location where the eclipse is total, you will see one of the most amazing sights in nature.

As the time when the Moon's disk completely covers the Sun (totality) approaches, the landscape takes on an eerie light. Shadows become incredibly sharp and black: even individual hairs on your head cast crisp shadows. Sunlight filtering through leaves creates tiny bright crescents on the ground. Seconds before totality, pale ripples of light sweep across the ground, and to the west the deep purple shadow of the Moon hurtles toward you at more than 1000 miles an hour. In one heartbeat you are plunged into darkness. Overhead, the sky is black, and stars become visible. Perhaps a solar prominence—a tiny, glowing, red flamelike cloud in the Sun's atmosphere—may protrude beyond the Moon's black disk (fig. 1.24C). The corona of the Sun—its outer

FIGURE 1.24
Pictures of a total solar eclipse in 2010. (A) One hour before totality, the Moon only partially eclipses the Sun. (B) About 5 minutes before totality. (C) With the bright part of the Sun covered, the Sun's glowing pink atmosphere becomes visible. (D) Faint hot gases form a corona around the Sun. (E) As the Moon slides off the Sun, the first glimpse of the bright portion of the Sun makes a "diamond ring," while thin clouds in Earth's atmosphere are colored by optical phenomena.

FIGURE 1.25
(A) When the Moon casts a shadow on the Earth, the Moon's orbit shifts it from west to east along a narrow line. (B) The locations of recent and upcoming total solar eclipses are shown through 2035. The paths show where totality can be observed. In regions outside of these paths, a partial eclipse may be visible.

Q: According to the map, when will the next total solar eclipse occur after 2017 in North America? In South America?

Edlipse paths generated based on data from NASA's website: http://eclipse.gsfc.nasa.gov

atmosphere—gleams with a steely light around the Moon's black disk (fig. 1.24D). Birds call as if it were evening. A deep chill descends, because for a few minutes the Sun's warmth is blocked by the Moon. The horizon takes on sunset colors: the deep blue of twilight with perhaps a distant cloud in our atmosphere glowing orange. As the Moon continues in its orbit, it begins to uncover the Sun, and in the first moments after totality, the partially eclipsed Sun looks a little like a diamond ring (fig. 1.24E). Now the cycle continues in reverse. The sky rapidly brightens, and the shadow of the Moon, racing away to the east, may be glimpsed on distant clouds or mountains.

Total solar eclipses can be seen only within a narrow path where the Moon's shadow crosses the Earth (fig. 1.25A). Because the Moon is physically smaller than the Sun, the Moon's shadow grows narrower farther from the Moon, as illustrated in figure 1.24A, and is at most a few hundred kilometers wide at the distance of the Earth. The locations of the paths of totality are shown for total eclipses from 2008 to 2035 in figure 1.25B. The first total solar eclipse visible in the continental United States since 1979 will occur in 2017, with a path crossing from the northwest to the southeast. If you have the chance to travel to the path of totality, do it!

FIGURE 1.26

An annular eclipse of the Sun in 1992 occurring near sunset. The Moon is at a distant point in its orbit, so it cannot block the Sun entirely.

Eclipses and the Moon's orbital inclination

Sometimes the Moon is so far away that its shadow does not reach the Earth. What we see when this happens is that the Moon does not completely cover the Sun, even though it is precisely in line with the Sun. An example is shown in figure 1.26, where a ring of sunlight is seen as the Sun is setting. This is called an **annular eclipse** because it leaves an *annulus* of the Sun's surface still visible.

Rarity of Eclipses

Given that the lunar cycle is about 29.5 days, you may wonder why we do not have eclipses every month. The answer is that the Moon's orbit is tipped with respect to the Earth's orbit (fig. 1.27). Because of this tip, even if the Moon is new, the Moon's shadow may pass above or below Earth, as you can see in figure 1.27A. As a result, no eclipse occurs. Similarly, when the Moon is full, the Earth's shadow may pass above or below the Moon so that again no eclipse occurs. Only a nearly exact alignment of the Earth, Moon, and Sun leads to eclipses, a point that is easier to appreciate if you look at figure 1.27B, which shows the Earth and Moon and their shadows drawn to scale.

The tilt of the Moon's orbit remains fixed—like that of the spinning Earth—by a gyroscopic effect or, more technically, by the conservation of angular momentum. The result is that twice each year, the Moon's orbital plane (if extended) passes through the Sun, as shown in figure 1.27A. At those times—**eclipse seasons**—eclipses will happen when the Moon crosses the Earth's orbital plane, the ecliptic. In 2012 the eclipse seasons were within about two weeks of the end of May and November. Only at those

FIGURE 1.27

(A) The Moon's orbit keeps approximately the same orientation as the Earth orbits the Sun. Because of its orbital tilt, the Moon generally is either above or below the Earth's orbit. Thus, the Moon's shadow rarely hits the Earth, and the Earth's shadow rarely hits the Moon. Eclipse seasons are when the Earth is in either of two places in its orbit, about 6 months apart, when the Moon's orbital plane, if extended, intersects the Sun. (B) The Earth and Moon are drawn to correct relative size and separation, with their orbits seen here edge on. Note how thin their shadows are.

Table 1.1	Some Upcoming Solar and Lunar Eclipses					
Solar Eclipses			**Lunar Eclipses**			
2016 March 9	Total	Sumatra, central Pacific	2017 August 7	Partial	Europe, Africa, Asia, Australia	
2016 September 1	Annular	Africa, Madagascar	2018 January 31	Total	Asia, Australia, Pacific, w. Americas	
2017 February 26	Annular	S. America, Atlantic, Africa	2018 July 27	Total	S. America, Europe, Africa, Asia, Aus.	
2017 August 21	Total	N. America	2019 January 21	Total	Asia, Australia, Pacific, N. America	
2019 July 2	Total	S. Pacific, S. America	2019 July 16	Partial	S. America, Europe, Africa, Asia, Aus.	
2019 December 26	Annular	Asia, Australia	2021 May 26	Total	Asia, Australia, Pacific, Americas	
2020 June 21	Annular	Africa, S. Europe	2021 November 19	Partial	Americas, Europe, Asia, Aus., Pacific	
2020 December 14	Total	Pacific, S. America	2022 May 16	Total	Americas, Europe, Africa	

Data from NASA's eclipse website: http://eclipse.gsfc.nasa.gov/. Partial solar eclipses and "penumbral" lunar eclipses, are not listed.

times could eclipses happen: at other times, the shadows of the Earth and Moon fall on empty space. You can also see from figure 1.27A that when a solar eclipse occurs at new moon, conditions are right for a lunar eclipse to happen at either the previous or the following full moon. Thus, eclipses can occur in pairs or triplets, with a solar eclipse followed approximately 14 days later by a lunar eclipse, or vice versa. This can be seen in table 1.1 where several upcoming solar and lunar eclipses are listed.

Precession of the Moon's Orbit

Eclipse seasons do not always remain in the same months, because the orientation of the Moon's orbit does not remain exactly the same over time. The plane of the orbit slowly changes orientation, as illustrated in figure 1.28. That is, the Moon's orbit *precesses,* swinging once around about every 18.6 years. This orbital **precession** makes the dates of the eclipse seasons shift by 1/18.6 of a year (about 20 days) each year. Thus, eclipses occurred about 3 weeks earlier in 2015, on average, than in 2014.

If one of the eclipse seasons occurs in early January with the next in June, a third eclipse season may sometimes happen in late December. As a result, as many as seven eclipses, solar and lunar combined, can occur each year. No matter when the eclipse season falls, at least two solar and two lunar eclipses must happen each year, but that does not mean they will be visible to an observer at a given location, since the eclipse may be visible only from another part of the Earth. Most of these eclipses are partial, only partially dimming the Sun or Moon, so they may go unnoticed even where they are visible.

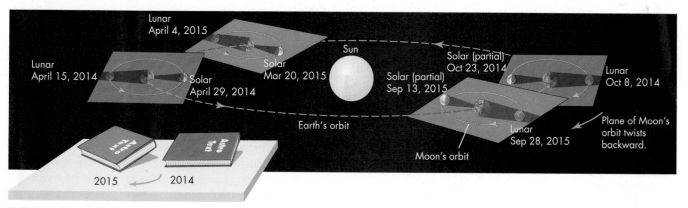

FIGURE 1.28
Precession of the Moon's orbit causes eclipses to come a few weeks earlier (on average) each year. The shift of the orbital plane is similar to twisting a tilted book that has one edge resting on a table, as illustrated in the inset diagram. (Sizes and separations are not to scale.)

SUMMARY

The night sky looks like a giant dome, which we model as part of a *celestial sphere*. Star patterns on the celestial sphere are called constellations. According to this model, stars rise in the east and set in the west as the celestial sphere rotates around the Earth. This apparent motion is actually caused by the Earth's spin.

The Sun's glare hides the stars behind it. However, as the Earth orbits the Sun, the Sun changes its position with respect to the stars, making different constellations visible at different times of year. The path that the Sun follows around the celestial sphere is called the ecliptic, and the 12 constellations close to the ecliptic are called the zodiac.

The ecliptic is tipped at an angle of 23.5° to the celestial equator because the Earth's rotation axis is tipped by that amount with respect to its orbit. The solstices and equinoxes mark when the Sun reaches its maximum distance from the celestial equator and when it crosses the equator, respectively. These dates define the onsets of the seasons.

The Earth's spin keeps its rotation axis pointing in nearly a fixed direction as we orbit the Sun. Because the axis is tipped, the Sun shines more directly on the Northern Hemisphere for half the year and on the Southern Hemisphere for the other half of the year. This difference in exposure to the Sun's light and warmth creates the seasons.

Ancient peoples built monuments to trace the motions of the Sun through the seasons. They also tracked the position of the Moon, which moves through a cycle of phases every 29.5 days. The plane of the Moon's orbit around the Earth is at a small angle to Earth's orbital plane around the Sun (the ecliptic). When a new or full Moon is close to the ecliptic, there can be a solar or lunar eclipse, respectively.

Because of the small size of the Moon relative to the Earth, the full Moon can be completely in the Earth's shadow during a lunar eclipse, but during a solar eclipse the Moon's shadow covers only a narrow path across the Earth. The dates when the orbital planes of the Moon and the Earth cross are called eclipse seasons, which gradually shift as the orientation of the Moon's orbit changes over time.

QUESTIONS FOR REVIEW

1. (1.1) What is the celestial sphere? What are the celestial equator and the ecliptic?
2. (1.1) What is the difference between rotation and revolution?
3. (1.1/1.2) What is a constellation, and what is special about the zodiac constellations?
4. (1.2) What causes the seasons?
5. (1.3) What causes the Moon's phases?
6. (1.3) How long does it take the Moon to go through a cycle of phases?
7. (1.4) What is the difference between lunar and solar eclipses?
8. (1.4) Why aren't there eclipses each month?

6. (1.2) If the shape of the Earth's orbit were unaltered but its rotation axis were shifted so that it had no tilt with respect to the orbit, how would seasons be affected?
7. (1.2) Why does the position of sunrise along the eastern horizon change during the year?
8. (1.2) Why do we have time zones? Sketch and label a diagram to justify your answer.
9. (1.3) Provide two or three pieces of evidence you could use to explain to someone that the Moon's phases are not caused by the Earth's shadow.
10. (1.3) If the Moon orbited the Earth in the opposite direction, but everything else remained the same, how would the sidereal and solar months change (if at all)? Create a drawing like figure 1.19 representing this situation.

THOUGHT QUESTIONS

1. (1.1) If you were standing on the Earth's equator, where would you look to see the north celestial pole? Could you see this pole from Australia?
2. (1.1) Draw a sketch of the Earth and a distant North Star, and show that your latitude is the angle of the north celestial pole above the northern horizon.
3. (1.1) Can you think of an astronomical reason why the zodiac may have been divided into 12 signs rather than 8 or 16?
4. (1.1) Draw sketches to show the angles setting stars would make relative to the horizon for someone watching at the equator, the north pole, and a midlatitude.
5. (1.1/1.2) When it is winter in New York, what season is it in Australia, and in Paris? If you see Orion in the evening in New York, would you see it in the evening in Australia or Paris?

PROBLEMS

1. (1.1) If the Earth turns one full rotation in approximately 24 hours, how many degrees per hour does the sky turn?
2. (1.2) From a latitude of 55°, what is the highest and lowest altitude above the horizon of the noon Sun? What will be the altitude on September 22?
3. (1.3) Make a sketch to calculate what times the waxing crescent moon will rise and set. Indicate the observer's location and lines of sight to the Moon for these times.
4. (1.3) Calculate how many degrees the Moon moves in its orbit in one day based on its 27.3 day period relative to the stars. Use this result and the answer to problem 1 to determine how much later the Moon rises each day.
5. (1.3/1.4) The Moon crosses down through the ecliptic every 27.21222 days ("draconic period"). Its synodic period, the

period of the phases, is 29.5306 days. Show that 242 draconic periods very nearly equals 223 synodic periods. How long is this in years? What does this suggest about eclipses and why? (This match of cycles is called the saros and was used by ancient astronomers to predict eclipses.)

6. (1.4) Find how many hours it takes the Moon to move in its orbit a distance equal to the Earth's diameter. (You will need to determine the speed of the Moon in its orbit. You can find values for the diameter of the Earth and the radius and period of the Moon's orbit in the appendix.) How does this relate to the time it takes for a lunar eclipse to occur?

7. (1.4) List some of the details left out of problem 5 that you would need to consider to exactly calculate the length of an eclipse. What effect would each have on the final answer?

8. (1.4) The Moon's shadow at the Earth is much smaller than the Moon's diameter—it is only a few hundred kilometers wide. Is the Moon's speed still a good estimate of how fast the shadow moves? Repeat problem 6 to estimate the duration of a solar eclipse.

TEST YOURSELF

1. (1.1) If you are standing at the Earth's North Pole, which of the following will be directly overhead?
   (a) The celestial equator   (d) The north celestial pole
   (b) The ecliptic   (e) The Sun
   (c) The zodiac

2. (1.1) If you observe Polaris to be 30° above the horizon, you are at a latitude of approximately
   (a) 6.5°   (b) 30°   (c) 53.5°   (d) 60°   (e) 83.5°

3. (1.1/1.2) For this question, choose as many answers as are correct. If the Earth reversed its direction of spin,
   (a) the Sun would rise in the west and set in the east.
   (b) the seasons would be reversed.
   (c) the stars would circle Polaris clockwise.
   (d) the Moon would rise in the west and set in the east.
   (e) the Moon would rise in the east and set in the west.

4. (1.2) In the Northern Hemisphere, summertime is warmer than wintertime because
   (a) the Earth's orbit is an ellipse.
   (b) the Sun is visible for more hours.
   (c) sunlight is more concentrated on the ground.
   (d) both b and c.
   (e) All the answers are true.

5. In which of the following locations can the length of daylight range from zero to 24 hours?
   (a) Only on the equator
   (b) At latitudes closer than 23.5° to the equator
   (c) At latitudes between 23.5° and 66.5° north or south
   (d) At latitudes greater than 66.5° north or south
   (e) Nowhere on Earth

6. (1.3) If the Moon is waning gibbous in Chicago, then that night in Australia the Moon will be
   (a) waxing crescent.   (c) waxing gibbous.
   (b) waning gibbous   (d) waning crescent.

7. (1.3) You observe the Moon rising at 6 P.M., around sunset. Its phase is
   (a) 1st quarter   (b) new   (c) full   (d) 3rd quarter

8. (1.3) You observe the Moon rising at 3 P.M., a few hours before sunset. Its phase is
   (a) between new and first quarter.
   (b) between first quarter and full.
   (c) between full and third quarter.
   (d) between third quarter and new.

9. (1.3) If you see a full moon at midnight, about how long will it be until there is a new moon?
   (a) 12 hours   (b) 3 days   (c) 2 weeks   (d) 6 months

10. (1.4) Figure 1.22 (right) shows an eclipse of the Sun. The black circle in the middle of the photo is
    (a) the Earth's shadow on the Sun.
    (b) the Sun's shadow on the Moon.
    (c) the Moon covering the Sun.
    (d) the Earth's shadow on the Moon.
    (e) a dark cloud in our atmosphere.

11. (1.4) If the Moon were to expand to twice its current diameter, we would have total solar eclipses
    (a) every month.
    (b) more often than now but less often than every month.
    (c) never.
    (d) occasionally, but less often than now.

KEY TERMS

annular eclipse, 32   lunar eclipse, 29
celestial equator, 18   model, 16
celestial poles, 17   phase, 26
celestial sphere, 16   precession, 33
constellation, 16   rotation axis, 20
eclipse season, 32   solar eclipse, 29
ecliptic, 19   solstice, 22
equinox, 22   zenith, 23
horizon, 15   zodiac, 19

𝒬: FIGURE QUESTION ANSWERS

WHAT IS THIS? (chapter opening): The figure at the start of the chapter shows the Moon's shadow on the Earth's surface. The shadow is usually a few hundred kilometers across. People within the region of the shadow would be able to see a total solar eclipse.

FIGURE 1.4: Counterclockwise.

FIGURE 1.23: In both cases they are lit by sunlight that has passed through our atmosphere, which has removed most of its blue light.

FIGURE 1.25: 2024 for North America; 2019 for South America.

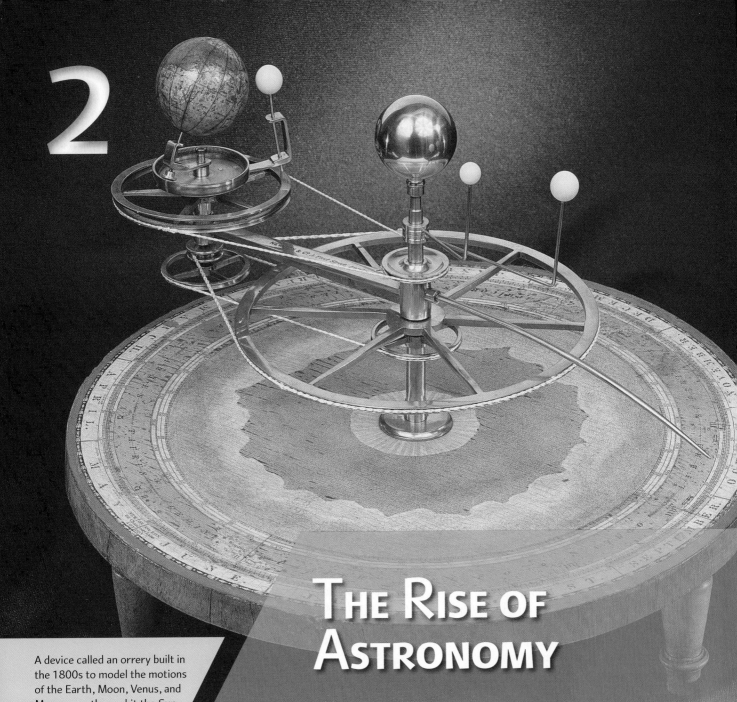

2

THE RISE OF ASTRONOMY

A device called an orrery built in the 1800s to model the motions of the Earth, Moon, Venus, and Mercury as they orbit the Sun.

LEARNING OBJECTIVES

Upon completing this chapter you should be able to:

- Explain the different lines of simple observational evidence that prove the Earth is round.
- Carry out the kind of calculation that Eratosthenes used to measure the size of the Earth.
- Show how the relative distances and sizes of the Moon and Sun can be estimated from basic observations.
- Explain why ancient astronomers thought the Earth was at the center of the Universe, and describe what they thought planets were and how they explained planets' motions.
- Explain Copernicus's arguments that the Earth is a planet orbiting the Sun, and explain how his reasoning accounts for planets' retrograde motion.
- Describe the characteristics of planetary orbits discovered by Kepler as given by his three laws.
- Calculate the period of a planet's orbit from its semimajor axis, or calculate its semimajor axis from its period.
- Describe Galileo's telescopic observations, and discuss why these were so upsetting to ancient beliefs about the nature of the Universe.
- Describe the general trends in the development of astrophysics in the centuries after Kepler and Galileo.

Our understanding of the Universe has been assembled bit by bit from many separate discoveries—discoveries made by scientists from many parts of the world, at many times in the past, and in many disciplines. How those discoveries led to our current knowledge is the subject of this chapter.

The astronomical phenomena that we discussed in chapter 1 (the rising and setting of Sun, Moon, and stars; the constellations; annual motion of the Sun; phases of the Moon and eclipses) were the basis of ancient knowledge of the heavens. With these observations, we can now describe people's early attempts to explain the heavens. We will see that some of their conclusions were incorrect, just as we today are probably in error about some aspects of modern astronomy. We study ancient ideas of the heavens not so much for what they tell us about the heavens but to learn how observation and reasoning can lead us to an understanding of the Universe.

Much of what we know about the Universe can be shown by carrying out simple observations and making a few logical deductions. For example, by observing the shape of the Earth's shadow during a lunar eclipse, it is possible to deduce the shape of the Earth and its size relative to the Moon. This was understood by ancient Greek philosophers more than 2000 years ago. It is only a myth that the Earth was widely believed to be flat until recent times.

Astronomers of classical times determined a remarkable amount about the Earth, Moon, Sun, and stars. However, they struggled to understand the motions of the planets. The puzzling motions of these objects in the sky finally forced humans to consider the possibility that they did not live at the center of the Universe. This revolution of thinking during the Renaissance led to the development of new mathematical and scientific ideas and the birth of astrophysics.

CONCEPTS AND SKILLS TO REVIEW

- The seasonal motion of the Sun (1.2)
- The Moon's motions and phases (1.3)
- Lunar eclipses (1.4)

2.1   EARLY IDEAS OF THE HEAVENS: CLASSICAL ASTRONOMY

The ancient Greek astronomers of classical times were some of the first to try to explain the workings of the heavens in a careful, systematic manner, using observations and models. Given the limitations of naked-eye observation, these astronomers were extraordinarily successful, and their use of logic, mathematics, and geometry as tools of inquiry created a method for studying the world around us that we continue to use even today. This method is in many ways as important as the discoveries themselves.

The Shape of the Earth

The ancient Greeks knew that the Earth is round. As long ago as about 500 B.C., the mathematician Pythagoras (about 560–480 B.C.) was teaching that the Earth is spherical, but the reason for his belief was as much mystical as rational. He, like many of the ancient philosophers, believed that the sphere was the perfect shape and that the gods would therefore have utilized that perfect form in the creation of the Earth.

By the fourth century B.C., however, Aristotle (384–322 B.C.) was presenting arguments for the Earth's spherical shape that were based on simple naked-eye observations that anyone could make. Such reliance on careful, firsthand observation was the first step toward acquiring scientifically valid knowledge of the contents and workings

FIGURE 2.1

(A) A sequence of photographs during a partial lunar eclipse. The edge of the Earth's shadow on the Moon is always a portion of a circle, showing that the Earth must be round. (B) As a traveler moves from north to south on the Earth, different stars become visible. Some stars that were previously hidden become visible above the southern horizon. This variation would not occur on a flat Earth.

FIGURE 2.2

A sequence of photos taken from a boat traveling away from Boston. Note that the tops of the tallest buildings remain visible as the bottom parts and shorter buildings disappear over the curved ocean surface.

of the Universe. For instance, Aristotle noted that if you look at an eclipse of the Moon when the Earth's shadow falls upon the Moon, the shadow can be clearly seen as curved, as figure 2.1A shows. He wrote in his treatise "On the Heavens":

> The shapes that the Moon itself each month shows are of every kind—straight, gibbous, and concave—but in eclipses the outline is always curved: and, since it is the interposition of the Earth that makes the eclipse, the form of this line will be caused by the form of the Earth's surface, which is therefore spherical.

Another of Aristotle's arguments that the Earth is spherical was based on the observation that a traveler who moves south will see stars that were previously hidden below the southern horizon, as illustrated in figure 2.1B. For example, the bright star Canopus is easily seen in Miami but is invisible in Boston. This could not happen on a flat Earth.

It was also observed that as ships sailed away from port, the lower parts of the ships would disappear below the horizon while the sails remained visible. Today you can see this phenomenon if you travel away from a city across the ocean: the bottoms of buildings disappear below the horizon, while the tops remain visible (fig. 2.2). If the surface of the ocean were flat, the bottom of a building (or a ship) would remain visible at any distance. Therefore the surface of the ocean must be curved.

Distances and Sizes of the Sun and Moon

About a century after Aristotle, Aristarchus of Samos (an island in the Mediterranean) used geometric methods to estimate the relative sizes of the Earth, Moon, and Sun, and the relative distances to the Moon and Sun. His values for these numbers were not very accurate, but they were the best estimates for almost 2000 years, and gave at least the correct sense of the order of sizes and distances of these bodies compared to the Earth.

Aristarchus estimated the relative distances of the Moon and the Sun through a clever bit of reasoning. He realized that when the Moon appears exactly half lit (first or third quarter), as shown in figure 2.3, the Sun must be shining down on the Moon at an angle exactly 90° to our line of sight. However, if the Sun were only a few times farther away than the Moon, as sketched in figure 2.3, we would observe an angle between the Sun and the Moon much less than 90° at these phases. What Aristarchus found is that the half-lit Moon is only slightly less than 90° from the Sun, so the Sun

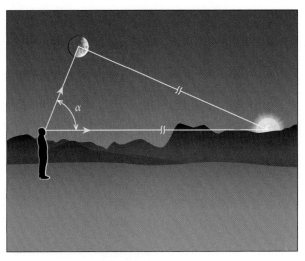

FIGURE 2.3

Aristarchus estimated the relative distance of the Sun and Moon by observing the angle between the Sun and the Moon (α in the diagram) when the Moon is exactly half lit. Angle β must be 90° for the Moon to be half lit. By observing the angle α, he could then set the scale of the triangle and thus the relative lengths of the sides. (Sizes and distances are not to scale.)

must be much farther away than the Moon. He estimated 20 times farther away. Today we know the Sun is much farther away than that, about 400 times the Moon's distance. The problem is not with the method but with the difficulty in making measurements that are accurate enough with just the unaided eye. The important thing was that Aristarchus showed that the Sun is much more distant than had been previously suspected.

If we know the relative distances of the Sun and Moon, we can also determine their relative sizes. Recall that the Moon just barely covers the Sun during a total solar eclipse (chapter 1), so the two orbs appear to be about the same size in the sky. Astronomers call this apparent size of an object its **angular size,** as is illustrated in figure 2.4. The Sun and the Moon both have an angular size of about ½°. (Note that our perception of angular sizes is not always reliable, as discussed in Extending Our Reach: "The Moon Illusion," so it is important to measure them with appropriate instruments.) If the Sun were 20 times farther from us than the Moon, for example, to have the same angular size, it would have to be 20 times bigger than the Moon (fig. 2.4).

Aristarchus further realized that he could estimate the Moon's size by comparing it to the size of the Earth's shadow during a lunar eclipse, as illustrated in figure 2.5. He carried out his measurement by timing how long the Moon took to cross Earth's shadow, and estimated that the Moon's diameter is about 0.35 times the Earth's. This

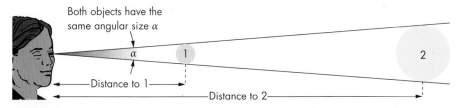

FIGURE 2.4

The angle that an object covers from an observer's point of view is called its angular size. Note that a larger object at a larger distance may have the same angular size as a nearer, smaller object.

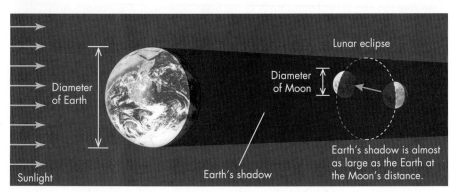

FIGURE 2.5

Aristarchus used the size of the Earth's shadow on the Moon during a lunar eclipse to estimate the relative size of the Earth and Moon.

EXTENDING *our reach*   THE MOON ILLUSION

The Moon sometimes appears to be huge when you see it rising. In fact, if you measure the Moon's angular diameter carefully, you will find it to be smaller when it is near the horizon than when it is overhead, regardless of how huge it looks. This misperception, known as the **Moon illusion,** is still not completely understood but is an optical illusion caused, at least in part, by the observer's comparing the Moon with objects seen near it on the horizon, such as distant hills and buildings. You know those objects are big even though their distance makes them appear small. Therefore, you unconsciously magnify both them and the Moon, making the Moon seem larger. You can verify this

sense of illusory magnification by looking at the Moon through a narrow tube that blocks out objects near it on the sky line. Seen through such a tube, the Moon appears to be its usual size.

Figure 2.6 shows a similar effect. Because you know that the rails are really parallel, your brain ignores the apparent convergence of the railroad tracks and mentally spreads the rails apart. That is, your brain provides the same kind of enlargement to the circle near the rails' convergence point as it does to the rails, causing you to perceive the middle circle as larger than the lower one, even though they are the same size.

FIGURE 2.6
Circles beside converging rails illustrate how your perception may be fooled. The bottom circle looks smaller than the circle on the horizon but is in fact the same size. Similarly, the circle high in the sky looks smaller than the circle on the horizon.

It is also possible to estimate the size of the Earth's shadow during a lunar eclipse by looking at the curvature of the Earth's shadow on the Moon. Look again at the opening "What is this?" picture at the start of the chapter.

is a slight overestimate, because at the distance of the Moon the Earth's shadow is actually a little smaller than the Earth itself. We now know that the correct ratio of the bodies' diameters is about 0.27, so the Moon's diameter is about 1/4 that of the Earth.

These observations equipped Aristarchus with enough information to estimate the size of the Sun relative to the Earth. By his measurements, the Moon was 0.35 times as big as the Earth, but the Sun was 20 times bigger than the Moon. This meant that the Sun was about 7 (= 20 × 0.35) times larger than the Earth. Today we know the Sun is even bigger, more than 100 times Earth's diameter, but Aristarchus was the first to show that the Sun is the largest body in the Solar System. It was perhaps his recognition of the vast size of the Sun that led Aristarchus to the idea that the Earth orbits the Sun. Aristarchus was right, of course, but his idea was too revolutionary, and another 2000 years passed before scientists became convinced of its correctness.

Arguments for an Earth-Centered Universe

In ancient Greek times there was a good reason for not believing that the Earth moves around the Sun. If it did, argued the critics of Aristarchus, the positions of stars should change during the course of the year. Looking at figure 2.7, you can see two examples of why they expected to see effects of the Earth's motion.

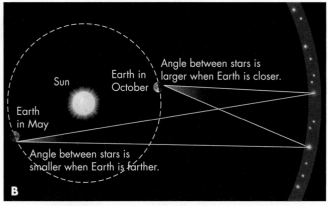

FIGURE 2.7

Most ancient astronomers argued against the idea of the Earth revolving around the Sun because: (A) if some stars are nearer than other stars, we would see their positions appearing to shift relative to their neighbors (stellar parallax) as the Earth moved around the Sun; and (B) even if all the stars lay at the same distance (on the celestial sphere), as Earth orbited we would sometimes be closer to the stars and sometimes farther, so the angular size of constellations would change. Neither effect was seen because stars are so tremendously distant.

If some stars are nearer than others, they would shift against background stars due to Earth's changing perspective (fig. 2.7A). This apparent shift in position of a foreground star relative to the background is called the star's **parallax.** Even if all stars lay at the same distance on the celestial sphere, as the Earth moved closer and farther from stars forming a constellation on one part of the celestial sphere, the angular size of constellation would appear to change (fig. 2.7B).

Aristarchus's critics were absolutely right in supposing that these shifts in stars' positions should occur. So, when they did not observe any effects caused by the Earth's motion, they concluded that Aristarchus's Sun-centered system must be wrong. But what no one appreciated at the time was how tiny these shifts would be.

The size of the parallax shift grows smaller the farther away a star is, but the ancient Greeks did not imagine that stars could be so enormously far away that their parallaxes would be imperceptible to the human eye. In Aristarchus's time, about 2000 years before the telescope was invented, there was no hope of detecting the parallax of stars. It was not until 1838 that astronomers had telescopes of sufficient accuracy to measure the nearest stars' parallaxes. Thus Aristarchus's idea was rejected for reasons that were logically correct but were based on inaccurate data.

The Size of the Earth

Even though Aristarchus had established a great deal about the *relative* sizes and distances of astronomical bodies, his methods could not say whether the Moon was a thousand or a million miles across. All the sizes were related to Earth's diameter, so if that could be measured, then the other sizes would be known. But how was it possible to find the diameter of the Earth in an age long before there were means of circling the globe? It required another remarkable piece of geometry and deduction to reveal the true physical dimensions of the cosmos.

Eratosthenes (276–195 B.C.), head of the famous Library of Alexandria in Egypt, succeeding in making the first measurement of the Earth's size. He obtained a value for its circumference of about 25,000 miles, remarkably close to its actual value. Eratosthenes's demonstration is one of the most beautiful ever performed. Because it so superbly illustrates how science links observation and logic, the demonstration is worth describing in some detail.

By ancient Greek times, astronomers were very well acquainted with the yearly movement of the Sun and could predict accurately the times of the solstices and

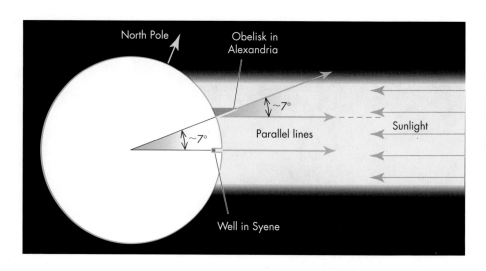

FIGURE 2.8
Eratosthenes's calculation of the circumference of the Earth. The Sun is directly overhead at local noon on the summer solstice at Syene, in southern Egypt. On that same day and time, Eratosthenes found the Sun to be 1/50th of a circle (about 7°) from the vertical in Alexandria, in northern Egypt. Eratosthenes deduced that the angle between two verticals placed in northern and southern Egypt must be 1/50th of the circumference of the Earth.

equinoxes (chapter 1). The summer solstice marked the day of the year in Alexandria when the Sun would reach its highest point in the sky at noon. However, the Sun was not straight overhead but still cast a shadow at noon. Eratosthenes, a geographer as well as an astronomer, heard that lying to the south, in the Egyptian town of Syene (the present city of Aswan), the Sun would be directly overhead at noon and cast no shadow. Proof of this was the fact that at that time the Sun shone straight down a deep well near there.

Appreciating the power of geometry, Eratosthenes realized he could deduce the circumference of the Earth. He analyzed the problem as follows: Because the Sun is far away from the Earth and much larger, as shown by Aristarchus, its light travels in nearly parallel rays toward the Earth. Thus, two rays of sunlight, one hitting Alexandria and the other shining down the well, are parallel lines, as depicted in figure 2.8, and the ray hitting the well in Syene would be aimed directly toward the center of the Earth.

Now imagine drawing a straight line from the center of the Earth outward so that it passes vertically through the Earth's surface in Alexandria. The angle between that line and the direction of the Sun's rays in Alexandria is the same as the angle between that line and the line from the center of the Earth up through the well in southern Egypt (fig. 2.8). The reason is that a single line crossing two parallel lines forms the same angle to both (a geometric theorem).

The angle between sunlight and vertical directions in Alexandria can be measured with sticks and a protractor (or its ancient equivalent) and is the angle between the direction to the Sun and the vertical to the ground (fig. 2.8). Eratosthenes found this angle to be about 1/50th of a circle. Therefore the angle formed by a line from Alexandria to the Earth's center and a line from the well to the Earth's center must also be 1/50th of a circle.

To find the circumference of the Earth, all that is needed is to find the distance between Alexandria and the well, which represents 1/50th of the distance around the Earth. Soldiers marching between Alexandria and Syene estimated the distance to be 5000 stadia (where a stadium is about 0.1 mile), so the distance around the entire Earth is 50 × 5000 stadia, or 250,000, stadia. When expressed in miles, this is roughly 25,000 miles, close to the circumference of the Earth as we know it today.

Eratosthenes's measurement of the Earth's size was a triumph of logic and the scientific technique, and with it we have the key to the sizes of the Moon and the Sun. Furthermore, because there is a relationship between angular size, physical size, and distance, this measurement provides enough information to determine the Moon's and Sun's immense distances. This is worked out in detail in Astronomy by the Numbers: "The Diameter–Distance Relation of Astronomical Objects."

You can use Eratosthenes's technique yourself to measure the size of the Earth by collaborating with someone at a known distance north or south of you, and comparing the difference in angle of the noontime Sun.

ASTRONOMY
by the numbers

THE DIAMETER–DISTANCE RELATION OF ASTRONOMICAL OBJECTS

We can find an astronomical body's true diameter from its angular diameter if we know its distance, or its distance if we know its diameter. We need either the body's distance or its diameter because angular size changes with both. For example, a building looks big when it is near us and small when it is far away, as shown in figure 2.9. And, of course, a larger building also appears bigger. Furthermore, it is easy to verify that the angular size of a distant object changes inversely with the object's distance. That is, if we double the distance to an object, its angular size is halved.

FIGURE 2.9
How angular size varies with distance.

To find an object's true diameter from its angular diameter and distance, imagine we are at the center of a circle passing through the object, as illustrated in figure 2.10. Let ℓ be the diameter of the body and d the distance to the body, which is the *radius* of the circle in the figure. Next draw lines from the center to each end of ℓ, letting the angle between the lines be α, the object's angular diameter.

We now determine the object's true size, ℓ, by forming the following proportion: ℓ is to the circumference of the circle as α is to the total number of degrees around the circle, which we know is 360°. Thus,

$$\frac{\text{Object's diameter}}{\text{Circumference}} = \frac{\text{Angle between lines}}{360°}$$

$$\frac{\ell}{\text{Circumference}} = \frac{\alpha}{360°}$$

However, we know from geometry that the circle's circumference is $2\pi d$, so

$$\frac{\ell}{2\pi d} = \frac{\alpha}{360°}$$

Multiplying both sides of the equation by $(360°d/\alpha)$, we can now solve for d, and find that

$$d = \frac{360°\ell}{2\pi\alpha}$$

Thus, given a body's actual and angular diameters, we can calculate its distance. For example, suppose we apply this method to measure the Moon's distance from the ancient Greek measurements. We stated previously that the Moon's angular diameter is about 1/2°, while its diameter is 0.27

the Earth's, or about 2100 miles. Therefore its distance is about

$$d = \frac{(360°)(2100 \text{ miles})}{2\pi(0.5°)} = \text{about 240,000 miles}$$

or about 380,000 kilometers.

We can work another example to find a diameter from a distance. We know the angular diameter of the Sun is also about 1/2°, and the Sun's distance is today known to be about 150 million kilometers. The Sun's diameter must therefore be

$$\ell = \frac{2\pi d\alpha}{360°} = \frac{2\pi(150{,}000{,}000 \text{ km})(0.5°)}{360°}$$

$$= \text{about } 1{,}300{,}000 \text{ km}$$

The Sun is more than a million kilometers across!

$$\frac{\ell}{\text{Circumference}} = \frac{\ell}{2\pi d} = \frac{\alpha}{360°}$$

therefore, $\ell = 2\pi d \times \dfrac{\alpha}{360°}$

FIGURE 2.10
How to determine linear size from angular size.

2.2 THE PLANETS

Many ancient cultures noted that there are five bright "stars" visible in the night sky that do not stay fixed relative to the rest of the stars in the sky. The Greeks called them *plane¯tai,* meaning "wanderers," from which our word *planet* comes.

Because these wandering stars seemed to have a will of their own, many cultures named the planets after divine beings. The names we use for them today come from Greco-Roman mythology. Mercury, named for the fleet-footed messenger god, is seen always near the Sun, switching back and forth between the evening and morning skies half a dozen times each year. Venus, named for the goddess of love and beauty, spends about 9 months as the brightest star gracing the evening sky, then 9 months in the predawn sky, then back again. Mars is probably named for the god of war because of its blood-red color. Jupiter, named for the king of the gods, shines steadily as one of the brightest stars, moving at a stately pace among the stars. Saturn, usually the faintest and slowest-moving of the planets, was Jupiter's father in mythology, cast into the deepest recesses when his son overthrew him.

Today we know that the planets move across the background stars because of a combination of the Earth's and their own orbital motion around the Sun. One of the more striking features of this motion is that the planets always remain close to the ecliptic, within the constellations of the zodiac. The motion of the planets lies in the same narrow zone as the Sun because their orbits, like that of the Earth, all lie in nearly the same plane, as illustrated in figure 2.11. Thus, like the path of the Sun through the stars, the paths of the planets are tilted by about 23.5° to the celestial equator, moving into our northern and southern skies depending on their position in their orbits.

The motions of the planets relative to the stars are gradual, detectable only through observations over many nights. Therefore, like the Sun, the planets rise and

FIGURE 2.11
To the naked eye, the planets look like bright stars that "wander" through the sky. Although they move, they always remain near the ecliptic in the constellations of the zodiac, like the Sun and the Moon.

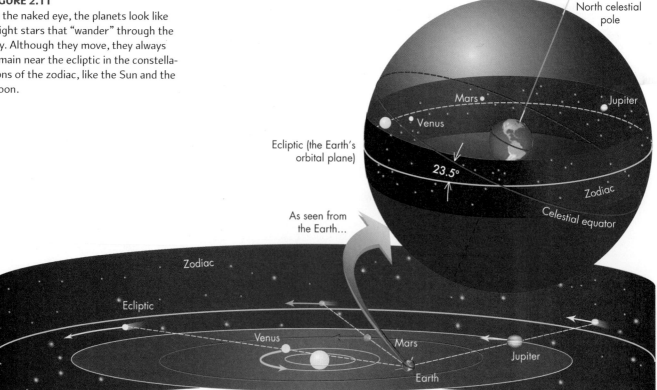

set each day—reflecting, of course, the rotation of the Earth. The motion of a planet through the zodiac can be seen by marking off its position on the celestial sphere over a period of a week or more. Figure 2.12 illustrates such a plot and shows that planets normally move eastward through the stars as a result of their orbital motion around the Sun.

Although planets usually move from west to east through the stars, this does *not* mean that they rise in the west and set in the east. As seen from Earth, planets *always* rise in the east and set in the west because they are carried across the sky—just as the stars are—by the Earth's rotation. However, the motion of the planets is usually slower than that of the stars because their orbital motion partly offsets the rotation of the Earth that causes this apparent motion of the stars. Generally, when we observe a star and a planet rising side by side, at some later time that evening the planet will not be as far above the horizon as the star. Therefore, *with respect to the stars,* the planet has moved to the east because of its orbital motion around the Sun.

This simple pattern of movement is sometimes interrupted. Occasionally a planet will move west with respect to the stars, a condition known as **retrograde motion** and shown in figure 2.13. The word *retrograde* means "backward," and when a planet is in retrograde motion, its path through the stars bends backward, sometimes even forming a loop, for a few months. All planets undergo retrograde motion for a portion of their paths around the sky. This motion greatly complicates the otherwise straightforward idea that the celestial sphere and its bodies rotate around the Earth. In fact, the search for a simple, plausible explanation of retrograde motion was what led astronomers ultimately to reject models of the Solar System with the Earth at the center.

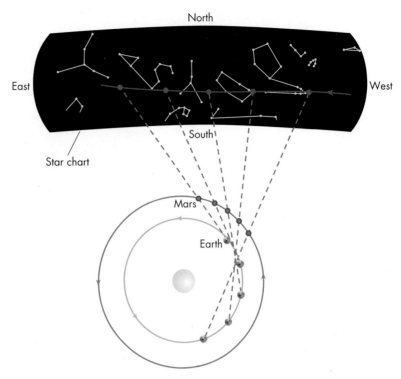

FIGURE 2.12

A planet's eastward drift against the background stars plotted on the celestial sphere. Note: Star maps usually have east on the left and west on the right, so that they depict the sky when looking south.

Retrograde motion

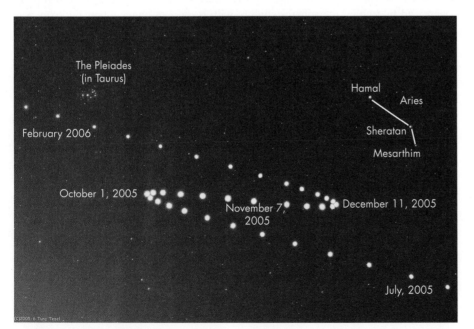

FIGURE 2.13

A sequence of images of Mars made in late 2005, showing its motion relative to the background stars. The pictures were taken roughly a week apart. Mars underwent retrograde motion in October and November of that year.

Q: Why does the brightness of Mars change in the image? (Hint: Draw a sketch of the positions of Mars and the Earth as Mars undergoes retrograde motion.)

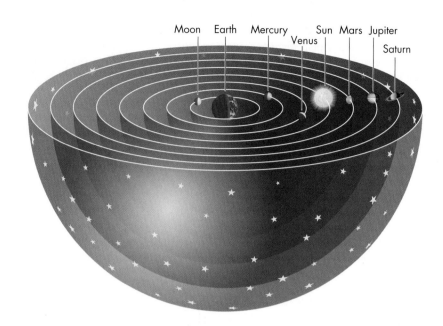

Moon   Earth   Mercury   Venus   Sun   Mars   Jupiter   Saturn

FIGURE 2.14

A cutaway view of the geocentric model of the Solar System according to Eudoxus.

Explaining the Motion of the Planets

Following the basic discoveries about the size and distance of the Sun and Moon, the main thread of astronomical research for almost the next 2000 years centered on the motion of the planets. The Sun, Moon, stars, and planets appear to move around the Earth, rising in the east and setting in the west once a day with slight differences in timing. The earliest models placed the Earth at the center of the Universe with all other bodies revolving around it. Descriptions of the Universe of this type are called **geocentric** models.

Figure 2.14 shows an early geocentric model based on the work of the Greek astronomer Eudoxus, who lived about 400–347 B.C. In this model, the celestial bodies all lie on transparent spheres that revolve around the Earth. The bodies that move fastest across the sky are those that are nearest to the Earth. Thus, the Moon, whose path through the stars takes only about 27 days, is nearest to the Earth, whereas Saturn, whose path through the stars takes roughly 29 years, is located the farthest out of the planets known then. By assuming that each body was mounted on its own revolving sphere and by tipping the spheres slightly with respect to one another, Eudoxus was able to explain most of the motions of the heavenly bodies.

Unfortunately, such a model does not explain retrograde motion, unless one believes that the giant spheres sometimes stop, reverse direction, stop again, and then resume their original motion. This idea is clumsy and unappealing. Eudoxus explained retrograde motion by requiring that each planet moved on *two* interconnected spheres, one inside the other. By adjusting their rotation rates and axes, he was able to get rough agreement with the observed positions of the planets as they shifted across the sky.

Ptolemy

By about A.D. 150, the great astronomer of the Roman Empire, Claudius Ptolemy, developed a more elaborate model that could predict the planets' motions with much better accuracy. Ptolemy lived in Alexandria, Egypt, which at that time was one of the intellectual centers of the world, in part because of its magnificent library. Ptolemy's era was one of social and political instability for the Roman Empire, which accounts for our uncertainty about the year of his birth or death. We know of him mainly through his great book, the *Almagest,* a compendium of the astronomical knowledge of the ancient Greeks. The book includes tables of star positions and brightnesses and is the source of much of our knowledge of ancient Greek astronomy.

ANIMATION ▶

Ptolemy's model of motion of a planet

In the *Almagest* Ptolemy fashioned a model of planetary motions in which each planet moved on one small circle, which in turn moved on a larger one (fig. 2.15). The small circle, called an **epicycle,** was supposedly carried along on the large circle like a Frisbee spinning on the rim of a bicycle wheel. Ptolemy probably developed his model of epicycles based on the writings of Hipparchus, who lived about 150 B.C.*

According to Ptolemy's model, the motion of a planet from east to west across the night sky is caused by the rotation of the large circle (the bicycle wheel, in our analogy). Retrograde motion occurs when the epicycle carries the planet in a reverse direction (caused by the rotation of the Frisbee, in our model). By choosing epicycles of the right size and spin rate, Ptolemy's model was able to account for retrograde motion, and predict planetary positions with reasonable accuracy.

Unfortunately, discrepancies remained between the predicted and true positions of the planets. This led to further modifications of the model, each of which led to slightly better agreement but at the cost of adding greater complexity. Ptolemy's model remained dominant until the 1500s, when its inability to make precise predictions despite a steadily growing complexity led astronomers to look for better, simpler models. Simplicity is an important element of scientific theory. As the medieval British philosopher William of Ockham wrote in the 1300s, "Entities must not be unnecessarily multiplied," a principle known as "Ockham's razor."

Islamic Astronomy

A great deal of what we know of Ptolemy, and of Greek and Roman astronomy (and their civilizations more broadly), we owe to the Islamic civilization that flourished around the southern edge of the Mediterranean from about 700 to 1200. Islamic scholars preserved, studied, and expanded upon ancient texts while most of Europe struggled through the Middle Ages.

Islamic civilization, like so many others, relied on celestial phenomena to set its religious calendar, and Islamic astronomers made many detailed studies of the sky and the motions of Sun, Moon, and planets. Islam's influence is very evident in astronomy through Arabic words such as *zenith* and the names of nearly all the bright stars—Betelgeuse, Aldebaran, and so on. In addition, Islamic scholars revolutionized mathematical techniques through innovations such as algebra (another Arabic word) and Arabic numerals.

Asian Astronomy

The early people of Asia, like their contemporaries to the west, studied the heavens. They too devised constellations, but based on their own mythologies, and they too made maps of the sky. Although the ancient astronomers of East Asia did not devise elaborate geometric models of the heavens, their careful observations of celestial events nevertheless prove useful to astronomers even today. For example, Chinese, Japanese, and Korean astronomers kept detailed records of unusual celestial events, such as eclipses, comets, and exploding stars.

Based on their records, Chinese astronomers devised ways to predict eclipses. They even noted dark spots on the Sun (sunspots) that they could occasionally see with the naked eye when the Sun was low in the sky and its glare was dimmed by dust or haze. These records have allowed astronomers to discover ancient patterns of variation in the Sun's behavior. Their records of exploding stars also allow today's astronomers to determine the dates of many of these celestial outbursts.

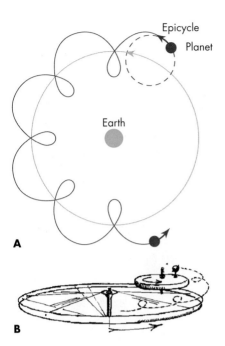

FIGURE 2.15
Epicycles are a bit like a bicycle wheel with a Frisbee bolted onto its rim.

---

*Hipparchus is best known to astronomers for his invention of the magnitude system for measuring stellar brightness (see chapter 13), and for his discovery of precession (see chapter 6). The latter was made possible by his meticulous observations of star positions and the care with which he compared his data to those of his predecessors.

2.3 ASTRONOMY IN THE RENAISSANCE

FIGURE 2.16
Nicolaus Copernicus (1473–1543)

Nicolaus Copernicus

The person who began the demolition of the geocentric model and the revolution in astronomical ideas that continues to this day was a Polish physician and lawyer by the name of Nicolaus Copernicus (fig. 2.16). During the early 1500s Copernicus made many attempts to reconcile Ptolemy's geocentric model with the centuries of data on planetary positions that had been collected, but all such attempts failed. Thus, he was led to reconsider Aristarchus's ancient idea that the Earth moves around the Sun.

A **heliocentric model** in which the Sun (*helios,* in Greek) is the center of the planets' motion had been proposed nearly 2000 years earlier by Aristarchus, but it had been rejected partly because the observational tools available at that time were inadequate to detect stellar parallax. Nevertheless, such models offer an enormously simpler explanation of retrograde motion. In fact, if the planets orbit the Sun, retrograde motion becomes a simple consequence of one planet overtaking and passing another, as Copernicus was able to show.

To see why retrograde motion occurs, examine figure 2.17. Here we see the Earth and Mars moving around the Sun. The Earth completes its orbit around the Sun in 1 year, whereas Mars takes 1.88 years to complete an orbit, with the Earth overtaking and passing Mars every 780 days. If we draw lines from Earth through Mars, we see that Mars appears to change its direction of motion against the background stars as the Earth overtakes and passes it. A similar phenomenon occurs when you drive on a highway and pass a slower car. Both cars are moving in the same direction, but as you pass the slower car, it *looks* as if it shifts backward relative to stationary objects beyond it.

Copernicus described his model of a Sun-centered Universe in one of the most influential scientific books of all time, *De revolutionibus orbium coelestium* (*On the*

The retrograde motion of Mars according to the heliocentric model

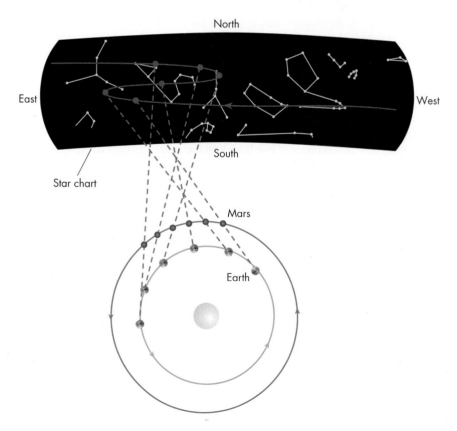

FIGURE 2.17
Why we see retrograde motion.
(Object sizes, positions, and distances are exaggerated for clarity.)

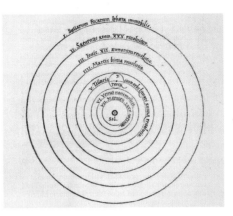

Table 2.1	Planetary Distances According to Copernicus	
Planet	Copernicus's Distance	Actual Distance
Mercury	0.38 AU	0.39 AU
Venus	0.72 AU	0.72 AU
Earth	1.00 AU	1.00 AU
Mars	1.52 AU	1.52 AU
Jupiter	5.22 AU	5.20 AU
Saturn	9.17 AU	9.54 AU

FIGURE 2.18

The title page and a diagram showing the heliocentric system of the planets from the first edition of *De revolutionibus orbium coelestium*, published in 1543.

Revolutions of the Celestial Spheres, fig. 2.18). Because his ideas were counter to the teaching of the Catholic Church, they were met with hostility and skepticism. The book itself was not published until shortly before Copernicus's death (which was perhaps just as well for him), and according to legend he saw the first copy while on his deathbed.

With his heliocentric model, Copernicus not only could give a simple explanation of retrograde motion but also could also explain why Venus and Mercury never move very far from the Sun. In Ptolemy's geocentric model this was caused by a coincidence in the rotation rates of the planetary cycles and epicycles. In the Copernican model these two planets have orbits smaller than the Earth's, so their angle from the Sun is limited by the size of their orbits (fig. 2.19). As shown in Astronomy by the Numbers: "How Copernicus Calculated the Distances to the Planets," Copernicus was able to use geometry to determine each planet's distance from the Sun. The distances found in this manner must be expressed in terms of the Earth's distance from the Sun, the astronomical unit or AU (whose value was not known accurately until several hundred years later), but table 2.1 illustrates that they agree well with modern values.

Ironically, some of the criticism of Copernicus's work was justified. Although his model was basically correct, it did not account for the observed positions of the planets any more accurately than did Ptolemy's more complicated but incorrect model.

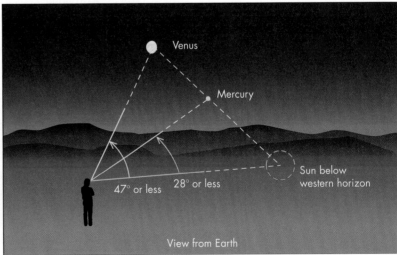

FIGURE 2.19

The greatest elongations of Mercury and Venus and the Evening Star phenomenon. The left-hand diagram also shows that Mercury and Venus can never appear more than 28° and 47°, respectively, from the Sun.

ASTRONOMY
by the numbers

HOW COPERNICUS CALCULATED THE DISTANCES TO THE PLANETS

When an inner planet appears farthest from the Sun, the planet's angle on the sky away from the Sun, α, can be measured as illustrated in figure 2.20A. You can see from the figure that the planet makes an angle of 90° with the Sun. The planet's distance from the Sun can then be calculated with geometry, if one knows the value of the angle α and the fact that the Earth–Sun distance is 1 AU.

Finding the distance to an outer planet is more complicated. First you must determine when the outer planet is directly opposite the Sun (rising when the Sun sets, for example). Then you must count the number of days until the planet is 90° away from the Sun in the sky. From that time interval we can determine the fraction of their orbits that the Earth and planet moved in that time. Multiplying those fractions by 360° gives the angles for those movements; we then take their difference to find the angle β in figure 2.20B.

For example, in 2012 Mars was opposite the Sun on March 3, and then at right angles from the Sun on June 8. During those 97 days, the Earth moved through approximately $(97/365) \times 360° \approx 96°$. Because Mars takes 687

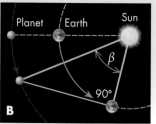

FIGURE 2.20
Finding the size of orbits for (A) planets closer to the Sun than the Earth, and (B) planets farther from the Sun.

days to complete its orbit, it has moved through an angle of about $(97/687) \times 360° \approx 51°$. The difference between those angles gives the angle $\beta \approx 45°$. We could then construct a triangle with this shape, and compare the sides, or use trigonometry, to find that Mars is approximately 1.4 AU from the Sun. Mars actually varies between 1.38 and 1.67 AU from the Sun, so many measurements around its orbit are necessary to give the correct mean value.

This lack of complete agreement between model and observation arose at least in part because Copernicus insisted that the planetary orbits were circles. Furthermore, his model again raised the question of why no stellar parallax could be seen. Finally, his views of planetary motion ran counter to the teachings of Aristotle, views supported both by "common sense" and by the Catholic Church at that time. After all, when we observe the sky, it *looks* as if it moves around us. Moreover, we do not detect any sensations caused by the Earth's motion—it feels at rest. This mixture of rational and irrational objections made even scientists slow to accept the Copernican view.

However, by this time there was a growing recognition of the immensity of the Universe. Astronomers such as the Englishman Thomas Digges and the Italian Giordano Bruno went so far as to claim that the stars were other suns, perhaps with other worlds around them. This new scientific open-mindedness, coupled with the aesthetic appeal of the simpler system, led to a growing belief in the Copernican system.

Tycho Brahe

Copernicus's model, although not the only stimulus, marked the opening of a new era in the history of astronomy. Conditions were favorable for new ideas: the cultural renaissance in Europe was at its height; the Protestant Reformation had just begun; the New World was being settled. In such an environment, new ideas found a more receptive climate than in earlier times, at least among scientists.

One scientist whose ideas flourished in this more intellectually open environment was the sixteenth-century Danish astronomer Tycho Brahe (fig. 2.21). Born into the Danish nobility, Tycho utilized his position and wealth to indulge his passion for study of the heavens, a passion based in part on his professed belief that God placed the planets in the heavens to be used as signs to mankind of events on Earth. Driven by this interest in the skies, Tycho designed and had built instruments of far greater accuracy than any yet devised in Europe. Tycho then used these devices to make precise measurements of planetary positions. His meticulous observations turned out to be crucial

FIGURE 2.21
Tycho Brahe (1546–1601)

not only for showing the superiority of the heliocentric over the geocentric system but also for revealing the true shape of planetary orbits.

Tycho did more than just record planetary positions; he recognized opportunity when he saw it. In 1572, when an exploding star (what we today call a supernova) became visible, Tycho demonstrated from its lack of motion with respect to other stars that it was far beyond the supposed spheres on which planets move. Likewise, when a bright comet appeared in 1577, he showed that it lay far beyond the Moon, not within the Earth's atmosphere, as taught by the ancients. These observations suggested that the heavens were both changeable and more complex than was previously believed.

Although Tycho could appreciate the simplicity of the Copernican model, he remained unconvinced of its validity because he could not detect any stellar parallax. Instead, he offered a compromise model in which all of the planets except the Earth went around the Sun, while the Sun, as in earlier models, circled the Earth. Tycho was the last major astronomer to hold that the Earth was at the center of the Universe.

Johannes Kepler

After Tycho Brahe's death, his young assistant, Johannes Kepler (fig. 2.22), was able to derive from Tycho's huge set of precise information a detailed picture of the path of the planet Mars. Whereas all previous investigators had struggled to fit the planetary paths to circles, by using Tycho's superb data Kepler was able to show that the path of Mars was not circular but elliptical.

An **ellipse** can be drawn with a pencil inserted in a loop of string that is hooked around two thumbtacks. If you move the pencil while keeping it tight against the string, as shown in figure 2.23A, you will draw an ellipse. Each point marked by a tack is called a **focus** of the ellipse. Not only was Mars's orbit elliptical, Kepler determined that the Sun was located at a spot that was *not* the center of the ellipse but was off center at a focus. Using an elliptical shape for the orbit, he was able to obtain excellent agreement between the calculated and the observed positions of the other planets as well. Kepler's discovery that planetary orbits are ellipses and not circles was a critical step in understanding planetary motion.

Along with discovering the shape of planetary orbits, Kepler also measured the relative sizes of the orbits. Because an orbit is elliptical, its size cannot be described by a single number. The shape of an ellipse is instead given by its long and short dimensions, called its major and minor axes, respectively (fig. 2.23B). Astronomers usually use the orbit's **semimajor axis**—half the major axis, analogous to a circle's radius. To describe the ellipse's shape, astronomers usually report its *eccentricity*, which indicates how far from the center of the ellipse each focus is located. The eccentricity of a circle is 0, but approaches 1 as the ellipse becomes more stretched out. Several ellipses with the same semimajor axis but different eccentricities are displayed in figure 2.23C.

FIGURE 2.22
Johannes Kepler (1571-1630)

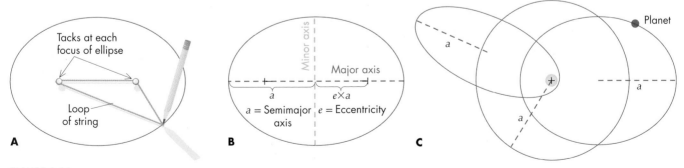

FIGURE 2.23
(A) Drawing an ellipse. (B) The major and minor axes. The semimajor axis, *a*, is half of the major axis. The distance that each focus is off-center in the ellipse determines the eccentricity, *e*, of the ellipse. (C) Three orbits are shown that have the same size semimajor axis but differing eccentricities. The Sun lies at one focus of the ellipse.

Table 2.2	Table Illustrating Kepler's Third Law for the Planets Known at His Time			
Planet	Distance from Sun (a) (in Astronomical Units)	Orbital Period (P) (in Years)	a^3	P^2
Mercury	0.387	0.241	0.058	0.058
Venus	0.723	0.615	0.378	0.378
Earth	1.0	1.0	1.0	1.0
Mars	1.524	1.881	3.54	3.54
Jupiter	5.20	11.86	141.0	141.0
Saturn	9.54	29.46	868.0	868.0

Based on Tycho's measurements, Kepler could measure not only the shape of a planet's path but also its speed as it changes distance from the Sun. And when Kepler compared the size of a planet's semimajor axes with how long the planet takes to orbit the Sun—its orbital **period**—Kepler discovered the relationship shown in table 2.2: the square of the period is proportional to the cube of the orbital size, as measured by the semimajor axis.

Kepler's discoveries of the nature of planetary motions are expressed in what are known today as **Kepler's three laws:**

ANIMATION ▶

Kepler's laws

I. Planets move in *elliptical* orbits with the Sun at one focus of the ellipse (see fig. 2.24-I).
II. The orbital speed of a planet varies so that a line joining the Sun and the planet will sweep over equal areas in equal time intervals (see fig. 2.24-II).
III. The amount of time a planet takes to orbit the Sun is related to its orbit's size, such that the period, P, squared is proportional to the semimajor axis, a, cubed (fig. 2.24-III). Mathematically,

$$P^2 = a^3$$

where P is measured in years and a is measured in astronomical units.

These three laws describe the essential features of planetary motion around our Sun.

The second law—in its statement that a line from the planet to the Sun sweeps out equal areas in equal times—implies that when a planet is near the Sun, it moves more rapidly than when it is farther away. We can see this by considering the shaded areas in figure 2.24-II. For the areas to be equal, the distance traveled along the orbit in a

INTERACTIVE ⚙

Kepler's second law

FIGURE 2.24

Kepler's three laws. (I) A planet moves in an elliptical orbit with the Sun at one focus. (II) A planet moves so that a line from it to the Sun sweeps out equal areas in equal times. Thus, the planet moves fastest when nearest the Sun. For purposes of the drawings a two-month interval is chosen. (III) The square of a planet's orbital period (in years) equals the cube of the semimajor axis of its orbit (in AU), the planet's distance from the Sun if the orbit is a circle.

given time must be larger when the planet is near the Sun. Thus, according to Kepler's second law, as a planet moves along its elliptical orbit, its speed changes, increasing as it nears the Sun and decreasing as it moves away from the Sun.

The third law also has implications for planetary speeds, but it deals with the relative speeds of planets whose orbits are at different distances from the Sun, not the speed of a given planet. Because the third law states that $P^2 = a^3$, a planet far from the Sun (larger a) has a longer orbital period (P) than one near the Sun (see table 2.2). For example, the Earth takes 1 year to complete its orbit, but Jupiter, whose distance from the Sun is slightly more than 5 times Earth's distance, takes about 12 years. Thus, a planet orbiting near the Sun overtakes and passes a planet orbiting farther out, leading to the phenomenon of retrograde motion, as discussed earlier in this section.

Kepler's third law has other implications. For example, we shall see in chapter 3 that the law gives information about the nature of the force holding the planets in orbit. Also, it implies that a planet close to the Sun moves along its orbit faster than a planet far from the Sun. Finally, the third law allows us to calculate the distance from the Sun of any body orbiting it if we measure the body's orbital period. (See Astronomy by the Numbers: "Using Kepler's Third Law for Orbit Calculations.") The distance we obtain will only be relative to the Earth's distance, but the law thereby gives us at least the relative scale of the Solar System.

Apart from such astronomical applications, Kepler's laws have an additional significance. Kepler's laws are the first mathematical formulas to describe the heavens correctly, and as such they revolutionized our way of thinking about the Universe. Without such mathematical formulations of physical laws, much of our technological society would be impossible. These laws are therefore a major breakthrough in our quest to understand the world around us.

It is perhaps ironic that such mathematical laws should come from Kepler, because so much of his work is tinged with mysticism. For example, as a young man he sought to explain the spacing of the planets as described in Copernicus's work in terms of nested geometrical figures, the sphere, the cube, and so on. In fact, it was Tycho's notice of this work that led to his association with Kepler. Moreover, Kepler's third law evolved from his attempts to link planetary motion to music, using the mathematical relations known to exist between different notes of the musical scale. Kepler even attempted to compose "music of the spheres" based upon such a supposed link. Nevertheless, despite such excursions into these nonastronomical matters, Kepler's discoveries remain the foundation for our understanding of how planets move. The work of Tycho Brahe and Johannes Kepler was the pinnacle of pre-telescopic astronomy. However, even as Kepler was developing his geometric and mathematical laws describing the motion of the planets, the nature of astronomy was about to change dramatically.

Kepler's third law

ASTRONOMY *by the numbers* — USING KEPLER'S THIRD LAW FOR ORBIT CALCULATIONS

Kepler's third law can be used to calculate the period or size of orbits around the Sun. Here are two examples:

Example 1 – The period of Pluto's orbit. To find how long Pluto takes to orbit, we use its distance from the Sun, which is about 39.5 AU. Putting this into Kepler's third law, we have
$$P^2 = a^3 = 39.5^3 = 61630.$$
Taking the square root of both sides, we have
$$P = \sqrt{61630} = 248 \text{ yrs.}$$
So, since its discovery in 1930, Pluto has completed only about 1/3rd of an orbit.

Example 2 – Asteroids in resonance with Jupiter. An asteroid with an orbital period half as long as Jupiter's (11.86 years) will suffer repeated gravitational deflections that might send it into a collision course with Earth. At what distance would such an asteroid orbit? Using Kepler's third law, we solve for the semimajor axis of an orbit with $P = 5.93$ years. We set
$$a^3 = P^2 = 5.93^2 = 35.2.$$
Taking the cube root of each side
$$a = \sqrt[3]{35.2} = 3.28.$$
So these dangerous asteroids orbit at 3.28 AU (chapter 11).

2.4 THE BIRTH OF ASTROPHYSICS

Galileo Galilei

At about the same time that Tycho Brahe and Johannes Kepler were striving to understand the motion of heavenly bodies, the Italian scientist Galileo Galilei (fig. 2.25) was also trying to understand the heavens. However, his approach was entirely different.

Galileo was interested not just in celestial motion but in all aspects of motion. He studied falling bodies and swinging weights hung on strings, and tried to find universal laws of motion. In addition, he used the newly invented telescope to study astronomical objects. Galileo did not invent the telescope himself. That invention seems to have been the work of the Dutch spectacle-maker Johannes (Hans) Lippershey. However, Galileo was the first person we know of who used the telescope to study the heavens and published his interpretations of his findings.* His book *Starry Messenger* was published in 1610. What he found was astonishing.

In looking at the Moon (fig. 2.26A), Galileo saw that its surface had mountains and was in that sense similar to the surface of the Earth. Therefore, he concluded that the Moon was not some mysterious ethereal body but a ball of rock. He looked (without taking adequate precaution) at the Sun and saw dark spots (now known as sunspots) on its surface. He noticed that the position of the spots changed from day to day, showing not only that the Sun had blemishes and was not a perfect celestial orb but that it also changed. Both these observations were in disagreement with previously held conceptions of the heavens as perfect and unchangeable. In fact, by observing the changing position of the spots from day to day, Galileo deduced that the Sun rotated.

Galileo looked at Jupiter and saw four smaller objects orbiting it, which he concluded were moons of the planet (fig. 2.26B). When Galileo's contemporary, Johannes Kepler, saw these moon's through a small telescope, he gave them the name **satellites** because their motion around the planet made him think of attendants or bodyguards—*satelles,* in Latin. These four moons of Jupiter are known today as the Galilean satellites

FIGURE 2.25
Galileo Galilei (1564–1642).

---

* Thomas Harriott (1560–1621), an English mathematician-scientist, appears to have used a telescope to study the heavens a little before Galileo. He too saw sunspots and the moons of Jupiter, but he failed to publish his discoveries at the time.

FIGURE 2.26
Drawings from Galileo's 1610 book *Sidereus Nuncius (Starry Messenger)*. (A) A sketch of the Moon seen through his telescope, showing mountainous features. (B) A series of diagrams of Jupiter and its moons, seen shifting from night to night as they orbited Jupiter. (C) Numerous faint stars near the belt and sword of the constellation Orion, illustrating the existence of stars too faint to be seen with the unaided eye.

in honor of Galileo's discovery. They proved unambiguously that there were at least *some* bodies in the heavens that did not orbit the Earth, and they raised the fundamental question of what force held them in orbit around Jupiter.

Galileo also discovered that the sky was populated with an uncountable number of stars everywhere he looked (fig. 2.26C). This single observation, by demonstrating that there were far more stars than previously thought, shook the complacency of those who believed in the simple Earth-centered Universe.

When Galileo looked at Saturn, he discovered that it did not appear as a perfectly round disk but that it had blobs off the edge. However, his telescope was too small and too crudely made (inferior to inexpensive modern binoculars) to show these as rings. That discovery that had to wait until 1656, when they were first recognized by the Dutch scientist Christiaan Huygens as features that were detached from the planet.

Galileo observed that Venus went through a cycle of phases, like the Moon, as shown in figure 2.27. The relation between the phase of the planet and its position with respect to the Sun left absolutely no doubt that Venus must be in orbit around the Sun, because if it orbited the Earth it would always remain in a crescent phase (fig. 2.28). Perhaps more than any other observation, this one dealt the death blow to the old geocentric model of planetary motion.

Galileo's contributions to science would be honored even had he not made all these important observational discoveries, for he is often credited with originating the experimental method for studying scientific problems. From his experiments on the manner in which bodies move and fall, Galileo deduced the first correct "laws of motion," laws that ultimately led Newton to his explanation of why the planets obey the laws of planetary motion that Kepler discovered.

Galileo's probings into the laws of nature led him into trouble with religious "law." He was a vocal supporter of the Copernican view of a Sun-centered Universe and wrote and circulated his views widely and somewhat tactlessly. His exposition followed the style of Plato, presenting his arguments as a dialog between a wise teacher (patterned after himself) and an unbeliever in the Copernican system named Simplicio who, according to his detractors, was patterned after the pope. Although the pope was actually a friend of Galileo, more conservative churchmen urged that Galileo be brought before the Inquisition because his views that the Earth moved were counter to the teachings of the Catholic Church. Considering that his trial took place at a time when the papacy was attempting to stamp out heresy, Galileo escaped lightly. He was made to recant his "heresy" and was put under house arrest for the remainder of his life. Only in 1992 did the Catholic Church admit it had erred in condemning Galileo for his ideas.

FIGURE 2.27
Images of Venus made with a small telescope in 2004 show it changing from a gibbous phase on the far side of the Sun to a crescent phase as it passes between the Earth and Sun.

The phases of Venus according to the Ptolemaic and Copernican systems

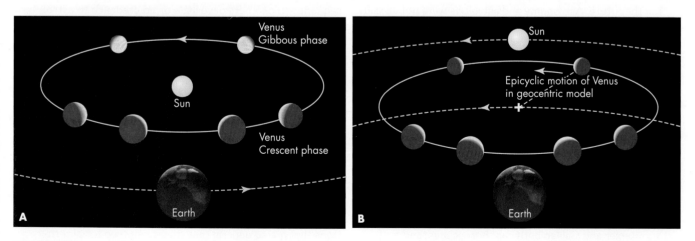

FIGURE 2.28
As Venus orbits the Sun, it goes through a cycle of phases (A). The relation between phase and the planet's position with respect to the Sun shows conclusively that Venus cannot be orbiting the Earth. The gibbous phases Galileo observed occur for the heliocentric model but cannot happen in the Earth-centered Ptolemaic model (B), where Venus is shown on its epicycle.

FIGURE 2.29
Isaac Newton (1642–1727)

Isaac Newton

Isaac Newton (fig. 2.29), who was born the year Galileo died, is arguably the greatest scientist of all time. Newton's contributions span mathematics, physics, and astronomy. Moreover, Newton pioneered the modern studies of motion, optics, and gravity. In his attempts to understand the motion of the Moon, Newton not only deduced the law of gravity but also discovered that he needed mathematical methods for calculating the gravitational force of a spherical body and that no such methods were then available. This realization led him to invent what we now know as calculus.

What is especially remarkable about Newton's work is that the discoveries he made in the seventeenth century still form the core for most of our understanding of gravity and the motion of bodies, discoveries we will discuss in more detail in chapter 3. In chapter 4 we will discuss some of Newton's ideas and discoveries about light, ideas that are also still in use.

Newton was a fascinating individual. He came from very modest origins and rose to high positions not only in academia but also in the government. He was Warden of the Mint and is alleged to have invented milling, the process whereby grooves are cut in the edge of coins to detect metal being pared off them, which would debase their value. He was also a deeply religious man and wrote prolifically on theological matters as well as science.

Newton's laws of motion, when combined with his law of gravity, were successfully applied for the next 200 years to essentially all problems of the motion of astronomical bodies. They still form the foundation for space flight today. These laws allow one to predict all future astronomical motions from a detailed knowledge of current motions, positions, and forces. Such a "clockwork universe" had no room for mystical effects of celestial bodies on human affairs such as had been part of the belief system of astrology, which had been part of the subject of astronomy until the seventeenth century. See Extending Our Reach: "Astronomy and Astrology."

EXTENDING *our reach*   ASTRONOMY AND ASTROLOGY

Astrology is an ancient belief, thousands of years old, that the positions and patterns of celestial bodies in the sky exert an influence on the course of human events, or foretell the future. Astronomy and astrology were not considered separate subjects before the seventeenth century.

Actually, one motivation behind the astronomical discoveries of the Renaissance was the hope of better understanding the motions of celestial bodies in order to cast more accurate horoscopes. The idea of a horoscope is that the positions of celestial bodies at the time of a person's birth (particularly the position of the Sun in the zodiac) along with their current positions could provide predictive power over human events. It might be believed, for example, that when the planet Mars is in the birth constellation of the leader of a country, then war is likely or even advised, so predicting the position of the planets accurately would be a critical ability of an astronomer.

Kepler and Galileo both cast horoscopes, and both pondered whether their new discoveries, such as the existence of satellites around the planets, might provide new insights into astrology. However, Newton's discoveries of the laws of motion and gravity removed the mystery of the motions of the Sun, Moon, and planets. He and subsequent astronomers gave little or no credence to astrology, and it was dropped from studies of astronomy.

Carefully conducted studies show that astrology has no predictive power. A simple test can be done in a classroom by passing out horoscopes from the previous day with all indication of the astrological "signs" removed. On average about 1 in 12 students—which is what is expected by random chance—will pick the horoscope that was intended for them. This is not to say that astrologers never offer useful advice or even cast horoscopes that seem to be accurate. In fact, students trying to select from anonymized horoscopes often express surprise that so many of the horoscopes seem appropriate. A skillfully written horoscope apparently offers advice and predictions that are so general that they seem true for almost anyone. This makes astrology a belief system rather than a predictive science.

New Discoveries

Newton's enormous contributions tend to overshadow other advances in astronomy during the eighteenth and nineteenth centuries. That period began with observational discoveries that increased astronomers' confidence in using physical laws to understand the structure and workings of astronomical bodies. However, by the end of the period, newly found physical laws gave astronomers totally new tools for studying the heavens. In fact, the increasing use of the word *astrophysics* describes that shift well.

The shift of stars due to parallax as the Earth orbits the Sun was not detected until 1838, but proof of the Earth's motion was discovered in 1729. The motion of the Earth actually causes the observed positions of all stars to shift throughout the year because as the Earth moves through space, the angle of the light entering a telescope changes. This is the same effect that causes droplets to hit the front of your body more than your back as you run through falling rain. Your motion causes the rain to appear to fall at an angle toward your front side, and likewise the Earth's motion makes the light appear to come in at an angle toward the direction of the Earth's motion as it orbits the Sun.

Unexpected discoveries play a major role today in expanding our knowledge of the heavens, no less so than in the time immediately after Newton's death. For example, in 1781 the English astronomer Sir William Herschel discovered the planet Uranus. He also discovered that some stars have companion stars in orbit around them. The motion of such double stars offered additional tests of Newton's laws, but the most striking triumph of these laws of motion was their explanation of irregularities in the orbital motion of Uranus. Such irregularities hinted that another body was exerting a gravitational force on Uranus, and from Newton's laws, astronomers could calculate the position of the unseen body. As we will discuss further in chapter 10, a search of the sky near the calculated position revealed the planet Neptune.

New Technologies

Steady improvements in telescopes played an important role during this period. For example, refinements in optics allowed astronomers to build bigger telescopes and thereby observe much fainter objects. Among these objects, astronomers found dim, fuzzy patches of light—the so-called *nebulae* (fig. 2.30). Some of these were gas clouds within the Milky Way; others turned out to be external star systems similar to the Milky Way.

Another important technological advance was the application of photography to astronomy, starting in the middle of the nineteenth century. Photographic film gave astronomers permanent records of what they saw, and because film could store light during long exposures, astronomers were now able to detect objects much fainter than the eye could see in a single moment.

The scientific and technical advances described here have a direct bearing on astronomy, but scientific discoveries often influence totally unconnected areas. For example, during the eighteenth and nineteenth centuries, many scientists were studying the nature of matter and heat. The study of heat was prompted, at least in part, by a desire to improve the newly invented steam engine. Understanding the generation of heat and energy in turn gave new insights into how stars work, but it also presented a mystery—stars were generating far more power than could be explained by any known source of energy. This conundrum was finally resolved with the discovery of nuclear energy in the twentieth century.

It was also not until the twentieth century that the discovery of a tiny discrepancy in the motion of Mercury, as calculated using Newton's work, showed scientists that Newton's laws were not the last word on planetary motion. His descriptions of motion require modification if we are to correctly describe motion at speeds near that of light or where gravitational fields are very intense. These modifications are incorporated in Einstein's theories of relativity, described in essay 2.

FIGURE 2.30
Sketches of nebulae as seen by Sir John Herschel in the early 1800s.

SUMMARY

Ancient peoples noted the basic patterns of the night sky, but the Greeks appear to have been the first to give explanations of planetary motion based on a combination of observations and geometric analysis. The Greeks pictured the planets, Sun, and Moon all orbiting the Earth on crystalline spheres.

Through the work of Aristotle and Eratosthenes, respectively, the Greeks determined the shape and size of the Earth. Aristarchus measured the relative size and distance of the Moon and Sun and about 300 B.C. proposed that the Earth orbited the Sun. However, his model was rejected because the expected shift in star positions (parallax) was unobservable at that time.

Planets look like bright stars that move with respect to the constellations, but always within the narrow band of the zodiac, like the Sun and the Moon. The usual direction of planetary motion is from west to east with respect to the stars, also like the Sun and the Moon. However, during several months of

each orbit, planets shifts in the other direction, undergoing apparent retrograde motion.

Based on earlier Greek models, Ptolemy (about A.D. 150) developed a complex model of planetary motion with the Earth at the center (geocentric) and with retrograde motion explained by planets moving on epicycles. This model was widely used for more than a millennium.

The geocentric model began to crumble in the 1500s with Copernicus's revival of the heliocentric model. Better observations by Tycho Brahe and detailed mathematical models by Kepler based on those observations placed the heliocentric model on a firmer basis. Galileo's observations with the recently invented telescope helped prove the heliocentric model.

Newton's discovery in the 1600s of the law of gravity and the laws of motion allowed him to explain why Kepler's laws worked, thereby completing the understanding of planetary motions.

QUESTIONS FOR REVIEW

1. (2.1) List some observational evidence that the Earth is round.
2. (2.1) What is meant by the phrase *angular diameter?*
3. (2.1) If you triple your distance from an object, what happens to its angular size?
4. (2.1) What is parallax and how is it measured?
5. (2.2) Where on the celestial sphere would you look for the planets?
6. (2.2) Sketch the path on the sky that a planet makes when undergoing retrograde motion.
7. (2.2) Will a planet in retrograde motion rise in the east or west?
8. (2.2/2.3) Contrast the geocentric and heliocentric models.
9. (2.3) What are the three laws of planetary motion?
10. (2.4) How does astrology differ from astronomy?
11. (2.1–2.4) Describe the major astronomical contribution(s) of the following in a sentence or two for each: Eudoxus, Aristotle, Aristarchus, Eratosthenes, Ptolemy, Copernicus, Tycho, Kepler, Galileo, and Newton.

THOUGHT QUESTIONS

1. (2.1) Explain why the Moon's angular size is largest when it is directly overhead. (A sketch or two may help.)
2. (2.1) Suppose the stars were very much closer than they really are. How might that have made it easier for Aristarchus to persuade people that the Earth moves around the Sun?
3. (2.2/2.3) Tycho argued that the Sun orbits the Earth but that the other planets orbit the Sun. Could Tycho's model

explain the phases of Venus as observed by Galileo? Why?
4. (2.3) Which of Kepler's laws explains why the Sun has a slightly larger angular diameter in January than in July?
5. (2.3) We know from Kepler's laws that the periods of the outer planets are very long. Jupiter, for example, has a period of almost 12 years. How then is it that, over a matter of months, Jupiter's position on the sky moves from one side of the Sun, to closer to the Sun, then to past the Sun to the other side? (Drawing a sketch might be helpful).
6. (2.3) You may have noticed that although every 10 years or so there is a comet visible in the night sky, the same comet is seen only once or twice during a human lifetime. Use this fact and Kepler's third law to deduce how the semimajor axis and shape of a comet's orbit must compare to the Earth's orbit.
7. (2.4) Describe how modern astrophysics differs from ancient astronomy, with examples based on the work of specific astronomers or astrophysicists.
8. (2.1–2.4) Make a table listing the astronomers named in review question 11 above, and then add the approximate dates of their births and deaths. Then add a few historic events of each period, as well as names of famous artists, writers, musicians, or politicians who lived at about the same time.

PROBLEMS

1. (2.1) A small probe is exploring a spherical asteroid. As the probe creeps over the surface, it drills holes to take soil samples. Scientists on Earth notice that the Sun shines straight

down into one of the holes. At the same time, 10 kilometers due "north," the shadow of the vertical antenna on the main landing craft allows the scientists to deduce that the Sun is 15° from directly overhead. What is the radius of the asteroid? How many times smaller or bigger than Earth's is its radius?

2. (2.1) If the distance between Alexandria and Syene had been 15,000 stadia, what would Eratosthenes have calculated for the circumference and diameter of the Earth?

3. (2.1) On average, Mercury is 0.387 times Earth's distance from the Sun, and Pluto is 39.53 times Earth's distance from the Sun. If the Sun has an angular diameter of 0.5° as seen from Earth, what is the Sun's angular diameter as seen from Mercury? From Pluto?

4. (2.1) The great galaxy in Andromeda has an angular diameter along its long axis of about 5°. Its distance is about 2.2 million light-years. What is its linear diameter?

5. (2.3) Suppose a planet is found with an orbital period of 64 years. How might you estimate its distance from the Sun? If its orbit is circular, what is its radius?

6. (2.3) In 2003, astronomers discovered Sedna, an object in the outer Solar System with a semimajor axis of 526 AU. What is its orbital period?

7. (2.3) Suppose a planet orbits a nearby star once every 125 years. If the star is identical to the Sun, how could you find the planet's distance from its star? If the planet's orbit is a perfect circle, how far from the star is the planet in AU?

8. (2.3) Suppose that future observations with a new telescope reveal a planet about 16 AU from a star whose mass is the same as our Sun's. How long does it take the planet to orbit the star?

TEST YOURSELF

1. (2.1) A total solar eclipse demonstrates that the Moon and Sun are very nearly the same angular size. If the Sun is 400 times farther from us than the Moon, then the radius of the Moon must be _____ the radius of the Sun.
   (a) 1600 times   (c) the same as   (e) 1/1600th of
   (b) 400 times   (d) 1/400th of

2. (2.2) A planet in retrograde motion
   (a) rises in the west and sets in the east.
   (b) shifts westward with respect to the stars.
   (c) shifts eastward with respect to the stars.
   (d) will be at the north celestial pole.
   (e) will be exactly overhead no matter where you are on Earth.

3. (2.2) "Ockham's razor" refers to
   (a) a device used by the ancient Greeks to measure the angle between the Sun and planets.
   (b) a metaphor for the process of discriminating between models based on their simplicity.
   (c) another term to describe the heliocentric model.
   (d) a method used to execute heretics.
   (e) a description of retrograde motion of planets.

4. (2.3) If an asteroid has an average distance from the Sun of 4 AU, what is its orbital period?
   (a) 1 year   (c) 4 years   (e) 16 years
   (b) 2 years   (d) 8 years

5. (2.3) Kepler's third law
   (a) relates a planet's orbital period to the size of its orbit around the Sun.
   (b) relates a body's mass to its gravitational attraction.
   (c) allowed him to predict when eclipses would occur.
   (d) allowed him to measure the distance to nearby stars.
   (e) showed that the Sun is much farther away than the Moon.

6. (2.4) Galileo used his observations of the changing phases of Venus to demonstrate that
   (a) the Sun moves around the Earth.
   (b) the Universe is infinite in size.
   (c) the Earth is a sphere.
   (d) the Moon orbits the Earth.
   (e) Venus follows an orbit around the Sun rather than around the Earth.

7. (2.4) A major objection to the heliocentric model not resolved until the development of high-quality telescopes was that
   (a) the speed of light had been thought to be infinite.
   (b) the Moon was believed to shine by its own light, not reflected light from the Sun.
   (c) the stars did not exhibit parallax.
   (d) Jupiter did not show a crescent phase.
   (e) Earth's gravitational pull was originally estimated to be stronger than the Sun's.

KEY TERMS

angular size, 39
ellipse, 51
epicycle, 47
focus, 51
geocentric model, 46
heliocentric model, 48
Kepler's three laws, 52

Moon illusion, 40
parallax, 41
period, 52
retrograde motion, 45
satellite, 54
semimajor axis, 51

Q: FIGURE QUESTION ANSWERS

WHAT IS THIS? (chapter-opening): Photograph of a partial lunar eclipse. The person in the photo is holding up a hoop, which has about the same angular diameter as the shadow of the Earth as seen by the photographer. Early astronomers recognized that the shape of Earth's shadow is always part of a circle, so they realized that the Earth must be spherical.

FIGURE 2.13: Mars appears to move backward when Earth is passing it at the same time as Earth is closest to Mars, so Mars appears its brightest.

Backyard Astronomy

You can learn many of the same things that classical astronomers did by simply watching the night sky. But there is a bonus as well. Backyard astronomy is just plain fun, as evidenced by the many thousands of amateur astronomers who in their spare time pursue activities ranging from simply stargazing to searching for new comets.

This essay is intended to give you some hints on how to become an amateur astronomer, beginning with learning the constellations, using star charts, and locating the planets. We will then discuss how to improve your observing experience with your eyes and cameras. Finally, we offer some suggestions about purchasing a small telescope, and taking your amateur astronomy to the next level.

LEARNING THE CONSTELLATIONS

One of the best ways to get started as an amateur astronomer is to learn the constellations. All it takes is a star chart (such as the ones provided at the back of this book), a dim flashlight, and a place that is dark and has an unobstructed view of the night sky. The star chart will tell you how to hold it so that it matches the sky for the date and time that you are observing.

Start by determining which way is north, using a compass if necessary. Then try to locate a few of the brighter stars,

matching them up with your star chart. This will give you some sense of how big a piece of the sky the chart corresponds to. Next, try to identify a few of the constellations. Focus at first on just a few of the brighter ones.

Probably the most familiar star grouping for people in the Northern Hemisphere is the Big Dipper. It is not a constellation but rather is called an **asterism.** An asterism is an easily recognized grouping of stars that may be part of one constellation or may incorporate pieces of several. For example, the Big Dipper is part of the constellation Ursa Major, the Great Bear. If you live northward of latitude 35°N, the Big Dipper is always visible in the northern part of the sky.

As you attempt to find and identify stars, your spread hand held at arm's length makes a useful scale. For most people, a fully spread hand at arm's length covers about 20° of sky, or about the length of the Big Dipper, as shown in figure E1.1A. For smaller angles, you can use your thumb's width, which is about 2°, or your little finger's width, which is about 1° wide (fig. E1.1B).

The Big Dipper is an excellent signpost to other asterisms and stars. For example, the two stars at the end of its "bowl" away from the "handle" (fig. E1.1A) are called the "pointers" because they point, roughly, to the North Star, Polaris, about 30° or 1½ handspreads away. Because Polaris lies nearly above

FIGURE E1.1

(A) The Big Dipper, part of the constellation Ursa Major, the Great Bear. A line through the two pointer stars points toward Polaris. The Big Dipper spans about 20° of the sky. The sky is shown approximately as it looks in mid-September at about 8 p.m. from midnorthern latitudes. (B) You can estimate angular separations on the sky using your hand stretched out at arm's length in front of you. Your handspread is about 20°, your thumb is about 2° wide, and the tip of your little finger is about 1° wide.

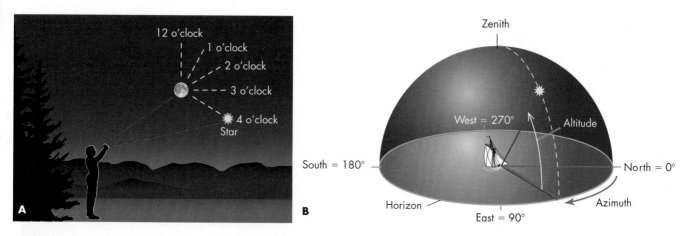

FIGURE E1.2

(A) Describing the location of stars by clock position. The star is half a handspread from the Moon and at the 4 o'clock position. (B) A star's position can be indicated by its altitude above the horizon and its azimuth measured eastward around the horizon from true north.

the Earth's North Pole, it is useful in orienting yourself to compass directions. Polaris marks the end of the handle of the Little Dipper, an asterism that is part of the constellation Ursa Minor, the Little Bear. If you extend the arc formed by the stars in the handle of the Big Dipper, you will find a path that curves to the bright star Arcturus ("follow the arc to Arcturus"). Arcturus is also about 1½ handspreads away from the Big Dipper in the constellation Boötes.

Estimating angles with your hand makes it easy to point out stars to other people. For example, you might say that a star is half a handspread away from the Moon and at the 4 o'clock position, as illustrated in figure E1.2A. A more general method for locating a star is to measure its altitude and azimuth, as shown in figure E1.2B. The star's **altitude** is its angle above the horizon, while its **azimuth** is defined as the angle measured eastward from due north to the point on the horizon below the star. Due east is at azimuth 90°, south at 180°, and west at 270°.

Learning to recognize some of the brightest stars is another way to locate constellations. A good example is the asterism known as the Summer Triangle, which spans three constellations. It consists of three bright stars conspicuous in evenings most places from July to November: Deneb (in Cygnus, the Swan), Altair (in Aquila, the Eagle), and Vega (in Lyra, the Harp), shown in figure E1.3. The three stars almost form an isosceles triangle, with Deneb and Altair 38° apart, while Deneb and Vega are 23° apart. To Polynesians, these were known as the Navigator's Triangle because of their importance for traveling between Pacific islands. A modern invention is that they mark out a "V" for summer Vacation.

Once you recognize a few constellations, you may find that learning the stories behind them will help you remember their shapes and locations. It has been suggested that many such stories were created as aids to memory, especially important when familiarity with the stars could be literally a matter of life or death to a farmer or a navigator. Scientists have even shown that baby birds learn to recognize star patterns and movements

and use them to navigate safely—unguided by their parents—across thousands of miles of ocean to their winter homes.

The native inhabitants of North America had a story about the Big Dipper. Its bowl represented a huge bear, and the handle represented three warriors in pursuit of the bear. They had wounded it, and it was bleeding. The red color of autumn leaves was said to be caused by the bear's blood dripping on them when the constellation lies low in the sky during the evening hours of the autumn months.

Stories are also told that connect multiple constellations. For example, if you follow the pointer stars in the Big Dipper

FIGURE E1.3

Dominating the night sky in July, August, and September are the three bright stars Vega, Altair, and Deneb, which form the Summer Triangle. This sketch shows how the sky looks looking south (from midnorthern latitudes) at about 9 P.M. in early September.

past the Little Dipper and Polaris, you will come to a set of constellations tied together by an ancient Greek myth. The constellations are shown in figure E1.4 , as they might be seen in a northern autumn sky. Their story goes as follows:

In ancient days there lived a queen of Ethiopia, Cassiopeia, who was very beautiful but also very vain. She and king Cepheus, her husband, and their daughter, Andromeda, lived happily until one day the queen boasted that she was more beautiful than the daughters of Nereus, a sea god. In punishment for such pride, the sea god Neptune sent a sea monster, Cetus, to ravage the kingdom. To save his people and appease the gods, Cepheus was instructed to tie his daughter, Andromeda, to a rock for the monster to devour. Meanwhile, Perseus was returning home from a quest in which he slew the snakehaired Gorgon, Medusa. Upon Medusa's death, her blood dripped into the sea and turned into the flying winged horse, Pegasus. Perseus saw the maiden's peril and flew to her rescue, slaying the monster. They all lived as happily ever after as most mythological families.

There are many other stories about constellations, but the one just described may give you some sense of those that have been handed down over thousands of years of written and oral history. Explore these stories as you learn the constellations: They will help you remember the relative locations in the sky of the various constellations.

CELESTIAL MAPPING

Star charts show the pattern of stars on the sky, usually using larger dots to indicate brighter stars. Lines may be drawn between stars to help suggest the shape of what the constellation is supposed to represent. For example, figure E1.5 shows a portion of the foldout star chart at the back of the book centered on the constellation Orion. Canis Major (the large dog) and Gemini (the twins) are also fairly recognizable based on the connecting lines, but other constellations are not as easily deciphered.

In the early 1900s the International Astronomical Union divided up the sky into 88 official constellations based primarily on Western cultural traditions. Some constellations have names that go back at least several thousand years, while others are much more recent. In particular, constellations near the south celestial pole were only added to star charts in the last few centuries by European navigators, who named them for practical items on a ship such as a pump (Antlia), chisel (Caelum), table (Mensa), or even a telescope (Telescopium).

Several hundred of the brightest stars have proper names, usually given by Arab astronomers, which often are descriptive of their location in a constellation. For example, Betelgeuse derives from Arabic words meaning "hand of the central one," while Rigel means "foot." To make a more complete naming system that included fainter stars, astronomers use Greek letters then numbers, beginning with alpha (α) for the brightest star and after omega (ω) continuing with numbers 25 on up. However, there were inconsistencies and variant numbering schemes, and sometimes disagreements about which constellation a star belonged to, so a more general system is used today.

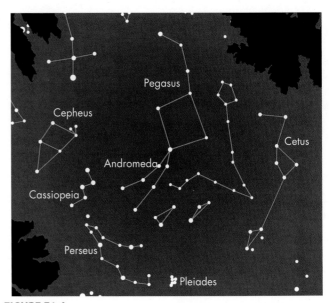

FIGURE E1.4
Perseus, Andromeda, Cassiopeia, Cepheus, Cetus, and Pegasus. The sky is drawn as it looks in mid-November at about 8 P.M., looking approximately straight overhead from midnorthern latitudes.

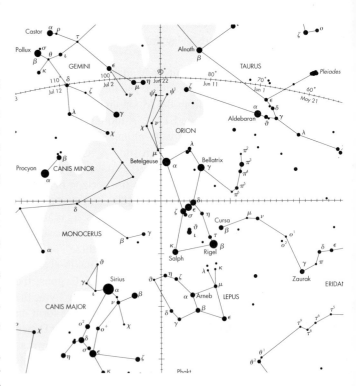

FIGURE E1.5
The region around the constellation Orion from the foldout star chart in the back of the book. This part of the sky is visible in the evening from December to February. Orion (the hunter) is easily recognized from the three stars of his belt. (Compare this chart of Orion to the photographs in figures E1.1B and E1.15A.) In the chart, brighter stars are shown by larger dots, and they are labeled by Greek letters that go approximately in order from brightest to faintest. The Milky Way is illustrated by a pale blue band.

Astronomers define locations of stars on the sky much as navigators define locations on Earth. Astronomers use a grid of lines running east–west on the celestial sphere, parallel to the celestial equator, and another set running north–south, connecting one celestial pole to the other. The east–west lines play the same role as latitude on the Earth, but to avoid confusion with terrestrial coordinates, they are called lines of **declination,** or "dec" for short. The north–south lines play the same role as longitude on the Earth and are called lines of **right ascension,** or "RA" for short.

As illustrated by the diagram of the celestial sphere in figure E1.6, declination values run from +90° to –90°, the north and south celestial poles respectively. The celestial equator is at declination 0°. Right-ascension lines divide the celestial sphere into 24 equal zones that are labeled not in degrees but in units of time because the sky rotates overhead once every 24 hours. Thus, the right ascension of an object is given in hours (h), minutes (m), and seconds (s), and the RA difference between objects indicates how much later one will cross the meridian than the other. Because the 360° around the sky is divided into 24 segments, each hour of RA equals 15°; that is, 360° / 24 = 15°. The point $0^h\ 0^m\ 0^s$ of RA is arbitrarily chosen to be where the Sun's path, the ecliptic, crosses the celestial equator as the Sun moves north, marking the vernal equinox.

One of the pleasures of studying the night sky is to look at objects too faint to be seen by the naked eye with binoculars or a small telescope. If we know the right ascension and declination of the object, we can use a star chart to locate it.

Star charts are designed much like maps of the Earth, representing on a flat surface a map of something curved. In the one case we depict the "interior surface" of the celestial sphere; in the other the surface of the Earth. These maps' coordinate grids are also very similar. A typical star chart (fig. E1.7) shows the location of the constellations, the stars, and other objects. Detailed star charts can be found in book form, generated with software, or even displayed using apps on a smartphone. It is helpful when working with any star chart to be able to rotate it as you compare it to the sky, because the orientation of constellations changes throughout the night and depends on your location.

Using a star chart to find a faint object requires some practice. For example, M31 is a galaxy in the constellation Andromeda, at right ascension 0 hours 43 minutes ($0^h\ 43^m$), declination + 41°16′, as illustrated by the large red ellipse in the star chart in figure E1.7. If you were trying to find this object in the night sky, you will first want to locate the constellation Andromeda, then orient the chart to the match the positions of the bright stars in the constellation. With binoculars or a small telescope you will be able to identify fainter stars in the chart, and step your way from the naked-eye stars toward M31. Actually, M31 is just barely visible with the naked eye in a dark sky away from city lights, so it is a good test of using a star chart. With binoculars, dozens of other galaxies, faint star clusters, and remnants of dying stars are visible. With a small telescope you can see hundreds of these objects.

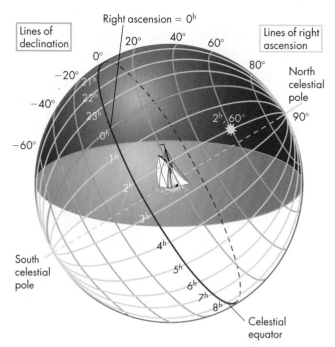

FIGURE E1.6

Locating a star according to right ascension and declination.

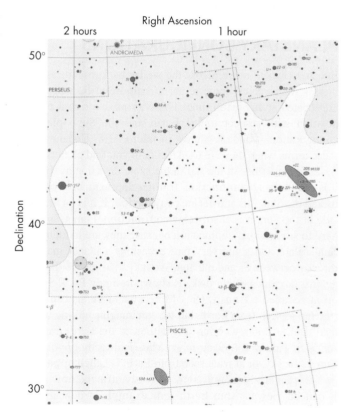

FIGURE E1.7

A star chart showing stars, galaxies, and coordinates. Black circles are stars. Their size indicates their brightness—larger are brighter. Red ellipses are galaxies; the large one at right is M31, the nearest large galaxy to our own. Blue shading indicates the brightness of the Milky Way.

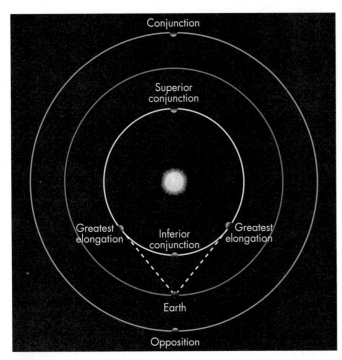

FIGURE E1.8
Planetary configurations: opposition, superior conjunction, and inferior conjunction.

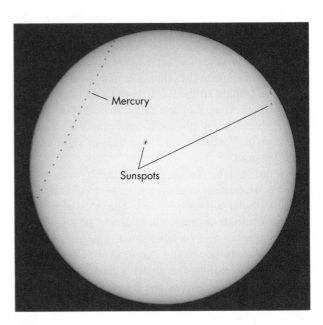

FIGURE E1.9
Transit of Mercury, May 7, 2003, photographed by Dominique Dierick of Belgium. Twenty-three separate images, taken about 15 minutes apart, were combined to form this picture. The next three transits will occur on May 9, 2016; Nov. 11, 2019; and Nov. 13, 2032.

PLANETARY CONFIGURATIONS

Because planets move across the stellar background, astronomers have invented some terms to help describe where they are located at any given time. These terms describe a planet's position with respect to the Earth and the Sun—planetary configurations—and are shown in figure E1.8. Understanding these terms when they are used can help you find planets.

If a planet lies in the sky in the same direction as the Sun, it is said to be at **conjunction.** If it lies approximately between us and the Sun, it is at **inferior conjunction.** If it is on the other side of the Sun, it is at **superior conjunction.** Planets are very hard to see at either conjunction because they are hidden in the Sun's glare. On some very rare occasions a planet may pass directly between us and the Sun. We may then see it silhouetted against the Sun's bright disk, as shown in figure E1.9. Such an event is called a **transit.** Only Mercury and Venus can transit the Sun as seen from Earth, but we can imagine talking with an astronaut on Mars who has just witnessed the Earth transiting the Sun. This would occur when Mars is directly opposite the Sun in the sky, or at what seen from Earth is called **opposition.**

A planet's configuration strongly affects how easily it can be viewed from Earth. For example, when an outer planet is at opposition, it is at its nearest to the Earth. The planet is therefore also at its brightest. Being opposite the Sun, a planet at opposition rises at sunset.

Inner planets, on the other hand, are easiest to see when they lie far from the Sun in our sky, so that they are not lost in the Sun's glare. However, there is a limit to how far from the Sun in our sky an inner planet can be, because their orbits are smaller than the Earth's. When Venus or Mercury is at its largest angular separation from the Sun (fig. E1.8), it is said to be at **greatest elongation**—which can be either western (morning) or eastern (evening). It is for this reason that Mercury and Venus are usually visible only in the morning or evening sky when the Sun is just below the horizon. Venus is so bright in the dawn or dusk sky that is often called the **Morning Star** or **Evening Star.**

The foldout chart at the back of the book includes "Moon and Planet Finder" tables that identify when these planetary configurations occur. Table E1.1 shows a subset of these tables covering 3 years. For Venus, Mars, Jupiter, and Saturn, the constellation in which each appears each month is listed. The constellations are printed in green in months when the planet rises before sunrise, blue when the planet is already up in the sky at sunset, and black when the planet is in opposition. When a planet is in conjunction, "Sun" is listed for that month instead.

Mercury remains so close to the Sun that the constellations are difficult to see, so just the date of greatest elongation is given. The best opportunity for seeing Mercury is within about a week of the date of greatest elongation, just before sunrise if the date is printed in green, and just after sunset if the date is printed in blue. Months when Venus reaches greatest elongation are marked with an asterisk.

In addition to planetary configurations, the tables give the dates of the new moon each month. Dates when the Moon causes a total solar eclipse are noted with a black circle; total lunar eclipses (when the Moon is full) are indicated with a

Table E1.1   Moon and Planet Finder

2016

	New Moon & Eclipses	Mercury	Venus	Mars	Jupiter	Saturn
Jan	10	Sun	Oph	Vir	Leo	Oph
Feb	8	7	Sgr	Lib	Leo	Oph
Mar	**9**	Sun	Aqr	Sco	Leo	Oph
Apr	7	18	Psc	Oph	Leo	Oph
May	6	Sun	Ari	Sco	Leo	Oph
Jun	5	5	Sun	Lib	Leo	Oph
Jul	4	Sun	Cnc	Lib	Leo	Oph
Aug	2	16	Leo	Sco	Vir	Oph
Sep	1	28	Vir	Oph	Sun	Oph
Oct	1, 30	Sun	Lib	Sgr	Vir	Oph
Nov	29	Sun	Sgr	Cap	Vir	Oph
Dec	29	11	Cap	Aqr	Vir	Sun

2017

	New Moon & Eclipses	Mercury	Venus	Mars	Jupiter	Saturn
Jan	28	19	Aqr*	Aqr	Vir	Oph
Feb	26	Sun	Psc	Psc	Vir	Oph
Mar	28	Sun	Sun	Ari	Vir	Sgr
Apr	26	1	Psc	Tau	Vir	Sgr
May	25	17	Psc	Tau	Vir	Sgr
Jun	24	Sun	Ari*	Gem	Vir	Oph
Jul	23	30	Tau	Sun	Vir	Oph
Aug	**21**	Sun	Gem	Cnc	Vir	Oph
Sep	20	12	Leo	Leo	Vir	Oph
Oct	19	Sun	Vir	Vir	Sun	Oph
Nov	18	24	Lib	Vir	Lib	Oph
Dec	18	Sun	Cap	Vir	Lib	Sun

2018

	New Moon & Eclipses	Mercury	Venus	Mars	Jupiter	Saturn
Jan	17, **31**	1	Sun	Lib	Lib	Sgr
Feb	15	Sun	Aqr	Oph	Lib	Sgr
Mar	17	15	Psc	Sgr	Lib	Sgr
Apr	16	29	Ari	Sgr	Lib	Sgr
May	15	Sun	Tau	Cap	Lib	Sgr
Jun	13	Sun	Cnc	Cap	Lib	Sgr
Jul	13, **27**	12	Leo	Cap	Lib	Sgr
Aug	11	26	Vir*	Cap	Lib	Sgr
Sep	9	Sun	Vir	Cap	Lib	Sgr
Oct	9	Sun	Sun	Cap	Lib	Sgr
Nov	7	6	Vir	Aqr	Sun	Sgr
Dec	7	15	Lib	Aqr	Oph	Sgr

red circle. Tables for additional years can be found on the foldout star chart at the back of the book.

The time interval between successive planetary configurations of the same type is called the **synodic period.** The synodic period differs from the planet's orbital period because both the Earth and the other planets move around the Sun. Thus, the interval between oppositions is neither an Earth year nor another planet's orbital period. For example, the Earth takes about 2 years to catch up to and overtake Mars after an opposition. The Earth overtakes the slower moving, more distant planets more quickly, and the interval between oppositions is close to a year. Thus, the Martian synodic period is about 780 days, whereas the Saturnian synodic period is 378 days, just slightly more than 1 year.

YOUR EYES AT NIGHT

Your eyes are marvelous instruments for studying the night sky, but you need to give them an opportunity to adjust to the low levels of the light. In fact, the longer you stay outside in dim light, the more sensitive your eyes will become and the fainter the stars you will be able to see. This is the result of physiological changes in your eye referred to as **dark adaption.**

The simplest change in your eye occurring in dim light is that the pupil (fig. E1.10) opens wider. In bright light your pupil typically has a diameter of just 2 millimeters, but in total darkness its diameter may expand to about 7 or 8 millimeters, thereby allowing more light to enter your eye.

Your eyes undergo another change in the dark. Chemical changes make the dark-adapted retina about 1 million times more sensitive to light than under full daylight conditions. The process takes about 20 minutes to get well established but is undone by even a few seconds' exposure to bright light. Thus,

once you are dark-adapted, you should stay away from bright lights for as long as you intend to observe.

In dark adaption, your eye changes its sensitivity to color slightly. In daylight, the eye responds best to greenish colors. At low light levels, it responds best to bluer colors. This is probably the result of natural selection, because starlight is bluer than sunlight and eyes responsive to blue will therefore aid survival. It is certainly the case that night-flying insects see blue light better than yellow. That is why using yellow light-bulbs for outdoor night lighting attracts fewer insects, and why astronomers use red lights when they look at star charts.

You may also notice that it is easier to see very faint objects if you don't look directly at them but instead look a little to one side. The greater sensitivity you enjoy from this so-called **averted vision** arises because the center of your field of view (the fovea) is densely packed with receptors that allow you to see fine details in bright light. Looking slightly to one side of a faint object allows more sensitive parts of your eye to see it.

FIGURE E1.10

A diagram illustrating the structure of the eye.

FIGURE E1.11

(A) Early morning image of clouds and stars over Bora Bora. The Large Magellanic Cloud is visible in the right half of the image. This picture was made with a 30-second exposure with the camera sitting on a railing. About 10 attempts yielded one sharp picture. (B) A picture toward the center of the Milky Way. This is another 30-second exposure, with the sensitivity (ISO) setting of the camera set to maximum.

IMAGING THE SKY

With a digital camera, even a cell-phone camera, it is possible to make quite lovely images of the night sky. To do astrophotography, you will need a camera that allows you to control the length of the exposure. Cameras that automatically adjust the focus and other settings are convenient, but they rarely produce good pictures of astronomical objects. Fortunately, many digital cameras have a manual mode that allows you to choose your exposure settings. There are even apps for most smart phones that allow you to adjust the exposure.

For almost all astrophotography, you should also set the camera's lens as wide open as possible (small "f-ratio") and the focus distance at infinity, so manual controls are usually more successful. In addition, if you can change the ISO setting, make it at least 400, preferably higher (higher ISO values allow you to image dimmer objects at the expense of a "noisier" image). Your camera's instruction guide will tell you how to make these changes.

Because the light is dim, the camera will usually need to take a long exposure, collecting light for several seconds or minutes, so it is important to use something to keep the camera steady, such as a tripod. Two examples of images made with 30-second exposures are shown in figure E1.11. For these images, the camera was simply set on a solid surface, and images were repeated several times until an acceptable image was achieved. It also helps to adjust the contrast to bring out dimmer objects using imaging software.

If you expose for more than about 30 seconds, the Earth's rotation will smear the star image into a noticeable streak. It is possible to "stack," or average together, many short-exposure images to bring out fainter detail. This is closer to how astronomers usually build up images of faint objects. Some image software programs have functions for combining multiple exposures and creating images with excellent sensitivity.

It can be quite dramatic to leave a camera open for a long exposure, deliberately allowing the stars to smear into long "star trails." This is particularly interesting looking toward the celestial poles as shown in figure E1.12. In this image the star Polaris is visible close to the north celestial pole, but it is not exactly at the pole, so it still makes a short trail. To make star-trail pictures, place the camera on a tripod and leave the shutter open for 20 minutes or more. To make such long exposures you need the capability of starting and stopping the exposure, which is possible for most higher-end cameras.

FIGURE E1.12

A time exposure showing how Polaris remains essentially fixed while the sky pivots around it.

FIGURE E1.13

Two pictures of the Moon made with a handheld digital camera. (A) The automatic features of the camera here overexposed the Moon, and the camera could not be kept steady during the shot. (B) The length of the exposure was forced to be just 1/1000 of a second.

Another popular object for astrophotography is the Moon. Even though the Moon is quite bright, it presents challenges for photography. In automatic mode, most cameras try to balance the exposure between the dark sky and the bright Moon, resulting in an overexposed Moon. And because it is difficult to hold a camera still for more than a small fraction of a second, the Moon becomes smeared out (fig. E1.13A). Recall that the sunlit side of the Moon is similar in brightness to daytime on Earth, so the exposure should actually be very short (fig. E1.13B). Even if you have a fully automatic camera, it may be tricked into taking a shorter exposure by taking the picture with other bright objects in the frame, or sometimes by turning on the flash.

SMALL TELESCOPES

Anyone with access to even very modest equipment, such as binoculars or a low-powered telescope, has better equipment than Galileo ever had. With such equipment and access to a dark sky, you can explore a wide variety of objects too faint to be seen with your unaided eye, and you will gain far better and more interesting views of the Moon and planets. With practice and dedication, you may even discover a new comet or an exploding star. There is an enormous variety of telescope designs available for amateur astronomy with a wide range of prices. Selecting the best one for your needs can be confusing, so we give a few pointers here.

A good starting size for a telescope is one with a 100- to 150-mm (4- to 6-inch) diameter *aperture*. The aperture is the opening of the telescope where light enters—usually the diameter of the main tube. Telescope sizes refer to the aperture diameter, because that indicates how much light the telescope collects. When astronomer speak of a 4-inch (or 100-mm) telescope, this means it has a main lens or mirror that gathers light over a circle with that diameter. Two telescopes with the same aperture may vary enormously in their other dimensions and cost, but both will deliver approximately the same amount of light to your eye. With a 100-mm telescope, you can easily study our Moon, the moons of Jupiter, the rings of Saturn, and most of the star clusters, nebulae, and galaxies in the Messier Catalog (Appendix table 13).

Perhaps the most familiar type of telescope is called a *refractor*. It has a lens at the front of the telescope that focuses the light to the point where you observe it at the back (fig. E1.14A). Glass bends light of different colors by different amounts like a prism (chapter 4), so you should avoid inexpensive refractors that have a single-element primary lens. Achieving a quality image requires the use of two, three, or more lenses designed to correct each other's color problems and provide an undistorted image. Good-quality refractors have what are called "achromatic" lenses and the best have "apochromatic" lenses—indicating that the color problems of the lens have been corrected to different degrees.

The color problems caused by lenses can be avoided by using a mirror to focus the light. A relatively inexpensive design is called a Newtonian reflector, named for its inventor, Isaac Newton. It uses a mirror at the back of the telescope to reflect and focus the light back up the telescope tube, and then a second mirror to reflect the light out to the side near the top (fig. E1.14B). These telescopes tend to be a bit more bulky, but a high-quality mirror is easier to produce, and it does not suffer the color problems of a lens.

FIGURE E1.14

Telescope designs. Both the optical design of a telescope and the design of its mount are important. (A) A refracting telescope (using a lens as the main focusing element) is illustrated on an azimuth altitude mount. This type of mount is convenient for aiming and is fairly easy to balanced, but it cannot track the stars as smoothly. (B) A Newtonian reflecting telescope (using a mirror as the main focusing element) is shown on an "equatorial" mount. An equatorial mount can more smoothly track stars because the whole assembly can rotate to cancel out the Earth's rotation, but these mounts tend to be heavier and more difficult to balance. The mounts are mostly interchangeable between reflectors and refractors.

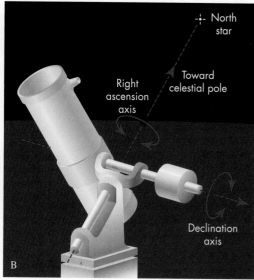

If you plan to take your telescope on the road, perhaps to enjoy dark skies in the wilderness, you will probably want it to be compact. A popular portable design is called a "Cassegrain" telescope, which is a hybrid between a reflector and a refractor. This type of telescope uses both a large mirror and a large lens, called a corrector plate, to focus the light. A Cassegrain folds the path of the light back on itself, sending the light through a hole in the primary mirror, making the telescope more compact. For the same size aperture, this type of telescope tends to be intermediate in cost between a reflector and a refractor.

Refractors are particularly favored for bright objects such as the Moon and planets, because they can give the crispest and highest-contrast images. Their high contrast is a result of having a clear aperture without secondary mirrors or internal support structures that scatter light inside the telescope. The "spikes" coming out of stars in many images are caused by the internal structures in most reflectors. Cassegrain designs can use the corrector plate to support the secondary mirror, but that mirror blocks some of the aperture, resulting in a lower-contrast image.

To see very faint objects, you will need a telescope with an aperture of 20 cm or more, but at increasing expense and decreasing portability. Keep in mind that using your eyes well can have more effect on what you will be able to see than increasing your telescope's size. By shielding your eyes from bright lights for at least half an hour before observing and using averted vision, you will be able to see things through the telescope that would otherwise be invisible to you.

Notice that we have said nothing of magnifying power. You can change the magnification of any telescope by using different eyepieces. (The magnification can be found by dividing the *focal length* of the telescope by the focal length of the eyepiece—consult your telescope's manual.) For most amateur telescopes, the maximum useful power is limited by distortions to the light as it passes through our atmosphere. These distortions make a magnification of about 100 to 200 the useful limit, so even a high-quality eyepiece is unlikely to provide a sharp image at higher magnifications. To get the most out of a telescope, you should expect to spend about a quarter or more of the cost of the telescope on good eyepieces.

Whatever type of telescope you choose, be sure to get a sturdy mount for it. A photograph made with a camera looking through a telescope, as in figure E1.15B, requires a long exposure and a heavy-duty mount. Even at 100 power, tiny vibrations of the telescope caused by wind or the touch of your hand will make the image jiggle, hopelessly blurring it. There are two basic styles of telescope mounts. The simpler design rotates in azimuth and altitude (fig. E1.14A). Many modern telescopes offer computer tracking on an azimuth design. However, because the sky rotates around the north celestial pole, it is useful for the telescope to rotate around an axis parallel to the Earth's axis (fig. E1.14B). This second type is especially useful for sky photography. These telescopes can keep a star centered in your images without producing trails.

Before you actually buy a telescope, you might look for a local astronomy club, where you may get an opportunity to try out different kinds of telescopes, or even purchase one secondhand. There are also online communities, such as the American Association of Variable Star Observers (www.aavso.org), where you can take your observing to another level and make scientifically valuable observations.

FIGURE E1.15
(A) A photograph of the constellation Orion made with an ordinary camera. (B) A picture of the Orion Nebula taken with a small backyard telescope.

SUMMARY

Not only is looking at the night sky fun, it will help you understand some of the phenomena described in chapters 1 and 2. Star maps will help you identify constellations and bright stars, and by learning the mythology of the stars, you will be able to find your way around the night sky more easily. In addition, you will forge a link to distant and ancient cultures. For many people, backyard astronomy—even with simple equipment—is an enjoyable and exciting hobby. Perhaps you will discover a comet and have it named for you!

QUESTIONS FOR REVIEW

1. Do we see the same constellations today as ancient cultures saw?
2. What are right ascension and declination?
3. What are altitude and azimuth?
4. What is the main advantage of the celestial coordinate system over altitude-azimuth coordinates?
5. What is a transit?
6. Approximately where would you look for Mercury in the sky at about sunset?
7. Is magnification or aperture diameter more important when selecting a telescope?
8. What is meant by "Morning Star"?
9. What is meant by "dark adaption"? What is averted vision?
10. Why does the pupil of your eye grow wider in dim light?

THOUGHT QUESTIONS

1. If a planet is at opposition and you see it high in the sky, about what time of night must it be?
2. Could you ever see Mercury in the western sky at dawn?
3. Considering the orbits in figure E1.8, where would Venus and Mercury be when they appear closest together on the sky? What is the name for the alignment when planets are in that position compared to Earth? Would they be easy to see from Earth when closest together? Why?
4. Pirates are often depicted wearing an eye patch. Can you think of any reason a pirate with two good eyes might wear an eye patch while on a ship?

TEST YOURSELF

1. If Venus is 42° away from the Moon, then it should be about _____ away from it.
   (a) two hands
   (b) a hand and three fingers
   (c) two and a half hands
   (d) two hands and a thumb

2. As a star rises and moves across the sky, which of the following change? (More than one may be correct.)
   (a) Its right ascension
   (b) Its declination
   (c) Its azimuth
   (d) Its altitude
   (e) None of the above

3. Imagine that you are lost, but you see two bright stars, both initially at an altitude of 20°, one about 90° to the right of the other. After about half an hour, the star to the right is still at the same altitude, but the star on the left is about 5° higher. What direction is north?
   (a) Toward the star on the left
   (b) Toward the star on the right
   (c) Opposite the star on the left
   (d) Opposite the star on the right

4. A planet is at inferior conjunction. It therefore rises at approximately
   (a) sunset.
   (b) sunrise.
   (c) midnight.
   (d) 2 hours before the Sun.
   (e) You can't tell from the available information.

5. If Mercury is at greatest elongation to the west of the Sun, it will be easiest to see
   (a) just before dawn.
   (b) just after sunset.
   (c) about midnight.
   (d) just before sunset.
   (e) None of the above

6. Which of the following planets can be at inferior conjunction?
   (a) Jupiter   (c) Uranus   (e) All of them
   (b) Mars   (d) Venus

7. When your eye is dark-adapted, (select all that apply)
   (a) your pupils are smallest.
   (b) your pupils are biggest.
   (c) your color vision is at its most sensitive.
   (d) your eyes are most sensitive to light.

KEY TERMS

3

GRAVITY AND MOTION

The space shuttle *Endeavour* launched at night in 2008.

LEARNING OBJECTIVES

Upon completing this chapter you should be able to:

- Define the concept of inertia, and illustrate it with examples from your everyday experiences.
- Explain why either a changing speed or a changing direction implies the presence of a force.
- Estimate the relative size and direction of forces at work on an object, given information about its motion and mass.
- Define Newton's three laws of motion, and give examples of how they are used.
- Explain what objects produce a gravitational force and how that force depends on mass and varies with distance.

- Describe how a body's mass can be calculated from the motion of an object orbiting it.
- Contrast mass and weight, and illustrate the differences with examples.
- Explain why objects of different masses fall at the same rate, and define the term *surface gravity*.
- Define what is meant by an escape velocity and what factors it depends upon.
- Carry out calculations to find the mass of a body, its surface gravity, or its escape velocity.

Gravity gives the Universe its structure. It is a universal force that acts on all the objects in the Universe so that every particle is drawn toward every other particle by its pull. Gravity holds together astronomical bodies of all sizes, from the Earth to the Universe itself. But the role of gravity extends beyond giving structure to astronomical bodies. Gravity also controls their motions, holding the Earth in orbit around the Sun, the Sun in orbit within the Milky Way, and the Milky Way within the Local Group. Thus, gravity and motion are tightly connected in the Universe. This connection is the theme of this chapter.

Astronomers of antiquity did not make the connection between gravity and astronomical motion that we recognize today. They were puzzled about what kept the planets moving and why. They could not understand why, if the Earth moved, objects thrown into the air were not left behind.

The solutions to these mysteries began with a series of careful experiments conducted by Galileo in the 1600s. Apart from his famous—and perhaps fictitious—demonstration of weights dropped from the Leaning Tower of Pisa, Galileo experimented with projectiles and with balls rolling down planks. Such experiments led him to propose several laws of motion. More significant, perhaps, these experiments demonstrated the power and importance of the experimental method for verifying scientific conjectures.

Perhaps the most remarkable discovery of these investigations is that the same laws that govern the motions of objects here on Earth also govern the motions of objects in space. If we really understand what makes objects move or change direction at home, we have the keys to understanding the motions of planets, stars, and galaxies throughout the Universe.

See end of chapter for the answer.

Q: WHAT IS THIS?

CONCEPTS AND SKILLS TO REVIEW

- Kepler's laws (2.3)
- Forces (Preview, p. 8)

3.1 INERTIA

Central to an understanding of motion is the concept of **inertia.** Inertia is the tendency of a body at rest to remain at rest or of a body in motion to keep moving in a straight line at a constant speed. Aristotle noted that bodies at rest resist being moved, but he failed to link this property to the tendency of objects to keep moving once they are set in motion. Instead, he thought some force must be steadily applied to an object to ensure that it would keep moving. Kepler recognized inertia's importance and in fact was first to use that term. However, it was Galileo who provided insight into inertia through careful experiments.

In one such set of experiments, Galileo rolled a ball down a sloping board and observed that it always sped up as it rolled down the slope (fig. 3.1). When the ball rolled up a slope, he observed, it always slowed down as it approached the top. He observed that on a flat surface the speed remained nearly constant, and hypothesized that if the flat surface produced no forces on the ball—such as friction—the ball's speed would

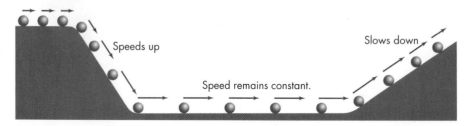

Speeds up

Slows down

Speed remains constant.

FIGURE 3.1
A ball rolling down a slope speeds up. A ball rolling up a slope slows down. A ball rolling on a flat surface rolls at a constant speed if no forces (such as friction) act on it.

neither increase nor decrease but remain constant. That is, in the absence of forces, inertia keeps an object already in motion moving at a fixed speed. Inertia is familiar to us all in everyday life. Apply the brakes of your car suddenly, and the inertia of the bag of groceries beside you keeps the bag moving forward at its previous speed until it hits the dashboard or spills onto the floor.

Newton recognized the special importance of inertia. He described it in what is now called **Newton's first law of motion** (sometimes referred to simply as the law of inertia). The law can be stated as follows:

> **A body continues in a state of rest or of motion in a straight line at a constant speed unless made to change that state by forces acting on it.**

In applying Newton's first law, we should note two important points. First, we have not defined force yet but have relied on our intuitive feeling that a force is anything that pushes or pulls. Second, we need to note that when we use the term *force,* we are talking about *net* force; that is, the total of all forces acting on a body. For example, if a box at rest is pushed equally by two opposing forces, the forces are balanced. Therefore, the box experiences no net force and accordingly does not move (fig. 3.2).

Newton's first law may not sound impressive at first, but it carries an idea that is crucial in astronomy: that if a body is *not* moving in a straight line at a constant speed, some net force *must* be acting on it.

Actually, Newton was preceded in stating this law by the seventeenth-century Dutch scientist Christiaan Huygens. However, Newton went on to develop additional physical laws and—more important for astronomy—showed how to apply them to the Universe. For example, let us look at what happens if we swing a mass tied to a string in a circle (fig. 3.3). Newton's first law tells us that the mass's inertia will carry it in a straight line if no forces act upon it. What force, then, is acting on the circling mass? The force is the one exerted by the string, preventing the mass from moving in a straight line and keeping it moving in a circle. We can feel that force as an outward pull on the string as the mass swings around.

We can see the importance of the force if the string suddenly breaks. Many people expect the mass to retain some "memory" of its former circular motion and travel along the path labeled B in figure 3.3 at least a little while longer. Or sometimes they think it should move outward along the path labeled C because of the sudden loss of the inward force. However, with the force of the string no longer acting on it, the mass flies off in

Balanced forces = no change in motion

FIGURE 3.2
Balanced forces lead to no acceleration.

FIGURE 3.3
For a mass on a string to travel in a circle, a force (green arrow) must act along the string to overcome inertia. Without that force, inertia makes the mass move in a straight line.

Side view

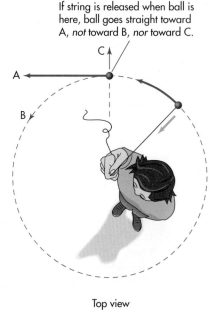

If string is released when ball is here, ball goes straight toward A, *not* toward B, nor toward C.

C

A

B

Top view

a straight line in the same direction it was moving at the moment the string broke, as illustrated by the arrow labeled A in figure 3.3, demonstrating Newton's first law.

We can translate this example to an astronomical setting and apply it to the orbit of the Moon around the Earth, or of the Earth around the Sun, or of the Sun within the Milky Way. Each of these bodies follows a curved path. Therefore, each must have a force acting on it, the origin and nature of which we will now describe.

3.2   ORBITAL MOTION AND GRAVITY

Newton was not the first person to attempt to discover and define the force that holds planets in orbit around the Sun. Nearly 100 years before Newton, Kepler recognized that some force must hold the planets in their orbits and proposed that something similar to magnetism might be responsible. Newton was not even the first person to suggest that gravity was responsible. That honor belongs to Robert Hooke, another Englishman, who noted gravity's role in celestial motions several years before Newton published his law of gravity in 1687. Newton's contribution is nevertheless special because he demonstrated the *properties* that gravity must have if it is to control planetary motion. Moreover, Newton went on to derive equations that describe not only gravity but also its effects on motion. The solution of these equations allowed astronomers to predict the positions and motions of the planets and other astronomical bodies.

According to legend, Newton realized gravity's role when he saw an apple falling from a tree. The falling apple drawn downward to the Earth's surface presumably made him speculate whether Earth's gravity might extend to the Moon. Influenced by an apple or not, Newton correctly deduced that Earth's gravity, if weakened by distance, could explain the Moon's motion.

Most of Newton's work is highly mathematical, but as part of his discussion of orbital motion, he described a thought experiment to demonstrate how an object can move in orbit. Thought experiments are not actually performed; rather, they serve as a way to think about problems. In Newton's thought experiment, we imagine a cannon on a mountain peak firing a projectile (fig. 3.4A). From our everyday experience, we know that whenever an object is thrown horizontally, gravity pulls it downward so that its path is an arc. Moreover, the faster we throw, the farther the object travels before striking the ground.

Now imagine increasing the projectile's speed more and more, allowing the projectile to travel ever farther. As the distance traveled becomes very large, we see that the Earth's surface curves away below the projectile (fig. 3.4B). Therefore, if the projectile moves sufficiently fast, the Earth's surface may curve away from it in such a way that the projectile will never hit the ground. Such is the nature of orbital motion and how the Moon orbits the Earth. The balance between inertia and the force of gravity maintains the orbit.

ANIMATION ▶

Newton's cannon

Q: Why are space vehicles normally launched from regions near the equator? Why are space vehicles normally launched to the east?

FIGURE 3.4
(A) A cannon on a mountain peak fires a projectile. If the projectile is fired faster, it travels farther before hitting the ground. (B) At a sufficiently high speed, the projectile travels so far before it drops that the Earth's surface curves out from under it, and the projectile is in orbit.

Slow   Fast

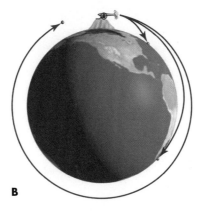

A   **B**

We can analyze this thought experiment more specifically with Newton's first law of motion. According to that law, in the absence of forces the projectile will travel in a straight line at constant speed. But because a force, gravity, is acting on the projectile, its path is not straight but curved. Moreover, the law helps us understand that the projectile does not stop, but continues moving, because its inertia carries it forward.

If we apply the same reasoning to the Moon's motion, we conclude that

- the Moon moves along its orbit because its inertia carries it forward, and
- its path is curved (not a straight line) because the force of gravity deflects it.

Notice that in the above discussion we used *no* formulas. All we needed was Newton's first law and the idea that gravity supplies the deflecting force. However, if we are to understand the particulars of orbital motion, we require additional laws. For example, to determine how rapidly the projectile must move to be in orbit, we need laws that have a mathematical formulation.

3.3   NEWTON'S SECOND LAW OF MOTION

We stated that an object's inertia causes it to move at a constant speed in a straight line in the absence of forces. However, suppose forces do act on the object. How much deviation from constant straight-line motion will such forces produce? To answer that question, we first need to define more carefully what we mean by motion.

Motion of an object is a change in its position, which we can characterize in two ways: by the direction the object is moving and by the object's speed. For example, suppose a car is moving east at 30 miles per hour. If the car's speed and direction remain constant, we say it is in uniform motion (fig. 3.5A). If the car changes either its speed or its direction, it is no longer moving uniformly, as depicted in figures 3.5B and C. Such nonuniform motion is defined as **acceleration.**

Acceleration

We are all familiar with acceleration as a change in speed. For example, when we step on the accelerator in a car and it speeds up from 30 to 40 miles per hour, we say the car is accelerating. Its speed has changed, and its motion is therefore nonuniform. Although in everyday usage acceleration implies an increase in speed, scientifically

FIGURE 3.5

Views of a car in uniform motion and accelerating. (A) Uniform motion implies no change in speed or direction. The car moves in a straight line at a constant speed. If an object's (B) speed or (C) direction changes, the object undergoes an acceleration.

Q: Is it possible for an object to travel with constant speed and still accelerate?

Same force, *F*

Empty cart—large acceleration, *a*.

Cart coasts

Same force, *F*

Full cart—small acceleration, *a*.

Cart coasts

FIGURE 3.6
A loaded cart will not accelerate as easily as an empty cart.

any change in speed is an acceleration. Thus, technically, a car "accelerates" when we apply the brakes and it slows down.

In this first example, we produced an acceleration by changing the car's speed. We can also produce an acceleration by changing an object's direction of motion. For example, suppose we drive a car around a circular track at a steady speed of 30 miles per hour. At each moment, the car's direction of travel is changing, and therefore it is not in uniform motion. Similarly, a mass swung on a string or a planet orbiting the Sun is experiencing nonuniform motion and is therefore accelerating. In fact, a body moving in a circular orbit constantly accelerates, even though its speed is not changing.

How do we produce an acceleration? Newton realized that for an object to accelerate, a force must act on it. For example, to accelerate—change the direction of—the mass whirling on a string, we must constantly exert a pull on the string. Similarly, to accelerate a shopping cart, we must exert a force on it. In addition, experiments show that the acceleration we get is proportional to the force we apply. That is, a larger force produces a larger acceleration. For example, if we push a shopping cart gently, its acceleration is slight. If we push harder, its acceleration is greater.

While pushing harder produces a larger acceleration, experience shows us that more than just force is at work here. For a given push, the amount of acceleration also depends on how heavy the object is. A lightly loaded shopping cart may scoot away under a slight push, but a push of the same strength on a heavily loaded cart gives it a much smaller speed, as illustrated in figure 3.6. Thus, the acceleration produced by a given force also depends on the amount of matter being accelerated.

Strictly speaking, the word *speed* indicates the rate of motion, irrespective of its direction. The term *velocity* refers to a speed in a given direction. Thus, a body's velocity changes if either its speed or its direction changes. With velocity so defined, acceleration is simply the change in velocity over some interval of time.

Mass

The amount of matter an object contains is measured by a quantity that scientists call mass. Technically, **mass** measures an object's inertia. The more inertia, the more mass, and vice versa. Scientists generally measure the mass of ordinary objects in kilograms (abbreviated kg). One kilogram equals 1000 grams or about 2.2 pounds of mass. Under normal conditions, a liter of water (roughly a quart) has a mass of 1 kilogram, but it is important to remember that mass is not the same as weight.

Because an object's mass describes the amount of matter in it, its mass in kilograms is a fixed quantity. An object's weight, however, measures the force of gravity on it, a point we will explore further in section 3.7. Thus, although an object's mass is fixed, its weight changes if the local gravity changes. For example, on Earth we have one weight, on the Moon, where gravity is less, we have a lesser weight. However, no matter where we are, we have the same mass. When measuring masses it is therefore best to use a balance scale (fig. 3.7) rather than a spring scale, which uses springs and measures the force compressing the spring.

FIGURE 3.7
A balance determines an object's mass by comparing it to objects of known mass.

Mass is the final quantity needed to understand **Newton's second law of motion.** This law states that when a force, *F*, acts on an object whose mass is *m*, it produces an acceleration, *a*, according to the equation $F = ma$. However, the second law's meaning is clearer if we write:

$$a = F/m$$

or, stated in words:

F = Net force
m = Mass of the body being accelerated
a = Amount of acceleration

The acceleration of a body is proportional to the net force exerted on it, but is inversely proportional to the mass of the body.

This astonishingly simple equation allows scientists to predict virtually all features of a body's motion. With $a = F/m$ and with knowledge of the masses and the forces in action, engineers and scientists can, for example, drop a spacecraft safely between Saturn and its rings or use a computer to design an airplane that will fly successfully the first time it is tested.

3.4   NEWTON'S THIRD LAW OF MOTION

Newton's studies of motion and gravity led him to yet another critical law, which relates the forces that bodies exert on each other. This additional relation, **Newton's third law of motion,** is sometimes called the law of action–reaction. This law states:

When two objects interact, they create equal and opposite forces on each other.

Two skateboarders side by side may serve as a simple example of the third law (fig. 3.8A). When one skateboarder pushes on another, *both* move. According to Newton's third law, when X exerts a force on Y, Y exerts an equal force on X, so that both are accelerated.

Note that although the forces between the skateboarders are equal and opposite in direction, the resulting magnitude of the acceleration is not necessarily the same. Here we have to recall Newton's second law: $a = F/m$. If one of the skateboarders weighs much more than the other, the acceleration of the more massive of the two will be smaller by the same factor that the mass is bigger (fig. 3.8B).

FIGURE 3.8
Skateboarders illustrate Newton's third law of motion. (A) When X pushes on Y, an equal push is given to X by Y. Because both are the same mass, they both accelerate to the same speed. (B) When X pushes on Z, who is more massive, the resulting acceleration is smaller for Z in proportion to how much more massive he is.

Gases accelerated to high speed

Rocket accelerated to smaller speed

F F

Ignited rocket fuel pushes gases backward with same force that it pushes rocket forward.

FIGURE 3.9
A rocket accelerates according to Newton's third law. Ignited propellants are pushed out the back of the rocket at high speed. They apply an equal and opposite force on the rocket, accelerating the ship forward. Nothing needs to be behind the rocket to be "pushed on."

The same principle explains why a spacecraft accelerates in response to firing its rockets. It is often incorrectly believed that the propellant "pushes against" the ground or air, but if that were true, a spacecraft would be unable to accelerate once it was in space. By Newton's third law, pushing propellant out the rocket nozzle produces an equal push in the opposite direction on the ship (fig. 3.9). This also shows how an astronaut who accidentally floats free from a spacecraft can redirect her motion back to the spacecraft. By throwing an object in a direction away from the spacecraft, an equal and opposite force is exerted on the astronaut, who is thereby pushed back toward the ship.

3.5 | THE LAW OF GRAVITY

Using Newton's three laws, we can determine an object's motion if we know its initial state of motion and the forces acting on it. For astronomical bodies, that force is often limited to gravity, and so to predict their motion, we need to know how to calculate gravity's force. Once again we encounter Newton's work, for it was he who first worked out the **law of gravity.** On the basis of his study of the Moon's motion, Newton concluded the following:

Every mass exerts a force of attraction on every other mass. The strength of the force is directly proportional to the product of the masses divided by the square of their separation.

We can write this important result in a shorthand mathematical manner as follows: Let m and M be the masses of the two bodies (fig. 3.10) and let the distance between their centers be d. Then the strength of the force of gravity between them, F_G, is

$$F_G = \frac{GMm}{d^2}.$$

The factor G is a constant, a conversion factor that allows us to get out the units of force when we put the masses and distance into the equation. The resulting value for G depends on the units chosen to measure M, m, d, and F, but once determined, G is the same for any pair of objects anywhere in the Universe whenever the same units are used. Scientists usually use meters (m), kilograms (kg), and seconds (sec) to measure physical properties, so these are sometimes called MKS units. These are the basis of the *Système International* (SI) set of units, and in this system forces are measured in units of kg·m/sec², which has been named the "newton," appropriately enough.

ANIMATION ▶

Gravity produces a force of attraction between bodies.

F_G = Strength of the gravitational force between two bodies
M = Mass of one body
m = Mass of second body
d = Separation between the bodies' centers
G = Gravitational constant

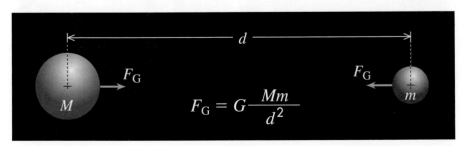

d

F_G

F_G

M

m

$$F_G = G\frac{Mm}{d^2}$$

FIGURE 3.10
Gravity produces a force of attraction (green arrows) between bodies. The strength of the force depends on the product of their masses, m and M, and the square of their separation, d^2. G is the universal gravitational constant.

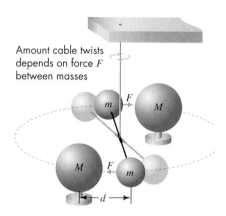

FIGURE 3.11
Cavendish's experiment that measured the gravitational constant G, by measuring the twist caused by a pair of large masses.

Determining the value of G is not easy because the gravitational force is so very weak for any object whose mass we can measure in a laboratory. It was first measured by the British scientist Henry Cavendish in 1798 by measuring the effect of large lead weights placed close to two smaller weights suspended by a thin wire (fig. 3.11). Placing the large weights on either side and at different distances caused the suspended weights to twist toward them by a small amount. From these kinds of experiments, the value of the gravitational constant has been found to be

$$G = 6.67 \times 10^{-11} \text{ meter}^3/(\text{kilogram} \cdot \text{second}^2) = 6.67 \times 10^{-11} \text{ m}^3 \cdot \text{kg}^{-1} \cdot \text{sec}^{-2}.$$

This can equivalently be written as:

$$G = 6.67 \times 10^{-11} \text{ newton} \cdot \text{m}^2 \cdot \text{kg}^{-2},$$

which makes it clear how G converts separations and masses to give a force in newtons.

An important reason for writing the law of gravity as an equation is that it helps us see more clearly how the force works. If either M or m increases and the other factors remain the same, the force increases. If d (the distance between two objects) increases, the force gets weaker. Moreover, the force weakens as the square of the distance. That is, if the distance between two masses is doubled, the gravitational force between them decreases by a factor of four, not two. Finally, although one object's gravitational force on another weakens with increasing distance, the gravitational force never completely disappears. Thus, the gravitational attraction of an object reaches across the entire Universe, so the Earth's gravity not only holds you on to Earth's surface but also extends to the Moon and exerts the force that holds the Moon in orbit around the Earth.

The gravitational force between the Earth and the Moon provides another astronomical example of Newton's third law and takes us a step closer to understanding orbital motion. The gravitational force of the Earth on the Moon is exactly equal to the gravitational force of the Moon on the Earth. We can see this from Newton's law of gravity, where the gravitational force between two objects depends on the product of their masses, so we get the same force regardless of whether we let the Earth act on the Moon or vice versa. Why, then, does the Moon orbit the Earth and not the other way around?

Newton's second law supplies the answer. Written in the form $a = F/m$, we see that the acceleration an object experiences is inversely proportional to its mass; that is, the more massive it is, the more force is required to accelerate it. Thus, even though the forces acting on the Earth and Moon are precisely equal, the Moon accelerates 81 times *more* because it is 81 times *less* massive than the Earth. Because the Moon's acceleration is so much larger than the Earth's, the Moon does most of the moving (fig. 3.12A). In fact, however, the Earth does move a little bit as the Moon orbits it, much as you must move if you swing a child or a heavy object around you (fig. 3.12B).

FIGURE 3.12
(A) The Earth and Moon exert equally strong gravitational forces on each other, but the Moon is 81 times less massive, so that force produces an acceleration (and resulting velocity) that is 81 times larger. (B) You feel a similar "reaction" force as the Earth when swinging a heavy object around.

Knowledge of orbital motion is important for more than just understanding the paths of astronomical objects. From the orbit's properties (such as size and orbital period), astronomers can deduce physical properties of the orbiting objects, such as their mass.

The basic method for determining an astronomical object's mass uses a modified form of Kepler's third law and was first worked out by Newton using his laws of motion and gravity. The underlying idea is very simple:

- The masses of the orbiting objects determine the gravitational force between them.
- The gravitational force in turn sets the properties of the orbit.

Thus, from knowledge of the orbit, astronomers can work backward to find the masses of the objects.

To see how this can be done, we consider a simple case: orbital motion in a perfect circle with the orbiting object having a mass so small that its gravitational effect on the central body can be ignored. These restrictions are met to high precision in many astronomical systems, such as the Earth's motion around the Sun and the Sun's motion within the Milky Way. By assuming that the mass being orbited is very large compared with the orbiting body, we can ignore the acceleration of the central body (as we discussed in section 3.5) and assume it is at rest. These assumptions simplify the problem but the solution turns out to be the same even when the problem is solved in the more general case.

To work out the orbital properties of one object moving around another, we use Newton's laws of motion and his law of gravity. From the first law, we know that if an object moves along a circular path, there is a net force (an unbalanced force) acting on it because balanced forces give straight-line motion. This force (known as a centripetal force) must be applied to any object moving in a circle, whether it is a car rounding a curve, a mass swung on a string, or the Earth orbiting the Sun.

Using Newton's second law of motion together with some algebra and geometry gives us an equation that shows that if a mass m moves with a velocity V around a circle at a distance d from the center, the centripetal force F_C needed to hold it in a circular orbit (fig. 3.13A) is

$$F_C = \frac{mV^2}{d}.$$

For example, the gravitational force the Sun exerts on a planet to make it move in a circle must equal the centripetal force. Therefore, we can equate the force in Newton's law of gravity (fig. 3.13 B) to the centripetal force corresponding to the orbital velocity and distance of the planet from the Sun to find a new equation for the central mass.

Newton drew upon Kepler's third law in deriving the law of gravity. Although this is not simple to see, it turns out that the exponents in Kepler's 3rd law (the square of P and the cube of a) are set by the power (square of d) in the law of gravity.

INTERACTIVE

Orbital velocity

F_C = Force needed to hold a body in orbit
m = Mass of the body
d = Radius of the orbit
V = Velocity of the body

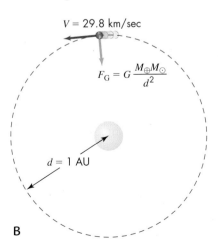

A   B

FIGURE 3.13
(A) The centripetal force, F_C, depends on the mass and speed at which an object swings in a circle as well as the object's distance from the center of the circle. (B) The gravitational force between the Sun and Earth holds the Earth in its nearly circular orbit.

ASTRONOMY by the numbers — WEIGHING THE SUN

To find the mass of the Sun, we need the distance of a planet orbiting the Sun, and the speed with which the planet is moving in its orbit. We will use the Earth's orbital parameters, but any planet would do.

The Earth–Sun distance, d, is 1 astronomical unit (1 AU), which in meters is 1.50×10^{11} m. (Fundamental values like this can be found on the inside cover of the book as well as in the appendix.)

To find the velocity V of Earth's orbit we can divide the circumference of Earth's orbit by the time it takes the Earth to complete one orbit, a length of time we call the orbital period P. The circumference of a circle of radius d is $2\pi d$ (note that in this case the radius is d). Therefore,

$$V = \frac{2\pi d}{P}.$$

Now we can solve for the mass of the Sun, $M_\odot$. The symbol $\odot$ is used by astronomers to represent the Sun.

Putting the expression we derived for V into the orbital equation for mass gives

$$M_\odot = \frac{V^2 d}{G} = \left(\frac{2\pi d}{P}\right)^2 \frac{d}{G} = \frac{4\pi^2 d^3}{GP^2}.$$

This is a modified form of Kepler's third law, and it is important because we can use it to measure the mass of any object given the orbital period, P, and orbital distance, d, of a much smaller object moving around it.

To calculate the Sun's mass, we need P and d for the Earth's orbit. If we measure P in seconds and d in meters, we can use the SI value of the gravitational constant $G = 6.67 \times 10^{-11}$ m$^3 \cdot$ kg$^{-1} \cdot$ s^{-2}.

To find P in seconds, we remember that it takes the Earth one year to orbit the Sun. Thus, P is one year. We can express P in seconds by multiplying the number of seconds in a minute (60), times the number of minutes in an hour (60), times the number of hours in a day (24), times the number of days in a year (365.25). The result of that calculation, rounded off to three significant figures, is

$$P = 3.16 \times 10^7 \text{ seconds}$$

Putting these values and the value of π and G into the expression for $M_\odot$ we find

$$M_\odot = \frac{4(3.14)^2 \times (1.50 \times 10^{11} \text{ m})^3}{6.67 \times 10^{-11} \text{ m}^3 \cdot \text{kg}^{-1} \cdot \text{sec}^{-2} \times (3.16 \times 10^7 \text{ sec})^2}$$

$$= \frac{4 \times 9.86 \times 3.38 \times 10^{33} \text{ m}^3}{6.67 \times 9.99 \times 10^{(14-11)} \text{ m}^3 \cdot \text{kg}^{-1} \text{ sec}^{-2} \text{ sec}^2}$$

$$\approx 2 \times 10^{30} \text{ kg}.$$

This is more than 300,000 times the Earth's mass.

Let the mass of the large central body be M and the orbiting object's mass be m, where the mass m is assumed to be much smaller than M. Then if the orbiting body moves in a circular orbit at distance d from the center of the large mass, the gravitational force between them is given by the gravitational force GMm/d^2, and this must match the centripetal force mV^2/d to maintain the orbit.

Setting these two forces equal to each other, we have

$$\frac{GMm}{d^2} = \frac{mV^2}{d}.$$

Multiplying both sides of the equation by d^2 and dividing both by Gm, we obtain

$$M = \frac{dV^2}{G}.$$

M = Mass of an orbited body
d = Radius of the orbit
V = Velocity of the orbit
G = Gravitational constant

This orbital equation for mass gives the mass of the central object if the orbiting object's velocity and distance from it are known. Note that the mass of the orbiting body cancels out of the equation, so we do not need to know the mass m of the smaller body. This means that all orbiting bodies will have the same orbit, so long as their mass is small compared to the object they are orbiting.

The orbital equation for mass can be used to find the mass of any object around which a smaller object orbits—a satellite orbiting a planet, a planet orbiting a star, or a star orbiting the center of a galaxy. This is illustrated in Astronomy by the Numbers: "Weighing the Sun." Thus, gravity becomes a tool for determining the mass of astronomical objects, and we shall use this method many times throughout our study of the Universe.

3.7   SURFACE GRAVITY

One of Galileo's famous discoveries was that objects of different masses all fall at the same rate. Many people incorrectly believe that lighter objects fall slower, but that is just a result of air friction. In a vacuum all objects accelerate downward at the same rate. Astronomers call this acceleration **surface gravity,** which gives a measure of the gravitational attraction at a planet's or star's surface. This acceleration determines not only how fast objects fall, but because an object's weight depends on its mass and the acceleration of gravity, surface gravity also determines what a mass weighs.

We can determine the strength of a planet's surface gravity as follows. The law of gravity states that a planet of mass M exerts a gravitational force F on a body of mass m at a distance d from its *center* given by GMm/d^2. However, at the planet's surface, $d = R$, the planet's radius, so the force on the body—in other words its weight—is

$$F = \frac{GMm}{R^2} = ma.$$

The second equality in this equation is just Newton's second law, which tells us the acceleration a mass m experiences when the force F is applied. We can cancel out the m's to give us the resulting acceleration, in other words the surface gravity, which astronomers usually denote by the letter g:

$$g = \text{gravitational acceleration at surface} = \frac{GM}{R^2}.$$

Note that the surface gravity does not depend on the mass of the object that is falling, which explains Galileo's discovery about falling objects.

However, the surface gravity of other astronomical bodies is usually different from Earth's, depending on their mass and radius. For example, the Moon's surface gravity is about 6 times smaller than the Earth's, as illustrated in figure 3.14 and calculated in Astronomy by the Numbers: "The Surface Gravity of the Earth and Moon" below.

FIGURE 3.14
Apollo 16 astronaut John Young easily leaps high in the Moon's low gravity despite a massive space suit.

g = Surface gravity
G = Gravitational constant
M = Mass of the attracting body
R = Radius of the attracting body

ASTRONOMY *by the numbers*   THE SURFACE GRAVITY OF THE EARTH AND MOON

To compare the surface gravity of the Moon with that of the Earth, we need to know the mass and radius of both. Those numbers are given in table 3.1. We can make this calculation two ways. The first way is to plug into the equation for g the values of M and R appropriate for the Earth (astronomical symbol $\oplus$) and then repeat the process with values appropriate for the Moon ($\leftmoon$).

An easier way to make the comparison is to work with ratios. To do that, we write out on one line the expression for $g_\oplus$ on the Earth and then write out on the line below it the value for $g_\leftmoon$ on the Moon. Next we draw a horizontal line between them to indicate division, which gives the following equation:

$$\frac{g_\oplus}{g_\leftmoon} = \frac{GM_\oplus/(R_\oplus)^2}{GM_\leftmoon/(R_\leftmoon)^2}.$$

Notice that by doing this, the value of G, the gravitational

Table 3.1	Mass and Radius of the Earth and Moon	
	Mass	**Radius**
Earth ($\oplus$)	6.0×10^{24} kg	6.4×10^6 m
Moon ($\leftmoon$)	7.3×10^{22} kg	1.7×10^6 m
Earth/Moon ratio	81	3.8

constant, cancels out. If we then group the terms in M and the terms in R separately, we find

$$\frac{g_\oplus}{g_\leftmoon} = \frac{M_\oplus/M_\leftmoon}{(R_\oplus/R_\leftmoon)^2} = \frac{(81)}{(3.8)^2} = 5.6 \approx 6.$$

This shows that the ratio of g on the Earth to g on the Moon is about 6:1, so that you weigh about 6 times more on the Earth than you would on the Moon.

The surface gravity formula tells us that two planets with the same radius but different masses will have different surface gravities—the planet with the larger mass will have the larger surface gravity. Similarly, if two planets have the same mass but different radii, the planet with the larger radius will have a smaller surface gravity.

Besides telling us how much more or less we would weigh, surface gravity affects the steepness of geological features, the degree of compression of a planet's atmosphere, and even the overall shape of an astronomical body. Small objects such as asteroids are not spherical because the gravitational pull at their surface is too weak to crush them into round shapes.

Gravity variations

3.8   ESCAPE VELOCITY

To overcome a planet's gravitational force and escape into space, a rocket must achieve a critical speed known as the **escape velocity.** The escape velocity, V_{esc}, for a spherical body, such as a planet or star, can be found from the law of gravity and Newton's laws of motion and is given by the following formula:

$$V_{esc} = \sqrt{2GM/R}.$$

V_{esc} = Escape velocity
G = Gravitational constant
M = Mass of the body to be escaped from
R = Radius of the body to be escaped from

Here, M stands for the mass of the body from which we are attempting to escape, and R is its radius, as shown in figure 3.15.

The escape velocity is the speed an object must have to be able to move away from a body and not be drawn back by its gravity. We can understand how such a speed might exist if we think about throwing an object upward. The faster the object is tossed

FIGURE 3.15
Escape velocity is the speed an object must have to overcome the gravitational force of a body such as a planet or star and not fall back. In general, the larger the mass of the planet or star and the smaller the distance from its center, the greater the escape velocity will be.

Escape velocity

Escape velocity

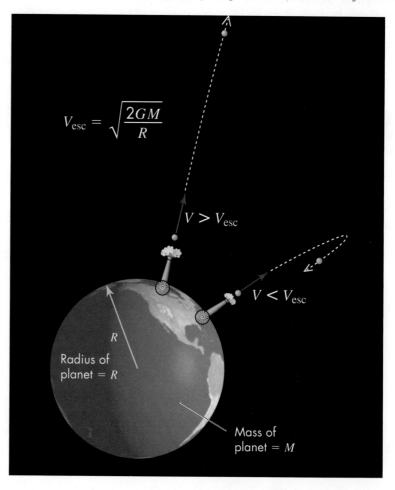

$$V_{esc} = \sqrt{\frac{2GM}{R}}$$

$V > V_{esc}$

$V < V_{esc}$

R

Radius of planet = R

Mass of planet = M

ASTRONOMY *by the numbers*   THE ESCAPE VELOCITY FROM THE MOON

To illustrate the use of the formula for escape velocity, we apply it to the Moon using the data in table 3.1. We insert the Moon's mass and radius in the formula and find

$$V_{esc}(\leftmoon) = \sqrt{2G\,M_{\leftmoon}/R_{\leftmoon}}$$

$$= \sqrt{\frac{2 \times 6.7 \times 10^{-11}\,\mathrm{m^3\,sec^{-2}\,kg^{-1}} \times 7.3 \times 10^{22}\,\mathrm{kg}}{1.7 \times 10^{6}\,\mathrm{m}}}$$

$$= \sqrt{5.8 \times 10^{5}\,\mathrm{m^2/sec^2}}.$$

Note that when we take the square root, we take the square root of both the number and the units. This is easy to do. Just divide the exponents by 2. Thus $\sqrt{\mathrm{m^2/sec^2}} =$ m/sec. Carrying out the calculation gives

$$V_{esc}(\leftmoon) = 2.4 \times 10^{3}\,\mathrm{m/sec} = 2.4\,\mathrm{km/sec}.$$

This escape velocity is about 5000 mph, which is about five times smaller than Earth's.

upward, the higher it goes and the longer it takes to fall back. Escape velocity is the speed an object needs so that it will *never* fall back, as depicted in figure 3.15. Thus, escape velocity is of great importance in space travel if craft are to move away from one object and not be drawn back to it.

Notice in the equation for V_{esc} that if two objects of the same radius are compared, the larger mass will have the larger escape velocity. Likewise, if two objects of the same mass are compared, the one with the smaller radius will have the greater escape velocity. The escape velocity from the Moon's surface is calculated in the Astronomy by the Numbers box above. A similar calculation shows that the escape velocity from the Earth is 11 kilometers per second. Thus, it is much easier to blast a rocket off the Moon than off the Earth. In chapter 7 we will see that this low escape velocity is partly responsible for the Moon's lack of an atmosphere, as illustrated in figure 3.16.

Although escape velocity is usually calculated from the surface of a body, where R is the body's radius, the escape velocity can also be found at any larger distance. For example, at Earth's distance from the Sun (1 AU), we would find that a spacecraft would need to be launched at 42 km/sec to escape the Sun's gravity. Thus a spacecraft launched just fast enough to escape the Earth would not have enough speed to leave the Solar System.

Escape velocity is particularly important in understanding the nature of black holes. In chapter 15 we will see that a black hole is an object whose escape velocity exceeds the speed of light, so no light can escape it. The escape velocity equation helps us to understand that a black hole has such a huge escape velocity, not necessarily because it has an unusually large mass, but because it has an abnormally small radius.

A large mass body (such as the Earth) has a large escape velocity, so it is difficult for gas molecules to escape into space. Thus, *has* an atmosphere.

A small mass body (such as our Moon), other things being equal, has a lower escape velocity, so gas molecules rapidly escape into space. Thus, *no* atmosphere.

$V < V_{esc}(\text{Earth}) = 11.2$ km/sec

$V > V_{esc}(\text{Moon}) = 2.4$ km/sec

FIGURE 3.16
The Moon's escape velocity is almost five times smaller than the Earth's. A low escape velocity, in general, leads to the absence of an atmosphere on a planet or satellite.

SUMMARY

An object's inertia makes it remain at rest or move in a straight line at a constant speed unless the object is acted on by a net force. Thus, for example, for a planet to orbit the Sun, the Sun must exert a force on it.

Newton formulated a gravitational force that exists between any two objects in the Universe. The strength of this force depends on the masses of the bodies and their separation. Gravitational forces hold astronomical bodies together and in orbit about one another. The law of inertia and Newton's other laws of motion, when combined with the law of gravity, allow us to relate the size and speed of orbital motion to the mass of the central object.

The gravitational force exerted by a planet determines its surface gravity and escape velocity. The former determines your weight on a planet. The latter is the speed necessary to leave the surface and escape without falling back.

QUESTIONS FOR REVIEW

1. (3.1) What is meant by inertia?
2. (3.1) What does Newton's first law of motion tell you about the difference between motion in a straight line and motion along a curve?
3. (3.2) Explain how inertia and gravity are both involved in an orbit.
4. (3.3) How does mass differ from weight?
5. (3.3) If your mass is 70 kg on Earth, what is it on the Moon?
6. (3.5) What is Newton's law of gravity?
7. (3.6) How can you measure the mass of a planet by studying the motion of a moon orbiting it?
8. (3.7) If you weigh 110 pounds on the Earth, do you weigh 110 pounds on the Moon? Why?
9. (3.7) What does surface gravity measure?
10. (3.8) What is meant by escape velocity?

THOUGHT QUESTIONS

1. (3.1/3.3) A cinder block can be weightless in space. Would you want to kick it with your bare foot? Even if it is weightless, does it have inertia?
2. (3.1/3.2) In some amusement park rides, you are spun in a cylinder and are pressed against the wall as a result of the spin. People sometimes describe that effect as being due to "centrifugal force." What is really holding you against the wall of the spinning cylinder? Drawing a sketch and using Newton's first law may help you answer the question.
3. (3.2) Is there a force of gravity between the orbiting International Space Station (ISS) and the Earth? If so, is it large enough that the ISS is affected by it? Why do the astronauts in the ISS float freely?
4. (3.3) Use Newton's second law of motion to explain why smaller cars tend to get better mileage than larger ones.
5. (3.4) How many times greater is the Earth's gravitational force on the Moon than the Moon's gravitational force on the Earth? Think about Newton's third law of motion before answering this.
6. (3.4) When you walk, does the ground push on you, or do you push on the ground? Explain clearly how you are accelerated forward. Why is it harder to walk across the beach than down a road?

7. (3.5) Is there a force of gravity between you and this textbook? What is the dependence of the force of gravity on distance? If the book is much closer to you than the center of the Earth, why does it accelerate to the ground instead of toward you when you drop it?
8. (3.5) Using $F = ma$ and $F = GMm/d^2$, deduce the units of Newton's gravitational constant G if masses are measured in kilograms, times in seconds, and distances in meters.
9. (3.5) How many times greater is the Sun's gravitational force on the Earth than the Earth's gravitational force on the Sun? (Consider Newton's third law and the law of gravity.)
10. (3.6) Consider a binary star system in which one star has ten times the mass of the other. Which star's orbit would you expect to have a larger circumference? Why?
11. (3.7) Explain how a larger planet could have a smaller surface gravity than a smaller planet, or why it could not.

PROBLEMS

1. (3.3) If you apply a force F to a mass m, it results in an acceleration a. What acceleration would result if you applied a force of (a) $2F$ to m, (b) $2F$ to $2m$, (c) $10F$ to m, and (d) $10F$ to $3m$?
2. (3.3) You are working at the hockey rink, and your resurfacing machine breaks down in the middle of the ice. Assuming you can get it moving, how much force will you need to apply to accelerate it to 2 m/sec in 25 seconds if it has a mass of 2500 kg?
3. (3.3/3.7) Calculate your weight on the Moon.
4. (3.6) Given that Neptune is about 30 times farther from the Sun than the Earth, calculate its orbital speed if the mass of the Sun is 2×10^{30} kg. How many years does it take Neptune to complete one orbit?
5. (3.6) Assuming that the mass of the Milky Way Galaxy is 10^{11} times that of the Sun and that the Sun is 2.6×10^{20} m from its center, what is the Sun's orbital speed around the center of the Galaxy? How long does it take the Sun to orbit the center of the Milky Way? (In this problem, we assume that the Galaxy can be treated as a single body. Strictly speaking, this isn't correct, but the more elaborate math needed to calculate the problem properly ends up giving almost the same answer.)
6. (3.6) Gliese 581e is an exoplanet with a mass of 1.9 Earths that orbits a red dwarf star at a distance of 5×10^{10} m

(0.33 AU). If its orbital period is 124 days, find the mass of the star in kg. Divide your answer by the Sun's mass to see how much more or less massive the star is than our Sun.

7. (3.7) Using the method of section 3.7, compare the surface gravity of the Earth with the surface gravities of Jupiter and Pluto.

8. (3.8) Calculate the escape velocity from the Earth, given that the mass of the Earth is 6×10^{24} kilograms and its radius 6×10^6 meters. In this problem, round off G to 7×10^{-11} m^3/(kg·sec^2).

9. (3.8) Convert the escape velocity of Earth (problem 8) into miles per hour.

10. (3.8) Calculate the escape velocity from the Sun, given that its mass is 2×10^{30} kg and its radius is 7×10^8 meters.

11. (3.7/3.8) Which body has a larger escape velocity, Mars or Saturn? Solve this problem using ratios in a way similar to the comparison of the Earth's and the Moon's gravity in section 3.7. Show your work. In the appendix you can find values for Mars's and Saturn's radii and masses in terms of Earth's.

12. (3.8) Calculate the ratio of the escape velocities from the Moon and Earth.

13. (3.8) A good baseball pitcher can throw a ball at 100 miles/hour (about 45 m/sec). If the pitcher were on Sinope, one of Jupiter's smaller moons, could the pitcher throw the ball fast enough to escape Sinope's gravity? Sinope is roughly spherical with a radius of 18,000 m and a mass of 6×10^{16} kg.

TEST YOURSELF

1. (3.1) Which of the following demonstrate the property of inertia? Select all that apply.
   (a) A car skidding on a slippery road
   (b) The oil tanker *Exxon Valdez* not being able to stop and running aground
   (c) A brick sitting on a tabletop
   (d) Whipping a tablecloth out from under the dishes set on a table

2. (3.1–3.3) If an object moves along a curved path at a constant speed, you can infer that
   (a) a force is acting on it.   (c) it is in uniform motion.
   (b) it is accelerating.   (d) both (a) and (b) are true.

3. (3.2) Newton's work added _____ to Kepler's laws.
   (a) the concept of elliptical orbits
   (b) an equation for ellipses
   (c) a physical reason
   (d) planetary data
   (e) an experimental verification

4. (3.3) An astronaut has a mass of 60 kilograms before she takes off in her ship. When she reaches Earth orbit, her mass is _____; when she lands on the Moon, her mass is _____.
   (a) zero; the same as on Earth

(b) much smaller than on Earth; smaller than on Earth
(c) smaller than on Earth; larger than on Earth
(d) zero; smaller than on Earth
(e) the same as on Earth; the same as on Earth

5. (3.4) A rocket blasts propellant out of its thrusters and "lifts off," heading into space. What provided the force to lift the rocket?
   (a) The propellant pushing against air molecules in the atmosphere
   (b) The propellant heating and expanding the air beneath the rocket, and so pushing the rocket up
   (c) The action of the propellant accelerating down, giving a reaction force to the rocket
   (d) The propellant reversing direction as it strikes the ground below the rocket, then bouncing back and pushing the rocket up

6. (3.4/3.5) The strength of the gravitational force exerted by the the Earth on you is the same as the strength of the gravitational force exerted by you on the Earth.
   (a) True   (b) False

7. (3.5) If the distance between two bodies is increased by a factor of 4, the gravitational force between them is_____ by a factor of ___.
   (a) increased; 4   (c) decreased; 8   (e) decreased; 64
   (b) decreased; 4   (d) decreased; 16

8. (3.7) If the Earth were 16 times as far from the Sun as it is, the Earth's orbital velocity would be about ___ times slower.
   (a) 4   (b) 8   (c) 16   (d) 64   (e) 256

9. (3.8) Two planets have identical diameters but differ in mass by a factor of 36. The more massive planet therefore has an escape velocity _____ than the other.
   (a) 36 times larger   (d) 36 times smaller
   (b) 6 times larger   (e) 6 times smaller
   (c) 1296 times larger

KEY TERMS

acceleration, 74	Newton's second law of
escape velocity, 82	motion, 76
inertia, 71	Newton's third law of motion,
law of gravity, 77	76
mass, 75	surface gravity, 81
Newton's first law of motion, 72	

ⓐ: FIGURE QUESTION ANSWERS

WHAT IS THIS? (chapter opening): This is a time-lapse photo of the launch of the space shuttle *Atlantis* on July 12, 2001.

FIGURE 3.4: To take advantage of the Earth's spin in helping the vehicle reach orbital velocity.

FIGURE 3.5: Yes—by traveling along a curved path.

4

LIGHT AND ATOMS

Light reflects inside raindrops, and the paths of different colors (different wavelengths) are bent by different amounts. This produces a rainbow (spectrum) of the Sun's light.

LEARNING OBJECTIVES

Upon completing this chapter you should be able to:

- Identify the basic properties of light, contrasting its wave-like and particle-like characteristics.
- Describe the meanings of the wavelength and frequency of light and how they relate to its color.
- Explain what white light is.
- Relate the different bands of electromagnetic radiation to visible light and to each other.
- Calculate the energy of photons from their wavelength.
- Compare the different temperature scales and explain why the Kelvin scale is more useful for scientific comparisons than the Celsius or Fahrenheit scales.
- Describe how and under what conditions the color of an object changes with temperature, and use Wien's law to calculate its temperature.

- Describe how electrons and photons interact to produce emission or absorption lines, and explain the resulting spectrum for hydrogen gas.
- Explain how spectra can be used to determine the chemical composition of an astronomical source, and identify the main elements that are encountered astronomically.
- Classify the physical conditions in objects according to whether they produce a continuum, an absorption, or an emission spectrum.
- Indicate the ways Earth's atmosphere interacts with light across the electromagnetic spectrum.
- Describe the conditions that produce a Doppler shift, and carry out a calculation using the Doppler formula to measure astronomical motions.

Q: WHAT IS THIS?

See end of chapter for the answer.

Our home planet is separated from other astronomical bodies by such vast distances and extremely harsh conditions that, with few exceptions, we cannot learn about them by direct measurements of their properties. If we want to know how hot the Sun is, we cannot stick a thermometer into it. We cannot directly sample the composition of a distant star or a planet that orbits it. However, we can sample such remote bodies indirectly by analyzing the light they emit or reflect.

Whenever light interacts with matter, an imprint is left on the light that tells us something about the matter's chemical and physical properties. Light from a distant star or planet can tell us what the body is made of, its temperature, its motion, and many of its other properties. Light, therefore, is our key to studying the Universe. To use the key, however, we need to understand the nature of light.

In this chapter, we will discover that light is a form of energy that can be thought of either as a wave or as a stream of particles. Furthermore, we will discover that the light we see with our eyes is just a small part of the radiation emitted by astronomical objects. We will also learn that light can be produced within an atom by changes in its electrons' energies. These changes leave a precise "fingerprint" on the light that we can use to identify the atoms involved.

By examining light, we might learn about the characteristics of a star in which the light was generated. We may also learn about the properties of an interstellar cloud that the light passed through on its journey to Earth. However, the light may also bear unwanted messages. For example, when light reaches our atmosphere, gases there alter its properties, blocking some rays and bending and blurring others. These distortions limit what astronomers can learn from the ground, so we sometimes need to carry out observations from space to interpret the light correctly.

The goal of this chapter is to explain the nature of light, how it is produced, and how it interacts with matter. Our first step toward this goal is to better understand what light is.

CONCEPTS AND SKILLS TO REVIEW

- Structure of atoms (Preview, p. 8)
- Forces (Preview, p. 8)

4.1 PROPERTIES OF LIGHT

Light is radiant energy; that is, it is energy that can travel through space from one point to another without the need of a direct physical link. Therefore, light is very different in its basic nature from, for example, sound. Sound can reach us only if it is carried by a medium such as air or water, whereas light can reach us even across empty space. In empty space, we can see the burst of light of an explosion, but we will hear no sound from it at all.

Light's capacity to travel through the vacuum of space is paralleled by another very special property: its high speed. In fact, the speed of light is an upper limit to all motion. In empty space, light travels at the incredible speed of about 300,000 kilometers per second. An object traveling that fast could circle the Earth seven times in under a second.

The speed of light in empty space is a constant and is denoted by "*c*." However, in transparent materials, such as glass, water, and gases, the speed of light is reduced. Furthermore, different colors of light are slowed differently. For example, in nearly all materials, blue light travels slightly more slowly than red light. As we will see in chapter 5, lenses and prisms work because they slow the light as it travels through them.

To be more precise, light travels at 299,792.458 km/sec in a vacuum. It travels just 90 km/sec slower through air, but its speed is only 3/4 as large through water, and even lower through glass.

The Nature of Light—Waves or Particles?

Observation and experimentation on light throughout the last few centuries have produced two very different models of what light is and how it works. Light can be modeled as waves or as particles. Each model captures some of the characteristics of light, but neither fully describes all of light's properties.

According to the wave model, light is a mix of electric and magnetic energy, swinging up and down in intensity, as depicted in figure 4.1A. Because light is a mix of electric and magnetic energy, it is often called an **electromagnetic wave** or **electromagnetic radiation.** The ability of such radiation to travel through empty space comes from the interrelatedness of electricity and magnetism.

You can see this relationship between electricity and magnetism in everyday life. For example, when you start your car, turning the ignition key sends an electric current from the battery to the starter. There, the electric current generates a magnetic force that turns over the engine. Similarly, when you pull the cord on a lawn mower, you spin a magnet that generates an electric current that creates the spark to start its engine.

This ability of electricity and magnetism to generate each other is what leads to a wave. A small disturbance of an electric field creates a magnetic disturbance in the adjacent space, which in turn creates a new electric disturbance in the space adjacent to it, and so on. Thus, a fluctuation of electric and magnetic field spreads out from its source carried by the fields. In this fashion, light can move through empty space "carrying itself by its own bootstraps."

As the electromagnetic wave travels through matter, it may disturb the atoms, causing them to vibrate the way a water wave makes a boat rock. It is from such disturbances in our eyes, a piece of film, or an electronic sensor that we detect the light.

The model of light as a wave works well to explain many phenomena, but it fails to explain some of light's other properties when it interacts with matter. In those circumstances, it is necessary (and easier!) to use a model in which light is thought of as a stream of particles called **photons** (fig. 4.1B) The photons are individual packets of energy moving through space in a straight line at the speed of light. When photons strike "photoreceptive" chemicals in your eye, they produce the sensation of light.

Although light is described as both particles and waves, it is not unique in this respect. According to the laws of quantum physics, subatomic particles such as electrons and protons can also behave like waves. For this reason, scientists often speak of light and subatomic particles as having a **wave–particle duality,** and they use whichever

 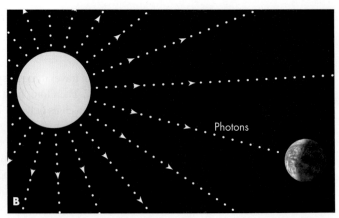

FIGURE 4.1
(A) A wave of electromagnetic energy moves through empty space at the speed of light, about 300,000 kilometers per second. The wave carries itself along by continually changing its electric energy into magnetic energy and vice versa. The curving lines illustrate the changing strength of electric and magnetic energy as the light travels through space. (B) Photons—particles of energy—stream away from a light source at the speed of light.

model—wave or particle—best describes a particular phenomenon. For example, reflection of light off a mirror is easily understood if you imagine photons striking the mirror and bouncing back just the way a ball rebounds when thrown at a wall. On the other hand, the focusing of light by a lens is best explained by the wave model.

Usually we will discuss light using the wave model, so that we do not have to constantly refer first to photons and then to waves. But regardless of which model we use, light has two important properties that we need to describe. The first of these properties is its brightness (or intensity). Brightness is a measure of the total amount of energy carried by the light. In the wave picture, brightness is related to the height of the wave. In the photon picture, brightness is related to the number of photons traveling in a given direction. The second important property of light is its color.

Light and Color

Human beings can see colors ranging from deep red through orange and yellow into green, blue, and violet, and we call these colors the **visible spectrum.** But what property of photons or electromagnetic waves corresponds to light's different colors?

According to the wave theory, the color of light is determined by the light's **wavelength,** which is the spacing between wave crests (fig. 4.2). That is, instead of describing a quality (color), we specify a quantity (wavelength), which we usually denote by the Greek letter lambda, λ. For example, the wavelength of deep red light is about 7×10^{-7} meters. The wavelength of violet light is about 4×10^{-7} meters. Intermediate colors have intermediate wavelengths. The wave–particle duality model allows us to make a similar connection between wavelength and color for photons. Thus, we can also characterize photons by their wavelength.

The wavelengths of visible light are very small (roughly the size of a bacterium). They are therefore usually measured not in meters but in billionths of a meter, a unit called the **nanometer,** abbreviated nm. Thus, the wavelength of red light is about 700 nanometers and that of violet light is about 400 nanometers. Table 4.1 lists the wavelengths of the primary colors in nanometers as well as micrometers, which are sometimes used to measure wavelengths.

The relation between wavelength and color is important, and we will refer to it repeatedly in later chapters. In doing so, we will often mention only the red and blue colors of visible light. The other colors are not missing: they are just not referred to explicitly. Given this simplification, you should be sure to remember that *red* colors refer to *long* wavelengths of visible light and *blue* colors refer to *short* wavelengths.

Note that the wavelength of light is independent of the intensity, or *amplitude,* of the electromagnetic wave (fig. 4.2). Thus, the strength of the variations of the electromagnetic radiation do not change its color. In the particle description of light, we might say that more-intense red light contains a larger number of red photons.

ANIMATION ▶

Photons stream away from a light source at the speed of light.

Table 4.1	Colors and Wavelengths*	
Red	700 nm	0.70 μm
Yellow	580	0.58
Blue	480	0.48
Violet	400	0.40

* Wavelengths listed are only approximate. The micrometer, also called a micron and abbreviated μm, equals 10^{-6} meters. Note that in older texts astronomers often used the unit "angstroms" to measure wavelength. Angstroms are 1/10th of a nanometer, so yellow light would be at 5800 angstroms, for example.

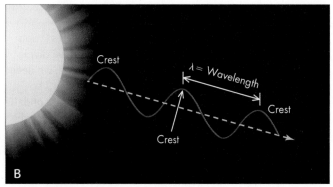

FIGURE 4.2
The distance between crests defines the wavelength, λ, for any kind of wave, be it (A) water or (B) electromagnetic.

wavelength

A   **B**   **C**

FIGURE 4.3
The frequency of a wave is determined from the time it takes for a full wavelength of the wave to pass you. If you are bobbing up and down as a wave passes you, and it takes 2 seconds to move from the trough of the wave (A) through one crest (B), and back into the next trough (C), the frequency of the wave would be said to be 0.5 per second or 0.5 hertz. A longer-wavelength wave moves by at about the same speed, so the time between crests and troughs is longer, so you bob up and down at a lower frequency—if at half the rate, then 0.25 hertz.

Characterizing Electromagnetic Waves by Their Frequency

The unit "hertz" (cycles per second) is named for the German physicist Heinrich Hertz who first demonstrated the existence of electromagnetic waves.

Sometimes it is useful to describe electromagnetic waves by their frequency rather than their wavelength. **Frequency** is the number of wave crests that pass a given point in 1 second. It is usually measured in hertz (abbreviated as Hz) and is usually denoted by the Greek letter nu, ν. The number of hertz indicate the number of waves passing by a point each second as illustrated in figure 4.3. You can see an everyday example of this on a radio dial, where you tune in a station by its frequency rather than its wavelength.

For all kinds of waves, the frequency and wavelength are related to the wave speed, because in one vibration a wave must travel a distance equal to one wavelength. This implies that for light, the product of the wavelength (λ) and the wave frequency (ν) equals the speed of light (c): that is, $\lambda\nu = c$.* Because all light travels at the same speed (in empty space), we can treat c as a constant. Thus, specifying λ determines ν and vice versa. Examples of converting between the two are given in Astronomy by the Numbers: "Wavelength and Frequency." While we will generally use λ to characterize electromagnetic waves, ν is just as good.

---

* In using this equation, if ν is in hertz, the units of λ will be set by the units you choose for c. Thus, if c is in meters per second, λ will be in meters.

ASTRONOMY *by the numbers* WAVELENGTH AND FREQUENCY

Because the product of wavelength and frequency equals the speed of light, you can always calculate light's wavelength from its frequency and vice versa.

For example, the wavelength of an FM radio station at 88.5 MHz can be calculated as follows. First we divide both sides of $\lambda\nu = c$ by ν to give us: $\lambda = c/\nu$. Next, we need to remember that the M in MHz indicates "million," so there are 88.5 million or 88,500,000 waves per second. If we take the speed of light in meters per second, we find:

$$\lambda = \frac{3.00 \times 10^8 \text{ m/sec}}{8.85 \times 10^7 \text{ waves/sec}}$$

$$= 0.339 \times 10^1 \text{ m/wave}$$

$$= 3.39 \text{ meters per wave.}$$

If instead we ask what the frequency of blue light is, we would divide both sides of $\lambda\nu = c$ by λ to give us: $\nu = c/\lambda$. Taking the wavelength of blue light to be 480 nm from table 4.1 (and recalling that a nm is 10^{-9} m), we find:

$$\nu = \frac{3.00 \times 10^8 \text{ m/sec}}{4.8 \times 10^{-7} \text{ m/wave}}$$

$$= 0.625 \times 10^{15} \text{ wave/sec}$$

$$= 6.25 \times 10^{14} \text{ Hz}$$

Almost a million-trillion waves of blue light pass by every second!

White Light

Although wavelength is an excellent way to specify most colors of light, some light seems to have no color. For example, the Sun when it is seen high in the sky and an ordinary lightbulb appear to have no dominant color. Light from such sources is called **white light.**

White light is not a special color of light; rather it is a mixture of all colors. That is, the sunlight we see is made up of all the wavelengths of visible light—a blend of red, yellow, green, blue, and so on—and our eyes perceive this blend as white. Newton demonstrated this property of sunlight by a very simple but elegant experiment. He passed sunlight through a prism (fig. 4.4) so that the light was spread out into the visible spectrum (or rainbow of colors). He then recombined the separated colors to reform the beam of white light.

You can see how colors of light mix if you look at a television or computer screen with a magnifying glass. You will notice that the screen is covered with tiny red, green, and blue dots. In a red object, only the red spots are lit. In a blue one, only the blue spots are lit. In a white object, all three are lit, and the brain mixes these three colors to form white. Other colors are made by appropriate blending of red, green, and blue.

The way light mixes is very different from the way pigments of paint mix. Red, green, and blue paint when mixed give a dark brownish color. Paints work by absorbing all but the colors they reflect, so a red paint absorbs blue and green light. When we mix together several colors of paint, we produce a mix that absorbs most colors and reflects relatively little light.

Why do we see sunlight as white? It is actually quite blue compared to most indoor lighting, as you may find comparing photographs taken outdoors versus those taken indoors under incandescent lights. Both look "white" because our senses make us aware of *changes* in our surroundings. Thus, we ignore the ambient "color" of sunlight just as we come in time to ignore a steady background sound or smell.

While white light is a mix of all the wavelengths of light that our eyes can sense, there is more to light than what meets the eye. Just as red is but one part of the visible spectrum, so too the visible spectrum itself is but one part of a much wider spectrum of electromagnetic waves.

A

B

FIGURE 4.4

(A) A prism spreads "white" light into its component colors (a spectrum). Combining the colors again with a lens makes the light "white" again. (B) Spectra in everyday life.

| **4.2** | # THE ELECTROMAGNETIC SPECTRUM: BEYOND VISIBLE LIGHT |

Visible light is just one of the many kinds of electromagnetic waves that exist. For example, radio waves, X rays, and ultraviolet radiation are also electromagnetic waves. They differ from visible light only in their wavelengths, and to indicate this fundamental unity, scientists refer to them collectively as the **electromagnetic spectrum.**

The electromagnetic spectrum has been studied over a huge range of wavelengths. The longest electromagnetic waves yet detected have wavelengths thousands of kilometers long.* The shortest have wavelengths of 10^{-18} meters or less. Ordinary visible light falls in a very narrow section in about the middle of the known spectral range (see fig. 4.5 and table 4.2).

Although our eyes can detect only a tiny portion of the electromagnetic spectrum—namely, the part we call visible light—various instruments allow us to explore most of the other wavelength regions, too. In fact, new instruments allow astronomers to "see" such astronomically important events as the formation of stars, the remnants left behind when stars die, and, indirectly, black holes. Some examples of astronomical imaging in different portions of the electromagnetic spectrum are shown in figure 4.5. In the following subsections, we discuss these different wavelength regions in the order in which they were discovered.

---

* Such long waves have not been detected from astronomical sources, however, and cannot pass through our atmosphere.

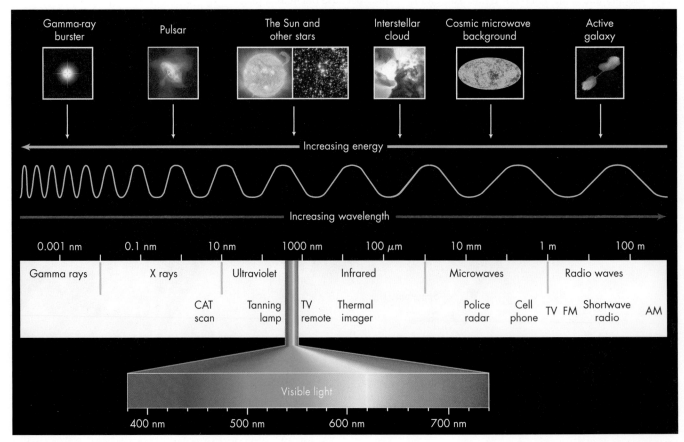

FIGURE 4.5
The electromagnetic spectrum.

Table 4.2	Electromagnetic Spectrum	
Wavelength	**Kind of Radiation**	**Astronomical Sources**
100–500 meters	Radio (AM broadcast)	Pulsars (remnants of exploded stars—also emit X rays)
10–100 meters	Shortwave radio	Active galaxies
1–10 meters	TV, FM radio	Solar radio outbursts, interstellar gas
10–100 centimeters	Radar, cell phones	Planets, active galaxies
1–100 millimeters	Microwaves	Interstellar clouds, cosmic background radiation
700 nm–1000 μm	Infrared (heat)	Young stars, planets, interstellar dust
400–700 nanometers	Visible light	Stars, Sun
10–400 nanometers	Ultraviolet	Stars
0.01–10 nanometers	X rays	Collapsed stars, hot gas in galaxy clusters
10^{-7}–0.01 nanometers	Gamma rays	Active galaxies and gamma-ray bursters

Note: 1 mm = 1000 μm; 1 μm = 1000 nm

Infrared Radiation

The exploration of the electromagnetic spectrum began in 1800, when Sir William Herschel (discoverer of the planet Uranus) showed that heat radiation, such as you feel from the Sun or from a warm radiator, though invisible, is related to visible light.

Herschel was trying to measure heat radiated by astronomical sources. He projected a spectrum of sunlight onto a tabletop and placed a thermometer in each color to measure its energy. He was surprised that when he put a thermometer just off the red end of the visible spectrum, the thermometer registered an elevated temperature there just as it did in the red part of the spectrum. He concluded that some form of invisible energy perceptible as heat existed beyond the red end of the spectrum and he therefore called it **infrared.**

Even though our eyes cannot see infrared light, nerves in our skin can feel it as heat. Several kinds of snakes, including the rattlesnake, have special infrared sensors located just below their eyes. These allow the snake to "see" in total darkness, helping it to find warm-blooded prey such as rats.

A camera equipped with sensors that react to infrared photons allows us to image the intensity of the infrared radiation and determine the relative warmth of different objects. The infrared image of a dog in figure 4.6 shows the strongest infrared radiation coming from the dog's warm eyes and the least from its cold paws and nose. Infrared imagers are also used medically to identify and screen people who have a fever.

FIGURE 4.6

An infrared image of a dog. This is a "false-color image," with colors used to help represent the intensity of the infrared radiation.

Ultraviolet Light

Another important part of the electromagnetic spectrum, **ultraviolet** radiation, was discovered in 1801 by J. Ritter while he was experimenting with chemicals that might be sensitive to light. Ritter noted that when he shone a spectrum of sunlight on a layer of silver chloride, the chemical blackened most strongly in the region just beyond the violet end of the spectrum, implying the presence there of some invisible radiation.

Infrared and ultraviolet radiation differ in no physical way from visible light except in their wavelengths. Infrared has longer wavelengths and ultraviolet shorter wavelengths than visible light (see table 4.2). Exploration of those parts of the electromagnetic spectrum with wavelengths much longer and much shorter than visible light had to await the growth of new technology, as the development of radio astronomy demonstrates.

Radio Waves and Microwaves

James Clerk Maxwell, a Scottish physicist, predicted the existence of radio waves in the mid-1800s. It was some 20 years later, however, before Heinrich Hertz produced them experimentally in 1888, and another 50 years had to pass before Karl Jansky discovered naturally occurring radio waves coming from cosmic sources. Jansky's discovery in the 1930s that the center of the Milky Way was a strong source of radio emission was the birth of radio astronomy.

Radio waves have wavelengths of about 1 meter and longer, making them much longer than visible and infrared waves. Today we can generate radio waves and use them in many ways, ranging from communication to radar. Astronomers detect radio waves using radio telescopes. Radio signals, generated by natural processes, allow astronomers to obtain radio "views" of forming stars, exploding stars, active galaxies, and interstellar gas clouds. Radio wavelengths are even being searched for signals that might reveal the existence of extraterrestrial civilizations.

Originally, astronomers drew no clear line between the radio and infrared parts of the spectrum. It is difficult to observe from the ground between radio and infrared wavelengths, from about 1 mm to 1 m (see section 4.5). This range is now usually called microwaves. Besides being useful for cooking food, radiation at these wavelengths also allows astronomers to study molecules in interstellar clouds, and it proves to be particularly important for studying light from the Universe when it was very young.

X Rays and Gamma Rays

X rays were discovered by Wilhelm Roentgen in 1895, but many decades passed before their first astronomical detection in the late 1940s when the Sun was found to emit them. X ray wavelengths are far shorter than those of visible light, typically between 0.01 and 10 nanometers, but they too are important. Doctors and dentists use X rays to probe our bones and organs. Astronomers use X ray telescopes to detect X rays emitted by the hot gas surrounding black holes and the tenuous gas in distant groups of galaxies.

Even more extreme are gamma rays, the shortest wavelengths known, which are associated with some of the most violent events in the Universe, such as supernova explosions. Both of these wavelength regions are difficult to study from the ground because, as we will see in section 4.5, they fall in wavelength bands that are strongly blocked by the Earth's atmosphere. However, orbiting telescopes have now given astronomers preliminary views of the sky in these wavelengths.

Energy Carried by Electromagnetic Radiation

Despite the enormous variety of electromagnetic radiation, it is all the same physical phenomenon: the vibration of electric and magnetic energy traveling at the speed of light. Equivalently it can be described as streams of photons, whether radio photons, visible photons, or gamma-ray photons. The essential difference between these many kinds of electromagnetic radiation is their wavelength (or frequency). This difference alters not only how we perceive the light but also how much energy each photon can carry.

The warmth we feel on our face from a beam of sunlight demonstrates that light carries energy, but not all wavelengths carry the same amount of energy. The amount of energy, E, carried by a photon depends on its wavelength, λ. Each photon is an energy packet that carries an amount of energy given by

$$E = \frac{hc}{\lambda}.$$

E = Energy carried by a photon of wavelength = λ
h = Planck's constant
c = Speed of light (constant)

The term h is known as Planck's constant. If E is measured in joules and λ in meters, then $h = 6.63 \times 10^{-34}$ joule · second, and $hc = 1.99 \times 10^{-25}$ joule · meter. The speed

| Radio - Effelsburg | Infrared - *Herschel* | Visible - Kitt Peak | Ultraviolet - *Swift* | X ray - *ROSAT* |
| 6 cm | 70–250 µm | 400–700 nm | 160–330 nm | 0.8–3 nm |

FIGURE 4.7
A series of images of the Andromeda galaxy (M31) at wavelengths from radio waves to X rays.

of light c and Planck's constant h are unchanging, so the equation tells us that if the wavelength of the light decreases, the energy it carries increases:

Short-wavelength photons carry more energy than long-wavelength photons in inverse proportion to their wavelengths.

An ultraviolet photons therefore carries more energy than an infrared photon. In fact, an ultraviolet photon of sufficiently short wavelength carries so much energy that it can break apart molecular bonds. As we will see in chapter 14, this can cause intense heating of gas near stars. Nearer to home, it is the reason ultraviolet light gives you a sunburn but an infrared heat lamp does not.

The differing appearance of the galaxy M31 in different wavelengths shown in figure 4.7 gives an idea of what we can see in different wavebands. While visible light shows the overall distribution of stars in the galaxy, the infrared image shows where dust particles in interstellar space have been heated by the stars. In ultraviolet light we see mostly hot young stars and gas clouds that surround them. At X ray wavelengths we can pick out some of the regions where stars have recently exploded, and radio emission is also generated in the gas clouds around both young and exploding stars.

4.3   THE NATURE OF MATTER AND HEAT

Like so many of our ideas about the nature of the Universe, our ideas of matter date back to the ancient Greeks. For example, Leucippus, who lived about the fifth century B.C. in Greece, and his student Democritus taught that matter is composed of tiny indivisible particles. They called these particles *atoms,* which means "uncuttable" in Greek.

Our current model for the nature of atoms dates back to the early 1900s, with the work of the British physicist Ernest Rutherford. Rutherford showed with a series of experiments that atoms have a tiny core, the nucleus, around which yet smaller particles, called electrons, orbit (fig. 4.8). Electrons have a negative electric charge while the nucleus of an atom has a positive charge; atoms are held together by the electrical attraction between oppositely charged particles. This electrical attraction is what causes clothes to stick together in a dryer—electrons can rub off from one garment to another, building up static electricity. The attraction between the nucleus of one atom and the electrons of a neighboring atom also can link atoms together to form *molecules.*

The presence of electrical charges in atoms allows them to generate electromagnetic radiation and to interact with photons. These interactions leave an "imprint" on electromagnetic radiation that allows us to determine many properties of a material, including its temperature and the kinds of atoms and molecules out of which it is made.

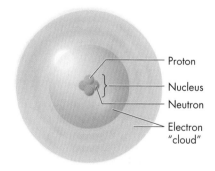

Proton
Nucleus
Neutron
Electron "cloud"

FIGURE 4.8
A schematic diagram of a helium atom. Two electrons orbit in a wave–particle cloud surrounding a nucleus that contains two protons and two neutrons.

FIGURE 4.9

Temperatures on the Kelvin, Celsius, and Fahrenheit scales.

The Kelvin Temperature Scale

Before we examine the interactions between matter and light, we need to introduce the scale that astronomers and other scientists use to measure the temperature of materials.

One of the most important contributors to the understanding of heat and molecular motion was the English physicist Lord Kelvin. Kelvin studied numerous problems in physics and astronomy, ranging from the motion of fluids to the properties of gases. Much of this latter work was motivated by his attempts to improve the energy efficiency of steam engines. In the course of studying the energy content of gases, Kelvin devised a temperature scale that is used today in virtually all the physical sciences. The reason for this wide usage is that on the Kelvin scale, a body's temperature is directly related to its energy content and to the speed of its molecular motion. That is, the greater a body's Kelvin temperature, the more rapidly its atoms move and the more energy it possesses. Similarly, if the body is cooled toward a temperature of zero on the Kelvin scale, molecular motion within it slows to a virtual halt and its energy approaches zero. Partly as a result of this, the Kelvin scale has no negative temperatures.

Temperatures on the Kelvin scale are not given in degrees but are simply called "kelvin." For example, the freezing and boiling points of water are very nearly 273 and 373 kelvin, respectively. Room temperature is about 300 kelvin. Relatively simple formulas allow conversion between the Kelvin scale and the more familiar Fahrenheit and Celsius degree scales. Celsius temperatures, $°C$, are simply the number of kelvins minus 273. Fahrenheit temperatures, $°F$, can be calculated by using the formula $°F = 9/5 \, K - 459.4$, where K is the temperature in kelvins. Figure 4.9 shows how the Kelvin, Celsius, and Fahrenheit temperature scales compare. Because of its direct relation to so many physical processes, we will use the Kelvin scale in most of the remainder of this book.

Temperature and Radiation

Hot objects emit light, as you can easily demonstrate if you turn on a burner on an electric stove. As the stove's burner warms up, it begins to glow. Initially the burner emits only a dim, deep red color of light. However, as the burner grows hotter, the light it emits becomes brighter and its color changes (fig. 4.10), becoming brighter red and eventually yellow. If we could make the burner even hotter, it would glow blue.

Recalling that blue light has a shorter wavelength than red light, you can see from this simple demonstration an important relation between an object's temperature and the color of the light it emits:

As an object's temperature increases, the object radiates light more strongly at shorter wavelengths.

That is, as an object heats up, the color of the light it emits shifts from red (long) wavelengths toward yellow (medium) wavelengths and, if hot enough, to blue (short) wavelengths. This connection between an object's temperature and the color of the light it emits applies to more than just electric stove burners. It is a general property of many hot objects, including stars, and it allows astronomers to measure the temperature of stars (and many other astronomical objects) from their light.

The hotter burner glows more orange than the cooler burner.

FIGURE 4.10

A hotter burner glows brighter and at a shorter wavelength than a cooler burner.

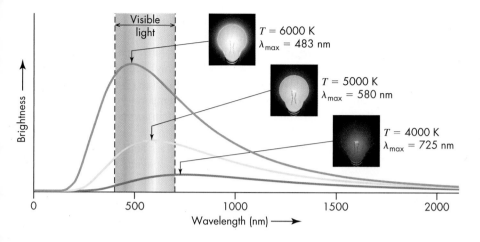

FIGURE 4.11
As an object is heated, the wavelength at which it radiates most strongly, λ_{max}, shifts to shorter wavelengths, a relation known as "Wien's law." Note also that as the object's temperature rises, the amount of energy radiated increases at *all* wavelengths.

In this discussion we have not given a value to the temperature but merely said "hot" or "hotter." However, we can find a numerical value for the temperature using a relation first worked out by the German physicist Wilhelm Wien (pronounced "veen") about 1900. **Wien's law** states that the wavelength (color) at which an object radiates most strongly is inversely proportional to the object's temperature.

In figure 4.11 we illustrate this principle by plotting the amount of energy radiated at each wavelength (color) for an object heated to different temperatures. You can see from the curves for these objects that the hottest one is brightest (emits its greatest amount of energy) at 483 nanometers. That is, its curve is highest at that wavelength. On the other hand, the coolest one emits its greatest amount of energy at 725 nanometers. Finally, the object of intermediate temperature has the peak of its curve at an intermediate wavelength (580 nm). You can therefore see in this figure the relation between temperature and wavelength we mentioned above—namely, that hotter objects emit more strongly at shorter wavelengths. We show how to use this relation in Astronomy by the Numbers: "Taking the Temperature of the Sun."

You might note that the wavelength at which the Sun radiates most strongly corresponds to a blue-green color, yet the Sun looks yellow-white to us. The reason we see it as whitish is related to how our eyes perceive color. Physiologists have found that the human eye interprets sunlight (and light from all extremely hot bodies) as whitish, with only tints of color. Keep in mind that although hot bodies emit most strongly at

Blackbody radiation and stellar luminosity

ASTRONOMY *by the numbers*   TAKING THE TEMPERATURE OF THE SUN

To measure the temperature of an object using Wien's law, we proceed as follows. First we measure the object's brightness at many different wavelengths to find at which particular wavelength it is brightest (that is, its wavelength of maximum emission). Then we use the law to calculate the object's temperature. To see how this is done, however, we need a mathematical expression for the law.

If we let T be the object's temperature measured in kelvin, and λ_{max} be the wavelength in nanometers at which it radiates most strongly (fig. 4.11), Wien's law can be written in the form

$$T = \frac{2.9 \times 10^6 \, K \cdot nm}{\lambda_{max}}.$$

The subscript "max" on λ is to remind us that it is the wavelength of maximum emission. The constant 2.9×10^6 is more accurately $2.898 \times 10^6 \, K \cdot nm$. We round it off here to make calculations easier. The error this creates is small.

As an example, let's measure the Sun's temperature. The Sun turns out to radiate most strongly at a wavelength of about 500 nanometers. That is, its $\lambda_{max} = 500$ nm. Then, substituting that value for λ_{max} in our expression for T, we find

$$T = \frac{2.9 \times 10^6 \, K \cdot nm}{500 \, nm} = \frac{2.9 \times 10^6}{5 \times 10^2} \, K$$

$$= 5.8 \times 10^3 \, K = 5800 \text{ kelvin.}$$

This is within a hundred degrees of the actual value.

a wavelength indicated by Wien's law, they emit at all other wavelengths too. Thus, cool stars look white tinged with red, while very hot stars look white tinged with blue.

The difference between regions of different temperatures can sometimes be seen on the surface of the Sun. Sunspots (fig. 4.12) are "stormy" regions on the surface of the Sun that are cooler than surrounding regions as discussed further in chapter 12. Sunspots are actually very bright, but because their temperature is typically about 4500 kelvin, they look dark and somewhat reddish in color in contrast to the 6000 kelvin surrounding regions.

Although Wien's law works accurately for most stars and planets, it has some important exceptions. For example, the red color of an apple and the green color of a lime come from the light they *reflect* and have nothing to do with their temperature. The apple does emit some radiation, but if it is at normal room temperature, its radiation will be mostly in the infrared.

Wien's law makes good sense if you think about the relation between energy and temperature. Hotter things carry more energy (other quantities being equal) than cooler things. Also, bluer light carries more energy than red. Thus, it is reasonable to expect that hotter bodies emit bluer light.

Our discussion above has been qualified several times by terms such as *usually* and *most*. The reason for these qualifications is that Wien's law is exact only for a class of objects known as blackbodies. A **blackbody** is an object that absorbs all the radiation falling upon it. Because such a body reflects no light, it looks black to us when it is cold; hence its name. Experiments show that when blackbodies are heated, they radiate more efficiently than other kinds of objects. Thus, they are both excellent absorbers and excellent emitters. Moreover, the intensity of their radiation changes smoothly from one wavelength to the next with no gaps or narrow peaks of brightness, as illustrated by the curves in figure 4.11. Very few objects are perfect blackbodies, but many of the objects we will study are near enough to being blackbodies that we can use Wien's law with little fear of its being in error. For example, the electric stove burner, the Sun, and the Earth all obey Wien's law quite satisfactorily.

FIGURE 4.12
Sunspots. The darker spots in the picture are cooler than surrounding regions, so they look dark by contrast. However, they are generating light.

4.4   RADIATION FROM INDIVIDUAL ATOMS

Both solid matter and dense collections of atoms emit blackbody radiation, but gases generally behave quite differently. We see these contrasting kinds of emission in ordinary lightbulbs. Incandescent lightbulbs emit light by heating a solid filament to high temperature, which emits light according to Wien's law. However, fluorescent lights and neon signs are not blackbodies. They instead produce light by first pulling electrons free from the atoms in the gas, which then emit light when the electrons recombine with the atoms.

This same difference is found in nature. Interstellar clouds, for example, radiate strongly only at specific wavelengths, such as a narrow wavelength range in the red part of the visible spectrum or the millimeter wavelength part of the radio spectrum. The clouds' colors are determined by characteristics of the individual atoms in the gas more than by temperature.

You can demonstrate the importance of composition in determining color with a gas flame on a stove or Bunsen burner. Normally the flame has a blue part and a yellow part. The yellow part is blackbody radiation from very hot specks of carbon soot. However, the blue part is caused by non-blackbody emission from carbon atoms. If you add chemicals to the flame, the flame's color may change dramatically. For example, if you hold some copper sulfate crystals in the flame with a pair of pliers, the flame will take on a greenish-blue color caused by the emission wavelengths of copper (fig. 4.13A). Likewise, the strontium in a highway emergency flare gives its light a strong red color (fig. 4.13B).

A   B

FIGURE 4.13
These photographs show the effect of (A) copper (green) and (B) strontium (red) on the burner's flame.

The structure of atoms determines both their chemical properties and their light-emitting and light-absorbing properties. For example, iron and hydrogen not only have very different atomic structures but also emit very different wavelengths of light. From those differences astronomers can deduce whether an astronomical body—a star or a planet—contains iron, hydrogen, or whatever chemicals happen to be present. Therefore, an understanding of the structure of atoms ultimately leads us to an understanding of the nature of stars.

The Chemical Elements

Iron and hydrogen are examples of what are called chemical **elements.** A chemical element is a substance composed only of atoms that all have the same electrical charge in their nucleus. We described in section 4.3 how an atom has a dense core called a *nucleus* around which particles called *electrons* orbit. The nucleus is in turn composed of particles called *protons* and *neutrons;* the protons have a positive charge, whereas the neutrons have no charge. The number of protons therefore determines the kind of chemical element the atom is.

For example, hydrogen consists exclusively of atoms that contain 1 proton; helium, of atoms that contain 2 protons; carbon, 6; oxygen, 8; and so forth. Although the identity of an element is determined by the number of protons in its nucleus, the chemical properties of each element are determined by the electrons orbiting its nucleus. However, the number of electrons normally equals the number of protons. The protons attract an equal number of the oppositely charged electrons until the atom is electrically neutral.[†]

Table 4.3 lists some of the more important elements we will discuss during our exploration of the Universe and the number of protons each contains. Most elements can have various forms with different numbers of neutrons, called *isotopes*. Isotopes have the same chemical properties, but different masses. The table lists stable isotopes of each kind of atom. Other numbers of neutrons are possible, but the resulting nucleus is unstable, or *radioactive*.

Table 4.3	Astronomically Important Elements	
Element	Number of Protons	Number of Neutrons*
Hydrogen	1	0,1
Helium	2	2,1
Carbon	6	6,7
Nitrogen	7	7,8
Oxygen	8	8,10,9
Neon	10	10,11,12
Silicon	14	14,15,16
Iron	26	30,31,32

* The number of neutrons listed is the number found in stable forms of the element, the most abundant listed first. Different neutron numbers lead to what are called isotopes of the element. Isotopes with different numbers of neutrons than those listed are unstable.

Electron Orbitals

The orbits of electrons in an atom are generally extremely small. For example, the diameter of the smallest electron orbit in a hydrogen atom is only about 10^{-10} (1 ten-billionth) meter. This infinitesimal size leads to effects that operate at an atomic level and have no counterpart in larger systems. The most important of these effects is that the electron orbits may have only certain prescribed sizes.

Although a planet may orbit the Sun at any distance, an electron may orbit an atomic nucleus at only certain distances, much as when you climb a set of stairs, you can be only at certain discrete heights. For example, in a hydrogen atom the electron must have an orbital radius of 0.053 n^2 nanometers, where $n = 1$, or 2, or 3, or ..., etc. That is, the radius can be 0.053, 0.21, 0.48, etc. nanometers, but it cannot have intermediate values. We describe this restriction on the allowable sizes of orbits by saying that they are **quantized.**

The restriction on orbital sizes results from the electron's acting not just as a particle but also as a wave. That is, just as light itself has a wave–particle duality, so too does an electron. The electron's wave nature forces the electron to move only in orbits whose circumference is a whole number of wavelengths. If it were to move in other orbits, the electron's wave nature would "cancel" it out. Figure 4.14 illustrates this property of electron orbits and compares it to painters being confined to only certain levels when they work on a scaffold.

[†] Under some circumstances, an atom may lose or gain one or more electrons. Such atoms are said to be ionized, as we will discuss in chapters 12 and 16.

Level '3'

Level '2'

Ground level = '1'

Hydrogen atom

Electron cloud
orbital $n = 3$

Electron cloud
orbital $n = 2$

Electron cloud
orbital $n = 1$

Proton in
nucleus

$r = 0.053$ nm

$r = 0.21$ nm

FIGURE 4.14

Just as the painters can only be at levels 1, 2, 3, . . . of the scaffold (and cannot "float in between"), so too an electron must be in orbital 1, 2, 3, ..., etc.

The wave nature of the electron has another important effect. It "smears out" the location of the electrons. As a result, although we have described the electrons as orbiting like tiny planets around the nucleus, most scientists prefer to think of them as existing in an electron cloud, which is called an **orbital.** The shape of the orbital describes the probability of finding the electron at different positions. Simplified depictions of several orbitals are illustrated in figure 4.14.

Electrons in orbitals have another property totally unlike those of planets in orbit: they can routinely shift from one orbital to another. This shifting occurs when there is a change in their energy, as can be understood by a simple analogy. The electrical attraction between the nucleus and the electron creates a force between them like a spring. If the electron increases its distance from the nucleus, it is like stretching the spring. This requires giving energy to the atom. Likewise, if the electron moves closer to the nucleus, it is as if the spring relaxes and the atom must give up, or emit, energy. We perceive that emitted energy as light or, more generally, electromagnetic radiation. The wavelength of that radiation is precisely the same for atoms of the same type, but it is not the same for other kinds of atoms because of the different charges of the nucleus and the interactions between electrons. Therefore the electrons in hydrogen atoms behave in one way, but the electrons in iron atoms have a very different pattern of behaviors.

In summary, then, atoms consist of a nucleus containing protons and neutrons surrounded by electrons in orbitals. The identity of the atom—the element—is determined by the number of protons in its nucleus. The electrons are bound to the nucleus by the electric attraction between the protons and electrons. Electrons may shift from one orbital to another accompanied by a change in the atom's energy. With this picture of the atom in mind, we can now turn to how light is generated within atoms.

The Generation of Light by Atoms

We saw above that when an electron moves from one orbital to another, the energy of the atom changes. If the atom's energy is increased, the electron moves outward from an inner orbital. Such an atom is said to be **excited.** On the other hand, if the electron moves inward toward the nucleus, the atom's energy is decreased.

Although the energy of an atom may change, the energy cannot just disappear. One of the fundamental laws of nature is the **conservation of energy.** This law states that energy can never be created or destroyed, it can only be changed in form. According to this principle, if an atom loses energy, that energy *must reappear in some other form.* One important form in which the energy reappears is light, or, more generally, electromagnetic radiation.

How is the electromagnetic radiation created? When the electron drops from one orbital to another, it alters the electric energy of the atom. As we described in section 4.1, such an electrical disturbance generates a magnetic disturbance, which in turn generates a new electrical disturbance. Thus, the energy released when an electron drops from a higher to a lower orbital becomes an electromagnetic wave, a process called **emission** (fig. 4.15). Emission plays an important role in many astronomical phenomena. The aurora borealis (northern lights) is an example of emission by atoms in the Earth's upper atmosphere, and sunlight and starlight are examples of emission in those bodies.

The reverse process, in which light is stored in an atom as energy, is called **absorption** (fig. 4.16). Absorption lifts an electron from a lower to a higher orbital and excites the atom by increasing the electron's energy. Absorption is important in understanding such diverse phenomena as the temperature of a planet and the identification of star types, as we will discover in later chapters.

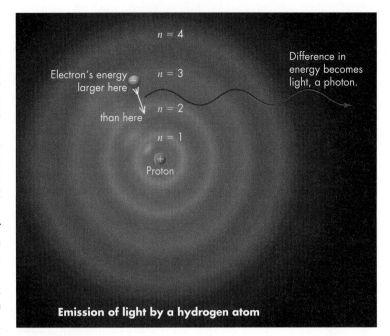

Emission of light by a hydrogen atom

FIGURE 4.15

Energy is released when an electron drops from an upper to a lower orbital, causing the atom to emit electromagnetic radiation.

Energy is released when an electron drops from an upper to a lower orbital.

Absorption

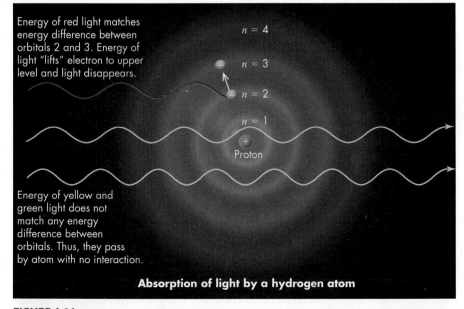

Energy of red light matches energy difference between orbitals 2 and 3. Energy of light "lifts" electron to upper level and light disappears.

Energy of yellow and green light does not match any energy difference between orbitals. Thus, they pass by atom with no interaction.

Absorption of light by a hydrogen atom

FIGURE 4.16

An atom can absorb light if its energy matches the energy difference between two orbitals.

Q: If the light reflecting off or passing through a leaf is green, what does that tell us about the wavelengths used for photosynthesis?

Emission and absorption are particularly easy to understand if we use the photon model of light. According to this model, an atom emits a photon when one of its electrons drops from an upper to a lower orbital. Similarly, an atom absorbs light when a photon of the right energy collides with it and "knocks" one of its electrons into an upper level.

You may find it helpful in understanding emission and absorption if you think of an analogy. Absorption is a bit like drawing an arrow back preparatory to shooting it from a bow. Emission is like the arrow being shot. In one case, energy of your muscles is transferred to and stored in the flexed bow. In the other, it is released as the arrow takes flight.

4.5 FORMATION OF A SPECTRUM

The key to determining the composition and conditions of an astronomical body is its spectrum. The technique used to capture and analyze such a spectrum is called **spectroscopy.** In spectroscopy, the light (or more generally the electromagnetic radiation) emitted or reflected by the object being studied is collected with a telescope and spread into its component colors to form a spectrum by passing it through a prism or a grating consisting of numerous, tiny, parallel lines.

Figure 4.17 shows not only how to form a spectrum but also what the spectrum looks like—a band of rainbow colors, in the case of visible light. In general, we can show a spectrum as it would look to us or as a plot of the light's brightness at each color. In the case shown in figure 4.17, all colors are present and are more or less equally bright. As we will discover later, not all spectra look like this. Sometimes only a few colors are present.

Because light is emitted from atoms as electrons shift between orbitals, we might expect that the light will bear some imprint of the kind of atom that creates it. That is usually the case, and astronomers can search for the atom's "signature" by measuring how much light is present at each wavelength.

Spectroscopy is such an important tool for astronomers that we will look in greater detail at how it works. Specifically, why does an atom produce a unique spectral signature? To understand that, we need to recall how light is produced.

FIGURE 4.17

Sketch of a spectroscope and how it forms a spectrum. Either a prism or a grating may be used to spread the light into its component colors.

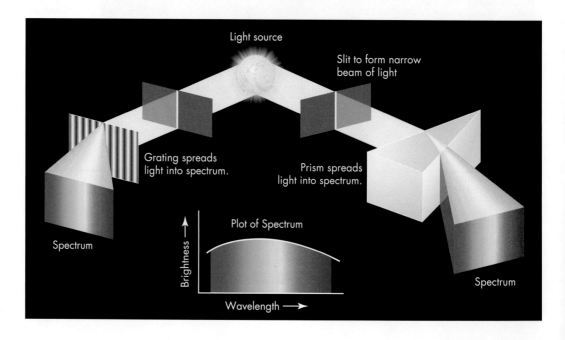

How a Spectrum Is Formed

We saw earlier that each kind of atom has a different number of electrons. This means that each kind of atom has a different set of electron orbitals. We also saw that the orbital in which an electron is located at any given moment sets the atom's energy. For this reason scientists sometimes refer to the orbitals as **energy levels.**

When an electron moves from one energy level (orbital) to another, the atom's energy changes by an amount equal to the *difference* in the energy between the two levels. As an example, suppose we look at light from heated hydrogen. Heating speeds up the atoms, causing more forceful and frequent collisions, knocking each excited atom's electron to outer orbitals. However, the electrical attraction between the nucleus and the electron draws the electron back almost at once. Suppose we look at an electron shifting from orbital 3 to orbital 2, as shown in figure 4.18. As the electron shifts downward, the atom's energy decreases, and the energy lost appears as light.

The wavelength of the emitted light can be calculated from the energy difference of the levels and the relation between energy (E) and wavelength (λ): $E = hc/\lambda$. If we evaluate the wavelength of this light, we find that it is 656 nanometers, a bright red color. An electron dropping from orbital 3 to orbital 2 in a hydrogen atom will always produce light of this wavelength.

If, instead, the electron moves between orbital 4 and orbital 2, there will be a different change in energy because orbital 4 has a different energy from that of orbital 3. That different energy will have a wavelength different from 656 nanometers. A calculation of its energy change leads in this case to a wavelength of 486 nanometers, a turquoise blue color.

Similar calculations show that when the electron jumps from orbital 5 to orbital 2 or from orbital 6 to orbital 2, other spectrum lines are emitted. However, we will see no light at most other wavelengths because hydrogen has no electron orbitals corresponding to those energies. Therefore, the hydrogen spectrum shows a set of brightly colored lines separated by wide, dark gaps. This is how an emission-line spectrum is formed.

The model for the atom that we have used here and that so successfully explains its spectrum was developed by Danish physicist Niels Bohr. He won the 1922 Nobel Prize for Physics for his work.

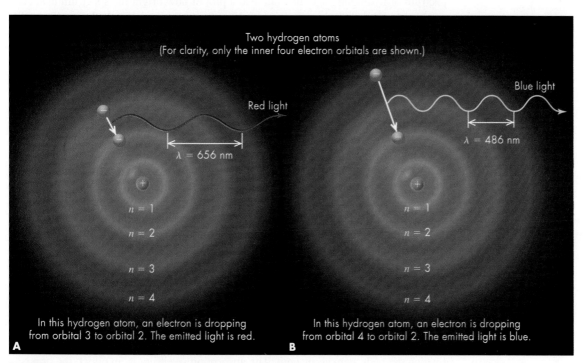

Two hydrogen atoms
(For clarity, only the inner four electron orbitals are shown.)

Red light

$\lambda = 656$ nm

Blue light

$\lambda = 486$ nm

$n = 1$
$n = 2$
$n = 3$
$n = 4$

$n = 1$
$n = 2$
$n = 3$
$n = 4$

In this hydrogen atom, an electron is dropping from orbital 3 to orbital 2. The emitted light is red.

In this hydrogen atom, an electron is dropping from orbital 4 to orbital 2. The emitted light is blue.

A

B

FIGURE 4.18

Emission of light from a hydrogen atom. The energy of an electron dropping from an upper to a lower orbital is converted to light. The light's color depends on the orbitals involved.

FIGURE 4.19

The spectrum of hydrogen in the visible wavelength range.

ANIMATION ▶

Atomic emission and absorption

You can see these emission lines in figure 4.19, which shows not only what the spectrum of hydrogen looks like but also a plot of how bright the spectrum is at each wavelength. We might now imagine making the same kind of calculations for a different chemical element, such as helium. If we did, we would discover that its wavelengths are in general different from those of hydrogen (fig. 4.20). Thus, hydrogen's signature is its red 656-nanometer and blue 486-nanometer lines, and that signature offers astronomers a way to determine what astronomical objects are made of.

Identifying Atoms by Their Light

In the previous paragraphs we have described how atoms emit light. Moreover, we have seen that each chemical element emits a particular set of spectrum lines and that these emission lines provide a way to identify the presence of that element in a hot gas.

It is also possible to identify atoms in a gas from the way they absorb light. Light is absorbed if the energy of its wavelength corresponds to an energy that matches the difference between two energy levels in the atom. If the wavelength *does not* match, the light will not be absorbed, and it will simply move past the atom, leaving itself and the atom unaffected.

For example, suppose we shine a beam of light through a box full of hydrogen atoms. The light initially contains all the colors of the visible spectrum, but after it has passed through the box, we will find that certain wavelengths of the light are missing from the spectrum (fig. 4.21). In particular, the spectrum will contain gaps that appear as dark lines at 656 nanometers and 486 nanometers, precisely the wavelengths at which the hydrogen atoms emit. The absorption spectrum is, in effect, the opposite of the emission spectrum and an atom's absorption lines have exactly the same wavelengths as its emission lines.

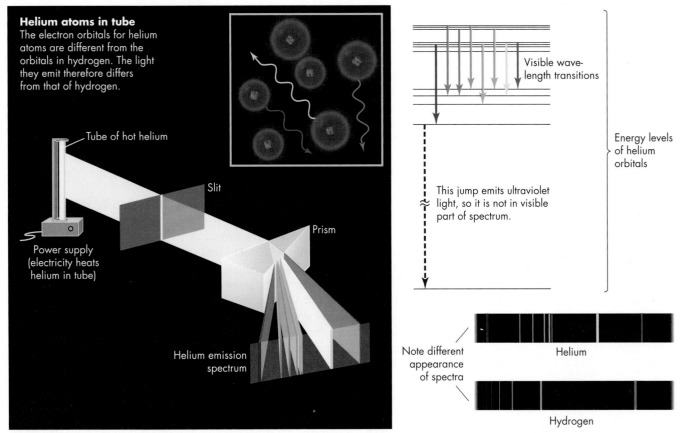

Helium atoms in tube
The electron orbitals for helium atoms are different from the orbitals in hydrogen. The light they emit therefore differs from that of hydrogen.

Tube of hot helium

Slit

Prism

Power supply (electricity heats helium in tube)

Helium emission spectrum

Visible wavelength transitions

This jump emits ultraviolet light, so it is not in visible part of spectrum.

Energy levels of helium orbitals

Note different appearance of spectra

Helium

Hydrogen

FIGURE 4.20
The spectrum of helium in the visible wavelength range.

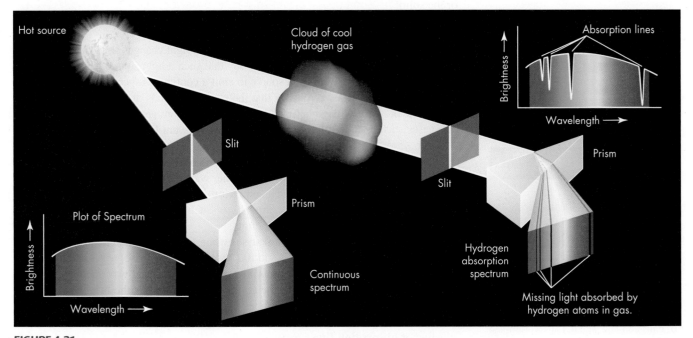

Hot source

Cloud of cool hydrogen gas

Absorption lines

Brightness

Wavelength

Slit

Prism

Slit

Prism

Plot of Spectrum

Brightness

Wavelength

Continuous spectrum

Hydrogen absorption spectrum

Missing light absorbed by hydrogen atoms in gas.

FIGURE 4.21
A hot, dense substance produces a continuous spectrum. Atoms in a gas cloud between an observer and the source of continuous emission absorb only those wavelengths whose energies equal the energy difference between their electron orbitals. The absorbed energy lifts the electrons to upper orbitals. The lost light makes the spectrum darker at the wavelengths where it is absorbed.

These gaps are created by the light with a wavelength of 656 nanometers interacting with the hydrogen and lifting the electron in some atoms from orbital 2 to orbital 3 while the light at 486 nanometers lifts the electron in other hydrogen atoms from orbital 2 to orbital 4. Light at other wavelengths in this range has no effect on the atom. Thus, we can tell that hydrogen is present from either its emission or its absorption spectral lines.

Light is not only emitted and absorbed by individual atoms in a gas. If the atoms are linked to one another to form molecules, such as water or carbon dioxide, the molecules too produce emission and absorption lines. The spectra of molecules are generally quite complex, including not just transitions of electrons between orbitals but a variety of low-energy transitions involving vibration or rotation of the molecular structure.

Even solid objects may imprint spectral lines on light that reflects off them. For example, when light from the Sun reflects from an asteroid, some wavelengths are absorbed by materials on the asteroid's surface. This gives astronomers information about the composition of bodies too cool to emit significant light of their own.

We conclude that in general we can identify the kind of atoms or molecules that are present by examining either the bright or the dark spectrum lines. Gaps in the spectrum at 656 nanometers and 486 nanometers imply that hydrogen is present. Similar gaps at other wavelengths would show that other elements are present. By matching the observed gaps to a directory of absorption lines, we can identify the atoms that are present. *This is the fundamental way astronomers determine the chemical composition of astronomical bodies.*

Types of Spectra

Whether a spectrum will have emission lines or absorption lines depends on certain general properties having to do with the density and temperature of the source of light and any intervening material. For example, the spectrum of a hot, tenuous gas is different from that of a hot, dense solid, regardless of the composition of either the gas or the solid.

Spectra have the following three basic forms:

A. For some sources, the brightness of the emitted light changes smoothly with wavelength and all colors are present. We say that such a light source has a **continuous spectrum** (fig. 4.22A). For a source to emit a continuous spectrum, its atoms must in general be packed so closely that the electron orbitals of one atom are distorted by the presence of neighboring atoms. Such conditions are typical of solids or dense gases such as the heated filament of an incandescent lightbulb or the interior of a star.

B. Some heated objects have a spectrum in which light is emitted at only a few particular wavelengths while most of the other wavelengths remain dark (fig. 4.22B). This type of spectrum is called an **emission-line spectrum.** Emission-line spectra are usually produced by *hot, tenuous gas,* such as that in a fluorescent tube, the aurora, and many interstellar gas clouds.

C. A still different type of spectrum arises when light from a hot, dense body passes through cooler gas between it and the observer. In this case, nearly all the colors are present, but light is either missing or much dimmer at wavelengths absorbed by the atoms or molecules in the cooler gas. This causes the bright background to be crossed with narrow dark lines where the light of some colors is fainter or absent altogether (fig. 4.22C). The resulting spectrum is therefore called a dark-line or **absorption-line spectrum.**

Absorption-line spectra were first detected astronomically in 1802, when the English scientist William H. Wollaston viewed sunlight through a prism and a narrow slit

Continuous spectrum

A

Emission-line spectrum (hydrogen gas)

B

Absorption-line spectrum (hydrogen gas)

C

FIGURE 4.22
Types of spectra: (A) continuous, (B) emission-line, and (C) absorption-line.

FIGURE 4.23
(A) The spectrum of the Sun. Note the narrow dark absorption lines. (B) A graphical representation of the spectrum.

(fig. 4.23). He noticed dark lines between some of the colors but paid little attention to them. These dark lines in the Sun's spectrum were independently discovered a few years later by the German scientist Joseph Fraunhofer, who catalogued them and discovered similar lines in other stars. In fact, because nearly all stars have absorption-line spectra, this spectrum type is especially important in astronomy. Considering the physical process that gives rise to absorption-line spectra, this observation reveals that a continuous spectrum from the dense interior of the star shines through cooler gas in layers above it.

Astronomical Spectra

The first step facing an astronomer who wants to analyze a spectrum is to identify the spectral lines. This is done by measuring the wavelengths of the lines and then consulting a directory of spectrum lines. By matching the wavelength of the line of interest to a line in the table, astronomers can determine what kind of atom or molecule created the line. A look at a typical spectrum will show you that some lines are hard to see, being faint and weak. On the other hand, some lines may be very obvious and strong. The strength or weakness of a given line turns out to depend on the number of atoms or molecules absorbing (or emitting, if we are looking at an emission line) at that wavelength. Unfortunately, the number of atoms or molecules that can absorb or emit depends not just on how many of them are present but also on their temperature, as we will discuss more fully in chapter 13. Nevertheless, astronomers can deduce from the strength of emission and absorption lines the relative quantity of each atom producing a line and thereby deduce the composition of the material in the light source. Table 4.4 shows the result of such an analysis for our Sun, a typical star.

Let us now apply what we know about spectra to astronomical bodies. We begin by using a telescope to obtain a spectrum of the object of interest. Next we measure the wavelengths and identify the lines. As an example, consider the spectrum of the Sun in figure 4.23.

We can see from the spectral lines that the Sun contains hydrogen. In fact, when a detailed calculation is made of the strength of the lines, it turns out that about 71% of the Sun's mass is hydrogen. (This is about 90% of the atoms, because hydrogen is so light.) Similar observations

Table 4.4	Composition of a Typical Star, Our Sun*	
Element	**Relative Number of Atoms**	**Percent by Mass**
Hydrogen	10^{12}	71.1%
Helium	9.64×10^{10}	27.4%
Oxygen	5.75×10^{8}	0.65%
Carbon	2.88×10^{8}	0.25%
Neon	8.91×10^{7}	0.13%
Nitrogen	7.94×10^{7}	0.08%
Silicon	4.07×10^{7}	0.06%
Iron	3.47×10^{7}	0.14%
Gold	8	0.00000011%
Uranium	0.4	0.000000007%

* The table lists eight of the most common elements along with gold and uranium to illustrate how extremely rare they are. Data on relative number of atoms drawn from Lodders (2003) *The Astrophysical Journal*, vol. 591, pp. 1220–1247.

FIGURE 4.24

Emission-line spectra at a variety of wavelengths. (A) A spectrum of a comet at visible and ultraviolet wavelengths. (B) A microwave spectrum of a cold interstellar cloud. (C) An X ray spectrum of hot gas from an exploding star.

show that the spectrum of a comet consists mainly of emission lines from such substances as the molecules carbon dioxide (CO_2) and cyanogen (CN). Thus, we know that comets contain these substances. Moreover, recalling how the different types of spectra (continuous, emission-line, or absorption-line) form, we can tell that the CN and CO_2 must be gaseous, because the spectrum consists of emission lines. There may be other gases present too, but without seeing their spectral features, we cannot tell for sure.

Although the examples we have used above involve spectra of visible light, one of the most useful features of spectroscopy is that it can be used in *any* wavelength region where an atom or molecule emits or absorbs electromagnetic radiation. For example, figure 4.24A shows ultraviolet emission lines from the gas cloud around a comet. Figure 4.24B shows microwave emission lines from molecules inside a cold interstellar cloud. X ray emission lines from ionized atoms in a very hot region are shown in figure 4.24C. Regardless of the wavelength region we use, the spectrum allows us to determine what kind of atoms and molecules are present. In addition, by detailed analysis of the exact wavelength and shape of the spectral lines, we can sometimes determine in what direction and how fast that material is moving.

Absorption in the Atmosphere

Gases in the Earth's atmosphere absorb electromagnetic radiation, affecting the flow of heat and light through it. The amount of this absorption depends strongly on wavelength. For example, the gases affect visible light hardly at all, and so our atmosphere is nearly completely transparent at the wavelengths we see with our eyes. On the other hand, some of the gases strongly absorb infrared radiation while others strongly block ultraviolet radiation.

This nearly total blockage of infrared and ultraviolet radiation results from the ability of molecules such as carbon dioxide, water, and ozone to absorb at a wide range of wavelengths. For example, carbon dioxide and water molecules strongly absorb infrared wavelengths. Likewise, ozone (O_3) and ordinary oxygen (O_2) strongly absorb ultraviolet radiation, while oxygen and nitrogen absorb X rays and gamma radiation. As a result of this absorption by molecules, virtually no infrared, ultraviolet, X ray, or gamma-ray radiation can pass through our atmosphere.

Molecules in general are excellent absorbers (and emitters) because they can store energy in more ways than isolated atoms can. Individual atoms can store energy by exciting electrons into higher-energy orbitals. Molecules can store energy not only by exciting electrons but also by the spinning and vibrating motions of the molecule as a whole. These added ways to store energy are what make molecules such powerful blockers of radiation at many wavelengths. The transparency of the atmosphere to visible light compared to its opacity (nontransparency) to infrared and ultraviolet

ANIMATION ▶

Absorption by atmosphere

The addition of infrared-absorbing gases to our atmosphere can contribute to the greenhouse effect. Some pollutants can also cause depletion of Earth's ozone layer. These problems are discussed further in chapter 6.

EXTENDING
our reach

OBSERVING THE CRAB NEBULA AT MANY WAVELENGTHS

In midsummer A.D. 1054 just after sunset, astronomers in China and other East Asian countries noticed a brilliant star near the crescent moon in a part of the sky where no bright star had previously been seen. They wrote of this event: "In the last year of the period Chih-ho,… a guest star appeared. … After more than a year it became invisible."

We know today that these astronomers of long ago witnessed a supernova explosion, the violent event that marks the death of a massive star. Their record—nearly 1000 years old—begins a story that continues today as astronomers try to understand what causes a star explosion. Although the story began with naked-eye observations, it continued with observations made with telescopes on the ground and in space. Moreover, the story illustrates how astronomers have come to rely on observing radiation at many wavelengths, not just visible light.

Despite its initial brilliance, the dying star seen so long ago faded and disappeared from the sky and astronomical records. Then, in 1731, John Bevis, a British physician and amateur astronomer, noticed with his telescope a faint dim patch of light in the constellation Taurus. (You can see Taurus and where the Crab Nebula lies in Looking Up #5 at the front of the book.) Twenty-seven years later, Charles Messier, a French astronomer and comet hunter, rediscovered the glowing cloud and made it the first entry in his catalog of fuzzy patches of light that were not comets. In 1844, Lord Rosse, a British astronomer and telescope builder, noticed that the fuzzy patch contained filaments (fig. 4.25A) that to his eye resembled a crab. He therefore named it the Crab Nebula.

In 1921, John Duncan, an American astronomer, compared two photographs of the nebula taken 12 years apart and noticed that it had increased slightly in diameter.

He therefore deduced that the nebula was expanding. At the same time, several other astronomers came across the ancient Chinese records and noticed the coincidence in position of the nebula with the report of the exploding star. Then, 7 years later, Edwin Hubble, at Mount Wilson Observatory in California, measured the increase of size more accurately and calculated from the rate of expansion that the nebula was about 900 years old—roughly the same age as the dying star seen nearly a millennium earlier by the Chinese astronomers.

Since then astronomers have examined the Crab Nebula at virtually all wavelengths and, in doing so, have added still more to their understanding of a star's demise. For example, in 1948, Australian astronomers discovered that the Crab Nebula is a powerful source of radio waves (fig. 4.25B). In 1968, further observations at radio wavelengths revealed that a faint, peculiar star near the center of the nebula is spinning about 30 times per second and that it is the core of the star whose explosion created the Crab Nebula. Astronomers have discovered that it is also a source of X ray radiation (fig. 4.25C).

What have all these observations shown? They have given astronomers their best view yet of the last moments of a star's life. From visible-wavelength observations, astronomers measure that the gas ejected when the star exploded is expanding with a speed of about 1000 km/sec. From radio-wavelength observations, they deduce that the nebula contains charged particles moving at nearly the speed of light and that the central star pulses on and off about 30 times per second. The X ray observations confirm this picture. Thus, by observing the Crab Nebula and its stellar corpse at a variety of wavelengths, astronomers have shown that it is a far richer and more mysterious object than could be deduced from observations at one wavelength alone.

A   B   C

FIGURE 4.25
(A) Visible-light photograph of the Crab Nebula. (B) Radio image of the Crab Nebula. (C) X ray image, in false color, of the core of the Crab Nebula.

Q: What does the flattened round shape of the glowing gas in panel C suggest about the gas motion?

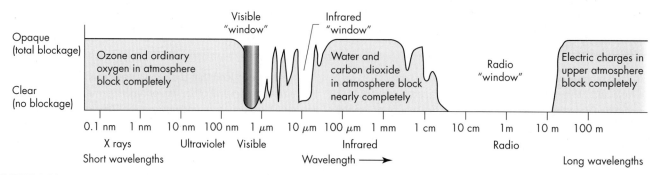

FIGURE 4.26

Atmospheric absorption. Wavelength regions where the atmosphere is essentially transparent, such as the visible spectrum, are called "atmospheric windows." Wavelengths and atmosphere are not drawn to scale.

radiation creates what is called an **atmospheric window.** An atmospheric window is a wavelength region in which energy comes through easily, compared to other wavelengths (fig. 4.26).

Without atmospheric windows, it would be impossible for us to study astronomical objects from the ground. As it is, the visible window allows us to study stars and galaxies (which radiate lots of visible energy), but the lack of ultraviolet windows and the rarity of infrared windows makes it very difficult to observe objects that radiate strongly in those spectral regions. This is one of several reasons why astronomers so badly need telescopes in space, where there is no absorption by our atmosphere. Today telescopes of many varieties orbit Earth so that we can study the whole electromagnetic spectrum. These observations have given us a much more complete picture of the objects in the Universe, and the technologies are steadily improving so that we continue to make new discoveries. We will discuss telescopes further in chapter 5.

We have seen how important a source of information electromagnetic radiation is. It provides information about the temperature, composition, and physical conditions of astronomical objects. The box Extending Our Reach: "Observing the Crab Nebula at Many Wavelengths" describes one example of how astronomers used light from across the electromagnetic spectrum to understand an unusual object.

4.6 THE DOPPLER SHIFT: DETECTING MOTION

Doppler shift

The Doppler effect

If we observe light from a source that is moving toward or away from us, we will find that the wavelengths we receive from it are altered by the motion. If the source moves toward us, the wavelengths of its light will be shorter. If it moves away from us, the wavelengths will be longer, as illustrated in figure 4.27A. Furthermore, the faster the source moves, the greater those changes in wavelength will be. This change in wavelength caused by motion is called the **Doppler shift,** and it is a powerful tool for measuring the speed and direction of motion of astronomical objects.

The Doppler shift occurs for all kinds of waves. You have perhaps heard the Doppler shift of sound waves from the siren of an emergency vehicle as it passes you and moves away: the siren's pitch drops as the wavelength of its sound increases (fig. 4.27B). Likewise, the Doppler shift of a radar beam that bounces off your car reveals to a law enforcement officer how fast your car is moving (fig. 4.27C).

It is easy to see why the Doppler shift occurs. Imagine that the waves from the moving source are a wiggly line. If the source is moving away from you, the wiggles get stretched so the spacing of the waves increases. If the source is moving toward you, the wiggles are scrunched up so the spacing of the waves decreases (fig. 4.27D).

Mathematically, the Doppler shift arises because the wavelength we observe (λ) is the original wavelength (λ_0) plus the distance the source travels during the time a

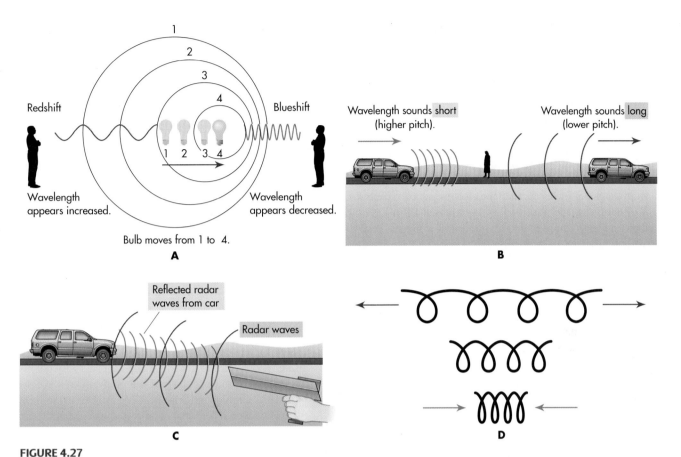

FIGURE 4.27
(A) The Doppler shift: waves appear to shorten as a source approaches and to lengthen as it recedes. (B) The Doppler shift of sound waves from a passing car. (C) The Doppler shift of radar waves in a speed trap. (D) A Slinky illustrates the shortening of the space between its coils as its ends move toward each other and a lengthening of the space as the ends move apart.

single wave is emitted. That distance depends on the speed, V, of the source along the line from the source to the observer, a speed astronomers call the radial velocity. Some mathematics (omitted here) then leads to the Doppler shift formula

$$V = c(\lambda - \lambda_0)/\lambda_0 = c(\Delta\lambda/\lambda_0)$$

where c is the speed of the wave and $\Delta\lambda = (\lambda - \lambda_0)$ is simply shorthand for the change in wavelength.* We will see applications of this law in later chapters. Here, our goal is simply to indicate that the Doppler shift allows us to find out how fast a source is moving away from us (positive V) or toward us (negative V).

Doppler shift measurements can be made at any wavelength of the electromagnetic spectrum. However, regardless of the wavelength region observed or whether the waves are of visible light, astronomers generally refer to any shift that increases the measured wavelength as a redshift. Any shift that decreases the measured wavelength is referred to as a blueshift. Thus, even though we may be describing radio waves, we will say that an approaching source is blueshifted and a receding one is redshifted.

Doppler shift measurements are very powerful for understanding the nature of astronomical objects. They allow us to study the shapes of rotating asteroids, using the different wavelengths of reflected radar signals from a rotating asteroid to determine its shape (fig. 4.28), to determine how fast stars are rotating and moving through our Galaxy, and to study the motions of galaxies as they speed away from each other in the expanding Universe.

$\Delta\lambda$ = Wavelength shift
λ = Measured wavelength (what we observe)
λ_0 = Emitted wavelength
c = Speed of light
V = Velocity of source along the line of sight (radial velocity)

FIGURE 4.28
Doppler image of a bowling-pin-shaped asteroid made by measuring the time delay and Doppler shift of a radar signal.

---

*The Greek letter Δ, or delta, is widely used to stand for "the change in quantity."

SUMMARY

Light can be described in two complementary ways: as a stream of particles called photons, or as electromagnetic waves. In the wave picture, the energy increases as the wavelength decreases. The wavelength of light determines its color. Red light has a longer wavelength than blue light. In addition to the electromagnetic radiation that we see as visible light, many electromagnetic waves, such as infrared, ultraviolet, radio, X, and gamma rays, are invisible to the eye. The entire assemblage of electromagnetic waves is called the electromagnetic spectrum.

Energy can be absorbed by or released from an atom when an electron moves to a higher or lower orbital, respectively. If an electron drops from an upper to a lower orbital, the energy appears as light. If light of the appropriate energy (wavelength) hits an atom, it may lift an electron in the atom from a lower to an upper orbital. The generation of light is called emission. The removal of light from a beam of radiation is called absorption.

Each kind of atom has a unique set of wavelengths at which it emits and absorbs, allowing us to determining the composition of many objects. However, atoms packed close together (in a dense gas or solid) will generally emit and absorb over a wide range of wavelengths, producing a continuous blackbody spectrum that shifts to shorter wavelengths as the temperature of the material grows hotter.

Gases in Earth's atmosphere absorb light outside of visible and radio wavelength "windows." Telescopes in space are used to observe astronomical objects at other wavelengths.

Motion of atoms alters the wavelengths we observe, creating a Doppler shift from which we can deduce their speed toward or away from us.

QUESTIONS FOR REVIEW

1. (4.1) Why is light called electromagnetic radiation?
2. (4.1) What is a photon? How fast can photons travel?
3. (4.1/4.3) How are color and wavelength related? What about temperature and wavelength of a glowing body?
4. (4.2) What is meant by the electromagnetic spectrum?
5. (4.2) Name the regions of the electromagnetic spectrum from short to long wavelengths.
6. (4.3) Describe the Kelvin temperature scale.
7. (4.3) How does the color of dense materials change with temperature? How does this relate to the idea of a blackbody?
8. (4.4) What makes elements different from each other? What is the arrangement of the parts of an atom?
9. (4.5) What is the difference between emission and absorption in terms of what happens to an electron in an atom?
10. (4.5/4.6) What are some of the things astronomers can learn about astronomical objects from their spectra?
11. (4.5) Which gases in the atmosphere absorb infrared radiation? Which gases absorb ultraviolet radiation?
12. (4.6) Explain how the Doppler shift affects waves reflected by or emitted from a moving body.

THOUGHT QUESTIONS

1. (4.1–3) If red stars are cooler than blue stars, and red light has less energy than blue light, why do you suppose we associate the color red with hot and the color blue with cold in everyday life?
2. (4.2) Why do night-vision cameras use infrared detectors?
3. (4.3/4.4) Through a telescope, you see a red object. Is that enough information to tell what temperature it is? Explain.
4. (4.3–4.5) Why don't atoms emit a continuous spectrum?

5. (4.3–4.5) Given that water absorbs microwaves very strongly, can you explain why a Pop-Tart gets very hot inside while its crust stays cool if you heat it in a microwave oven?
6. (4.5) How can you tell what sort of gas is emitting light?
7. (4.5) How would a spectrum help you learn what the atmosphere of Venus is made of?
8. (4.5) Review the types of spectra. What kinds of spectra do an incandescent lightbulb and a compact fluorescent bulb produce? Make an argument why a 23-W fluorescent bulb can light up a room as effectively as a 100-W incandescent. Where does the other energy from the incandescent go?
9. (4.5) If you added more water or carbon dioxide to our atmosphere, how would it alter the loss of heat from our planet? Would the Earth get warmer or colder? Why?
10. (4.3–4.5) Can you explain why the atmospheric layer containing ozone is much warmer than the levels above and below it?

PROBLEMS

1. (4.1) Use the Sun's distance of 150 million kilometers to calculate how long light takes to travel from the Sun to the Earth.
2. (4.1) Suppose you are operating a remote-controlled spacecraft on Mars from a station here on Earth. How long will it take the craft to respond to your command if Mars is at its nearest point to Earth? Use data in the appendix for your calculations.
3. (4.1) A frequency commonly used for cell phones, wireless Internet, and even in microwave ovens is 2.4 GHz. What is the wavelength of this radiation?
4. (4.2) A solar flare emits X rays and radio waves simultaneously. Which reaches the Earth first? If the X rays have a wavelength of 0.25 nanometers and the radio waves have

a wavelength of 6 cm, how many times larger is the frequency of the X rays than the radio waves?

5. (4.3) Your body temperature is about 300 K. At what wavelength do you radiate most strongly? What region of the electromagnetic spectrum is this? Do you understand now how a rattlesnake can bite you in the dark?

6. (4.3) An electric stove burner on "high" radiates most strongly at about 2000 nanometers. What is its temperature?

7. (4.3) The Earth's temperature averaged over the year is about 300 kelvin. At what wavelength does it radiate most strongly? In what part of the electromagnetic spectrum does this wavelength lie? Can you see it?

8. (4.4) Sketch an atom emitting light. Does the electron end up in a higher or lower orbit? Repeat for an atom absorbing light.

9. (4.6) Calculate the Doppler shift for blue light (wavelength of 500 nanometers) reflected off a sports car traveling away from you at 150 km/hr. What is $\Delta\lambda$? What is the wavelength you see? Could we see the shift in color with our eyes?

10. (4.6) You are analyzing a radio spectrum of an outer part of a distant spiral galaxy. A spectral line expected to be at 21 centimeters is instead measured to be at 21.010 centimeters. Is the outer part of the galaxy rotating toward or away from you? How fast is that part of the galaxy moving?

TEST YOURSELF

1. (4.2) Which kind of light travels fastest?
(a) Ultraviolet (b) Visible (c) Gamma rays (d) Radio waves
(e) They all travel at the same speed.

2. (4.2) Which of these types of electromagnetic radiation has the shortest wavelength?
(a) Ultraviolet   (c) Radio   (e) Visible
(b) Gamma ray   (d) X ray

3. (4.2) Which of these photons has the lowest energy?
(a) Ultraviolet   (c) X ray   (e) Radio
(b) Visible   (d) Infrared

4. (4.3) If we doubled the thermal energy of a rock that had a temperature of 7°C = 45°F = 280 K, the new temperature would be
(a) 14°C.   (c) 560 K.
(b) 90°F.   (d) all of the above.

5. (4.3–4.5) Suppose we detect red photons at 656 nanometers emitted by electrons dropping from the $n = 3$ to $n = 2$ orbital in hydrogen. The hydrogen is in an interstellar cloud at 5000 K. If the cloud were heated to 10,000 K, what would be the wavelength (in nanometers) of the photons emitted by the transition?
(a) 328   (b) 656   (c) 1312   (d) 658   (e) 654

6. (4.5) An astronomer finds that the visible spectrum of a mysterious object shows bright emission lines. What can she conclude about the source?
(a) It contains cold gas.

(b) It is an incandescent solid body.
(c) It is rotating very fast.
(d) It contains hot, relatively tenuous gas.
(e) It is moving toward Earth at high speed.

7. (4.5) Most stars have spectra showing dark lines against a continuous background of color. This observation indicates that these stars
(a) are made almost entirely of hot, low-density gas.
(b) have a warm interior that shines through hotter, high-density gas.
(c) have a hot interior that shines through cooler, low-density gas.
(d) are made almost entirely of cool, low-density gas.

8. (4.6) If an object's spectral lines are shifted to longer wavelengths, the object is
(a) moving away from us.   (c) very hot.
(b) moving toward us.   (d) very cold.

KEY TERMS

: FIGURE QUESTION ANSWERS

WHAT IS THIS? (chapter opening): The ring of light around the Sun is called a halo and is caused by sunlight refracted in tiny atmospheric ice crystals. Haloes are quite common (perhaps one a week) and may be seen around both the Sun and the Moon. They are easier to see around the Sun if you cover the Sun with your hand or block its direct light with a building or tree, as shown here.

FIGURE 4.16: Chlorophyll absorbs blue and red wavelengths, but not colors in between.

FIGURE 4.25: It is spinning.

Special and General Relativity

A favorite theme in science fiction is human space travel. At present such travel is limited to flights orbiting the Earth, although in the 1960s and 1970s astronauts traveled to the Moon, landed there, and explored some of its surface. The Moon is figuratively on our doorstep, however. Can we expect to ever be able to travel the vastly longer distances to other stars or galaxies? In science fiction stories, these immense interstellar distances are crossed by craft using faster-than-light travel (fig. E2.1).

Scientists have excellent reasons for concluding that faster-than-light travel is impossible and that travel even at near-light speed requires long spans of time. For example, it takes light more than 4 years to reach us from even the nearest star beyond the Sun, and tens to hundreds of thousands of years to reach us from nearby galaxies. This seems to mean that for astronauts to conquer interstellar distances, they would have to be prepared to live in space for decades, perhaps having the journey completed by their descendants. But it might surprise you to learn that astronauts could in principle travel *millions* of light-years to another galaxy and return to Earth in their own lifetime! The science that explains how such an immense journey could be made is one theme of this essay. Along the way, we will learn a little about two of science's most important and intriguing theories, the theories of special and general relativity. Before we deal with travel at near the speed of light, however, we need to look at how we describe motion.

REST FRAMES

Astronomical objects are in constant motion: the Earth moves through space around the Sun; the Sun moves through space within the Milky Way Galaxy; and so forth. To describe such motions we need a frame of reference, or **rest frame.** In everyday life we often use the ground as our rest frame. For example, we drive a car at 60 mph (~100 km/hr) along the freeway or walk to class across the campus at 4 mph (~6 km/hr) with respect to the ground.

Suppose, however, we are traveling in an airplane and we walk down the aisle of the plane from the back to the front. How fast are we moving? With respect to the plane, we are moving at a walking pace, say a few miles per hour. However, we could also measure our speed in the plane with respect to the ground, in which case our speed would be that of the plane plus our walking speed with respect to the plane, or hundreds of miles per hour. In other words, our measurement of an object's motion depends on the rest frame that we use for our observation.

Describing an object's motion in one rest frame when it is viewed from another rest frame is not difficult. If we consider again

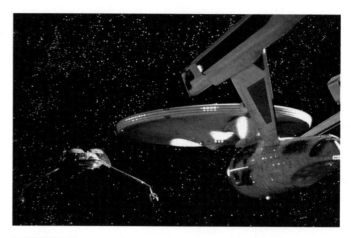

FIGURE E2.1

Spaceships that can travel interstellar distances in a short period of time are popular in science fiction. The realities of such travel are very different, but no less strange.

a person walking in an airplane, we simply add the speed of the person to the speed of the plane. For another example, suppose a person is running at 10 mph and throws a javelin forward at 40 mph in the direction of his or her motion—the javelin will move across the ground at the speed 10 mph + 40 mph = 50 mph. Addition of speeds in this fashion (thrower plus javelin) (fig. E2.2) is an example of what is sometimes called **Galilean relativity,** in honor of Galileo, one of the first scientists to recognize how motions combine with one another. But Galilean relativity fails for light.

FIGURE E2.2

The speed of a javelin with respect to the ground is found by adding the speed of the thrower across the ground to the speed with which the javelin is thrown. That is, the velocities add.

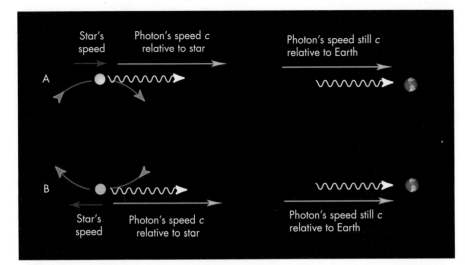

FIGURE E2.3
Light from orbiting stars reaches Earth at the same speed whether the star is moving toward or away from us.

THE SPEED OF LIGHT FROM MOVING OBJECTS

Suppose we have two stars, A and B, in orbit around each other (fig. E2.3). At some point in the orbit, one of the stars, say star A, is moving toward us while the other, say star B, is moving away from us. Light leaves the surface of each star at the speed of light, c. We might then expect that light from the approaching star A will travel toward us a little faster than c, while light from star B, the one swinging away from us, will travel toward us a little slower than c. However, when scientists tried to measure this effect, they found no change in the light's speed: the light arrived at the same speed, c, regardless of the motion of the stars. This is not so odd as it might first seem. We see a similar effect in waves generated by a boat moving across a lake. The waves travel through the water at the same speed irrespective of what speed or direction the boat moves. Such arguments led scientists in the 1800s to conclude that light moves through space like the waves on the water. Waves move at a speed relative to the water they move through. What substance plays the role of the water for light? Scientists at that time concluded that space must be filled with a transparent substance that conducted the light, which they call the **æther.**

Calculations showed that the æther must have very special properties. It must be very rigid for light waves to have their high observed velocity (the stiffer a substance, the faster waves travel in it). At the same time, it must freely allow astronomical objects to plow through it with no resistance. In particular, it must flow past the Earth as we orbit the Sun.

THE MICHELSON-MORLEY EXPERIMENT

In 1887 the American scientists Albert Michelson and Edward Morley designed an apparatus to search for the æther. They predicted that light traveling along the direction of the Earth's motion should move at a different speed from that of light traveling perpendicular to the Earth's motion. But when Michelson and Morley conducted their experiment, they detected no difference in the speed of light along the perpendicular paths. At first they thought this might just be bad luck: that during the experiment, the combination of the Earth's motion around the Sun and the Sun's motion through space might just have happened to make the Earth stationary relative to the æther at that point in its orbit. So they repeated their experiment many times over the course of a year. At some point they should have easily detected the motion of Earth relative to the æther, but they found no sign of the Earth's motion at all.

The Michelson-Morley experiment has been called the most famous "failed" experiment in history because it led to a revolution in physics. The results implied that there was no æther regulating the speed of light. However, if light was not moving relative to an æther, then physicists had no explanation for the constancy of the speed of light reaching us from sources moving at different speeds toward or away from us.

All experiments made then and now show that no matter how fast the source generating the light is moving, and no matter how fast the observer measuring the light is moving, the speed of light through space is always measured to be exactly the same: $c = 299{,}792{,}458$ meters per second. How can this be?

One explanation offered in the late 1800s was that motion through space somehow caused matter to contract* in the direction of motion. If matter contracts when it is moving, this would change our perception of length so that we might be tricked into thinking that the speed of light had not changed. For example, Michelson and Morley were searching for a difference in the speed of light in two perpendicular directions. If the apparatus were compressed in the direction it was moving through space but not in the perpendicular direction, the path the light traveled would be shorter. Such a contraction could potentially cancel out the effect that Michelson and Morley were searching for. The factor by which the apparatus would need to contract

---

* The idea of a contraction caused by motion was first proposed by Irish physicist George Fitzgerald. The Dutch physicist Hendrik Lorentz developed a model for explaining how the contraction might arise.

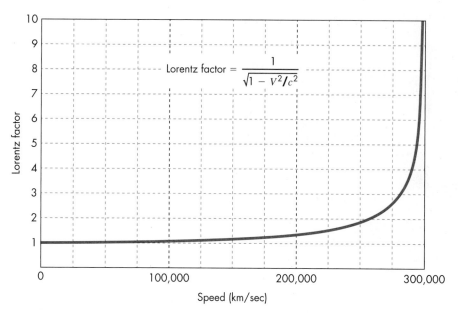

FIGURE E2.4
The Lorentz factor expresses the amount by which objects appear to compress in the direction of their motion. It is also the factor by which a moving clock appears to slow down, and the factor by which a moving object's mass appears to increase.

Table E2.1	The Lorentz Factor at High Speeds
Speed	**Lorentz Factor**
$0.87\,c$	2.0
$0.97\,c$	4.1
$0.99\,c$	7.1
$0.999\,c$	22.4
$0.9999\,c$	70.7
$0.99999\,c$	223.6

is known today as the **Lorentz factor,** usually denoted by the Greek letter gamma (γ). Lorentz hypothesized that an object's length shortens in the direction it is moving by a factor equal to

$$\gamma = \frac{1}{\sqrt{1 - V^2/c^2}}$$

where V is the speed of the object and c is the speed of light. The value of γ for different speeds V is plotted in figure E2.4, and values at high speeds are given in table E2.1.

The Lorentz factor is close to 1 at small speeds. For example, at the speed at which the Earth is orbiting the Sun, 30 kilometers per second, the Earth would contract by only a few centimeters in its direction of motion. At much higher speeds the contraction factor becomes very large, growing to infinity if the speed V were to reach the speed of light. The Lorentz contraction factor can explain the Michelson-Morley experiment, but it cannot explain a number of other conflicting results that were found. Still, it contains an important idea that grew into a whole new concept of the nature of space and time.

EINSTEIN'S THEORY OF SPECIAL RELATIVITY

In 1905 a 26-year-old graduate student named Albert Einstein (fig. E2.5) took on the problem of the seemingly inexplicable measurements of the speed of light. He was completing his physics degree while working in the Swiss patent office and supporting a family. In that one year alone, he completed his doctorate degree and wrote four papers in several areas of physics. Physicists widely agree that three of these papers were each worthy of a Nobel Prize! He was little known at the time and had few colleagues with whom to discuss his ideas; but nonetheless in one of these papers he came up with a brilliant new approach to the question of the motion of light.

Einstein began by concentrating on the finding that light travels at the same speed no matter what the speed of its source or of the observer measuring the light. Even though many experiments had come to this conclusion, most physicists had assumed it was an impossibility and so were seeking other explanations—such as errors in the experiments or the Lorentz contraction. Einstein, instead of thinking of a constant speed of light for all observers as an impossibility, accepted this notion as correct, and proceeded to work out its consequences. He found that this led inevitably not only to a Lorentz-like contraction of space, but also to a stretching of time, or **time dilation,** by the same factor. Even more important, he found that this contraction was not relative to some imagined æther filling space, but that these effects depended only on the relative motions of any two objects.

FIGURE E2.5
Albert Einstein (1879–1955).

FIGURE E2.6
When a spacecraft travels by us, its length appears to be contracted by the Lorentz factor, and clocks on board run more slowly by the same factor. From on board the ship it appears that the Earth is compressed and time is running slowly on Earth by the same factor.

These alterations of our perception of space and time affect everything we see that has any speed relative to us, forming the foundation of what is known as the theory of **special relativity.** According to this theory, if a star or a rocket moves by us at high speed, we will see it squashed in its direction of motion by the Lorentz factor (fig. E2.6). If we could watch a clock tick or measure the speed of an astronaut's heart in the spaceship, we would discover that all these processes occur more slowly by the Lorentz factor.

What is remarkable is that the mathematics Einstein worked out showed that the situation is exactly symmetrical for the astronaut moving by us at high speed. That astronaut will sense herself as being the one who is stationary and will see us as moving by her at high speed. She will measure us and the Earth as being contracted in the direction of "our" motion, and she will measure our clocks and our hearts as running slowly (fig. E2.6). So what two observers in relative motion see is parallel—each would find that the other was the one undergoing the distortions of space and time. Thus, there is no preferred rest frame.

An especially important feature of Einstein's work is the behavior of light. Suppose we and the passing astronaut shine a light beam at each other. We will each measure that the light is moving past us at the same speed, c. However, as we watch each other making these measurements, we will each think that the other is measuring a shorter distance and using a clock that runs too slow.

This theory of special relativity is far-reaching in its implications. The theory is the basis for Einstein's discovery of the relationship between energy and mass: $E = mc^2$. It also predicts that other fundamental quantities change for moving objects. For instance, a moving object grows more massive by the same Lorentz factor that describes how lengths grow shorter and time slows.

These effects mean that nothing can reach the speed of light in our rest frame, let alone exceed it. The rocket ship traveling by us can fire its rockets and accelerate forever, but even though the ship goes ever faster, its mass grows ever greater, making it harder and harder to accelerate. The ship may approach the speed of light, but because the Lorentz factor goes to infinity, it would require infinite energy to reach the speed of light.

SPECIAL RELATIVITY AND SPACE TRAVEL

The theory of special relativity may seem strange, but it has been tested by more than a century of high-precision experiments. There is not a single verified contradiction to it, and its predictions about such things as the slowing of time have been verified directly.

Special relativity is also about more than just perceived differences in space, time, and mass. For example, when atomic clocks (the most accurate clocks, used for establishing time worldwide) are flown between locations, it is found that their travel in airplanes leaves them a little bit slow relative to the network of fixed clocks maintained around the world. The moving clocks must be readjusted after each trip. The Lorentz factor for traveling at airplane speeds of about 1000 km/hour (600 mph) is just $\gamma = 1.0000000000004$, so every tick of the clock on an airplane takes about 4 ten-trillionths of a second longer than a tick of the clock in the ground-based network; but after several hours of flying the effect is measurable.

More intriguing is what happens at such high speeds that the Lorentz factor is large. Experiments have demonstrated, for example, that a subatomic particle called a *muon* normally has a lifetime of about 2 millionths of a second before it decays. However, when muons are traveling at 99% of the speed of light (and therefore have a Lorentz factor of about 7), they live about 14 millionths of a second. This longer lifetime allows them to travel distances that would be impossible for them within their normal lifetimes.

When speaking of such brief times, this change in the rate at which time passes seems minor; but the same slowing factor would apply for a human traveling at 99% of the speed of light.

FIGURE E2.7

(A) From the rest frame of Earth and a star 70 light-years away, it appears that a spaceship traveling at 99% of c is shortened to one-seventh of its original length. (B) On the spaceship, once it is up to speed, it appears that the Earth and star are both moving at 99% of c in the opposite direction. They and the distance between them are shortened to one-seventh of their original length.

If a spaceship could be built that traveled that fast, time would effectively run seven times more slowly on board compared with time here on Earth. If astronauts had food and air supplies for a 10-year trip, the Lorentz factor of $\gamma \approx 7$ (corresponding to $V = 99\%$ the speed of light) would mean they could travel for $\gamma \times 10$ years ≈ 70 years in Earth time. At their speed relative to Earth of $0.99\,c$, they would be able to reach a distance of almost 70 light-years (fig. E2.7A) before they ran out of supplies. At speeds even closer to c the Lorentz factor becomes even larger (see table E2.1) and the potential distances greater.

From the perspective of astronauts on a craft traveling at $0.99\,c$, it would not seem that time was passing any more slowly than normal. Nor would they feel that they or their ship was foreshortened or more massive. From their perspective, the Earth and the star they are visiting *and the distance between the two* are contracted by the Lorentz factor (fig. E2.7B). The distance that looked like 70 light-years when they were stationary (in the rest frame of Earth), would shrink to one-seventh the distance once they reached their high speed!

These marvelous "tricks" of relativity open up possibilities for traveling distances far greater than we might once have imagined. Theoretically, one could travel a million light-years within a human lifetime—although it is far beyond current technologies. (See Astronomy by the Numbers: "A Lorentz Factor of a Million" on page 120.) It is also important to realize that such travel would have major challenges beyond simply reaching such high speeds. From the perspective of the spaceship traveling among the stars at near the speed of light, every atom and every dust particle in space along the ship's path has a mass increased by the Lorentz factor, and it is heading toward the ship at nearly the speed of light! This can give a pebble the impacting force of a ship-destroying asteroid.

Also, as intriguing as these possibilities are, they offer no time savings for the rest of us back on Earth. Consider again the astronauts traveling at 99% of the speed of light for 10 years to visit a distant star. If they then turned around and came home in another 10 years (by their reckoning), they would find that 140 years had passed on Earth. Everyone they knew would have grown old and died. They might even be younger than their great-great-grandchildren!

THE TWIN PARADOX

Something may seem wrong with the description of space travel in the previous section. When we first introduced special relativity, we noted that the time stretching appears to be symmetrical. That is, from the rest frame of the Earth it appears that the astronauts' time is running slowly, while from the rest frame of the astronauts it appears that time on Earth is running slowly. This is sometimes presented as the **twin paradox:** If one of the astronauts left a twin back on Earth, how can we say that one ages more than the other?

The explanation of this seeming paradox is that the situations are *not* in fact symmetrical. The astronaut twin experiences accelerations as her ship speeds up and then slows down at the destination. Furthermore, when she turns her ship around to return to Earth, she experiences yet more accelerations as the ship again speeds up and later slows down as it reaches Earth again. But people remaining back on Earth experience none of these accelerations because they always remain in the same rest frame. Thus, their progression of time remains constant. By contrast, an astronaut's rest frame keeps changing, and in the end the astronaut returns to the rest frame of the Earth.

Imagine, for example, that one of the astronaut twins sends messages once each day to her twin back on Earth, and meanwhile the twin on Earth sends messages once each day to his astronaut sister on the ship. As they part, they each receive the other's messages much more slowly—both because of the Lorentz factor *and* because the separation between the ship and Earth is growing larger and the messages take longer to reach each other. When the astronaut twin reaches the distant star, she sends her 10th-year message. Until this time, the situations are symmetrical. Both have received only a small fraction of each other's messages because of the Lorentz factor and growing separation.

The astronaut twin turns around and starts back, receiving the messages from her brother that were sent years earlier and have been on their way to her across the 70-light-year gap. They come more quickly now because she is approaching Earth, cutting the distance each message has to travel. She reads about her brother getting older and older, now seemingly very rapidly. In the meantime, her brother back on Earth is still reading her messages from the outgoing trip. By the time he dies, he would not even have received the 10th-year message announcing that the ship had reached the other star. At the speed of light in Earth's rest frame, that 10th-year message would take 70 years to reach Earth.

In fact, the spaceship, traveling at 99% of the speed of light, nearly keeps up with the messages sent during its return trip. The ship will reach the Earth in a time just 1% longer than it takes the first message on the return journey announcing that the astronauts were beginning their journey back to Earth.

In the movies, space travel is fast and everyone ages at the same rate. In reality, traveling at high speeds means that the travelers must leave behind not only their homes but their own times and the people in them.

RETHINKING GRAVITY

The theory of special relativity has been demonstrated in experiment after experiment, yet despite its success, Einstein saw that the theory was not yet complete. The ideas of motion and different rest frames did not really account for the effects of gravity. For example, the International Space Station orbits the Earth under the pull of Earth's gravity. Astronauts inside the station feel weightless, as though no forces are acting on them, even though they are constantly changing direction. If there were no windows, the astronauts could easily believe that they were traveling in a straight line, far from any massive objects.

Gravity has the unique ability to change the direction or speed of objects without them feeling a force being applied. Imagine that you had the bad fortune to be inside an elevator whose cable broke. As it was falling, you would find yourself floating inside the elevator, like an astronaut in a space capsule far from any massive objects (fig. E2.8). On the other hand, if you were on a stationary elevator in Earth's gravitational field, the acceleration you feel is no different from what you would experience in a spaceship that was accelerating at 9.8 meters per second per second, or 1 g (fig. E2.9).

Einstein had the remarkable insight that gravity actually alters the nature of space. When you are near a massive object, it causes space to, in effect, flow past you. Thus, when you let go of an apple, it moves downward as if you were moving upward through space. What you observe is the same as letting go of an apple in an accelerating spaceship (fig. E2.9). In the spaceship we would say that the apple remains stationary as the floor of the ship accelerates upward to meet it. This idea that gravity is the same as being in an accelerating frame of reference is known as the **principle of equivalence.**

Newton described gravity as a force that causes objects to accelerate or their paths to curve. By contrast, Einstein described space itself as having motion or as being curved so that an object would follow a curved path in much the way a golf ball follows a curved path or accelerates as it rolls along the hills and slopes of a putting green. At first this idea may sound like a semantic difference, but it proves to have fundamental consequences for the nature of space and time.

FIGURE E2.8
If you are inside a free-falling elevator, the elevator and everything inside it all fall at the same rate so you feel weightless, just as you would feel in a spaceship that is drifting in deep space.

FIGURE E2.9
The principle of equivalence. If you are stationary near the surface of the Earth, objects fall in a way that is equivalent to being aboard a spaceship accelerating at 1 g.

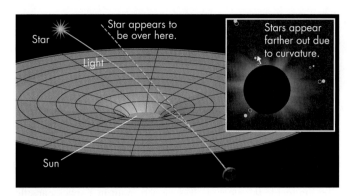

FIGURE E2.10
The "flow" of space toward a massive object like the Sun deflects the path of starlight passing near it. This prediction of general relativity was confirmed in 1919.

GENERAL RELATIVITY

In 1915, a decade after publishing the theory of special relativity, Einstein arrived at his theory of **general relativity.** General relativity is a theory of gravity in that it replaces Newton's law with a mathematical description of how space is curved by mass. This curvature affects the trajectories of objects moving through space. General relativity is broader than Newton's theory—it also describes how energy as well as mass can affect space and how mass and energy alter the flow of time.

While it is mathematically complex to carry out the calculations of how mass and energy "curve" space and time, a simple analogy illustrates many of the important features of general relativity. Imagine a large artificial lake, in the middle of which there is a drain that draws out water at a steady rate. If you sit in a boat on the lake, the outflow of water draws you toward the drain, with a stronger pull the closer you get to the drain. This is analogous to how gravity exerts a stronger pull the closer

you get to an object. If you were to travel across the lake in a motorboat, holding your steering wheel in a fixed direction, you would discover that your path was curved by the flow of water toward the drain. This is similar again to the trajectory of a spacecraft passing by a planet.

The water in this analogy represents space, and it illustrates an important idea of general relativity: space itself can have motion. While the notion of "nothing" having motion seems strange, it is demonstrated in a number of surprising ways. For example, clocks on the surface of the Earth run slower than clocks in deep space. In fact, standing still on the surface of the Earth, we experience a gravitational time dilation the same as the Lorentz factor for an object moving at the escape velocity (11 km/sec) from the Earth's surface.*

The "flow" of space also bends the paths of photons. For example, as starlight passes close to the Sun, where its gravity is strongest, it is deflected like the boat in our analogy, even though photons have no mass (fig. E2.10). This was confirmed in 1919 during a total eclipse of the Sun, where the positions of stars showed a shift that agreed exactly with Einstein's prediction. Einstein further showed that the changes in space and time close to the Sun explained some peculiarities of Mercury's orbit that had long puzzled astronomers when using the Newtonian understanding of gravity.

General relativity has become a cornerstone of modern physics. It is more than a hypothetical curiosity. For instance, if we did not correct for the effects of general relativity between our position on the Earth's surface and satellites orbiting in space, global positioning system (GPS) units would not work. Furthermore, the flow of space and time in general relativity are central to our understanding of black holes and the expansion of the entire Universe, as we shall discover in later chapters.

---

* The gravitational time dilation factor at any point is equal to $1/\sqrt{1 - V_{esc}^2/c^2}$ where V_{esc} is the escape velocity at that position.

ASTRONOMY *by the numbers* — A LORENTZ FACTOR OF A MILLION

At the opening of this Essay, the comment was made that astronauts could (in principle) travel millions of light-years within their own lifetime. To do that they would have to be traveling so fast that the Lorentz factor was about a million so time would slow relative to the rest of the Universe by about that factor. Just how fast is that?

Here we want to solve for V assuming that the Lorentz factor $\gamma = 1,000,000$. Let's begin by writing out the Lorentz formula:

$$10^6 = \frac{1}{\sqrt{1 - V^2/c^2}}$$

Next we can invert and square both sides of the equation, giving:

$$(1/10^6)^2 = 1 - V^2/c^2$$

The term on the left equals 10^{-12}, or 0.000000000001.

We can rearrange the terms to provide a calculation for V/c by adding and subtracting terms from both sides of the equation, then taking the square root, giving us:

$$V/c = \sqrt{1 - 10^{-12}} = \sqrt{0.999999999999}$$

$$\approx 0.9999999999995$$

This is less than one part in a trillion away from the speed of light! Note that many calculators cannot handle this many digits, so you might need to use a calculator program on a computer to complete the calculation.

SUMMARY

Experiment shows that the speed of light is a constant and is not affected by the motion of its source or the observer. This puzzling result became the basis for Einstein's theory of special relativity. A consequence of that theory is that an observer watching an object move past him sees its length shrunk (the so-called Lorentz contraction) and the rate of passage of its time slowed (time dilation).

These changes in space and time cause someone in a non-accelerating rest frame to observe anyone moving relative to them as having clocks that run slower than their own. On the other hand, anyone undergoing acceleration finds that their clocks have in fact run slower, even if the acceleration occurred as a result of being in a gravitational field. Einstein developed a new theory of gravity, general relativity, that showed how mass and energy "curve" space and time, changing the flow of both and producing the effects of gravity described by Newton's laws.

QUESTIONS FOR REVIEW

1. What is Galilean relativity? Give an example of how it is used.
2. Describe what the Michelson-Morley experiment was trying to detect, and how it failed.
3. What is the Lorentz factor?
4. How are length, time, and mass altered according to special relativity?
5. What is the twin paradox, and how is it resolved?
6. What is general relativity?
7. How does gravity affect space and time?

THOUGHT QUESTIONS

1. Would you be willing to travel to a nearby galaxy if it meant you would return to Earth one million years in the future?
2. Given the Lorentz factor, does time pass for a photon? What about for a place where gravity is so strong that the escape velocity equals the speed of light?
3. With everyone's time running at different speeds, can you imagine a story line that would make a good movie if it portrayed space travel accurately?

PROBLEMS

1. To travel 100,000 light-years in 10 years of your own time, at what fraction of the speed of light would you have to travel?
2. Mercury orbits the Sun at speeds ranging from 59 km/sec to 39 km/sec when it is nearest and farthest from the Sun, respectively. What are the Lorentz factors for these two speeds?
3. The escape velocity from the Sun is 76 km/sec at Mercury's closest distance to the Sun and 62 km/sec at Mercury's farthest distance. Using the gravitational time dilation formula $1/\sqrt{1 - V_{esc}^2/c^2}$, find how slowly time runs in these two places. Compare your results to the previous question, and discuss how time running at different speeds in the orbit might affect Mercury's orbit.

TEST YOURSELF

1. When a spaceship is traveling at 99% of the speed of light (Lorentz factor = 7), an astronaut on board the ship will find that

   (a) everything in the ship weighs 7 times more.
   (b) the ship is very cramped—only 1/7th its original length.
   (c) everyone onboard talks 7 times more slowly than normal.
   (d) All of the above.
   (e) None of the above. Everything seems normal to the astronaut on board.

2. Suppose Tom and Molly are both flying in spaceships toward each other at half the speed of light (0.5 c). If Tom shines a light toward Molly, what speed will Molly measure for the light coming toward her?

   (a) 0.25 c (c) 1.0 c (e) 2.0 c
   (b) 0.5 c (d) 1.5 c

3. If Bob travels at close to the speed of light to another star and then returns, he will find that his twin sister Alice, who remained on Earth, is

   (a) younger than him.
   (b) older than him.
   (c) the same age as him.
   (d) He cannot return to Earth because it would violate special relativity.

4. What is "equivalent" in the principle of equivalence?

   (a) Space and time
   (b) Matter and energy
   (c) Gravity and acceleration
   (d) Forward and reverse directions of time
   (e) All frames of reference

5. Sort the following in order of where time runs slowest to where it runs fastest.

   (a) On the Earth's surface
   (b) On the Moon's surface
   (c) In deep space
   (d) On the Sun's surface

KEY TERMS

5

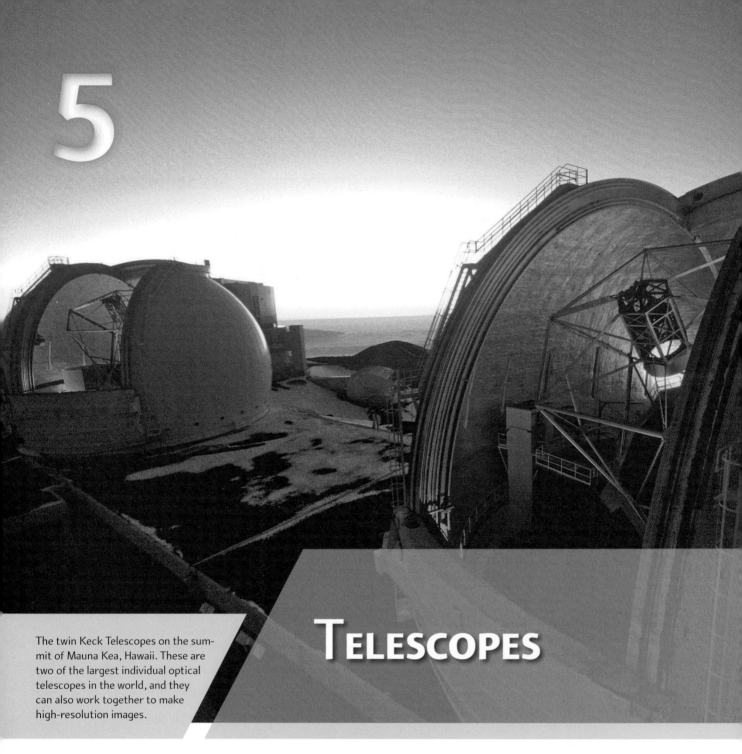

The twin Keck Telescopes on the summit of Mauna Kea, Hawaii. These are two of the largest individual optical telescopes in the world, and they can also work together to make high-resolution images.

TELESCOPES

LEARNING OBJECTIVES

Upon completing this chapter you should be able to:

- Classify the common types of telescope designs.
- Compare the advantages and disadvantages of reflectors and refractors.
- Describe what causes refraction and how lenses focus light.
- Identify the important aspects for determining a telescope's sensitivity.
- Compare the light-gathering power of different telescopes.
- Describe the factors affecting telescope resolution, and calculate the diffraction limit for a telescope.

- Describe the idea of interferometry and how astronomers use it to improve resolution.
- Describe the methods used for detecting visible light and other wavelengths of electromagnetic radiation.
- Discuss the problems caused by observing through the Earth's atmosphere, and describe the methods astronomers use to overcome these problems.
- Identify the wavelength ranges in which telescopes cannot operate from the ground and the reasons for this.
- Describe the causes of and remedies for light pollution.

Astronomers, like all scientists, rely heavily on observations to guide them in theorizing and in testing theories already developed. Unlike most scientists, however, astronomers cannot directly probe the objects they study. Rather, they must perform their observations from vast distances and can only passively collect radiation emitted by the bodies they seek to study.

The extraordinary advances in astronomy since Galileo's first telescopic observations have been made possible by the development of larger and more precise telescopes since the early 1600s. Just as critical has been the development of new techniques for making images since the mid 1800s, and the invention of new kinds of detectors to study light in different electromagnetic bands. Collecting enough radiation to be useful in studying astronomical objects is difficult because most objects are so remote that their radiation is extremely faint by the time it reaches Earth. Moreover, extracting the desired information from the radiation requires special instruments—instruments that can measure to high precision the brightness, spectrum, and position of objects. For example, to collect enough light to detect remote galaxies, astronomers use telescopes with mirrors the size of a small swimming pool, radio telescopes the size of a city block, and arrays of telescopes spread thousands of miles apart. To avoid the blurring and blocking effects of our atmosphere, they use orbiting observatories. To analyze and display the observations, they use high-speed computers.

This chapter describes some of the more important devices and how they work. We will see that modern telescopes bear little resemblance to the long tubes depicted in cartoons. Moreover, modern astronomers rarely sit at the eyepiece of a telescope. They are more likely to be sitting at a computer terminal operating a telescope remotely, examining the data collected, or solving equations that describe such things as the paths stars take when two galaxies collide.

CONCEPTS AND SKILLS TO REVIEW

- Light as a wave or photons (4.1)
- Red light has a longer wavelength than blue (4.1)
- The electromagnetic spectrum (4.2)
- Absorption in Earth's atmosphere (4.5)

5.1 TELESCOPE FUNDAMENTALS

A telescope enables an astronomer to observe things not visible to the naked eye. Although our eyes are superb detectors, they cannot see extremely faint objects or fine details on distant sources. For example, the Andromeda galaxy M31 (fig. 5.1) is visible as a faint patch of light, barely visible with the unaided eye (fig. 5.1A). Through a telescope, much more becomes visible. First, the telescope gathers much more light than enters your eye's pupil, so many more stars become visible. The telescope can also magnify details too small to see directly with your eye, but the eye is still only able to examine the light each moment as it arrives. Telescopes are made much more powerful by instruments that can collect the light for a long period of time. This allows them to record far fainter signals, revealing structures not otherwise visible (fig. 5.1B). In this section we will examine how telescopes are designed to gather light and form images. In the following sections we will examine the physical properties of light that limit the detail visible through a telescope, and the kinds of modern detectors that permit us to examine light in far greater detail than in the past.

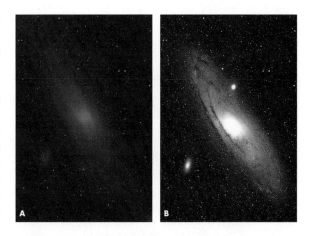

FIGURE 5.1
The galaxy M31 as seen (A) by the unaided eye, and (B) through a telescope with an imaging device to collect light.

FIGURE 5.2

A large lens collects more light (photons) than a small one, leading to a brighter image. We therefore say that the larger lens has a greater "light-gathering power."

Light-Gathering Power

For our eyes to see an object, photons (light) from it must strike the retina in large enough numbers to stimulate chemical reactions in nerve cells. How bright an object appears to us depends on the number of its photons that enter our eye per second, a number limited by the size of our eyes' pupils. Astronomers overcome that limit by collecting photons with a telescope that is much larger than our eyes, which then "funnels" the photons to our eyes. The bigger the telescope's collecting area, the more photons it collects, as shown in figure 5.2. Thus, a larger-diameter mirror or lens gives a telescope a greater **light-gathering power.**

A larger telescope produces a brighter image, which allows us to see dim stars that are invisible in telescopes with smaller gathering areas. Astronomers usually describe a telescope by the diameter of its lens or mirror. Thus, the 10-meter Keck Telescopes in Hawaii have mirrors spanning 10 meters (roughly 30 feet) in diameter. Because the gathering area of a circular lens or mirror of radius R is πR^2, increasing the radius produces a rapid increase in the number of photons caught. For example, doubling the radius of a lens or mirror increases its light-gathering area by a factor of 4, and a large telescope will have enormously more light-gathering power than your eye, as shown in Astronomy by the Numbers box below.

ASTRONOMY *by the numbers*   LIGHT-GATHERING POWER OF A TELESCOPE

The diameter of a person's pupil when fully adapted to the dark is typically about 8 mm. This is the diameter of the "aperture" of the eye. By contrast, some of the largest telescopes have aperture diameters of about 8 m. How does their light-gathering power compare?

To determine this, we need to calculate the collecting area of each. The eye collects light over an area equal to

$\pi R^2 = \pi (8 \text{ mm})^2 = \pi (0.008 \text{ m})^2 = 2.0 \times 10^{-4} \text{ m}^2$. By contrast, the telescope collecting area is $\pi R^2 = \pi (8 \text{ m})^2 = 2.0 \times 10^2 \text{ m}^2$ —which is a million (10^6) times larger.

Combined with technologies that can collect light for long periods of time, large telescope can detect objects much more than a million times fainter than what your eye can see.

FIGURE 5.3
(A) A lens focuses light by bending (or refracting) the path of light to a point. (B) A curved mirror focuses light by reflecting it to a point.

FIGURE 5.4
A refracting telescope. Completed in 1897 for the University of Chicago's Yerkes Observatory in Williams Bay, Wisconsin, this refractor has a lens approximately 1 meter (40 inches) in diameter, making it the world's largest refracting telescope.

Focusing the Light

Once light has been gathered, it must be focused to form an image or to concentrate it on a detector. A telescopes in which light is gathered and focused by a lens is called a "refracting telescope," or **refractor** for short. The lens of a refractor focuses the light by bending the rays, as shown in figure 5.3A. This bending is called **refraction,** and it happens when light moves from one substance (such as air) into a different substance (such as glass), as discussed in Extending Our Reach: "Refraction." Figure 5.4 shows a photograph of the world's largest refractor, the 1-meter (40-inch) diameter Yerkes Telescope of the University of Chicago.

Lenses have many serious disadvantages in large telescopes, however. First, large-diameter lenses are very expensive to fabricate. Moreover, a lens must be supported at its edges so as not to block light passing through it. This makes the lens "sag" in the middle (though by only tiny amounts), distorting its images. A third difficulty with lenses is that most transparent materials bring light of different colors to a focus at slightly different distances from the lens. This creates images fringed with color, a flaw called "chromatic aberration." Finally, many lens materials completely absorb short-wavelength light, making them, for some purposes, as useless in a telescope as a chunk of concrete.

To avoid such difficulties with lenses, most modern telescopes use mirrors to gather and focus light, and they are therefore called **reflectors.** The mirrors are made of glass that has been shaped to a smooth curve, polished, and then coated with a thin layer of aluminum or some other highly reflective material. As figure 5.3B shows, such a curved mirror can focus light rays reflected from it, creating an image just as well as a lens can. Moreover, because the light does not pass through glass, it focuses all colors equally well and does not absorb short-wavelength light. Furthermore, because the light does not have to pass through the mirror, the mirror can be supported from

EXTENDING
our reach
REFRACTION

When light moves at an angle from one material into another (for example, from air into water), its direction of travel generally bends. This phenomenon is called refraction. Refraction is the principle by which our eyes and eyeglasses focus light. You can easily see its effects by sticking a ruler in water and noticing that the ruler appears bent, as seen in figure 5.5. The ruler in water also illustrates an important property of refraction. If you change the ruler's tilt, you will see that the amount of bending (refraction) changes. Exactly vertical rays are not bent at all, nearly vertical rays are bent only a little, and rays entering at a grazing angle are bent most.

Refraction occurs because light changes speed as it enters matter, generally becoming slower in denser material. This decrease in the speed of light arises from its interaction with the atoms through which it moves. To understand how this reduction in light's speed makes it bend, imagine a light wave approaching a slab of material. The part of the wave that enters the material first is slowed while the part remaining outside is unaffected, as depicted in figure 5.6A. To see why slowing part of the wave makes it bend, imagine what would happen if the wheels on the right-hand side of your car turned more slowly than those on the left. Your car would swerve to the right, a result that lies behind the reason cars have a differential. By allowing one wheel to turn faster than the other, the differential "swings" your car smoothly around corners. A similar effect occurs if you walk hand-in-hand with a friend, and your friend walks more slowly than you do. You will soon find yourself traveling in a curve (see fig. 5.6B). So, too, if one portion of a light wave moves more slowly than another, the light's path will bend.

FIGURE 5.5
Refraction of light in water. Note how the ruler appears to be bent where it crosses from the air into the water.

Refraction not only bends light but also generally spreads the light into its component colors, breaking white light into a spectrum, or rainbow. This spreading occurs because different colors of light travel at different speeds in most materials and are therefore bent by different amounts in a process called **dispersion**. Thus, if light consisting of a mix of colors enters a block of glass, each color is slowed to a different speed and is therefore deflected differently. The result is that colors initially traveling together separate into different beams. This is how a prism creates a spectrum.

FIGURE 5.6
Cause of refraction.
(A) Light entering the denser medium is slowed, while the portion still in the less dense medium proceeds at its original speed.
(B) A similar effect occurs when you walk hand-in-hand with someone who walks slower than you do.

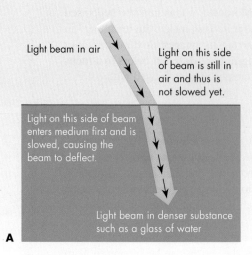

Light beam in air

Light on this side of beam is still in air and thus is not slowed yet.

Light on this side of beam enters medium first and is slowed, causing the beam to deflect.

Light beam in denser substance such as a glass of water

A

Fast walker

Slow walker

B

FIGURE 5.7
The largest telescope mirrors yet built are 8.4 meters (about 27 feet) in diameter. Two of these huge mirrors are used together on the Large Binocular Telescope on Mount Graham, Arizona.

behind, thereby reducing the sagging problem that affects large lenses. For these and other reasons, major observatories now use reflecting telescopes almost exclusively. Figure 5.7 shows a photograph of the largest telescope mirrors yet built—a pair of 8.4-meter diameter reflectors in Arizona. The mirrors can work together to give the same collecting area as a single 11.8-meter diameter mirror.

Light striking a mirror is focused in front of the mirror. Thus, to see the image, the observer or camera would ordinarily have to be positioned in front of the mirror, thereby blocking some of the light (fig. 5.8A). To overcome this difficulty, a secondary mirror is often used to deflect the light off to the side (fig. 5.8B) or back toward the mirror and out through a hole in its center (fig. 5.8C).

Most telescopes are mounted on huge pivots that allow them to follow astronomical objects as they move across the sky. Swinging the many tons of metal and glass smoothly and with precision requires great care in construction and design. Moreover, as the telescope moves, its lenses or mirrors must keep their same precise shapes and relative positions, if the images are to be sharp. This is one of the

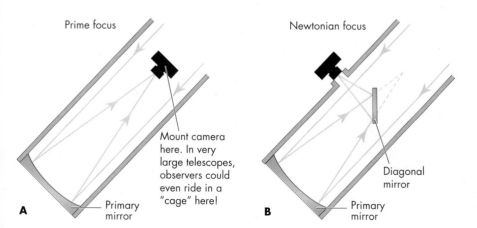

FIGURE 5.8
Sketches of different focus arrangements for reflectors.

FIGURE 5.9

The Gran Telescopio Canarias, currently the largest visual-wavelength telescope in the world. (A) View of the telescope dome, with clouds visible below the mountain top. (B) View of the telescope while it was under construction shows part of the 10.4-meter diameter multimirror as well as the support structure for the hexagonal mirrors, which allows their precise positioning to create the overall large mirror size.

most technically demanding parts of building a large telescope, because the large pieces of glass used in lenses and mirrors bend slightly when their positions are shifted.

In the past, astronomers made mirrors thick to make them less susceptible to deforming as the telescope was tilted in different orientations. Large pieces of glass, however, weigh more than smaller pieces and thus can cause the whole telescope to sag more, just as a smaller gob of whipped cream will keep its shape on a tilted plate whereas a large gob will slump under its own weight. As a result, a 5-meter diameter telescope completed in 1949 on Mount Palomar, California, remained the largest telescope for decades. The structural limits of glass required astronomers to develop a different approach. They discovered that a thin piece of glass, if properly supported, could keep its shape better than a thick piece. Thus, astronomers have sought ways to make thin mirrors that are then kept precisely shaped by alignment systems on the back side of the mirror.

While thin mirrors have allowed astronomers to build larger telescopes than in the past, another approach shows promise for much larger telescopes. Instead of using a single mirror, telescopes are being designed with many smaller mirrors aligned to collect and focus the light as if they were a single mirror. These are called segmented mirrors. Currently, the largest segmented mirror is a 10.4-meter reflector in the Canary Islands (fig. 5.9). The mirror consists of 36 separate mirrors that are kept aligned by lasers that measure precisely the tilt and position of each mirror. If any misalignment is detected, tiny motors shift the offending mirror segment to keep the image sharply in focus. Astronomers think that this method will permit the building of telescopes perhaps 30, 50, or even 100 meters in diameter in the future, as illustrated in figure 5.10.

The light-gathering power of telescopes makes dim objects bright enough to see. Telescopes, however, serve another important function—they increase our ability to see fine detail. This also depends on the size of the telescope.

FIGURE 5.10

Design for a 39-meter segmented-mirror telescope planned for completion by the European Southern Observatory in the 2020s.

5.2   RESOLVING POWER

If you mark two black dots close together on a piece of paper and look at them from the other side of the room, your eye may see them as a single dark mark, not as separate spots. Likewise, stars that lie close together or markings on planets may not be distinguishable. A telescope's ability to discern such detail depends on its **resolving power.**

Resolving power is limited by the wave nature of light. For example, suppose two stars are separated by a very tiny angle. For them to be discernible as separate images, their light waves must not be mixed up. Such mixing, however, always occurs when waves pass through an opening, because as each wave passes the opening, weaker secondary waves are produced in a phenomenon called **diffraction.** Figure 5.11A shows how water waves are diffracted as they pass through an opening.

Light waves are similarly diffracted as they enter a telescope. The result of diffraction is that point sources of light become surrounded by faint patterns of light. One way to see the effects of diffraction is by looking at a light source, such as a small bright lightbulb, through a piece of cloth, such as a shirt. The light will be surrounded by diffraction rings produced as the light waves pass through the tiny openings in the weave of the fabric. The Hubble Space Telescope likewise sees faint diffraction patterns, as figure 5.11B demonstrates. In addition to the diffraction caused by the aperture of a telescope, internal support structures in the path of the light also produce diffraction effects. This is the source of the spikes seen coming from the star in figure 5.11B as well as in other astronomical images.

Diffraction presents a fundamental limit to the detail visible through a telescope. This is governed by both the diameter of the telescope and the wavelength of light being observed. In fact, diffraction theory shows that for a telescope of diameter D observing at a wavelength λ, two points of light can only be distinguished if they are separated by an angle α greater than $2.5 \times 10^5\ \lambda/D$, where the angle α is measured in arc seconds*.

We can rewrite this relationship to find the minimum size a telescope needs to be to resolve detail of a particular angular extent. If D is expressed in centimeters, λ in nanometers, and α in arc seconds, then

$$D > 0.025\ \lambda/\alpha .$$

For example, to resolve two stars separated by 0.1 arc seconds when observing in visible light ($\lambda \approx 500$ nanometers), you need a telescope whose mirror has a diameter greater than 125 centimeters (about 50 inches). We can also compare resolving power as shown in Astronomy by the Numbers: "Resolving Power of a Telescope."

---

* An arc second is a unit of angle and is equal to 1/3600 of a degree.

As we will discover later in this chapter, our atmosphere seriously blurs fine details in astronomical objects, degrading the resolving power of large ground-based telescopes to far below their diffraction limits.

D = Diameter of telescope in centimeters
λ = Wavelength of light in nanometers
α = Separation angle in arc seconds

FIGURE 5.11
(A) Water waves diffracted as they pass through a narrow opening. (B) A highly magnified image of a star made with the Hubble Space Telescope. The light from a single point is spread out by diffraction at the edges of the mirror and structures within the telescope. (These diffraction features are very faint relative to the star, but are amplified in this image.)

Let's compare again a person's pupil when dilated to 8 mm to a giant telescope with an aperture diameter of 8 m, this time asking how their resolving power compares.

We will assume both are observing at 500 nm, in the middle of the visible range. To compare their resolving power, we need to determine α in the equation $D > 0.025\,\lambda/\alpha$. Multiplying both sides of the equation by α and dividing both by D, we find that $\alpha > 0.025\,\lambda/D$. With $\lambda = 500$ nm, and for the pupil diameter of $D = 0.8$ cm, we have $\alpha > 0.025\,(500/0.8) \approx 16$ arcsec.

The telescope's aperture diameter is $D = 800$ cm, so we have $\alpha > 0.025\,(500/800) = 0.016$ arcsec. The telescope can therefore detect objects as much as a thousand times smaller than your eye can discern.

Interferometers

The limit to resolution caused by diffraction can never be eliminated, but it can be improved by enlarging the area over which light is collected beyond the size of any single telescope that can feasibly be built. Astronomers accomplish this with a device called an **interferometer.** With an interferometer, observations are made simultaneously through two or more widely spaced telescopes (fig. 5.12) that direct the light to a common detector, which combines the separate light beams.

The interferometer is so-named because when it mixes the separate beams, the light waves of one "interfere" with the waves from the other. Where the crests of two waves arrive together, they create a bright region. Where the crest of one wave arrives simultaneously with the trough of another, they cancel and create a dark patch. The

FIGURE 5.12

Photograph of an infrared and optical wavelength interferometer (IOTA). Light from the object of interest is collected by the three telescopes and sent to a control room. Computers there combine the light and reconstruct an image of the object.

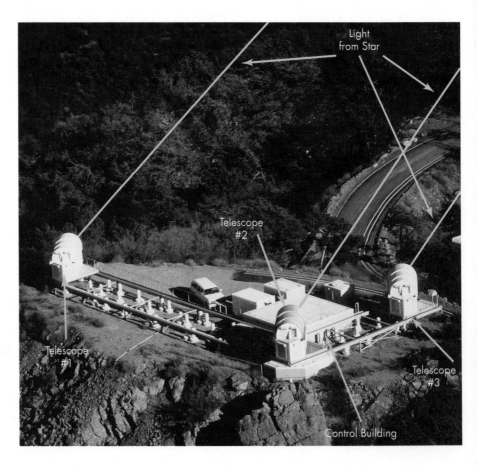

result is a complex pattern of alternating light and dark regions, which can be analyzed by a computer to create an image of the object observed.

The result of this process is an image in which the resolution is set not by the size of the individual mirrors but rather by their separation. If the mirrors are 100 meters apart, for instance, the interferometer has the same resolving power as a telescope 100 meters in diameter. The resulting fine resolution is far beyond what can be obtained in other ways. For example, figure 5.13A shows a view of two closely spaced stars as observed with a small telescope. Their images are severely blended as a result of diffraction. Figure 5.13B shows the same stars observed with an interferometer and after the image has been processed by a computer. The two stars can now be easily distinguished: the two separate mirrors produce the resolving power of a single mirror whose diameter equals the spacing between them.

FIGURE 5.13
A young star observed with an ordinary telescope (A) appears to be a single star. An interferometer image (B) reveals that the "star" is actually two stars in orbit around one another.

5.3 DETECTING LIGHT

Visible Light

Once light has been collected, it must be detected and recorded. In olden days, the detector was the eye of an astronomer who sat at the telescope eyepiece and wrote down data or made sketches of the object being observed. The human eye, marvel that it is, has difficulty seeing very faint light. Many astronomical bodies are too distant or too dim for their few photons to create a sensible effect on the eye. For example, if you were to look at any but the nearest galaxies through even the Gran Telescopio Canarias, the galaxies would appear merely as dim smudges. Only by storing up their light, sometimes for hours, can a quality picture of them be made. Thus, to see very faint objects, astronomers use detectors that can store light in some manner. Such storage can be done chemically with film or electronically with detectors similar to those used in digital cameras.

From the late 1800s until the 1980s, astronomers generally used photographic film to record the light from the bodies they were studying. Film forms an image by absorbing photons that cause a chemical change, making the film dark where light has fallen on it. This process, however, is very inefficient: fewer than 4% of the photons striking the film produce a useful image. The result of such low efficiency is that many hours are needed to accumulate enough light to create an image of faint objects. Moreover, the film must be developed, thereby delaying the observing process even further.

Astronomers today use many kinds of electronic detectors. One of the main types is the **CCD** (charge-coupled device). Modern CCDs can make pictures with a sensitivity to faint light approximately 20 times greater than photographic film. In these devices, the incoming light strikes a semiconductor surface, allowing electrons to move within the material as shown in figure 5.14. The surface is divided into millions of little squares called pixels, in which the electrons are temporarily stored. The number of electrons in each pixel is proportional to the number of photons hitting it (that is, proportional to the intensity of the light). An electronic device coupled to a computer then scans the detector, counting the number of electrons in each pixel and generating a picture, in much the same way as a TV screen or newsprint photo creates a picture made up of many separate tiny dots.

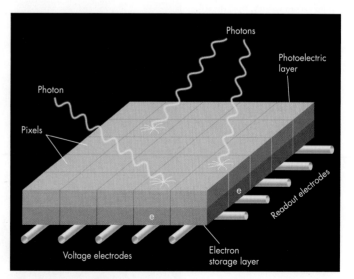

FIGURE 5.14
Simplified diagram of a CCD. Photons striking the upper layer free an electron (e). A positive voltage applied to one set of electrodes attracts the electrons and holds them in place under each pixel. During readout, voltages are changed to move out the collected electrons.

Such electronic devices are extremely efficient, recording 75% (or more) of the photons striking them, allowing astronomers to record images much faster than with film. Electronic detectors have other advantages as well. For example, they record the signal digitally, essentially counting every photon that falls on each part of the detector. Such digital images can be processed by computers to sharpen them, remove extraneous light, and enhance contrast.

Detecting Other Wavelengths

Visible light, which we can see because its wavelengths are detectable by our eyes, is just one of many wave bands of electromagnetic radiation, as discussed in chapter 4. Many astronomical objects, however, radiate at wavelengths that our eyes cannot see, and so astronomers have devised ways to observe such objects. For example, cold clouds of gas in interstellar space emit little visible light but large amounts of radio energy. To observe them, astronomers use radio telescopes. Radio-wavelength detectors are similar to radio receivers used for man-made broadcasts but are much more sensitive. They are also made highly directional by building huge radio "mirrors" (fig. 5.15), just as we build large mirrors to make higher-resolution observations with visible light. Radio telescopes can also be joined together to form interferometers for even higher resolutions. Some radio interferometers use telescopes in different continents to form a telescope nearly as large as the whole Earth.

Different challenges face astronomers when building telescopes in different wave bands. For example, dust clouds in space are too cold to emit visible light, but they do radiate infrared energy, which astronomers observe with infrared telescopes. One of the challenges for infrared telescopes is that the telescope itself may emit infrared radiation that can mask the objects being observed. These telescopes must be carefully designed to prevent that local radiation from entering the detectors, and parts that cannot be shielded are kept at extremely low temperatures.

FIGURE 5.15

Photograph of the radio telescope at the Owens Valley Radio Observatory, operated by the California Institute of Technology. In the background you can see the Sierras. The "telescope" is an array of six separate dishes that collect the radio waves. The captured radiation is then combined by computer to increase the resolution of the instrument.

Radio waves from space

Signals are focused here and carried by cable to the control room.

An antenna 10.4 meters (about 34 feet) in diameter collects radio waves and reflects them to focus.

Mounting allows telescope to track sources.

Instrument room

FIGURE 5.16

(A) A false-color picture of a radio galaxy. We can't see radio waves, so colors are used to represent their brightness—red brightest, blue dimmest. (B) A false-color X ray picture of Cas A, an exploding star. In this case colors represent different wavelengths of X ray photons (bluer colors corresponding to more-energetic photons).

Q. What kind of common false-color map is the color scheme in figure 5.16A based upon? Where else do we commonly use false-color maps?

Designing a telescope for observing X rays presents different challenges. X rays entering a normal telescope would strike the mirror surface and be absorbed, making the telescope no more effective for observing than a slab of concrete. Astronomers have found, however, that X rays can be reflected if they strike a smooth surface at a very shallow angle, somewhat as a rock can skip over the surface of water if thrown nearly horizontally. X ray telescopes are like curved funnels, gradually redirecting the X ray photons toward the detector. As for visual photons, CCDs are again used as detectors, but because X ray photons carry so much energy, a single photon frees up many electrons when it strikes a CCD pixel. This allows astronomers to measure the energy of each X ray photon by reading out the CCD quickly enough to avoid multiple-photon hits.

Because our eyes cannot see these other wavelengths, astronomers must devise ways to depict what such instruments record. The most common way to illustrate the radiation is with false-color pictures, as shown in figure 5.16. In a false-color picture, the colors represent different properties of the radiation. For example, in figure 5.16A (a radio "picture" of a radio galaxy and the jet of hot gas spurting from its core), astronomers color the regions emitting the most intense radio emission red; they color areas emitting somewhat weaker emission yellow and the faintest areas blue. Thus, if we could "see" radio waves, the red areas would look brightest and the blue areas dimmest. Another approach is to "translate" the energies of the photons into colors. For example, figure 5.16B shows a false-color X ray "photograph" of the gas shell ejected by an exploding star. In this image the highest-energy photons are colored blue, intermediate ones yellow, and the lowest-energy photons red. In this case, if our eyes were sensitive to the X ray band instead of the visible band, this is what X rays might look like to us. Astronomers sometimes use false-color images to bring out particular features or even to depict calculated quantities, such as magnetic field strength or pressure—quantities that we could never directly see with our eyes.

Telescopes operating at infrared, ultraviolet, and X ray wavelengths face an additional obstacle: most of the radiation they seek to measure cannot penetrate the Earth's atmosphere. If astronomers want to view an object in a blocked wavelength, they must use a telescope in space, operated remotely from the ground or, more rarely, by a scientist-astronaut in space.

5.4 OBSERVATORIES ON THE GROUND AND IN SPACE

The large telescopes and the associated equipment astronomers use are extremely expensive. Therefore, the largest telescopes are often national or international facilities, such as the National Optical Astronomical Observatory of the United States and the Anglo-Australian Telescope. Despite the expense of such facilities, many colleges and universities have their own large research telescopes (in addition to smaller ones near campus for instructional purposes). In addition, some large private groups, such as the Carnegie Institution, operate observatories. Altogether, several thousand observatories exist around the world, on every continent. There are even telescopes at the South Pole in Antarctica to take advantage of the extreme dryness of the bitterly cold Antarctic air.

The largest optical telescopes in the United States at this time are the twin 10-meter Keck Telescopes pictured in the chapter-opening image. These telescopes pioneered the use of segmented mirrors, and the slightly larger Gran Telescopio Canarias is based on their design. The two Keck Telescopes can be operated individually or as an interferometer with double the collecting area.

The optical telescope with the largest collecting area in the world is the VLT (for "very large telescope"), a group of four 8.2-meter telescopes that work individually or as an interferometer (fig. 5.17). The VLT is operated by a consortium of European countries and Chile and is located in the extremely dry northern part of Chile. Several other large telescopes have begun operation recently, such as the 8.3-meter Subaru Telescope operated by Japan and located in Hawaii. Others are the Gemini telescopes—two identical 8.1-meter instruments. One is located in Hawaii; the other is in Chile. These twin telescopes are run by a consortium consisting of the United States, the United Kingdom, Canada, Chile, Australia, Brazil, and Argentina.

Visible light is special not only because our eyes can detect it, but also because it is one of the few wavelength regions, called **atmospheric windows,** in which it is possible to peer out into space from the ground (see section 4.5). Gases in our atmosphere such as ozone, carbon dioxide, and water strongly absorb infrared, ultraviolet, and shorter wavelengths, as illustrated in figure 5.18. For example, infrared radiation with a wavelength of 50 micrometers is strongly absorbed by water and carbon dioxide in our atmosphere. Astronomers can make some observations from high-flying airplanes or balloons, but other wavelength ranges are so strongly absorbed that it is necessary to launch telescopes into space.

Figure 5.18 depicts a few of the many telescopes astronomers have launched into space. One of the challenges of space observatories is maintaining them for many years. For example, the Spitzer Space Telescope is an infrared observatory that used

> Subaru is the Japanese name for the Pleiades star cluster. Look carefully at the logo on a Subaru car and you will see the stars in the design.

FIGURE 5.17
The four 8.2-meter diameter telescopes of the VLT in Cerro Paranal, Chile, can work in unison or independently.

Opening is about 1.2 m (3.9 feet) in diameter

X rays enter here

~12.2 meters (40 feet) long

Chandra X ray Telescope Observatory

Extreme Ultraviolet Explorer – EUVE

Infrared radiation enters here

~4 meters (15 feet) long

Spitzer Infrared Space Telescope

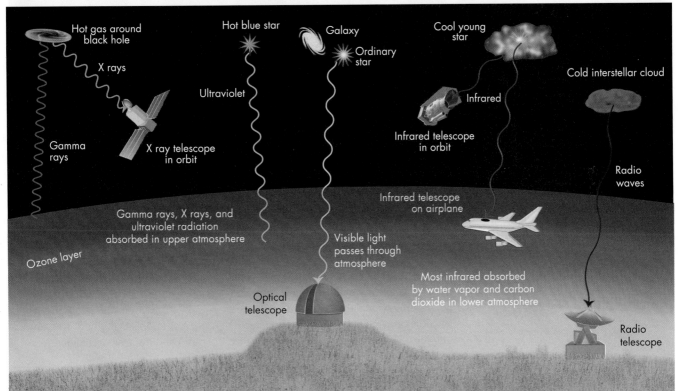

Hot gas around black hole

Hot blue star

Galaxy

Cool young star

X rays

Ultraviolet

Ordinary star

Cold interstellar cloud

Infrared

Gamma rays

X ray telescope in orbit

Infrared telescope in orbit

Radio waves

Infrared telescope on airplane

Gamma rays, X rays, and ultraviolet radiation absorbed in upper atmosphere

Visible light passes through atmosphere

Ozone layer

Most infrared absorbed by water vapor and carbon dioxide in lower atmosphere

Optical telescope

Radio telescope

FIGURE 5.18

Diagram illustrating how light of different wavelengths passes through or is blocked by the Earth's atmosphere. Along the top there are drawings of several space observatories: Chandra, EUVE, and Spitzer. Chandra (short for Chandrasekhar) and Spitzer are named for important astronomers of the twentieth century.

liquid helium to keep its instruments cold for maximum sensitivity. Once the helium on board the craft was used up in 2009, the telescope could no longer operate with full sensitivity, but some instruments continue to collect data.

Probably the best known of all space telescopes is the Hubble Space Telescope (HST). It can observe parts of the infrared and ultraviolet bands, but it is best known for its visible-wavelength images. Even though there is an atmospheric window available, HST was placed in orbit because of the blurring that our atmosphere causes.

Although the HST initially had a number of problems, astronauts repaired the major defects, and now the clarity of its images is excellent. These images reveal details never before seen by telescopes on the ground because such telescopes must peer through the blurring effects of our atmosphere. The HST has several different instruments, including cameras for wide-field views and for detailed images, as well as spectrographs for analyzing light from stars and galaxies.

2.4 m diameter aperture

Hubble Space Telescope

Hourglass Nebula—a dying star

Sombrero Galaxy—system of billions of stars and dusty interstellar clouds

FIGURE 5.19
The Hubble Space Telescope and two of the remarkable images it has collected.

The HST's mirror, 2.4 meters (about 8 feet) in diameter, produces strikingly sharp images, as shown by two examples in figure 5.19. This leads many people to imagine that it must be the largest telescope ever built, but it is actually quite modest in size compared to most ground-based research telescopes. The HST has remained an exceptional instrument for so long because it was designed to be serviced by astronauts aboard the space shuttle. They have replaced and repaired instruments, allowing the HST to remain a state-of-the-art instrument. A final servicing mission in 2009 will hopefully keep the HST running for another decade until a new, larger space telescope can take its place.

Dozens of other space telescopes operate at wavelength regions not observable from the ground. Some of the most exciting discoveries have been made as astronomers explore new wavelength regions. Extending Our Reach: "Exploring New Wavelengths: Gamma Rays" describes a phenomenon that was never previously seen before it was discovered from a space observatory.

Radio astronomy can be conducted from the ground thanks to another atmospheric window at these long wavelengths. This is fortunate since radio telescopes have to be extremely large to observe fine detail, given the dependence of resolving power on λ/D (section 5.2). The 300-meter Arecibo telescope (fig. 5.20) observing at 20 cm wavelength, and the 50-meter LMT (fig. 5.21) observing at 3 mm, are examples of large telescopes built to explore the radio and microwave bands.

FIGURE 5.20
The 300-meter diameter Arecibo radio telescope in Puerto Rico.

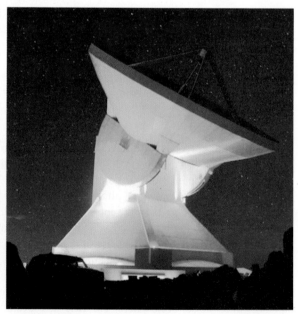

FIGURE 5.21
The 50-meter Large Millimeter Telescope in Mexico.

FIGURE 5.22
The Atacama Large Millimeter Array (ALMA) is a microwave interferometer in the high desert of Chile. It is being built by a collaboration of many countries to fund the cost of over one billion dollars. When completed, the array will comprise 66 antennas and will be able to study objects at millimeter and submillimeter wavelengths with unmatched sensitivity and resolution.

Most modern observatories, built to detect anything from radio waves to gamma rays, are also designed and operated by international consortiums. For instance, the X ray space telescope *Chandra* has detectors designed by teams in Germany, the United Kingdom, and the United States. New ground-based telescopes increasingly rely on international collaborations to share the expense of building the best possible instruments. A recent example of this is the Atacama Large Millimeter Array (ALMA), a large interferometer being built in the high desert of Chile, jointly funded by agencies in Canada, Chile, Europe, Japan, Taiwan, and the United States (fig. 5.22). Among other objects, ALMA will be able to study galaxies forming shortly after the Universe began. By pooling their resources, astronomers can build a far more powerful instrument than would be possible if each country built a separate instrument.

EXTENDING *our reach*   EXPLORING NEW WAVELENGTHS: GAMMA RAYS

Astronomers have made many of their most exciting discoveries when new telescopes allowed them to observe the sky at wavelengths not previously detectable. Gamma-ray wavelengths are among the last of the wavelength regions to be explored, and astronomers are still trying to interpret what they see.

Gamma-ray astronomy began in 1965 when a small and (by modern standards) primitive satellite detected cosmic gamma rays. A few years later, a slightly more advanced satellite detected gamma rays coming from the center of our Galaxy, the Milky Way. By the 1970s, astronomers had discovered that many familiar sources, such as the Crab Nebula and the remnants of other exploded stars, emit gamma rays. However, the most interesting gamma-ray sources were discovered earlier by accident.

In 1967 the United States placed several military surveillance satellites in orbit to watch for the gamma rays produced when a nuclear bomb explodes. The satellites were designed to monitor the United States–Soviet Union ban on nuclear bomb tests in the atmosphere. Curiously, on a number of occasions the satellites detected gamma-ray bursts coming not from the Earth but from space. Unfortunately for astronomers, the discovery of the bursts was top secret at the time and was not made public until 1973.

Astronomers' thirst for more information about these high-energy sources was unsatisfied for many years because our atmosphere absorbs gamma rays and ordinary telescopes cannot focus gamma rays. Nevertheless, with ever more complex instruments in satellites, astronomers discovered that gamma-ray sources—apart from the bursts—coincided with known astronomical objects such as dying stars and some peculiar galaxies. Gamma-ray bursts, on the other hand, would appear suddenly in otherwise blank areas of the sky, flare in intensity for a few seconds, and then fade to invisibility. It has taken nearly 30 years of study to answer even the simple question "Are they near or far?" The breakthrough came in December 1997, when astronomers detected a gamma-ray burst that coincided with a distant galaxy. This solved the mystery of the bursters' distance but leaves unanswered what they are. Theories to explain the bursts abound. According to a favored hypothesis, the bursts are "hypernovas," stellar explosions caused when massive stars run out of fuel and collapse to form black holes. This latter proposal gained support when, in 2002, astronomers obtained spectra of a burst that showed emission lines suggestive of the explosion of a massive star. Today, the source of some gamma-ray bursts is still mysterious, but mysteries are what make doing science exciting.

Going Observing

When astronomers want to carry out observations with a telescope, they do not just run off to the observatory. Generally, they submit an observing proposal that describes what objects they wish to look at and why such observations are important. They must also show that all necessary equipment will be available and that it has the sensitivity needed to complete the proposed observations. At all major observatories and space facilities, proposals are invited from astronomers from around the world to encourage the best science possible. Proposals are screened by a committee that then allocates telescope time according to the scientific merits of the proposals.

For ground-based telescopes, observing sessions are typically assigned in blocks of several nights, and the astronomer travels to the site to set up the equipment and monitor the observations. If the astronomer is using an optical telescope and must therefore observe at night, he or she must also become accustomed to switching schedules to be awake all night and sleep during the day. Most observatories have small dorm rooms with special shades to make the room dark and quiet during daylight hours for astronomers trying to catch up on sleep. Many also have small cafeterias where food and coffee are available at odd hours. Sometimes observing runs go smoothly and the astronomers can return to their home institutions laden down with data. At other times weather or equipment will not cooperate, and the runs are a total loss.

For space-based telescopes, astronomers have developed techniques for operating the telescope remotely. These techniques are increasingly being applied to ground-based telescopes as well, so that astronomers are not required to travel to the observatory, but instead may operate the telescope from their desk or prepare instruction files that allow the telescope to run on its own. While this style of observing lacks much of the adventure of traveling to remote mountain peaks, it is generally more efficient.

In the last few decades the computer has become one of the astronomer's most important tools. In fact, for many astronomers today, operating a computer and being able to program is more important than knowing how to use a telescope. Astronomers use computers not only to solve equations but also to move the telescope, store the information gathered by the detectors, convert the data obtained to a useful form, and communicate with other astronomers.

The developments in computers and the Internet have also opened up new "observing" possibilities. A number of projects have systematically observed the sky at one or more wavelengths, then stored the data in large archives that can be accessed over the Internet. Instead of gathering data on a few particular objects, such maps record information about *every* astronomical object they can detect. For example, astronomers working on 2MASS (the Two Micron All Sky Survey) mapped the entire sky at short infrared wavelengths that get through our atmosphere, compiling a database of several million images. You can see these images of gas clouds, galaxies, star clusters, and so on at the relevant websites, where they can be downloaded, then carry out analyses of the data. Similarly, astronomers working on the Sloan Digital Sky Survey have mapped large portions of the sky at visible wavelengths and carried out spectroscopic observations of more than a million objects. The use of archives does not make direct observing obsolete, but it can give a first look at a problem so that when an astronomer later goes to a telescope, he or she can use the time there more efficiently.

Modern computers have additionally made possible a new age in modeling astronomical objects. When astronomers attempt to interpret observational data, they now use computers to examine the physical processes that they suspect are taking place. This may involve simulating the gravitational interactions between two galaxies that appear to be colliding, calculating what kinds of nuclear interactions must be occurring in the core of a star, or perhaps tracing how photons are expected to move through an interstellar cloud of a particular molecular composition. These computer simulations allow astronomers to test and refine their hypotheses, and make predictions for what they expect to see when making future observations.

The Infrared Processing and Analysis Center (IPAC) archives infrared data from 2MASS as well as many other projects at www.ipac.caltech.edu.

The Sloan Digital Sky Survey can be accessed at www.sdss.org.

5.5 CHALLENGES AND NEW DIRECTIONS IN GROUND-BASED OBSERVING

Atmospheric Blurring

Anyone who has ever watched the stars flicker and "twinkle" on a clear night has seen the blurring that our atmosphere creates. Twinkling, more properly called **scintillation,** is caused by atmospheric irregularities refracting the star's light. These irregularities result from slight variations in the air's density caused by small temperature differences. As light moves through these irregularities, the light is slightly refracted.

If there were no irregularities, a star's light would reach your eye along a single path and be steady. However, the many irregularities create many paths by which the light reaches you. Moreover, these paths change direction rapidly and erratically as the irregularities move, carried by the wind. As a result, the starlight you see at any instant is a blend of light from many slightly different directions, a blend that smears the star's image and makes it dance (fig. 5.23). Atmospheric irregularities also slightly disperse the light, making the star's color dance, too. Such refractive twinkling, though very pretty to watch, seriously limits the ability of Earthbound observers to see fine details in astronomical objects. The dancing image of a star or planet distorts its picture when recorded by a camera or other device. This distortion is called **seeing** by astronomers. The atmosphere also bends the light by larger amounts near the horizon, which shifts the apparent positions of stars upward and distorts the shape of objects, as explored in Extending Our Reach: "Distortion of the Sun's Shape."

Until recently, ground-based astronomers had to simply accept the distortions of seeing, but now they can partially compensate for such seeing in several ways. One technique involves observing a known "reference" star simultaneously with the object of interest. By measuring carefully how the atmosphere distorts the known star's image, corrections can be made in the pictures of other objects. Unfortunately, it is rare that there is a bright enough star close enough to an object of interest for this technique to work. Astronomers have therefore developed a technique using a powerful laser beam to create an artificial star where they need it.

The laser beam is projected on the atmosphere, as shown in figure 5.24A. The distortions of the artificial star image are recorded by a computer that then triggers tiny actuators

FIGURE 5.23
Twinkling of stars ("seeing") is caused by moving atmospheric irregularities that refract light in random directions.

> **Q:** Why do stars twinkle more when they are low in the sky and close to the horizon than when they are nearly overhead?

FIGURE 5.24
(A) A laser beam creates an artificial star whose image serves as a reference to eliminate the atmosphere's distortion of real stars (Starfire Optical Range of the Phillips Laboratory at Kirtland Air Force Base in New Mexico). (B) A pair of images made at the Canada-France-Hawaii Telescope on Mauna Kea illustrating the difference in detail seen in a galaxy with adaptive optics turned on (left) and off (right).

on a correcting mirror placed in the telescope's light beam. The actuators create tiny adjustments in the correcting mirror that cancel out those created by the atmosphere. This technique, called **adaptive optics,** has already given astronomers dramatically improved views through the turbulence of our atmosphere, as illustrated by the inset images in figure 5.24B.

The elimination of seeing problems caused by the turbulence of our atmosphere is one reason astronomers have launched visible-wavelength telescopes into space. However, even though these space observatories can produce sharper images, the great expense of launching a telescope into space guarantees that much astronomical work, and the largest telescopes, will be ground-based for the foreseeable future. Moreover,

EXTENDING
our reach

DISTORTION OF THE SUN'S SHAPE

Refraction is easy to see in the Earth's atmosphere when it distorts the shape of the Sun as it rises and sets.

Refraction distorts the shape of the rising or setting Sun because when sunlight enters the Earth's atmosphere from space, it is refracted and bent slightly toward the ground. Thus, if you are on the ground and look back along the light ray, the light seems to come from slightly higher than it would with no atmosphere(fig. 5.25A). That is, refraction makes astronomical objects look higher in the sky than they actually are. This effect is greatest when objects are near the horizon, as shown in figure 5.25B. The result is that light from the lower edge of the Sun is refracted more than light from its upper edge, which "lifts" the lower edge more than the upper and makes the Sun look flattened (fig. 5.25C).

Refraction also slightly alters the time at which the Sun seems to rise or set. When it is at the horizon, the Sun is "lifted" by about the height of its diameter. Thus, at the moment when we see the Sun touch the horizon, it has actually already set.

By "lifting" the Sun's image above the horizon, even though it has not yet risen or has already set, refraction slightly affects the length of the daylight hours. As a result, the day of the year the Sun is above the horizon for exactly 12 hours is not the equinox, but instead several days before the spring equinox and several days after the autumnal equinox. It turns out that near latitude 40° N, St. Patrick's Day (March 17) is the day with almost exactly 12 hours between sunrise and sunset in the spring.

FIGURE 5.25
Atmospheric refraction makes stars look slightly higher in the sky than they really are. (B) Refraction is stronger for objects nearer the horizon. Objects seen at the horizon would actually be about half a degree below the horizon but for the atmosphere. (C) The Sun is flattened because refraction "lifts" its lower edge more than its upper edge. The red line near the bottom of these images is the Sun's reflection in water, showing the position of the horizon.

with ground-based equipment, problems can be corrected easily without the expense, delay, danger, and complexity of sending humans into space.

Because huge telescopes in space or even on the Moon will remain dreams for years to come, astronomers choose with care the location of ground-based telescopes. Sites are picked to minimize clouds and the inevitable distortions and absorption of even clear air. Nearly all observatories are built in dry, relatively cloud-free regions of the world, such as the American Southwest, the Chilean desert, Australia, and a few islands such as Hawaii and the Canaries. Moreover, astronomers try to locate observatories on mountain peaks to get them above the haze that often develops close to the ground in such dry locales.

While the properties of the atmosphere pose a number of challenges to ground-based observatories, another problem is posing a threat not only to astronomical observatories, but to everyone's enjoyment of the night sky, and even to our health.

Light Pollution

Recently astronomers have had to contend with another factor that affects the location of observatories: light pollution. Most inhabited areas are peppered with nighttime lighting such as street lights, advertising displays, and automobile headlights (fig. 5.26A and B). Although some lighting may increase safety, much of it is wasted energy, illuminating unessential areas and spilling light upward into the sky, where it serves no purpose at all. Figure 5.26C shows a satellite view of North America at night and illustrates the waste of energy involved in light pollution. Such stray light can seriously interfere with astronomical observations. Some observatories have been essentially shut down because of light pollution.

Light pollution not only interferes with astronomy, it wastes energy, disrupts ecosystems, and interferes with sleep cycles. While some regional planning bodies have been persuaded to implement lighting codes to minimize light pollution, the problem continues to grow worse. Light pollution can be minimized with proper light fixtures, saving money in energy costs in the long term, and preserving a part of our heritage—the ability to see stars at night.

> More information about light pollution and approaches to solving it are available from the International Dark-Sky Association (www.darksky.org).

FIGURE 5.26
Photographs illustrating light pollution. (A) Los Angeles basin viewed from Mount Wilson Observatory in 1908. (B) Los Angeles at night in 1988. (C) Notice the pattern of the interstate highway system visible in the satellite picture of North America at night.

SUMMARY

Astronomers use telescopes to collect radiation from astronomical sources. Telescopes generally have large-diameter mirrors or lenses to collect as much radiation as possible and allow faint objects to be seen. This gives them large light-gathering power.

A large mirror or lens also increases a telescope's ability to resolve detail, giving sharper images, but such gains are seriously limited for ground-based visible-light telescopes by the blurring effects of our atmosphere. Interferometers give the resolving power of a single large area by combining radiation detected with two or more small but widely separated collectors. Adaptive optics also provides a way of overcoming the atmosphere's blurring effects.

Astronomers use special telescopes to observe bodies that radiate in the radio, infrared, X ray, or gamma-ray regions of the electromagnetic spectrum. Because many of these wavelengths do not penetrate our atmosphere, telescopes have been put in space to observe them.

QUESTIONS FOR REVIEW

1. (5.1) What is light-gathering power? How does it affect the ability to see faint objects?
2. (5.1) What is the difference between a reflecting and a refracting telescope? What are some advantages of a reflecting telescope? Which type are the biggest telescopes?
3. (5.2) What is resolution of a telescope? What physical process limits it?
4. (5.2) How is resolution affected by the size of a telescope's mirror or lens?
5. (5.2) What is the purpose of an interferometer?
6. (5.3) What is a CCD and how is it better than film?
7. (5.4) Why do astronomers put X ray observatories in space rather than just on high mountains?
8. (5.4) What kinds of astronomy can be done from the ground?
9. (5.4) What is meant by adaptive optics?
10. (5.5) How do astronomers get to use observatories?

THOUGHT QUESTIONS

1. (5.1) Apart from magnification, how do binoculars help you see better? All else being equal, what difference will you see with 50-millimeter lenses versus 25-millimeter lenses?
2. (5.1) Put a pencil straight down into a glass of water. Notice whether the pencil looks bent. Now tilt the pencil and note how its apparent bending changes. How does this illustrate why light from objects near the horizon is refracted more strongly?
3. (5.1) Why isn't there a hole in the image from a reflecting telescope because the secondary mirror blocks some light? Is your answer true even if there is a hole in the center of the primary mirror, like in the Cassegrain focus shown in figure 5.8C?
4. (5.2) Is it better to have a telescope with a high resolving power or a high magnification? Explain why.
5. (5.2/5.4) The Very Large Array is a radio interferometer observatory in New Mexico with twenty-seven 25-meter telescopes. In its widest arrangement, it has the resolving power of a telescope 36 kilometers in diameter. What is better about this arrangement than a single 36-kilometer diameter telescope? What is lacking compared to a single 36-kilometer dish?
6. (5.2/5.4) Why does the useful resolving power of a ground-based telescope with a 2-meter diameter mirror not match its theoretical value?
7. (5.4) It is difficult to observe 1-nanometer, 1-millimeter, and 100-meter radiation with ground-based telescopes. What are the reasons, for each?
8. (5.4) If you look with binoculars down a beach on a hot day, you will see that distant objects appear shimmery. How is this related to astronomical "seeing"?

PROBLEMS

1. (5.1) Compare the light-gathering power of a telescope with a 10-centimeter (about 4-inch) diameter mirror to that of the human eye. (Take the diameter of the pupil of the eye to be about 5 millimeters.)
2. (5.2) Estimate your eye's resolving power by drawing two lines 1 millimeter apart on a piece of paper. Put the paper on the wall and then step back until the two lines appear as one, measuring that distance. From the distance and the separation of the lines (1 millimeter), estimate their angular separation. How does your result for the eye's resolving power compare with that calculated from the resolving-power formula, using a pupil diameter of 5 millimeters and a wavelength of 500 nanometers?
3. (5.2) Can the unaided human eye resolve a crater on the Moon whose angular diameter is 2 minutes of arc (= 120 seconds of arc)? (Take the diameter of the pupil of the eye to be about 5 millimeters and the wavelength of the light to be 500 nanometers.)
4. (5.2) Determine the resolving power of a 25-meter radio telescope observing 10-centimeter radio waves. Compare this to its resolving power for 1-meter radio waves. (Remember to convert units for the equation in section 5.2.)
5. (5.2) Using ratios or proportionalities, determine how large a diameter "eye" a person would need to see as well (1) in the infrared at wavelength of 12 micrometers, and (2) in the radio at a wavelength of 10 centimeters, as we can in the visible at 500 nm with a 5-mm pupil.

6. (5.1/5.3) Compute the collecting area of the 27 telescopes in the Very Large Array radio interferometer if each has a diameter of 25 meters. If this were the collecting area of a single dish, what would be its diameter?

7. (5.1/5.3) If a CCD could record 80% of the photons striking it, and a photograph about 4%, how many times larger in diameter would a telescope have to be, to take a photograph equal in sensitivity to a CCD image in the same amount of time?

8. (5.4) The altitude of Hubble's orbit is about 569 km above the surface of the Earth. Calculate the circumference of the orbit, the orbital velocity, and the period of the orbit (see section 3.6). How does this period affect observations?

TEST YOURSELF

1. (5.1) Telescope A's mirror has four times the diameter of telescope B's. How much greater is A's light-gathering power?
   (a) 4 times   (c) 16 times   (e) 2 times
   (b) 8 times   (d) 64 times

2. (5.2) A telescope's resolving power measures its ability to see
   (a) fainter sources.
   (b) more distant sources.
   (c) finer details in sources.
   (d) larger sources.
   (e) more rapidly moving sources.

3. (5.2) One way to increase the resolving power of a telescope is to
   (a) make its mirror bigger.
   (b) make its mirror smaller.
   (c) replace its mirror with a lens of the same diameter.
   (d) use a mirror made of gold.
   (e) observe objects using longer wavelengths.

4. (5.2) Astronomers use interferometers to
   (a) observe extremely dim sources.
   (b) measure the speed of remote objects.
   (c) detect radiation that otherwise cannot pass through our atmosphere.
   (d) enhance the resolving power (see fine details) in sources.
   (e) measure accurately the composition of sources.

5. (5.3) Which of the following are advantages of a CCD over photographic film? (Select all that are correct.)
   (a) CCDs can collect light for a long time.
   (b) CCDs do not need to be changed out for each exposure.
   (c) CCDs are not affected by blurring of the Earth's atmosphere.
   (d) CCDs do not need to be corrected for instrumental effects.
   (e) CCDs record a greater percentage of the photons striking them.

6. (5.1/5.3) Suppose you were examining a pulsing radio signal from a stellar remnant in a distant part of the Milky Way. Knowing that ionized gas in interstellar space causes dispersion of radio waves, what effect would you expect this to have on the signal?
   (a) The time when the pulses arrived would be different for different wavelengths.
   (b) The path would be bent so the signal would come from a different direction than it started from.
   (c) The wavelengths would all grow longer as they ran out of energy.
   (d) The signal would be slowed down—stretched out to fill a much longer time.

7. (5.4) To use ground-based optical telescopes to their theoretical specifications, astronomers must use
   (a) much larger mirrors than we have today.
   (b) adjustable mirrors that can adapt to correct for the atmosphere.
   (c) far more sensitive detectors.
   (d) space satellite surveys to plan observations.

8. (5.5) Atmospheric refraction makes the Sun look
   (a) smaller in diameter.   (c) bluer in color.
   (b) more luminous.   (d) higher in the sky.

9. (5.5) The purpose of adaptive optics is to make telescopes
   (a) more flexible so they can fit in smaller buildings.
   (b) look in several directions without having to move the primary mirror.
   (c) capture more photons within their aperture.
   (d) adjust for the distortions caused by the Earth's atmosphere.
   (e) All of the above.

KEY TERMS

adaptive optics, 140
atmospheric window, 134
CCD, 131
diffraction, 129
dispersion, 126
interferometer, 130
light-gathering power, 124

reflector, 125
refraction, 125
refractor, 125
resolving power, 129
scintillation, 139
seeing, 139

: FIGURE QUESTION ANSWERS

WHAT IS THIS? (chapter opening): This is believed to be Galileo's telescope. It is in the Galileo Museum in Florence, Italy.

FIGURE 5.16: Astronomers frequently use the same coloring as is used in topographic maps of the Earth, with low elevations blue as in the oceans, and mountain peaks red or even white. Weather maps, for example, use false colors.

FIGURE 5.23: Their light passes through much more atmosphere and thus is more likely to encounter atmospheric irregularities. (Fig. 5.25B shows the greater path length through the atmosphere for a star that is low in the sky.)

6

THE EARTH

The Earth seen from *Apollo 11* as it traveled to the Moon. East Africa is visible near the center of the crescent.

LEARNING OBJECTIVES

Upon completing this chapter you should be able to:

- Explain how density and seismic waves can be used to determine the Earth's internal composition.
- Identify the main elements that make up the Earth's crust and interior, and list the main regions beneath the crust.
- Describe the difference between P and S waves and how they show that part of the Earth's interior is molten.
- Explain the major factors that made and keep the Earth's interior so hot.
- Describe the process of differentiation and how it relates to the Earth's internal composition.
- Explain how radioactivity is used to find the Earth's age.
- Describe how plate tectonics produces surface features.
- Describe the source of the Earth's magnetic field, how

it changes over time, and how it affects cosmic particles.
- Identify the major layers of the Earth's atmosphere, and the main gases composing it.
- Explain why pressure declines at higher altitudes.
- Identify common greenhouse gases and explain the mechanism by which they can cause global warming.
- Identify the location of most of Earth's ozone, why it is important to life on Earth, and the causes of ozone loss.
- Describe different ways the Earth's atmosphere may have originated and the ways in which it has changed over time.
- Describe the Coriolis effect and its effect on surface motions in both hemispheres.
- Describe precession, how it alters apparent star position, and how it may affect climate.

Earth is a beautiful planet. Even from space we can see its beauty—blue seas, green jungles, red deserts, white clouds. Much of our appreciation of the Earth comes from knowing that it is home for us and the billions of other living things that share this special and precious corner of the Universe. But why study Earth in an astronomy course? The reason is that we know Earth better than any other astronomical body, and from it we can learn about many of the properties that shape other worlds.

In the simplest terms, Earth is a huge, rocky sphere spinning in space and hurtling around the Sun. In the time it took you to read that sentence, the Earth carried you about 100 miles through the black hostile space around the Sun. But you were protected by a blanket of air, a screen of magnetism, and filters of molecules that blocked most of the hazards of interplanetary space. Other planets share many of these properties but not in the right mix to make it possible for life as we know it to live on them.

Earth's special characteristics result in large measure from its dynamic nature. The Earth is *not* a dead ball of rock; both its surface and its atmosphere have changed greatly during its vast lifetime. Even today, the ground below our feet sometimes trembles and wrenches in response to dynamic forces, crumpling our planet's crust into mountains, stretching it, and tearing it open to form new ocean basins.

These slow but violent motions within the Earth arise as heat generated deep within flows toward the surface. That heat also drives volcanic eruptions, which vent gases and molten rock. Over billions of years, such gases accumulated, in part creating our atmosphere—an atmosphere that has itself been changed by the presence of abundant water and life.

CONCEPTS AND SKILLS TO REVIEW

- Determining the shape and size of the Earth (2.1)
- Newton's law of gravity (3.5)
- The law of inertia (3.1)
- Absorption in the Earth's atmosphere (4.5)

6.1 THE EARTH AS A PLANET

Stripped of its thin layer of water and atmosphere, the Earth might look something like figure 6.1—a large ball of rock, not much more colorful than the other rocky planets. Yet even without its colorful skin, there are many features of the solid Earth that clearly distinguish it from other rocky bodies in the Solar System. Understanding those differences require us to understand a variety of physical processes that are taking place on Earth and to understand its deep interior and deeper past.

Shape and Size of the Earth

Astronomers in ancient Greece knew that the Earth is a sphere with a radius of about 6400 kilometers or 4000 miles (chapter 2). Aristotle even argued that the Earth formed by material being pulled together in such a way that the jostling of the pieces led to a spherical shape. He didn't refer to gravity specifically, but we know today that it is gravity, in fact, that makes the Earth round.

Gravity is the great leveler. Over millions of years, the force of gravity crushes and deforms rock, pulling high points down and rounding large bodies off. However, this shaping process is effective only if the body exceeds a critical size that depends on the object's composition. For bodies made of rock, the critical radius is about 350 kilometers. An object with a radius larger than this has strong enough gravity that it can pull

FIGURE 6.1
The solid Earth. A reconstructed image, with the atmosphere and oceans removed.

145

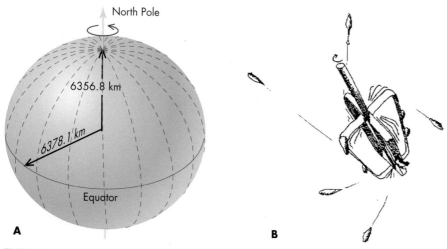

FIGURE 6.2

(A) Rotation makes the Earth's equator bulge. (B) Inertia makes material move away from the axis of a spinning object.

You can demonstrate rotationally caused bulges with a water balloon. If you toss a water balloon (gingerly!) into the air, it will take on an almost spherical shape. If you set it spinning as you toss it up, it will become noticeably bulged. This is a great demonstration, but choose an appropriate place.

itself into a sphere even if initially it was irregular. Smaller objects retain their irregular shape, as we will see when we examine asteroids in chapter 11.

Although the Earth is approximately a sphere, it is not a perfect one. It bulges out at the equator, as illustrated in figure 6.2A. The existence of this bulge was first demonstrated in the eighteenth century, when detailed mappings of the Earth showed that its equatorial radius is about 21 kilometers (about 13 miles) greater than its polar radius. This discovery was not a complete surprise, because astronomers such as Newton and Hooke had already suggested that the Earth's spinning motion might make its equator bulge into a shape technically known as an oblate spheroid. Furthermore, some other planets, such as Jupiter and Saturn, spin faster than Earth and have such large equatorial bulges that they can be easily seen through a small telescope.

To understand why the Earth's equator bulges, think of what happens when you lift a spinning electric beater from a bowl of cake batter (fig. 6.2B). Particles on such a spinning object fly outward as a result of their inertia, the tendency of all moving objects to keep moving in a straight line, as described by Newton's first law of motion. This same tendency moves matter away from the Earth's rotation axis and is strongest at the equator because the Earth rotates fastest there.

All points on the Earth take the same time (one day) to rotate once around its axis, but because points near the equator travel farther than points near the pole, they must travel faster. At the equator, a point on the Earth's surface moves at about 1000 miles per hour, while at middle latitudes, such a point moves at about 700 miles per hour. The greater speed of the equator is harder for the Earth's gravity to overcome, so the equator bulges outward.

Table 6.1	Composition of Earth's Crust	
Chemical Element	**Mass in Crust**	**Density (g/cm³)**
Oxygen (O)	46%	*
Silicon (Si)	28%	2.42
Aluminum (Al)	8%	2.70
Iron (Fe)	6%	7.9
Calcium (Ca)	4%	1.55
Magnesium (Mg)	2%	1.74
Sodium (Na)	2%	0.97
Potassium (K)	2%	0.87
Titanium (Ti)	0.6%	4.5
Hydrogen (H)	0.1%	*
Others	1%	

* Oxygen and hydrogen in crustal rock are not gases, but are part of various minerals.

Composition of the Earth

Although we may call the Earth a ball of rock, the statement is not very informative because so many different kinds of rock exist. Rocks are composed of minerals, and minerals in turn are composed of chemical elements. Analysis of the surface rocks of the Earth shows that the most common elements in them are oxygen, silicon, aluminum, magnesium, and iron. Furthermore, silicon and oxygen usually occur together as **silicates.** For example, ordinary sand (particles of the silicate mineral quartz) is nearly pure silicon dioxide (SiO_2). Table 6.1 lists a few of the most abundant elements in the Earth's crust.

Other kinds of minerals are more complicated, with atoms of calcium, magnesium, or iron included with the silicates. For example, much of the rock deep below the Earth's crust is composed of the mineral olivine, which is an iron-magnesium silicate. It gets its name from its olive-green color. Pieces of olivine are sometimes carried to the surface in lava flows (fig. 6.3).

At this point you might ask how we can tell what the interior of the Earth is made of. We can infer what lies deep inside our planet in several ways. One is by studying earthquake waves, a point we will take up in more detail in section 6.2. Another way is by analyzing the Earth's density.

Density of the Earth

Density is a measure of how much material is packed into a given volume. It is defined as an object's mass divided by its volume and is usually measured in terms of the mass in grams of 1 cubic centimeter of the substance. In these units, for example, water has a density of 1 gram per cubic centimeter. Ordinary rocks, on the other hand, have a density of about 3 grams per cubic centimeter, and iron has a density of almost 8 grams per cubic centimeter. In other words, a volume of iron has about 8 times as much mass as a similar volume of water; that is, iron is much denser than water.

We see from this that density gives some clue to an object's composition. For instance, it would be easy to tell whether a closed box contained a block of iron or a block of wood because an equal mass of a dense substance takes up a smaller volume. This is the basis of the famous story of the ancient scientist Archimedes leaping from his bath and shouting "Eureka!" Seeing how his body displaced the water, he realized that was how he could test the density of the king's crown to determine if it was pure gold. Likewise, we can use the density of the Earth to estimate *its* composition.

We find the Earth's density by dividing its mass by its volume. Its mass is 6.0×10^{27} grams (see box below), and its radius is about 6400 kilometers, or 6.4×10^8 centimeters. To actually make the calculation, we divide the mass of the Earth by its volume, $\frac{4}{3}\pi R^3$, assuming it is a sphere. Thus, the density is

FIGURE 6.3
A fist-sized piece of olivine (the greenish rock) in a lava sample.

$$\frac{M}{\left(\frac{4}{3}\right)\pi R^3} = \frac{6.0 \times 10^{27} \text{gm}}{\left(\frac{4}{3}\right)\pi (6.4 \times 10^8 \text{ cm})^3} = \frac{6.0}{\left(\frac{4}{3}\right)\pi \times 6.4^3} \times \frac{10^{27} \text{ gm}}{10^{24} \text{ cm}^3} = 5.5 \text{ gm/cm}^3$$

M = Mass of the Earth
R = Radius of the Earth
$\left(\frac{4}{3}\right)\pi R^3$ = Volume of a sphere

That is about twice the density of ordinary rock.

The density as defined here is really an *average density* over the whole planet. Because we can measure directly that the average density of surface rocks is much less than 5.5 grams per cubic centimeter, we can therefore infer that other parts of the Earth must have a density much greater than 5.5 grams per cubic centimeter. That by itself does not tell us what lies inside the Earth, but if we ask ourselves what substances are both dense and abundant in nature, we find that iron is a likely choice, as table 6.1 shows. We therefore deduce that the Earth has an iron core. But we can do better than merely deduce. We can test that hypothesis by taking advantage of one of nature's most violent phenomena: earthquakes.

Note that the number of grams per cubic centimeter is the same as the number of kilograms per liter or the number of metric tons per cubic meter. For example, water's density can also be written as 1 kg/L.

ASTRONOMY *by the numbers* — DETERMINING THE INTERNAL COMPOSITION OF THE EARTH

If a planet is made up of just rock and iron, it is possible to find the fraction of each from the planet's overall density. For example, mixing together a volume 90% of rock (3 gm/cm^3) with a volume 10% of iron (8 gm/cm^3), one cubic centimeter of the mixture contains 90% of 3 gm of rock (or 2.7 gm) plus 10% of 8 gm of iron (or 0.8 gm) for a total of 3.5 gm/cm^3.

We might write this as a formula as follows. If there is a fraction f of iron, then the overall density is:

$$(1 - f) \times 3 \text{ gm/cm}^3 + f \times 8 \text{ gm/cm}^3$$
$$= 3 \text{ gm/cm}^3 + f \times (8 - 3) \text{ gm/cm}^3$$
$$= (3 + 5f) \text{ gm/cm}^3$$

The final line of this formula can be understood as follows: As the fraction of iron goes from 0 to 1, the density goes from that of pure rock to that of pure iron.

Because the Earth has an overall density of 5.5 gm/cm^3, we can solve for the volume percentage of iron as follows:

$$(3 + 5f) \text{ gm/cm}^3 = 5.5 \text{ gm/cm}^3$$

Subtracting 3 from both sides gives:

$$5f = 5.5 - 3 = 2.5$$

Finally, divide both sides by 5 and convert to a percentage:

$$f = 2.5/5 = 0.5 = 50\%$$

So approximately half of Earth's volume is iron. A more detailed calculation accounts for the higher density of rock in the mantle and the higher density of strongly compressed iron in the core, but our simple calculation provides a reasonable first estimate.

FIGURE 6.4
A sonogram allows a doctor to "see" inside a patient. Here, a developing fetus in the womb is visible.

If we ask how the Earth's interior can be studied, your first reaction might be to say, "Why not drill a very deep hole and take a look or pull out samples?" Unfortunately, the deepest hole yet drilled in the Earth penetrates only 12 kilometers, a mere scratch when compared to the Earth's 6400-kilometer radius. If the Earth were an apple, the deepest holes yet drilled would not have broken the apple's skin.

To study the Earth's interior, we rely on indirect means somewhat like how doctors use sound waves to "see" inside our bodies. To make a picture of your internal organs, sound waves are sent through your body and are then picked up with a sensitive microphone. Because the sound travels at different speeds in bone, tissue, cartilage, and so forth, a medical technician can analyze the signals with a computer to make a picture of your anatomy or of an unborn child (fig. 6.4). The powerful waves produced by earthquakes similarly allow us to make images of the Earth's interior.

Probing the Interior with Earthquake Waves

When earthquakes shake and shatter rock within the Earth, they generate **seismic waves** that travel outward from the location of the quake through the body of the Earth. Seismic waves slightly compress rock or cause it to vibrate side to side. The speed of the waves depends on the properties of the material through which they move. A wave's speed can be determined by carefully timing its arrival at remote points of the world. From that speed, scientists can deduce a picture of the Earth's interior along the path of that wave. Thus, seismic waves allow us to "see" inside the Earth. You can use a similar, though obviously much cruder, technique to locate wall studs by thumping areas of a plaster wall with your knuckle, and petrogeologists use small explosions in a similar way to hunt for underground oil deposits.

Seismic waves in the Earth are of two main types: S waves and P waves.* P waves form as matter in one place pushes against adjacent matter—whether solid or liquid—compressing it. They travel easily through both solids and liquids. By contrast, S waves form as matter "jerks" adjacent material from side to side, like a wriggle in a shaken rope (fig. 6.5). In a liquid, material easily slips past adjacent matter, preventing S waves from spreading. Thus, S waves can travel only through solids. Therefore, if a laboratory on the far side of the Earth from an earthquake detects P waves but no S waves, the seismic waves must have encountered a region of liquid on their way from the earthquake to the detecting station (fig. 6.6), an indication that the Earth has a liquid interior. Observations show precisely this effect, from which we infer that the Earth has a liquid core. More complicated analyses can then reveal the density of the material and give clues to its composition.

Seismic studies show that the Earth's interior has four distinct regions. The surface layer is a solid, low-density **crust** about 20 to 70 kilometers (12 to 43 miles) thick and composed of rocks that are mainly silicates. Beneath the crust is a region of hot, essentially solid rock called the **mantle.** This region is also composed of silicates, the most common of which is the mineral olivine. The mantle extends roughly halfway to the Earth's center and, despite being basically solid, is capable of slow flow when stressed, much the way a wax candle can be bent by a steady pressure.

Beneath the mantle is a region of dense liquid material, probably a mixture of iron, nickel, and perhaps sulfur, called the **liquid** (or **outer**) **core.** At the very center is a **solid** (or **inner**) **core,** probably also composed of iron and nickel. Figure 6.7 illustrates these different layers and their relative sizes.

A **B**

FIGURE 6.5
(A) A P wave can be made when you push and pull a spring, compressing and stretching it. (B) An S wave can be made in a spring when you shake it side to side.

---

* P and S stand for "primary" and "secondary" and refer to the waves' arrival time at a distant site. That is, the primary waves arrive first.

ANIMATION ▶

S and P waves generated by earthquakes

Q: P-type seismic waves have a velocity of ~10 km/sec. About how long does it take such a wave to travel straight through the Earth from surface to core to the opposite surface?

FIGURE 6.6
P and S waves move through the Earth, but the S waves cannot travel through the liquid core. The speed of the waves depends on the density and composition of the material they pass through, and this causes the waves to curve, much like the refraction of light waves.

You may be puzzled as to why there is a solid core inside a liquid core at the Earth's center. If it is hot enough to melt part of the interior of the Earth, why is the very center not liquid as well? The solid core is not cooler, but rather it has a higher melting point because it is under greater pressure. At very high pressures, a previously melted material may resolidify. You can understand why this happens in the following way. For a solid to form, the atoms composing it must be able to link up to their neighbors to form rigid bonds. Heating makes the atoms move faster and breaks the bonds between neighboring atoms. With the bonds broken, the solid has nothing to hold it together, and so it becomes liquid. However, if the material is highly compressed, the atoms may be forced so close together that, despite the high temperature, bonds to neighbors may hold and keep the substance solid.

The compression needed to solidify the Earth's inner core comes from the weight of the overlying material. The thousands of miles of rock above the Earth's deep interior generate an enormous pressure there. To help visualize that pressure, imagine what it would feel like to have a pile of cinder blocks a mile high put on your stomach. In the Earth's core, the huge pressure squeezes what would otherwise be molten iron into a solid.

FIGURE 6.7
The Earth's internal structure.

Heating and Differentiation of the Earth's Core

The Earth's interior is much hotter than its surface (a fact that figures in folklore and theology!). Anyone who has ever seen a volcano erupt can hardly doubt this. In fact, the rise in temperature as you move deeper into the Earth can be measured easily in deep mines where air conditioning must be used to create a tolerable working environment. Just below the surface, the temperature rises about 2 K every 100 meters you descend. If the temperature increase continued at this same rate all the way to the center, the Earth's core would be a torrid 120,000 K. However, by measuring the amount of heat escaping from the deep interior and from laboratory studies of the properties of heated rock, geologists estimate that the temperature in the core of the Earth is "only" about 6500 K—hotter than the Sun's surface! What makes our planet's core so hot?

Scientists think that the Earth was born hot. According to this theory, the Earth formed from many smaller bodies drawn together by their mutual gravity. As each body hit the accumulating young Earth, the impact generated heat. You can demonstrate that impact generates heat by hitting a small piece of metal repeatedly with a hammer and then feeling the metal. It will be warmer than it was before you began hitting it. For the forming Earth, the impacting objects act as the hammer, and gravity is the force that drives them onto the young Earth. Technically, therefore, impact heating is release of gravitational potential energy. When the bombardment stopped, the Earth's surface cooled, but its interior has remained hot. That is, the Earth has behaved much like a baked potato taken from the oven, cooling on the outside but remaining hot inside because heat leaks only slowly from its interior to its surface.

Even if the Earth had been cold at its birth, another source of heat is important: the natural radioactivity in the Earth's interior. That is, the Earth generates heat much as a nuclear reactor does. All rock contains trace amounts of naturally occurring **radioactive elements** such as uranium. A radioactive element is one that breaks down into another element by ejecting a subatomic particle from its nucleus, a process called **radioactive decay.** Radioactive decay releases energy, generating heat. If that heat is created in a small piece of rock at the Earth's surface, it simply escapes into the surroundings and the rock's temperature barely increases. However, in the Earth's deep interior the heat is trapped by the outer rocky layers, slowing its escape. The amount of heat lost depends on the surface area, but the amount of heat contained depends on the volume. Because a smaller body has proportionately a much larger surface area compared to its volume than a larger body does, the smaller body cools quickly while the larger one remains hot (fig. 6.8). Thus, with the heat trapped, the temperature of the Earth's interior would gradually rise and the rock eventually melts.

Scientists estimate that somewhat more than half of the Earth's current heat loss is supplied by radioactive decays, as determined from measurements of subatomic particles coming from these decays. Thus, radioactive heating is slowing our planet's cooling from its original high temperature, but the present levels are not sufficient to maintain the core at its current temperature. Because radioactivity steadily depletes the radioactive elements, the amount of radioactive heating must have been larger in the past, and the planet's interior must have been much hotter.

The high temperature of the Earth's interior in the past also helps to explain the layered structure of the Earth's interior. From seismic studies we saw that dense materials, such as iron and nickel, are at the center of the Earth, and the lower-density materials, such as silicates, are found nearer the surface in the crust and mantle. They presumably sorted themselves in a process that scientists call **differentiation.** Differentiation can occur if a mixture of high- and low-density material melts, allowing the dense substances to sink and the lighter ones to rise.

You have seen differentiation at work if you have ever had the misfortune to have a half gallon of mint chocolate chip ice cream melt. When you open the carton, you find all the chips have sunk to the bottom and the air in the ice cream has risen to the top as foam. Because the Earth is differentiated, we can infer that it must have been almost

Surface area =
6 square meters

Volume =
1 cubic meter

$\dfrac{\text{Surface}}{\text{Volume}} = \dfrac{6}{1}$

1 meter cube

Surface area =
5.1×10^{14} square meters

Volume =
1.1×10^{21} square meters

12,800 km

$\dfrac{\text{Surface}}{\text{Volume}} = \dfrac{5.1 \times 10^{14}}{1.1 \times 10^{21}} \approx \dfrac{5}{10^{7}}$

FIGURE 6.8
Heat readily escapes from small rocks but is retained in larger bodies.

ANIMATION ▶

The differentiation of the Earth's core

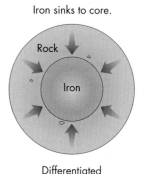

Mint chocolate chip foam

Gravity

Ice cream

Chips

Iron and rock mixed

Heating melts rock and iron.

Iron sinks to core.

Rock

Iron

Undifferentiated

Differentiated

FIGURE 6.9

Melting ice cream "differentiates" as the dense chocolate chips sink to the bottom of the carton. So, too, melting has made much of the Earth's iron sink to its core.

entirely melted at some time in the past (fig. 6.9). Differentiation also generates heating as the dense iron and nickel are dragged through the lower-density silicates, and the friction generates heat. Indeed, astronomers think that differentiation is currently helping to heat the interior of Saturn as gases with higher and lower density differentiate. Like the initial impact heating, the ultimate source of this energy is gravity.

6.3   THE AGE OF THE EARTH

Earth's radioactivity gives us a powerful tool for measuring our planet's age by measuring the amount of radioactive material that rocks contain. As a rock ages, its radioactive atoms decay into so-called **daughter atoms.** For example, uranium decays into lead, and radioactive potassium decays into calcium and the gas argon. The more daughter atoms a rock contains relative to the original number of radioactive atoms, the older the rock must be.

For instance, suppose a rock crystal formed with 100,000 atoms of radioactive potassium when it solidified, as illustrated in figure 6.10. No argon would be present when the rock first crystallized, because argon is a gas that does not combine chemically. Laboratory studies of radioactive potassium show that it decays into calcium and argon at a steady rate such that half the potassium present at a particular time will decay within the next 1.28 billion years. About 90% of the decaying potassium atoms turn into calcium, the other 10% turn into argon. Neither the argon nor the calcium is radioactive. Although argon is the rarer of the daughter atoms, it is easier to identify as a decay product, so we will ignore the calcium from now on in this discussion.

The argon is trapped within the rock unless the rock melts, and so at the end of 1.28 billion years, the sample will contain 50,000 potassium atoms. The other 50,000 potassium atoms have decayed into 45,000 calcium atoms plus 5,000 argon atoms (10% of the 50,000). The argon/potassium ratio at this point will be 5,000/50,000 = 0.1. In another 1.28 billion years, half the surviving potassium will decay, leaving

Technically, the time it takes for half of a radioactive element's atoms to decay is called its half-life. Thus, 1.28 billion years is the half-life of this form of potassium.

FIGURE 6.10

The amount of argon compared to potassium in a sample of rock gives information about the rock's age.

● Potassium
● Calcium
● Argon

100,000 Potassium	50,000 Potassium 45,000 Calcium 5,000 Argon	25,000 Potassium 67,500 Calcium 7,500 Argon	12,500 Potassium 78,750 Calcium 8,750 Argon	6,250 Potassium 84,375 Calcium 9,375 Argon
Time = 0	1.28 billion yrs	2.56 billion yrs	3.84 billion yrs	5.12 billion yrs
Argon/Potassium = 0.0	0.1	0.3	0.7	1.5

25,000 potassium atoms and creating another 2,500 argon atoms. The argon/potassium ratio will now be $(2,500 + 5,000)/25,000 = 0.3$. The ratio of argon to potassium in the rock therefore changes with time from 0, when it solidified, to 0.1, after 1.28 billion years, and so on. This ratio gives us the age of the rock sample, as figure 6.10 shows.

From such studies, scientists have found that the oldest rocks on the Earth have an age of over 4 billion years. These ancient rocks are found in such diverse places as northern Canada, southern Africa, and Australia. Thus, the Earth must be at least 4 billion years old. However, scientists think that the Earth is even older, because rock samples from other Solar System bodies, such as the Moon and asteroids (fragments of which fall to the Earth as meteorites), have ages of up to 4.55 billion years. Moreover, some small mineral crystals within old Earth rocks can be individually dated to about 4.4 billion years old. In addition, other lines of evidence indicate that the Sun is also about this old, as we will see in chapter 14. All of this evidence suggests a common age for the Solar System of a little over 4.5 billion years, an age that we will also take to be the Earth's. Why, then, are there no rocks this old on Earth? They were probably destroyed by processes we will discuss in the next section.

The age of the Earth is immense. To illustrate, if 4.5 billion years were compressed into a single year, all of *Homo sapiens* existence would take place during the last half hour, all human recorded history would have happened during the last minute of the year, and a human life span would be less than 1 second.

6.4 MOTIONS IN THE EARTH'S INTERIOR

The brevity of human life compared to the vast age of the Earth prevents us from seeing how dynamic our planet is. Mountains and seas appear to us permanent and unchanging, but even they change over the vast epochs that Earth has existed. Such changes have their ultimate cause in the heat slowly flowing from Earth's interior, a flow that creates motions in the Earth's interior and crust.

Convection in the Earth's Interior

Heat in the Earth's interior, whether left over from our planet's birth or generated by radioactive decay, creates movement of the material inside the Earth. Heating often causes motion, as you can see by watching a pan of soup on a hot stove. If you look into the pan as it heats, you will see some of the soup, usually right over the burner, slowly rising from the bottom to the top, while some will be sinking again (fig. 6.11A). Such circulating movement of a heated liquid or gas is called **convection.**

Convection occurs because heated matter expands and becomes slightly less dense than the cooler material around it. Being less dense, it rises, the basis on which a hot-air balloon operates. As the hotter material flows upward, it carries heat along with it. Thus, convection not only causes motion but also carries heat.

Convective motions in a pot of soup are easy to see—here a lima bean rises, there a noodle sinks. Such motions are less obvious in the Earth. Our planet's crust and mantle are not bubbling and heaving like the soup; rather, they are solid rock. Nevertheless, when rock is heated, it too may develop convective motions, though they are very, very slow and therefore difficult to observe. Deep within the Earth, hot molten material rises in great, slow plumes. When such a plume nears the surface, it flows parallel to the surface below the crust (fig. 6.11B), cooling and gradually growing more dense before cycling back down toward the Earth's core. Despite its slowness, the results of convective motion are evident all around us. These motions create such diverse phenomena as earthquakes, volcanoes, the Earth's magnetic field, and perhaps even the atmosphere itself.

Hot liquid rises.

A

Crust

Mantle

Hot core

B

FIGURE 6.11

Examples of convection. (A) When the soup in a pan is heated on a stove, the heated liquid drops in density, rises, cools at the surface, then sinks. (B) Convection occurs in the Earth's mantle, but over vastly longer times.

FIGURE 6.12
(A) Rifting may occur where rising material in the mantle nears a planet's surface. (B) Subduction builds mountains where material sinks back toward the interior of the Earth.

Plate Tectonics

Where hot mantle material rises then spreads beneath the surface, it drags sections of the crust apart in a process called **rifting** (fig. 6.12A). In other places the convection currents force one piece of crust to collide with another, driving one piece of crust beneath the other in a process called **subduction** (fig. 6.12B). These two processes sculpt Earth's unique landscape and cause much of Earth's seismic activity.

Rifting takes place along long breaks in the crust where convection in the mantle pulls it apart. Molten rock rises into the rifts, producing long ridges on the ocean floors, almost resembling the seams on a baseball as can be seen in the topographic map of the Earth shown in figure 6.13. Where sections of the crust are driven into each other, subduction pushes large sections of the crust down, often producing long deep trenches in the ocean floor.

The motions of the Earth's crust do not happen smoothly. The rocks stick, then suddenly break, generating a sudden lurch in the crust—an earthquake. Heat generated

FIGURE 6.13
A topographic map of the Earth's surface, including the ocean floors. Rifting occurs along long ridges on the ocean floor, where plates are moving apart, such as the Mid-Atlantic Ridge. Deep trenches occur where one plate is forced down under another, which can be seen around much of the rim of the Pacific Ocean.

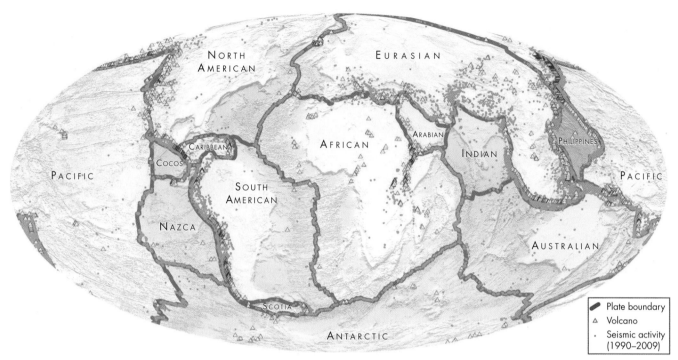

FIGURE 6.14

Locations of earthquakes and volcanoes identify plate boundaries.

by these forces melts the rock and produces volcanoes. A map of the Earth showing the locations of earthquakes and volcanoes (fig. 6.14) outlines large stable sections of crust covering Earth's surface.

The large stable sections of the crust are known as plates, and the geologic processes shifting them around on the planet's surface is called **plate tectonics.** This shifting of large blocks of the Earth's surface used to be called *continental drift;* however, it is not just the continents that move, but whole large sections of the crust. The term *plate* is used because the pieces of the Earth's crust that move are only about 50 kilometers deep but many thousands of kilometers across. Plate motion is a little like the movement of the crust on bubbling oatmeal. The crust breaks apart where hot oatmeal bubbles up and collects where the slightly cooler oatmeal sinks.

Many of the plates are named for the continents they contain, such as the North American and Australian plates (fig. 6.14). Some plates are entirely made up of ocean floor, such as the Pacific and Caribbean plates. The continents are made up of lower-density kinds of rock, such as granite, that "float" above denser rock making up the ocean floor and still denser types of rock below the surface. Most of the rifts are found in the oceans, but in some places, such as between the African and Arabian plates, stretching breaks the continental material apart. Where plates collide, the continental material buckles upward to form mountain ranges such as the Rockies and Andes along the western coasts of North and South America.

Today we can directly measure the shifting positions of the continents using global positioning system (GPS) satellites. The continents can be seen to move up to about 10 centimeters per year relative to each other. Our world is literally changing beneath our feet, growing new crust at mid-oceanic ridges and devouring it at subduction zones. This devoured rock is not lost. As it is carried downward, it is heated, and eventually its lower density causes it to rise again toward the surface.

Some of the highly eroded mountain ranges we find in scattered locations today were produced by the collision of ancient plates with completely different configurations of the material that makes up the continents. Geologists have pieced together a portion of the history of plate motion as shown in figure 6.15. The low-density rock that makes up the continents has broken apart and reassembled in many different ways far back into the remote past.

Plate motion over time

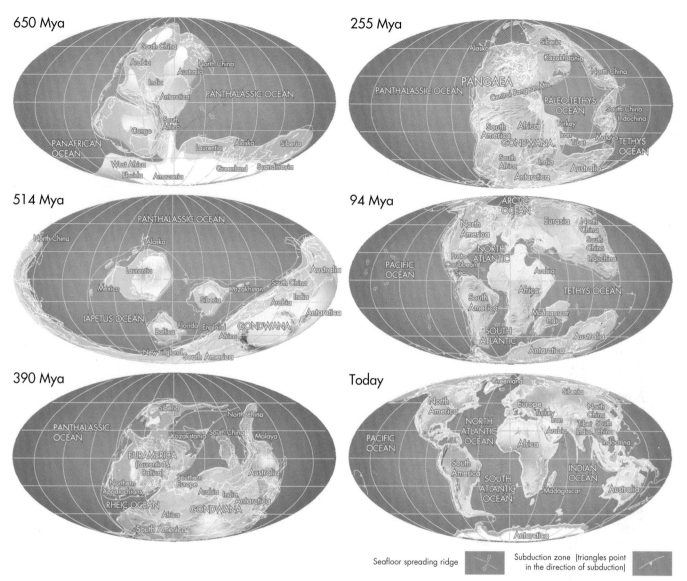

FIGURE 6.15
Continental masses have shifted throughout the Earth's history because of plate tectonics. The changes that geologists can trace over the last 650 million years ago (Mya), just 14% of Earth's history, are illustrated above. Several hundred million years ago, most of the continental material was joined in a single large continent called Pangaea, but about 250 million years ago this began to break apart to form the modern continental features.

Q: 94 million years ago, large portions of North America were under water. What else back then was different from today that could explain this?

The development of plate tectonic theory is an interesting example of how science works. As early as 1596, Abraham Ortelius, a Flemish cartographer, noticed that the newly mapped coast of South America matched the coastline of Africa like two pieces of a giant jigsaw puzzle. In 1858 a French scientist, Antonio Snider-Pellegrini, also remarked how similar the coastlines were and noted that fossils found at matching locales on both sides of the Atlantic were also very alike. He conjectured that the continents had broken apart, creating the Atlantic Ocean in the opening rift between, but apart from the similarities in fossils, he offered little in the way of supporting evidence for his idea. Similarly, in 1910 the American geologist F. B. Taylor published a paper proposing that South America and Africa had once been joined, but he too offered only slight evidence supporting his hypothesis.

The scientist who began to develop the modern theory was the German meteorologist Alfred Wegener. In 1912 he published a paper called "The Origin of the

FIGURE 6.16
(A) The idea that continents broke apart and separated was first suggested by how their coastlines fit together. Later, geological features were found that match across the dividing line. (B) Investigations of fossils reveal that they are found in bands that would have fit together if the continents were once assembled into a single large continent dubbed Pangaea.

Continents," in which he amassed fossil and geological evidence to support his hypothesis (fig. 6.16). He made a strong case for the continents having shifted position, rather than having been joined by land bridges that were later destroyed. Wegener proposed that all the continents were originally assembled in a single supercontinent that he called Pangaea (literally, "all-Earth"). Pangaea began to split into smaller plates that became the familiar continents of today (fig. 6.15), taking about 250 million years for the plates to move into their present locations. Wegener's ideas were not well received at first, and it was not until the 1960s that the accumulating evidence led to wide acceptance of the idea. Today it has become a central part of our understanding of Earth's geology. The internal motions of the Earth even provide a new understanding of another interesting feature of the Earth, its magnetic field.

6.5 THE EARTH'S MAGNETIC FIELD

The English natural philosopher William Gilbert (1540–1603) was perhaps the first to appreciate that the Earth acts like a magnet, though the ancient Chinese had used the Earth's magnetism in their invention of the compass many hundreds of years earlier.

Magnetic forces are communicated by what is called a **magnetic field.** Although some forces are transmitted directly from one body to another (as when two billiard balls collide), other forces, such as gravity or magnetism, need no direct physical contact. Magnetic fields are often depicted by a diagram showing magnetic lines of force, where each line represents the direction in which a tiny compass would point in response to the field. For example, the field lines of an ordinary toy magnet emanate from one end of the magnet, loop out into the space around it, and return to the other end, as can be illustrated by sprinkling iron particles around it (fig. 6.17). The Earth's magnetic field has a similar shape, as illustrated in figure 6.17.

Magnetic fields have an important property called **polarity.** All magnets have both a north pole and a south pole. The existence of north and south poles allows magnets to either attract or repel. Two north poles or two south poles repel each other, but a north and a south pole attract. A compass works on this principle. Its needle is a magnet, and its north pole is attracted to the Earth's magnetic south pole, and its south pole is attracted to the Earth's magnetic north pole.

The Earth's magnetic poles do not align exactly with its rotation axis (true north and south). Therefore, in general a compass needle points, not to true north, but instead several degrees away to what is called "magnetic north." Both the position and the strength of the Earth's magnetic poles change slightly but measurably, from year to year, even reversing their polarity about every 250,000 years on average. Thus, at some time in the future, a compass that now points north will point south. These changes in our planet's magnetic field are recorded in the rocks on the ocean floor (see Extending Our Reach: "Measuring Reversals of the Earth's Magnetic Field").

FIGURE 6.17
Schematic view of Earth's magnetic field lines and photograph of iron filings sprinkled on a toy magnet, revealing its magnetic field lines. "N" indicates the Earth's rotational North Pole. The magnetic north pole lies fairly close to the rotation axis, but its orientation changes over time.

When molten rock cools, magnetic minerals within the rock align to the direction of the Earth's magnetic field much as a compass needle does. This leaves rocks slightly magnetic after they solidify, and the orientation of Earth's magnetic field at the time the lava solidified is frozen into them.

During the last century, geologists discovered that Earth's magnetic field in some lava flows from millions of years ago was oriented in the opposite direction—its polarity was reversed. A compass needle would have pointed in the opposite direction in those ancient times.

In the 1960s, geologists discovered that the ocean floor showed stripes of opposite magnetic polarity running parallel to the mid-ocean ridges (see figure 6.18). The theory of plate tectonics explains how this occurs. New crust is created at the ridge and spreads away from it like paper peeling off a roll. The molten rock cooling along the ridge becomes magnetically aligned with the Earth's magnetic field at the time it emerges. This preserves a recording of Earth's magnetic field going back millions of years.

By dating the rock on the ocean floor, geologists can determine how fast the crust has moved and how often the magnetic field has reversed. One can calculate the speed of the spreading motion by dividing a rock's distance from the ridge by its age. For example, if rocks 50 kilometers from the ridge are 5 million years old, the plates must have shifted at an average speed of 50 kilometers per 5 million years, or 1 centimeter per year, a fairly typical plate speed.

The magnetic reversals occur erratically. Sometimes the orientation remains the same for tens of millions of years, but at other times for just thousands of years. On average, it reverses once every 250,000 years. The current orientation has persisted for 780,000 years. It is

suspected that during a reversal, we would lose much of our shielding from cosmic rays. Our ancestors have survived this many times before. It may, however, be a bigger problem for our current satellites and technologies.

Rift valley at ridge crest

Normal magnetic direction

Reversed magnetic direction

5 4 3 2 1   1 2 3 4 5

|— 10 kilometers —|

Age (millions of years ago)

FIGURE 6.18

Magnetic reversals recorded in the ocean floor. As plates spread apart, the upwelling lava cools and records the direction of the magnetic field at that time. Portions of the ocean floor having the same magnetic orientation as we have currently are shown in black, while those with reversed orientation are in white.

Origin of the Earth's Magnetic Field

An important question about any field is, How is it generated? Gravitational fields are generated by masses. Magnetic fields are generated by electric currents. You can easily demonstrate this by wrapping a few coils of insulated wire around an iron nail and attaching the wire ends to a battery. The nail will now act like a magnet and be able to pick up small pieces of iron or deflect a compass needle. Microscopic electric currents on the atomic scale create the magnetism of toy magnets.

The magnetic field of the Earth is generated by electric currents flowing in the Earth's molten iron core. Scientists hypothesize that these currents originate from a combination of Earth's rotational motion and convection in what is called the **dynamo model.** Studies of the magnetic fields of other Solar System bodies support this model. Planets with weak or no magnetic fields, such as Mars and Venus, are either too cold to have a large convecting core or rotate very slowly. On the other hand, bodies with large magnetic fields, such as Jupiter and Saturn, rotate rapidly and have very hot cores.

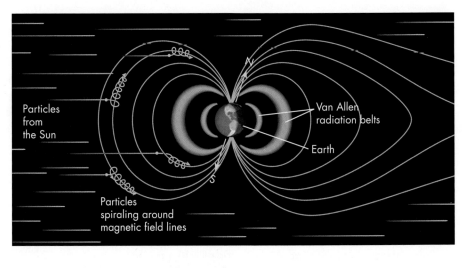

FIGURE 6.19
Electrically charged particles from the Sun spiral in the Earth's magnetic field. Some of these particles become trapped in the Van Allen radiation belts, two regions ranging out to many times the Earth's radius. "N" and "S" indicate the Earth's North and South Poles based on its rotation.

Planetary variations

Magnetic Effects on Cosmic Particles

The Earth's magnetic field does more than make a compass work. It partially screens us from electrically charged particles emitted by the Sun and even more energetic particles called *cosmic rays* produced during violent cosmic events, such as the explosion of a dying star. Many of these particles are energetic enough to damage living cells and are therefore potentially harmful to us. The Earth's magnetic field protects us because when charged particles encounter it, they are deflected into a spiraling motion around the field lines (fig. 6.19). This diverts the particles streaming from the Sun and deep space, causing many of them to flow along the field lines toward the polar regions.

The magnetic field traps some of these charged particles in two doughnut-shaped regions called the **Van Allen radiation belts** (fig. 6.19). The particles trapped in the Van Allen belts are energetic enough to penetrate spacecraft and could be a hazard to space travelers, damaging their genetic material or other tissue as well as sensitive electronic equipment. Astronauts therefore try to either avoid passing through the belts or go through them as quickly as possible.

As the charged particles flow toward the magnetic poles, they generate electric currents in the upper atmosphere. These currents, circulating around the magnetic poles, drive electrons along the magnetic field lines. The moving electrons spiral around the field lines, colliding with molecules of nitrogen and oxygen. Such collisions excite atmospheric gases, lifting their electrons to higher energy orbitals. As the electrons drop back to lower orbitals, they emit the lovely light we see as the **aurora** (fig. 6.20). The exact process by which the aurora forms is still not completely understood, but there is no doubt that its beautiful streamers are shaped by the Earth's magnetic field.

The most powerful cosmic rays can penetrate the Earth's magnetic field, but fortunately for life on Earth's surface, we have a second line of defense, the atmosphere. Most of a cosmic ray's energy is spent as it slams into molecules in the upper atmosphere. The atmosphere also blocks certain types of electromagnetic radiation, and this too plays a critical role in making the Earth a hospitable planet for life.

FIGURE 6.20
Photographs of an aurora (A) from the ground and (B) from the International Space Station.

6.6 THE EARTH'S ATMOSPHERE

Structure of the Atmosphere

Surrounding the solid body of the Earth is a veil of gases that constitutes our atmosphere. Most planets in the Solar System have an atmosphere, but the Earth's has many unique features as well as similarities that help us understand atmospheres in general.

Our atmosphere extends from the ground to an elevation of hundreds of kilometers, but at the highest altitudes the air is extremely thin. In fact, the density of the atmospheric gases decreases steadily with height, as shown in figure 6.21. Gases near the ground are compressed by the weight of gases above them. Thus, the atmosphere is a little like a tremendous pile of pillows. The pillow at the bottom is squashed by the weight of all those above it. Likewise, a block of air near sea level is more compressed and therefore has a greater density than a block of air near the top of the atmosphere. That is why it is so difficult to breathe on a mountaintop, where the air is far less dense.

The lowest layer of the atmosphere is known as the **troposphere,** which extends up to about 12 kilometers. This is the part of the atmosphere most familiar to us, where clouds and airplanes are generally found. Because of compression by the overlying layers, roughly three-quarters of the mass of the atmosphere is within the troposphere. Above the troposphere is the **stratosphere,** a region particularly important to us because of the protective ozone layer located there. Above this are other, even less dense layers gradually merging with the near vacuum of interplanetary space.

The small drag caused by rarified gas located far above Earth's surface gradually slows spacecraft in low orbits, making them spiral into the denser lower regions where they burn up. Even the International Space Station, orbiting more than 300 kilometers (about 180 miles) above the surface requires periodic boosts from docked space shuttles to keep from spiraling back down to the surface.

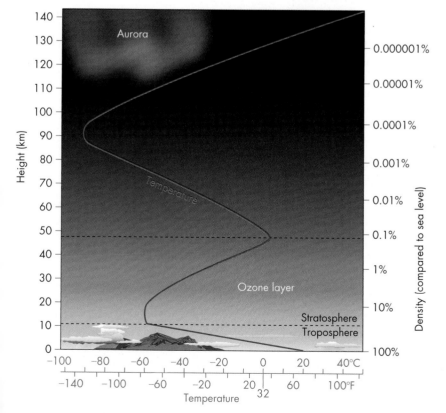

FIGURE 6.21

The Earth's atmosphere becomes steadily less dense with height, but it has a number of layers and varying temperatures, as illustrated in this diagram. The weather we experience is located in the lowest layer, known as the troposphere. Above this is the stratosphere, where most of the ozone is located. The charged particles associated with the aurora occur far above even the stratosphere, where the gas is less than one-millionth as dense as at sea level, while the International Space Station orbits about 300 kilometers above the surface.

One cubic centimeter (a volume roughly the size of the end of your little finger) of the air around you contains about 10^{19} molecules. When you take even a tiny sniff, you inhale a number of molecules roughly comparable to the number of grains of sand in a pile the size of the Astrodome.

FIGURE 6.22

In our atmosphere, puffy cumulus clouds form when the Sun heats the ground, moisture evaporates, and humid air rises. When the air reaches a low enough temperature, water vapor condenses into small droplets, forming a visible cloud, and the water precipitates back to the surface.

Table 6.2	Atmospheric Gases
Gas	% of Molecules by Number (excluding water vapor)
Nitrogen (N_2)	78.08
Oxygen (O_2)	20.95
Argon (Ar)	0.93
Water (H_2O)	Variable—typically between about 0.1 and 3.0
Carbon dioxide (CO_2)	0.039
Trace gases (less than 0.01%) Neon, helium, ozone, krypton, hydrogen, methane, carbon monoxide, and many pollutants both natural and human-made.	

Composition of the Atmosphere

One of the most striking differences between the atmosphere of the Earth and that of other planets is its composition. For example, the atmospheres of Mars and Venus are nearly completely carbon dioxide, while the atmospheres of Jupiter and Saturn are mostly hydrogen and helium. The atmosphere of the Earth is primarily a mixture of nitrogen and oxygen. Nitrogen molecules make up about 78% of our atmosphere's gas and oxygen about 21%. The remaining 1% includes carbon dioxide, ozone, and water, gases crucial for protecting us and making life possible. Table 6.2 lists the main gases in our atmosphere.

It may surprise you to see water listed as a gas. But even at low temperatures liquid water evaporates into individual water molecules that mingle with the other gases in our atmosphere. We call such free water molecules *water vapor*. Water vapor is almost completely concentrated in the troposphere. Because water molecules are lightweight, air rich with water vapor—humid air—tends to rise until its temperature gets so low that the water condenses, forming clouds of tiny ice crystals and water droplets, and rain that falls back to the surface (fig. 6.22). The circulation of water from liquid to gaseous form and back to liquid drives much of our weather. It has also played an important role in removing carbon dioxide from our atmosphere. Because carbon dioxide dissolves in water, rain scrubs the air of carbon dioxide. In fact, there is about 50 times more carbon dioxide dissolved in Earth's oceans than is present in the atmosphere.

The circulation of our atmosphere is similar to the circulation of rock in the Earth's mantle. Both are convection processes driven by heating from below. However, in the atmosphere, the heating originates not from the interior of the Earth, but from sunlight.

The Greenhouse Effect

The transparency of the Earth's atmosphere to visible radiation allows sunlight to enter the atmosphere and reach the Earth's surface. The energy of the photons is absorbed by surface materials, heating them. The warmed surface radiates infrared energy, but the atmosphere is not very transparent at infrared wavelengths, largely due to water vapor and carbon dioxide. This reduces the heat loss and makes the surface warmer than it would be if the infrared energy could escape freely, a phenomenon illustrated in figure 6.23 and known as the **greenhouse effect.**

You can get some idea of how effectively water vapor traps heat by noticing how the temperature drops dramatically at night in desert regions or on clear nights. All gardeners know that it is clear nights (with no clouds and little water vapor) that are most likely to have frost. On humid or cloudy nights, heat is retained.

It is important to recognize that the greenhouse effect does not generate heat; rather, it limits the heat loss to space. The greenhouse effect therefore warms the Earth

The air inside a greenhouse is warmer than air outside because the greenhouse confines the air within it, preventing it from rising and cooling the way outside air can. This is different from the atmosphere, so some scientists prefer the term *atmosphere effect* rather than greenhouse effect.

UV radiation is absorbed by O_3 and O_2, breaking them apart.

Some visible radiation reflects back to space.

IR radiation may be "recycled" between atmosphere and ground several times before escaping to space.

Stratosphere

Troposphere

CO_2 and H_2O absorb and reradiate IR radiation.

CO_2

H_2O

Visible radiation is absorbed by the ground and heats it. Ground then emits IR radiation.

Q: The red light waves in this figure are drawn with longer "wiggles" than the violet light waves. Why might this have been done?

FIGURE 6.23
The greenhouse effect. Radiation at visible wavelengths passes freely through the atmosphere and is absorbed at the ground. The ground heats up and emits infrared radiation. Atmospheric gases absorb the infrared radiation and warm the atmosphere, which in turn warms the ground.

the same way a blanket warms you. The blanket doesn't make you generate more heat; it simply slows down the loss of heat already there. Likewise, the water and carbon dioxide do not create heat of their own; they simply slow down the loss of heat from the ground by absorbing the infrared radiation. Eventually they re-emit it, but much of it is re-emitted back down toward the ground so it is not lost to space as quickly. That extra infrared energy reradiated to the ground helps keep the surface warm at night.

We can see how important our atmosphere is as a heat blanket by comparing the temperature of the Earth and the Moon. Although their distance from the Sun is the same, the average temperature of the Earth is much higher than the average temperature of the Moon. Averaged over seasons and latitude, the Earth's average temperature is 59°F (15°C), whereas the Moon's average temperature is a frigid −4°F (−20°C). Furthermore, tracers of the Earth's temperature and atmospheric carbon dioxide content for the past 800,000 years show a clear correlation (fig. 6.24A). Some greenhouse warming is critical for making Earth a habitable planet, but many scientists are concerned that humans are adding carbon dioxide to the atmosphere so rapidly that we might drive Earth's temperature up to problematic levels—a process called **global warming.** Rising CO_2 levels and surface temperatures over recent decades suggest this is happening (fig. 6.24B).

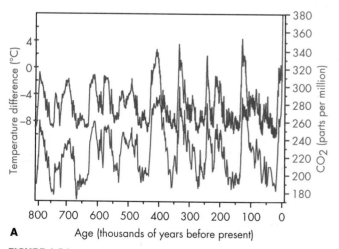

A   Age (thousands of years before present)

B   Year

FIGURE 6.24
(A) By examining trapped air in ice core samples, scientists can determine the carbon dioxide levels in the Earth's atmosphere for the past hundreds of thousands of years (red line). Isotopes also provide clues to the Earth's temperature (blue line). The two are closely coupled, as shown here. (B) Direct measurements of carbon dioxide levels and average global atmospheric temperatures since 1880.

FIGURE 6.25

The Antarctic ozone hole in September 2012. The colors in this image indicate the amount of ozone above the Earth's surface. The violet colors over Antarctica indicate only about one-third the normal amount of ozone. Chlorine (Cl) atoms released by pollutants appear to be one of the main causes of the ozone hole. As illustrated in the three small panels at top, chlorine can eliminate ozone and oxygen atoms.

The origin of Earth's atmosphere by volcanoes, comet impacts, and planetesimal collision

The Ozone Layer

The oxygen in our atmosphere is important to us not only for breathing but also as a vitally protective blanket that shields us from harsh solar ultraviolet radiation. Some of that shielding is provided by O_2 (the normal form of oxygen), but much of it comes from another molecular form of oxygen, O_3, or **ozone.**

Most of the ozone in our atmosphere is located in the ozone layer at an altitude of about 25 kilometers (80,000 feet) within the stratosphere. Ozone is formed when O_2 molecules absorb solar ultraviolet photons that have enough energy to split O_2 into individual oxygen atoms. The splitting occurs because the ultraviolet radiation makes the O_2 molecule vibrate so energetically that it flies apart. These individual oxygen atoms then combine with other O_2 molecules to form O_3.

Ozone is important because it is a strong absorber of ultraviolet radiation; without the ozone layer, solar ultraviolet radiation would pour into the lower atmosphere. The short wavelength (and therefore the high energy) of the radiation would damage living organisms. Without the protective ozone layer, you would get a severe sunburn on exposed skin simply by stepping outside. In fact, it is doubtful that life could exist on the Earth's surface without the ozone layer to shield us.

Concerns about the ozone layer were triggered by satellite studies of the atmosphere over Antarctica. During the long polar winter, atmospheric pollutants build up over Antarctica. Some pollutants destroy ozone, creating an "ozone hole" (fig. 6.25). Ozone levels in the Antarctic spring have been seen to decline by more than half over the last few decades, providing us with a warning of what might happen to the rest of the ozone layer if the amount of ozone-destroying pollutants continues to increase. For this reason, governments around the world are cooperating to limit the use of chemicals that can rise into the stratosphere where they can chemically combine with ozone and destroy it.

Oxygen and ozone are similar to water vapor and carbon dioxide in the sense that all are gases that can absorb radiation that might otherwise pass through our atmosphere. This may explain some of the common confusion between the problems of global warming and ozone depletion. Both problems are related to how humans are changing the atmosphere, but they are caused by different gases in different layers of the atmosphere. It is important to recognize that ozone blocks ultraviolet light from the Sun as it is entering the Earth's atmosphere. By contrast, greenhouse gases block infrared radiation from the Earth on its way out through the atmosphere.

Origin of the Atmosphere

Astronomers have proposed several ideas to explain how our atmosphere formed. According to one hypothesis, the gases of our atmosphere were originally trapped inside the solid material that eventually became the Earth. When that material was heated—either by volcanic activity (fig. 6.26A) or by the violent impact of asteroids hitting the surface of the young Earth (fig. 6.26B)—the gases escaped and formed our atmosphere.

Some astronomers have proposed a very different hypothesis to account for our atmosphere. They suggest that the gases were not originally part of the Earth but were brought here by comets. As we will see in chapter 11, comets are made mostly of a mixture of frozen water and gases. When a comet strikes the Earth, the impact melts the ices and vaporizes the frozen gas. Given a large enough number of impacts, comets could have delivered enough gas to form the atmosphere (fig. 6.26C). We know from the collision of Comet Shoemaker-Levy 9 with Jupiter in July 1994 that comets collide with planets even today. But such collisions were almost certainly far more common billions of years ago, when the Earth was young, because at that remote time the Solar System was full of smaller objects—the pieces from which the planets themselves grew.

FIGURE 6.26

Sources of our atmosphere. (A) Volcanic gas venting from ancient eruptions built some of our atmosphere. (B) Asteroids collide with young Earth and release gas—another source of our atmosphere. (C) Comets striking young Earth and vaporizing. The released gases also contributed to our atmosphere.

In both of these hypotheses, the early atmosphere had a very different composition than the air we breathe today. For example, our planet's ancient atmosphere probably contained far more methane (CH_4) and ammonia (NH_3) than it does now. Although these gases are still abundant in the giant planets such as Jupiter and Saturn, they have all but disappeared from our atmosphere, which is fortunate because both methane and ammonia are poisonous.

What has rid Earth of these noxious gases? Astronomers think that sunlight is responsible. Solar ultraviolet radiation is intense enough at Earth's distance from the Sun to break the hydrogen atoms out of both methane and ammonia, leaving carbon and nitrogen atoms, respectively. The nitrogen and carbon remain behind, supplying at least some of the nitrogen in our atmosphere. The hydrogen, however, gradually escapes into space because Earth's gravity is too weak to hold it. Only huge planets such as Jupiter and Saturn have strong enough gravities to retain their hydrogen and thus preserve the large amounts of methane and ammonia we see there today.

Which of these hypotheses about the origin of our atmosphere—delivered by comet or liberated from Earth's own material—is correct? Scientists have tried hard to test these very different ideas. For example, they have studied whether erupting volcanoes produce enough gas to have supplied our atmosphere and whether the composition of these gases can explain the mix of molecules in the air around us.

It requires a certain nonchalance to walk up to the lip of a bubbling volcano and hold a collecting tube over the edge to sample the foul-smelling exhalation, but when geologists make such tests, they find that nitrogen, water, and carbon dioxide are added to the atmosphere even now by volcanic eruptions. Moreover, the eruptions can account for the amount of these gases we see today if we assume that eruptions have been about as frequent in the past as they are now and that they ejected comparable amounts of gas. Thus, these tests confirm that the gases of our atmosphere might have been released by heating the material from which our planet formed.

Testing the comet delivery idea is difficult because of our lack of precise knowledge about the number of them that impacted the young Earth. Spectroscopic studies of gases that "boil" off comets reveal that heavy isotopes of hydrogen and nitrogen are about twice as abundant in comets as on Earth. These results suggest that water from comets could have contributed at most a few percent of the water in Earth's oceans. However, some models suggest that the Earth formed with such low abundances of these heavy isotopes that to raise the abundances to their present level, enough comets would have had to strike the Earth to provide most of its atmosphere. Thus, scientists remain divided about which theory is correct. Perhaps, as with so many differences of opinion, each side is partly right.

None of the hypotheses for the origin of our atmosphere—volcanic exhalations, asteroids, or comets—can account for the large amount of oxygen in our atmosphere. Where, therefore, did that vital ingredient originate? Chemical analysis of ancient rocks, particularly those rich in iron compounds that react with oxygen, shows that our atmosphere once contained much less oxygen than it does today. In fact, over the past 3 billion years, the amount of oxygen in our atmosphere has steadily increased, a rise paralleled by the spread of plant life across our planet. Most scientists therefore agree that the bulk of the free oxygen, which we breathe, was created from H_2O and CO_2 by ancient photosynthesizing microorganisms and plants. This intimate connection between life and the environment of our planet is a fact that we ignore at our peril.

Microorganisms and plants have created most of our oxygen by photosynthesis, but not all of it. Some has come from water molecules split by solar ultraviolet radiation into hydrogen and oxygen. The lighter hydrogen slowly drifts to the top of the atmosphere and escapes, leaving oxygen behind. This mechanism for adding oxygen to the atmosphere was probably the dominant source of that gas in the early history of the Earth.

6.7   THE SPIN OF THE EARTH

Considering what happens to many of us on amusement park rides, it is just as well that we are unaware of the Earth's many motions. Our planet spins on its axis, orbits the Sun, orbits along with the Sun around the Galaxy, and moves through the Universe with the Milky Way. We have already discussed how the Earth's rotational and orbital motions define the day and year, and cause seasons. But our planet's motions have other effects. Earth's spin strongly influences winds and ocean currents, and over thousands of years a slow "wobble" of its rotation axis plays a role in the onset of ice ages.

Air and Ocean Circulation: The Coriolis Effect

The Coriolis effect

If you sit with a friend on a rotating school yard merry-go-round and toss a ball back and forth, you will discover that the ball does not travel in the direction in which you aim but instead seems to curve off to the side. Similarly, ocean and air currents sweeping across a spinning planet like Earth are deflected from their original direction of motion. This phenomenon is called the **Coriolis effect.**

The Coriolis effect, named for the French engineer who first studied it, alters the path of objects moving over a rotating body, such as the Earth, other planets, or stars. To understand why the Coriolis effect occurs, imagine standing at the North Pole and throwing a rock as far as you can toward the equator (fig. 6.27). As the rock arcs through the air, the Earth rotates under it. Thus, if you were aiming at a particular point on the equator, you will miss because the surface has turned beneath the rock's path,

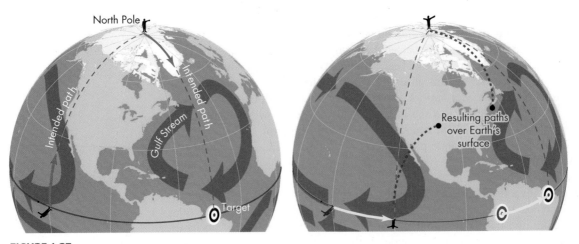

FIGURE 6.27
Coriolis effect on a rock thrown toward the equator from the North Pole or vice versa. The deflection also gives rise to the direction of ocean currents (shown by large blue arrows).

making the rock appear to have been pushed to the right. A rock thrown in the opposite direction will also curve to the right because initially it has a very large speed to the east, thanks to the Earth's rotation, but as it moves northward, the land underneath it has a smaller speed to the east because it is closer to the Earth's axis. Air, water, rockets, and all other things moving across the rotating Earth in any direction are affected similarly. In the Southern Hemisphere the effect causes paths to curve to the left.

You can clearly see the results of the Coriolis effect in the pattern of ocean currents, as illustrated in figure 6.27. The Coriolis effect deflects ocean currents to the right in the Northern Hemisphere, creating a clockwise oceanic circulation. It also creates atmospheric currents such as the trade winds. The trade winds form in response to the Sun's heating of equatorial air, which rises and expands, flowing away from the equator toward the poles at high elevations. However, before this air can reach the poles, it cools and sinks toward the surface and flows back toward the equator. As the air approaches the equator, the Coriolis force deflects it to the west (right in the North, left in the South), creating a steady surface wind that traders in sailing ships relied upon.

The Coriolis effect also causes storm systems, hurricanes, and tornadoes to spin in a counterclockwise direction in the Northern Hemisphere, and clockwise in the Southern Hemisphere. This might at first seem to be backward from what you would expect, but these storms arise in low pressure regions, which draw in air from surrounding regions. As the air is pulled in, it is deflected to the right (in the Northern Hemisphere), giving the air a counterclockwise spin, as illustrated in figure 6.28, which shows a weather satellite pictures of a large storm system. Contrary to urban myth, the Coriolis effect does not determine the direction in which water spirals down a drain or toilet! Air or water must travel over a sizable portion of the Earth's surface to receive any significant spin from the Coriolis effect. A tiny splash has far more power to set the water spiraling down a drain in either direction than the Coriolis effect has.

At one time the Coriolis effect was of special interest because it is an indirect proof that the Earth rotates. This was demonstrated with a Foucault pendulum, a massive swinging ball on a long wire that you may have seen at a science museum (fig. 6.29). If you watch the pendulum for half an hour or so, you will notice that its swing changes direction. On each swing it is deflected very slightly to the right (in the North), causing its direction to slowly shift. This can be made more evident by setting pegs in a circle around the pendulum; the pendulum will knock the pegs over, one by one. This provided one of the first clear demonstrations that the Earth must be rotating.

FIGURE 6.28
Weather satellite pictures show clearly the spiral pattern of spinning air around a storm that results from the Coriolis effect.

The Foucault pendulum

FIGURE 6.29
A Foucault pendulum.

FIGURE 6.30
Cloud bands on Jupiter created in part by the Coriolis effect.

Looking Up #1 shows part of the North Celestial Pole region, including the star Thuban in the constellation Draco.

The Coriolis force also establishes the direction of the **jet streams,** which are narrow bands of rapid, high-altitude winds. Jet streams are an important feature of the Earth's weather and are found on other planets as well. On rapidly rotating planets like Jupiter, Saturn, Uranus, and Neptune, the Coriolis effect is much stronger than on the Earth, creating extremely fast jet streams. The striking cloud bands we see on Jupiter, for example, are partly caused by this effect (fig. 6.30).

Precession

As the Earth moves around the Sun over long periods of time, the direction in which its rotation axis points changes very slowly. This motion, similar to the wobble that occurs when a spinning coin or toy top begins to slow down, is called **precession.** If the Earth were perfectly spherical, precession would not occur. But the Earth's spin makes its equator bulge slightly, so the Sun and Moon exert an unbalanced gravitational attraction on our planet, twisting it slightly. That twisting makes the Earth's rotation axis slowly change direction, completing one swing in about 26,000 years (fig. 6.31A).

Currently the North Pole points almost at the star Polaris. In about A.D. 14,000 the North Pole will point instead nearly at the bright star Vega (fig. 6.31B). Thirteen thousand years later the North Pole will again point nearly at Polaris. Thuban is of interest because it was the pole star at the time the pyramids were built in ancient Egypt, and the passage into the Great Pyramid pointed at that star.

Precession is of minor importance in day-to-day life, but over long periods of time it appears to alter the Earth's climate as well as the climate of other planets, such as Mars. As noted in chapter 1, the Earth is nearest the Sun when it is winter in the Northern Hemisphere. This changes as the Earth precesses, however. In 13,000 years, the Earth will be farthest from the Sun in Northern Hemisphere winter (fig. 6.32A). That will make winters slightly colder in the Northern Hemisphere, altering the balance of seasonal heating and cooling. And because the continents are concentrated in the Northern Hemisphere, this may help to trigger ice ages.

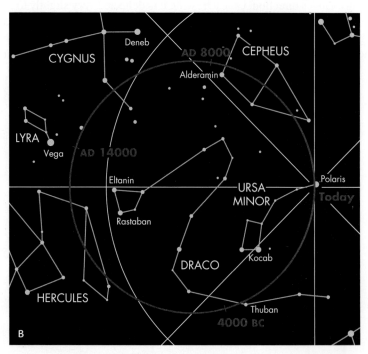

FIGURE 6.31
(A) Precession makes the Earth's rotation axis swing slowly in a circle, similar to the "wobble" of a spinning top. (B) The direction that the Earth's North Pole points changes as the Earth precesses over a 26,000-year period, so the "North Star" is steadily changing.

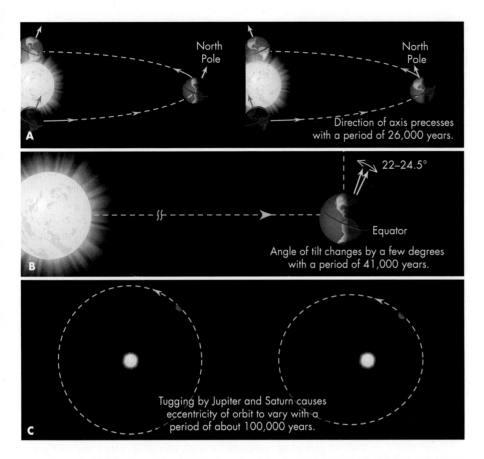

FIGURE 6.32
Long-period changes that affect Earth's climate: (A) precession of its axis; (B) the angle of its tilt; and (C) the eccentricity of its orbit.

The tug of Jupiter and the other planets also gradually alters the tilt of the Earth's axis (fig. 6.32B) and the eccentricity of its orbit (fig. 6.32C). All three of these gradual changes appear to contribute to periodic changes in Earth's climate, perhaps accounting for some of the changes seen in figure 6.24A.

SUMMARY

The Earth is roughly spherical, and its radius is about 6400 kilometers (4000 miles). Trace amounts of radioactivity in rocks here and elsewhere in the Solar System reveal that the Earth formed about 4.6 billion years ago. Radioactive material also adds heat to its interior. The flow of heat to the Earth's surface stirs slow convective motions, which shift the Earth's crust (plate tectonics), creating mountains, volcanoes, ocean basins, and earthquakes.

The waves generated by earthquakes (seismic waves) allow us to study the Earth's interior. They show it is stratified into four distinct regions: a very thin crust of ordinary rock, a mantle of hot but essentially solid silicates, an outer core of liquid iron and nickel, and an inner core of solid iron and nickel.

Currents created by motions in the Earth's core generate the Earth's magnetic field. That field in turn affects the motion of charged particles in the upper atmosphere. Such particles may create auroral displays when they collide with oxygen and nitrogen in the upper atmosphere.

The nitrogen, carbon dioxide, and water of our atmosphere may have come from volcanic gases vented over the Earth's history. Alternatively, these atmospheric gases may be the evaporated remains of comets that hit the Earth in its infancy. Plant life has created the atmosphere's oxygen by photosynthesis. Oxygen, and the ozone created when oxygen is broken apart, absorbs ultraviolet radiation from the Sun, thereby protecting us from its biologically harmful effects.

Carbon dioxide and water vapor absorb infrared radiation, trapping heat radiated from the Earth's surface. By slowing heat loss from our planet into space, these gases create the greenhouse effect and make Earth slightly warmer than it would be if the infrared radiation could escape freely.

The Earth's spinning motion creates a Coriolis effect that deflects objects moving over its surface. The Coriolis effect makes large storm systems rotate and is essential for driving the circulation of the atmosphere and the oceans. The Earth's axis also wobbles (precesses) over a period of 26,000 years.

QUESTIONS FOR REVIEW

1. (6.1) Why is the Earth round? Is it perfectly spherical?
2. (6.1) What are some of the most common elements composing the Earth's crust, mantle, and core?
3. (6.2) How do we know that the Earth has a liquid core? Why is the inner core solid even though it is hotter than the outer liquid core?
4. (6.2) What are two explanations that scientists offer for why the interior of the Earth is hot? How hot is it?
5. (6.3) How can scientists determine the age of the Earth?
6. (6.4) What is convection? What are some other examples of convection besides hot soup?
7. (6.4) What is the relation between rising and sinking material in the Earth's interior and subduction and rifting?
8. (6.4) On what plate of the crust are you located? Which way is it taking you?
9. (6.4) What is happening where one tectonic plate is smashing into another?
10. (6.5) What factors are thought to be responsible for the Earth's magnetic field?
11. (6.5) How is the aurora related to the Earth's magnetic field?
12. (6.5) How does the fact that the Earth has a magnetic field help provide evidence for the theory of plate tectonics?
13. (6.6) What were the main components of the atmosphere when the Earth formed, and what are the main components today? How and why did they change?
14. (6.6) Explain how the greenhouse effect works and how it relates to global warming.
15. (6.6) What is ozone? Why is it important?
16. (6.7) What is the Coriolis effect? How does it affect life on Earth?
17. (6.7) What is precession? What are some of its possible consequences?

THOUGHT QUESTIONS

1. (6.1) When you choose fruit at a supermarket, you might heft the fruit in your hand to test its weight. How does this tell you whether the fruit is dried out inside? How is that similar to using mean density as an indicator of the composition of the Earth's interior?
2. (6.1) Submarines contain "ballast tanks" that can take on or expel seawater. Explain how these tanks allow a submarine, which is largely constructed of steel with a density much higher than that of water, to rise and submerge at will.
3. (6.1) According to the *Guinness Book of Mountains and Mountaineering,* the summit of the volcano Chimborazo in Ecuador is the point on the Earth's surface farthest from the center. However, the book also states that the summit of Mount Everest is the highest point above sea level. Are these claims inconsistent? Why?
4. (6.2) Flicking your finger against your cheek makes a different sound from flicking it against your forehead. How

is that similar to studying the interior of the Earth with seismic waves?
5. (6.4) How does the eventual acceptance of the plate tectonic theory illustrate some aspects of the scientific method?
6. (6.5) If the Earth rotated more slowly, would you expect it to have as strong a magnetic field?
7. (6.6) Compute the mass of one molecule of H_2O, N_2, and O_2, in atomic mass units, and use this information to explain why humid air rises.
8. (6.7) Think about a stone thrown from the pole toward the equator. If the Earth rotated faster, would the Coriolis effect be larger?
9. (6.7) As seen from above the North Pole, the Earth rotates counterclockwise. Using the thrown-stone argument of question 8, explain why the Coriolis effect deflects objects to the right of their motion in the Northern Hemisphere.

PROBLEMS

1. (6.1) Suppose the Earth's radius were only half of its real value. If the Earth's mass remained the same, what would be the average density? What if the Earth's mass were twice its real value, but the Earth's radius remained the same?
2. (6.1) Using the periodic table in the back of the book, determine the astrophysical source(s) of the elements in the Earth's crust listed in table 6.1. Compare this source to the source of "precious" metals like gold, silver, and platinum.
3. (6.1) What is the average density of an alloy made of 40% titanium, 30% iron, and 10% each of calcium, magnesium, and aluminum?
4. (6.2) Seismic waves are partly reflected when they cross a boundary such as that between the mantle and the liquid core. Suppose that a P wave has a constant velocity of 8.0 km/sec. Suppose further that 700 seconds (about 12 minutes) after an earthquake near the surface, a seismometer detects a reflected P wave. How far below the surface is the liquid core–mantle boundary? Compare your answer to the distance in figure 6.7. *Hint:* Remember that this is an echo.
5. (6.3) The half-life of carbon-14, which is commonly used to date organic materials, is 5700 years. Make a graph of the percentage of original carbon present versus the age of a sample. What is the minimum age of a sample in which less than 13% of the original carbon-14 is left? In which less than 2% is left?
6. (6.4) Studies of the South American and African plates indicate that for tens of millions of years they have spread apart at a roughly constant rate of approximately 4 centimeters per year. How many kilometers farther apart are the two continents now than they were 80 million years ago? How does this compare to 6000 kilometers, approximately the distance between some matching parts of the South American and African coastlines?
7. (6.6) The total mass of the Earth's atmosphere is about 5.1×10^{18} kg. If you assume it is entirely made of nitrogen (N_2) and oxygen (O_2) gas molecules, what is the mass of

oxygen gas in the atmosphere? The mass of one molecule of gas is equal to the sum of the masses of each atom (its atomic mass times 1.66×10^{-27} kg) in the molecule. Consult table 6.1 and the periodic table in the back of the book.

8. (6.7) Use the principle of the Coriolis effect to draw a diagram to predict the air circulation flowing away from a "high-pressure system."

TEST YOURSELF

1. (6.1) Scientists think the Earth's core is composed mainly of
   (a) silicate rocks.   (c) lead.   (e) iron.
   (b) uranium.   (d) sulfur.
2. (6.2) What evidence indicates that part of the Earth's interior is liquid?
   (a) With sensitive microphones, sloshing sounds can be heard.
   (b) We know the core is lead, and we know the core's temperature is far above lead's melting point.
   (c) Deep bore holes have brought up liquid from a depth of about 4000 kilometers.
   (d) No S-type seismic waves are detectable at some locations after an earthquake.
   (e) S-type waves are especially pronounced at all locations around the Earth after an earthquake.
3. (6.3) Scientists use radioactivity in rock samples to measure
   (a) the temperature in the Earth's core.
   (b) the depth of the oceans.
   (c) the Earth's age.
   (d) the composition of the mantle.
   (e) the composition of the inner core.
4. (6.4) The slow shifts of our planet's crust are believed to arise from
   (a) the gravitational force of the Moon pulling on the crust.
   (b) the gravitational force of the Sun pulling on our planet's crust.
   (c) the Earth's magnetic field drawing iron in crustal rocks toward the poles.
   (d) heat from the interior causing convective motion, which pushes on the crust.
   (e) the great weight of mountain ranges forcing the crust down and outward from their bases.
5. (6.4) Plate motion at subduction zones can cause (more than one answer may be correct)
   (a) earthquakes.
   (b) convection currents in the Earth's mantle.
   (c) plates to grow larger.
   (d) volcanic activity.
   (e) the creation of mountains.
6. (6.5) The presence of a strong magnetic field around a planet like the Earth is evidence for
   (a) rotational and convective motion in a liquid core.
   (b) the presence of an atmosphere.
   (c) a slow rotational period.
   (d) intense heat in the core.

7. (6.6) Why is carbon dioxide called a "greenhouse gas"?
   (a) It is generated when plants are burned.
   (b) It is needed by plants to grow.
   (c) It absorbs infrared light.
   (d) It appears greenish when concentrated.
   (e) All of the above.
8. (6.6) The layer of the Earth's atmosphere in which weather occurs is the
   (a) stratosphere.   (c) ionosphere.
   (b) troposphere.   (d) hydrosphere.
9. (6.7) At what location would a pendulum's direction appear to change the most over a day?
   (a) On the ice at the North Pole
   (b) On a high mountain at midlatitude
   (c) On an island at the equator
   (d) On the ice at the north magnetic pole
   (e) It would change the same amount at all locations.

KEY TERMS

aurora, 158
convection, 152
Coriolis effect, 164
crust, 148
daughter atoms, 151
density, 147
differentiation, 150
dynamo model, 157
global warming, 161
greenhouse effect, 160
jet streams, 166
liquid or outer core, 148
magnetic field, 156
mantle, 148

ozone, 162
plate tectonics, 154
polarity, 156
precession, 166
radioactive decay, 150
radioactive elements, 150
rifting, 153
seismic waves, 148
silicates, 146
solid or inner core, 148
stratosphere, 159
subduction, 153
troposphere, 159
Van Allen radiation belts, 158

ⓠ: FIGURE QUESTION ANSWERS

WHAT IS THIS? (chapter opening): This is a satellite picture of the volcanic island Santorini in the Mediterranean Sea. The central bay in the island was created when the volcano exploded violently about 3600 years ago, possibly giving rise to the legend of Atlantis. A new volcanic cone can be seen growing in the middle of the bay.

FIGURE 6.6: The time, t, for something moving at a speed V to travel a distance d is given by $t = d/V$. The Earth's diameter is about 12,000 km. Therefore $t = 12,000$ km/(10 km/sec) = 1200 seconds, or about 20 minutes.

FIGURE 6.15: There were no polar ice caps, so sea level was much higher.

FIGURE 6.23: Red light has a longer wavelength than violet light.

Keeping Time

From before recorded history, people have used events in the heavens to mark the passage of time. The day, the month, and the year were all originally defined in terms of obvious astronomical phenomena. The day was the time interval from sunrise to sunrise. The month was the interval from new moon to new moon. The year was the time it takes for the Sun to complete one circle of the zodiac. Astronomical events are not perfect time markers, however: even the day and year need to be defined with care. It requires some careful astronomical study for clocks and calendars to work as they are intended to work.

THE DAY

The length of the day is set by the Earth's rotation speed on its axis. One day is defined as one rotation. However, we must be careful how we measure our planet's rotation. For example, we might use the time from one sunrise to the next to define a day. That is how many ancient civilizations measured time, but there is a problem with this: The time from sunrise to sunrise changes steadily throughout the year as a result of the seasonal change in the number of daylight hours.

A better time marker is the time it takes the Sun to move from its highest point in the sky on one day (what we technically call apparent noon) to its highest point in the sky on the next day—a time interval that we call the **solar day.** If we measure the length of the solar day, however, we will discover that it too is not exactly 24 hours. Its length changes by almost 1 minute over the course of the year. As we will discuss below, this variation arises from the Earth's motion around the Sun. Thus, although the Sun's motion across the sky determines the

day–night cycle, the Sun's motion is *not* a good reference for the actual time it takes our planet to complete one spin.

We can avoid most of this variation in the day's length if, instead of using the Sun, we use a star as our reference. For example, if we pick a star that lies exactly overhead at a given moment and measure the time it takes for that same star to return to exactly overhead, we will find that the time interval is an unchanging 23 hours 56 minutes 4 seconds. We call this day length, measured with respect to the stars, a **sidereal day.**

Why do the solar and sidereal day differ in length? We can see the reason by looking at figure E3.1, where we measure the interval between successive apparent noons—a solar day. Let us imagine that while we are watching the Sun, we can also watch a star, and that we measure the time interval between the star's passages overhead, a sidereal day.

As we wait for the Sun and star to move back overhead, the Earth moves along its orbit. The distance the Earth moves in one day is so small compared with the star's distance that we see the star in essentially the same direction as on the previous day. However, we see the Sun in a measurably different direction, as figure E3.1 shows. The Earth must therefore rotate a bit more before the Sun is again overhead. That extra rotation, needed to compensate for the Earth's orbital motion, makes the solar day slightly longer than the sidereal day.

It is easy to figure out how much longer, on average, the solar day must be. Because it takes us 365¼ days to orbit the Sun and because a circle has by definition 360°, the Earth moves approximately 1° per day in its orbit around the Sun. That means that for the Sun to reach its noon position, the Earth must rotate approximately 1° past its position at the previous

Day 1, Noon—Sun and star are both overhead.

North Pole

Sun

Day 2, 11:56 A.M.—Earth has turned once with respect to the star, so star is back overhead, but the Sun is not.

To star

Day 2, Noon—Earth has now turned once with respect to the Sun but has made *more* than one full turn with respect to the star.

FIGURE E3.1

The length of the day is shorter measured with respect to the stars than with respect to the Sun. The Earth's orbital motion around the Sun makes it necessary for the Earth to rotate a tiny bit more before the Sun will be back overhead. (Motion and sizes are exaggerated for clarity.)

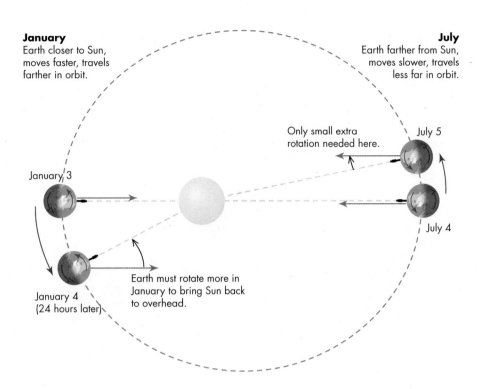

January
Earth closer to Sun, moves faster, travels farther in orbit.

July
Earth farther from Sun, moves slower, travels less far in orbit.

Only small extra rotation needed here.

July 5

January 3

July 4

Earth must rotate more in January to bring Sun back to overhead.

January 4 (24 hours later)

FIGURE E3.2
As the Earth moves around the Sun, its orbital speed changes in accordance with Kepler's second law of motion. For example, the Earth moves faster in January when it is near the Sun than in July when it is far from the Sun. Thus, in 24 hours the Earth moves farther along its orbit in January than in July. As a result, the Earth must turn slightly more in January to bring the Sun back to overhead. This makes the interval between successive noons longer in January than in July and means they are not exactly 24 hours. For that reason, time is kept using a "mean Sun" that moves across the sky at the real Sun's average rate. (Note that the ellipticity of the Earth's orbit has been exaggerated to make the differences clearer in this figure.)

noon. In 24 hours = 24 × 60 = 1440 minutes, the Earth rotates 360°. Therefore, to rotate 1° takes 1440/360 minutes, or about 4 minutes. The solar day is therefore about 4 minutes (3 minutes 55.9 seconds, to be precise) longer than the sidereal day.

The motion of the Earth around the Sun not only makes the solar day longer than the sidereal day, it also makes the length of the solar day vary. If you measure carefully the time interval from one apparent noon to the next, it ranges from 29.8 seconds more than 24 hours (about December 23) to 21.4 seconds less (about September 17). This variation arises because of two factors: the Earth's axis is tilted, which makes the Sun's daily "longitudinal" shift smaller around the equinoxes; and its orbit is not circular, which makes our orbital velocity change according to Kepler's second law.

The Earth moves along its orbit faster when it is near the Sun and slower when it is farther away. When the Earth is moving rapidly in its orbit, it takes a little longer for a point on the Earth to swing around to face the Sun than when it is moving slowly (fig. E3.2). Hence, the solar day is longer when we are near the Sun and shorter when we are farther away. The variations are small, but they must be accounted for if our clocks are to always read about noon when the Sun is highest in the sky.

It would be complicated and confusing to design clocks that had hours of different lengths at different times of the year. Instead, it is much easier to define the length of the day differently, using not the true interval from one apparent noon to the next, but the *average* value of that interval over the year. That average day length is called the **mean solar day,** and it has, by definition, 24 hours of clock time. We therefore use mean solar time in our daily timekeeping.

Over the course of the year, the differences in length between the mean solar day and the true solar day accumulate

and lead to a difference of up to 16½ minutes between clock time and time based on the position of the Sun. This difference is called the "equation of time" and is shown graphically in figure E3.3. The equation of time gives the correction needed on a sundial if it is to give the same time as your watch.

Although we use solar time in regulating our daily activities, astronomers find sidereal time more useful. This is because at a given location, a given star always rises at the same sidereal time. To avoid the nuisance of A.M. and P.M., sidereal time is measured on a 24-hour basis. At a particular location, we might find that the star Betelgeuse in the constellation Orion rises at about 10 P.M. in November, at about 8 P.M. in December, but always at about 23:50 sidereal time.

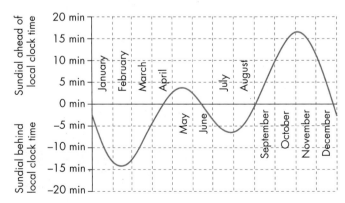

FIGURE E3.3
The equation of time is the correction that must be applied to the true Sun to determine mean solar time. It can be shown as a graph (as here) or as a figure-8 shape called an "analemma," often seen on globes of the world.

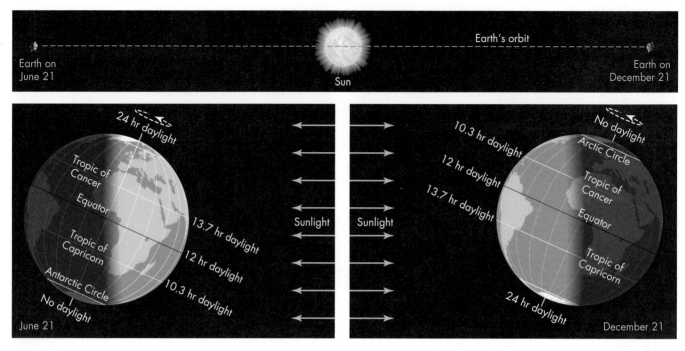

FIGURE E3.4

The tilt of the Earth affects the number of daylight hours. Locations near the equator always receive about 12 hours of daylight, but locations toward the poles have more hours of dark in winter than in summer. In fact, poleward of latitude 66.5° north or south, the Sun never sets for part of the year and never rises for another part of the year (the midnight sun phenomena). At the equinoxes, all parts of the Earth receive the same number of hours of light and dark. (Sizes and separation of the Earth and Sun are *not* drawn to scale.)

HOURS OF DAYLIGHT

Although each day has almost exactly 24 hours, the number of hours of daylight (the amount of time the Sun is above the horizon) changes greatly throughout the year unless you are close to the equator. At a latitude of 40° (approximately that of New York, Rome, or Beijing), summer has about 15 hours of daylight and only 9 hours of night. In the winter, the reverse is true.

This variation in the number of daylight hours is caused by the Earth's tilted rotation axis. Remember that as the Earth moves around the Sun, its rotation axis points in roughly a fixed direction. At the start of summer in the Northern Hemisphere, the North Pole is tilted toward the Sun, and six months later it is tilted away. The result (as you can see in fig. E3.4) is that only a small part of the Northern Hemisphere is unlit in the summer, but a large part is unlit in the winter. Thus, as the Earth's rotation carries us around, only a relatively few hours of a summer day are unlit, but a relatively large number of winter hours are dark. On the first day of spring and of autumn (the equinoxes), the hemispheres are equally lit, so that day and night are of equal length everywhere on Earth.

If we change our perspective and look out from the Earth, we see that during the summer, the Sun's path is high in the sky, so that the Sun spends a larger portion of the day above the horizon. This gives us not only more heat (because the sunlight falls more directly on the ground) but also more hours of daylight. On the other hand, in winter the Sun's path across the sky is much shorter, giving us less heat (because the sunlight falls less directly on the ground) and fewer hours of light.

From the arctic and antarctic, there are days when the Sun will be up for all 24 hours of a day. This is illustrated by the sequence of photographs in figure E3.5, taken near the Arctic Circle at the beginning of summer. Six months later the Sun remains below the horizon for the entire day.

ANIMATION ▶

The change in number of hours of daylight as seasons change

FIGURE E3.5

Sequence of 24 pictures of the Sun taken each hour from a spot close to the Arctic Circle near the beginning of summer. At "midnight" the Sun is just above the horizon.

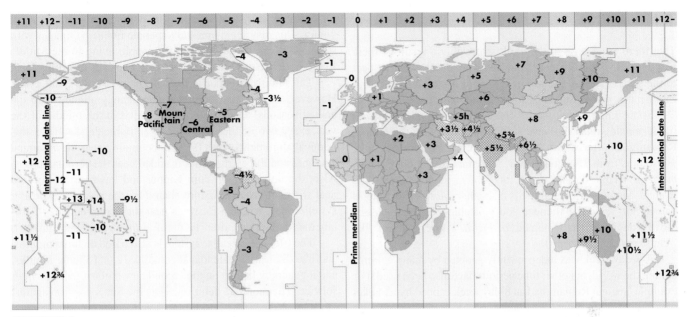

FIGURE E3.6

Time zones of the world and the international date line. Local time = Universal time + numbers on top of chart. Many regions also add an hour for daylight saving time during the summer and surrounding months. Regions with nonstandard times have checkerboard shading.

TIME ZONES

Throughout history, towns set their clocks according to the local solar time. This meant that the clocks in towns east or west of each other would differ by some minutes depending on their distance and latitude. This did not cause confusion until the mid-1800s with the development of high-speed travel and communications. Then it became difficult to set up, for example, train timetables between cities that kept different times.

The solution eventually developed was to divide the Earth into a set of 24 major **time zones** separated by 15° in longitude in which the time differs by one hour from one zone to the next. Within each time zone, the Sun is close to "overhead" at noon, yet the times are the same or differ by a whole number of hours.

Across the contiguous 48 United States, four time zones were adopted, Eastern, Central, Mountain, and Pacific time (see fig. E3.6). The common time within each time zone is called **standard time.** Thus, in the eastern zone, the time is denoted Eastern Standard Time (EST). The time zone boundaries have often been modified to follow political or natural boundaries. Some countries, such as China, have adopted a single time zone for the whole country, so the Sun is highest in the sky about 4 P.M. in the western part of the country.

As you travel westward, you need to reset your watch one hour earlier for each time zone you enter. It might seem that if you went through 24 time zones, you would end up a day earlier, but of course you cannot actually gain a day this way. The time zone system is arranged so that when you cross the **international date line** (fig. E3.6), you add a day to the calendar if you are traveling west and subtract a day if you are traveling east. The international date line generally follows 180° longitude (roughly down the middle of the Pacific Ocean), but

bends around extreme eastern Siberia and some island groups to ensure they keep the same date as their neighbors.

UNIVERSAL TIME

The nuisance of having different times at different locations can be avoided by using **Universal time,** abbreviated as UT. Universal time is the time kept in the time zone containing the longitude zero, which passes through Greenwich, England. By using UT, which is based on a 24-hour system to avoid confusion between A.M. and P.M., two people at remote locations can decide to do something at the same time without worrying about what time zone they are in.

DAYLIGHT SAVING TIME

In many parts of the world, people set clocks ahead of standard time during the summer months and then back again to standard time during the winter months. This has the effect of shifting sunrise and sunset to later hours during the day, thereby creating more hours of daylight during the time most people are awake. Time kept in this fashion is therefore called **daylight saving time** in the United States. In other parts of the world, it is called "Summer Time."

Daylight saving time was originally established during World War I as a way to save energy. With clocks set ahead, less artificial light was needed during work hours late in the day. Nowadays, it allows us more daylight hours for recreation after work during the summer. As of March 2007, daylight saving time in the United States runs from the second Sunday in March to the first Sunday in November.

THE WEEK

That there are 7 days in the week is possibly a result of there being seven visible objects that move across the sky with respect to the stars: the Sun, the Moon, and Saturn are obvious in our English day names Sunday, Monday, and Saturday.

Some English day names come to us through the names of Germanic gods who have a direct parallel with the Greco-Roman gods after whom the planets are named. For example, Tuesday is from *Tīw,* god of war, like Mars (matching Spanish *martes*). Wednesday is named for *Wōden,* identified with the Roman Mercury (matching Spanish *miércoles*). Thursday is named for *Thor* the thunder god (matching Spanish *jueves,* "Jove's day"). Friday is named for *Freya,* a love goddess, like Venus (matching *viernes*).

A 7-day week was independently adopted by many cultures around the world, but other time periods have also been used. For example, ancient Romans had an 8-day week capped by a market day, and after the French Revolution people experimented with a 10-day week. Other collections of days, such as sets of 20 days used by ancient Mayans (fig. E3.7), are also known.

THE MONTH AND LUNAR CALENDARS

While the week is somewhat arbitrary, almost all cultures have marked intervals of time based on the Moon. The month, whose name derives from "Moon," is a period of time reflecting the duration of the lunar cycle of phases. The time interval between full moons is about 29.5 days, so 12 lunar cycles adds up to about 354 days. This is about 11 days short of a full year, which is about 365 days. This makes it complicated to design a calendar that keeps track of both the phase of the Moon and the time of year.

The Islamic calendar is based solely on the Moon, with a year consisting of 12 months of either 29 or 30 days. The resulting year of 354 days therefore shifts about 11 days earlier compared to seasonal markers such as the solstices or equinoxes each year. This means that the holy month of Ramadan,

FIGURE E3.7
Portion of a Mayan calendar, which is broken up into 20-day "weeks."

for example, can sometimes fall in summer and sometimes in winter. This may seem unusual to people living in climates with strong seasonal variations, but for people living in the Middle East where seasons are not so extreme, it makes less difference.

The Jewish calendar is likewise based on the lunar cycle. However, to keep the months approximately aligned with the annual seasons, an extra month is added in the middle of the year every two or three years. The Jewish calendar has astronomical alignments, with the year beginning near the autumnal equinox, and the extra month is added near the vernal equinox. Also, the holy days of Yom Kippur and Passover are located near the equinoxes, a feature shared by the common calendar system today, wherein Easter is near the vernal equinox (actually the first Sunday after the first full moon after the equinox).

The Chinese calendar system is also based on lunar months, and like the Jewish calendar it contains 12 or 13 lunar months, and it generally begins on the second new moon after the winter solstice. The month is determined by the Sun's position against the stars at the start of each lunar cycle, according to 12 equal intervals along the ecliptic (see chapter 1). In some years the Sun may be at the beginning of an interval at the start of one month, and still within the same interval at the start of the next lunar cycle. In that case, there is a second month of the same name—for example, a "second August."

The Chinese calendar has some other unusual features. The years are grouped into 60-year cycles composed of 5 cycles repeating every 12 years. The years in each 12-year cycle are given names such as the Year of the Rat, the Year of the Dog, and so forth.

THE MAYAN CALENDAR

One of the more unusual calendar systems was developed by the Maya. They tracked 260-day periods along with the approximate year length of 365 days, which would match up every 52 years (which is the same length as 73 cycles of 260 days). It is unclear why the Maya kept track of a 260-day cycle, although some hypothesize that it was used to keep track of the changing apparent position of the planet Venus, which figured prominently in Mayan beliefs.

To measure longer periods of time, the Maya combined time intervals that added up to about 400 years, and placed those into groups of 13 (about 5000 years). In the Mayan calendar, at the end of one of these 5000-year periods, the count would roll over, like a car odometer reaching 99999 miles and returning to 00000. The fact that the Mayan long count cycle has a "rollover" in December 2012 prompted many silly claims that this was a prediction by the Maya of catastrophes such as a collision with a "rogue planet," an outburst from the Sun, or a peculiar alignment with the Milky Way. There was no astronomical basis to any of these claims. The Maya themselves wrote of dates beyond this, so clearly they did not believe that a catastrophe was coming. The obsession with apocalyptic events as calendars reach a millennium or other "special" date seems to be a question for sociology, not astronomy.

THE COMMON CALENDAR

Although lunar calendars are still used to mark many religious occasions, the calendar in most widespread use is based only on the solar year. This calendar is based on one initially developed about 200 B.C. by the Romans. In fact, the word *calendar* is itself of Roman origin. There is some controversy about how the original Roman calendar was organized. It may have had only 10 months, and it probably began on the first day of spring (the vernal equinox) rather than in January.

The names of our months date from that calendar and its modifications. For example, if the year began in March, then September, October, November, and December were the 7th (Sept.), 8th (Oct.), 9th (Nov.), and 10th (Dec.) months, respectively. The 5th and 6th months (*Quintilis* and *Sextilis* in ancient Rome) were renamed later in honor of Roman emperors. Possible origins of the names of other months are listed in table E3.1.

Because it did not contain the right number of months and days to match the astronomical phenomena, this original calendar became a form of political patronage. The priests who regulated the calendar would add days and even months to please one group, and take days off to punish another. So much confusion resulted from these abuses that in 46 B.C. Julius Caesar asked the astronomer Sosigenes to design a calendar that would fit the astronomical events better and give less room for the priests and politicians to tinker with it. The resulting calendar, known as the **Julian calendar,** consisted of 12 months, which with the exception of February alternated between 31 and 30 days in length.

For that reason, some of the months are made 31 days long. In fact, if every other month, starting with January, were a 31-day month, the year would total 366 days. To make the days add up to 365, February was trimmed 1 day, to 29.

Table E3.1	Origin of the Names of the Months
January	Janus, god of gates, a two-faced god looking to the past and future
February	Februa, Roman festival of purification
March	Mars, god of war, month to resume wars
April	Etruscan *apru*, from the Greek Aphrodite, goddess of love
May	Maia, the eldest of the Pleiades, mother of Hermes, and Roman goddess of spring
June	Juno, principal Roman goddess, wife and sister of Jupiter
July	Julius Caesar, Roman emperor who reformed the calendar
August	Augustus Caesar, successor to Julius
September–December	Seventh- to tenth-month. The *-ember* may come from the same root as *month*.

But, you protest, that is not the way the calendar looks. Although January, March, May, and July have 31 days, the sequence is broken in the later months. You see at work there the politics of ancient Rome. The Julian calendar barely survived Caesar before the politicians were at it again. First, the name of the seventh month was changed to *Julio* to honor Julius Caesar—hence our *July*. Next, on the death of Julius Caesar's successor, Augustus Caesar, a very able and highly respected leader, it was decided to name the eighth month in *his* honor—hence, the name *August*. However, because it would have been impolitic to have his month a day shorter than Julius's, August became a 31-day month, and all the following months had the number of their days changed to maintain the alternation. Unfortunately, this led to using up one more day than there were days in the year. Thus, poor February, already one day short, was trimmed a second day, leaving it with only 28 days. With only minor modifications, this is the calendar we use today. However, those modifications are important, as we discuss next.

LEAP YEAR

The ancient Egyptians knew that the year is not exactly 365 days long. It turns out that it takes about 365 and ¼ days for the Earth to complete an orbit around the Sun, which is how we measure a year. Because we can't have fractions of a day in the calendar, a calendar based on a year of 365 days will come up 1 day short every 4 years.

Your first reaction might be, So what? However, the seasons are set by the orientation of the Earth's rotation axis with respect to the Sun, not by how many days have elapsed. We therefore want to make sure that we start each year with the Earth having the proper orientation. Otherwise, the seasons get out of step with the calendar. For example, because in 4 years you will lose 1 day, in 120 years you will lose a month, and in 360 years, you will lose an entire season. With a 365-day year, in a little over three centuries April would be coming in what is now January.

This problem is corrected by the leap year, a device implemented by the Julian calendar to keep the calendar in step with the seasons. The leap year correction adds a day to the calendar every fourth year. The extra day is traditionally added to February because it is the shortest month.

Unfortunately, the year is actually a little bit shorter than 365¼ days. Thus, having leap year every four years corrects a tiny bit too much. For nearly 1600 years after the adoption of the Julian calendar, the small errors accumulated, adding up to about 10 days, making it obvious that the calendar was out of sync with the seasons. To prevent further accumulation of errors, three leap years needed to be dropped every four centuries. Therefore it was decided that centuries not divisible by 400 would be eliminated as leap years. Thus, 1900 was not a leap year, but 2000 was. This modification of omitting leap year for all century years not divisible by 400 was added in 1582 at the direction of Pope Gregory XIII. The calendar we use today is thus known as the **Gregorian calendar.**

The inauguration of the Gregorian calendar in 1582 was not a peaceful affair. To bring the calendar back into synchrony with the seasons, Pope Gregory simply eliminated 10 days from the year 1582 so that the day after October 4 became October 15. Although the changeover went smoothly in most places, non-Catholic countries such as Protestant England refused to abide by the Pope's edict. The calendar in England and in a few other northern European countries was not altered. This made commerce between Catholic and non-Catholic countries very difficult because the day and sometimes even the month and year were different from one country to the next.

Eventually the Gregorian calendar was adopted essentially worldwide, but the change was not made in England until 1752. This elimination of by then 11 days from the calendar supposedly led to riots by people fearing they would be charged a full month's rent for only 20 days. In Russia the change was not made until the revolution in the early part of the twentieth century. Other countries (Greece and Turkey, for example) changed in the 1920s.

MOON LORE

The Moon figures prominently in folklore around the world. Most stories concerning its powers are false. For example, people often claim that the full moon triggers antisocial behavior, hence the term *lunatic*. All studies to look for such effects have found nothing. Automobile accidents, murders, admissions to clinics, and so forth show no increase when the Moon is full.

On the other hand, "once in a blue moon," indicating a rare event, is a phrase with a basis in fact, because on rare occasions the Moon may look blue. This odd coloration comes from particles in the Earth's atmosphere. Normally our atmosphere filters the blue colors from light better than it filters the red ones. For example, light from the rising or setting Sun passes through so much atmosphere that little blue light remains by the time it reaches us. Therefore, the Sun looks red when it is low in the sky. However, if the atmosphere contains particles whose size falls within a very narrow range, the reverse may occur. Dust from volcanic eruptions or smoke from forest fires may have just the right size to filter out the red light, allowing mainly the blue colors to pass through. Under these unusual circumstances, we may therefore see a "blue Moon."

A different meaning for "blue moon" has appeared more recently, referring to months with two full moons. Because the cycle of phases is 29.5 days, it is rare for a second full moon to occur in the same month, and some calendars have printed the second full moon in blue.

Another well-known phrase is "harvest moon," the full moon nearest the time of the autumn equinox. As it rises in the east at sunset, the light from the harvest moon helps farmers see to get in the crops. Full moons in other months also have popular names, but only the harvest and hunter's moon are widely known. Other names occasionally used in American folklore are listed in table E3.2.

Table E3.2 **Names Used for Full Moons***

January	old moon	July	thunder or hay
February	hunger	August	grain or green corn
March	sap or crow	September	harvest
April	egg or grass	October	hunter's
May	planting	November	frost or beaver
June	rose or flower	December	long night

*Most of these names derive from Native American usage.

In the last few decades the term "supermoon" has become popular to describe the Moon when it is particularly large in angular size when it is full or new. This occurs at least twice every year, and is not astronomically significant, because it simply reflects the fact that the Moon's orbit is elliptical. Details of the Moon's orbit are examined further in chapter 7.

THE ABBREVIATIONS A.M., P.M., B.C., A.D., B.C.E., AND C.E.

Four abbreviations are used frequently in the measure of time and calendars. They are the familiar letters A.M., P.M., B.C., and A.D. The first two have specific astronomical meaning. The last two have cultural meaning.

A.M. and P.M. stand for "ante meridian" and "post meridian," respectively. The meridian is the line passing from due north to due south and passing through the point exactly overhead (the zenith), dividing the eastern and western halves of the sky. As the Sun moves across the sky, it crosses the meridian at the time called apparent noon. Before noon, it lies before (ante) the meridian. After noon, it lies past (post) the meridian. Hence, A.M. and P.M.

B.C. stands for "before Christ," referring to the year of his birth. Oddly, by convention 1 B.C. refers to the year before his birth, to avoid having a year "0." Most historians believe this chronology was inaccurate and that Jesus would have been born about 5 B.C.

A.D. stands, not for "after death," but for *anno Domini*, meaning "in the year of the Lord." The term A.D. was introduced by the sixth-century monk Dionysius Exiguus, about A.D. 528, in his attempts to trace the chronology of the Bible.

Recently, two different abbreviations have begun to replace A.D. and B.C. They are B.C.E. and C.E., which stand for, respectively, "before the common era" and "common era." "Common era" refers to our present calendar, which is used nearly worldwide for most business purposes and thereby avoids reference to a particular religion. Yet another abbreviation—B.P.—is used, especially in anthropological and geological works. B.P. stands for "before present (era)" and is used for dates determined by analyzing the radioactive carbon in the object of interest. It takes 1950 C.E. as its base year.

SUMMARY

Our system for keeping time is based on the motions of the Earth, Moon, and Sun. The day is determined by the Earth's spin, the month by the Moon's orbital motion around the Earth, and the year by the Earth's orbital motion around the Sun.

The solar day is based on the time interval between one apparent noon and the next. The sidereal day, or the interval between the time of star-rise for a given star and the time of its next rising, is about 4 minutes shorter than the solar day. This difference arises because as the Earth moves along its orbit, the direction to the Sun shifts slightly. We must therefore wait a little longer to allow the Earth's rotation to carry us into the same position with respect to the Sun.

Time zones divide the Earth into regions such that the time differs by 1 hour (in general) from zone to zone. The resulting time difference allows the Sun to be approximately at its highest point above the horizon at noon in each zone.

The Earth makes approximately 365.25 rotations in the time it takes it to complete one orbit around the Sun. Thus, every 4 years an extra day accumulates, which in leap years we add to the calendar as February 29.

QUESTIONS FOR REVIEW

1. How is the solar day defined? How is the sidereal day defined?
2. Why do the sidereal and solar days differ in length?
3. How does the number of hours of daylight vary with location and time of year?
4. What are time zones? Why are they useful?
5. What is the month based on? What is a lunar calendar?
6. Why do we need a leap year?
7. What do A.M., P.M., B.C., A.D., B.C.E., and C.E. stand for?

THOUGHT QUESTIONS

1. Why do you suppose that ancient mathematicians chose to divide a circle into 360°?
2. Suppose the Earth's spin slowed down until there were just 180 days in a year. Compare the length of a sidereal and a solar day in this new situation. (Do not redefine units of time—just express them in terms of our current hours, minutes, and seconds.)
3. One might logically conclude that it would make the most sense to add a leap day at the end of the calendar year. Thinking about the history of the calendar, reconcile this idea with the fact that February has 28 days.
4. Speculate about what factors may have caused some cultures to base their calendars more strongly on the Moon's cycles instead of the cycle of seasons.
5. Suppose you were asked to revise the calendar. What changes would you make?

PROBLEMS

1. Your friend lives in a town at a longitude 5° to the east of you. Both of you define "noon" as when the Sun reaches its highest point in the sky. How do your clocks differ from each other?
2. How long are nights at the equator? Justify your answer with diagrams.
3. Compare the 7-day week to the Roman 8-day week or the French Revolution experiment with a 10-day week. In trying to match lunar and solar cycles with a whole number of "weeks," which works best? Is there another number of days that works better?
4. If there were no leap days, after how many years would the seasons align correctly with the calendar again?

TEST YOURSELF

1. What effect does the Earth's orbit around the Sun have on stars' rising?

   (a) They rise only once a year.
   (b) They rise several minutes earlier each night.
   (c) They rise a little farther south or north each night, depending on the season.
   (d) Nothing—only the Sun rises, not stars.
   (e) Nothing—they rise the same time every night.

2. Suppose the Earth's rotation axis were not tilted with respect to its orbit. How would the number of daylight hours change throughout the year?

   (a) The number would be no different.
   (b) Days would be longer and nights shorter all year.
   (c) Days and nights would be of equal length all year.
   (d) Days would be shorter and nights longer all year.
   (e) None of the above

3. If on a given date there are 24 hours of night at the North Pole, how many hours of night are there at the South Pole?

   (a) 12 hours   (d) 48 hours
   (b) 24 hours   (e) There is no night then.
   (c) 36 hours

4. Suppose that the length of the year were 365.2 days instead of 365.25 days. How often would we have leap year? Every

   (a) 2 years.   (c) 10 years.   (e) 50 years.
   (b) 5 years.   (d) 20 years.

KEY TERMS

daylight saving time, 173
Gregorian calendar, 175
international date line, 173
Julian calendar, 175
mean solar day, 171

sidereal day, 170
solar day, 170
standard time, 173
time zone, 173
Universal time, 173

7

THE MOON

Apollo 17 astronaut Harrison Schmitt unloads the lunar rover next to Shorty crater.

LEARNING OBJECTIVES

Upon completing this chapter you should be able to:

- Describe the primary surface features of the Moon and explain why they are different from Earth's.
- Identify what features are produced by impacts and explain the processes that form them.
- Describe the primary features of the Moon's internal structure, how that structure has been determined, and how it relates to external features of the Moon.
- Explain the factors causing the Moon to have so little internal activity or atmosphere.

- Describe the Moon's orbit and spin, and how they are similar to those of other satellites and how they differ.
- Discuss the different hypotheses for the Moon's formation and summarize the evidence for the collision model.
- Describe how gravity causes ocean tides, and explain why it produces two tidal bulges.
- Use geometric principles to estimate when tides will occur, and explain why their strength varies with lunar phase.
- Describe how tides cause the Earth's spin to slow and to cause the Moon's distance to grow.

The Moon is our nearest neighbor in space, a natural satellite orbiting the Earth. It is a barren ball of rock, with about one-fourth the diameter of Earth, possessing no air, water, or life. In the words of lunar astronaut Buzz Aldrin, the Moon is a place of "magnificent desolation."

The Moon is the frontier of direct human exploration, an outpost that we reached more than 40 years ago but from which we have since drawn back. Despite our retreat from its surface, the Moon remains of great interest to astronomers. Although originally it was molten, its small mass and radius made the Moon unable to generate or retain any appreciable internal heat. It is therefore a dead world, with neither plate tectonic nor volcanic activity.

The Moon has not always been inactive. Shortly after its formation, it was pelted with a hail of rocky fragments up to 200 kilometers (about 120 miles) in diameter. The smaller fragments made craters, and the big fragments made huge basins. The basins subsequently flooded with lava (long since congealed) to create several dark, nearly circular plains easily visible to the naked eye. The Earth probably once bore such features, but erosion and plate motions have erased them. On the Moon's windless, rainless, airless surface, they remain as a record of events in the early Solar System, giving clues to the birth of not only the Moon but also the Solar System.

The Moon has much to teach us about the early Solar System. One intriguing possibility is that we might find pieces of Earth blasted out by impacts when our planet was young, which could teach us much about the early stages of Earth's formation—perhaps even providing us with fossil remains of the earliest life. There also may be useful resources that can be excavated on the Moon, and it may make a good launch platform for missions elsewhere in the Solar System. Both the United States and China have expressed interest in sending new missions to the Moon, and both are developing heavy-lift rockets that could permit major new missions.

In this chapter, we will describe the Moon's surface and how impacts created most of its features. We will see that lunar rocks differ significantly from terrestrial rocks and how they point to the Moon's having been born in a cataclysmic event early in the Earth's history. Finally, we will discuss how the Moon affects Earth today through tides, which gradually alter the Moon's orbit and the Earth's spin.

Q: WHAT IS THIS?

See end of chapter for the answer.

CONCEPTS AND SKILLS TO REVIEW

- Lunar phases (1.3)
- Density (6.1)
- Law of gravity (3.4)
- Escape velocity (3.8)

7.1 THE SURFACE OF THE MOON

Surface Features

To the naked eye, the Moon is a world of grays. Some patches are darker than others, creating a vague impression of what some see as a face ("the man in the moon") or maybe a rabbit. Closer examination reveals that the dark patches are in fact quite different from their lighter surroundings. Through a small telescope or even a pair of binoculars, you can see that the dark areas are smooth while the bright areas are covered with numerous large circular pits called **craters,** such as the one explored by *Apollo 17* astronauts shown on the opening page of the chapter. The near vacuum of the Moon's surface and lack of geologic activity has preserved the Moon's surface features, which were formed over billions of years.

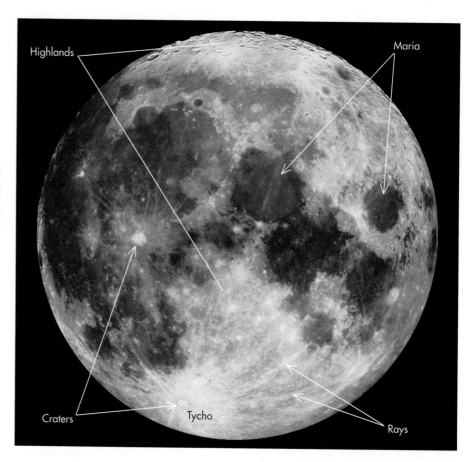

Q: Some rays cross maria. What does this imply about the relative age of the rays and the maria?

FIGURE 7.1

Photograph showing the different appearance of the lunar highlands and maria. The highlands are heavily cratered and rough. The maria are smooth and dark and have few craters. The long, narrow, white streaks radiating away from some of the craters are lunar rays.

The large, smooth, dark areas of the Moon's surface seen in figure 7.1 are called **maria** (pronounced MAR-ee-ah), from the Latin word for "seas." However, these regions, like the rest of the Moon, are essentially devoid of water. This usage comes from early observers who believed the maria looked like oceans and who gave them poetic names such as Mare (pronounced MAR-ay) Serenitatis (Sea of Serenity), and Mare Tranquillitatis (Sea of Tranquility), the site where astronauts first landed on the Moon.

The bright areas that surround the maria are called **highlands.** The highlands and maria differ in brightness because they are composed of different rock types. The maria are basalt, a dark, congealed lava rich in iron, magnesium, and titanium silicates, like the rock that makes up most of the ocean floors on Earth. The highlands, on the other hand, are mainly anorthosite, a rock type rich in calcium and aluminum silicates. This difference has been verified from rock samples obtained by Apollo astronauts. Moreover, the samples also show that the highland material is generally less dense than mare rock and considerably older.

The highlands are not only brighter and their rocks less dense than the maria, they are also more rugged, being heavily pitted with craters. Highland craters are so abundant that many overlap, as shown in figure 7.2A. Contrast this picture with the mare region shown in figure 7.2B, in which only a few, small craters are visible.

Craters can be found all over the Moon, but very few appear to be volcanic in origin. Most lunar craters have a raised rim and range in size from tiny holes less than a centimeter across to gaping scars more than 200 kilometers (120 miles) across. Many appear somewhat softer in outline, probably because of the accumulation of dust and debris over billions of years. Other craters look relatively recent, suggesting that cratering has taken place over an extremely long period of time. From some craters, long, light streaks of pulverized rock called **rays** radiate outward, as can be seen in figure 7.1. A particularly bright set spreads out from the crater Tycho near the Moon's south pole and can be seen easily with a pair of binoculars when the Moon is full.

By convention, most lunar craters are named for famous scientists..

A
B

FIGURE 7.2
(A) Overlapping craters in the Moon's highlands. (B) Isolated craters in the smooth mare.

Q: In (A) a small crater lies at the edge of a larger one. Which formed more recently: the small one or the large one?

Origin of Lunar Surface Features

Though telescopic images have taught us a great deal about the Moon, space probes and lunar landings provide far more detail than we can achieve from Earth (see Astronomy by the Numbers: "The Limits of Telescopic Observations of the Moon"). Close study shows that nearly all the surface features we see on the Moon—craters, maria, and lunar rays—were made by the impacts of solid bodies on its surface. When an object hits a solid surface at high speed, it disintegrates into a cloud of vaporized rock and fragments. The blast wave from the impact makes a hole whose diameter depends on the mass and velocity of the impacting object.

As the vaporized rock expands from the point of impact, it forces surrounding rock outward, piling it into a raised circular rim. Pulverized rock spatters in all directions, forming rays. Sometimes the impact compresses the rock below the crater sufficiently that it rebounds upward, creating a central peak, as shown in figure 7.3A. Figure 7.3B shows a similar process happening in a high-speed picture of a raindrop falling into water.

A
B

FIGURE 7.3
(A) Central peak in a crater and slumped inner walls. Apollo astronauts took this photograph of the crater Eratosthenes on the last manned flight to the Moon. This crater is 58 kilometers (approximately 36 miles) in diameter. (B) A drop of rain falling into water produces an effect similar to the one that creates central peaks in lunar craters. The drop falls into the water but is then pushed up again as the water rebounds.

FIGURE 7.4

(A) Mountains along the edge of a mare were probably thrown up by the impact that created the mare. (B) Mare Orientale shows the multiple ring structure from a major impact. The central area of the impact crater was flooded by lava after the impact.

FIGURE 7.5

Euler Crater and a close-up view of the wall of the crater made by the *Lunar Reconnaissance Orbiter*. The impact blasted out a section of Mare Imbrium, exposing the layers of lava that built up the mare's floor. Layers are measured to be 3 to 12 meters thick.

Astronomers think that the maria are also impact features, but to understand their formation we must briefly describe the early history of the Moon. From the great age of the highland rocks (in some cases as old as 4.5 billion years), astronomers deduce that these rugged uplands formed shortly after the Moon's birth. At its birth the Moon was probably molten, allowing dense, iron-rich material to sink to its interior while less-dense material floated to the lunar surface. On reaching the surface, the less-dense rock cooled and congealed, forming the Moon's crust. A similar process probably formed the Earth's continents. The highlands were then heavily bombarded by solid bodies from space, forming the numerous craters we see there.

Before the Moon's interior solidified completely, a small number of exceptionally large bodies (over 100 kilometers or 60 miles in diameter) struck the surface, blasting huge craters and pushing up mountain chains along their edges (fig. 7.4A). The impact that formed Mare Orientale (fig. 7.4B) sent shock waves out into the surrounding crust that formed multiple rings around the mare. These impacts probably produced some melting during the impact, but radiometric dating of the dark lava that flooded these low areas indicates that the flooding occurred up to 1.4 billion years after the Moon formed, much later than the period of heavy bombardment. In fact, molten material from within the Moon appears to have flooded the vast craters repeatedly, as suggested by the layering seen in the wall of a deep crater in one of the maria (fig. 7.5).

ASTRONOMY *by the numbers*   THE LIMITS OF TELESCOPIC OBSERVATIONS OF THE MOON

Even though the Moon is very close to the Earth (astronomically speaking), telescopes can show only a limited amount of its surface detail. In chapter 5 we saw that a telescope's resolving power is fundamentally limited by the telescope's diameter and the wavelength of light being observed.

For the Hubble Space Telescope (HST), this angle is about $\alpha = 0.05$ arcsec $= 0.000014°$ at visible wavelengths.

At the distance of the Moon, $d = 380,000$ km, the formula for angular size vs. distance (chapter 2) tells us that the smallest object the HST can discern has a size

$$\ell = \frac{2\pi d\alpha}{360°} = 380,000 \text{ km} \frac{0.000014}{57.3} = 0.09 \text{ km}.$$

Therefore, anything smaller than about 90 meters across (the size of a football field) cannot be seen by the HST. To study the Moon in greater detail, astronomers must send probes such as the *Lunar Reconnaissance Orbiter*.

FIGURE 7.6
Large impacts late in the process of the Moon's formation formed huge basins. Lava flooded the basins to make the maria.

A possible explanation for this was provided by the *Lunar Prospector* satellite. From gamma-ray emissions it found that the regions of the maria have a higher abundance of radioactive elements. Radioactive heating in the upper mantle may have melted rock there, and the deep basins carved by the impacts became filled in with basalt in successive volcanic eruptions (fig. 7.6). Thus, rather than producing the vast seas of basalt in a single episode, the major impacts appear to have provided a pathway for lava to flow to the surface at later times. Because the maria formed after most of the impacting bodies were gone, few bodies remained to crater the maria. The maria therefore remain relatively smooth to this day.

The Moon has a variety of other surface features with more uncertain origins. There are lunar canyons known as **rilles,** some of which can be seen through a small telescope. Some look like river valleys (fig. 7.7A) but were probably carved by ancient lava flows. Elsewhere, straight rilles gouge the surface, probably the result of crustal cracking (fig. 7.7B). Drying mud and chocolate pudding left too long in the refrigerator show similar cracks. There are also some long linear cliffs where sections of the crust appears to have been pushed together. These features might indicate that the Moon shrank as it cooled, or perhaps they arose from something similar to the early stages of plate tectonics on Earth.

Presumably Earth was also battered by impacts in its youth. Although the vast multitude of these craters have been obliterated by erosion and plate tectonics, a few remain in ancient rock whose measured age is typically hundreds of millions of years. From the scarcity of such craters, astronomers can deduce that the main bombardment must have ended billions of years earlier. On the other hand, land forms common on Earth, such as folded mountain ranges and volcanic peaks, are rare on the Moon.

FIGURE 7.7
Photographs of lunar rilles. (A) The sinuous rilles in this region are thought to have formed from lava flows. (B) Linear rilles may have formed from shifting blocks of crust.

The Moon's small size relative to the Earth explains most of the differences between the two bodies. Because its volume compared to its surface area is small relative to the Earth's, heat escapes far more easily from the Moon. Thus, the Moon has cooled far more than the Earth has. (Think of how a french fry cools much faster than a baked potato.) Thus, having a much cooler interior, the Moon lacks the convection currents that drive plate tectonic activity on the Earth.

Without tectonic activity to recycle the debris from impacts that cratered the Moon, the surface has become covered with a **regolith**—meaning "blanket of rock"—tens of meters deep. The regolith consists of both rock chunks and fine powder, the result of successive impacts breaking rock into smaller and smaller pieces. This powdery nature is easily seen in the crispness of the astronauts' footprints (fig. 7.8). Samples of the regolith picked up by astronauts show that these surface rocks are typically the same type as the underlying rock. That is, the regolith on maria is generally broken-up basalt, whereas that on the highlands is broken-up highland material.

FIGURE 7.8
Footprint of an astronaut on the Moon.

Crust and Interior

The Moon's low overall density (3.3 grams per cubic centimeter) tells us its interior contains little iron. Recall that in chapter 6 we saw that the Earth's high density (about 5.5 grams per cubic centimeter) is an indication that it has a large iron core. In addition, the Moon lacks a magnetic field, suggesting that the core is at most partially molten.

The Moon's interior can be studied by seismic waves just as the Earth's can. Apollo astronauts set up seismic detectors on the Moon that showed that the Moon's interior is essentially inactive and has a simpler structure than the Earth's. Below the surface layer of rocky rubble is the Moon's crust, about 100 kilometers (60 miles) thick, on average. The Moon's crust, like the Earth's, is composed of silicate rocks relatively rich in aluminum and poor in iron. Beneath the crust is a thick mantle of solid rock, extending down a little more than 1000 kilometers (600 miles). The Moon's mantle is probably rich in olivine, the same type of dense, greenish rock that composes most of the Earth's mantle. Unlike the Earth's mantle, however, it appears to be too cold and rigid to be stirred by the Moon's feeble heat.

The crust is much thinner (about 65 kilometers) on the side of the Moon that faces the Earth than on the far side (about 150 kilometers), as shown in figure 7.9. The reason for this difference is not clear, but it may have resulted from the Earth's gravity shifting the Moon's core slightly toward Earth billions of years ago, when the Moon's interior was molten. The crust on the near side—being slightly closer to the Moon's core because of that shift—might therefore have become hotter and as a result thinner

FIGURE 7.9
Structure of the Moon's interior. Notice the thinner near-side crust, so the Moon's center of mass is displaced by a few kilometers toward the Earth. The Moon's iron core is small, with an outer liquid part, like the Earth's, and the surrounding mantle rock may be partially molten too.

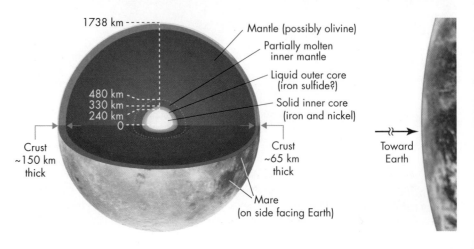

1738 km
Mantle (possibly olivine)
Partially molten inner mantle
Liquid outer core (iron sulfide?)
Solid inner core (iron and nickel)
480 km
330 km
240 km
0
Crust ~150 km thick
Crust ~65 km thick
Mare (on side facing Earth)
Toward Earth

than that on the far side. Subsequently, the Moon cooled, leaving the crust thinner on one side than on the other.

The internal asymmetry of the Moon helps to explain some of the differences between the Moon's near and far sides. The far side was first seen in 1959 by a Soviet space probe that passed by the Moon and sent back pictures. It is distinctly different from the near side, consisting entirely of very rough terrain with no maria. Figure 7.10 shows an image made by Apollo astronauts. The Moon's far side is heavily cratered like the highlands on the near side. The thinner crust on the near side made it much easier for impact craters to be flooded by basalt, forming the maria. In fact, the largest impact feature, Aitken Basin, is on the far side of the Moon as shown in the figure 7.11, a topographic map of the Moon. Although this basin is more than 10 kilometers beneath the height of the surrounding terrain, it was not flooded with basalt because of the thick crust there.

The Absence of a Lunar Atmosphere

Lunar scientists have detected only tiny quantities of gas above the Moon's surface— less than one-quadrillionth the density of our atmosphere. Most of this gas is helium, probably a by-product of radioactive decay in the Moon's interior. There are also traces of hydrogen near the Moon's poles. Lunar scientists suspected this came from the breakdown of frozen water mixed with rock in craters that remain perpetually in shadow. A NASA space probe confirmed this in 2009 when it was crashed into a crater near the south pole, raising a debris cloud containing water. The water may have originally come from comets striking the Moon and vaporizing. The water vapor then condensed in the coldest places on the Moon (the polar craters into which sunlight never shines). You may have seen this tendency for frost to form in cold spots if you have taken something out of a freezer and left it for a while on a table.

FIGURE 7.10

An image made by *Apollo 16* astronauts showing some of the heavily cratered far side of the Moon. Part of the near side is visible in the left portion of the image. Mare Crisium is on the edge of the picture at the 9:30 position.

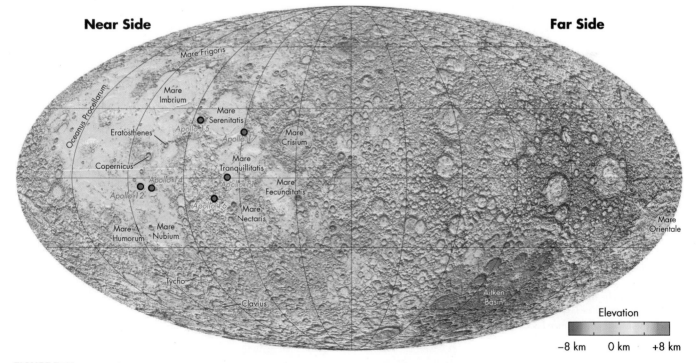

FIGURE 7.11

Topographic map showing the near (left half) and far (right half) sides of the Moon. The elevations were mapped by the *Clementine* satellite and are shown in different colors. The maria are generally at lower elevations, but the largest impact feature, the Aitken Basin, is at an even lower elevation. Several major features and the locations of the six Apollo landing sites are labeled.

Q. Why don't we ever see one side of the Moon from the Earth?

EXTENDING
our reach

IS THE MOON COMPLETELY DEAD?

The *Lunar Reconnaissance Orbiter*, placed in orbit around the Moon in 2009, has made highly detailed images of the Moon's surface that have revealed several unusual features, such as the "crater" Ina (fig. 7.12).

Ina is about 3 kilometers wide, and appears to be a region where eruptions have blown away most of the surface regolith. The small number of impact craters within Ina suggest that it may have erupted under a million years ago, suggesting that there is still some activity in the Moon's interior.

FIGURE 7.12
The peculiar feature Ina, probably the site of volcanic outgassing. It appears that in the lighter-colored parts of this D-shaped region, the surface regolith has been blown away.

The Moon lacks a significant atmosphere for two reasons. First, its interior is too cool to cause much volcanic activity, which as we saw in chapter 6 was probably an important source of Earth's atmosphere (see Extending Our Reach: "Is the Moon Completely Dead?"). Second, and more important, even if volcanos or comets created an atmosphere in its youth, the Moon's gravity is too weak to retain gas for long. In chapter 3 we learned that the Moon's escape velocity is only about one-fourth that of the Earth's (2.4 kilometers per second versus 11 kilometers per second), and so atoms in the Moon's atmosphere would have found it easier to escape its gravity. With virtually no atmosphere to absorb and trap heat, temperatures on the Moon soar during the day and plummet at night, and no wind blows to stir the thick dust on its surface.

7.3   ORBIT AND MOTIONS OF THE MOON

By watching the Moon for a few successive nights, you can see it move against the background stars as it follows its orbit around the Earth. If you carefully measure the Moon's angular size, you will find that it varies by about 14% during its orbit. This is a consequence of the Moon's elliptical orbit and its changing distance (the relationship between angular size and distance is discussed in chapter 2). The Moon's orbit has an average distance from Earth of 380,000 kilometers (about 250,000 miles), but it varies from about 360,000 to 405,000 kilometers during its orbital period. As a result, a full moon may be significantly larger some months than others (fig. 7.13A).

FIGURE 7.13
(A) The Moon's distance from the Earth varies during its orbit, which can be seen in the changing angular size of the Moon. (B) The precise distance from the Earth to the Moon can be determined by measuring the time radar or laser signals take to reach the Moon and bounce back to Earth.

Moon closest to Earth

Moon farthest from Earth

A

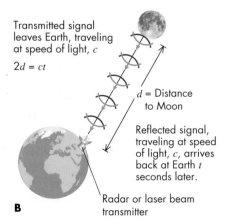

Transmitted signal leaves Earth, traveling at speed of light, c

$2d = ct$

d = Distance to Moon

Reflected signal, traveling at speed of light, c, arrives back at Earth t seconds later.

Radar or laser beam transmitter

B

The Moon's distance can today be measured very precisely, as sketched in figure 7.13B, by bouncing either a radar pulse or a laser beam off special reflectors that were placed on the Moon by the Apollo astronauts. Measuring the time it takes for a signal to travel to the Moon and back, multiplied by the speed of light, gives the round-trip distance. Halving that value provides distances to an accuracy of centimeters.

The Moon's Rotation

As it orbits, the Moon keeps the same side facing the Earth, as you can see by watching it through a cycle of its phases. You might think from this that the Moon does not rotate. Figure 7.14A shows, however, that the Moon must slowly rotate to keep the same features facing the Earth. Thus, the Moon *does* turn on its axis but with a rotation period exactly equal to its orbital period, a condition known as **synchronous rotation.** The Earth's gravity caused this locking of the Moon's spin to its orbital motion, as we will discuss in section 7.5. This is a common characteristic of satellites in the Solar System, which almost all keep the same face toward their planet.

The rotation of the Moon

Oddities of the Moon's Orbit

Unlike almost all other large moons, our Moon has an orbit with a large tilt with respect to its planet's equator. In discussing eclipses in chapter 1, we noted that the Moon's orbit is tilted by a little more than 5° with respect to the Earth's orbit around the Sun, and the Moon's orbit gradually precesses or "wobbles" over a period of 18 years. As a result, its orbit is tilted between 18° and 29° with respect to the Earth's equator, as shown in figure 7.14B. This is unlike all of the major moons of Jupiter, Saturn, and Uranus, which lie nearly exactly in their planet's equatorial plane. The only large satellite that has a larger discrepancy than the Moon is Neptune's satellite Triton, which may have been an outer Solar System body captured by Neptune, as we will discuss in chapter 10.

Our Moon is also exceptional in its size relative to its planet. Even the largest of the moons of Jupiter and Saturn have masses less than 1/1000th that of their planet. But our Moon's mass is 1/81 that of the Earth. These oddities suggest that our Moon formed differently from the moons of other planets, as we discuss next.

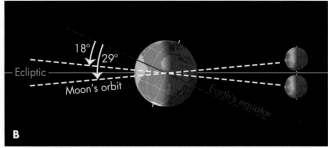

FIGURE 7.14
(A) The Moon rotates once each time it orbits the Earth, as can be seen from the changing position of the exaggerated lunar mountain. Notice that when the Moon is new (left side in the diagram), the lunar peak faces to the right, while when it is new it faces to the left. Thus, from the Earth, we always see the same side of the Moon even though it turns on its axis. (B) The Moon's orbit is nearly in the plane of the Earth's orbit (the ecliptic), but is quite tilted with respect to the Earth's equator, which is very unusual for large satellites of planets. (Separation of Earth and Moon not to scale.)

7.4   ORIGIN AND HISTORY OF THE MOON

ANIMATION ▶

The birth of the Moon

Lunar rocks brought back to Earth by the Apollo astronauts caused astronomers to radically revise their ideas of how the Moon formed. (Until recently, and despite hundreds of pounds of lunar rocks brought back to Earth by astronauts, there were a number of rather silly claims that the Moon landings never occurred—see Extending Our Reach: "The Moon Landing 'Hoax.'") Before the Apollo program, lunar scientists had three hypotheses regarding the Moon's origin:

- The Moon was originally a small planet orbiting the Sun; it approached the Earth and was captured by Earth's gravity (capture theory).
- The Moon and Earth were "twins," forming side by side from a common cloud of dust and gas (twin formation theory).
- The Earth initially spun enormously faster than now and formed a bulge that ripped away from the Earth to become the Moon (fission theory).

Each of these hypotheses led to different predictions about the composition of the Moon. For example, had the Moon been a captured planet, its composition might be very unlike the Earth's. If the Earth and Moon had formed as twins, their overall composition should be nearly identical. Finally, if the Moon was once part of the Earth, its composition should be the same as the Earth's crust. When the rock samples were analyzed, astronomers were surprised that for some elements the composition was the same, but for others it was very different. For example, the Moon has a relatively high abundance of high-melting-point materials such as titanium and an almost complete lack of low-melting-point materials such as water. It also has much less iron than the Earth, as we pointed out when discussing its interior and low density.

The failure of evidence based on lunar surface samples to confirm any of the three hypotheses led astronomers to consider alternatives, and now a completely different picture of the Moon's origin has emerged. According to the new hypothesis, the Moon formed from debris blasted out of the Earth by the impact of a Mars-sized body, as illustrated in figure 7.16. The great age of lunar rocks and the absence of any impact feature on the Earth indicate that this event must have occurred during the Earth's own formation, at least 4.5 billion years ago. The colliding body melted and vaporized millions of cubic kilometers of the Earth's surface rock and hurled it into space in an incandescent plume. As the debris cooled, its gravity drew it together into what we now see as the Moon.

EXTENDING
our reach   THE MOON LANDING "HOAX"

For years, a number of sensationalists claimed that the lunar landings were a hoax. Most of their arguments were silly, as can be determined by anyone who spends a few hours examining the voluminous image and video data freely available from NASA. These claims were profitable vehicles for their authors, suggesting the real hoax.

Because the landers are too small to be seen by Earth-bound telescopes, it was difficult to refute these claims directly. However, the *Lunar Reconnaissance Orbiter* has now imaged the Moon's surface with unprecedented detail, showing the landing sites and even the tracks left by the astronauts (fig. 7.15).

Apollo lunar surface experiments package

Astronauts' foot trails

Apollo 12 lander

FIGURE 7.15
An example of an LRO high-resolution image of one of the Apollo landing sites. In addition to the lander, even the tracks left behind by astronauts remain visible.

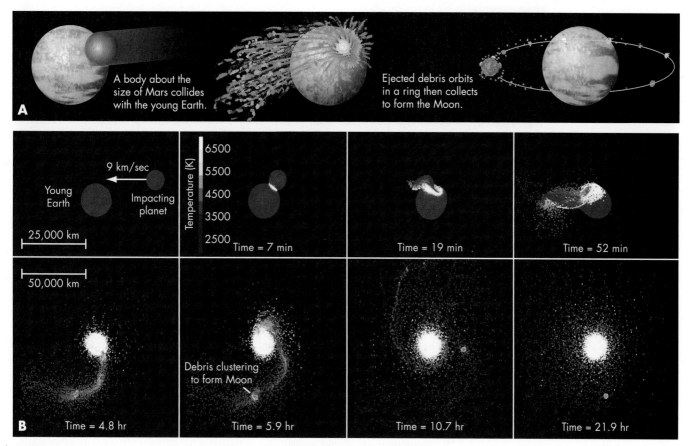

FIGURE 7.16

Origin of the Moon. (A) Sketch of the main stages in the birth of the Moon by a major collision. (B) This computer simulation shows how the Moon might have formed when a Mars-size object hit the young Earth and splashed out debris that later assembled into the Moon.

This violent-birth hypothesis explains many of the oddities of the Moon. The impact would vaporize low-melting-point materials and disperse them, leaving, for example, little water to be incorporated into the lunar body. Computer models of such an event (fig. 7.16) also show that only surface rock would be blasted out of the Earth, leaving our planet's iron core intact, thereby also explaining the low iron content of the Moon. The splashed-out rock would condense in an orbit whose shape and orientation were determined by the collision rather than by the orientation of the Earth's equator. Furthermore, we would expect both similarities and differences in composition between the Earth and Moon because the Moon was made partly from Earth rock and partly from rock of the impacting object.

After the Moon's birth, stray fragments of the ejected rock pelted its surface, creating many of the craters that blanket the highlands. A few huge fragments plummeting onto the Moon later in its formation process blasted enormous holes that later flooded with molten interior rock to become the maria. That rock was probably melted in the Moon's interior by radioactive decay, as happened in the Earth. During the time it took the rock to melt, about half a billion years, most of the debris remaining in the Moon's vicinity fell onto its surface.

A recent hypothesis proposes that instead of forming the Moon immediately, the ejecta from the massive collision with Earth coalesced into two moons at first. Sometime after the surfaces of these two moons solidified, the two bodies collided and merged into the present-day Moon, with the final collision giving rise to the asymmetry between the near and far sides of the Moon. Since these early times, the Moon has been almost a dead world geologically, although a few regions may still show some signs of life.

7.5 | TIDES

ANIMATION ▶

Tidal forces

Anyone who has spent even a few hours by the sea knows that the ocean's level rises and falls during the day. A blanket set on the sand a few feet from the water's edge may be inundated an hour later, or a boat pulled ashore may be left high and dry. This regular change in the height of the ocean is called the **tides** and is caused mainly by the Moon.

Cause of Tides

Just as the Earth exerts a gravitational pull on the Moon, so too the Moon exerts a gravitational attraction on the Earth and its oceans and draws material toward it. The attraction is stronger on the side of the Earth near the Moon and weaker on the far side (fig. 7.17) because the force of gravity weakens with distance (recall Newton's law of gravity, section 3.4). The difference between the strong force on one side and the weaker force on the other is called a **differential gravitational force.**

The differential gravity draws water in the oceans into a **tidal bulge** on the side of the Earth facing the Moon, as shown in figure 7.17.* It may seem surprising at first, but it creates an identical tidal bulge on the Earth's far side. This second tidal bulge can be viewed as a result of the Moon's gravity pulling the Earth "out from under" the water on the far side.

A better explanation can be obtained, however, by examining the Moon's gravitational forces on the Earth and its oceans as measured on Earth's surface, as shown in figure 7.18. The arrows in figure 7.18 (top) represent the Moon's gravitational pull

North Pole

To Moon

Moon's gravitational attraction creates tidal bulges.

FIGURE 7.17
Tides are caused by the Moon's gravity creating tidal bulges.

---

* Because of tidal braking, which we will discuss at the end of this section, the tidal bulges do not exactly align with the Moon.

FIGURE 7.18
(Top) Arrows schematically show the Moon's gravitational force at different points on the Earth. (Bottom) Tidal forces from the point of view of an observer on the Earth. These arrows represent the difference between the Moon's gravitational force at a given point and its force at the Earth's center (C). Graphically, you can find the tidal force by "adding" the arrows. The figure shows schematically how to do this, but details are omitted.

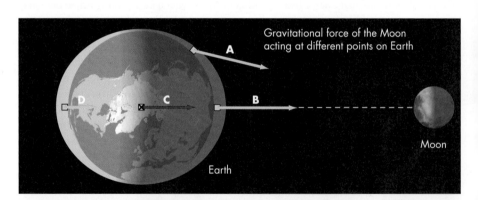

Gravitational force of the Moon acting at different points on Earth

Moon

Earth

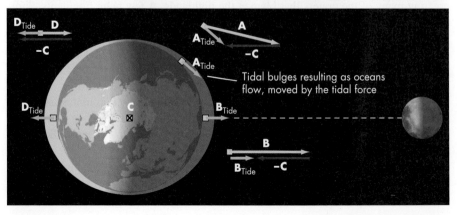

Tidal bulges resulting as oceans flow, moved by the tidal force

FIGURE 7.19
As the Earth rotates, it carries points along the coast through the tidal bulges. Because there are two bulges where the water is high and two regions where the water is low, we get two high tides and two low tides each day at most coastal locations.

at several points on the Earth. Points on the side of the Earth near the Moon undergo a stronger pull (B) toward the Moon than points on the far side (D), and so the arrow from point B is longer than the arrow from point D. Likewise, because point C is closer to the Moon than point D, the arrow from point C (which is the Moon's pull on the center of the Earth) is longer than the arrow from point D.

To see how the tidal bulges form, we need to look at the *difference* between the gravitational force at a given point and at the center of the Earth. For example, at point B the force is larger than at point C, and so matter at point B will be pulled *away* from point C. This creates one tidal bulge. But matter at point C is in turn pulled away from point D, which creates a second tidal bulge. If we now draw a second set of arrows to represent the *difference* between the force at C and at every other point (the differential gravitational force), we find the forces illustrated in figure 7.18 (bottom). These drive the oceans into the bulges* that we see.

Up to this point we have ignored the Earth's rotation. The tidal bulges point toward and away from the Moon, but the Earth spins. The Earth's rotation carries us first into one bulge and then the next. As we enter one of the bulges, the water level rises, and as we leave it, the level falls. Because there are two bulges, we are carried into high water twice a day, creating two high tides. Between the times of high water, as we move out of the bulge, the water level drops, making two low tides each day (fig. 7.19).

This simple picture must be modified to account for the ocean's inability to flow over land areas. Thus, water tends to pile up at coastlines when the tidal bulge reaches shore. In most locations the tidal bulge has a depth of about 2 meters (6 feet), but it may reach 10 meters (30 feet) or more in some long narrow bays (as you can see in the photographs of high and low tides along the Maine coast in fig. 7.19) and may even rush upriver as a tidal bore—a cresting wave that flows upstream. On some rivers, surfers ride the bore upstream on the rising tide.

The motion of the Moon in its orbit makes the tidal bulge shift slightly from day to day. Thus, high tides come almost 50 minutes later each day, the same delay as there is between the times when the Moon is highest in the sky from one day to the next, as we saw in chapter 1.

---

* Tides also occur in the atmosphere and solid ground, but tides in the ground are smaller because the ground is rigid and cannot move as easily as water or air.

FIGURE 7.20

The Sun's gravity creates tides, too, though its effect is only about one-half that of the Moon. (A) The Sun and Moon both create tidal bulges on the Earth. When the Sun and Moon are in line, their tidal forces add together to make larger-than-normal tides. (B) When the Sun and Moon are at 90° as seen from Earth, their tidal bulges are at right angles and partially nullify each other, creating smaller-than-normal tidal changes.

Solar Tides

The Sun creates tides on the Earth like the Moon does, but although the Sun is much more massive than the Moon, it is also much farther away. The result is that the Sun's tidal force on the Earth is only about one-half the Moon's. Nevertheless, it is easy to see the effect of their tidal cooperation in spring tides, which are abnormally large tides that occur at new and full moon. At those times, the lunar and solar tidal forces work together, adding their separate tidal bulges, as illustrated in figure 7.20A. Notice that spring tides have nothing to do with the seasons; rather, they refer to the "springing up" of the water at new and full moon.

It may seem odd that spring tides occur at both new and full moon, because the Moon and Sun pull together when the Moon is new but in opposite directions when it is full. However, the Sun and the Moon both create two tidal bulges, and the bulges add together regardless of whether the Sun and Moon are on the same or opposite sides of the Earth. On the other hand, at first and third quarters, the Sun and Moon's tidal forces work at cross-purposes, creating tidal bulges at right angles to one another, as shown in figure 7.20B. The so-called neap tides that result are therefore not as extreme as normal high and low tides.

Tidal Braking

Tides create forces on the Earth and Moon that slow their rotation, a phenomenon known as **tidal braking.** Figure 7.21 shows how the Moon tidally brakes the Earth. As the Earth spins, friction between the ocean and the solid Earth below drags the tidal bulge ahead of the imaginary line joining the Earth and Moon, as depicted in figure 7.21. The Moon's gravity pulls on the bulge, as shown by the long green arrow in the figure, and holds it back. The resulting drag is transmitted through the ocean to the Earth, slowing its rotation the way a brake shoe on a car or your hand placed on a spinning bicycle wheel slows the wheel.

As the Earth's rotation slows, the Moon experiences a force that causes it to move farther from the Earth, as is required by the need to conserve angular momentum. The Moon moves outward because the tidal bulge it raises on the Earth exerts a gravitational force back on the Moon (Newton's third law of motion), which pulls the Moon ahead in its orbit, as shown by the short green arrow at the right side of figure 7.21. That force makes the Moon move away from the Earth at about 4 centimeters (roughly 1½ inches) per year, a tiny increase in the Earth–Moon distance, but nevertheless detectable with the laser range finders placed on the Moon's surface by Apollo astronauts.

Thus, the Moon was once much closer to the Earth, as discussed in Astronomy by the Numbers: "The Distance of the Moon in the Past." The Earth must have spun much faster, perhaps as rapidly as once every 5 hours 4.5 billion years ago. Over that immense period of time, the Moon has receded to its present distance, and the Earth's

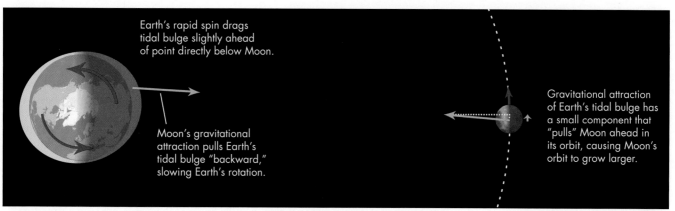

FIGURE 7.21
Tidal braking slows the Earth's rotation and speeds up the Moon's motion in its orbit. Friction between the oceans and Earth's solid crust "drags" the bulges of water "ahead" of the Earth–Moon line.

Q: Why doesn't the pull of the bulge on the far side of the Earth cancel the effects of the bulge closer to the Moon?

rotation has slowed to 24 hours. These processes occur even now: tidal braking lengthens the day by about 0.002 seconds each century.

Tidal braking is also the reason the Moon always keeps the same face to the Earth. Just as the Moon raises tides, which slow the Earth, the Earth raised tides on the Moon, which slowed its spin when it was young. These lunar tides distort the Moon's crust and have braked the Moon until it was locked it into synchronous rotation. The Moon's braking of the Earth will eventually make the Earth rotate synchronously with the Moon's orbital motion. Billions of years from now, the Earth and Moon will orbit so that each constantly presents the same face to the other: the Moon will then be visible only from one side of the Earth! Similar tidal effects have locked almost all of the moons of other planets into synchronous rotation, but the planets themselves have not been noticeably slowed, except the dwarf planet Pluto, whose large moon Charon has locked it into synchronous rotation. Similarly, tidal braking by the Sun appears to have slowed the rotation of Mercury and Venus.

The Moon's gravitational pull on the Earth may also stabilize our climate. Astronomers have recently found through computer simulations that the tilt of a planet's rotation axis may change erratically by many tens of degrees if the planet has no moon. Because the tilt causes seasons, changes in the tilt will alter the severity of the seasons. Our Moon is so large that its gravitational attraction on Earth's equatorial bulge helps hold the Earth's tilt relatively fixed, sparing us from catastrophic climate changes.

ASTRONOMY *by the numbers*   THE DISTANCE OF THE MOON IN THE PAST

The Moon is moving away from the Earth at about 4 centimeters per year at present. If it has moved at this rate for the last 4.5 billion years, how close was it to the Earth when they first formed?

This is a straightforward distance-velocity-time calculation (see appendix). Expressed in scientific notation the speed is $V = 4$ cm/yr $= 4 \times 10^{-2}$ m/yr, and the time is $t = 4.5 \times 10^9$ yr. We can therefore solve for the distance:

$$d = V \times t = 4 \times 10^{-2} \text{ m/yr} \times 4.5 \times 10^9 \text{ yr}$$
$$= 1.8 \times 10^8 \text{ m} = 180,000 \text{ km}$$

This is nearly half of the Moon's current distance of about 380,000 km from the Earth. In fact, the tidal effects would have been even larger in the past, making the rate of change even greater, so the Moon must have been *much* closer in the past.

It is quite reasonable to suppose that when the Moon first formed it was only 10,000 km above Earth's surface. The Moon then would have looked enormous in the sky—perhaps 20° across, and the tides would have been hundreds of times larger than they are today.

SUMMARY

The Moon is the Earth's satellite. It is much smaller than the Earth: it has about one-fourth the Earth's radius and about 1/81 its mass. Its small size has allowed its internal heat to escape, so its core cooled more rapidly than Earth's, thereby leaving the Moon much less geologically active than Earth. The Moon has no atmosphere because it is too cool to create one by volcanic outgassing and too small for its low gravity to retain gases that may have been present in the past.

With neither atmosphere nor geologic activity, the Moon's surface is largely unaltered except by impact features: craters, rays, and the maria. Maria are enormous lava flows that have flooded into basins made by large impacting bodies late in the Moon's formation.

The Moon is asymmetrical, internally and externally, and keeps the hemisphere containing its major maria always facing the Earth. The crust is thicker on the far side, where no maria are present even though the largest impact basin is located there.

The Moon probably formed when a Mars-size body collided with the Earth and splashed material from the Earth into orbit. That debris, drawn together by its own gravity, then reassembled into the Moon, explaining its composition.

The Moon's gravity creates tides, and as the Earth rotates beneath the tidal bulge of the ocean, our planet's rotation is slowed. Similar tidal braking exerted by the Earth on the Moon probably slowed the Moon's spin long ago, making its spin synchronous with its orbital motion around the Earth.

QUESTIONS FOR REVIEW

1. (7.1) Describe a crater and how it is formed. Why do some craters contain maria?
2. (7.1) How do the maria differ from the highlands?
3. (7.1) What are lunar rilles? What are rays?
4. (7.1) What formed the maria? Why are they smooth?
5. (7.2) What is regolith? How does it form?
6. (7.2) List the structure and composition of the Moon from surface to core. How is it different from Earth's?
7. (7.2) Why does the Moon lack an atmosphere?
8. (7.3) List two ways to measure the distance to the Moon.
9. (7.4) How do astronomers think the Moon formed? What supports this theory? How does the theory explain why the Earth and Moon have such different densities?
10. (7.4) Why is the Moon's surface heavily cratered but the Earth's is not?
11. (7.5) How are tides formed on the Earth?
12. (7.5) Why does the Moon form two tidal bulges on the Earth?
13. (7.3/7.5) Describe the Moon's rotation. How has it been affected by tidal interaction with the Earth?

THOUGHT QUESTIONS

1. (7.1) Highway surfaces develop "potholes" over time. How can you use the number of potholes as an indication of the "age" of the paving? How is this like using craters to estimate the age of the Moon's surface?
2. (7.2) Bergmann's rule states that individuals of a given species—for example, bears—will be larger in cold climates than in warmer climates. How is an explanation of this rule similar to an explanation of the temperature difference between the Earth's interior and the Moon's interior?
3. (7.2) How is the apparent lack of water on the lunar surface an argument against the idea that comets were a significant source of Earth's water and atmosphere? What is a good counterargument to your answer?
4. (7.2) Why will an astronaut's footprint on the Moon last so long?
5. (7.3/7.5) If the Moon were not in synchronous rotation with the Earth, would its phases be affected? What if both the Earth and the Moon were in synchronous rotation?
6. (7.4) How has our understanding of the Moon changed because of data available only from missions to the Moon? Make an argument for future missions based on results discussed in this chapter or that you look up.
7. (7.5) If the day were 12 hours long, what would be the approximate time interval between high and low tide?
8. (7.5) As the Moon recedes from the Earth, are the tides getting taller or shorter? If the Moon is someday twice as far from the Earth, how many high tides will there be each day?
9. (7.5) Why do tides happen about an hour later each day?

PROBLEMS

1. (7.1) Use data from the appendix to calculate the ratio of the Moon's mass to the Earth's, and the ratio the Moon's radius to the Earth's radius.
2. (7.1) Mare Serenitatis has an angular diameter of 5 minutes of arc. What is its diameter in kilometers? (See section 2.1.)
3. (7.1) The crater Tycho is 88 kilometers wide. What is its angular diameter from Earth? Could you see a crater this size with the naked eye?
4. (7.2) Calculate the Moon's density (see the end of section 6.1 for how to calculate density). The Moon's mass and radius can be found in the appendix. On the basis of your value for the density, what can you say about the amount of iron in the Moon? (See table 6.1 for iron's density.)

5. (7.2) The density of Swiss cheese is about 1.1 g/cm³. If the Moon were in fact made of (incompressible) cheese, what would be its mass?

6. (7.2) The *Lunar Reconnaissance Orbiter* orbits the Moon 50 kilometers above its surface. Its period is about 113 minutes. Use these values to find the Moon's mass.

7. (7.3) A laser pulse takes 2.56 seconds to travel from Earth to the Moon and return. Use this to calculate how far away the Moon is. How might this time delay affect conversations between an astronaut on the Moon and someone back on Earth?

8. (7.2/7.4) Because the Earth and Moon are both rocky spheres, we can make a crude estimate of how much faster the Moon cooled than the Earth. Compute the ratio of the surface area to the volume of the Moon, and compare it to the same ratio for the Earth (formulas for surface area and volume, and values of the radii, can be found in the appendix; also review fig. 6.8).

9. (7.5) If the Earth constantly slowed down at a rate of 0.002 seconds/century, how many years ago would the Earth's day have been only 5 hours long?

TEST YOURSELF

1. (7.1) The large number of craters on the lunar highlands compared to those on the maria is evidence that
   (a) the maria have a liquid surface, so craters disappear.
   (b) the highlands are composed of soft, easily cratered material.
   (c) the bodies that struck the Moon and made the craters were clumped, and missed hitting the maria.
   (d) the maria are much younger than the highlands.
   (e) the maria are much older than the highlands.

2. (7.2) What is the best explanation for the Moon's lack of an atmosphere?
   (a) The Moon never had any atmosphere.
   (b) The Moon has weak gravity and is close to the Sun.
   (c) All the comets that might have hit the Moon hit the Earth instead.
   (d) The Moon's atmosphere has frozen onto its surface.

3. (7.3) If the Moon did not rotate on its own axis, we would observe
   (a) both sides of the Moon.
   (b) the Moon remaining stationary against the stars.
   (c) a lack of tides on Earth.
   (d) the Moon from only one hemisphere of Earth.
   (e) everything the same as now—it doesn't rotate.

4. (7.4) The Moon does not undergo plate tectonics because
   (a) it has no areas of thin crust (like the Earth's ocean floors) where spreading ridges can form.
   (b) it does not have a substantial magnetic field.
   (c) its mantle is cold and rigid.
   (d) it has no active volcanoes.
   (e) its mantle is made of iron.

5. (7.5) If photographs are taken at high tide and the next low tide (as in figure 7.19), about how much time elapses between the pictures?
   (a) 3 hours   (c) 12 hours   (e) 1 month
   (b) 24 hours   (d) 6 hours

6. (7.5) As a result of the Moon's gravitational pull, when would you weigh the least?
   (a) When it is high tide locally
   (b) When it is low tide locally
   (c) When the Moon is overhead
   (d) When you are near one of the Earth's poles
   (e) Your weight is the same at all locations and times.

KEY TERMS

craters, 179
differential gravitational force, 190
highlands, 180
maria, 180
rays, 180

regolith, 184
rilles, 183
synchronous rotation, 187
tidal braking, 192
tidal bulge, 190
tides, 190

⊘: FIGURE QUESTION ANSWERS

WHAT IS THIS? (chapter opening): This picture was taken by *Apollo 15* astronauts orbiting the Moon shortly before their return to the Earth. The bright crescent is the Earth seen above the Moon's surface. Note that when the Moon is near its full phase, the Earth appears to be near its new phase from the perspective of the Moon.

FIGURE 7.1: The rays are younger than the maria.

FIGURE 7.2: The smaller one.

FIGURE 7.11: The Moon rotates on its axis so that it always keeps the same face toward the Earth. That is, it makes exactly one turn for each orbit around the Earth.

FIGURE 7.21: The bulge on the far side of Earth is farther away, so its gravitational pull is weaker.

8

SURVEY OF SOLAR SYSTEMS

Artist's depiction of a solar system in its early stages of formation.

LEARNING OBJECTIVES

Upon completing this chapter you should be able to:

- Identify the primary components of the Solar System, and describe their distinctive properties.
- Discuss the differences between terrestrial, Jovian, and dwarf planets, and their satellites, and recount why astronomers reclassified Pluto.
- Explain how astronomers measure masses and radii for bodies in the Solar System, and carry out a calculation to find a body's density from these measurements.
- Describe the densities of different classes of objects in the Solar System and relate this to their composition.
- Recall the age of the Solar System and explain how it is determined.
- Explain the various methods currently being used to detect exoplanets, and the information each provides.

- Discuss the limitations of each exoplanet detection method and what consequences this has for our understanding.
- Describe the ways in which exoplanetary systems differ from the Solar System, and in what ways the Solar System appears to be unusual.
- Describe the steps in the formation of the Solar System according to the nebular theory and relate these to the properties of the planets and other bodies.
- Explain why disks are expected to form around stars, and describe the observations that indicate disks are present around young stars.
- Describe the role of planetesimals in planet formation and modification, and where some can still be found.
- Discuss the roles of rocky, icy, and gaseous materials in the formation of planets and their atmospheres.

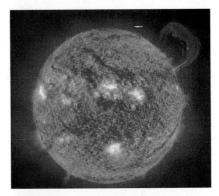

The Solar System consists of the Sun and the bodies in its gravitational domain: the eight planets, dozens of dwarf planets, and swarms of moons, asteroids, and comets. Although earthlings have not walked on any objects except the Earth and Moon, we have detailed pictures sent to us from spacecraft of most of the planets and their satellites. Some are naked spheres of rock; others are mostly ice. Some have thin, frigid atmospheres so cold that ordinary gases crystallize as snow on their cratered surfaces; others have thick atmospheres the consistency of molten lava and no solid surface at all. Despite such diversity, the Solar System possesses an underlying order, an order from which astronomers attempt to read the story of how our Solar System came to be.

The Solar System formed in the extremely remote past, over 4.5 billion years ago. Astronomers hypothesize that the Sun and planets formed from the collapse of a huge, slowly spinning cloud of gas and dust. Most of the cloud's material fell inward and ended up in the Sun, but in response to rotation, some settled into a swirling disk around it. Then, within that disk, dust particles coagulated—perhaps aided by electrostatic effects such as those that make lint cling to your clothes—to form pebble-size chunks of material, which in turn collided and sometimes stuck together, growing ever larger to become the planets we see today. The objects that formed in the disk retained the motion of the original gas and dust, and so we see them today, moving in a flattened system, all orbiting the Sun in the same direction.

Seeing planets around other stars is much more challenging. However, astronomers have developed an array of techniques that have revealed more than a thousand planets around other stars and even other "solar systems" in their first stages of formation. Many of the other systems detected so far look very different from our own, challenging our understanding of how solar systems form. In this chapter, we will survey the general properties of our Solar System and others. In later chapters we will explore the components of our Solar System in much more detail.

See end of chapter for the answer.

CONCEPTS AND SKILLS TO REVIEW

- Law of gravity (3.5)
- Density (6.1)
- Modified form of Kepler's third law (3.6)

8.1 COMPONENTS OF THE SOLAR SYSTEM

The Solar System is just one among billions of planetary systems in our galaxy. It is by far the best-studied, so we begin by studying our system to provide a context for interpreting the many other planetary systems now being discovered.

The Sun

The Sun is a star, a ball of incandescent gas (fig. 8.1) whose light and heat are generated by nuclear reactions in its core. It is by far the largest body in the Solar System—more than 700 times the mass of all the other bodies put together—and its gravitational force holds the planets and other bodies in the system in their orbital patterns about it. This gravitational domination of the planets by the Sun justifies our calling the Sun's family the **Solar System.**

The Sun is mostly hydrogen (about 71%) and helium (about 27%), but it also contains trace amounts of nearly all the other chemical elements (carbon, iron, uranium, and so forth) in vaporized form, as we can tell from the spectrum of the light it emits. Stars can have lower or higher amounts of the trace elements, although planets are more common when there are more of the trace elements.

FIGURE 8.1
Image of the Sun made with an ultraviolet telescope that reveals high-temperature gases in the Sun's atmosphere.

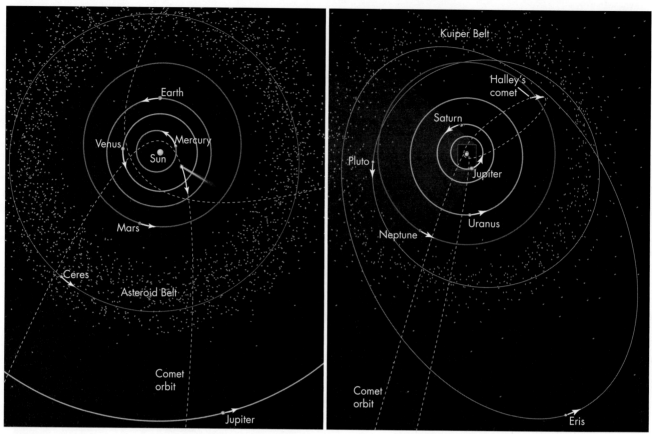

FIGURE 8.2

Diagrams of the Solar System from above. The orbits are shown in the correct relative scale in the two drawings. Because of the great difference in scale, the inner and outer Solar System are displayed separately.

The Planets

The planets are much smaller than the Sun and orbit about it in nearly circular orbits. They emit no visible light of their own but shine by reflected sunlight. In order of increasing distance from the Sun, they are Mercury, Venus, Earth, Mars, Jupiter, Saturn, Uranus, and Neptune, as shown in figure 8.2. The **inner planets**—Mercury, Venus, Earth, and Mars—are small rocky bodies with relatively thin or no atmospheres. The **outer planets**—Jupiter, Saturn, Uranus, and Neptune—are gaseous and liquid. They are much larger than the inner planets and have deep, hydrogen-rich atmospheres. For example, Jupiter is more than 10 times larger in diameter than the Earth and has 318 times its mass. These differences can be seen in figure 8.3, which also shows a small part of the edge of the Sun to illustrate how the Sun dwarfs even the large planets.

Instead of "inner" and "outer" planets, astronomers sometimes use "terrestrial" and "Jovian" to describe the two types of planets. The **terrestrial planets** (Mercury to Mars) are so-named because of their resemblance to the Earth. The **Jovian planets** (Jupiter to Neptune) are named for their resemblance to Jupiter.

Although the two categories of planets neatly describe the eight most massive objects that orbit the Sun, astronomers have found many smaller objects that fit neither category. Pluto has long failed to fit, because of its small size, composition of ice and rock, and odd orbit. Not only is its orbit highly tilted with respect to the other planets, it also crosses Neptune's orbit. Moreover, in the last decade astronomers have discovered more than a thousand icy objects orbiting at similar distances from the Sun as Pluto. In 2005 it was discovered that one of these objects, named Eris, is an icy world more massive than Pluto that orbits about 68 AU from the Sun, roughly half again Pluto's distance from our star.

FIGURE 8.3

The planets and the Sun to scale.

In response to the discovery of Eris and half a dozen other objects similar in size to Pluto, in 2006 the International Astronomical Union introduced a new category of Solar System objects called **dwarf planets.** A dwarf planet is an object that orbits the Sun, is massive enough that its gravity compresses it into an approximately spherical shape, but has not swept its orbital region clear of other objects that add up to a mass comparable to its own mass. To recognize Pluto as the first of these objects discovered, the IAU decided in 2008 to call dwarf planets that orbit beyond Neptune *plutoids.*

Most planets are themselves orbited by satellites. Jupiter, Saturn, Uranus, and Neptune have large families of 67, 62, 27, and 14 moons, respectively, discovered to date. Mars has 2, Earth has 1, while Venus and Mercury have none. Some of the dwarf planets also have moons: Pluto has 5 and Eris has 1. Many of these satellites are just a few kilometers in size and very difficult to detect, but others are so large that they would be termed planets or dwarf planets if they orbited the Sun themselves.

Asteroids and Comets

The Solar System is filled with millions of objects far smaller than planetary bodies. The **asteroids** are rocky or metallic bodies, the largest of which is the dwarf planet Ceres with a diameter of about 970 kilometers (600 miles). Smaller asteroids do not qualify as dwarf planets because their gravity is not strong enough to have pulled them into a spherical shape; this includes the next most massive asteroid, Vesta (fig. 8.4). Most asteroids orbit the Sun in the large gap between the orbits of Mars and Jupiter, a region called the **asteroid belt** (fig. 8.2). They are probably material that failed—perhaps as a result of disturbances by Jupiter's gravity—to aggregate into a planet.

Beyond Neptune, extending to perhaps 50 AU from the Sun, is a region called the **Kuiper** *(KY-per)* **belt.** As seen in figure 8.2, the Kuiper belt looks similar to the asteroid belt, but the objects here are made mostly of ice. Pluto and dozens of other dwarf planet candidates orbit in the Kuiper belt, and uncounted icy bodies such as Eris orbit even farther from the Sun in a scattered region whose extent is not well known.

Objects at such large distances from the Sun are so dimly illuminated that they are very difficult to detect. Our main clue to what bodies orbit the Sun at even larger distances are the **comets.** These are icy bodies typically about 10 km (about 6 miles) in diameter that enter the inner Solar System on highly elongated orbits (fig. 8.2). When they approach the Sun, they grow huge tails of gas and dust as their ices are partially vaporized. Most comets orbit far beyond Neptune in a region of the Solar System called the **Oort cloud,** which may extend 100,000 AU from the Sun. Although the

Oort cloud and Kuiper belt

FIGURE 8.4
Photographs show that the Earth is round but the asteroids Gaspra and Vesta are not. Gaspra is too small for its gravity to make it spherical, but Vesta is nearly big enough.

majority of comets probably originate in the Oort cloud, some come from the Kuiper belt. We will discuss more details of the Oort cloud and Kuiper belt in chapter 11, but for now we simply note that together they probably contain more than 1 trillion (10^{12}) comet nuclei, only a few of which get close enough to the Sun to be detected.

The Orbits and Spins of the Planets

When several planets are visible in the evening sky, we can see that they lie along a linear band extending away from the Sun (fig. 8.5). The planets appear to lie along a line because their orbits, as well as the Earth's, all lie in nearly the same plane, as shown in the side view of the Solar System in figure 8.6. Mercury's orbit has the largest tilt, just 7° from the average of the rest of the planets. The planetary orbits out to Neptune have about the same relative thickness as 3 CDs stacked together. The planets also all travel around the Sun in the same direction: counterclockwise as seen from above the Earth's North Pole, and this is the same direction in which the Sun itself spins.

As the planets orbit the Sun, each also spins on its rotation axis. The spin is generally in the same direction as the planets' orbital motion around the Sun (again, counterclockwise as seen from above the Earth's North Pole), and the tilt of the rotation axes relative to the plane of planetary orbits is generally not far from the perpendicular. However, there are two exceptions: Venus and Uranus. Uranus has an extremely large tilt to its rotation axis, which lies nearly in its orbital plane (fig. 8.7). Venus's rotation axis has such a large tilt that it spins backward, a motion technically called "retrograde rotation." However, despite this backward spin, Venus orbits the Sun in the same direction as the rest of the planets.

Many dwarf planets and other small bodies in the Solar System have highly inclined orbits and randomly oriented spins. Although the objects in the asteroid belt and Kuiper belt may have orbits tilted by up to about 45°, on average they are very close to same plane as the planets. On the other hand, the comets that arrive from the largest distances may have orbits oriented in any direction. This leads astronomers to conclude that the Oort cloud surrounds the Solar System in a roughly spherical shape.

Like the planets orbiting the Sun, most of the moons orbiting the planets move along approximately circular paths that are roughly in the planet's equatorial plane, their orbits tilted like the planets themselves. Thus, each planet and its moons resemble a miniature Solar System—an important clue to the origin of these satellites. Some large moons and many of the smaller moons have much more irregular orbits, suggesting that they may have been captured.

FIGURE 8.5
Sunset view of four planets strung along the zodiac on March 1, 1999. Their straight-line arrangement results from the flatness of the Solar System. From top to bottom, you can see Saturn, Venus, Jupiter, and Mercury (nearly lost in the twilight).

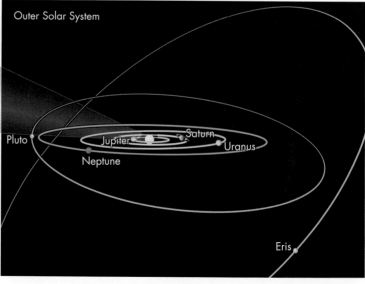

FIGURE 8.6
The planets' orbits from the side. The dwarf planets Ceres, Pluto, and Eris are also shown, illustrating their highly inclined orbits. This view also shows that although Pluto sometimes gets closer to the Sun than Neptune, its orbit actually remains well separated from Neptune's.

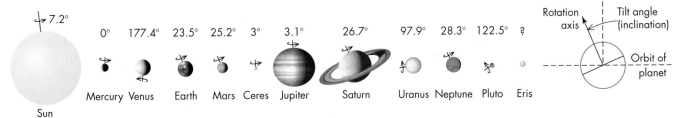

FIGURE 8.7
Sketches showing the orientation of the rotation axes of the planets and Sun. The figure illustrates that most of them spin in the same direction, counterclockwise as seen from above the Earth's North Pole. The dwarf planets Ceres, Pluto, and Eris are also shown. The bodies are not shown to the same scale.

The significance of another feature of the planets' orbits is a matter of some debate. In the 1700s astronomers noticed that the spacing between the orbits of the planets seems to follow a fairly regular progression. This mathematical progression may indicate something about the natural spacing between orbits of large bodies, or it may be a chance pattern as discussed in Astronomy by the Numbers: "Bode's Rule: The Search for Order."

Its flattened structure, and the orderly orbital and spin properties of its planets, are two of the most fundamental features of the Solar System, but a third and equally important feature is that the planets fall into two families, called inner and outer planets, based on their size, composition, and location in the Solar System, as we discuss next.

ASTRONOMY *by the numbers*   BODE'S RULE: THE SEARCH FOR ORDER

A curious—and as yet unexplained—feature of the orbits of the planets is their regular spacing. Very roughly, each planet is about twice as far from the Sun as its inner neighbor. This progression of distance from the Sun can be expressed by a simple mathematical relation known as **Bode's rule,** which works as follows: write down 0, 3, and then successive numbers by doubling the preceding number until you have nine numbers. That is, 0, 3, 6, 12, 24, and so on. Next, add 4 to each, and divide the result by 10, as shown in table 8.1. The resulting numbers, with two exceptions, are very close to the distances of the planets from the Sun in astronomical units.

Bode's rule was worked out before the discovery of Uranus, Neptune, and Pluto, and when Uranus was discovered and found to fit the law, interest was focused on the "gap" at 2.8 AU. Astronomers therefore began to search for a body in the gap, and, as we will see in chapter 10, Giuseppi Piazzi, a Sicilian astronomer, soon discovered the dwarf planet Ceres, which fit the rule splendidly.

The next planet to be found, Neptune, did not fit the rule at all, nor did the dwarf planet Pluto. These irregularities show that Bode's rule is not a law like the "law of gravity," which is why we prefer to call it "rule" to emphasize this difference. It is not based on any (known) physical principles, but computer simulations of planet formation sometimes produce planets at similar spacing patterns. It may tell us that systems of planets are not likely to remain in stable orbits for billions of years unless their orbits are a factor of 1.5 to 2 times larger than the next planet interior to them. Or perhaps it merely shows the human fascination with patterns and our tendency to see order where none may actually exist.

Table 8.1	Bode's Rule		
Bode's Rule	**Number**	**Object**	**True Distance**
$(0 + 4)/10 =$	0.4	Mercury	0.39
$(3 + 4)/10 =$	0.7	Venus	0.72
$(6 + 4)/10 =$	1.0	Earth	1.00
$(12 + 4)/10 =$	1.6	Mars	1.52
$(24 + 4)/10 =$	2.8	Ceres	2.77
$(48 + 4)/10 =$	5.2	Jupiter	5.2
$(96 + 4)/10 =$	10.0	Saturn	9.5
$(192 + 4)/10 =$	19.6	Uranus	19.2
$(384 + 4)/10 =$	38.8	Neptune	30.1
$(768 + 4)/10 =$	77.2	Pluto	39.5

Composition Differences Between the Inner and Outer Planets

The composition differences between the rocky inner planets and the hydrogen-rich outer planets are critcally important to our understanding of the history of the Solar System. Therefore we will look more closely at how we determine these properties.

Astronomers can deduce a planet's composition in several ways. From its spectrum, they can measure its atmospheric composition and get some information about the nature of its surface rocks. However, spectra give no clue as to what lies deep inside a planet where light cannot penetrate. To learn about the interior, astronomers must therefore use indirect methods.

We saw in chapter 6 that we can examine Earth's internal structure by studying seismic waves, but the only other planet where it has been possible to use this method is Mars, and only to a limited extent. Our main clue to a planet's composition is its density. The average density of a planet is its mass divided by its volume. Both of these quantities can be measured relatively easily. For example, we showed in chapter 3 how to determine a body's mass from its gravitational attraction on a second body orbiting around it by applying Newton's modification of Kepler's third law. Thus, from this law, we can calculate a planet's mass by observing the orbital motion of one of its moons or a passing spacecraft. We can determine a planet's volume ($\mathscr{V}$) from the formula $\mathscr{V} = \left(\frac{4}{3}\right)\pi R^3$, where R is the planet's radius. We can measure R in several ways—for example, from its angular size and distance, a technique we used in chapter 2 to measure the radius of the Moon. With the planet's mass, M, and volume, $\mathscr{V}$, known, we can calculate its average density straightforwardly by dividing M by $\mathscr{V}$ (fig. 8.8).

Once the planet's average density is known, we can compare it with the density of common candidate materials to find a likely match. For example, we saw in chapter 6 that the average density of the Earth (5.5 grams per cubic centimeter) was intermediate between silicate rock (about 3 grams per cubic centimeter) and iron (8 grams per cubic centimeter). Therefore, we inferred that the Earth has an iron core beneath its rocky crust, a supposition that was verified from studies using earthquake waves.

Although density comparison is a powerful tool for studying planetary composition, it also has drawbacks. First, there may be several different substances that will produce an equally good match to the observed density. Second, the density of a given material can be affected by the planet's gravitational force. For example, a massive planet may crush rock whose normal density is 3 grams per cubic centimeter to a density of 7 or 8 grams per cubic centimeter. Thus, in making a match to determine the composition, we must take into account the compression by gravity.

> **Q:** Suppose you are given a tiny box that has a volume of 10 cubic centimeters and a mass of 30 grams. What is its density? Is it more likely to contain solid iron or rock?

Volume

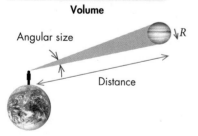

Measure angular size of planet, and use relation between angular size and distance to solve for planet's radius, R. Calculate volume, $\mathscr{V}$, of planet:

$$\mathscr{V} = \frac{4\pi R^3}{3}$$

for a spherical body of radius R.

Finding the density of a planet

Average density

Average density, ρ, equals mass, M, divided by volume, $\mathscr{V}$:

$$\rho = \frac{M}{\mathscr{V}}$$

Mass

Observe motion of a satellite orbiting planet. Determine satellite's distance, d, from planet and orbital period, P. Use Newton's form of Kepler's third law:

$$M = \frac{4\pi^2 d^3}{GP^2}$$

Insert measured values of d and P, and value of constant G. Solve for M.

FIGURE 8.8

Measuring a planet's mass, radius, and average density. Volume can be determined from the radius of a planet, which in turn is found from its distance and angular size (chapter 2). Mass can be determined from the orbit of a satellite (chapter 3).

The terrestrial planets have densities similar to the Earth's (3.9 to 5.5 grams per cm^3). On the other hand, the Jovian planets have much lower densities (0.7 to 1.7 grams per cm^3). After correcting for gravitational compression, we conclude that all the inner planets contain primarily rock and iron and that the iron has sunk to the core, as shown in figure 8.9. The outer planets contain mainly light materials, as borne out by their spectra, which show them to be mostly hydrogen, helium, and hydrogen-rich molecules such as methane (CH_4), ammonia (NH_3), and water (H_2O).

When we speak of rock, we mean material composed of silicates, which are composed of silicon (Si) and oxygen (O) with an admixture of other heavy elements such as aluminum (Al), magnesium (Mg), sulfur (S), and iron (Fe). If we consider the Solar System as a whole, rock is rare, because the silicon atoms that compose it are outnumbered more than 25,000 to 1 by hydrogen. However, in the warmth of the inner Solar System, rock dominates because intrinsically more abundant materials such as hydrogen, water, methane, and ammonia cannot condense to mingle with it.

The outer planets probably have cores of iron and rock roughly the size of the Earth beneath their deep atmosphere, as illustrated in figure 8.9. Astronomers deduce the existence of these cores in two ways. First, if the outer planets have the same relative amount of heavy elements as the Sun, they should contain several Earth masses of iron and silicates, and because these substances are much denser than hydrogen, they must sink to the planet's core. Secondly, detailed analyses of these planets' gravitational fields, determined from their effect on space probes, are best explained by dense

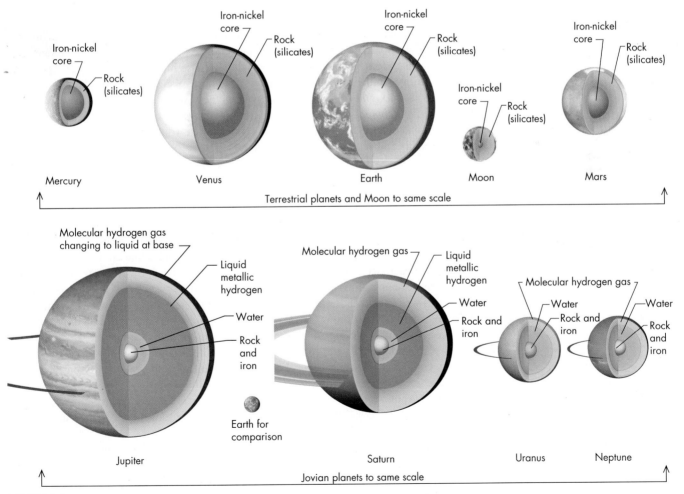

FIGURE 8.9
Sketches of the interiors of the planets. Details of sizes and composition of inner regions are uncertain for many of the planets.

cores. In the case of Jupiter, a core of roughly 7 times the Earth's mass is estimated. However, there is a large uncertainty in the exact value, with some recent models estimating twice as much.

The Jovian planets have no true "surface"; rather, their atmospheres thicken with depth and eventually compress to liquid form despite high temperatures. They have no distinct boundary between a thin "atmosphere" and a solid "crust" as we have on the Earth. Thus, we can never "land" on any of the Jovian planets because we would simply sink ever deeper into their atmospheres into an interior that contains extremely hot gas/liquid that is denser than rock.

Models of Uranus and Neptune suggest that their interiors may contain large amounts of liquid water, and other molecules that would be ices if they were not in the hot core of these planets. This contrasts with Jupiter and Saturn, which are primarily composed of hydrogen and have interiors composed of hydrogen gases so highly compressed that they take on a liquid form. Accordingly, planets like Neptune and Uranus are referred to as **ice giants,** while planets as large as Jupiter and Saturn are called **gas giants.**

The satellites of the Jovian planets and the objects orbiting beyond Neptune are similar in density to the Jovian planets, typically about 1.5 grams per cm^3 (fig. 8.10). These bodies do not have atmospheres, for the most part, but are instead made up of ice and rock. By ice, we mean frozen liquids and gases such as ordinary water ice (H_2O), frozen carbon dioxide (CO_2), frozen ammonia (NH_3), and frozen methane (CH_4). Some of these satellites have diameters comparable to Mercury's, although they are much less massive because they are built from lower-density materials. Asteroids and comets show the same split into two families; that is, rocky bodies and icy bodies.

The composition of bodies in the inner and outer Solar System furnishes another clue to the Solar System's origin: the planets and Sun were all made from the same material. Astronomers come to this conclusion because Jupiter and Saturn have a composition very similar to that of the Sun, and the inner planets have a similar composition *if we were to remove the Sun's hydrogen, helium, and other elements normally found in gaseous compounds.* Thus, we can explain the compositional difference between the inner and outer planets by proposing a process that would keep the inner planets from collecting and capturing large amounts of gas.

Age of the Solar System

Another important clue to how the Solar System formed comes from its age. The best evidence implies that, despite their great differences in size, structure, and composition, the Sun, planets, asteroids, and other bodies all formed at nearly the same time.

We can estimate the date when the Earth, Moon, and some asteroids formed from the radioactivity of their rocks. As discussed in chapter 6, by-products of radioactive decay remain trapped in rocks until they melt and recrystallize. Therefore, the oldest rocks we can find on Earth give us a lower limit to the planet's overall age. Some rocks are dated to over 4 billion years old; some individual crystals embedded in old rocks are arguably as old as 4.4 billion years. The Earth has had such an active geology that it is not easy to find rocks that have remained unaltered since the Earth formed. By contrast, because there has been relatively little geological activity on the Moon, lunar rock samples tend to be at least several billion years old; some rocks returned by the Apollo missions are as old as 4.5 billion years. Meteorites, which are pieces of asteroids that have fallen to Earth (chapter 11), have radioactive ages of up to about 4.6 billion years.

The radioactive dates are all consistent with the planets beginning to form about 4.6 billion years ago, with the smaller bodies cooling first. We find a similar age for the Sun, based on its current brightness and temperature and its rate of nuclear fuel consumption. Thus, it appears that the Solar System formed between about 4.6 and 4.5 billion years ago, creating most of the bodies that we still see today.

FIGURE 8.10
Galileo spacecraft image of Ganymede, the largest satellite of Jupiter, and even larger than Mercury, but with a density of only about 1.9 grams per cm^3.

OTHER PLANETARY SYSTEMS

The Discovery of Planets Beyond the Solar System

Astronomers have long searched for planets orbiting stars other than the Sun. Their interest in such **exoplanets*** (as these distant worlds are called) is motivated not merely by the wish to detect other planets. Equally important is the hope that study of such systems will help us better understand our own planetary system.

Spotting a planet orbiting a star is a little like trying to spot a gnat flying near a lightbulb from miles away. Planets are so very small that the light they reflect is almost completely drowned by the light of their star. Although direct imaging of exoplanets is very difficult, astronomers have been able to image a few exoplanets at infrared wavelengths. This has proved more successful than visible-wavelength observations because the star is dimmer in the infrared, and large planets often remain quite warm long after their formation. We will see in chapter 9 that most of the large planets in the Solar System radiate more energy than they receive from the Sun, even though it has been billions of years since they formed. Two examples of planets detected around stars in the infrared are shown in figure 8.11. The red object in figure 8.11A appears to be a young gas giant planet orbiting its star at a distance even farther than Neptune is from the Sun. Figures 8.11B and C show an infrared image of several gas giants orbiting a star and illustrate the difficult problem astronomers face in removing scattered light from the star to reveal the planets.

Even though their reflected light is difficult to see, astronomers have developed a variety of other techniques for detecting exoplanets. In fact, the first direct evidence for exoplanets came in the 1990s not from imaging but by observing how an exoplanet's gravitational pull affects the star it orbits. When a planet orbits its star, the planet

INTERACTIVE

Exoplanets

---

* A number of astronomers use the term *extra-solar planets*. However, this is a bit peculiar because, after all, Earth is extra-solar too, in the sense that it is orbiting outside the Sun.

Brown dwarf
2MASSWJ1207334-393254

Probable planet
glowing in infrared

├────── 50 AU ──────┤

A

Scattered light from star HR 8799

B

After subtraction of scattered starlight,
3 of 4 planets around HR 8799 visible

c   d

b

├──────100 AU──────┤

C

FIGURE 8.11

(A) The first image of an exoplanet was made at infrared wavelengths with the 8-meter Very Large Telescope (VLT) in Chile using adaptive optics. The exoplanet, which is seen glowing red from its infrared emission, is about 50 AU from the star it orbits, which is a low-mass "brown dwarf." (B) A Hubble Space Telescope infrared image of the star HR 8799 shows the challenge for finding planets produced by scattered light from the star. (C) Careful analysis of the scattered light allows most of it to be removed, revealing the presence of three giant planets orbiting the star. A fourth planet is known from other observations.

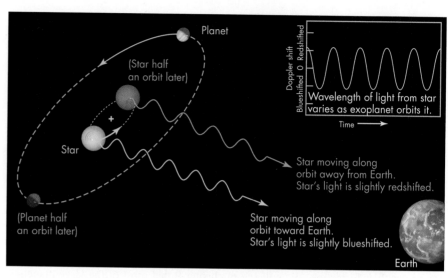

FIGURE 8.12

As an unseen planet orbits a star, the star's position "wobbles." This produces a changing Doppler shift. From the period of the wobble, the planet's distance from the star can be found. From the amplitude of the wobble, the planet's mass can be estimated.

exerts a gravitational force back on the star as a result of Newton's third law—the law of action–reaction. That force makes the star's position wobble slightly, just as you wobble a little if you swing a heavy weight around you. The wobble creates a Doppler shift in the star's light that astronomers can measure (fig. 8.12). From that shift and its change in time, astronomers can deduce the planet's orbital period, mass, and distance from the star. Using this Doppler method, astronomers had discovered nearly 600 exoplanets by early 2015.

It is even possible to detect multiple planets orbiting a star because each planet's pull produces a wobble with a different period as illustrated in figure 8.13. Astronomers have found no system of exoplanets yet that looks very much like our own, but some show similarities. For example, 55 Cancri is Sun-like star that has five planets orbiting within 6 AU of the star, just as in the Solar System (fig. 8.14). However, all of these planets are massive, at least 10 times Earth's mass, and three of the planets orbit at distances much closer than Mercury's distance from the Sun.

ANIMATION ▶

The position and Doppler shift of a star orbiting its common center of mass with a planet

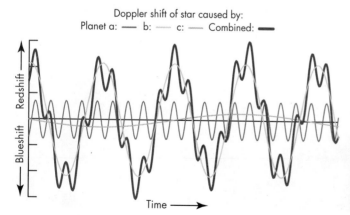

Doppler shift of star caused by:
Planet a: —— b: —— c: —— Combined: ——

FIGURE 8.13

Astronomers can detect more than one planet by determining the combination of orbits that produce the star's overall pattern of Doppler shifts. In the example illustrated here, planet a is closest and planet c is farthest from the star; a and c have the same mass, and b is 5 times more massive.

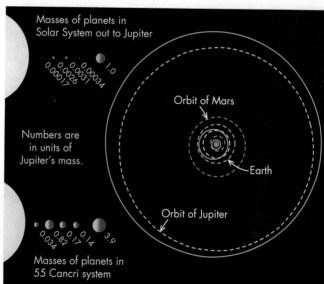

Masses of planets in Solar System out to Jupiter

0.0017 0.0026 0.0031 0.000034 1.0

Orbit of Mars

Numbers are in units of Jupiter's mass.

Earth

0.034 0.82 0.17 0.14 3.9

Orbit of Jupiter

Masses of planets in 55 Cancri system

FIGURE 8.14

The 55 Cancri system contains five known planets around a star that is very similar to the Sun. The estimated masses (compared to Jupiter) for these planets and their orbits are compared with the five innermost planets in the Solar System.

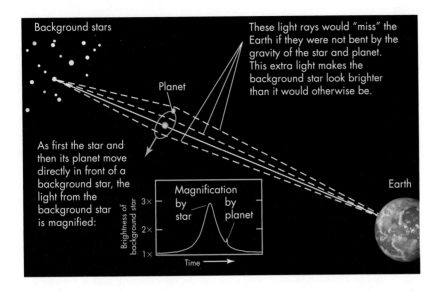

FIGURE 8.15
Detecting a planet by the slight bending of light from a background star caused by the planet's gravity.

The Doppler method for detecting exoplanets works best for massive planets that orbit near their star. This makes the star's wobble larger and faster. At present we would probably not be able to detect the planets in a system that was just like our Solar System, except possibly for Jupiter. Given that limitation, is there some way we can search for planets that more closely resemble our Earth?

One way is to use a discovery made by Einstein in the early 1900s. Einstein showed, as part of his general theory of relativity, that a mass bends space in its vicinity and that this bending creates the mass's gravity. As a result, if a ray of light passes near a mass, the bent space around the mass deflects the light and can bring it to a focus, as figure 8.15 schematically shows. As long ago as 1916, astronomers, following Einstein's suggestion, detected the bending of light from a distant star as the star's light traveled past our Sun (see essay 2). By analogy with the focusing ability of an ordinary lens, astronomers call such deflection of light **gravitational lensing.**

Gravitational lensing has proved to be a powerful tool for detecting low-mass planets. The method works approximately as follows. Suppose we look at some distant star and measure its brightness. Suppose further that, by chance, a star at an intermediate distance moves between us and the distant star. Rays of the distant star's light that would have traveled past us in the absence of the intermediate star are now bent so that they reach us (see fig. 8.15). Thus, we observe *more* light from the distant star when an intermediate-distance star is present. Moreover, we receive even more light (although only a very tiny amount more) if a planet is orbiting the intermediate-distance star.

It is very rare to find an intermediate star with an orbiting planet that is properly positioned. Thus, to search for planets by this method, astronomers monitor the brightness of millions of stars, and computers scan millions of bits of data for the tiny increase in brightness of a lensing event. The technique has successfully detected some planets with masses comparable to the Earth's. For example, in 2014 astronomers detected a brightening event in a star, caused by an exoplanet with a mass about 60% greater than the Earth's orbiting its star at distance of about 0.7 AU. The star is also orbited by a second star at a distance of about 15 AU. Both stars in this system are much smaller than the Sun, so the light the exoplanet receives from the nearer star is less than 1% as much light as the Earth receives from the Sun. This exoplanet is probably an icy object little resembling the Earth despite its similar orbital distance.

More than 30 exoplanets have been discovered by the gravitational lensing method, covering a wide range of masses and separations. However, there is little opportunity for follow-up observations because the method relies on rare chance alignments. Thus we cannot go back to search for other planets or determine other properties of the exoplanets found.

FIGURE 8.16

Detecting an exoplanet by the transit method. If a planet passes in front of (transits) its star, it diminishes the light we see. A planet with 10% the radius of its star, as shown, blocks 1% of its area.

Transiting Exoplanets

Another method for detecting exoplanets has become the major source of new discoveries in recent years. This method works for exoplanets whose orbits are aligned so the planets pass in front of, or **transit,** their star. As the planet transits in front of the star, the light is slightly dimmed, as illustrated in figure 8.16. At first this method was successfully applied only to a few relatively large planets, but the *Kepler* satellite launched by NASA in 2009 has much better precision than any Earth-based telescope, and it monitored nearly 150,000 stars for about 4 years. It identified more than 4000 candidate planets transiting their stars, and over 1000 have been confirmed.

One of the interesting features of the transit method is that it gives us information about the radius of the exoplanet from the amount of dimming that is observed. A larger planet blocks more of the light, so larger planets are easier to detect. For example, a planet the size of Jupiter can block about 1% of the light from a star like the Sun. Terrestrial planets block less than 0.01%, which is challenging to detect because stars often vary in brightness by much more than this on their own. With repeated transits, the period of the exoplanet's orbit can also be determined, and if we can estimate the mass of the star, we can then calculate the radius of the exoplanet's orbit.

More than 50 exoplanets the size of Earth or smaller have been confirmed so far. The smallest, orbiting Kepler-37, is not much bigger than the Moon (fig. 8.17). Astronomers expect that there are many more this size that have gone undetected because of the difficulty in measuring such a small change in the light from a star. More than 800 of the confirmed exoplanets discovered by *Kepler* are larger than Earth but smaller than Neptune. It is curious that this is the most common size of exoplanet found by *Kepler,* but there are no planets this size in the Solar System. It seems likely that planets will be rocky terrestrial bodies if they are slightly larger than Earth, and probably like the ice giants even if they are somewhat smaller than Neptune, but what happens in the range from 1.25 to 2 times Earth's radius (probably about 2 to 10 times the Earth's mass) is quite uncertain. They might be large terrestrial planets or mini ice giants, or perhaps something altogether different. Astronomers have begun calling this size of planet a **super-earth.** The star Kepler-37 has a super-earth in addition to its Moon-sized exoplanet as depicted in figure 8.17.

The chance of *Kepler* detecting exoplanets around any particular star is very low because their orbits have to be almost precisely edge-on for a transit event to occur. Presumably the orbits are randomly oriented, so for every star where an exoplanet is detected, there must be a hundred other stars whose planets' orbits are not aligned correctly. Smaller exoplanets are also missed when the dip in brightness is too small to be detected. Finally, because *Kepler* was only able to collect data for under 4 years, and several transits are necessary to definitely establish the orbit of an exoplanet, the data are only relatively certain for exoplanets with orbits taking under a year.

We can make adjustments for these limitations to the observations to extrapolate how many exoplanets are likely to be orbiting around other stars in orbits that are relatively close to the star—roughly 0.75 AU or less. The approximate distribution of

FIGURE 8.17

Artist's impression of the smallest exoplanet found by the *Kepler* spacecraft around a star that also hosts a nearly Earth-sized exoplanet and a planet nearly twice the Earth's diameter—a super-earth—unlike any of the planets in the Solar System. Solar System bodies are shown for comparison.

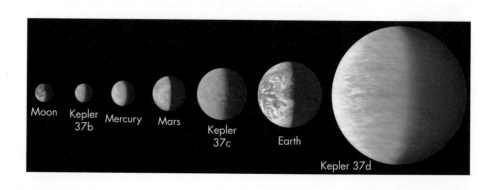

exoplanet sizes after making adjustments for the probability of detection is shown in figure 8.18. The results are remarkable. They indicate that 1 in 6 stars has a terrestrial planet orbiting so close, 1 in 5 has a super-earth, and 1 in 4 has an ice giant.

The census shows relatively few gas giants orbiting within this small distance of their star. This contrasts with the early results from the Doppler studies, which detected almost nothing other than massive exoplanets orbiting near their stars. The Doppler method has grown more sensitive, but still it has detected only two exoplanets that are in the terrestrial planet mass range. It seems likely that as studies explore larger distances from stars, many more gas giants will be detected, but other methods may be needed to find these more remotely orbiting exoplanets.

The great majority of the exoplanets *Kepler* has detected are in systems of two or more exoplanets. This tells us that the planets must be orbiting in very nearly the same orbital plane, like the planets in the Solar System. Figure 8.19 shows all of the systems where five or more planets have so far been detected orbiting their star. The diagram also shows the Solar System and systems discovered by the Doppler method, and some of these have additional known planets orbiting farther out than shown in the diagram.

Looking at figure 8.19, you can see that none of these exoplanetary systems looks very much like our own. All of these systems have three or more planets orbiting closer to their star than Mercury's distance from the Sun. In fact, nearly three-quarters of all the exoplanets detected to date orbit closer than Mercury, but this is probably less a reflection of the actual range of planetary orbits than it is of our detection methods, which favor finding large planets that orbit close to their stars.

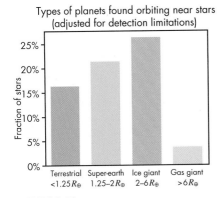

FIGURE 8.18

An estimate of the percentage of stars that have planets of various sizes orbiting them within about 0.75 AU.

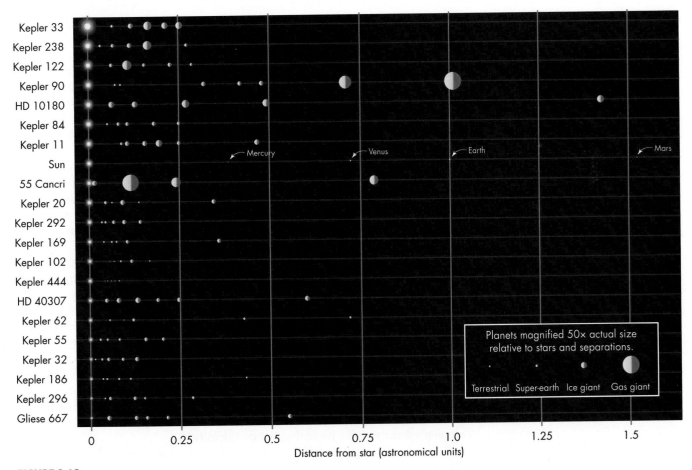

FIGURE 8.19

Comparison of the orbital radii and relative sizes of exoplanetary systems with the Solar System. Systems with four or more known exoplanets are shown, organized according to the mass of the star that they orbit. The sizes of the dots are based on the mass of each exoplanet, and approximately indicate their true relative size. The numbers indicate the mass of each planet in units of Jupiter's mass.

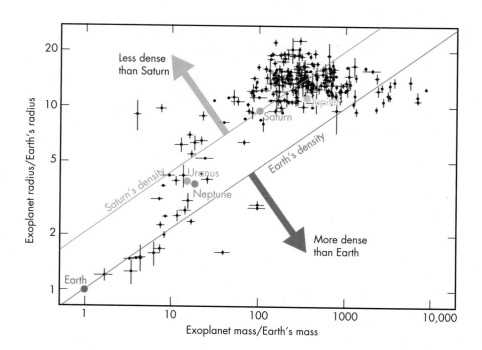

FIGURE 8.20

Plot comparing exoplanet masses to their radii. The values for the five largest planets in the Solar System are shown for comparison.

Composition of Exoplanets

The presence of giant planets so near their stars is distinctly different from the Solar System. From their radii alone we cannot determine if they are gaseous or rocky, so we might ask if these large-diameter bodies are like the ice and gas giants we are familiar with, or could they be giant rocky planets? We can answer this question if we can determine their densities. To do this we need to measure the exoplanet's mass, and then using the radius determined by *Kepler,* we can estimate its overall density. Astronomers are therefore striving to measure Doppler data on the stars that *Kepler* has identified. While this is not possible for every star, enough have been measured that we can begin to see some trends among the exoplanets.

Figure 8.20 presents a graph of the better radius and mass measurements made so far. While there is a good deal of individual variation, it appears that most of the exoplanets with masses up to ten times Earth's (and radii up to about twice as big) have densities fairly similar to Earth's, and probably contain a large fraction of rock like the terrestrial planets. Bodies larger than this, up to the mass of Saturn, generally have lower densities, and they are probably similar to our ice and gas giants.

At about Jupiter's mass, an interesting thing happens. Many exoplanets have substantially larger masses than Jupiter, but their radii are all fairly similar. This can only happen if they grow steadily more dense as their mass increases. This suggests that the gas becomes more and more compressed as mass is added to a planet. Indeed, if a planet gains enough mass, its interior will grow so dense and hot that nuclear reactions will begin, and it will become a star.

Transiting exoplanets offer us additional opportunities to study their composition. For example, a planet of the Sun-like star HD-209458 orbits so that it passes between us and the star every 3.5 days. The exoplanet's mass is about 70% of Jupiter's mass, and from the amount of the star's light it blocks, astronomers deduce that its diameter is about 1.3 times Jupiter's. This tells us that its density is less than one-third of Jupiter's density—even lower than Saturn's density.

When a planet passes in front of a star, a fraction of the star's light passes through the planet's atmosphere and the gas in the planet's atmosphere imprints very weak absorption lines on the spectrum. For the exoplanet orbiting HD-209458, the lines are from hydrogen, sodium, carbon, oxygen, and even water vapor. Analysis of the line strengths suggests that the planet is a gas giant planet similar to Jupiter. Notice, however, the extremely short orbital period of 3.5 days. Using Kepler's third law, astronomers deduce that this planet orbits a mere 0.05 AU from its star, roughly one-tenth the distance at which Mercury orbits our Sun. From the extent of the absorption lines seen, it appears that the planet is surrounded by a cloud of evaporating gas (fig. 8.21). The planet may have lost as much as a quarter of its mass over several billion years, according to some estimates.

Not only are many of the giant exoplanets extremely close to their stars, some also are in very elliptical orbits (rather than the nearly circular ones in our own system). An exoplanetary system with a massive planet on an elliptical orbit would likely create havoc for smaller planets in these systems. As a massive planet sweeps into the inner portion of a star system, it will, over time, disturb the orbits of smaller planets, either ejecting those from the system or causing them to fall into their star. Some evidence suggests this fate may have befallen planets in a few of these remote systems. A number of the stars with exoplanets are appreciably richer in iron than our Sun. One suggestion for why these stars are so iron-rich is that they have swallowed Earth-like planets and vaporized them. The iron from the vaporized planet's core then enriches the star, making its spectrum lines of iron stronger. This is not the only interpretation, however. Perhaps it is easier to make planets in the first place if a star has a higher-than-average concentration of iron. Which interpretation is correct? We do not yet know, but our rapidly expanding knowledge about exoplanetary systems is beginning to shed light on just how different they may be from our Solar System.

FIGURE 8.21
Artist's sketch of the evaporating gas giant orbiting close to the star HD-209458.

8.3 FORMATION OF PLANETARY SYSTEMS

How did the Solar System and other planetary systems form? What processes give them the features discussed in the previous sections? Our most detailed evidence comes from the Solar System, but the exoplanetary systems give us examples of how the process might vary. Our theory needs to explain why:

1. Planetary systems mostly appear flat, and in the Solar System all the planets orbit in the same direction.
2. In the Solar system there are rocky planets near the Sun and the gaseous or icy ones farther out, but some gaseous exoplanets are very close to their stars.
3. The composition of the planets and the Sun in the Solar System bear a close resemblance, differing primarily in the amount of gases present.
4. All the bodies in the Solar System whose ages have so far been determined appear to be about 4.6 billion years old.

We have listed only the major observed features that our theory must explain. There are many additional clues from the structure of asteroids, the number of craters on planetary and satellite surfaces, and the detailed chemical composition of surface rocks and atmospheres.

The modern theory for the origin of the Solar System derives from theories proposed in the eighteenth century by Immanuel Kant, the great German philosopher, and Pierre-Simon Laplace, a French mathematician. Kant and Laplace independently proposed what is now called the **solar nebula theory** that the Solar System originated from a rotating cloud (Latin *nebula*).

The solar nebula theory offers a natural explanation for the flattened shape of a planetary system as the cloud contracts under its own gravity. It explains the common direction of motion of the planets around the Sun and their compositions. There is nothing about these processes that is unique to the Solar System, so we expect to find evidence of similar processes occurring as other stars form. Therefore we can test this idea by searching for stars at various stages of this process.

Solar System builder

FIGURE 8.22
Hubble Space Telescope image of an interstellar cloud in which dust blocks background light. This may be similar to the one from which the Solar System formed.

ANIMATION ▶

Solar System formation from an interstellar cloud

Interstellar Clouds

The modern form of the solar nebula theory proposes that the Solar System was born 4.6 billion years ago from an **interstellar cloud,** an enormous rotating aggregate of gas and dust like the one shown in figure 8.22. The dust in this interstellar cloud blocks the light of background stars and glowing gas. Such clouds are common between the stars in our Galaxy, and astronomers now think all stars formed from them. Thus, although we are focusing on the birth of the Solar System, the solar nebula theory applies more broadly and implies that *most* stars could have planets, or at least disks of dust and gas surrounding them from which planets might form.

Because interstellar clouds are the raw material of the Solar System, we need to describe them more fully. Although such clouds are found in many shapes and sizes, the one that became the Sun and planets probably was a few light-years in diameter and contained at least twice the present mass of the Sun. If it was like typical clouds we see today, it was made mostly of hydrogen (71%) and helium (27%) gas, with traces of other chemical elements, such as gaseous carbon, oxygen, and silicon. In addition to the gases, interstellar clouds also contain tiny dust particles called **interstellar grains.**

Interstellar grains range in size from large molecules to micrometers or larger and are probably made of a mixture of silicates, iron compounds, carbon compounds, and water frozen into ice. Astronomers deduce the presence of these substances from their spectral lines, which are seen in starlight that has passed through dense dust clouds. Moreover, a few hardy interstellar dust grains, including tiny diamonds, have been found in ancient meteorites. This direct evidence from grains and the data from spectral lines shows that the elements occur in proportions similar to those we observe in the Sun. This is additional evidence that the Sun and its planets could have formed from an interstellar cloud.

The cloud began its transformation into the Sun and planets when the gravitational attraction between the particles in the densest parts of the cloud caused it to collapse inward, as illustrated in figure 8.23. The collapse may have been triggered by a star exploding nearby or by a collision with another cloud. But regardless of its initial cause, the infall was not directly to the center. Instead, because the cloud was rotating, it flattened. Flattening occurred because rotation retarded the collapse perpendicular to the cloud's rotation axis. A similar effect happens in an old-fashioned pizza parlor where the chef flattens the dough by tossing it into the air with a spin.

It takes million of years for an interstellar cloud to collapse and become a rotating disk called a **protoplanetary disk,** where the planets form. Gravity pulls most of

FIGURE 8.23
A sketch illustrating the collapse of an interstellar cloud to form a rapidly spinning disk. Note that the final size of the disk is not shown to scale—in actuality it would be thousands of times smaller than the cloud from which it formed.

the mass to the center, where the star forms. This explains the first obvious property of the Solar System—its flattened structure—which we noted at the beginning of this section. The protoplanetary disk that became the Solar System was probably about 200 AU in diameter and perhaps 10 AU thick. Its inner parts were hot, heated by the young Sun and the impact of gas falling on the disk during its collapse, but the outer parts were cold, far below the freezing point of water.

We are fairly certain of these dimensions and temperatures because we can observe protoplanetary disks around other stars. For example, figure 8.24 shows several examples of gas and dust disks imaged with the Hubble Space Telescope near the Orion Nebula, a star-forming cloud thought to be less than a million years old. The stars at the centers of these disks are only beginning to glow brightly enough to be visible through the dark, dusty disks that surround them. The disks are visible in silhouette against the glowing gas of the nebula.

Condensation in the Solar Nebula

Condensation occurs when a gas cools and its molecules stick together to form liquid or solid particles.* For condensation to happen, the gas must cool below a critical temperature (the value of which depends on the substance condensing and the surrounding pressure). For example, suppose we start with a cloud of vaporized iron at a temperature of 2000 K. If we cool the iron vapor to about 1300 K, tiny flakes of iron will condense from it. Likewise, if we cool a gas of silicates to about 1200 K, flakes of rocky material will condense.

At lower temperatures, other substances will condense. Water, for instance, condenses at room temperature, as you see when steam leaves a boiling kettle (fig. 8.25). Here, water molecules in the hot steam come into contact with the cooler air of the room. As the gaseous water cools, its molecules move more slowly, so that when they collide, electrical forces can bind them together, first into pairs, then into small clumps, and eventually into the tiny droplets that make up the cloud we see at the spout.

An important feature of condensation is that when a mixture of vaporized materials cools, the materials with the highest vaporization temperatures condense first. Thus, as a mixture of gaseous iron, silicate, and water cools, it will make iron grit when its temperature reaches 1300 K, silicate grit when it reaches 1200 K, and finally water droplets when it cools to only a few hundred degrees K. It is a bit like putting a jar of chicken soup in the freezer. First the fat freezes, then the broth, and finally the bits of chicken and celery. The condensation process in a protoplanetary disk stops if the temperature never drops sufficiently low. Thus, in the example above, if the temperature never cools below 500 K, water will not condense and the only solid material that forms from the gaseous mixture will be iron and silicates.

This kind of condensation sequence occurred in the solar nebula as it cooled after its collapse to a disk. Because the Sun heated the inner part of the disk, the temperature from the Sun to almost the orbit of Jupiter never dropped low enough for water and other substances with similar condensation temperatures to condense there. On the other hand, iron and silicate, which condense even at relatively high temperatures, could condense everywhere within the disk. Thus, the nebula became divided into two regions: an inner zone of silicate-iron particles, and an outer zone of similar particles on which ices also condensed, as illustrated schematically in figure 8.26A. Water, hydrogen, and other easily vaporized substances were present as gases in the inner solar nebula, but they could not form solid particles there. However, some of these substances combined chemically with silicate grains so that the rocky material from which the inner planets formed contained within it small quantities of water and other gases.

---

* Technically, condensation is the change from gas to liquid, and deposition is the change from gas to solid. However, we will not make that distinction here.

FIGURE 8.24
Young stars (the glowing red spot at the center of each image) surrounded by a dark disk of gas and dust. The dust in the disks blocks background light from glowing gas in the Orion Nebula.

500 AU   Diameter of Neptune's orbit

How does the process illustrated in figure 8.25 explain why you can see your breath on a cold morning?

FIGURE 8.25
Water vapor cools as it leaves the kettle. The cooling makes the vapor condense into tiny liquid water droplets, which we see as the "steam."

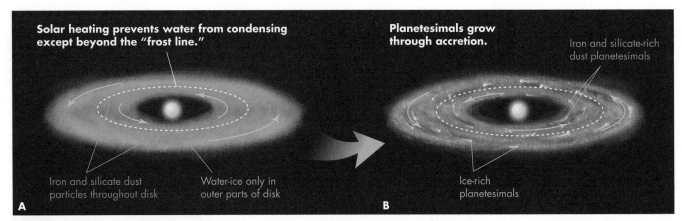

FIGURE 8.26

Depiction of the growth of planetesimals in the solar nebula. (A) Solid particles condensed in the solar nebula where the temperature was low enough, so water and other ices condensed only in outer regions beyond the frost line. (B) The particles stuck together, growing to kilometer-size planetesimals.

Accretion and Planetesimals

In the next stage of planet formation, the tiny particles that condensed from the nebula began to stick together, creating bigger pieces in a process called **accretion.** The process of accretion is a bit like building a snowman. You begin with a handful of loose snowflakes and squeeze them together to make a snowball. Then you add more snow by rolling the ball on the ground. As the ball gets bigger, it is easier for snow to stick to it, and it rapidly grows in size.

Similarly in the solar nebula, tiny grains stuck together and formed bigger grains that grew into clumps, perhaps held together by electrical forces similar to those that make lint stick to your clothes. Subsequent collisions, if not too violent, allowed these smaller particles to grow into objects ranging in size from millimeters to kilometers as illustrated in figure 8.26B. These larger objects are called **planetesimals**—that is, small, planetlike bodies.

Because the planetesimals near the Sun formed from silicate and iron particles, while those farther out were cold enough that they could incorporate ice and frozen gases as well, there were two main types of planetesimals: rocky-iron ones near the Sun and icy-rocky-iron ones farther out. This then explains the second observation we described at the beginning of this section—that there are different types of planets in the inner and outer Solar System.

Formation of the Planets

As planetesimals moved within the disk and collided with one another, planets grew. Computer simulations show that some collisions lead to the shattering of both bodies, but gentler collisions lead to merging, and the planetary orbits gradually become approximately circular. Merging of the planetesimals increased their mass and thus their gravitational attraction. That, in turn, helped them grow even more massive by drawing planetesimals into clumps or rings orbiting the Sun. Within these clumps, growth went even faster, so that over several million years, larger and larger objects formed.

In a few cases we can see indirect evidence of planet formation in a protoplanetary disk. Figure 8.27 shows Hubble Space Telescope images of material surrounding three stars with ages ranging from about 10 to 400 million years. The rings of material seen around the stars are perhaps similar to the Solar System's Kuiper belt. Material in the inner part of the disk has probably already accreted into planets or planetesimals, leaving a "hole" at the center (fig. 8.27 A and C). In figure 8.27B, we see an edge-on disk and a secondary disk at a slightly different angle. Astronomical models suggest that the best explanation for the secondary disk is that there is a planet orbiting at that angle and disturbing the debris.

FIGURE 8.27

Disks of dust and perhaps larger debris around three stars. The stars were observed with the Hubble Space Telescope, but the light from the central region was blocked out in order to see the extremely faint light reflected from the disks. (A) HR 4796A is estimated to be less than 10 million years old, and it is surrounded by a ring about the size of the Kuiper belt. (B) Beta Pictoris is somewhat older, perhaps as much as 20 million years old, and is surrounded by a disk seen nearly edge-on to us. A secondary disk around Beta Pictoris probably indicates the presence of a planet orbiting at a slight angle to the primary disk. (C) Fomalhaut is much older at more than 400 million years, and it is surrounded by a large ring of dust and debris. Astronomers believed they had detected a planet orbiting just inside the ring, but subsequent observations suggest it is an orbiting cloud of debris.

Planetesimal growth was especially rapid beyond 4 or 5 AU from the Sun. Planetesimals there had more material from which to grow, because the ices that could condense there are about 10 times more abundant than the silicate and iron compounds that were the only materials condensing in the inner Solar System.

Additionally, once a planet grew somewhat larger than the diameter and mass of the Earth, it was able to attract and retain gas by its own gravity. Because hydrogen was overwhelmingly the most abundant material in the solar nebula, planets large enough to tap that reservoir could grow vastly larger than those that formed only from solid material. In particular, Jupiter and Saturn may have begun as Earth-size bodies of ice and rock, but their gravitational attraction resulted in their becoming surrounded by the huge envelopes of hydrogen-rich gases that we see today. As discussed in the Extending Our Reach box below, some scientists think that there may have been enough gas for Jupiter to have formed directly from the gas, skipping the planetesimal phase.

The ice giants Neptune and Uranus have compositions that can mostly be explained by the accumulation of icy planetesimals that heated up as they collided to create their large atmospheres and liquid interiors. They may also have attracted gas from the surrounding nebula, but the addition of major amounts of hydrogen and helium is not required to explain the properties of ice giants. The smaller and warmer bodies of the inner Solar System could not capture hydrogen and therefore remained small and lack

EXTENDING
our reach   DIRECT FORMATION OF GAS GIANTS

Because astronomers have no direct way to observe how the Solar System formed, they rely heavily on computer simulations to study that remote time. Computer simulations try to solve Newton's laws of motion for the complex mix of dust and gas that we believe made up the solar nebula. The solutions then can reveal what might have happened as the dust particles stuck together to form planetesimals and how the planetesimals then drew together under the influence of their gravity to form planets.

One of the more interesting findings of such calculations is that Jupiter may have formed directly from

slightly denser regions of gas in the disk. Far from the Sun, where the gas is cold, gravity can more easily overcome warmer gas's resistance to being squeezed into a smaller region. (Think of how a balloon resists being squeezed.) This may have allowed gravity to pull gas together to make a giant planet without the need to first form cores from planetesimals.

Does this make the planetesimal theory wrong? No, just incomplete. Moreover, because this is an area of active research, astronomers are still searching for other evidence and performing more-detailed simulations.

that gas. This explains the third observation we mentioned at the beginning of this section—that the outer planets have a composition similar to the Sun's.

An important question raised by exoplanet discoveries is how there can be so many giant planets close to their stars. Perhaps they are objects that grew massive enough to overcome the higher temperatures close to the star and accumulated gas from the nebula. Or they may have formed far from the star, and their orbits may have been altered by gravitational interactions with other planets and the protoplanetary disk. The latter explanation is appealing because it may help to explain some additional features of our Solar System that we will discuss at the end of this chapter.

As planetesimals struck the growing planets, their impact released gravitational energy that heated both the planetesimal and the planet. Gravitational energy is liberated whenever something falls. For example, when a cinder block falls onto a box of tennis balls, the impact scatters the balls in all directions, giving them kinetic energy—energy of motion. In much the same manner, planetesimals falling onto a planet's surface give energy to the atoms in the crustal layers, energy that appears as heating.

You can easily demonstrate that motion can generate heat by hitting a steel nail a dozen or so times with a hammer and then touching the nail: the metal will feel distinctly hot. Imagine now the vastly greater heating created as mountain-size masses of rock plummet onto a planet. The heat so liberated, in combination with radioactive heating, melted the planets and allowed matter with high density (such as iron) to sink to their cores, while matter with lower density (such as silicate rock) "floated" to their surfaces. We saw in chapter 6 that the Earth's iron core probably formed by this process, and astronomers think that the other terrestrial planets formed their iron cores and rocky crusts and mantles the same way. A similar process probably occurred for the outer planets when rock and iron material sank to their cores.

Final Stages of Planet Formation

As the planetesimals were used up, the surfaces of the terrestrial bodies and icy bodies in the outer Solar System began to cool and solidify. There was still a steady rain of planetesimals, but the rate of collisions was slow enough that the surfaces could solidify between impacts (fig. 8.28). The record of these collisions is preserved on many of the solid surfaces in the Solar System where there has not been geological activity or erosion to erase them. Photographs of the surface of Mercury and one of the moons of Saturn in figure 8.29 show surfaces that are peppered with craters, demonstrating that bodies in both the inner and the outer Solar System were brutally battered by these violent collisions.

As more and more large bodies built up from smaller planetesimals, there were occasional impacts between bodies so large that they did not just leave a crater. For example, we saw in chapter 7 that the Moon was probably formed when the Earth

FIGURE 8.28

Final stages of planet formation. (A) Gravitational attraction between planetesimals causes them to grow in size, although many are left behind in the asteroid belt and the Kuiper belt. (B) Planet-sized bodies "sweep up" most of the remaining material orbiting at their distance. The high-speed collisions heat and crater the surface of the body.

FIGURE 8.29
Pictures taken by spacecraft showing craters on Mercury and Saturn's moon Dione.

was struck by a Mars-size body. Likewise, as we will discuss in chapters 9 and 10, Mercury may have suffered a massive impact that blasted away much of its crust, and the peculiar rotation of Uranus and Venus may also have arisen from major planetesimal collisions.

Although planet building consumed most of the planetesimals, some survived to form dwarf planets, small moons, the asteroids, and comets. Rocky planetesimals and their fragments remained between Mars and Jupiter, where, stirred by Jupiter's gravitational force, they were unable to assemble into a planet. We see them today as the asteroid belt.

Formation of Atmospheres

Atmospheres were the last part of the planet-forming process. The formation of an atmosphere proceeded quite differently in the inner and outer parts of the Solar System, which helps to explain the very different atmospheric compositions we find. As we discussed earlier, the extensive atmospheres of the outer planets probably grew through the process of planetesimal collisions and gravitational attraction of gas remaining in the protoplanetary disk. Some of the smaller icy satellites and dwarf planets in the outer Solar System also have atmospheres, and these are thought to originate from heating of the ices out of which these bodies formed.

The inner Solar System is quite different because the planets were not massive enough to capture gas from the solar nebula, so they are deficient in hydrogen, helium, and other substances that could not condense so close to the Sun.

Some ices were probably incorporated in the planetesimals that built the inner planets, so as the planets heated and differentiated, they probably had an early atmosphere spewed out in volcanic eruptions. The eccentric orbits of some asteroids and comets could have carried in more ices, which would vaporize upon impact. And if this happened after the planet's surface had cooled, the planet would have a better chance of retaining the gas. In fact, as a general rule, bodies too small to have captured atmospheres directly but that show clear signs of extensive volcanic activity (now or in the past) have atmospheres. More quiescent ones do not. Moreover, small bodies such as Mercury and our Moon keep essentially no atmosphere at all because their weak gravitational force means that their escape velocity is rather small, and atmospheric gases tend to escape easily from them.

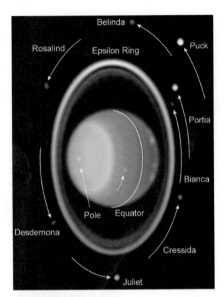

FIGURE 8.30

Hubble Space Telescope false-color infrared image of Uranus. The rings and satellites orbit around Uranus's equator even though the planet is tilted almost perpendicular to the plane of the Solar System, implying that these bodies all formed in place from a disk of material around the planet as it formed, like a miniature Solar System.

Formation of Satellite Systems

All four giant planets have flattened satellite systems in which the larger satellites (with few exceptions) orbit in the same direction as the planet spins. Several of these satellites are about as large as Mercury, and they would be considered full-fledged planets were they orbiting the Sun along an isolated orbit. A few of these bodies even have atmospheres, but they have too little mass (and thus too weak a gravitational attraction) to have accumulated large quantities of hydrogen and other gases from the solar nebula as their parent planets did. Thus, these moons are composed mainly of rock and ice, giving them solid surfaces, many of which record heavy cratering from the early history of the Solar System. Not all of them preserve this record, because volcanic activity and processes unique to their icy composition have sometimes melted their surfaces, erasing the craters.

The large systems of satellites around the outer planets probably were formed from the same set of planetesimals that was building the planets themselves. Once a body grew massive enough that its gravitational force could draw in additional material, it became surrounded with debris that settled into a flat disk or rings. Thus, moon formation was a scaled-down version of planet formation, and so the satellites of the outer planets have the same regularities as the planets around the Sun. This is particularly striking around Uranus, which has a system of rings and moons that orbit around its equator despite its nearly perpendicular tilt to the ecliptic (fig. 8.30). For the satellites to have ended up in orbit with this orientation, they must have formed while the planet itself was forming.

Cleaning Up the Solar System

Only a few million years were needed to assemble most of the mass of the planets from the solar nebula, though the rain of infalling planetesimals lasted several hundred million years. Such a time is long in the human time frame but short in the Solar System's. All the objects within the Solar System are about the same age—the fourth property of the Solar System mentioned at the beginning of this section.

One process still had to occur before the Solar System became what we see today: the residual gas and dust must have been removed. Just as a finished house is swept clean of the debris of construction, so too was the Solar System. In the sweeping process, the Sun was probably the cosmic broom, with its intense heat driving a flow of tenuous gas outward from its atmosphere. As that flow impinges on the remnant gas and dust around the Sun, the debris is pushed away from the Sun to the fringes of the Solar System. Such gas flows are seen in most young stars, and astronomers are confident the Sun was no exception. Even today some gas flows out from the Sun, but in youthful stars the flow is much more vigorous.

Migrating Planets and the Late Heavy Bombardment

Up to this point our exploration of the formation of the Solar System has followed the basic idea of the solar nebula theory, although obviously incorporating a great amount of detail that Kant and Laplace could not have known about. The theory is highly successful in explaining much about our Solar System and other planetary systems. However, some details are not well explained by the theory.

Until recently, most astronomers assumed that the Solar System's planets move along orbits that lie close to where the planets originally formed. But the discovery of ice and gas giants orbiting close to their star has led astronomers to examine whether planets may form at one distance from a star and then "migrate" to a new distance. Although we cannot watch real planets changing orbits over millions of years, computer simulations (fig. 8.31) suggest that in fact it should be common for a young planet to interact with the material in the protoplanetary disk in ways that will lead to a changing

FIGURE 8.31

Sketch of a large young planet migrating inward due to gravitational torques between it and the protoplanetary disk.

orbit. This has in turn inspired new studies of the young Solar System that shed light on some previously puzzling results.

The simulations show that interactions between a young planet and leftover material in the protoplanetary disk can gradually shift the planets' orbits either inward or outward, depending on whether there is more material orbiting inside or outside the planet's orbit. We saw earlier that Jupiter's gravity has interacted with the asteroid belt to keep a planet from forming there. Those same interactions slowly shift Jupiter's orbit inward. If the asteroid belt were much more massive, Jupiter could actually be drawn inward, "migrating" to the inner Solar System. This could explain the presence of giant exoplanets close to their stars.

Astronomers have attempted to model conditions in the young Solar System to see how this process might have operated when there was still much more debris leftover in the protoplanetary disk. According to one model (fig. 8.32) Jupiter shifted inward by less than 1 AU over several hundred million years. Meanwhile, Neptune shifted outward. It formed less than 20 AU from the Sun and as it migrated outward, it passed close to Uranus and underwent a dramatic gravitational encounter that sent it out to its present distance of about 30 AU. As the planets migrate, their gravitational interactions with small planetesimals send them careening into new orbits. The orbits of objects in the Kuiper belt are well explained by Neptune's flinging them outward in this model.

Planet migration may help explain a peculiarity found in the lunar rocks retrieved by Apollo astronauts. Based on the rocks' ages, it appears that about 600 million years after the Moon formed there was a second wave of cratering on the Moon, known as the **late heavy bombardment.** This is difficult to explain in the solar nebula theory, but planet migration can produce a delayed event like this. Planet migration may also help explain how the terrestrial planets gained water when comets were sent on new orbits, as well as through Jupiter's interactions with ice-rich asteroids in the outer asteroid belt.

Migration of planets has important consequences for planetary systems. For example, if a giant planet migrates inward toward its star, it might capture or fling smaller, Earth-size planets into wild new orbits as it passes them. Thus, small planets, suitable for life as we know it, may form but fail to survive in such systems.

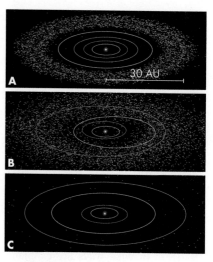

FIGURE 8.32

Illustration of a possible evolution of the young Solar System, based on a simulation carried out by astronomers in Nice, France. (A) Shortly after the major planet formed in a compact configuration, Neptune orbited closer than Uranus. (B) After about 800 million years, the orbits grow quite elliptical. (C) In another 100 million years, the orbits recircularize as they scatter most of the remaining disk material into collisions with the planets or into the far outer reaches of the Solar System.

SUMMARY

The Solar System consists of a star (the Sun) and planets, asteroids, and comets, which orbit it in a broad, flat disk. All the planets circle the Sun in the same direction, and most of them spin in the same direction. Their moons also form flattened systems, generally orbiting in the same direction. The planets fall into two main categories: small, high-density bodies (the inner, or terrestrial, planets) and large, low-density bodies (the outer, or Jovian, planets). The former are rich in rock and iron; the latter are rich in hydrogen and ice.

Astronomers have found many planets orbiting other stars. Study of these exoplanets helps us better understand the origin of planetary systems. Many of the systems found exhibit major differences from the Solar System, but current detection methods do not work well for exoplanets orbiting more than about 1 AU from their stars. Nevertheless, the results indicate that most stars probably have planets, and planets with sizes between Earth and Neptune are very common.

The features of most planetary systems can be explained by the solar nebula theory. In this theory, the Solar System was born from a cloud of interstellar gas that collapsed to a disk called the solar nebula. The center of the nebula became the Sun, and the disk became the planets. This explains the compositional similarities and the common age of the bodies.

The flat shape of the system and the common direction of motion around the Sun arose because the planets condensed within the nebula's rotating disk. Planet growth occurred in two stages: dust condensed and clumped to form planetesimals, and then later the planetesimals aggregated to form planets and satellites. Two kinds of planets formed because lighter gases and ice could condense easily in the cold outer parts of the nebula but only rocky and metallic material could condense in the hot inner parts. Impacts of surviving planetesimals late in the formation stages cratered the surfaces and may have tilted the rotation axes of some planets. Migration of planets over hundreds of millions of years may explain giant exoplanets near stars as well as orbits in the Kuiper disk and a period of heavy bombardment of planetary surfaces after the Solar System formed.

QUESTIONS FOR REVIEW

1. (8.1) Name the eight planets in order of increasing distance from the Sun. Which are inner and outer planets?
2. (8.1) What is Pluto, and why isn't it a planet?
3. (8.1) Where are the asteroid belt, the Kuiper belt, and the Oort cloud? What kind of objects are in or come from them?
4. (8.1) Make a sketch of the Solar System showing top and side views.
5. (8.1) What is Bode's rule?
6. (8.1) How do we know the composition of Jupiter?
7. (8.1) What properties, apart from position, distinguish the terrestrial and Jovian planets?
8. (8.1) How old is the Solar System? How do we know?
9. (8.2) In what ways are exoplanet systems similar to the Solar System? In what ways different?
10. (8.2) What methods are used to find exoplanets? What are their limitations?
11. (8.2) How do some exoplanets differ from what we have found in the Solar System? How common are exoplanets?
12. (8.2) What do we know about exoplanets' compositions?
13. (8.3) What is an interstellar cloud? What does it have to do with the Solar System?
14. (8.3) What is the solar nebula? What is its shape and why?
15. (8.3) Why are there two main types of planets?
16. (8.3) What is the difference between condensation and accretion?
17. (8.3) Describe the planetesimal theory of planet formation.
18. (8.3) How does the planetesimal theory of planet formation explain the asteroids?
19. (8.3) How did moons form around outer planets?
20. (8.3) How did the craters we see on many of the planets form?
21. (8.3) Describe a theory of how planets may have formed their atmospheres.
22. (8.3) What is planet migration, and how might it relate to the late heavy bombardment?

THOUGHT QUESTIONS

1. (8.1) Make arguments supporting the rules adopted for defining planets and dwarf planets, or create and justify your own set of rules.
2. (8.2) How are the kinds of exoplanets found by the Doppler method a biased sample of exoplanets? Give an example of a survey method that might give a biased result in everyday life.
3. (8.2) How do some exoplanets differ from what we might expect? Does this prove the nebula theory is wrong?
4. (8.3) What kinds of physics would be important to include in a computer simulation of solar system formation?

PROBLEMS

1. (8.1) (a) By what factor would the Sun be shrunk to be the size of a large beach ball, 1 meter in diameter? (b) Calculate the distances and diameters of Mercury, Earth, Ceres, Jupiter, Neptune, and Pluto if the whole Solar System were shrunk by the same amount. (b) Find their masses if their densities stayed the same.
2. (8.1) Calculate the densities of Venus and Jupiter (use the masses and radii given in the appendix). How do these numbers compare with the density of rock (about 3 grams per cm^3) and water (1 gram per cm^3)? (Note: Be sure to convert kilometers to centimeters and kilograms to grams if you are expressing your answer in grams per cm^3.)
3. (8.1/3.8) Look up the masses and radii of Mercury and Jupiter and calculate their escape velocities, using the equation in chapter 3. Does this help you see why the one body has an atmosphere but the other doesn't? (Note: Be sure to convert kilometers to meters as necessary.)
4. (8.1/3.8) Look up the masses and radii of Neptune and Mars and calculate their escape velocities, using the equation in chapter 3. Compare both with that of the Earth (see section 3.8). How do the atmospheres of these three planets differ? (Note: Be sure to convert kilometers to meters or the appropriate unit.)
5. (8.2/2.3) Kepler 30b is an exoplanet orbiting a star of about 1 solar mass. Its orbital period is nearly 29 days. Calculate the semimajor axis of its orbit in AU. (Use Kepler's laws.)
6. (8.2) Calculate the maximum Doppler shift that could be observed for the planet in question 5.
7. (8.2) Using the modified form of Kepler's laws given in figure 8.8, calculate the orbital period for Gliese 851d, an exoplanet with a mass about 7.1 times the mass of Earth. The star Gliese 851 is a red dwarf with a mass of 0.31 solar masses, and the planet orbits with a semimajor axis of 0.22 AU. (Remember to convert distances to meters and masses to kilograms when using the equation.)
8. (8.2) Imagine an alien is detecting the Earth as it transits our Sun. Compute the ratio of the areas of the Earth's disk and the Sun's to roughly estimate what percentage of the Sun's light the Earth blocks mid-transit. What percentage would Jupiter block?

TEST YOURSELF

1. (8.1) Which of the following objects are primarily rocky with iron cores?
   (a) Venus, Jupiter, and Neptune
   (b) Mercury, Venus, and Pluto
   (c) Mercury, Venus, and Earth
   (d) Jupiter, Uranus, and Neptune
   (e) Mercury, Saturn, and Eris

2. (8.1) Which of the following best describes the planets' spins?
   (a) All spin counterclockwise.
   (b) Very few spin at all.
   (c) The spins often reverse.
   (d) Most spin counterclockwise.
   (e) The spins are random.

3. (8.2) The Doppler-shift method for detecting the presence of exoplanets is best able to detect
   (a) massive planets near the star.
   (b) massive planets far from the star.
   (c) low-mass planets near the star.
   (d) low-mass planets far from the star.

4. (8.2) The transit method for detecting exoplanets works best for
   (a) very massive planets.
   (b) solar systems seen face-on.
   (c) planets very far from their stars
   (d) solar systems seen edge-on.
   (e) planets very close to their stars.

5. (8.3) Which of the following features of the Solar System does the solar nebula theory explain?
   (a) All the planets orbit the Sun in the same direction.
   (b) All the planets move in orbits that lie in nearly the same plane.
   (c) The planets nearest the Sun contain only small amounts of substances that condense at low temperatures.
   (d) All the planets and the Sun, to the extent that we know, are the same age.
   (e) All of the above

6. (8.3) The numerous craters we see on the solid surfaces of so many Solar System bodies are evidence that
   (a) they were so hot in their youth that volcanos were widespread.
   (b) the Sun was so hot that it melted all these bodies and made them boil.
   (c) these bodies were originally a mix of water and rock. As the young Sun heated up, the water boiled, creating hollow pockets in the rock.
   (d) they were bombarded in their youth by many solid objects.
   (e) all the planets were once part of a single, very large, and volcanically active mass that subsequently broke into many smaller pieces.

7. (8.3) Suppose a number of planets all have the same mass but different sizes and temperatures. Which of the following planets is most likely to retain a thick atmosphere?
   (a) Small, hot   (c) Large, hot
   (b) Small, cool   (d) Large, cool

KEY TERMS

accretion, 214	interstellar grain, 212
asteroid, 199	Jovian planet, 198
asteroid belt, 199	Kuiper belt, 199
Bode's rule, 201	late heavy bombardment, 219
comet, 199	Oort cloud, 199
condensation, 213	outer planet, 198
dwarf planet, 199	planetesimal, 214
exoplanet, 205	protoplanetary disk, 212
gas giant, 204	solar nebula theory, 211
gravitational lensing, 207	Solar System, 197
ice giant, 204	super-earth, 208
inner planet, 198	terrestrial planet, 198
interstellar cloud, 212	transit, 208

Q: FIGURE QUESTION ANSWERS

WHAT IS THIS? (chapter opening): This is a disk of gas and dust around a forming star. The dark disk is visible in silhouette against the glow of emission from the Orion nebula. The forming star glows red at its center.

FIGURE 8.8: The density is the mass (30 grams) divided by the volume (10 cm³). 30 grams/10 cm³ = 3 grams/cm³. Iron's density is about 8 gm/cm³, while a typical silicate rock's density is about 3 gm/cm³. It is thus more likely to be rock.

FIGURE 8.25: When you breathe out, warm moist air from your lungs comes in contact with the cold air outside. The moisture in your breath condenses and makes a tiny "cloud."

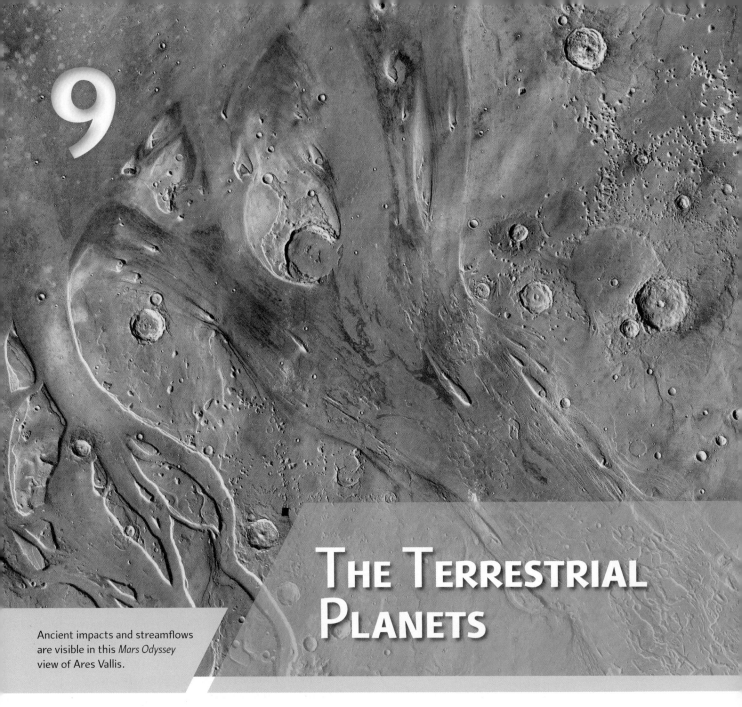

9

THE TERRESTRIAL PLANETS

Ancient impacts and streamflows are visible in this *Mars Odyssey* view of Ares Vallis.

LEARNING OBJECTIVES

Upon completing this chapter you should be able to:

- Describe and contrast the internal structure of the three other terrestrial planets with the Earth's.
- Describe the main surface features of Mercury and compare them to the surface of our Moon.
- Explain how Mercury's orbit and rotation are interlinked.
- Compare the landforms of Venus and the Earth and discuss hypotheses about why they differ.
- Describe the runaway greenhouse model and how the process is thought to have begun on Venus.
- Discuss the evidence that implies Venus's surface is about as young as the Earth's.
- Describe the major surface features present on Mars.

- Discuss the features that indicate Mars had geologic activity and how it differs from the activity on Earth or Venus.
- Describe the variety of evidence that implies Mars once had liquid water on its surface but does not at present.
- Discuss the evidence for frozen water on Mars at present and where it is located.
- Recall the history of the searches for life on Mars and the controversies about different claims for its existence.
- Describe the Martian moons and their probable origin.
- Discuss how the terrestrial planets' climates vary with distance from the Sun and orbital and rotational differences.
- Describe the effects of water and life on the atmosphere of a terrestrial planet.

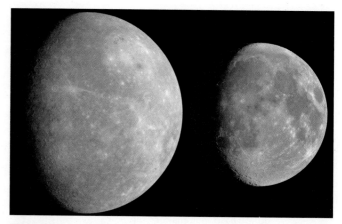
Terrestrial planets, as the name suggests, have a size and structure similar to Earth's. Within our Solar System, the planets Mercury, Venus, Earth, and Mars are terrestrial planets. Orbiting in the inner part of the Solar System, close to the Sun, these rocky worlds are too small and too warm to have captured massive gaseous envelopes like those that cloak the outer planets.

Despite their similarity in size and composition, the terrestrial planets differ greatly in their surface conditions. Our goal in this chapter is to better understand how these neighboring planets came to be so different from one another. We will discover that size plays a major role. For example, because Mercury is so small, it generates little internal heat to create surface activity, and so its crust is essentially unchanged from its birth. Venus, on the other hand, is large enough to have held in the heat from its formation and from the decay of the radioactive elements that are present in every planet. Its hot interior is rather like Earth's, and so it has a surface with mountains and volcanic peaks, but the character of those mountains is surprisingly different.

Size, coupled with distance from the Sun, creates the great atmospheric differences among these terrestrial worlds. For example, Mercury is too small, and its surface is too hot, to retain an atmosphere, while Mars, only slightly larger but farther from the Sun and therefore cooler, has retained one. Venus and Earth are both large enough to have sizable atmospheres, but Earth's slightly greater distance from the Sun has made it cool enough to have liquid water in its atmosphere. That simple fact has led to the profound difference between the atmospheres of Earth and Venus, because liquid water can remove carbon dioxide from the atmosphere. Moreover, liquid water has allowed life to form and flourish here, and life has not only removed additional carbon dioxide from our air but has also added oxygen.

Our study of the terrestrial planets provides us with clues about the nature of small exoplanets and perhaps allows us to extrapolate what super-earths may be like. Differences between the terrestrial planets can also tell us how our planet might change in the future as we alter its atmosphere and as our star evolves.

> **CONCEPTS AND SKILLS TO REVIEW**
>
> - Density (6.1)
> - Absorption in Earth's atmosphere (4.5)

9.1 MERCURY

Mercury is the smallest terrestrial planet and closest to the Sun. It is named for the Roman deity who was the speedy messenger of the gods, probably because it changes its position on the sky faster than any other planet, switching from morning to evening skies and back every few months. Relatively little was known about Mercury before 1974 when *Mariner 10* made its first flyby. *Mariner 10* photographed only one side of the planet, and it was not until more than 30 years later that the *Messenger* spacecraft provided a much more detailed view of all of Mercury's surface. These spacecraft revealed that Mercury resembles our Moon in both size and appearance (fig. 9.1), although Mercury lacks the dark maria. The flybys also allowed astronomers to make accurate measurements of the planet's mass from its gravitational pull on the spacecraft. Mercury's radius is about one-third and its mass about 1/28th that of the Earth.

FIGURE 9.1
Mercury (left) and the Moon (right) shown to the same scale.

FIGURE 9.2

Two views of Mercury from the *Messenger* spacecraft. (A) Some regions show a long history of overlapping craters. A recent impact produced a bright crater and splashed material outward to make rays across older craters. (B) A heavily cratered region in the foreground gives way to a smoother area in the background flooded by volcanic flows, somewhat like a lunar mare.

The Surface of Mercury

Mercury's surface is covered with craters and rays (fig. 9.2A), but it is not totally Moonlike. Congealed lava flows flood many of Mercury's old craters and pave much of its surface (fig. 9.2B). On our Moon, such flows are found almost exclusively within the maria, which contrast markedly with the heavily cratered terrain.

Among the more unusual features on Mercury are enormous **scarps**—cliffs formed where the crust has shifted—that run for hundreds of kilometers across Mercury's surface, as seen in figure 9.3A. Similar features have been found on the Moon, but they are not nearly as prominent or extensive. The scarps may have formed as the planet cooled and shrank, the crust wrinkling like the skin of a dried apple. *Messenger* has also identified some evidence of relatively recent volcanic activity (fig. 9.3B).

The largest crater on Mercury by far is the vast Caloris Basin, shown in figure 9.4. With a diameter of 1550 kilometers (about 950 miles), this mountain-ringed feature

FIGURE 9.3

(A) *Mariner 10* imaged many scarps. This scarp cuts across several older craters. (B) *Messenger* imaged this peculiar light-colored region from two angles. The irregularly shaped depression is thought to be a volcanic vent from which the lighter-colored material erupted.

Q: Which formed first, the scarp or the craters it passes through? On what do you base your answer?

FIGURE 9.4

(A) The edge of the Caloris Basin is seen in the semicircular set of rings surrounding Mercury's largest impact feature off the left side of this *Mariner 10* image. (B) The Caloris Basin appears light orange in this false-color image made by *Messenger,* enhancing color differences to show different mineral compositions. (C) The center of the Caloris Basin has a strange spidery pattern of troughs surrounding a 40-kilometer-diameter crater near the center of the basin.

Q: Does the Caloris Basin look like a lunar mare? Compare figures 9.4A and 7.4B, for example. How are they similar? How are they different?

is reminiscent of lunar maria. Moreover, its circular shape and surrounding hills indicate that, like the maria, it was formed by impact. The impact spawned volcanoes around the edge of the basin, several of which are visible as bright orange spots in the false-color image shown in figure 9.4B. There is also a peculiar set of radial cracks (fig. 9.4C) near the center of the basin that are still not well understood.

In a global topographic map of Mercury (fig. 9.5), the Caloris Basin does not stand out as a low region. The basin appears to have been largely filled by lava flows, and additional uplift occurred later, raising portions of its center—which may explain some

FIGURE 9.5

Topographic map of Mercury.

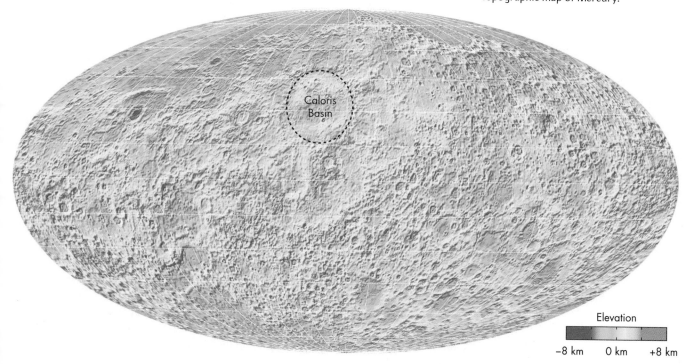

Caloris Basin

Elevation

−8 km   0 km   +8 km

FIGURE 9.6
Image of the chaotic terrain that lies on the side of Mercury opposite the Caloris Basin.

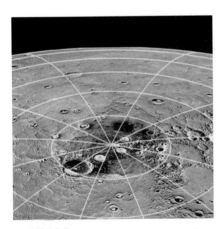

FIGURE 9.7
Mercury's north pole as imaged by *Messenger*. Yellow regions in craters show where the Arecibo radio telescope mapped probable ice, based on radar reflectivity.

of the cracking and troughs seen in its central volcanic fields. In fact the differences in elevation overall on Mercury are much less than on the Moon. This may be explained by Mercury's stronger gravity and hotter, more-molten interior, which allowed the planet to be pulled into a more spherical shape. The elevation differences are also less than on Earth, probably because while there was some geologic shifting, extensive plate tectonic motions never developed.

By counting craters within the Caloris Basin and comparing the number to other regions of Mercury, it appears that the basin formed during a collision about 3.8 billion years ago. This is about the time when some of the last major collisions occurred on the Moon, producing somewhat similar-looking features such as Mare Orientale (see fig. 7.4B).

Mercury also possesses the curious landscape illustrated in figure 9.6, which lies on the far side of the planet exactly opposite the Caloris Basin. Astronomers call this hummocky, jumbled surface "chaotic terrain" and think it was churned up by earthquake waves generated by the impact that created the basin. As the waves traveled around Mercury, they converged on its far side, heaving up the rock, much as dropping cream into coffee creates a tiny splash as the ripples reconverge.

Mercury's Temperature and Atmosphere

Mercury's surface is one of the hottest and coldest places in the Solar System. At its equator, noon temperatures reach approximately 710 K (about 820°F). On the other hand, nighttime temperatures drop to approximately 80 K (about −320°F). These extremes result from Mercury's closeness to the Sun, its slow rotation, and its lack of atmosphere. The slow rotation allows the surface to heat up strongly during the Mercurian day and to cool down greatly at night. The absence of an atmosphere means that nothing moderates the inflow of sunlight during the day or retains heat at night.

Mercury has almost no atmosphere for the same reason as the Moon. Mercury's small mass makes its gravitational attraction too small to retain much gas around it. Moreover, its high temperature causes molecules to move so fast that they escape into space. Traces of gas have been detected spectroscopically. Some of this gas is probably captured temporarily from gas flowing away from the Sun, but some appears to be escaping from the interior of the planet. The latter is indicated by higher gas levels in the vicinity of the Caloris Basin and the chaotic terrain, where the impact that formed these features fractured the crust and would have provided pathways for gas to escape.

Radar observations from Earth in 1992 detected spots in the polar regions with high reflectivity, possibly indicating ice (fig. 9.7). The presence of ice was confirmed by *Messenger* through studies of neutron emission from surface layers, which is strongly affected by the presence of the hydrogen atoms, most likely found in ice. This is surprising, because Mercury formed so close to the Sun that no ice would have condensed before the planet formed, and liquid water could never have survived on the surface. Where, then, did the ice at its poles come from?

In Mercury's polar regions sunlight shines at such a shallow angle that there is very little heating, and there is no atmosphere to transfer heat from the equatorial regions. Astronomers suspect the ice came from comets that struck the surface and vaporized, or volcanic outgassing. The gases released, which would include water vapor, are destined to gradually escape to space. However, while Mercury's gravity still hung onto them, they might drift toward the cold polar regions and freeze there, much as frost condenses on automobile windshields on a subfreezing morning. Over billions of years, frost deposits might have built up the ice seen on this otherwise hot planet.

Mercury's Interior

Mercury probably has an iron core beneath its silicate crust and mantle, but astronomers have little proof because no spacecraft has landed there to deploy seismic (earthquake) detectors. Their conclusion is therefore based on Mercury's density and gravitational field. A massive planet's gravity can compress its interior to high density, but Mercury is too small for this effect to be significant. Thus, its high density (5.4 grams per cm³) indicates an iron-rich interior with only a thin rock (silicate) mantle, as depicted in figure 9.8.

Why Mercury is so relatively rich in iron but poor in silicates is unclear. One possibility is that silicates did not condense as easily as iron compounds in the hot, inner solar nebula where Mercury formed. Another possibility is that Mercury once had a thicker rocky crust, which was blasted off by the impact of an enormous planetesimal, as the computer simulation in figure 9.9 illustrates. We saw in section 7.4 that a similar collision may have happened to Earth and formed our Moon.

Astronomers had not expected Mercury's core to be molten, because Mercury is not much bigger than the Moon, whose interior appears to have solidified relatively early in its history. A small radius allows heat from the interior to escape more readily. However, Mercury has at least a partially liquid (molten) core like the Earth. The existence of a liquid core was shown in 2007 by the way Mercury "wobbles." A similar effect can be seen by spinning hard-boiled and raw eggs on a tabletop. When you stop the hard-boiled egg from spinning, it remains stopped, but if you briefly stop the raw egg then let it go, it will begin spinning again. This is because the liquid material in its interior continues to rotate even after the shell is stopped. Precise measurements of the way Mercury's rotation responds to gravitational tugs by the Sun show a similar effect.

Mercury's magnetic field is only about 1% as strong as Earth's. As we discussed in chapter 6, the Earth's magnetic field is probably generated by circulating motions in the Earth's molten iron core combined with our planet's spin. Given that Mercury has a molten core, the weakness of its magnetic field is probably a consequence of the slowness of the planet's rotation.

FIGURE 9.8
Artist's depiction of Mercury's interior.

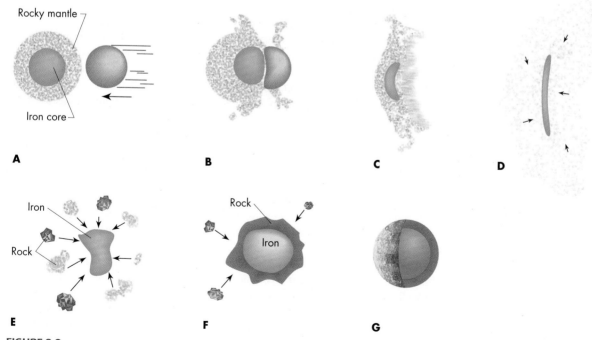

FIGURE 9.9
Computer simulation of a collision between Mercury and a large planetesimal. The impact strips away most of the outer rocky layers and leaves a highly distorted iron core surrounded by rocky debris. Gravity eventually reshapes the planet into a sphere.

FIGURE 9.10
Mercury's odd rotation. The planet spins three times for each two orbits it makes around the Sun.

: What resemblance do you see between Mercury's motions and those of our Moon?

ANIMATION ▶

The rotation of Mercury during 2 Mercury years

Mercury's Rotation

Mercury spins very slowly. Its rotation period is 58.646 Earth days, which is exactly two-thirds of its orbital period around the Sun of 87.969 Earth days. This means that it spins exactly three times for each two trips it makes around the Sun.

This unusual relationship between Mercury's rotation and revolution is almost certainly caused by the Sun's tidal forces. We might have expected the Sun to make Mercury rotate with the same face toward it all the time, just as the Earth has affected the Moon's spin. Mercury's orbit, however, is very elliptical. Thus, in accordance with Kepler's second law of planetary motion, Mercury's orbital speed changes as it moves around the Sun. Because of that changing speed, the Sun cannot lock Mercury into a purely synchronous spin; the closest to synchrony it can get is three spins for each two orbits (fig. 9.10). Such an integer ratio of periods is called a *resonance*.

Resonance occurs when a force that acts repeatedly on a body causes its motion to grow ever larger. For example, pushing a child on a swing can create a resonance. If you push just as the swing starts to move forward, the child will swing higher and higher, and the pushing force will be in resonance with the motion of the swing. On the other hand, if you push before the backward motion is stopped, the swinging motion will decrease, and no resonance occurs. Likewise, applying power resonantly to a car stuck in a ditch may "rock" it out. A similar resonance exists between the Sun's changing gravitational tug on Mercury as it moves along its elongated orbit and its rotation. The result is the 2:3 relation between its orbital and spin periods.

Mercury's odd rotation gives it an extremely long solar day (the time between successive sunrises) of 176 Earth days. During that time, the Sun sometimes changes its direction of motion across the sky. For example, if sunset occurs when Mercury is at the point of its orbit nearest the Sun, the Sun will set and then briefly rise again before setting a second time!

Venus is named for the Roman goddess of beauty and love, perhaps because it is the brightest object in the sky after the Sun and Moon, and it is almost always seen near dawn or dusk. Venus is so bright that if the air is very clear, you can sometimes find Venus in broad daylight. Of all the planets, Venus is most like the Earth in diameter and mass. We might therefore expect it to be like the Earth in other ways; however, over the past several decades, observations from Earth and with spacecraft and landers have revealed that Venus and the Earth have radically different surfaces and atmospheres.

The Venusian Atmosphere

Venus's atmosphere is so thick that it completely hides its surface (fig. 9.11). The atmosphere is 96% carbon dioxide as determined from its spectrum and from measurements with space probes. The kinds of gases in the atmosphere determine what wavelengths of sunlight are absorbed, so astronomers examine absorption lines in the reflected sunlight to find the composition and density of the gas. Moreover, spacecraft have descended through the atmosphere to the surface and have sampled its atmosphere. In addition to carbon dioxide, Venus's atmosphere contains about 3.5% nitrogen and very small amounts of water vapor and other gases.

Spectra also reveal the nature of the Venusian clouds: they are composed of sulfuric acid droplets that formed when sulfur compounds—perhaps ejected from volcanoes—combined with the traces of water in the atmosphere. These clouds permanently cover the planet and are very high and thick, beginning at about 30 kilometers (19 miles) above the surface and extending upward to about 60 kilometers (37 miles). Below the clouds, the Venusian atmosphere is relatively clear, and some sunlight penetrates to the surface. The light is tinged orange, however, because the blue wavelengths are absorbed in the deep cloud layer.

In Venus's upper atmosphere, wind speeds can exceed 350 kilometers per hour (210 mph), but near the surface the winds move much more slowly, just a few kilometers per hour. The motion of the atmosphere is driven by the Sun's heating near the equator, which causes the gas to expand most there. Its upper layers then flow toward the cooler polar regions, where they sink and flow back toward the equatorial regions. This produces a huge vortex near each pole, like the water running down a drain. The rapidly changing shape of these vortices has been studied by the European Space Agency's *Venus Express* mission (fig. 9.12).

Venus's atmosphere is extremely dense. It exerts a pressure roughly 100 times that of the Earth's, equivalent to the pressure you would feel under 1000 meters (3000 feet) of water. We will discuss in section 9.4 why Venus has such a massive atmosphere, but there are other features of the planet that we need to understand first. One of the most important of these features is that its lower atmosphere is extremely hot, more than 750 K (about 900°F)—hot enough to melt lead!

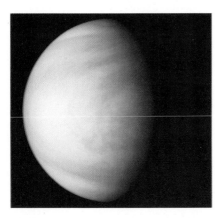

FIGURE 9.11
Photograph through an ultraviolet filter of the clouds of Venus. The picture is artificially colored and enhanced to show the clouds clearly.

Q: If this were the view of Venus from Earth, what would it imply about the position of Venus in its orbit?

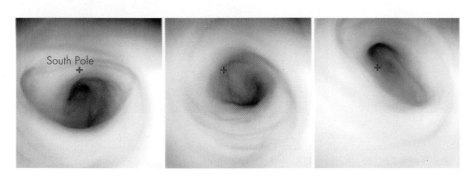

South Pole

FIGURE 9.12
Series of images of Venus's southern polar vortex made by ESA's *Venus Express* spacecraft at 24-hour intervals. The images are taken in infrared light, and darker parts of the image correspond to where the cloud tops are deeper. Each image is about 4000 km (about 2500 miles) across.

The Runaway Greenhouse Effect

Venus's surface is hotter than Mercury's even though it is almost twice as far from the Sun. It remains hotter than the hottest parts of Mercury even at night, and in fact there is very little change in temperature between day and night. What makes the surface of a planet so similar to the Earth, and only slightly nearer the Sun, so very hot?

Venus's carbon dioxide atmosphere creates an extremely strong **runaway greenhouse effect.** We discussed in chapter 6 how gases in Earth's atmosphere allow sunlight to enter and warm the surface but retard the heat so generated from escaping to space. That is, certain gases trap a planet's heat by hindering infrared radiation from escaping to space (see fig. 6.23). Carbon dioxide in the Earth's atmosphere traps heat, creating a weak greenhouse effect that keeps the Earth warmer than it would be otherwise. Venus, however, has about 300,000 times more carbon dioxide than the Earth, and so its greenhouse effect is correspondingly much stronger.

Measurements of hydrogen escaping to space made by the *Venus Express* spacecraft suggest that Venus used to have much more water, with oceans probably covering much of its surface billions of years ago. This suggests a dire scenario with cautionary implications for Earth. Venus may once have had an environment similar to Earth's, but greenhouse warming caused its oceans to begin to evaporate. This added further to the greenhouse effect because water vapor also absorbs infrared light. As the heat trapping grew stronger, the oceans completely boiled away. Over time sunlight broke down the water molecules in the atmosphere; the lightweight hydrogen atoms escaped, leaving behind the baked dry surface we see today.

The Surface of Venus

Despite the extremely hostile conditions on Venus's surface, several Russian *Venera* spacecraft landed there in the 1970s and 1980s and transmitted pictures back to Earth. These robotic spacecraft made a variety of measurements and sampled the rocks, showing them to be of volcanic origin. The landers lasted at most about two hours before succumbing to the high temperatures and atmospheric pressure. The pictures (fig. 9.13) show a barren surface covered with flat, broken rocks and lit by the pale orange glow of sunlight diffusing through the deep clouds. Two cameras on each spacecraft scanned a narrow strip from horizon to horizon. Figure 9.13 combines portions

FIGURE 9.13

Images from *Venera 13* (left) and *Venera 14* (right). The two cameras on each spacecraft scanned narrow strips that intersected at each end. Portions of each strip are combined to provide a clearer sense of the landscape. Distant hills are visible, as is nearby volcanic rock.

from each camera to show a partial view of the landscapes from the lander in the foreground to the horizon and the yellow sky.

The surface of Venus is hidden beneath its thick clouds, but planetary scientists can map its features with radar stations on Earth or with spacecraft orbiting Venus. Just as radar penetrates terrestrial clouds to locate a runway through fog, so too radar penetrates the Venusian clouds, revealing the planet's surface. Figure 9.14 shows a radar map of Venus made in the 1990s by *Magellan,* a U.S. spacecraft.

The radar map shows that Venus is less mountainous than Earth, with most of its surface being low, gently rolling lava fields. Only two major highland regions, Ishtar Terra and Aphrodite Terra, rise above the lowlands to form land masses similar to terrestrial continents. Ishtar, named for the Babylonian goddess of love, is about the size of Australia. Ishtar is studded with volcanic peaks, the highest of which, Maxwell Montes, rises more than 11 kilometers (about 7 miles) above the average level of the planet. (Notice that because no oceans exist on Venus, "sea level" has no meaning as a reference height.) The other major highland region, Aphrodite, bears the ancient Greek name for Venus and is about the size of South America. Together, Ishtar and Aphrodite compose only about 8% of Venus's surface, a far smaller fraction than for Earth, where continents and their submerged margins cover about 45% of the planet.

Recent measurements by the *Venus Express* spacecraft comparing the infrared radiation emitted by the highlands and lowlands regions suggest another similarity to Earth. The highlands rock has properties similar to the granite in Earth's continents. This is consistent with the hypothesis that Venus once had oceans, with volcanic activity forging granite when the water chemically combined with molten rock.

The many odd and unique structures seen in the radar maps have proved puzzling to planetary geologists. Venus is so similar in diameter and mass to the Earth that they expected to see landforms there similar to those on the Earth. However, the features seen bear little resemblance to the features that result from plate tectonics, such as continental blocks, crustal rifts, and trenches at plate boundaries.

Magellan met a deliberately engineered fiery doom in 1994. Its orbit was altered so that it plunged into Venus's atmosphere. Analysis of its final tumblings gave astronomers data on the density of Venus's upper atmosphere.

Venus

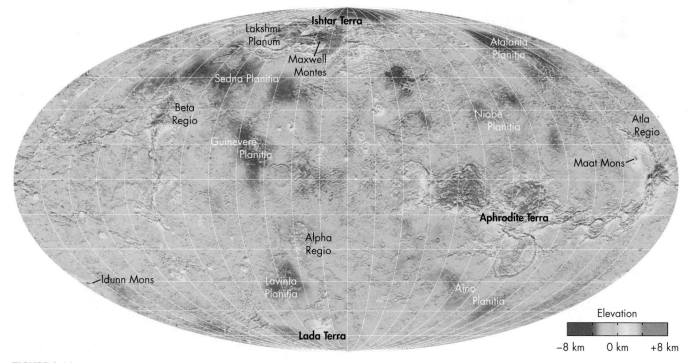

FIGURE 9.14

Global topographic map of Venus made by radar elevation measurements with the *Magellan* Venus-orbiting satellite. Colors indicate the relative height of surface features. Lowlands are blue; high elevations are orange.

Impact crater, Lavinia Planitia   Volcanoes, Atla Regio   Lava channels, Lada Terra   Lava domes, Alpha Regio

|← 100 km (about 60 miles) →|

FIGURE 9.15

Magellan radar images of a variety of features on Venus, all shown to the same scale. The views look straight down at the surface, with the radar illumination from the left. In addition to the individually noted features, most of these images show long cracks or faults in the bedrock.

FIGURE 9.16

Perspective image of Venus's volcano Idunn Mons based on *Magellan* radar data in the top image. The volcano appears to have relatively recent lava flows surrounding it, based on infrared observations made by the *Venus Express* satellite, shown by a false-color overlay of temperature differences in the bottom image.

The *Magellan* spacecraft was able to detect features as small as about 100 meters (about 300 feet) across. Figure 9.15 shows close-ups of some of the more intriguing images that *Magellan* transmitted. Brighter regions reflected radar more strongly, which relates to the angle and roughness of the surface.

Although Venus has some impact craters and crumpled mountains, volcanic landforms dominate. These include peaks with immense lava flows, "pancake-shaped" domes of uplifted rock, long narrow faults (cracks), and peculiar lumpy terrain. All these features indicate a young and active surface, a deduction borne out by the scarcity of impact craters. From the small number of craters, scientists have concluded that virtually all of Venus's original surface has been paved over by volcanic activity. The surface we see is probably at most half a billion years old, much younger than Earth's continental surface, and some regions may be less than 10 million years old. Such estimates of crustal age are difficult to make, however, because the Venusian atmosphere is so dense that all but the largest infalling bodies (bigger than a few hundred meters) are broken up in it.

Are the Venusian volcanoes still active? Eruptions have not been seen directly, but some lava flows appear very fresh. Differences in the infrared radiation from some volcanic peaks (fig. 9.16) seen by *Venus Express* suggest that these are relatively recent lava flows. In addition, electrical discharges, perhaps lightning, have been detected near some of the larger peaks. On Earth, volcanic eruptions frequently generate lightning, and some astronomers think the electrical activity indicates that Venus's volcanoes are still erupting. Such eruptions might also explain brief increases in sulfur content detected in the Venusian atmosphere, changes similar to those produced on Earth by eruptions here.

The numerous volcanic peaks, domes, and uplifted surface regions suggest to some scientists that heat flows less uniformly within Venus than within the Earth. Although some locations on Earth (Yellowstone Park and the Hawaiian Islands, for example) are heated anomalously by "plumes" of rising hot rock, such plumes seem to dominate on Venus. As hot rock wells upward, it bulges the crust, stretching and cracking it. We may be viewing on our sister planet what Earth looked like as its crust began to form and before smooth heat flows were established. Alternatively, Venus and Earth may differ for deeper reasons.

The Interior of Venus

The interior of Venus is probably similar to the Earth's, an iron core and rock mantle (fig. 9.17). Planetary geologists have no seismic information, so, as with Mercury, they must rely on deductions from Venus's gravity and density, which are similar to the Earth's. Why then are the surface features so different?

One important difference is the water content of the rock in these two worlds. Rocks that contain water trapped in their structure melt at a lower temperature than similar rocks that lack water. Moreover, when they become molten, they are "runnier," which makes it easier for the melted rock to flow. As a result, convection can be more vigorous in a planet whose rocks are rich in water compared to a planet with drier rocks. Furthermore, because the water-rich rock melts at a lower temperature, the solid crust of such a planet will be thinner. Other things being equal, we therefore expect wet Earth to have a thinner crust than dry Venus.

A thin crust, such as we have on Earth, breaks into plates more easily. This breaking allows crustal motion and activity to occur more or less continuously and at many places on the surface. For Venus, the thicker crust holds heat in, keeping the interior hot, but the convective motions that develop are unable to break up the thick crust.

Ultimately, however, the trapped heat must escape, and astronomers have proposed different hypotheses for what then occurs. According to one idea, at points where the hot, rising material reaches the crust, the surface bulges upward and weakens, and volcanoes may form. Where cooler material sinks, the thick crust crumples into a continent-like region. This creates a planet with few and small continents and whose surface is active but only in isolated spots.

Another idea is that the trapped heat gradually melts the bottom of the thick crust, thinning it and allowing it to break up. This may happen over widespread areas, flooding large portions of the planetary surface with lava in a brief time. The heat then rapidly escapes to space, the interior cools, and the crust again thickens. Surface activity therefore subsides, but heat is trapped once more, and the process may repeat, but only at intervals of hundreds of millions of years.

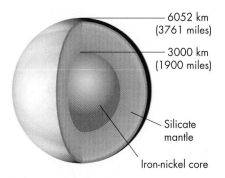

FIGURE 9.17
Artist's sketch of the interior of Venus.

6052 km
(3761 miles)

3000 km
(1900 miles)

Silicate
mantle

Iron-nickel core

Rotation of Venus

Venus spins on its axis more slowly than any other planet in the Solar System, taking 243 days to complete one rotation relative to the stars. Moreover, its spin is retrograde ("backward") compared with the direction of rotation of the other terrestrial planets. Because the planet's spin is retrograde, the Sun rises in the west and sets in the east. The slow rotation also makes the solar day there very long, approximately 117 Earth days (fig. 9.18).

This slow and retrograde spin has led some astronomers to hypothesize that Venus was struck shortly after its birth by a huge planetesimal; the impact set Venus spinning backward, and tidal forces exerted by the Sun have slowed it since. A less dramatic explanation of the spin is that Venus has been affected by a combination of tidal forces exerted by the Sun—and perhaps the Earth and Jupiter—so that the tilt angle of its rotation axis may have shifted over time. Venus probably has an interior similar to the Earth's, but it rotates so slowly that it cannot generate a strong magnetic field as the Earth does.

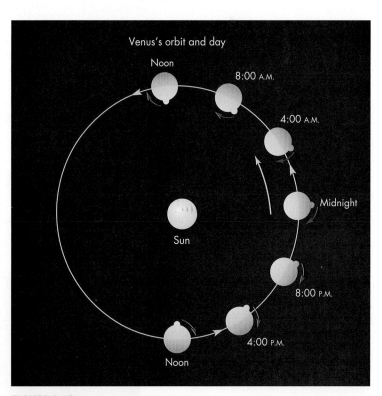

Venus's orbit and day

Noon

8:00 A.M.

4:00 A.M.

Midnight

Sun

8:00 P.M.

4:00 P.M.

Noon

FIGURE 9.18
Venus spins slowly backward, so that a day lasts a little more than half its orbit.

Mars is named for the Roman god of war, presumably because of its blood-red color. Compared to Mercury and Venus, Mars seems positively Earth-like. Although its diameter is only about half and its mass about one-tenth the Earth's, many other features are similar. The Martian day is just 39 minutes longer than an Earth day, and the tilt of its axis is almost the same as Earth's, so it experiences a similar sequence of seasons. On a warm day, the temperature at the Martian equator reaches about 50°F (10°C). Winds may sweep dust and patchy clouds of ice crystals through its sky, but generally the Martian atmosphere allows astronomers on Earth to view its surface clearly. Such views show other familiar features like polar caps of sparkling white, which contrast with the reddish color of most of the planet, as seen in figure 9.19.

The similarities of Mars to Earth have excited interest in the planet, perhaps even as a place we might someday inhabit. A series of spacecraft sent to explore the planet—*Mariner, Viking I* and *II, Mars Global Surveyor, Mars Odyssey, Mars Express, Mars Reconnaissance Orbiter,* and many others—have revealed the true marvels of the planet.

The Surface of Mars

Mars has some of the most dramatic surface features of any terrestrial planet. Along the equator runs a rift—Valles Marineris—that stretches 4000 kilometers (2500 miles) long, 100 kilometers (60 miles) wide, and 7 kilometers (4 miles) deep. This canyon, named for the *Mariner* spacecraft whose pictures led to its discovery, dwarfs the Grand Canyon and would span the continental United States. What would this immense canyon look like if we visited Mars? Planetary scientists used satellite data to construct an image of how Valles Marineris would look from a high-flying airplane (fig. 9.20).

At midlatitudes, a huge uplands called the Tharsis bulge (fig. 9.21) is dotted with enormous volcanic peaks. Another volcano at the edge of Tharsis, Olympus Mons, rises about 25 kilometers (about 16 miles) above its surroundings, nearly three times the height of Earth's highest peaks, and is illustrated in figure 9.22. If ever interplanetary parks are established, Olympus Mons should lead the list.

Planetary geologists think that the Tharsis region formed as hot material rose from the deep interior of the planet and forced the surface upward as it reached the crust. The hot matter then erupted through the crust to form the volcanoes, some of

FIGURE 9.19
Picture of Mars made by the Hubble Space Telescope.

Q: What season is it in this picture? (North is at top of image.)

Mars

FIGURE 9.20

A reconstructed view down Valles Marineris, the Grand Canyon of Mars. This image was constructed from hundreds of thousands of laser altimeter measurements made by the *Mars Odyssey* orbiter. This enormous gash may be a rift that began to split apart the Martian crust but failed to open farther.

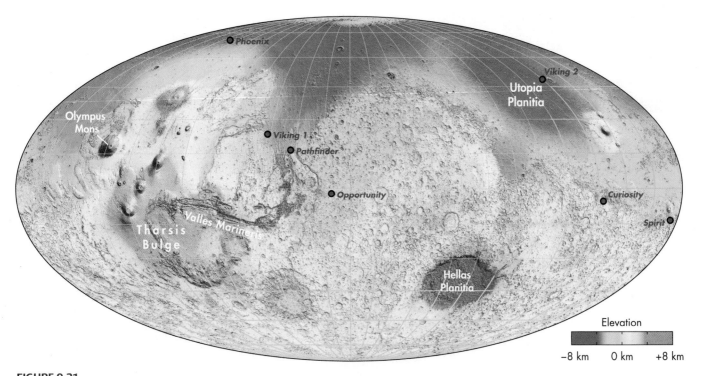

FIGURE 9.21

Topographic map of Mars showing its major features. The map is color coded according to elevations. Olympus Mons is the largest volcano in the Solar System, while Valles Marineris, the Grand Canyon of Mars, is an enormous gash in Mars's crust about 4000 kilometers (approximately 2500 miles) long. Were it on Earth, it would stretch from California to Florida.

which appear relatively young. For example, the small number of impact craters in its slopes implies that Olympus Mons is no older than 250 million years and that it may in fact have been active much more recently. Some planetary geologists think the Tharsis bulge may also have created the gigantic Valles Marineris, which lies to the southeast. According to this theory, Valles Marineris formed as the Tharsis region swelled, stretching and cracking the crust. Other planetary scientists think that this vast chasm is evidence for plate tectonic activity, like that of Earth, and that the Martian crust began to split, but the motion ceased as the planet aged and cooled.

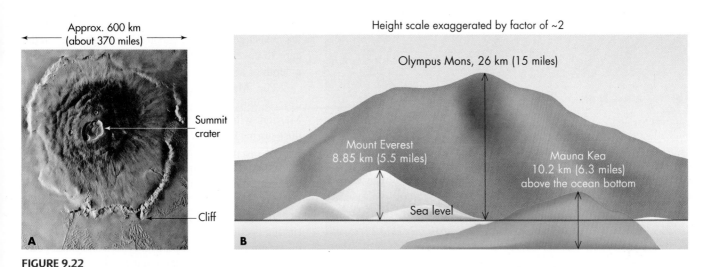

FIGURE 9.22

(A) Olympus Mons, the largest known volcano (probably inactive) in the Solar System. (B) Profile of Olympus Mons; Mount Everest, the highest mountain above sea level on Earth; and the Hawaiian volcano Mauna Kea, rising from the sea floor.

FIGURE 9.23

Pictures of (A) the south Martian polar cap and (B) the north Martian polar cap. (C) Note the layered structure of the north polar cap visible in this view from the *Mars Reconnaissance Orbiter*. This high-resolution image shows a portion of a chasm wall about 1.3 kilometers wide. Layers of dust and ice alternate in the chasm wall.

FIGURE 9.24

Mars Odyssey image of dunes that surround the northern polar cap. Color tinting has been added, ranging from blue for colder and yellow for warmer, based on the thermal imaging system.

Figure 9.23 shows the Martian polar caps. These frozen regions change in size during the cycle of the Martian seasons, which resemble Earth's seasons because of the similar tilt of Mars's rotation axis. Martian seasons are more extreme than terrestrial ones because the Martian atmosphere is much less dense than Earth's and therefore it does not retain heat as well.

Because Mars's seasonal changes are so extreme, its polar caps vary greatly in size, shrinking during the Martian summer and growing again during the winter. Much of the visible part of the southern cap is frozen carbon dioxide—dry ice—and in winter its frost extends in a thin layer across a region some 5900 kilometers (about 3700 miles) in diameter, from the south pole to latitude 40°, much as snow cover extends to middle latitudes such as New York in our winters. However, because the frost is very thin over most of this vast cap, it shrinks in the summer to a diameter of about 350 kilometers (220 miles).

The northern cap shrinks to a diameter of about 1000 kilometers (600 miles) in summer. Although the caps have a surface layer of CO_2, the bulk of the frozen material is ordinary water ice, as deduced from its temperature and radar studies. The northern cap consists of numerous separate layers, as can be seen in figure 9.21C. These strata indicate that the Martian climate changes cyclically. Thus, Mars may have "ice ages" similar to those on Earth.

Why do the Martian polar caps differ so? Altitude measurements made by the *Mars Global Surveyor* spacecraft show that Mars's south pole is considerably higher in elevation and is thus much colder than its north pole. This creates a strong wind pattern that carries water vapor and carbon dioxide away from the south pole toward the north pole. There, it precipitates out, leading to a larger north polar cap.

Satellites have been able to estimate the thickness of the ice caps using radar signals. From the depth of ice measured, there appears to be enough water in the caps to cover the entire surface of Mars with water to a depth of at least 10 meters (30 feet)

The Martian poles are bordered by immense deserts with dunes blown into parallel ridges by the Martian winds, as illustrated in figure 9.24. Huge dust storms blow the fine red dust over the entire surface of the planet, giving the planet its characteristic color. What makes it so red? The color comes from the iron minerals in its surface rocks. We know from everyday experience that a piece of iron will become rust-colored when exposed to air. Here on Earth, such rusting occurs because the iron combines with oxygen in our atmosphere to form iron oxides and other compounds. On Mars, even though there is little oxygen in its atmosphere, other chemical reactions with the iron in its surface minerals lead to the same effect.

FIGURE 9.25

Images from the *Viking* orbiter suggesting there was water on Mars in the past. All three images have the same scale and are oriented with north at top. (A) Picture of channels probably carved by running water on Mars. (B) Teardrop-shaped islands formed as water flowed (from bottom toward top of figure) around the rims of craters. (C) A Martian crater thought to have once been a "crater lake." Note the inflow channel at the south end of the crater, and the outflow channel to the northeast. The smooth floor (apart from a few small craters) suggests that the crater bottom is covered with sediment left behind as the lake dried out.

Water on Mars

Over the last several decades, scientists have sent dozens of spacecraft to Mars. Six have successfully landed on the surface and sent us back pictures and measurements of the Martian landscape and close-up images of rocks. These missions have many goals, but one of the major ones is searching for evidence that liquid water was once present on Mars. Why is liquid water of such interest? The answer is simple: scientists who study the possibility of life elsewhere in the Universe think that liquid water is a critical ingredient for living organisms of almost any type. Thus, the search for water is a first step in the search for life.

Perhaps the most intriguing features revealed by the two *Viking* orbiters in the 1970s were dry riverbeds, such as those seen in figure 9.25A. We infer that water once flowed on Mars along these winding channels, which look so similar to tributaries converging to make a large river channel on Earth. Some of the channels appear to have once had major flows that carved teardrop-shaped islands around crater rims (fig. 9.25B). In fact, many astronomers interpret the observed features to indicate that lakes and small oceans once existed on Mars. Evidence for these ancient bodies of water are seen in smooth terraces that look like old beaches around the inner edges of craters and basins, as you can see in the "crater lake" in figure 9.25C. Narrow canyons breach this crater's rim, showing where water flowed in from the south and drained out into lowland areas to the north, with the ancient shoreline still visible even though no surface liquid is present now.

The images from the *Viking* orbiters provided strong evidence that liquid water was once present on Mars, but it raised even more questions. How long ago was the water present? Was it ever a long-term feature of Mars, or did it occur in violent episodes of melting? For example, if there was ice buried under the soil, perhaps it was melted by an impact, with a sudden catastrophic flood. Furthermore, some features, such as the "crater lake" in figure 9.25C, could also be interpreted as arising from lava flows, so were there ever standing bodies of water on Mars? How much water remains today frozen in the ground in addition to the polar caps? And how much evaporated and dissociated, as on Venus, with the hydrogen lost into space?

FIGURE 9.26
A panoramic view of the Bonneville Crater, lying in the floor of the much larger Gusev Crater on Mars. This spot is near where the *Spirit* spacecraft landed.

FIGURE 9.27
A Martian rover. Cameras, a rock drill, and analysis instruments are powered by solar panels on the top of the rover.

The *Viking* landers provided our first view of the Martian surface in the 1970s, and in the 1990s the Mars *Pathfinder* mission provided a demonstration that we could operate a roving science vehicle across the surface of Mars. The next two NASA landers, *Opportunity* and *Spirit,* reached Mars in 2004 and carried out a remarkable set of explorations that far exceeded the original plan. These two missions were designed to explore Mars's surface, landing at sites that were chosen because pictures and spectral data taken from orbit suggested that water might have been present there long ago. *Spirit* landed in the center of the 150-km (90-mile) diameter Gusev Crater, a smooth-floored crater at the end of a narrow Martian valley that appeared to have once been flooded. Figure 9.26 shows a panorama from the *Spirit* lander site. *Opportunity* landed on the flat plains along the Martian equator on the opposite side of Mars from *Spirit*. Both craft deployed rovers—small, wheeled vehicles that can move away from the landing site and explore interesting features (fig. 9.27).

Both rovers were highly successful in their searches. For example, *Opportunity* took the pictures shown in figure 9.28, which both suggest processes that involved liquid water. Figure 9.28A shows small spheres (dubbed "Martian blueberries") that are made of hematite, which normally occurs from depositions of minerals in water. Figure 9.28B shows a rock outcropping thought to be material deposited in an ancient, now dried-up small sea. Examination of the rocks shows that they contain layers that are typical of sediment that sank to the bottom of a body of water and later was transformed into rock. This image also shows that the layers are wavy, similar to the ripple marks you see at the beach as water washes back and forth across the sand. Moreover, minerals in the rocks at the *Opportunity* site have a chemical makeup consistent with their having been deposited in a salty lake or small ocean. Half a world away *Spirit* found more layered rocks and other minerals that normally form in water. The evidence from the rovers has convinced many scientists that Mars once had large areas under liquid water. The original mission for the two rovers was planned for only 90 days, but the two rovers operated long past that, surviving cold winters and global dust storms. *Spirit* finally stopped responding in 2010, but *Opportunity* continues traveling over the Martian surface in 2015.

FIGURE 9.28
(A) Small spherical "blueberries" most likely formed from iron depositions in standing water. The lighter circular area was swept off by the *Opportunity* rover to study the rock type. (B) A close-up image of a rock outcropping at the landing site. The rocks show thin layers and contain minerals that suggest that they were formed on the bottom of a salty lake.

FIGURE 9.29

(A) Image from the *Mars Global Surveyor* of terraced features at the bottom of a Martian canyon. (B) *Mars Odyssey* image of former riverbed in which teardrop-shaped islands formed behind craters, and the surface layers were scoured by the stream flow. (C) *Mars Express* image of a crater with a lake of ice in its interior.

Meanwhile, a wide assortment of satellites have continued to send back detailed pictures of the surface of Mars with high resolution, allowing planetary scientists to build a more detailed understanding of Martian geology. The terraced layers in figure 9.29A are at the bottom of one of Mars's great canyons. Such features on Earth are usually laid down deep underwater. A detailed examination of the scouring of features in figure 9.29B implies that water carved these features with a flow exceeding that of the Mississippi River. Figure 9.29C shows a crater at a latitude of about 70° North that contains a large ice field, perhaps the remains of a lake.

In addition to being able to make highly detailed images, each orbiter includes a variety of other detectors. Some instruments can measure the spectrum of light reflected from the ground, whereas others collect radiation at other wavelengths. From these data, astronomers can deduce what minerals are present at different spots on the Martian surface. Matching the composition of those minerals with data on whether water is needed to produce those minerals gives additional evidence that Mars was once much wetter. Combining spectral information with the appearance of features can provide a much clearer picture of the geological origins of features, as illustrated in Extending Our Reach: "Analyzing Martian Geology."

EXTENDING *our reach*   ANALYZING MARTIAN GEOLOGY

Many features on Mars appear to be caused by water flow, but appearances can be deceiving. Similar looking features can often be produced by lava flows, for example. How then can we be sure? One way of deciding on the origin of a feature is to study the chemistry of the rocks left behind. We do not have to retrieve rock samples to do this. Visible and infrared spectroscopy can show what minerals are present. The image in figure 9.30 is an example of an outflow into what is thought to have been a crater lake. An imaging spectrometer on the *Mars Reconnaissance Orbiter* has identified clays and minerals that indicate that a body of water must have lasted for at least thousands of years.

FIGURE 9.30

Mars Reconnaissance Orbiter image of fanning outflow in a former crater lake. Spectral analysis of the surface features identified minerals normally found underwater, and different types of materials, such as clays, have been color coded in this image.

Mars Odyssey gamma ray data
Estimated hydrogen in top meter

Lower limit of water mass fraction

2% 5% 10% 20% 50%

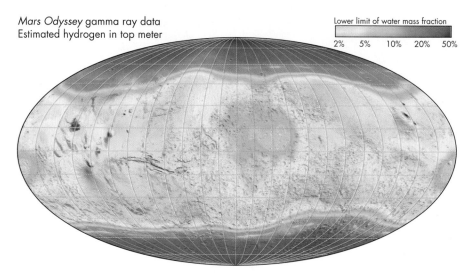

FIGURE 9.31
Global map of the likely percentage of water in Martian surface layers, made by *Mars Odyssey*.

FIGURE 9.32
The *Phoenix* lander scooped up soil samples, revealing ice under the polar dust.

FIGURE 9.33
Curiosity self-portrait made from a series of pictures by a camera on a robotic arm.

Other instruments on the *Mars Odyssey* orbiter can detect hydrogen atoms locked up in water in the Martian soil by studying gamma rays. This high-energy radiation is generated when cosmic rays from space strike the planet's surface. The radiation is reduced where there are hydrogen atoms, permitting scientists to estimate how much water is present. These measurements (fig. 9.31) have revealed what appears to be huge amounts of water, probably ice, in the upper meter of the Martian surface.

Another important experiment was carried out by the *Phoenix* lander, which reached the northern polar region of Mars in 2008. *Phoenix* carried out experiments on the soil, showing that it is quite different from Earth's—about as alkaline as baking soda with high levels of oxidizing chemicals. As the lander scooped up soil samples, it exposed ice below the surface (fig. 9.32).

The rover *Curiosity,* the size of a small car, reached Mars in 2012 (fig. 9.33). *Curiosity* has sophisticated instruments to analyze the geology and chemistry within Gale Crater. Planetary scientists suspect there was a crater lake that filled with sediment early in Mars's history, which was later eroded by streams and wind. By early 2015 *Curiosity* began to climb and analyze the layered central Mount Sharp (fig. 9.34A). Early in its mission it examined rocks probably laid down in the lake early in the crater's history, and it was able to do radioactive dating of the rock to establish that the earliest "mudstone" was laid down over 4 billion years ago.

Astronomers are both excited and puzzled by the possibility that liquid water was present on Mars. They are excited because of the belief that where there is water, there may be life—even if only microbes. They are puzzled because Mars does not have conditions that allow liquid water to be present on the planet today. To understand why, we need to look at the properties of the Martian atmosphere.

FIGURE 9.34
(A) The rover *Curiosity* landed in Gale crater, near the base of the central mountain, known as Mount Sharp, which it will climb and study. (B) Layered rocks in a region near the base of Mount Sharp appear to have been laid down about 4 billion years ago in a lake that lasted for perhaps 10 million years when Mars was young.

Curiosity's
A landing site

Proposed path to
climb Mount Sharp

~50 cm

B

FIGURE 9.35
(A) A "dust devil" (small tornado) spotted by the *Mars Reconnaissance Orbiter*. (B) Fog in Martian valleys seen by the *Viking* orbiter. (C) Frost on surface rocks near the *Viking* lander.

The Martian Atmosphere

Clouds and wind-blown dust are visible evidence that Mars has an atmosphere. Spectra and direct sampling by spacecraft landers confirm this and show that the atmosphere is mostly (95%) carbon dioxide with small amounts (3%) of nitrogen and traces of oxygen and water. From this, astronomers can determine the density of Mars's atmosphere, which turns out to be very low—only about 1% the density of Earth's.

This density is so low that, although the Martian atmosphere is mostly carbon dioxide, it creates only a very weak greenhouse effect. The consequent lack of heat trapping and Mars's greater distance from the Sun leaves the planet very cold. Temperatures at noon at the equator may reach a little above the freezing point of water, but at night they plummet to far below freezing. The resulting average temperature is a frigid 218 K (−67°F). Thus, although water exists on Mars, it is frozen solid, locked up either below the surface in the form of permafrost or in the polar caps as solid water ice.

Clouds of dry ice (frozen CO_2) and water-ice crystals (H_2O) drift through the Martian atmosphere carried by the Martian winds. These winds, like the large-scale winds on Earth, arise because air that is warmed near the equator rises and moves toward the poles. This flow from equator to poles, however, is deflected by the Coriolis effect arising from the planet's rotation. The result is winds that blow around the planet approximately parallel to its equator, and small tornadoes or "dust devils" (fig. 9.35A). The Martian winds are generally gentle, but seasonally and near the poles they become gales, which sometimes pick up large amounts of dust from the surface. The resulting vast dust storms occasionally cover the planet completely and turn its sky pink.

No rain falls from the Martian sky, despite its clouds, because the atmosphere is too cold and contains too little water. In fact, there is so little water in the Martian atmosphere that even if all of it were to fall as rain, it would make a layer only about 12 micrometers deep (less than 1/2000 inch). For comparison, Earth's atmosphere holds enough water to make a layer a few centimeters (inches) deep. Despite such dryness, however, fog sometimes forms in Martian valleys (fig. 9.35B), and frost condenses on the ground on cold nights (fig. 9.35C). In addition, during the Martian winter, CO_2 "snow" falls on the Martian poles.

Mars has not always been so dry, as we saw from the numerous channels in its highlands and other evidence from many Mars missions. But for a planet to have liquid water, it must have a temperature above freezing with a high enough pressure to keep the liquid from immediately boiling away. If the pressure on a liquid is very low, molecules can break free from its surface, evaporating easily because no external force

FIGURE 9.36

Sequence of images of the wall of a crater from early to mid summer. As the crater warms, dark streaks extend down the crater wall, resembling wet patches. It is possible that a briny solution, melting and seeping out from underground layers, might produce these features. Pure water would evaporate too quickly to keep the soil moist.

restrains them. On the other hand, if the pressure is high, molecules in a liquid must be heated strongly to turn them into gas. For example, at normal atmospheric pressure on Earth, water boils at 100° Celsius. If the pressure is reduced, however, the boiling point drops, an effect used by food producers to make "freeze-dried" foods, such as instant coffee. Coffee is brewed normally, then frozen and placed in a chamber from which the air is pumped out. The reduced pressure makes the liquid "boil" without heating and evaporate, leaving only a powder residue—instant coffee. Similarly, any liquid water on Mars's surface today would evaporate.

The existence of channels carved by liquid water on Mars is therefore strong evidence that in its past, Mars was warmer and had a denser atmosphere. However, that milder climate must have ended billions of years ago. How do we know? The large number of impact craters on Mars shows that its surface has not been significantly eroded by rain or flowing water for about 3 billion years. Why did Mars dry out, and where have its water and atmosphere gone?

Some water probably lies buried below the Martian surface as ice, as indicated by measurements from the *Mars Odyssey* orbiting spacecraft. If the Martian climate was once warmer and then cooled drastically, water would condense from its atmosphere and freeze, forming sheets of surface ice. Wind might then bury this ice under protective layers of dust, as happens in polar and high mountain regions of Earth. Figure 9.36 shows possible evidence that this buried ice continues to melt and trickle out of subsurface layers even today. The series of images made by the *Mars Reconnaissance Orbiter* shows dark lines growing down the inner wall of a crater over several months of the Martian summer. Scientists hypothesize that frozen groundwater is melting and seeping down the side of the crater.

If Mars had a denser atmosphere in the past, as deduced from the higher pressure needed to allow liquid water to exist, then the greenhouse effect might have made the planet significantly warmer than it is now. The loss of such an atmosphere would have weakened the greenhouse effect and plunged the planet into a permanent ice age. Such a loss could happen in at least two ways. According to one idea, repeated asteroid impacts on Mars when it was young may have blasted its original atmosphere off into space. Such impacts, although rare now, did occur for hundreds of millions of years after the Solar System began to form. Could there have been major impacts on Mars 3.9 billion years ago when the Moon appears to have experienced major impacts?

A less dramatic explanation for how Mars lost most of its atmosphere is that Mars's low gravity allowed gas molecules to escape over the first 1 to 2 billion years of the planet's history. Regardless of which explanation is correct, the loss of its atmosphere would have cooled the planet and locked up its remaining water as permafrost. But why have the Martian volcanoes not replenished its atmosphere, keeping the planet warm? Astronomers believe that the blame lies with Mars's low level of tectonic activity, a level set by conditions in its interior.

The Martian Interior

Astronomers think that the interior of Mars is differentiated like the Earth's into a crust, a mantle, and an iron core (fig. 9.37). However, Mars is so small compared with the Earth that its interior is cooler. Mars's smaller mass supplies less heat, and its smaller radius allows the heat to escape faster, as happens with the Moon (chapter 7). We have no direct confirmation of Mars's interior structure because no functioning seismic detectors have landed there yet. Thus, as is the case for Mercury and Venus, astronomers must rely on indirect evidence from its density and gravitational field to learn about the interior of Mars. Using the *Mars Global Surveyor* spacecraft in orbit around the planet, astronomers have measured Mars's magnetic field and internal structure. They concluded that Mars has a metallic core whose radius is approximately 1700 km, or about half of the overall radius of the planet. But Mars, unlike Earth, has no planetwide magnetic field, so its core is probably no longer molten.

Having a mass between that of dead Mercury and lively Earth and Venus implies that Mars should be intermediate in its tectonic activity. Such seems to be the case. Although it possesses numerous volcanic peaks and uplifted highlands, implying that it had an active crust, at least in the past, Mars bears no evidence of large-scale crustal motion like the Earth's. For example, it has no folded mountain ranges. Astronomers therefore think that Mars has cooled and its crust thickened to perhaps twice the thickness of the Earth's crust. As a result, the now-weak interior heat flow can no longer drive tectonic motions. A thick Martian crust may also explain why Mars has a small number of very large volcanoes, while the Earth has a large number of small ones. The volcanoes may have grown over hot spots in the core, and the crust did not shift to new positions. Mars's immense volcanoes are thus mute testimony to a more active past.

Mars's current low level of tectonic activity is also demonstrated by the many impact craters that cover its older terrain, far more than are seen on either Earth or Venus. The number of those craters implies that Mars has been geologically quiet for billions of years. Mars is probably not dead, however, because some regions (for example, the slopes of Olympus Mons and other volcanoes) are essentially free of craters. Thus, these immense peaks may still occasionally erupt. They do not erupt often enough, however, to replace the gas lost to space because of the planet's low gravity. It appears that Mars has entered a phase of planetary old age. Recently, however, using Earth-based telescopes, astronomers have detected methane in spectra of the Martian atmosphere. Because methane is rapidly destroyed in the Martian environment, this means that Mars must be producing it currently, indicating that there is still at least a low level of geological activity.

The Martian Moons

Mars has two tiny moons, Phobos and Deimos, which are named for the demigods of Fear and Panic (fig. 9.38). These bodies are only about 20 kilometers across and are probably captured asteroids. They are far too small for their gravity to have pulled them into spherical shapes. Both moons are cratered, implying bombardment by smaller objects. Phobos has cracks, suggesting that it may have been struck by a body large enough to split it nearly apart.

Phobos and Deimos were discovered in 1877, but by chance they appeared in literature nearly two centuries earlier in Jonathan Swift's book *Gulliver's Travels*. Gulliver stops at the imaginary country Laputa whose inhabitants include numerous astronomers. Among the accomplishments of these people is the discovery of two tiny moons of Mars. Even earlier, Kepler guessed that Mars might have two moons because the Earth has one moon and Jupiter, at least in Kepler's time, was known to have four. Mars, lying between these two bodies should therefore (according to Kepler's mystic argument) have a number of moons lying between 1 and 4, and he chose 2 as the more likely case.

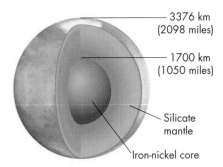

3376 km
(2098 miles)

1700 km
(1050 miles)

Silicate mantle

Iron-nickel core

FIGURE 9.37
Artist's sketch of the interior of Mars.

Q: What does the irregular shape of these bodies tell you about the strength of their surface gravity? Is it likely these moons have any atmosphere of their own?

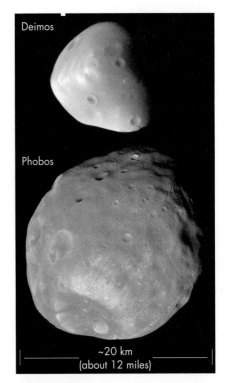

Deimos

Phobos

~20 km
(about 12 miles)

FIGURE 9.38
Images of Phobos and Deimos, the moons of Mars. These tiny bodies are probably captured asteroids.

FIGURE 9.39

Drawing of Mars made by Percival Lowell around 1900. Lowell thought he could see straight-line features that he believed were canals for irrigation or travel.

FIGURE 9.40

Fossils of ancient Martian life? The tiny rod-shaped structures look similar to primitive fossils found in ancient rocks on Earth. However, some scientists think these structures formed chemically.

Life on Mars?

Scientists have long wondered whether living organisms developed on Mars. Much of that interest grew from a misinterpretation of observations made in 1877 by the Italian astronomer Giovanni Schiaparelli. Schiaparelli saw what he took to be straight-line features on Mars and called them *canali,* by which he meant "channels." In English-speaking countries, the Martian *canali* became canals, with the implication that Mars must be inhabited by intelligent beings who built them. The interest in these canals had become so great by 1894 that the wealthy Bostonian Percival Lowell built an observatory in northern Arizona to study Mars and search for signs of life there (fig. 9.39).

Most astronomers could see no trace of the alleged canals, but they did note seasonal changes in the shape of dark regions, changes that some interpreted as the spread of plant life in the Martian spring. By the early 1970s scientists were excited by satellite photographs of water-carved canyons and old riverbeds, because water—at least on Earth—is so important for life. Therefore, to further the search for life on Mars, the United States landed two *Viking* spacecraft on the planet in 1976. These craft carried instruments to search for signs of carbon chemistry in the soil and to look for metabolic activity in soil samples that were put in a nutrient broth carried on the lander. All tests either were negative or ambiguous.

Then, in 1996, a group of American and English scientists reported possible signs of life in rocks from Mars. These were not samples returned to Earth by a spacecraft but samples of meteorites found in the Antarctic. They arrived here after being blasted off the surface of Mars, presumably by the impact of a small asteroid. Such impacts are not uncommon, but most fragments are scattered in space or fall back to Mars. Moreover, of those that are shot into space, only a tiny fraction have just the right combination of speed and direction to reach Earth.

How can astronomers tell if a meteorite has come from Mars? One way is to sample the gas trapped in tiny bubbles in the meteorite and see if it matches the composition of Mars's atmosphere as measured by the *Viking* Mars landers. For the meteorite in question, the match was excellent, assuring that it came from Mars. Scientists have even been able to match mineralogical details of the meteorite to a probable origin in the Valles Marineris region.

What was the evidence suggesting life? That turns out to be far more controversial. Microscopic examination of samples from the interior of the meteorite revealed many tiny, rod-shaped structures (fig. 9.40). These look very much like ancient terrestrial bacteria but are much smaller. Some scientists suggest they are fossilized primitive Martian life. The meteorite also contains traces of organic chemicals known as polycyclic aromatic hydrocarbons (PAHs, for short). Terrestrial bacteria make such chemicals when they die and decay, but PAHs can also form spontaneously, given the proper mix of chemicals. In fact, they have been found in a number of non-Martian meteorites and have also been detected by their spectrum lines in the radio emission from interstellar gas and dust clouds. Other structures in the meteorite can also be interpreted as having a biological origin. But other scientists have shown that ordinary chemical weathering can form very similar structures. As a result, most scientists today are unconvinced that any meteorite yet studied shows evidence of Martian life.

A more immediate opportunity to search for evidence of past life on Mars is provided by the rover *Curiosity. Curiosity* landed in a crater that is thought to have once been a lake or sea that filled with sediments for more than a billion years. Later erosion exposed the central mountain, which shows terraced layers. The rover is currently climbing the steep slopes of the mountain to investigate different geological layers. This may reveal how the Marian climate changed over time. The rover carries a variety of scientific instruments to study the chemical composition of the rocks and soil and to examine samples microscopically. It has already found that Mars was once probably habitable and that all the elements necessary for life are present. It may yet discover clearer signs that life arose when this cold dusty world was young.

9.4   WHY ARE THE TERRESTRIAL PLANETS SO DIFFERENT?

We have seen in sections 9.1–9.3 that the terrestrial planets have little in common with Earth, apart from being rocky spheres. They have different surfaces, atmospheres, and interiors. Astronomers think these differences arise from their different masses, radii, and distances from the Sun.

Role of Mass and Radius

A terrestrial planet's mass and radius affect its interior temperature and thus its level of tectonic activity, with low-mass, small-radius planets being cooler inside than larger bodies. For this discussion we can include the Moon as an example of an even smaller planetary body. We see a progression of activity from the relatively inert Moon and Mercury, to slightly larger and once-active Mars, to the larger and far more active surfaces of Venus and Earth, as illustrated in figure 9.41. The Moon's and Mercury's surfaces still bear the craters made as they were assembled from planetesimals. Mars has many craters, from which we infer that much of its surface is very old, but being larger and more tectonically active, it also has younger surface features such as volcanoes, canyons formed by surface cracking as hot material rose inside it, and erosional features such as canyons and riverbeds carved by running water. In contrast, Earth and Venus retain almost none of their original crust; their surfaces have been enormously modified by activity in their interiors over the lifetimes of these planets.

FIGURE 9.41

Gallery comparing the interiors of the terrestrial planets and the Moon, and indicating their relative stages of geological cooling.

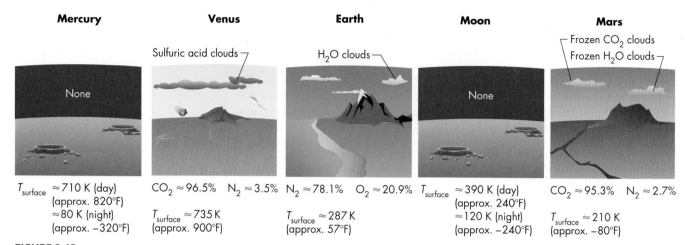

FIGURE 9.42

Gallery comparing interiors, atmospheres, and surfaces of the terrestrial planets.

Role of Internal Activity

Internal activity also affects a planet's atmosphere. In fact, the atmospheres of the terrestrial planets, though now greatly modified, are probably mostly volcanic gases vented as the result of their internal activity. Thus small, inactive Mercury probably never had much atmosphere, and Mars, active once but now geologically quiet, may have developed an atmosphere in its early history, but little gas has been added for billions of years. Moreover, these planets have so little mass and consequently such a small surface gravity relative to Venus and Earth that they have difficulty retaining what little gases they might once have had. As a result, Mercury is virtually without an atmosphere today, and Mars has only a small remnant of its original atmosphere. On the other hand, highly active Venus and Earth have extensive atmospheres. These and other properties are summarized in figure 9.42 as well as table 9.1.

Astronomers think that the compositions of the atmospheres of Venus, Earth, and Mars have changed appreciably over time through chemical processes. The original composition of all three bodies' atmospheres was probably nearly the same: primarily CO_2 with small amounts of nitrogen and water. But these original atmospheres have been modified by sunlight, tectonic activity, and, in the case of the Earth, life.

Role of Sunlight

Sunlight affects a planet's atmosphere in several ways. First, of course, it warms a planet by an amount that depends on the planet's distance from the Sun, and so we expect that Venus will be warmer than Earth and Earth will be warmer than Mars. These expected temperature differences are increased, however, by the atmospheres of these bodies. Moreover, even relatively small differences in temperature can lead to large differences in physical behavior and chemical reactions within an atmosphere. For example, Venus is just enough nearer the Sun than the Earth that even without a strong greenhouse effect, most of its atmosphere would be so warm that water would have difficulty condensing and turning to rain. Moreover, with the Venusian atmosphere being warmer throughout, water vapor can rise to great heights in the Venusian atmosphere, whereas on cooler Earth, water vapor condenses to ice at about 10,000 meters (about 30,000 feet), making our upper atmosphere almost totally devoid of water. You may have noticed this extreme dryness on a long plane trip, because after such a flight your skin may feel itchy and the inside of your nose may feel cracked or stuffy. That dryness is the result of bringing upper-atmospheric air into the cabin and compressing it to make it breathable but not adding moisture.

Table 9.1	Comparison of the Terrestrial Planets and the Moon				
	Mercury	**Venus**	**Earth**	**Mars**	**Moon**
Radius (Earth units) Radius (km)	0.383 2439	0.950 6052	1.000 6378	0.533 3398	0.273 1738
Mass (Earth units) Mass (kg)	0.055 3.30×10^{23}	0.815 4.87×10^{24}	1.000 5.98×10^{24}	0.107 6.42×10^{23}	0.012 7.35×10^{22}
Density (grams/cm^3)	5.43	5.25	5.52	3.94	3.34
Atmospheric composition	None	CO_2 (96.5%) N_2 (3.5%)	N_2 (78.1%) O_2 (20.9%)	CO_2 (95.3%) N_2 (2.7%)	None
Pressure (bars)	0.0	about 90	1.0	about 0.007	0.0

Surface features	Craters,
	   scarps
		Mountains,
		   volcanic peaks,
		   congealed lava plains
			Oceans,
			   mountains,
			   volcanic peaks
				Deserts, craters,
				   canyons,
				   volcanic peaks
					Craters,
					   maria

Sidereal day	58.65 Earth days	243.02 Earth days	23.9345 hours	24.62 hours	27.32 Earth days
Solar day	176 Earth days	116.8 Earth days	24 hours	24.66 hours	29.53 Earth days
Satellites	None	None	Moon	Phobos and Deimos	—
Distance from Sun	0.387 AU	0.723 AU	1 AU	1.524 AU	1 AU
Orbital period	87.969 Earth days	224.70 Earth days	365.26 Earth days	686.98 Earth days	—
Axial tilt	0°	177.4°	23.45°	23.98°	5.15°

Role of Water Content

Water alters a planet in many ways. We have seen that its cycling between gaseous and liquid states produces erosion and sedimentary rock, and water in surface rock alters the rock's chemistry and melting temperature. Both Earth and Mars have retained their water, in liquid or frozen states, but what became of the Venusian water?

We noted that water vapor can rise higher in Venus's atmosphere, and this has proved to have catastrophic consequences for Venus. At high altitudes, ultraviolet light from the Sun is intense enough to break apart any water molecules in that region into their component oxygen and hydrogen atoms, a process called **photo dissociation.** Being very light, the hydrogen atoms so liberated can escape into space, while the heavier oxygen atoms remain. Because water can rise to great heights in the Venusian atmosphere, over billions of years it has steadily been dissociated there and almost completely lost from our sister planet. In our atmosphere, however, water has survived, and that has altered our atmosphere even further.

Water makes possible chemical reactions that profoundly alter the composition of a planet's atmosphere. Most importantly, CO_2 dissolves in liquid water and creates carbonic acid, which in dilute form we drink as "soda water." In fact, the bubbles in soda water are just carbon dioxide that is coming out of solution. As rain falls through our atmosphere, it picks up CO_2, making the raindrops slightly acidic, even in unpolluted air. As the rain falls on the ground, the carbonic acid reacts chemically with rocks to form carbonates, locking some CO_2 into the rock. The amount of carbon dioxide that can be removed from an atmosphere by this process is impressive. If the CO_2 that is now bound into carbonates on Earth were to be freed again, our atmosphere's pressure would increase by a factor of about 60. This would make our atmosphere, apart from its oxygen, comparable in composition and mass to the existing atmosphere of Venus.

The role of liquid water is critical to the evolution of a planet's atmosphere. In the Solar System, the area where water can be liquid ranges from somewhere between Earth and Venus to about Mars's orbit. Around more luminous stars, water would be liquid on exoplanets that are farther out, and around less luminous stars, closer in.

Mercury
diameter = 4900 km

Venus
diameter = 12,100 km

Earth
diameter = 12,800 km

Moon
diameter = 3500 km

Mars
diameter = 6800 km

FIGURE 9.43

Images of the terrestrial planets and the Moon to the same scale (along top), with a sampling of pictures of their surface features below each.

Role of Biological Processes

The greens and blues of Earth's surface are what distinguish it most clearly from the other terrestrial planets (fig. 9.43). Life has given Earth its distinctive coloration, and it has shaped Earth in many other ways. Biological processes help remove CO_2 from the atmosphere. For example, plants use it to make the large organic molecules such as cellulose, of which they are composed. This CO_2 is usually stored for only short periods of time, however, before decay or burning releases it back into the atmosphere. More permanent removal occurs when rain carrying dissolved CO_2 runs off into the oceans, where sea creatures use it to make shells of calcium carbonate. As these creatures die, they sink to the bottom, where their shells form sediment that eventually is changed to rock. Thus, carbon dioxide is swept from our atmosphere and locked up both chemically and biologically in the crust of our planet. With most of the carbon dioxide removed from our atmosphere, mostly nitrogen is left. In fact, our atmosphere contains roughly the same total amount of nitrogen as the atmosphere of Venus.

Our atmosphere is also rich in oxygen, a gas found in such relative abundance nowhere else in the Solar System. Our planet's oxygen is almost certainly the product of plants and microorganisms breaking down H_2O molecules during photosynthesis, as we discussed in chapter 6. Thus, the great difference between the atmospheres of Earth and Venus may be largely a result of the fact that life was able to start on Earth. On our planet, liquid water removed much of the carbon dioxide from our atmosphere, and life locked it up in fossil forms, leaving only the tiny amount (0.03%) we have today. In removing the carbon dioxide, plants also produced the oxygen on which all animal life depends. An atmosphere rich in oxygen around a terrestrial exoplanet may provide our first clue that life has arisen there.

If water is so effective at removing carbon dioxide from our atmosphere, why does any CO_2 remain? We add small amounts of CO_2 by burning wood and fossil fuels, but the major contribution is from natural processes. Atmospheric chemists hypothesize that tectonic activity gradually releases CO_2 from rock back into our atmosphere. At plate boundaries, sedimentary rock is carried downward into the mantle, where it is melted. Heating breaks down the carbonate rock so that it releases carbon dioxide, which then rises with the heated rock to the surface and reenters the atmosphere. A similar process may once have occurred on Mars to remove carbon dioxide from its atmosphere, locking it up in rock there. Mars's lower level of tectonic activity, however, prevents its CO_2 from being recycled. Thus, with so little of its original carbon dioxide left, Mars has grown progressively colder. Our Earth, because it is active, has retained enough CO_2 in its atmosphere to maintain a moderate greenhouse effect, making our planet habitable. Thus, poised between one planet that is too hot and another that is too cold, Earth has been blessed with a relatively stable atmosphere, one factor in the complex web of our environment to which we owe our existence.

SUMMARY

The terrestrial planets—Mercury, Venus, Earth, and Mars—are alike in being rocky bodies comparable in size and internal structure, but they differ dramatically from one another. Their slight differences in mass and diameter have led to great differences in their internal heat and surface activity. In fact, when ordered by mass, the terrestrial planets are also ordered by activity. Mercury, the least massive, is least active and has the oldest surface. The many impact craters created late in its formation make Mercury look much like our Moon.

Venus is nearly as large as the Earth and probably just about as active geologically. Both have young surfaces, extensively altered by volcanic activity. On Earth, the flow of heat from its core has led to plate motion. On Venus, perhaps because of a thicker crust, the flow of heat to its surface is less uniform. As a result its surface is marked by isolated regions of intense volcanic uplift.

Mars, intermediate in mass and radius between Mercury and Venus, shows a surface with an intermediate level of activity. Much of it is cratered, implying a great age. Some regions, however, show volcanic uplift, implying that its interior was once much hotter than it is now. Moreover, its surface shows an amazing variety of landforms: polar caps of ordinary ice and frozen carbon dioxide, immense deserts with dune fields, and canyons created by both crustal cracking and erosion from running water, as well as features and minerals suggesting that there may have once been lakes or oceans.

Mercury, Venus, and Mars have very different atmospheres. Mercury was too small to either build up or retain an atmosphere and is now essentially airless. Venus and Mars both have carbon dioxide atmospheres, but Venus's is about 100 times denser than Earth's, while Mars's is about 100 times less dense than Earth's. The great thickness of the Venusian atmosphere creates a strong greenhouse effect, heating it to about 900°F. Thick clouds of sulfuric acid droplets high in the Venusian atmosphere block our view of the Venusian surface. Astronomers have nevertheless been able to study the Venusian surface with radar telescopes both on Earth and in orbit around Venus.

Mars's atmosphere is too thin now for a significant greenhouse effect, but the evidence of liquid water in its past implies that Mars once had a denser and warmer atmosphere, perhaps even resembling Earth's in its early history.

The great differences among the atmospheres of Venus, Earth, and Mars can probably be explained by the differences in their water content and the fact that life evolved on Earth.

QUESTIONS FOR REVIEW

1. (9.1–9.3) How do Mercury, Venus, and Mars each compare with the Earth in mass and radius?
2. (9.1–9.3) How do the interior structures of Mercury, Venus, and Mars compare with Earth's?
3. (9.1) What is the surface of Mercury like?
4. (9.1) Does Mercury have an atmosphere? Why or why not?
5. (9.1) What is peculiar about Mercury's rotation? What causes this oddity?
6. (9.2) What is the dominant gas in Venus's atmosphere? How do astronomers know this?
7. (9.2) What are the clouds of Venus made of?
8. (9.2) Why is Venus so hot?
9. (9.2) Can we see the surface of Venus? Why or why not? How do astronomers know what the surface of Venus is like?
10. (9.2) What sort of features are seen on Venus's surface? Is the surface young or old? Some of the youngest surfaces on Earth are from plate ridges—is the same true on Venus?
11. (9.3) Describe some of the surface features seen on Mars.
12. (9.3) What are the Martian polar caps composed of?
13. (9.3) What is the Martian atmosphere like?
14. (9.3) What is the evidence that Mars once had running water on its surface?
15. (9.3) What is the likely origin of Mars's two moons? What evidence supports this?
16. (9.3) What is the evidence that leads some scientists to believe life may have existed on Mars?
17. (9.4) What explanations have been offered for why the atmospheres of the terrestrial planets are so different?
18. (9.4) How do astronomers explain why the Earth's atmosphere ended up with so little CO_2, compared with that of Mars and Venus?

THOUGHT QUESTIONS

1. (9.1) What are some examples of resonances in everyday life?
2. (9.2) How does the surface temperature of Venus relate to concerns about global warming of the Earth?
3. (9.3) Suppose scientists discover simple life forms on Mars. Would that alter the way you look at life here on Earth?
4. (9.3) Would Olympus Mons be as large if it were on Earth? Why or why not?
5. (9.3) The average surface pressure on Mars is almost exactly equal to that below which water cannot be liquid. Review the topographic map of Mars in figure 9.21. If you wanted to search an area that had the greatest probability of any location on Mars for liquid water today, where would you go and why?
6. (9.2–9.4) Thinking about the relative roles of sunlight and greenhouse effect on temperature, volcanic activity and the gases it liberates, liquid water and a liquid water cycle, the presence or removal of CO_2 in the atmosphere, and the ability to retain an atmosphere, make an argument as to whether Mars or Venus would be easier to "terraform" into a condition suitable for colonization (by introducing organisms that would use carbon dioxide to photosynthesize oxygen).
7. (9.2–9.4) Reconcile the idea that Mars, Earth, and Venus all have approximately the same internal structure and composition, and the fact that the average density of Mars is nearly 1 gm/cm³ less than that of Earth or Venus.
8. (9.1–9.4) What evidence do we see from the terrestrial planets that supports the theory for the origin of the Solar System discussed in chapter 8?
9. (9.1–9.4) What role might planetesimal impacts have played in the history of the terrestrial planets?

PROBLEMS

1. (9.1/3.8) Calculate the escape velocity needed to launch a rocket from the surface of Mercury.
2. (9.2) Venus rotates (spins) clockwise while orbiting the Sun counterclockwise. Draw a diagram to show how, for a "retrograde" planet like Venus, the time from solar noon to solar noon is shorter than the time required to spin 360 degrees (the opposite is true for Earth—see Essay3 fig. E3.1). Try using a rotation period of about half of the orbital period.
3. (9.2) How long would it take to send a radio signal to a spacecraft at Venus when Venus is closest to Earth?
4. (9.3) Earth's atmospheric pressure drops by about a factor of 2 every 5 km higher up you go, so it is one-fourth the surface pressure at 10 km, for example. How high up in Earth's atmosphere would you have to go to reach a pressure similar to the 0.6% atmospheric pressure on Mars? (You can do this mathematically with logarithms or plot the changing pressure with elevation and estimate the height where this pressure is reached.)
5. (9.3/3.7) Olympus Mons is 26 kilometers above the surrounding surface of Mars. Calculate the surface gravity on Mars at the surface and on top of Olympus Mons (neglecting any effects on gravity from the mountain itself). What percentage of the surface value is the difference in the acceleration of gravity?
6. (9.2–9.4) In chapter 13, we will learn that the brightness of light decreases in proportion to the square of the distance from the source. We know sunlight plays an important role in the conditions on the planets and also is used to power equipment like the Mars rovers. How many times brighter is sunlight at Venus's distance from the Sun than at Earth? How many times dimmer is it at Mars than at Earth? (Distances from the Sun compared to Earth's can be found in table 9.1 or in the appendix.)

TEST YOURSELF

1. (9.1) Why does Mercury have so many craters and the Earth so few?
   (a) Mercury is far more volcanically active than the Earth.
   (b) Mercury is much more massive than the Earth and therefore attracts more impacting bodies.
   (c) The Sun has heated Mercury's surface to the boiling point of rock, and the resulting bubbles left craters.
   (d) Erosion and plate tectonic activity have destroyed most of the craters on the Earth.
   (e) Mercury's iron core and its resulting strong magnetic field have attracted impacting bodies.

2. (9.1) Mercury's average density is about 1.5 times greater than the Moon's, even though the two bodies have similar radii. What does this suggest about Mercury's composition?
   (a) Mercury's interior is much richer in iron than the Moon's.
   (b) Mercury contains proportionately far more rock than the Moon.
   (c) Mercury's greater mass has prevented its gravitational attraction from compressing it as much as the Moon is compressed.
   (d) Mercury must have a uranium core.
   (e) Mercury must have a liquid water core.

3. (9.2) Why is Venus's surface hotter than Mercury's?
   (a) Venus rotates more slowly, so it "bakes" more in the Sun's heat.
   (b) Clouds in Mercury's atmosphere reflect sunlight back into space and keep its surface cool.
   (c) Carbon dioxide in Venus's atmosphere traps heat radiating from its surface, thereby making it warmer.
   (d) Venus is closer to the Sun.
   (e) Venus's rapid rotation generates strong winds that heat the ground by friction as they blow.

4. (9.2) What are the Venusian clouds made of?
   (a) Mostly water droplets
   (b) Frozen carbon dioxide
   (c) Sodium chloride
   (d) Sulfuric acid droplets
   (e) Ice crystals

5. (9.3) Astronomers can measure the time since the last eruption of a volcano on Mars by
   (a) using radioactive dating.
   (b) counting the number of craters on its slopes.
   (c) measuring how much the lava has cooled.
   (d) searching ancient records of Chinese astronomers.
   (e) estimating based on the time since an eruption on Earth.

6. (9.3) What evidence suggests that liquid water was once present on the Martian surface? (Select all that are correct.)
   (a) Branching channels in the shape of riverbeds
   (b) High levels of humidity measured spectroscopically in the atmosphere
   (c) Teardrop-shaped islands behind craters.
   (d) Microscopic fossil creatures found by the Martian rovers
   (e) Compounds that form in water have been identified in rocks.

7. (9.4) Which body in the inner Solar System has the densest atmosphere?
   (a) Mercury   (b) Venus   (c) Earth   (d) Mars

8. (9.4) Which of the following features are shared by all of the terrestrial planets? (Select all that apply.)
   (a) A silicate mantle   (c) A strong magnetic field
   (b) An iron core   (d) Volcanic activity

KEY TERMS

photo dissociation, 247
resonance, 228
runaway greenhouse effect, 230
scarp, 224

: FIGURE QUESTION ANSWERS

WHAT IS THIS? (chapter opening): These are sand dunes on Mars imaged by the *Mars Global Surveyor.*

FIGURE 9.3: The scarp must be younger because it cuts the craters. If the craters had formed after the scarp, they would have destroyed the scarp within the craters.

FIGURE 9.4: The Caloris Basin is round and large, but its interior is rougher and lighter than lunar maria.

FIGURE 9.10: Mercury's rotation, like our Moon's, has been slowed by tidal interactions. Mercury tries to rotate so that its spin and orbital motions match, but because its orbit is so elliptical, and its orbital speed therefore changes, it can't make the match occur.

FIGURE 9.11: Venus does not look round in this image because we are seeing it with the Sun off to the left. Basically, we are seeing here a gibbous phase of Venus, which is visible from Earth only when Venus is on the far side of the Sun.

FIGURE 9.19: Because the south pole is illuminated, it must be spring or summer in the southern hemisphere. The polar cap has not yet shrunk to its smallest, so you are correct if you guessed spring.

FIGURE 9.38: Their irregular shape implies that they have only a weak surface gravity. If they had a stronger surface gravity, it would crush them into a rounder shape. Their weak surface gravity would make it virtually impossible for them to retain any atmosphere.

10

THE OUTER PLANETS

The satellite Io imaged in front of Jupiter's stormy atmosphere by the *Cassini* spacecraft.

LEARNING OBJECTIVES

Upon completing this chapter you should be able to:

- Describe and contrast the basic characteristics of the four outer planets, and compare them in size to the Earth.
- Explain the causes of Jupiter's banded atmosphere, and how it differs in appearance from Saturn.
- Explain the internal sources of heat in the Jovian planets.
- Summarize the properties of Jupiter's Great Red Spot, and explain what drives its rotation.
- Describe and contrast how Uranus and Neptune were discovered.
- Explain why Uranus and Neptune look blue.
- Describe the likely internal structures of Uranus and Neptune, and their similarities to and differences from Jupiter.

- Describe the major moons of the outer planets, and identify the unique geological features associated with several.
- Discuss what features of the satellite systems suggest that they formed along with the planet, except for Neptune.
- Explain why several of the outer planets' moons exhibit geological activity despite their small size and age.
- Contrast the properties of the ring systems around the Jovian planets.
- Discuss the ways in which rings can form and dissipate, and explain the relevance of the Roche limit.

Beyond Mars, the Solar System is a realm of ice and frozen gas. In this frigid zone, far from the Sun, where solar heat is only a vestige of what we receive on Earth, the giant planets formed. The low temperatures—about −100° Fahrenheit and colder—allowed objects condensing there, within the solar nebula, to capture hydrogen-rich ices such as methane, ammonia, and especially water. The largest of the bodies forming there had a strong enough gravitational attraction to pull in hydrogen and helium gas from the solar nebula.

Because hydrogen and hydrogen-rich molecules were far more abundant in the young Solar System than the silicate- and iron-rich material from which the terrestrial planets condensed, planets that formed in this cold environment had more material available for their growth. As a result, these cold planets became vastly larger than those near the Sun, and they developed very different structure and composition. The four largest planets—Jupiter, Saturn, Uranus, and Neptune—are composed mainly of gaseous and liquid hydrogen and its compounds. Although these giant bodies probably have cores of molten rocky matter, they lack solid surfaces and consequently have no surface features, such as mountains and valleys. Rather, it is their atmospheric features that give them such different appearances. Understanding these atmospheres is one of this chapter's goals. For example, it is popular to illustrate large exoplanets as looking like Jupiter and smaller ones as looking like Neptune, but we will see that distance from a star and the rate of rotation are probably more important in determining a giant planet's outward appearance. Mass is more important for determining the planet's internal structure.

The moons of these giant planets range in size from just a kilometer in diameter to larger than Mercury, and these planets all have ring systems made up of uncountable numbers of particles ranging from the size of buildings down to dust particles. As they orbit their parent planet, the moons form a family rather like a miniature Solar System around each of the giant planets. Some of the larger moons have brightly colored surfaces, others have numerous craters, and a few have surface features unlike anything seen elsewhere in the Solar System. A few moons even have active volcanoes or subsurface oceans. In fact, astronomers consider these diverse bodies, virtually unknown before the space age, to be some of the most interesting members of the Sun's family.

CONCEPTS AND SKILLS TO REVIEW

- Density (6.1)
- Coriolis force (6.7)

10.1   JUPITER

For the ancient Romans, Jupiter was the king of the gods. They did not know how immense the planet Jupiter is, so they probably named it for its steady brightness and stately motion around the ecliptic. Through even a small telescope, Jupiter is clearly different from the other giant planets. Richly colored parallel bands of clouds cloak the planet, as shown in figure 10.1.

Jupiter is the largest planet in our Solar System, both in radius and in mass. It is slightly more than 11 times the Earth's diameter and more than 300 times its mass. And while it is just slightly larger than Saturn, it is more than 3 times its mass. In fact, its mass is greater than that of all other planets in the Solar System combined.

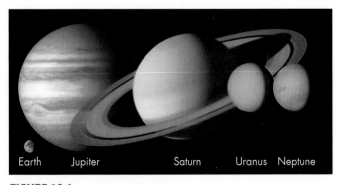

FIGURE 10.1
The giant outer planets compared to Earth.

Belts

Zones

Great Red Spot

FIGURE 10.2
Jupiter and its clouds.

Jupiter's Outer Atmosphere

Alternating dark **belts** and white **zones** that girdle Jupiter and its **"Great Red Spot"** (fig. 10.2) have been seen on the planet since some of the first telescopic observations made in the 1600s. These can shift in position and change in appearance over time, but the belts and zones appear to be semi-permanent features of the atmosphere.

Spectra of the sunlight reflected from these clouds show that Jupiter's atmosphere consists mostly of hydrogen, helium, and hydrogen-rich gases such as methane (CH_4), ammonia (NH_3), and water (H_2O). These gases were also directly detected in 1995 when a probe dropped from the *Galileo* spacecraft parachuted into Jupiter's atmosphere. The clouds themselves are more difficult to analyze, but theoretical calculations of the chemistry of Jupiter's atmosphere suggest they are particles of water ice and ammonia compounds. Their colors may come from complex organic molecules or compounds of sulfur or phosphorus. Time-lapse pictures show that Jupiter's clouds move swiftly, sweeping around the planet in jet streams that are far faster than those of Earth. Moreover, Jupiter itself rotates once every 10 hours, spinning so fast that its equator bulges significantly.

Jupiter's Interior

Astronomers cannot see through Jupiter's cloud layers to its interior, nor can they probe its interior with seismic detectors. Instead, they must rely on theory to tell them what lies inside this giant planet. For example, despite its great mass, Jupiter is much less dense than the Earth. To determine a planet's average density, astronomers divide its mass by its volume, using the method described in chapter 8 (see fig. 8.8). When such calculations are made for Jupiter, we find that its average density is only slightly greater than that of water—1.3 grams per cubic centimeter—showing that the bulk of the planet, not just its atmosphere, must be composed mainly of very light elements, such as hydrogen. Calculations based on Jupiter's low density, its shape, and measurement of its gravitational attraction on its satellites and passing spacecraft all give astronomers information on what lies below its clouds.

Jupiter's immense mass exerts a tremendous gravitational force on its atmosphere and interior, holding the planet together and compressing its gas. Near the cloud tops, this compression is only slight, so the gas density there is low. Deep in the interior, however, the weight of thousands of kilometers of gas compresses the matter to a high density (approximately 3 times the density of iron). Thus, the density of the gas increases with depth, as illustrated in figure 10.3A. Everywhere in the planet, however, the forces of gravity and pressure must balance, just as they do in our own atmosphere.

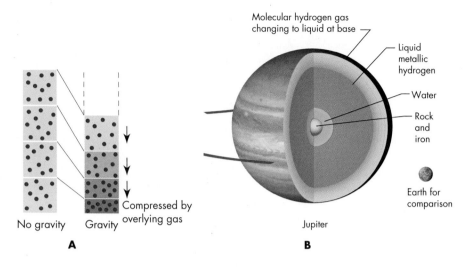

Molecular hydrogen gas
changing to liquid at base

Liquid
metallic
hydrogen

Water

Rock
and
iron

Earth for
comparison

Jupiter

No gravity   Gravity   Compressed by
overlying gas

A

B

Q: From the change in volume
shown here in (A), how
many times more dense is
the bottom, compressed
box than the top box?

FIGURE 10.3
(A) The density of gas increases with
depth because the overlying gas compresses the matter. (B) A sketch of what
astronomers think Jupiter's interior is like.

Deep within Jupiter, the compression created by its gravity presses molecules so close together that the gas changes to liquid. About 10,000 kilometers below the cloud tops—about one-sixth of the way into the planet—Jupiter's interior is a vast sea of liquid hydrogen. Deeper still, the weight of the liquid hydrogen compresses matter below it into a state known as liquid metallic hydrogen, a form of hydrogen that scientists have been trying to duplicate for decades in experiments on Earth, but the forces involved overwhelm almost any apparatus. Astronomers theorize that if the material from which Jupiter formed has the same overall composition as the Sun's, then Jupiter must also contain heavy elements such as silicon and iron. Because of their high density, these elements have probably sunk to Jupiter's center, forming a core of iron and rocky material a few times larger than that of the Earth (see fig. 10.3B). Detailed measurements of Jupiter's gravitational field made by orbiting satellite also indicate the presence of a core with a mass perhaps as high as 18 times the mass of the Earth.

Jupiter's interior is very hot, perhaps 30,000 K, which is about five times hotter than the Earth's core. This heat rises slowly to the planet's surface and escapes into space as low-energy infrared radiation. Astronomers have long known from measurements of this infrared radiation that Jupiter emits more energy than it receives from the Sun, but they are still uncertain what supplies that heat. Some of the energy is probably left over from Jupiter's formation. As we discussed in chapter 6, the formation of the Earth or any planet is a hot process. The hail of gas and planetesimals onto the forming world releases a huge amount of gravitational energy. Subsequently, additional heat may be generated by slow but steady shrinkage of the planet as it adjusts to the greater gravity created by the mass added in the last stages of its formation. In fact, giant gas planets such as Jupiter may still be shrinking slightly, which can continue to generate heat from gravitational energy. Heat is also generated as denser matter sinks toward the core. Whatever its source, the heat generates convection currents similar to those in the Earth's interior. These currents stir both Jupiter's deep interior and its atmosphere.

Circulation of Jupiter's Atmosphere

Heat within Jupiter generates convection currents in its outer layers that carry warm gas upward to the top of its atmosphere. There, the gas radiates its heat to space, becoming cooler and sinking again. As in Earth's atmosphere, when a rising mixture of gases cools, ice particles condense. On Jupiter it is primarily ammonia (NH_3) ice that condenses, forming the white clouds of the zones. As the gas sinks and warms, the cloud particles evaporate again, leaving the upper atmosphere more transparent. The darker belts are seen through these clearer regions, and the warmth of these deeper layers is visible in infrared images of the planet, as shown in figure 10.4. The structure of the high clouds varies over time, making some of the belts disappear on occasion (fig. 10.4), but generally the pattern remains similar over time.

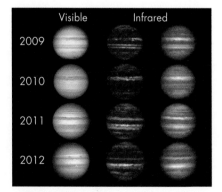

FIGURE 10.4
Series of visible and infrared images of
Jupiter showing the changing atmosphere.
The bright regions in the infrared images
show the warmer, deeper layer in the
belts, and dark cold regions in the zones.

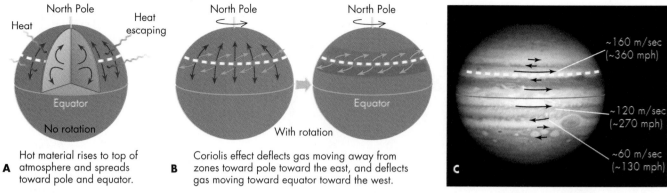

A Hot material rises to top of atmosphere and spreads toward pole and equator.

B Coriolis effect deflects gas moving away from zones toward pole toward the east, and deflects gas moving toward equator toward the west.

No rotation

With rotation

Heat

North Pole

Heat escaping

Equator

North Pole

North Pole

Equator

~160 m/sec (~360 mph)

~120 m/sec (~270 mph)

~60 m/sec (~130 mph)

C

FIGURE 10.5

(A) Rising gas from Jupiter's hot interior cools near the top of the atmosphere and sinks. (B) The Coriolis effect, arising from Jupiter's rotation, deflects the gas, creating winds that blow as narrow jet streams. (C) The high-speed winds reverse direction from region to region.

Gas rising to produce the high clouds in the white zones spreads out (fig. 10.5A). Because of this motion, the gas is subject to a Coriolis effect. We examined the Coriolis effect in chapter 6 when we looked at Earth's atmospheric circulation. As gas moves closer to the planet's rotation axis, it must speed up to conserve angular momentum, and when it moves farther from the rotation axis, it must slow down relative to the planet's rotation. If the gas moves toward the equator, the gas is deflected to the east, but if it moves toward a pole, it is deflected to the west (fig.10.5B). This creates a series of jet streams that can have velocities of up to 600 kilometers per hour (nearly 400 miles per hour) with respect to the planet's overall rotation. In adjacent regions they blow in opposite directions, as shown in figure 10.5C. Such reversals of wind direction (wind shear) from place to place also occur on Earth, where equatorial winds generally blow from east to west, while mid-latitude winds generally blow from west to east.

As the various jet streams circle Jupiter, gas between them rising from below may be spun into huge, whirling atmospheric vortices, much as a pencil between your palms twirls as you rub your hands together. Most of these spinning regions are shades of brown and white, but one exceptionally large vortex—bigger across than the Earth—is nearly brick red (fig. 10.6). Known as the Great Red Spot, this vortex was discovered in the seventeenth century. Since then, it has changed slightly in appearance and location, but both direct observations of the spot and theoretical models show that it may be an essentially permanent feature of Jupiter's atmospheric circulation.

ANIMATION ▶

The rotation of Jupiter

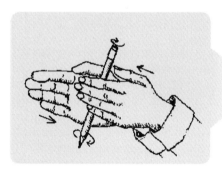

FIGURE 10.6

Vortices form between atmospheric streams of different velocities. Note the Great Red Spot and the white oval below it. The inset figure illustrates that the white zones move around Jupiter faster than the dark belts, creating a relative motion that can produce a vortex.

Great Red Spot

Matter also circulates in Jupiter's deep interior. There, convection in the metallic liquid hydrogen combines with the planet's rapid rotation to generate a magnetic field by a natural dynamo process similar to that which generates the Earth's magnetic field. Jupiter's dynamo process is far more powerful than Earth's, however, and creates the strongest magnetic field of any planet in the Solar System. Measurements taken by passing spacecraft show that the field near the top of Jupiter's cloud layer is about 20 times the strength of the magnetic field on the Earth's surface. When Jupiter's larger radius is taken into account, its field is some 20,000 times as powerful as Earth's. This was deduced even before it was directly measured both from Jupiter's auroral activity and from intense radio emission emitted by the giant planet.

The Earth's auroral activity is linked to its magnetic field, which steers incoming energetic particles from the Sun into our upper atmosphere, where the particles trigger the lovely pale glow of the northern and southern lights. Our planet's magnetic field also traps such particles in the Van Allen radiation belts. Jupiter's magnetic field does much the same: particles trapped magnetically in radiation belts far above the planet emit radio waves, making Jupiter a powerful source of radio emission. Some of those particles descend into Jupiter's upper atmosphere, where they create an aurora, as you can see in figure 10.7A, a picture made with the Hubble Space Telescope.

Jupiter's aurora also shows in figure 10.7B as the pale glow along the edge of the planet in a picture taken by the passing *Voyager* spacecraft. This figure also depicts another familiar phenomenon in Jupiter's atmosphere: lightning. Cloud particles carried up and down by the rising and sinking atmospheric motions collide and generate atmospheric electricity—thunderstorms—just as such motions do on Earth. In the picture, you can see many storms lighting up the Jovian night.

Jupiter's Rings

Jupiter is orbited by a system of very faint rings that are very difficult to see from Earth because the particles reflect back very little light. They were discovered after the *Voyager 1* spacecraft flew past Jupiter in 1979. Pictures taken from the craft looking back at the nightside of Jupiter clearly show rings, although very thin ones, as shown in figure 10.8A. More recently, the *New Horizons* spacecraft was able to image the rings from in front of Jupiter and from behind (fig. 10.8B). The inner parts of the ring are filled with small dust-size particles that show up much more clearly from behind as they scatter the Sun's light.

Jupiter's ring system is thought to be made of tiny dark particles of rock dust held in orbit by Jupiter's immense gravitational attraction. These particles are so tiny, however, that radiation from the Sun and collisions with gas trapped in Jupiter's magnetic field exert frictional forces on them, making them gradually drift down into Jupiter's atmosphere. There, they mingle with the swirling gas and are lost from sight. Thus, for the ring to persist, new dust particles must constantly be added to it. Where does this new ring material come from? It appears to come from two tiny satellites that orbit at the outer edge of these rings. The satellites suffer occasional impacts and the collision fragments contribute new dust to replenish the rings.

FIGURE 10.7
(A) Jupiter's auroral zone as seen by the Hubble Space Telescope. (B) Lightning on Jupiter, as observed by the *Voyager* spacecraft as it flew over Jupiter's nightside.

FIGURE 10.8
(A) Jupiter's rings shown in a montage of *Voyager 1* images from behind the planet, which show the small dusty particles more clearly. (B) The *New Horizons* spacecraft imaged the rings both from in front (top) and from behind (bottom) as it passed Jupiter on its way to Pluto.

Jupiter's Moons

When Galileo first viewed Jupiter with his telescope, he saw four moons orbiting the planet, bodies now called the "Galilean satellites" (fig. 10.9). The Galilean satellites—Io, Europa, Ganymede, and Callisto—are very large: all but Europa are larger than our Moon, and Ganymede is the largest moon in the Solar System, even larger in diameter than Mercury. Jupiter's fifth largest moon, Amalthea, is tiny by comparison, as can be seen in figure 10.10. Over the past three centuries, astronomers have found additional Jovian moons, with the total now reaching 67.

The Galilean satellites orbit approximately in Jupiter's equatorial plane, forming a flattened disk, rather like a miniature Solar System. They probably formed by a scaled-down version of the process that created the Solar System, aggregating from planetesimals that collected around Jupiter during its formation. Some astronomers think that Jupiter heated this orbiting debris, affecting the composition of the moons that formed by evaporating ices nearer the planet. The moons nearest Jupiter would then form with less ice and more rock, making them denser, much as the terrestrial planets near the Sun are denser than the outer planets. The densities of the Galilean satellites do in fact increase according to distance, with densities (in units of grams per cm³) of 3.53 for Io, 2.99 for Europa, 1.94 for Ganymede, and 1.85 for Callisto. Further evidence of heating comes from their strange, often colorful surface features, as illustrated in figure 10.10, and in the case of Io, from its active volcanoes.

Io (pronounced *eye-oh* or *ee-oh*) is named for a mythological maiden with whom Jupiter fell in love and whom he changed into a cow so that his wife, Hera, would not suspect his infidelity. Io is the nearest to Jupiter of the Galilean moons and therefore is subject to a strong tidal force created by Jupiter's gravity. That tidal force locks Io's spin to its orbital motion in the way our Moon's spin is locked to its motion around the Earth. But Io also undergoes a strong gravitational attraction from Europa, the Galilean satellite next closest to Jupiter, and Europa's gravitational tug twists Io from side to side. Moreover, Europa's pull forces Io into an orbit whose shape constantly changes. The outcome of these two effects is that Io is subject to a strong and changing gravitational force from Jupiter that distorts its shape. This deformation heats it by internal friction, much as bending a paperclip back and forth heats the wire (touch the bent spot to your lip and it will feel hot).

The heating drives eruptions on Io, creating volcanic plumes and lava flows, as shown in figure 10.11A. Sulfur, common in terrestrial volcanoes as well, is now the major component of Io's volcanic outpourings, and the erupted sulfur compounds give Io its rich red, yellow, and orange colors. A recent sulfur lava flow can be seen in figure 10.11B where the molten material has spread from the dark volcanic peak. (The peak's darker color indicates that the sulfur is molten there.) Over billions of years,

FIGURE 10.9
Jupiter and the Galilean satellites as seen through a small telescope. (This image was made by the Louisiana State Amateur Astronomy Club on March 19, 2004.) The starlike dots are the four large moons of Jupiter. From left to right: Europa, Io, Callisto, and Ganymede.

FIGURE 10.10
The four Galilean satellites—Io, Europa, Ganymede, and Callisto— and Jupiter's fifth largest satellite, Amalthea, compared to our Moon.

FIGURE 10.11

(A) *Galileo* image of Io. Note volcanic eruptions. (B) Close-up view of lava flow on Io. The colors are produced by sulfur compounds. (C) Model of Io's interior.

Q: What explanation has been offered for the source of heat in Io?

the heating has melted and driven out whatever ices were originally present, leaving a rocky interior and an iron core (fig. 10.11C).

Europa, the smallest of the Galilean moons, is named for another maiden whom Jupiter pursued. According to the legend, Jupiter disguised himself as a bull and carried her on his back across the Hellespont from Asia to Europe, thereby giving the continent its name. Europa looks rather like a cracked egg. Long, thin lines score its surface, as shown in figure 10.12A. The white material is probably a crust of ice, while the red material is probably mineral-rich water that oozed to the surface through the cracks and then froze. Close-up images made by the *Galileo* spacecraft show features reminiscent of ice floes (fig. 10.12B). The absence of large craters on its surface suggests that some process has eradicated them, for Europa's surface—like that of Ganymede and Callisto—must surely have been cratered during and following its formation. To explain the lack of craters, astronomers think that Europa may be heated by Jupiter's gravitational forces deforming it, although not as strongly as Io is. That heat, in combination with a small amount from radioactive decay of rocky material in its core, may be sufficient to keep a layer of water melted beneath Europa's crust, forming an ocean (fig. 10.12C). Moreover, the heat may soften the surface ice, allowing it to "flow," glacier-like, and obliterate craters as they form. Some astronomers speculate that Europa's ocean might harbor life forms, and the possibility of a mission that could send a probe through the ice crust to explore the ocean beneath it is being examined.

When the Galileo probe ran low on the fuel, NASA deliberately crashed it into Jupiter, where it burned up. This eliminated the chance of a crash into Europa. That is important because some scientists think that life might have developed in the liquid water below Europa's icy crust, and they don't want to risk contaminating it with terrestrial microorganisms.

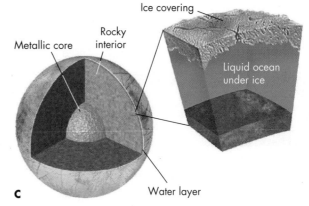

FIGURE 10.12

(A) Europa, as imaged by the *Galileo* spacecraft. (B) Close-up picture of Europa's surface showing fractures and spots where briny liquid from the interior has flowed onto the surface. (C) A model of Europa's interior with a liquid water ocean beneath the ice.

FIGURE 10.13
Images of Ganymede (top) and Callisto (bottom). (A) The surfaces of both are dark, but Ganymede has long linear features. (B) Close-up of regions about 70 km (45 miles) across. (C) Models of the interior of each satellite.

FIGURE 10.14
Two views of Jupiter's fifth largest satellite, Amalthea, which is about 250 kilometers (150 miles) in its longest dimension. It is an icy object coated with a layer of sulfur compounds from the moon Io's volcanoes.

Ganymede and Callisto, named for yet two more of Jupiter's loves, look somewhat like our own Moon. They are grayish brown and covered with craters made during the late stages of their formation, much as our own Moon was pockmarked by impacting debris (fig. 10.13). But these similarities to our Moon are only superficial, because the surfaces of Ganymede and Callisto are probably mostly ice, the white of the craters being similar to the white you see on ice cubes when they are shattered.

Ganymede's surface is less heavily cratered than Callisto's, implying that craters in the smoother areas have been destroyed, perhaps much as happened on our own Moon. That is, astronomers think that after Ganymede's surface solidified and was heavily cratered, a few large bodies may have hit it. These large bodies created basins that subsequently flooded with water and then froze into a relatively smooth surface with few craters. The smooth regions have curious parallel ridges (fig. 10.13B), possibly created by tectonic forces as the water in them froze.

From the average density of the Galilean satellites, astronomers deduce that their interiors are mostly a mix of icy and rocky material. Heating of these moons during their youth would have allowed rock and iron in them to sink to their centers and form cores, like the differentiation that occurred in our own planet. The internal distribution of mass has an effect on the trajectories of spacecraft passing near each moon, and scientists have used this information to show that all but Callisto appear to be clearly differentiated. Ganymede in particular appears most strongly differentiated, and it is the only moon with a strong magnetic field (several times Mercury's) suggesting that it has a partially molten iron core. Also, by studying how the moons interact with Jupiter's magnetic field, scientists have deduced from observations of Ganymede and Callisto that both these moons may have a liquid water ocean below their icy crusts, like Europa.

Jupiter's other moons are much smaller than the Galilean satellites. Amalthea (fig. 10.14) has the next largest extent at 250 km (150 miles) with a highly irregular shape and a density that suggests it is composed in large part of ice or is loosely assembled material. Most of rest are much smaller, with large orbits tilted steeply relative to Jupiter's equator, suggesting that these objects may be captured asteroids. Most of these are too small to be readily seen from Earth and were discovered by examining pictures taken by the *Voyager* and *Galileo* spacecraft.

10.2 SATURN

Saturn bears the name of an ancient Roman harvest god. Saturn came to be identified with Cronus (also spelled Kronos), who in Greek mythology was the last-born of the Titans and the father of the Olympian gods, including Jupiter. Saturn is the second largest planet in our Solar System and lies about twice as far as Jupiter from the Sun, at about 10 AU.

Saturn's Appearance and Structure

Saturn's diameter is about 9.5 times larger than the Earth's. Saturn's mass is about 95 times that of the Earth, and its average density is very small—only 0.7 grams per cubic centimeter, which is less than the density of water. Such a low density suggests that Saturn, like Jupiter, is composed mostly of hydrogen and hydrogen-rich compounds. Spectra of the planet bear this out, and astronomers think that Saturn is similar to Jupiter in its composition and internal structure, but not as compressed or as hot in its core, as depicted in figure 10.15.

Saturn radiates more energy than it gains from the Sun, implying that, like Jupiter, it has an internal heat source. Based on the predicted conditions inside Saturn, astronomers think that much of Saturn's heat comes from deep beneath Saturn's cold clouds, where helium droplets condense in its atmosphere, much as water droplets condense in Earth's atmosphere. As the helium droplets fall toward Saturn's core, they release gravitational energy, generating heat by friction as they press down through the compressed hydrogen.

If Jupiter and Saturn both have hot interiors and similar compositions, why does Saturn show only faint markings compared with the striking patterns seen on Jupiter? Saturn's greater distance from the Sun and its consequently lower temperature may provide an answer. Saturn's atmosphere is cold enough for ammonia gas to freeze into cloud particles that veil its atmosphere's deeper storms, much as the white zones on Jupiter cover the colorful belt layers below them. At near-infrared wavelengths (fig. 10.16A), we can see that Saturn has similar bands beneath its hazy outer layer. Every few decades, major storms do push up all the way to where they are visible near the surface layers, such as one the *Cassini* spacecraft observed in 2011 (fig. 10.16B). Saturn's north pole is surrounded by an unusual hexagonal jet stream (fig. 10.16C). Within it, clouds and storms can be seen, with a deep hurricane-like vortex over the pole where gas spirals downward. A similar vortex is located over the south pole.

FIGURE 10.15
Internal structure of Saturn.

FIGURE 10.16
(A) A false-color image of Saturn, made at infrared wavelengths by *Cassini*, reveals complex cloud belts and storms in Saturn's atmosphere. (B) A large storm erupted in the early spring of the northern hemisphere in 2011. (C) False-color image of a north polar vortex made at infrared wavelengths, with colors indicating relative heights of cloud layers—deep layers red and high layers green.

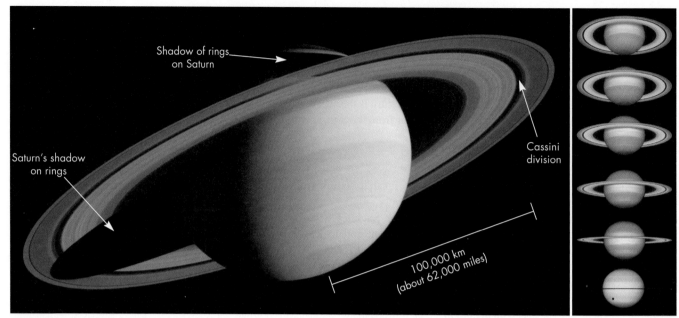

FIGURE 10.17

Saturn and its rings as imaged by the *Cassini* spacecraft are shown at left. At right, a sequence of images from the Hubble Space Telescope shows the changing appearance of the rings over a quarter of Saturn's orbit. These range from the south pole tilted most toward the Sun (top) to about 7 Earth years later when the Sun shines directly down on the equator and the rings are almost invisible (bottom).

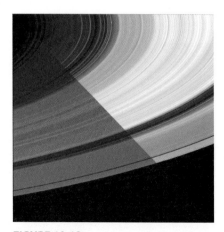

FIGURE 10.18

Cassini image detailing the substructure of Saturn's rings. Slight color differences have been greatly exaggerated over half the image to illustrate differences in the ringlets' composition.

Saturn's Rings

Saturn's spectacular rings, illustrated in figure 10.17, were first seen by the astronomer Galileo. Through his small, primitive telescope, however, they looked like "handles" on each side of the planet, and it was not until 1659 that Christiaan Huygens, a Dutch scientist, observed that the rings were detached from Saturn and encircled it.

The rings are very wide but very thin. The main band extends from about 30,000 km above the top of Saturn's atmosphere to a little more than twice the planet's radius (136,000 km), as illustrated in figure 10.17. Some faint inner rings can be seen even closer to Saturn, and faint outer rings extend considerably farther from the planet. Yet despite the rings' immense breadth, they are probably less than a few hundred meters thick—so thin that stars can be seen through them.

The British physicist James Clerk Maxwell, a pioneer in the study of electromagnetism, demonstrated that the rings must be a swarm of particles. He showed that no material could be strong enough to hold together in a solid sheet of such vast size. Spectra of the rings support Maxwell's theory: the inner and outer parts orbit Saturn at different velocities, obeying Kepler's third law, as shown by their Doppler shift. Thus, the rings must be a swarm of individual bodies.

Astronomers have since discovered that the ring particles are relatively small, only a few centimeters to a few meters across. Although these particles are far too small to be seen individually with telescopes, they reflect radar signals bounced off them. From the strength of the radar "echo," astronomers can estimate the particle sizes. More precise measurements were made with radio signals from the *Voyager* spacecraft. As *Voyager* transmitted data to Earth, the signals passed through the rings, scattering slightly from the particles. From the amount of scattering of the *Voyager* signal and the radar waves, astronomers can deduce not only the size of the ring particles but also their composition.

Better information about the composition of the rings, however, comes from analyzing the spectrum of sunlight reflected from them. Such analyses show that the rings are composed primarily of water-ice. However, the *Cassini* spacecraft pictures, such as figure 10.18, show that some parts of the rings have different colors than others, implying that the composition of the rings is not the same everywhere. Particles in the darker ring segments may be rich in carbon compounds similar to those found in some asteroidal material. Figure 10.18 also shows that the rings are not uniformly filled: they consist of numerous ringlets. Large gaps in the rings had been seen from Earth, but the many narrow gaps, forming "ringlets" in the rings, came as a surprise.

In 1866, Daniel Kirkwood, an astronomer who also discovered gaps in the distribution of asteroids (chapter 11), noticed that the largest gap—known as Cassini's division—occurs where ring particles orbit Saturn in exactly one-third the orbital period of its moon Enceladus. Thus, any ring particle that attempted to orbit in the gap would undergo a strong and repeated gravitational force from Enceladus every third orbit. Physicists call this sort of repetitive action that reinforces itself a **resonance.** Kirkwood concluded that over long periods of time the cumulative effect of Enceladus's force would pull particles from the gap, creating Cassini's division. A few years later, Kirkwood revised his theory of the gap to account for the action of the four largest Saturnian moons. Today astronomers think that Saturn's moon Mimas, whose period is twice that of particles in Cassini's division, primarily causes that large gap.

The many narrow gaps apparent in figure 10.18 have a different cause. Narrow gaps probably arise from a complex interaction between the ring particles and the tiny moons that orbit within the rings. As these moonlets—only tens of kilometers or less in size—orbit Saturn, their gravitational attraction on the ring particles generates waves. These waves spread through the rings much like ripples in a cup of coffee that is lightly tapped. Such ripples are circular in a cup or on a still pond, but in a planetary ring system they take a different form. Because the inner part of the ring orbits faster than the outer part, the spreading waves wrap into a tightly wound pattern called "spiral density waves." The crests of these density waves form the narrow rings. Moonlets may generate gaps within the rings in yet another way. If two moons move along orbits that lie very close together, their combined gravitational force may deflect ring particles into a narrow stream between them (fig. 10.19). Such **shepherding satellites** can also produce twists and knots as they interact gravitationally with the ring particles.

For centuries astronomers believed that Saturn was the only planet with rings, but in 1977 thin rings were detected around Uranus, and within a few years thin rings around Jupiter and Neptune were discovered as well. Rings made of small or dark particles are often easier to see when viewed from the side of the planet opposite the Sun, as shown in figure 10.20, where the rings around all four Jovian planets are shown, including faint outer rings of Saturn visible from "behind" the planet.

A 2:1 resonance

FIGURE 10.19
Two shepherding satellites (the bright dots) and a portion of a narrow outer ring around Saturn that they shape. Repeated gravitational interactions with the moons also produce twists in the ringlets.

FIGURE 10.20
Ring systems of the Jovian planets. The images of Jupiter's and Neptune's rings were made by the *Voyager 2* spacecraft from the nightside of the planets. Saturn's rings were also imaged from the nightside by *Cassini*. The Uranus image was made by the Hubble Space Telescope.

Origin of Planetary Rings

Astronomers have long debated how old Saturn's ring system might be. Some have argued that planetary rings were material left over from a planet's formation, perhaps matter that had failed to condense into a satellite. Others have argued that the rings must be recent additions, because the ring particles are subject to forces in addition to gravity. For example, gas trapped in a planet's magnetic field may exert a frictional force on the ring particles, gradually causing them to spiral into the planet's atmosphere, as we discussed earlier about the material in Jupiter's ring system. This suggests that new material must be added to the rings from time to time, for without such replenishment the rings would disappear in a few million years. One source of new material is the satellites orbiting the planet. A moon in a satellite system as complex as that of a Jovian planet is subject not only to the gravitational forces of the planet but also to those of the other moons. The cumulative effect of such forces alters the satellites' orbits and may lead to collisions between them.

Recent studies of Saturn's main rings suggest that they may be more stable than previously thought. The material in these rings appears to alternately clump together and break apart. These large clumps experience much less drag than tiny dust particles, so they can survive for long times. This also helps explain why Saturn's rings can remain bright even if they are very old. Older surfaces in the Solar System grow dark as they become covered with dust, but by clumping and breaking apart, fresh surfaces are exposed. Why these clumps never continued to combine to form a moon can be explained by a theory first suggested more than a century and a half ago.

The Roche Limit

In 1849 the French scientist M. E. Roche (pronounced *rohsh*), while studying the problem of a planet's gravitational effect on its moons, demonstrated mathematically that if a moon gets too close to its planet, the planet's gravity could rip the moon apart. This disruption occurs because the planet pulls harder on one side of the satellite than on the other. If the difference in this pull exceeds the moon's own internal gravitational force, the moon will be pulled apart, as shown in figure 10.21. Thus, if a moon—or any body held together by gravity—approaches a planet too closely, the planet raises a tide so large it pulls the encroaching object to pieces. Roche calculated the distance at which the tide becomes fatally large and showed that for a moon and planet of the same density, breakup occurs if the moon comes nearer to its planet than 2.44 planetary radii, a distance now called the **Roche limit.** All planetary rings lie near their planet's Roche limit, suggesting that rings might be caused by satellite disruption.

FIGURE 10.21
The Roche limit. A planet's gravity pulls more strongly on the near side of a satellite than on its far side, stretching it and, if strong enough, pulling it apart.

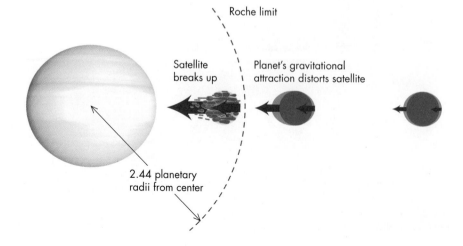

Roche limit

Satellite breaks up

Planet's gravitational attraction distorts satellite

2.44 planetary radii from center

Roche breakup of a moon

Moons are not the only bodies that can stray into the danger zone. Asteroids or comets may occasionally pass too close to a planet, and shattered fragments of such bodies may help keep rings filled. Figure 10.22 shows a comet that passed within Jupiter's Roche limit and was pulled apart into a string of comets. This string of comets eventually crashed into Jupiter in a series of spectacular collisions in 1994, but with a different trajectory it might have ended up as a ring of material. The existence side-by-side of ringlets around Saturn with different compositions (some rich in ice, others rich in carbon) is additional evidence that the rings were formed by the breakup of several different objects.

Finally, keep in mind that the Roche limit applies only to bodies held together by gravity. Artificial satellites or small bodies bonded together by chemical forces can pass safely through the Roche limit without effect.

Saturn's Moons

Saturn has one very large moon, Titan, and 61 smaller ones have been identified so far. The larger of these bodies, like Jupiter's moons, orbit in a flat "mini-Solar System" aligned with Saturn's equator. Saturn's moons have a lower average density than the Galilean satellites of Jupiter, from which astronomers deduce that their interiors must be mostly ice. Moreover, all have about the same density, implying that they were not strongly heated by Saturn as they formed or later on. This contrasts with the Galilean satellites nearest Jupiter, which have a higher density than those farther away, presumably a result of Jupiter's heating during the system's formation.

Figure 10.23 shows pictures of Saturn's seven largest moons taken by the *Cassini* spacecraft. The gravity of all seven is strong enough to have pulled them into round shapes. Saturn's smaller moons have irregular shapes, including the next largest, Hyperion, which has a highly unusual "spongy" appearance (fig. 10.24). The density of Hyperion is very low, under 0.6 grams per cm^3, which suggests that it is made of loosely packed ice. Most of these bodies are heavily cratered, implying that they have been extensively bombarded by infalling bodies.

FIGURE 10.22
Comet Shoemaker–Levy 9. This image shows 20 or so of the fragments into which it was broken by Jupiter's tidal force when it passed close to Jupiter in 1992. These pieces struck Jupiter in 1994.

FIGURE 10.23
Pictures of Saturn's seven largest moons shown to scale. Note the thin haze layer in Titan's upper atmosphere. All seven are large enough that their gravity has pulled them into a spherical shape.

Based on their surface features, which of these moons might be geologically active? Why?

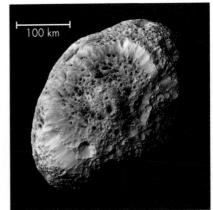

FIGURE 10.24
Saturn's satellite Hyperion is 360 kilometers (220 miles) across in its longest dimension. Hyperion has a heavily cratered surface and an extremely low density (under 0.6 gm/cm³), suggesting that it may be loosely packed material.

FIGURE 10.25

Three views of Saturn's moon Iapetus photographed by the *Cassini* spacecraft. Half of its surface is snow white as seen in figure 10.23, while the other half is coated with a layer of dark brown hydrocarbons. The first image is a view of the dark side. Note the strange 13-km-tall ridge along the moon's equator, which is shown in a close-up view in the second image. The third image shows spots of the dark material on the otherwise snow-white side of Iapetus.

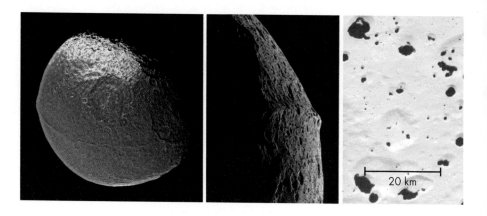

Many of Saturn's satellites have bright and dark streaks. Iapetus (*eye-AP-eh-tuss*) is particularly extreme, with the side of the moon that faces forward as it orbits Saturn covered with an extremely dark material (fig. 10.25). Scientists are unsure of the origin of this material, but it may have come from eruptions on other satellites or collisions that scattered fine debris in space. This material could then have been swept up on Iapetus's leading side, coating the surface with a thin layer of dark dust. Subsequently, solar heating of the darker side could have evaporated the remaining ice and further concentrated the dust and other dark material that was present in the surface ice.

The smooth surface seen on Enceladus (*en-SELL-ah-duss*) is the result of water that has erupted from its interior, flooding old craters and drowning them as it freezes. The *Cassini* spacecraft has observed these eruptions (fig. 10.26), which appear to be powered by tidal flexing (like the eruptions on Jupiter's satellite Io) and are probably the source of material in Saturn's faint outer ring (see fig. 10.20).

Titan, the largest Saturnian moon, has a diameter of about 5000 kilometers (3000 miles), making it slightly bigger in diameter than the planet Mercury and comparable in mass and radius to Jupiter's large moons Ganymede and Callisto. Because Titan is farther from the Sun than these bodies, it is much colder. Thus, as gas molecules leak from its interior, they move relatively slowly and are unable to escape Titan's gravitational attraction. This immense moon therefore possesses its own atmosphere, which spectra show to be mostly nitrogen. Clouds in Titan's atmosphere hide its surface, but during several flybys of Titan by NASA's *Cassini* spacecraft, astronomers have mapped Titan's surface with radar and infrared cameras (fig. 10.27).

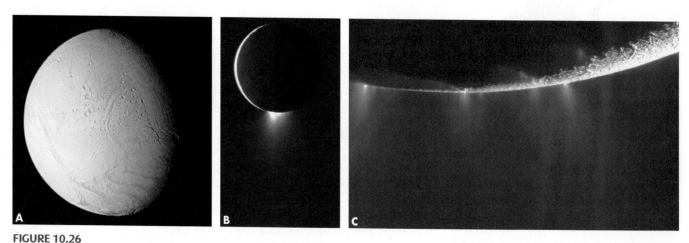

FIGURE 10.26

Saturn's moon Enceladus. (A) Long faults are visible near the south pole. Their blue color is enhanced in this image, making them stand out. (B) and (C) Views of the moon lit from behind by the Sun show eruptions of water from the faulted region.

FIGURE 10.27
Titan, as seen from the *Cassini* spacecraft. (A) Visible-light image shows Titan's thick clouds. (B)This false-color infrared image shows some surface features, glimpsed through the clouds. (C) Radar image reveals rivers and lake. (D) Radar image of impact crater and dunes.

Astronomers had shown on the basis of calculations that, because of Titan's extreme cold (94 K, or about −290°F), its clouds would be composed of methane (CH_4) and other hydrocarbons rather than water. Such models also predicted that liquid methane or ethane (C_2H_6) would "rain" from Titan's clouds and might even form oceans on this frigid moon. No oceans have been found, but radar maps of Titan's polar regions (fig. 10.27C) clearly show lakes, which appear to be liquid methane and ethane, and repeated imaging shows that they sometimes dry up.

In addition to the lakes and channels, other radar images show impact craters and rows of dunes, presumably built and aligned by winds (fig. 10.27D), and at least one "ice" volcano. But these seemingly familiar surface features differ in composition from similar looking features here on Earth. On Titan, the bitter cold makes ordinary water-ice as hard as rock, so Titan's dunes may be ice crystals (not particles of sand) and its river channels may have been cut by methane rain falling and eroding water-ice (not rock).

A small landing probe, carried by *Cassini* and built by the European Space Agency, parachuted into Titan's atmosphere in 2005. This probe, named *Huygens* in honor of Christiaan Huygens, descended through Titan's clouds and landed on the surface. As it drifted down to the surface, the probe took pictures of river networks (fig. 10.28A and B). On the surface it sent back pictures of a rock-strewn plain (fig. 10.28C), the "rocks" probably made of water-ice.

FIGURE 10.28
The landscape of Titan: (A) river networks carved by flowing liquid methane imaged by the *Huygens* lander from a height of 16 kilometers (10 miles); (B) panorama of the horizon from about 8 km (5 miles) above the surface; (C) these "rocks" of ice on the surface are about 10–15 cm (4–6 inches) across.

10.3 URANUS

FIGURE 10.29
Uranus as pictured from the *Voyager 2* spacecraft. Because of Uranus's odd tilt, we are viewing it nearly pole-on. Note the lack of clearly defined cloud belts.

Uranus, although small compared with Jupiter and Saturn, is much larger than the Earth. Its diameter is about 4 times that of the Earth, and its mass is about 15 Earth masses. Lying approximately 19 AU from the Sun (about twice Saturn's distance), Uranus is difficult to study from Earth, visible only as a blue but featureless disk. Even pictures of it taken by the *Voyager 2* spacecraft show few details (fig. 10.29), although computer processing of the images shows that it has faint cloud bands.

Uranus was unknown to the ancients, even though it is just visible to the naked eye. It was discovered by Sir William Herschel, a German émigré to England. Herschel, a musician, was at the time only an amateur astronomer interested in hunting comets, a task in which he collaborated with his sister Caroline. In 1781 he observed a pale blue object whose position in the sky changed from night to night. Herschel at first thought he had discovered a comet, but observations over several months showed that the body's orbit was nearly circular, and he therefore concluded that he had found a new planet. For this discovery, King George III named Herschel his personal astronomer, and to honor the king, Uranus was briefly known as "Georgium Sidus," or "George's Star." In ancient Greek mythology, Uranus was the father of Cronus and was identified as the God of the Heavens.

Uranus's Structure

Uranus, like Jupiter and Saturn, is rich in hydrogen and its compounds water, ammonia, and methane. We know this from spectra of its atmosphere, which show very strong absorption lines of methane. In fact, it is methane that gives the planet its deep blue color. When sunlight falls on Uranus's atmosphere, the methane gas strongly absorbs the red light. The remaining light, now blue, scatters from cloud particles in the Uranian atmosphere and is reflected into space, as depicted in figure 10.30. The cloud particles that cause the scattering are thought to be primarily crystals of frozen methane. Such crystals can form in Uranus's atmosphere because, being so far from the Sun, it is extremely cold.

Astronomers rely on indirect methods to study the interior of Uranus, using, for example, Uranus's density and shape. From its mass and radius, astronomers can calculate that Uranus has an average density of about 1.27 grams per cubic centimeter. This density is nearly twice that of Saturn and almost as large as Jupiter's. But a planet's density depends on both the planet's composition and the amount by which its gravity compresses it. Given that some part of the density of Jupiter and Saturn is simply a result of their greater mass, astronomers deduce that Uranus must contain proportionally fewer

FIGURE 10.30
Sketch illustrating why Uranus is blue. Methane absorbs red light, removing the red wavelengths from the sunlight that falls on the planet. The surviving light—now missing its red colors—is therefore predominantly blue. As that light scatters off cloud particles in the Uranian atmosphere and returns to space, it gives the planet its blue color.

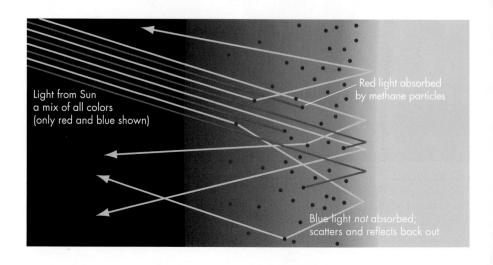

Light from Sun
a mix of all colors
(only red and blue shown)

Red light absorbed
by methane particles

Blue light *not* absorbed;
scatters and reflects back out

light elements, such as hydrogen, than those more massive worlds. On the other hand, the density is too low for Uranus to contain much rock or iron material. Astronomers therefore believe it must be composed of material that is light and abundant, such as ordinary water mixed with methane and ammonia. This mix satisfactorily explains both Uranus's density and spectrum.

Confirmation of the abundance of water also comes from studies of the planet's shape. Uranus, like Jupiter and Saturn, rotates moderately fast, with its equator rotating faster than its poles. At its equator, Uranus spins once every 17 hours, bulging the planet's equator. The size of such a bulge depends in part on the planet's gravitational attraction and therefore on how the mass generating that attraction is distributed inside the planet. Thus, astronomers can deduce the density and composition deep inside Uranus from the size of its bulge. Such studies are consistent with the hypothesis that Uranus is composed of mostly water and other hydrogen-rich molecules, and that it has a core of rock and iron-rich material, as illustrated in figure 10.31. It is not known whether the core formed first and then attracted the lighter gases and ices that condensed around it, or whether the core formed by heavy material sinking to the center after the planet formed. In fact, some astronomers think Uranus's core may be simply highly compressed water, ammonia, and methane with little rocky material. It appears that Uranus did not attract as much hydrogen and helium gas as Jupiter and Saturn when it formed. Its composition can be explained if it formed primarily from planetesimals rich in ices of water, methane, and ammonia. Astronomers consequently often describe Uranus (and Neptune as well) as ice giants, although it should be stressed that the central temperature is probably about 5000 K, so there is no ice inside them today.

Uranus's Odd Tilt

Uranus's rotation axis is tipped so that its equator is nearly perpendicular to its orbit. That is, it spins nearly on its side, as illustrated in figure 10.32. Moreover, the orbits of Uranus's moons are similarly tilted. They orbit Uranus in its equatorial plane, and as a result their orbits are also tilted at approximately 90° with respect to the planet's orbit. Some astronomers therefore hypothesize that during its formation, Uranus was struck by an enormous planetesimal whose impact tilted the planet and splashed out material to create its family of moons. Other astronomers think Uranus was tilted by gravitational tugs exerted on it by neighboring planets (especially Saturn).

Regardless of what caused it, the strong tilt of its axis gives Uranus an odd pattern of day and night. For part of its orbit, one pole is in "perpetual" day and the other pole is in "perpetual" night. Thus, sunlight heats the planet very unevenly, perhaps explaining why Uranus lacks the cloud bands seen on the other giant, gaseous planets. The lack of cloud bands may be temporary, however. Recent observations with an infrared telescope on Earth reveal an odd dark marking in Uranus's atmosphere, perhaps a feature that fades and reappears.

FIGURE 10.31

Artist's view of a recently suggested model for the interior of Uranus. Note how different it is from Jupiter and Saturn, which both contain large regions of liquid and metallic hydrogen.

The seasons of Uranus

FIGURE 10.32

Sketch of the odd tilt of Uranus and its satellites, perhaps caused by a huge impact early in the planet's history. Because of this tilt, when Uranus's north pole points toward the Sun (its northern summer), the Sun will be above the horizon for many Earth years. Likewise, the other pole will be in night for a corresponding period. On the other hand, during the Uranian spring and fall, the Sun rises and sets approximately every 17 hours. Size of bodies and orbits are not to scale.

FIGURE 10.33

The rings of Uranus were imaged by *Voyager 2* from behind the planet (left) and in front (right), illustrating that in addition to thin bright rings, there is dark dust spread between the rings. There is a mismatch between the two views of the outermost ring because it has an elliptical orbit.

FIGURE 10.34

The five largest Uranian moons. They are named for characters in English literature. Titania and Oberon are the queen and king of the fairies in Shakespeare's *A Midsummer Night's Dream*. Ariel and Miranda are characters in Shakespeare's *The Tempest*. Umbriel is a sprite in Alexander Pope's poem "The Rape of the Lock," in which Ariel also appears. The smaller Uranian moons are also named for characters from Shakespeare, such as Puck and Cordelia.

Uranus's Rings and Moons

Uranus is encircled by a set of narrow rings (fig. 10.33). The rings, like those of Saturn, are composed of particles ranging from dust grains to objects perhaps a meter or so in diameter, moving in individual orbits. The Uranian rings are very dark, more easily seen from scattered light from sunlight shining through the rings, implying that they are not made of, or coated with, ice like the bright rings of Saturn; instead, they may be rich in carbon particles or organic molecules. Some of Uranus's rings are very narrow and are held in place by "shepherding satellites," as occurs for some of Saturn's rings.

Uranus has five fairly large moons (fig. 10.34) and about 20 smaller ones. Like the moons of Jupiter and Saturn, they form a regular system and are probably composed mainly of ice and rock. Many of the Uranian moons are heavily cratered, but Miranda, the smallest of the five large moons, has a surface unlike that of any other known Solar System body (fig. 10.35). The surface is broken into distinct areas that seem to bear no relation to one another. One region is wrinkled, while an adjacent region has small hills and craters, rather like our Moon. Miranda's patchwork appearance leads some astronomers to think it may have been shattered by impact with another large body, with the pieces having been subsequently drawn back together by their mutual gravity, giving this peculiar moon its jumbled appearance. Alternatively, the curious surface might have been caused by rising and sinking motions driven by heat in Miranda's interior. Regardless of the cause, Miranda has some extremely curious and unexplained surface features, such as a set of cliffs, visible in figure 10.35, that are twice the height of Mount Everest.

Miranda

FIGURE 10.35

Miranda, an extremely puzzling moon, as observed by the *Voyager 2* spacecraft. Note the enormous cliffs glinting in the sunlight at the bottom of the picture.

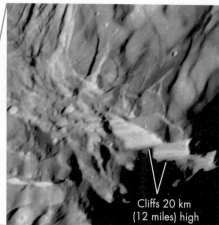

10.4 NEPTUNE

Neptune is the outermost of the major planets. It is very similar to Uranus in size, with a diameter about 3.9 times that of the Earth and a mass about 17 times the Earth's. Through a telescope on Earth, it looks like a chip of blue sapphire. Because of its great distance from the Sun—about 30 AU—it is difficult to study. Pictures taken by *Voyager 2* as it passed by in 1989 provide our most detailed pictures. They show it to be a deep blue world, with markings reminiscent of Jupiter's, as shown in figure 10.36. Cloud bands encircle it, and when *Voyager 2* passed by, it even had a large dark spot, a huge atmospheric vortex resembling Jupiter's Great Red Spot, but dark blue instead of red.

Neptune, named for the Roman god of the sea, was discovered in the 1840s from predictions made independently by a young English astronomer, John Couch Adams, and a French astronomer, Urbain Leverrier. Adams and Leverrier both noticed that Uranus was not precisely following its predicted orbit, and they therefore inferred that its motion was being disturbed by the gravitational force of an as yet unknown planet. From the size of these orbital disturbances, Adams and Leverrier predicted where the unseen body must lie.

Adams completed his calculations in 1845, but the Astronomer Royal, Sir George Airy, was unconvinced by his results and gave a low priority to the search for the unseen planet. In 1846, however, Airy was startled to read a paper by Leverrier detailing calculations nearly identical to those made by Adams. This spurred Airy to begin a search in earnest, but by then it was too late: Leverrier had given his predicted positions to Johann Galle, a German astronomer who that same night pointed his telescope to the predicted location and saw Neptune. Assignment of credit for the discovery of the new planet led to a rancorous dispute tinged with national pride that lasted decades. The discovery is now credited equally to Adams and Leverrier. Ironically, Galileo had seen Neptune in 1613 while observing Jupiter's moons. His observation notes record a dim object whose position changed with respect to the stars, as would be expected for a planet. Galileo failed, however, to appreciate the significance of that motion, so Neptune eluded discovery for another two centuries.

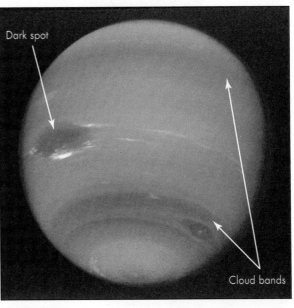

FIGURE 10.36

Image of Neptune taken by the *Voyager 2* spacecraft. Notice the dark blue oval, an atmospheric vortex whose origin is probably similar to that of Jupiter's Great Red Spot.

Q: Approximately how big is the dark spot compared to the Earth?

Neptune's Structure and Atmosphere

Neptune's structure (fig. 10.37) is probably similar to Uranus's. That is, the planet is likely composed mostly of a hot mixture of water, ammonia, and methane surrounded by a thin atmosphere rich in hydrogen and hydrogen compounds, such as methane. Our understanding of the atmosphere's composition is derived from the planet's spectrum. The composition of the interior is inferred from the planet's average density, which is 1.67 grams per cubic centimeter. Like with the other giant planets, such a low density implies that the planet must be composed mostly of light atoms. But during its formation, Neptune must also have collected heavy elements such as silicon and iron. These denser materials would have sunk to the planet's center, where they now form a core of rock and iron about as massive as the Earth.

Neptune's blue color is caused, like Uranus's, by methane in its atmosphere. Unlike Uranus, however, Neptune has distinctive cloud belts. Why do these two so similar bodies have such different cloud formations? Infrared observations offer a partial explanation. Neptune, like Jupiter and Saturn, radiates more energy than it gains from the Sun. That excess energy (also like Jupiter's and Saturn's) probably is left over from the planet's birth or is supplied by heavier material sinking even now to the

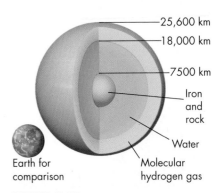

Earth for comparison

25,600 km
18,000 km
7500 km
Iron and rock
Water
Molecular hydrogen gas

FIGURE 10.37

Artist's view of the interior of Neptune.

ANIMATION ▶

The rotation of Neptune

planet's core. Whatever its source, the heat deep within Neptune generates convection currents that rise to its outer atmosphere. There, the rising gas is deflected into a system of winds by the Coriolis effect caused by Neptune's spin—one rotation every 16 hours. The resulting winds create cloud bands similar to those seen on Jupiter and Saturn but tinted deep blue by Neptune's methane-rich atmosphere.

Neptune's winds are extremely fast. For example, near its equator, strong easterly winds blow opposite to the planet's direction of rotation, much like the trade winds on Earth. But these Neptunian winds reach speeds of nearly 2200 kilometers per hour (about 1300 miles per hour). As these gales sweep around Neptune, gas sandwiched between adjacent streams spins, as described in our discussion of Jupiter's atmosphere. Such localized spinning motion created the "Great Dark Spot," as the large, dark blue spot illustrated in figure 10.36 was called. Curiously, pictures of Neptune taken by the Hubble Space Telescope in the mid-1990s showed that the spot was gone. Perhaps such spots grow and dissipate like terrestrial storms. Support for that possibility comes from recent Hubble Space Telescope observations of Uranus, which show that its previously bland appearing atmosphere had formed a small dark spot.

Neptune's Rings and Moons

Neptune, like the other giant planets, has rings, but they are very narrow, more like those of Uranus than those of Saturn, as shown in figure 10.38. They are probably composed of debris from small satellites or comets that have collided and broken up, and they contain proportionally more dust than the rings of either Saturn or Uranus. As a result, they are difficult to see from Earth, but sunlight scattered by the small dust particles is more easily visible from behind the planet. Moreover, photographs of the rings show that in some places the ring particles are not distributed uniformly around the ring but are gathered into arcs (fig. 10.38). This clumping may be the result of gravitational interactions with some of the satellites with eccentric orbits.

Neptune has six small moons orbiting close to the planet in roughly circular orbits and seven other moons at much greater distances, all named after mythological sea deities. One of these, Triton, is nearly as big as Jupiter's moon Europa, but it is highly unusual among large satellites in that it orbits "backward" (counter to Neptune's rotation) in a clockwise direction. This orbital peculiarity leads many astronomers to think that Triton came from the Kuiper belt. When Neptune captured it, the encounter destroyed or expelled most of the moons that Neptune originally possessed. In fact, Neptune's outermost moon, Nereid, follows such a highly elliptical path far beyond the other moons that it just barely missed escaping from its parent world.

Triton intrigues astronomers for more than its orbital oddity. It is massive enough that its gravity, in combination with its low temperature, allows it to retain gases around it. Triton is thus one of the few moons in the Solar System with an atmosphere.

Triton's surface also has many unusual features, as shown in figure 10.39. Its wrinkled surface has a texture much like that of a cantaloupe. Craters pock the surface elsewhere, and dark streaks extend from some of them. At least one of these streaks originates from a geyser caught in eruption by the passing *Voyager 2* spacecraft. Astronomers think that the matter ejected by the geysers is a mixture of nitrogen, ice, and carbon compounds. Sunlight, pale though it is at Triton's immense distance from the Sun, may warm such matter trapped below the surface and make it expand and burst through surface cracks. The erupted material cools and condenses in Triton's cold, thin atmosphere, where winds carry and deposit it as a black "soot."

FIGURE 10.38
Ring arcs around Neptune. This *Voyager 2* image shows dense regions along one of Neptune's rings, probably caused by gravitational interactions with one of its moons.

Wrinkled terrain

Streak of windblown volcanic debris

FIGURE 10.39
The surface of Triton.

Table 10.1   Comparison of the Four Giant Planets

	Jupiter	Saturn	Uranus	Neptune
$R_{equator}$ (Earth units) $R_{equator}$ (km)	11.19 71,492	9.46 60,268	3.98 25,559	3.81 24,764
M (Earth units) M (kg)	317.9 1.9×10^{27}	95.18 5.68×10^{26}	14.54 8.68×10^{25}	17.13 1.02×10^{26}
Density (gm/cm^3)	1.33	0.69	1.32	1.64
Sidereal day (hours)	9.924	10.656	17.24	16.11
Distance from Sun (AU)	5.203	9.539	19.19	30.06
Orbital period (years)	11.8622	29.4577	84.014	164.793
Axial tilt with respect to orbit (degrees)	3.12	26.73	97.86	29.56

SUMMARY

The outer planets formed in the solar nebula far from the Sun, where it was cold enough for them to accumulate hydrogen-rich material. Because of hydrogen's abundance relative to other elements, Jupiter, Saturn, Uranus, and Neptune grew much larger than the terrestrial planets in both diameter and mass (see table 10.1) and are composed mostly of hydrogen and its compounds, such as methane (CH_4), ammonia (NH_3), and water (H_2O). These gases form a deep atmosphere, often richly colored, that grows denser and hotter with depth and eventually is compressed into liquid form. The interiors of Jupiter and Saturn are composed mainly of hydrogen—liquid just below the atmosphere and metallic deeper down. Each is believed to have a rocky core whose diameter—within a factor of 2 or 3—is comparable to the Earth's. Uranus and Neptune have methane-rich atmospheres, but they are probably mostly water, ammonia, and methane surrounding a rocky core. Because they are smaller bodies, their weaker gravity has not allowed them to capture or retain as much hydrogen and helium, but they have retained the heavier gases.

The giant planets are probably heated by continued gravitational contraction and settling of heavier matter toward their cores. As this heat flows outward, it generates convective motions that stir them. The planets' rotation creates a Coriolis effect on the rising gas, drawing it into the cloud belts seen on Jupiter, Saturn, and Neptune.

Jupiter and Saturn spin rapidly, rotating in only about 10 hours, producing obvious equatorial bulges. The spinning motion, when combined with convection in their metallic hydrogen cores, generates powerful magnetic fields that in turn contribute to the formation of auroras and energetic, charged particle belts like the Van Allen belts of our Earth.

Each of the four giant planets has a ring system composed of small orbiting particles. The rings are probably debris from small objects that have broken up as the result of either collisions or tidal forces exerted by the planet. Collisions continue to break up small particles, replacing ring material that falls into the planets' atmospheres.

Many satellites orbit each of the four giant planets, resembling miniature Solar Systems. Some of these moons are comparable in diameter to Mercury. They are thought to be made mostly of water, ice, and rock, although Io, an exceptionally active moon of Jupiter, may be mostly sulfur and rock. Many of the moons are heavily cratered, but a few are smooth, indicating surface activity (Io's volcanos of sulfur, for example) that has filled in old craters. Saturn's moon Titan has an atmosphere of mainly nitrogen, but because of its great distance from the Sun, it is bitterly cold. "Rain" falling from Titan's clouds is not water, but liquid ethane, which flows across Titan's surface, carving rivers and collecting in lakes. Neptune's moon Triton has a backward orbit, suggesting that it may have been captured.

QUESTIONS FOR REVIEW
QUESTIONS FOR REVIEW

1. (10.1) How do Jupiter's mass and radius compare with the Earth's? How do they compare with those of the other outer planets?
2. (10.1) What does Jupiter look like?
3. (10.1) How do astronomers know what lies inside the outer planets?
4. (10.1–10.2) What are the major gaseous substances that make up Jupiter and Saturn?
5. (10.1–10.2) What is the interior structure of Jupiter and Saturn thought to be?
6. (10.1–10.2) Do Jupiter and Saturn have solid surfaces?
7. (10.1) What are Jupiter's internal heat sources?
8. (10.1) What sorts of atmospheric motion and activity are observed in Jupiter? What is the Great Red Spot?
9. (10.1)What sort of activity has been seen on Io? What is Io's heat source thought to be?
10. (10.2) What are the rings of Saturn made of? How do astronomers know this?
11. (10.2) What creates the gaps between the rings?
12. (10.2) How might the rings have formed?
13. (10.2) What is the Roche limit? Why does such a limit exist?
14. (10.3) What is unusual about Uranus's rotation axis? What might explain this peculiarity?
15. (10.3–10.4) How do Uranus and Neptune differ from Jupiter in their interiors?
16. (10.3–10.4) Why are Uranus and Neptune so blue?
17. (10.1–10.4) Why are the outer planets so large?
18. (10.1–10.4) What are the satellites of the outer planets thought to be composed of? Which have atmospheres? What might be special about Europa?

THOUGHT QUESTIONS
THOUGHT QUESTIONS

1. (10.1) If Jupiter were moved closer to the Sun, what do you think would happen to it?
2. (10.1–10.2) Why do the relatively uncratered surfaces of Europa and Enceladus imply that these moons may have been "active" recently?
3. (10.1–10.2) Ganymede and Callisto orbiting Jupiter and Tethys and Dione orbiting Saturn appear in many ways similar to our Moon. If one of these satellites were located close to the Earth, would the similarities remain? What might happen? How would the outer planets' satellites then differ from our Moon?
4. (10.2) Approximate the Roche limit as 2.44 times a planet's radius. Using appendix tables 3 and 5, calculate the Roche limit distances for Saturn and Uranus and make a list of the moons found inside these distances. How it is possible for these moons to orbit there? Would you expect these moons to be spherical or irregular? Why?
5. (10.2) Saturn's moon Titan exhibits an atmosphere, lakes, clouds, rocks, dunes, and at least one volcano. Explain how these features are substantially similar to or different from what we expect based on Earth's geology.
6. (10.3) Is Uranus's sky blue for the same reason our sky is blue? Compared to the yellow-white we see on Earth, would the Sun look redder or bluer from inside Uranus's atmosphere?
7. (10.3) How is the fact that Uranus's moons orbit its equator evidence that Uranus was in a collision with an asteroid or planetesimal? What if the moons had formed prior to the collision?
8. (10.3–10.4) Why might Uranus and Neptune have proportionally less hydrogen than their more massive companions, Jupiter and Saturn?
9. (10.3–10.4) Describe in detail why Neptune clearly shows bands of atmospheric circulation and Uranus does not. Explain when, if ever, Uranus could show such features.
10. (10.1–10.4) If only one new mission could be funded to visit or revisit one of the planets in this chapter, and land a rover on one of that planet's moons, make a case for which planet and moon would be most worth exploring.

PROBLEMS
PROBLEMS

1. (10.1–10.5) How long does it take sunlight to reach Jupiter? Uranus? How might this affect missions to these planets?
2. (10.1) Given its rate of rotation, at what speed does a point on the equator of Jupiter move? Give your result in km/hr.
3. (10.1) Use the modified form of Kepler's third law (illustrated in figure 8.8 and discussed in detail in chapter 3), to calculate Jupiter's mass using the orbital data for any of Jupiter's moons given in the appendix. Be sure to convert the orbital period to seconds and the orbital radius to meters before putting those numbers into the formula.
4. (10.2) If Saturn were compressed until it had the same density as Jupiter, what would Saturn's new radius be? Compare this to Jupiter's radius.
5. (10.2) At closest approach, Saturn is about 8.5 AU from the Earth. If the rings are 270,000 kilometers in diameter, what is their angular size seen from Earth?
6. (10.2) Use the modified form of Kepler's third law and Saturn's mass to calculate the period of the material at the inner and outer edges of Saturn's rings, with values for the semimajor axis a of 90,000 and 136,000 kilometers. Can you see why the rings cannot be solid?
7. (10.2) During a close flyby, the *Cassini* spacecraft yielded a new estimate of the mass of Enceladus of 1.08×10^{23} gm. Given that Enceladus's diameter is 499 km, calculate its density. What does this density suggest about Enceladus's composition? (Ice has a density of about 1 gm/cm^3 and rock about 3 gm/cm^3.)
8. (10.4) What is the "surface" gravity of Neptune (at the top of the clouds)?

TEST YOURSELF

1. (10.1) The low average densities of Jupiter and Saturn compared with the Earth's suggest that
   (a) Jupiter and Saturn are hollow.
   (b) the gravitational attraction of Jupiter and Saturn has compressed their cores into a rare form of iron.
   (c) Jupiter and Saturn contain large quantities of light elements, such as hydrogen and helium.
   (d) Jupiter and Saturn are very hot.
   (e) volcanic eruptions have ejected all the iron that was originally in Jupiter's and Saturn's cores.

2. (10.1) Astronomers think that the inner core of Jupiter is composed mainly of
   (a) hydrogen.   (d) rock and iron.
   (b) helium.   (e) water.
   (c) uranium.

3. (10.2) What is the Roche limit?
   (a) The mass a planet must exceed to have satellites
   (b) The smallest mass a planet can have and still be composed mainly of hydrogen
   (c) The greatest distance from a planet that its satellites can orbit without falling into the Sun
   (d) The distance at which a moon held together by gravity will be broken apart by the planet's gravitational attraction
   (e) The distance astronomers can see into a planet's clouds

4. (10.1–10.2) What source of energy allows Jupiter and Saturn to radiate more heat than they receive from the Sun?
   (a) A strong greenhouse effect
   (b) A high concentration of radioactive elements
   (c) A strong magnetic dynamo
   (d) Tidal stresses from their moons
   (e) Gravitational energy of sinking material

5. (10.3) What makes some astronomers think that Uranus was hit by a large body early in its history?
   (a) It goes around the Sun in a direction opposite to the other planets.
   (b) Its rotation axis has such a large tilt.
   (c) Its composition is so different from that of Neptune, Jupiter, and Saturn.
   (d) It has no moons.
   (e) All of the above.

6. (10.3) What is remarkable about Uranus's satellite Miranda?
   (a) It has several distinct types of surface terrain.
   (b) It is more massive than any other satellite of Uranus.
   (c) It is so dark that it is not visible in any photographs.
   (d) It has a satellite of its own.

7. (10.4) Neptune's rotational period is closest to two-thirds of an Earth _____.
   (a) day.   (b) month.   (c) year.   (d) decade.   (e) century.

KEY TERMS

belt, 254
Great Red Spot, 254
Roche limit, 264

resonance, 263
shepherding satellites, 263
zone, 254

Q: FIGURE QUESTION ANSWERS

WHAT IS THIS? (chapter opening): This is a picture made by the *Cassini* spacecraft of Titan in front of Saturn. The thin dark lines are Saturn's rings seen nearly edge-on.

FIGURE 10.3: The bottom box is about twice as dense as the upper box.

FIGURE 10.11: Io is heated by the gravitational (tidal) force exerted on it by Jupiter, which distorts and thereby heats it.

FIGURE 10.23: The moons completely covered with craters probably are inactive, or the craters would have been largely covered up or otherwise destroyed by eruptions or surface tectonics. Enceladus has some smooth areas, and it has been discovered to have eruptions.

FIGURE 10.36: Measure the spot and Neptune's disk, and divide the spot's size by Neptune's. The spot is not round, but its width is roughly 0.2 times Neptune's diameter. Neptune is about 3.8 times larger than Earth, so the spot's width is $0.2 \times 3.8 = 0.76$ times (about three-fourths) the Earth's diameter.

11

SMALL BODIES ORBITING THE SUN

Picture of Comet McNaught as seen from Australia in January 2007.

LEARNING OBJECTIVES

Upon completing this chapter you should be able to:

- Enumerate and describe the various classes of small bodies that orbit the Sun.
- Describe the different kinds of meteorites and what their different compositions indicate about their origins.
- Describe how asteroids were discovered, and retell the changing nomenclature used for some of them, from planet to asteroid to dwarf planet.
- Explain which objects are identified as dwarf planets and what distinguishes them from major planets.
- Describe what is meant by gaps in the orbits of asteroids and discuss ideas as to why asteroids did not form a planet.

- Explain how Pluto and other trans-Neptunian objects were discovered, and what is known about their properties.
- Describe the characteristics and structure of a comet.
- Calculate orbital properties of a comet from its period.
- Explain how comets' tails arise, in what directions they point, and comets' connection with meteor showers.
- Describe the effects of collisions with Earth, and explain why they can have lethal consequences.
- Recount records of some major and recent Earth impacts.
- Discuss the evidence for a major impact causing the demise of the dinosaurs and other species 65 million years ago.
- Calculate the energy released when a body strikes Earth.

Q: WHAT IS THIS?

See end of chapter for the answer.

Orbiting the Sun and scattered throughout the Solar System are numerous bodies much smaller than the planets—asteroids, comets, and dwarf planets. The asteroids are generally rocky objects in the inner Solar System. The comets are icy bodies and spend most of their time in the outer Solar System. Dwarf planets are larger versions of both of these kinds of objects, but too small to have established themselves as the dominant body orbiting at their orbital distance. These small members of the Sun's family, remnants from the formation of the Solar System, are of great interest to astronomers because they are our best source of information about how long ago and under what conditions the planets formed. In fact, some are planetesimals—the solid bodies from which the planets were assembled—that have survived nearly unchanged from the birth of the Solar System.

Apart from their scientific value, these smaller objects merit study because they can be both beautiful and deadly. A comet in the dawn sky with its tail a shining plume is a sight not to miss. But "miss" is precisely what we hope will happen if a large comet or asteroid is on a collision course with Earth. Such a collision in the past may have exterminated much of Earth's ancient life, and such an event in the future could well have equally disastrous effects on today's living things.

In this chapter, we will discover why meteorites (which are fragments of asteroids and comets that by chance fall into our atmosphere) are such important clues to the time of birth and the structure of the ancient Solar System. Likewise, we will study how a comet changes from a 10-kilometer diameter ball of ice into a beautiful banner of light in the night sky, and why an asteroid may be the reason you have hair rather than scales.

CONCEPTS AND SKILLS TO REVIEW

- Density (6.1)

11.1 METEORS, METEOROIDS, AND METEORITES

If you have spent even an hour looking at the night sky, you have probably seen a "shooting star," a streak of light that appears in a fraction of a second and as quickly fades (fig. 11.1). Astronomers call this brief but lovely phenomenon a **meteor.**

A meteor is the glowing trail of hot gas and vaporized debris left by a solid object heated by friction as it moves through the Earth's atmosphere at extremely high speed. Most of the heating occurs between about 100 and 50 kilometers in the outer fringes of our atmosphere. The solid body, while in space and before it reaches the atmosphere, is called a **meteoroid.**

Heating of Meteoroids

Meteoroids heat up on entering the atmosphere for the same reason a reentering spacecraft does. When an object plunges from space into the upper layers of our atmosphere, it collides with atmospheric molecules and atoms. These collisions convert some of the body's energy of motion (kinetic energy) into heat. In a matter of seconds, the outer layer of the meteor reaches thousands of degrees kelvin and glows. Given that reentry speeds are typically at least 10 kilometers per second and often 30 to 40 kilometers per second, the collisions with air molecules are extremely violent and tear atoms off the

Q: Why is the meteor's track straight, whereas the star images are curved?

FIGURE 11.1
A time-exposure photograph captures a "shooting star" (meteor) flashing overhead. The curved streaks are star trails.

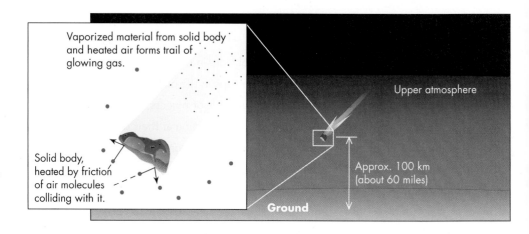

FIGURE 11.2

Sketch depicting how air friction heats an object entering our atmosphere, creating the glow we see as a meteor.

People are sometimes hit by meteorites, but only rarely. In one well-documented case, a meteor crashed through a roof and hit a woman, badly bruising her, and a number of cars and houses have also been hit.

body, vaporizing the surface layers. The trail of hot evaporated matter and atmospheric gas emits light, making the glow that we see, as shown in figure 11.2.

If the meteoroid is larger than a few centimeters, it creates a ball of incandescent gas around it and may leave a luminous or smoky trail. Such exceptional meteors, sometimes visible in daylight, are called "fireballs."

Meteoroids bombard the Earth continually: a hail of solid particles that astronomers estimate amounts to hundreds of *tons* of material each day. More strike between midnight and dawn than in the evening hours, and so it is best to watch for meteors in the early morning. This difference arises for the same reason that if you run through rain, your front will get wetter than your back. That is, the dawn side of our planet advances into the meteoritic debris near us in space, while the night side moves away from it.

Most meteors that we see last only a few seconds and are made by meteoroids the size of a raisin or smaller. These tiny objects are heated so strongly that they completely vaporize. Larger pieces, though heated and partially vaporized, are so drastically slowed by air resistance that they may survive the ordeal and reach the ground. We call these fragments found on the Earth **meteorites.**

Types of Meteorites

Astronomers classify meteorites into three broad categories based on their composition: stony (that is, composed mainly of silicate compounds), iron, and stony-iron. Most stony meteorites are composed of smaller rounded chunks of rocky material stuck together, as shown in figure 11.3A. The grains are called **chondrules** (*chondrule* is pronounced "KON-drool" and means "small grain" in Greek), and meteorites that have this lumpy structure are called **chondrites.** Chondrules appear to have been rapidly melted and cooled in the solar nebula. The cause of the heating is not known, although it may have been caused by outbursts from the Sun as it formed, collisions between planetesimals, or the explosion of a nearby star—perhaps an explosion that caused the solar nebula to begin collapsing.

Chondrules contain traces of radioactive material, which can be used to measure their age, as described in chapter 6. They are almost 4.6 billion years old, and are believed to be the first solid material that condensed within the solar nebula. Many chondrules contain even older material: dust grains that have survived from before the birth of the Solar System. Thus, chondritic meteorites offer us valuable information about the early history of the Solar System.

In some chondritic meteorites, the chondrules are embedded in a black, carbon-rich, coal-like substance and are therefore called **carbonaceous chondrites.** This carbonaceous matter contains organic compounds, including amino acids, the same complex molecules used by living things for the construction of the proteins on which

Chondrules

Ordinary chondrite

Carbonaceous chondrite

0   1   2 cm

A

Iron

Stony-iron

Achondrite

0   1   2   3   4 cm

B

FIGURE 11.3
Sliced samples of various kinds of meteorites. (A) Chondrites are stony meteorites made up of small round bits of rock called chondrules. Slices show the chondrules inside an ordinary chondrite and a carbonaceous chondrite, which has a matrix of carbon-rich material. (B) Iron and stony-iron meteorites as well as rocky achondrites come from fragments of differentiated asteroids. After the material has melted, the iron sinks to the core and the pattern of the chondrules is lost. Note the crystalline pattern in the iron meteorite. These patterns form only when the metal cools very gradually over millions of years.

they depend. Thus, the presence of amino acids in meteoritic matter indicates that the raw material of life can form in space and that it might therefore have been available right from the start within the Solar System. Irrespective of the existence of amino acids in meteorites, we can still ask how and where these bodies formed, and what brought them to Earth. Astronomers think that most of them are fragments of asteroids and comets. However, there is solid evidence that some come from the Moon and Mars, and perhaps Mercury as well. Chunks of rock from Venus are less likely to escape to space because of Venus's dense atmosphere, but meteorites from Earth may be found on the Moon and Mars.

Some stony meteorites have no chondrules and are therefore called **achondrites.** They appear to have melted and been transformed like the rock found on Earth, and therefore probably come from asteroids that were large enough to heat up and melt. In such bodies, iron would sink to the core, which would then explain the existence of iron meteorites. The unusual history of the iron in these meteorites is revealed when the iron is etched with acid. This shows a crystalline pattern that can form only when metals cool over millions of years (see fig. 11.3B). Finally, the rarest meteorites, stony-irons, contain silicate rock embedded in a matrix of iron, and may come from the transition region between the core and the mantle of a large asteroid.

FIGURE 11.4
Comparison of the Moon and Ceres, the fifth and sixth largest bodies in the inner Solar System.

Asteroids are small, generally rocky bodies that orbit the Sun. Giuseppe Piazzi discovered the first of this swarm of bodies in 1801 during his search for the "missing planet," which, according to Bode's rule (p. 201), should have been at 2.8 AU from the Sun. He named the asteroid that he found Ceres in honor of the patron goddess of Sicily, his home. It is the next largest object in the inner Solar System after the Moon, but it is less than half the Moon's diameter (fig. 11.4).

Several hundred thousand other asteroids have subsequently been found, but Ceres remains the largest by a substantial amount. Despite the huge number of asteroids, their combined mass is very small, amounting to less than 10% of the mass of the Moon. They are found throughout the Solar System, but most lie in the **asteroid belt**, a region between the orbits of Mars and Jupiter, stretching from about 2 to 4 AU from the Sun, as shown in figure 11.5.

Size and Shape

Most asteroids' diameters are difficult to measure because the asteroids are so small that they appear in ground-based telescopes merely as points of light. In fact, that is how they got their name, they look like stars (Greek *aster*) through a telescope. Even the largest, Ceres, was barely resolved before the *Dawn* spacecraft went into orbit around it in 2015. For the vast majority, we have only the amount of light they reflect. This is not always a good clue to an asteroid's size, because a large, poorly reflective object will look as bright as a small, highly reflective one. For this reason, the emitted infrared radiation is a better measure of diameter; bigger bodies emit more than smaller ones of the same temperature. From such measurements, astronomers have found that asteroids range tremendously in diameter, from Ceres, which is about 950 km (690 miles) across, down to bodies under a kilometer. For example, the tiny asteroid 1991 BA, which passed about 170,000 kilometers (less than half the distance to the Moon) from Earth in January 1991, is probably less than 9 meters (30 feet) across.

FIGURE 11.5
Diagram showing the distribution of more than 20,000 of the larger known asteroids. Notice that most lie between Mars and Jupiter, but the Trojan asteroids form two loose clumps located at Jupiter's orbital distance from the Sun. Making the plotted points large enough to see causes them to appear far more closely packed than they really are.

Q: With a ruler, estimate the distance (in AU) from the center of the Trojan swarms to the Sun and to Jupiter, and compare this with Jupiter's distance from the Sun.

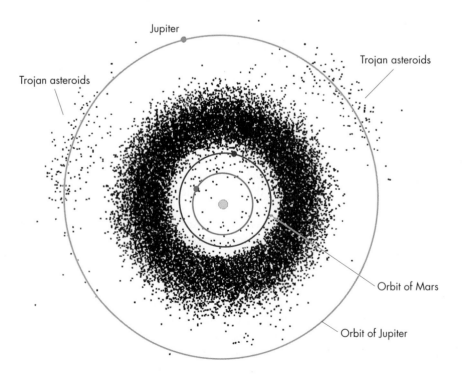

Jupiter

Trojan asteroids

Trojan asteroids

Orbit of Mars

Orbit of Jupiter

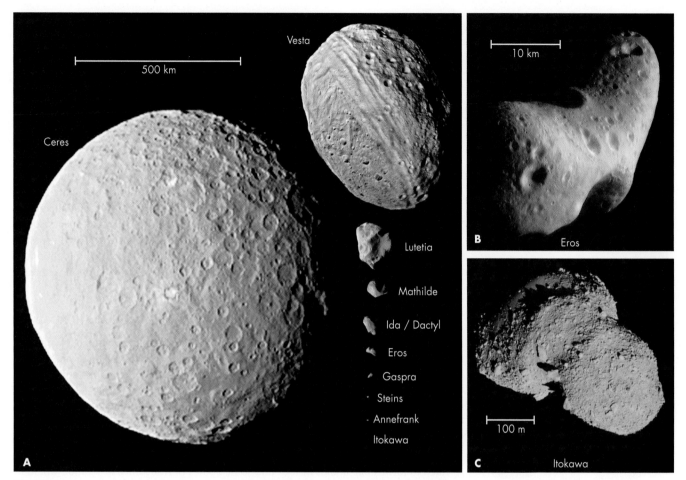

FIGURE 11.6

(A) Images of asteroids that have been visited by spacecraft compared to Ceres. Vesta is the second most massive asteroid, the others shown span a wide range of sizes. (B) Eros is very elongated, shaped like a fat banana. (C) Tiny Itokawa is only a few hundred meters across.

The smallest asteroids are far too tiny to observe unless they come extremely close to Earth or strike it. In that case, we see them as meteors. That is, meteoroids are in a real sense just very small (we hope!) asteroids that happen to reach our planet.

Most asteroids are irregularly shaped (fig. 11.6). Only the largest, Ceres, is nearly spherical. Because its gravitational force is strong enough to have crushed its material into a sphere, Ceres is now classified as a dwarf planet. Smaller asteroids with weak gravities remain irregular and are made more so by collisions blasting away pieces. Collisions leave the parent body pitted and lumpy, and the fragments become smaller asteroids in their own right, or sometimes even satellites of larger asteroids, such as Ida's moon Dactyl.

Although most meteorites are rocky or iron-rich, that does not mean the asteroids are all solid chunks of such material. Ceres has a density of 2.1 gm/cm^3, suggesting it contains a significant proportion of ice. Some small asteroids have even lower densities, such as Mathilde, which is just 1.4 gm/cm^3, suggesting that it is a loose pile of rubble, just barely held together by its gravity. Two smaller asteroids that pass near Earth are shown in figure 11.6B and C: Eros, which was visited by the *NEAR* spacecraft in 2000, and Itokawa, which was visited by the Japanese space probe *Hayabusa* (Japanese for "falcon"). The *Hayabusa* probe landed on Itokawa in 2005, and found its density to be 1.9 gm/cm^3. This density suggests that the asteroid is mainly loose rock lumps, weakly held together by its gravity. The probe collected dust particles that it returned to Earth in 2010, and these particles were found to be similar in composition to some types of chondritic meteorites.

FIGURE 11.7

Mineralogical map of Vesta made by the *Dawn* spacecraft. In this map green, red, and dark blue regions spectroscopically resemble a class of meteorites that were probably blasted from the asteroid by a major impact at Vesta's south pole. The edge of the giant impact is marked by a dashed line.

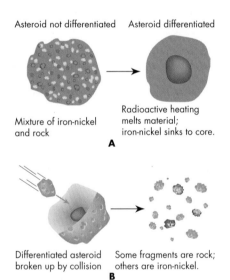

FIGURE 11.8

Sketch depicting (A) differentiation in asteroids and (B) their subsequent breakup by collision to form iron and stony bodies.

Differentiation in asteroids and their subsequent breakup by collisions to form iron and stony bodies

Composition

Most asteroids cannot be visited by space probes, so compositions are usually determined through spectroscopy. When sunlight falls on an asteroid, the minerals in its surface create absorption features in the spectrum of the reflected light, from which we can determine the asteroid's composition. Such spectra show that asteroids belong to three main compositional groups similar to those of meteorites: carbonaceous bodies, silicate bodies, and metallic iron-nickel bodies. The groups are not mixed randomly throughout the asteroid belt: inner-belt asteroids tend to be silicate-rich, and those in the outer belt tend to be carbon-rich.

Some asteroid spectra match meteorite samples, and many meteorites closely match spectra of Vesta. Some of these meteorites were types that would only form in the deep crust or mantle, which led astronomers to suspect that Vesta had undergone a major impact. The *Dawn* spacecraft was able to carry out more detailed spectroscopic analyses of Vesta's surface, confirming the association of the meteorites with specific regions on Vesta (fig. 11.7). Further, *Dawn* found clear evidence of the enormous impact that blasted away much of Vesta's south polar region about a billion years ago.

Origin of Asteroids

The properties of asteroids, such as composition, size, and their location between Mars and Jupiter, give us clues to their origin and support the solar nebula theory for the origin of the Solar System. According to the solar nebula theory, bodies that condensed in the inner asteroid belt should have a composition different from those that condensed farther out. The inner belt, being warmer, is richer in the easy-to-condense silicate and iron materials and contains less water or carbon-rich materials, as is indeed observed.

The existence of stony and iron meteorites might at first seem to be evidence *against* the solar nebula theory. How could a swirling mass of gas and dust have separated to form some bodies of rock and some bodies of iron? Such separation would be a bit like shaking a piece of cake and having it disintegrate into eggs, flour, sugar, and milk. We saw in chapter 6, however, that chemical elements *can* be separated by differentiation. That is, astronomers think that the Earth's rocky crust and iron core formed as a result of melting, with the iron then sinking to the core and the lighter rock floating. Larger asteroids are expected to become hot enough to differentiate (fig. 11.8A). There is good evidence that Vesta has had volcanic flows—this activity, now long dormant, is deduced from the spectroscopic detection of basalt, a volcanic rock.

After differentiation, collisions with neighboring asteroids broke up many of the large bodies (fig. 11.8B). The fragments are what we see today: pieces of crust became stony asteroids, while pieces of core became iron asteroids. But for a body to differentiate, it must be large enough to be able to melt from the heat liberated within it by radioactivity. Thus, the existence of stony and iron asteroids is strong evidence that the early Solar System contained intermediate-size bodies—planetesimals—like Vesta.

In the 1800s, astronomers speculated that the asteroid belt was created when a planet there was destroyed, but the solar nebula theory offers a more likely explanation. Planetesimals growing there would have to compete for material with Jupiter, whose immense gravity would disturb the accretion process, preventing the planetesimals' growth. Thus, asteroids are some of the oldest objects in the Solar System.

Asteroid Orbits

Even today Jupiter affects the asteroid belt. Figure 11.9 shows a partial census of the number of asteroids found at each distance from the Sun within the belt. Gaps can be seen at about 2.1, 2.5, and 2.8 AU, for example. The seemingly empty regions in the asteroid belt are called **Kirkwood gaps.**

FIGURE 11.9

The number of asteroids at each distance from the Sun within the asteroid belt. Notice the conspicuous gaps where there are few, if any, objects. These empty zones are called Kirkwood gaps.

The Kirkwood gaps are caused by the same process that creates the gaps in Saturn's rings: gravitational forces of an outlying body. Saturn's moons, particularly Mimas, create the gaps in the rings. Jupiter creates the Kirkwood gaps. As it orbits the Sun, Jupiter exerts a gravitational force on asteroids in the belt that slightly alters their orbits. If an asteroid has an orbital period that when multiplied by an integer equals Jupiter's period, the asteroid may be subject to a cumulative force that makes it drift to a new orbit. The gap at 2.5 AU arises because, according to Kepler's third law, an asteroid at that distance would have an orbital period exactly one-third that of Jupiter. Every third trip around the Sun, the asteroid would undergo exactly the same tug from Jupiter. Over time it will be shifted to a new orbit, leaving a gap.

Not all asteroids are found in the main belt. A few, the so-called Trojan asteroids, travel along Jupiter's orbit in two loose swarms, 60° ahead and 60° behind it, as illustrated in figure 11.5.

Nearer the Sun from the main belt are the **near-Earth objects,** whose orbits carry them into the inner Solar System across or close to the Earth's orbit. Fortunately for us, there are relatively few of these. The near-Earth objects are mostly small, with diameters of less than 1 kilometer, such as Itokawa (figure 11.6C). Over 10,000 have been identified, mostly the larger ones, and accurate orbits have been determined. None is likely to collide with the Earth over the next century, although several have been found that will approach to within one-tenth the distance of the Moon. On average, astronomers estimate that a body 100 meters in diameter or larger hits the Earth about every 10,000 years. The near-Earth objects have probably been shifted into their peculiar orbits by gravitational interactions with Jupiter and other planets. Some may be related to comets, but "dead" ones, stripped of ice and gas by their repeated passage around the Sun. We will describe how this happens in more detail later in section 11.4.

Chiron is another odd "asteroid." Its orbit stretches from just inside that of Saturn almost to Uranus, putting it far outside the main asteroid belt. Chiron changes brightness, sometimes flaring up as it ejects gas. Such behavior is more like that of a comet than a normal asteroid. From its infrared emission, astronomers deduce that Chiron has a diameter of about 230 kilometers (approximately 140 miles), making it much larger than most comets that enter the inner Solar System. Today, Chiron is thought to be an object that originally orbited beyond Neptune whose orbit has been altered by gravitational interactions with the giant planets. Chiron hints at a large population of objects in the dim recesses of the Solar System, which we examine next.

Q: What feature in Saturn's rings is similar to these gaps?

11.3 PLUTO, PLUTOIDS, AND BEYOND

Beyond Neptune no more major planets are known, but there are many smaller objects. Many of these **trans-Neptunian objects,** or **TNOs,** orbit the Sun in the Kuiper belt. Relatively little is known about them, although Neptune's moon Triton may be a captured TNO. The TNOs are so far from the Sun that they are difficult to detect from Earth. Most have been identified only in the last decade using large telescopes, though one was found more than 80 years ago.

Pluto

Pluto was discovered in 1930 by Clyde Tombaugh, an astronomer at Lowell Observatory in Flagstaff, Arizona. Tombaugh painstakingly examined pairs of photographs of the sky for over a year—scanning millions of star images—searching for objects whose position changed between the exposures, the telltale motion that distinguishes an orbiting body from a star (fig. 11.10A). Pluto is named for the Greek and Roman god of the underworld, and for more than 75 years it was considered a planet, although almost from the time of its discovery it was clear that it differed in many ways from the previously known planets.

Pluto's great distance from us and the Sun make it very dim and difficult to study. For years after its discovery, astronomers thought that Pluto might be even as large as the Earth. When Pluto's largest moon Charon was discovered in 1978, it finally became possible for astronomers to measure Pluto's mass using Newton's modified form of Kepler's third law of planetary motion (see chapter 8). Circling Pluto at a distance of just 19,600 kilometers, Charon takes only 6.4 days to complete a trip around Pluto. From this orbital data, Pluto's mass was found to be about 0.002 times the Earth's, or less than 1/20th that of Mercury, the smallest planet. Four other tiny moons have been discovered in the last decade orbiting about 2 to 4 times farther out than Charon.

In 2007 NASA launched the *New Horizons* spacecraft, which passed by Pluto in July 2015, giving us our first detailed pictures of this icy world and its moons. Pluto's diameter is a little less than one-fifth the Earth's, and Charon's diameter is slightly more than half Pluto's (fig. 11.10B). Charon is 12% as massive as Pluto—the largest relative mass of any known satellite—and its gravity is strong enough that both Pluto and Charon are both locked into rotation keeping the same face toward each

> Pluto's large moon Charon is named for the boatman who, in mythology, ferries dead souls across the river Styx to the underworld. The four smaller moons' names —Styx, Nix, Kerberos, and Hydra—also come from the mythology of Pluto.

FIGURE 11.10
(A) In this pair of photographs taken at Lick Observatory in California, Pluto looks like merely another dim star, distinguished only by its change in position from night to night. (B) Approximately true-color images of Pluto and its largest moon Charon obtained by the *New Horizons* probe. (C) False-color image of Pluto made by *New Horizons* indicates that the surface is colored by several different chemicals. (D) A close-up image of the region outlined in (C) shows old cratered dark regions, light-colored young plains, and mountains.

ASTRONOMY *by the numbers*   PLUTO'S ESCAPE VELOCITY

The escape velocity from Pluto can be calculated using the formula in chapter 3:

$$V_{esc} = \sqrt{\frac{2GM}{R}}$$

Pluto's mass is $M = 1.31 \times 10^{22}$ kg and its radius is $R = 1.16 \times 10^6$ m, so its escape velocity is:

$$V_{esc} = \sqrt{\frac{2 \times 6.67 \times 10^{-11} \text{m}^3\text{kg}^{-1}\text{sec}^{-2} \times 1.31 \times 10^{22}\text{kg}}{1.16 \times 10^6\text{m}}}$$

$$= \sqrt{1.51 \times 10^6 \text{ m}^2/\text{sec}^2}$$

$$= 1.23 \times 10^3 \text{ m/sec}$$

This is about 1/9th of Earth's escape velocity.

At Pluto's low temperature, gas molecules move about one-third as fast as they do in Earth's atmosphere, so with an escape velocity 1/9th of Earth's, Pluto is less able to hold on to an atmosphere than Earth. On the other hand, the low temperature makes Pluto better able to hold on to an atmosphere than the Moon, even though the Moon's escape velocity is twice Pluto's.

---

other. Because they orbit a point outside of Pluto, some astronomers think that Charon should be called a dwarf planet, too, making this a double dwarf planet system.

From their masses and radii, astronomers have determined that Pluto has a density of about 1.87 grams per cubic centimeter, while Charon's density is 1.65 gm/cm³. These values suggest that both are a mix of rock, ordinary water ice, and other ices, with Charon containing slightly less rocky material. Spectra of their surfaces made both from Earth and from *New Horizons* have identified ices of methane, nitrogen, and carbon monoxide as well as water. False-color images made by *New Horizons* show different color regions (fig. 11.10C). The darker regions are probably older, as indicated by the large number of craters in them, and appear to contain hydrocarbon compounds. The lighter regions are mainly nitrogen ice, and they appear to be much more geologically active based on the lack of impact craters (fig. 11.10D). Early analysis of these regions suggests that there are glacier-like flows of nitrogen ice, and mountainous regions may be "volcanic" in origin, made of water ice, which is as hard as rock at these temperatures. Spectra of the polar regions indicate that they contain frozen methane in addition to frozen nitrogen.

Astronomers have detected a thin atmosphere on Pluto, mostly nitrogen and carbon monoxide, with traces of methane. In the bitter cold (40 K, or about −87°F) of this remote world, molecules move so slowly that, despite its tiny mass, Pluto's gravity is strong enough to retain a thin atmosphere, as discussed in Astronomy by the Numbers: "Pluto's Escape Velocity." Because of the large 120° tilt of Pluto's axis, each hemisphere spends long periods of time in sunlight over the course of Pluto's 248-year orbit, so the ices that evaporate in the sunny hemisphere may flow to the other hemisphere and freeze back out. And because the orbit is quite elliptical, the amount of gas may vary dramatically as it travels closer to and farther from the Sun.

The Plutoids

Pluto's small size and peculiar orbit (it crosses Neptune's orbit, as shown in fig. 11.11) once led some astronomers to hypothesize that it was originally a satellite of Neptune that escaped and now orbits the Sun independently. Today, however, astronomers think almost the reverse—that Neptune has "captured" Pluto. Pluto's orbital period is 247.7 years, very close to one and one-half times Neptune's. Thus, Pluto makes two orbits around the Sun for every three made by Neptune. This match of orbital periods has created a cumulative gravitational attraction on Pluto that "tugged" it into its current orbit.

In fact, several hundred other objects a few hundred kilometers in diameter orbit at nearly the same distance from the Sun as Pluto. For example, Orcus, discovered in 2004, is about the size of Pluto's moon Charon and also orbits the Sun twice for every three orbits of Neptune. Even larger icy objects—presumably surviving

Q: If Pluto's orbit crosses Neptune's, why don't these planets collide?

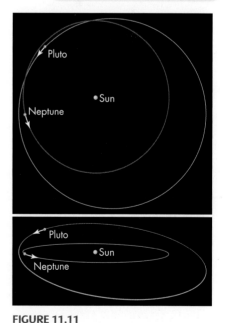

FIGURE 11.11
Pluto's and Neptune's orbits shown from "above" and edge-on. Pluto's odd orbit is highly tilted and highly eccentric—so much so that it comes closer to the Sun than Neptune for part of its orbit.

FIGURE 11.12
(A) Depictions of several of the largest trans-Neptunian objects shown to the same relative scale, along with Neptune's moon Triton, which may be a captured TNO. (B) Orbits of several of the TNOs.

planetesimals—orbit farther out but are more difficult to detect because of their large distance from the Sun. Some of the larger TNOs are illustrated in figure 11.12. The illustrations shown are mostly sketches, because astronomers can only estimate sizes and colors from the dim light we see. Astronomers think there are dozens more objects at least 1000 kilometers (600 miles) in diameter. Sedna is of particular interest because its orbit extends well beyond the standard Kuiper belt, and it has a strange red color that is not well understood. It currently happens to be relatively close to the Sun along its elliptical orbit, and it would not have been detectable through most of its orbit.

In 2005 astronomers reported the discovery of a dim object orbiting at about 97 AU from the Sun. Now officially named Eris (for the Greek goddess of discord), this remote world turns out to have a diameter of about 2400 kilometers (about 1500 miles), making it slightly larger than Pluto. Its orbit is both highly elliptical (ranging from 38 to 97 AU from the Sun) and highly inclined (about 44°) to the main disk of the Solar System. Because of its size, some astronomers initially described it as a tenth planet, despite its orbit being so different from that of the other planets. Eris is like Pluto in many ways besides its size. Its surface is covered with frozen methane, and it has a moon almost as large as Pluto's moon Charon.

Largely because of the discovery of an object more massive than Pluto, and faced with the prospect of dozens of more objects that were not significantly different from Pluto, astronomers decided to define a new category for Pluto and these other moderately large icy bodies. In 2006 the International Astronomical Union came up with the designation *dwarf planets* to describe bodies like Pluto and Eris, which are large enough that their gravity has pulled them into a roughly spherical shape, but which do not dominate all other masses orbiting at the same distance from the Sun. They also created the designation *plutoid* for dwarf planets orbiting beyond Neptune. Two more TNOs, Haumea and Makemake (fig. 11.12), have been identified as plutoids recently, but there will probably be many more. Terminology is not the only issue posed by the TNOs, though. Many of them are too large to have formed in the low-density regions of the outer Solar System where we see them now. Thus, astronomers hypothesize that the TNOs formed nearer the Sun, perhaps near Neptune, and have been ejected into their present orbits by successive gravitational tugs exerted by the massive outer planets as they settled into their current orbits. Many more small objects were ejected into these remote orbits, some of which come much closer to Earth, as we see next.

11.4 COMETS

A bright comet is a stunning sight, as illustrated by the chapter-opening figure. Sadly, such sights are now rare because light pollution from our cities drowns out the view for most people. Comets have long been held in fear and reverence, and their sudden appearance and equally sudden disappearance after a few days—or, in some cases, weeks—have added to their mystery. Today we recognize them as visitors from the most remote parts of the Solar System as well as being visitors from the remote past.

The Appearance and Structure of Comets

Comets consist of three main parts—the tail, the coma, and the nucleus (fig. 11.13). The largest part is the long **tail,** a narrow column of dust and gas that may stretch across the inner Solar System for as much as 100 million kilometers (nearly an AU!).

The tail emerges from a cloud of gas called the **coma,** which may be some 100,000 kilometers in diameter (10 times or so the size of the Earth). However, despite the great volume of the coma and the tail, these parts of the comet contain very little mass. The gas and dust are extremely tenuous, and so a cubic centimeter of the gas contains only a few thousand atoms and molecules. By terrestrial standards, this would be considered a superb vacuum. This extremely rarefied gas is matter that the Sun's heat has boiled off the heart of the comet, its nucleus.

The comet **nucleus** is a block of ice and gases that have frozen in the extreme cold of the outer Solar System into an irregular mass whose diameter is typically about 10 kilometers. The nucleus of a comet has been described as a giant "iceberg" or "dirty snowball," and it contains most of the comet's mass.

Most comets appear unexpectedly, often discovered by amateur astronomers with nothing but binoculars. They are too small to be seen in the outer Solar System, where they spend most of their time. They are named for their discoverer, though one of the most famous, Comet Halley, got its name because Sir Edmund Halley was the first to propose that it orbited the Sun like a planet. He predicted that the great comet of 1682 would reappear in 1759. It did, but Halley did not live to see his prediction verified.

Comet Halley also provided astronomers with their first close look at a comet nucleus when the *Giotto* spacecraft approached to within 600 km of it. Launched by the European Space Agency as part of an international study of Comet Halley during its

There is disagreement about how to pronounce *Halley*. Most astronomers say it as if it rhymes with *Sally*, but there is evidence that Halley himself pronounced his name as *haw-lee*.

FIGURE 11.13

Artist's depiction of the structure of a comet, showing the tiny nucleus, surrounding coma, and long tail.

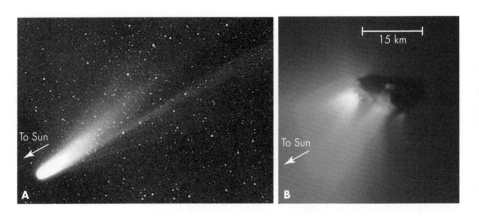

FIGURE 11.14
Comet Halley in March 1986: (A) photographed from Earth when the tails were more than 50 million kilometers long; and (B) its nucleus photographed from the *Giotto* spacecraft.

most recent return in 1986 (fig. 11.14A), *Giotto* was named after an Italian artist who in 1301 painted a Christmas scene with a comet, which may have been comet Halley.

The *Giotto* spacecraft revealed the size and appearance of a comet nucleus for the first time. Despite its icy composition, the nucleus of comet Halley is dark, as you can see in figure 11.14B, which is one of the pictures made by the *Giotto* spacecraft. The dark color comes from dust and carbon-rich material (similar to that of the carbonaceous chondritic meteorites) coating the surface of the nucleus. Other visible features of the nucleus are its irregular shape and the jets of gas erupting from the frozen surface. The jets form when sunlight heats and vaporizes the icy material. Because of the comet's weak gravity, the shape is not expected to be round, and uneven melting of the nucleus during previous passages by the Sun may also affect the shape.

Formation of the Comet's Tails

Most comets are visible only when they are less than about 5 AU from the Sun, where solar heating is able to vaporize the ices into gases that escape to make the coma. The comet then appears through a telescope as a dim, fuzzy ball. For most of their orbits we cannot see them, but we can extrapolate the rest of their orbits (see Astronomy by the Numbers: "Calculating Comet Halley's Orbit" on p. 290). As they get closer to the Sun, the tail forms. Actually, comets have not one but two tails (fig. 11.15), made of different materials that flow off the nucleus.

The gas evaporating off the nucleus of a comet is rapidly ionized by ultraviolet light from the Sun. It is then caught up in a flow of ions that streams from the Sun into space, a flow called the **solar wind.** The solar wind blows away from the Sun at about 400 kilometers per second. It is very tenuous, containing only a few atoms per cubic centimeter. However, the material in the comet's coma is tenuous too, and the solar wind is dense enough to blow it into a long plume. This is known as the **ion tail.** The ion tail usually has a bluish color and points directly away from the Sun. Magnetic fields in the solar wind enhance its effect on the comet's ion tail, helping to drag matter out of the coma and channel its flow, just as magnetic fields in the Earth's atmosphere channel particles to form the aurora.

The second tail is made of dust. Sunlight striking dust grains imparts a tiny force to them, a process known as **radiation pressure.** We don't feel radiation pressure when sunlight falls on us because the force is tiny and the human body is far too massive to be shoved around by solar photons. However, the microscopic dust grains in the coma do respond to radiation pressure and are pushed away from the Sun, as illustrated in

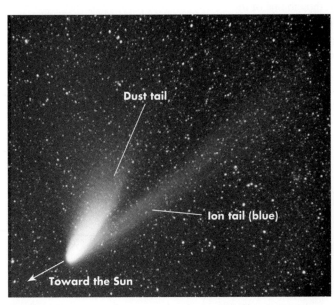

FIGURE 11.15
Comet Hale-Bopp in 1997, illustrating the dust and ion tails. The ion tail is affected almost exclusively by the solar wind, so it points nearly directly away from the Sun. By contrast, dust particles feel the effect of the Sun's gravity as well as its radiation pressure and continue on orbits around the Sun. The particles farther from the Sun orbit more slowly, so the dust tail arcs behind the comet.

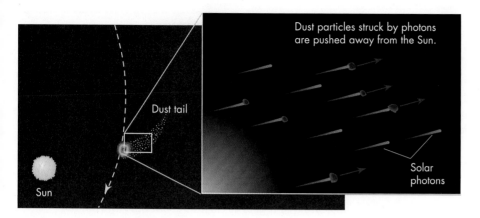

FIGURE 11.16
Sketch of how radiation pressure pushes on dust particles. Photons hit the dust, and their impact drives the dust away from the Sun, forming a dust tail. Sizes and distances are not to scale.

figure 11.16. Smaller grains experience a bigger acceleration, but all the grains move in the same direction, away from the Sun, forming a **dust tail.** Particles in the dust tail swing on a wider and therefore slower orbit than the comet nucleus, so the tail curves slightly behind the comet's direction, although still away from the Sun.

Thus, two forces, radiation pressure and the solar wind, act on the cometary material to drive out tails. Because those forces are directed away from the Sun, the comet's tails always point away from the Sun, and the tails even point out ahead of the comet as it moves away from the Sun (fig. 11.17). It might help you understand this seemingly odd phenomenon if you think of a runner carrying a torch. If the air is still, the smoke from the torch will, of course, trail behind the runner. However, if a strong wind is blowing at, say, 40 miles per hour, the smoke will be carried along in the direction of the wind regardless of which way the runner moves. Likewise, the high velocity of the solar wind (400 kilometers per second versus about 40 kilometers per second for the comet) carries the ion tail of a comet outward from the Sun regardless of the comet's motion, so it points ahead of the comet on the comet's outward path.

ANIMATION ▶

The orientation of comet tails

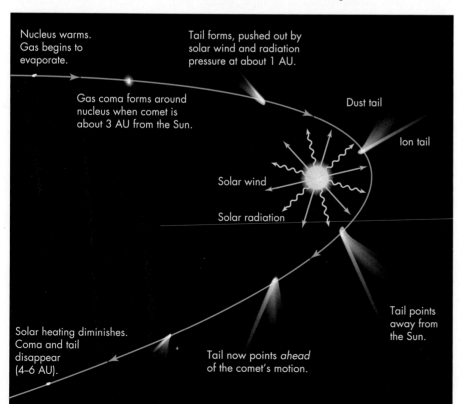

FIGURE 11.17
Sketch illustrating how radiation pressure and the solar wind make a comet's tail always point approximately away from the Sun. Sizes and distances are not to scale. Comets may orbit in any direction around the Sun.

ASTRONOMY *by the numbers* — CALCULATING COMET HALLEY'S ORBIT

Sir Edmund Halley's discovery that comets orbit the Sun just like planets do means that it is possible to apply Kepler's laws to comet orbits. As an example, consider Comet Halley, which Sir Edmund estimated returns every 76 years to the inner Solar System.

First, we can use Kepler's third law to predict the semimajor axis of Comet Halley's orbit. We set $P^2 = a^3$, and because $P = 76$ years, we know that

$$a^3 = 76^2 = 5776.$$

Taking the cube root of each side, we have

$$a = \sqrt[3]{5776} = 18.$$

So the semimajor axis of Comet Halley's orbit is 18 AU.

For most of its orbit, Comet Halley is invisible to us. Only when it comes within a few AU of the Sun is it visible. How far out does it go?

Recall that the semimajor axis is half of the long axis of the orbital ellipse (Kepler's first law), so the sum of Comet Halley's smallest distance and largest distance must be 2×18 AU $= 36$ AU. Observations of the comet indicate that it gets as close as about 0.5 AU from the Sun, so at its farthest point it must be $36 - 0.5 = 35.5$ AU from the Sun. Thus, it travels out beyond Neptune's orbit into the Kuiper belt.

Composition of Comets

The escaped gas from the comet offers astronomers a way to probe the comet's composition. The gases emit light of their own by a process called **fluorescence.** Fluorescence is produced when light at one wavelength is converted to light at another wavelength. A familiar example is the so-called black light that you may have seen for illuminating posters. Black light is really ultraviolet radiation, which we have difficulty seeing because of its short wavelength. When this light falls on certain paints or dyes, the chemicals in the pigment absorb the radiation and convert it into visible light, so the poster appears to glow.

Similarly, a photon of ultraviolet radiation from the Sun can lift an electron in one of the comet's gas molecules to an upper, excited level in a single leap. The electron then returns to its original level in two or more steps, emitting a photon each time it drops. The combined energy of these photons must equal that of the absorbed ultraviolet photon to conserve energy. The smaller energy of each photon gives them a longer wavelength, which, if they are in the visible range, we can see with our eyes. Thus, fluorescence creates the soft glow of the comet's ion tail. In addition, the spectrum of the fluorescing gas tells us what the comet is made of.

Spectra of gas in the coma and tail show that comets are rich in water, CO_2, CO, and small amounts of other gases that condensed from the primordial solar nebula. Evaporating water is broken up by solar ultraviolet radiation to create oxygen and hydrogen gas, so most comets are surrounded by a large cloud of hydrogen.

Astronomers have recently gotten even closer looks at comets. The *Stardust* probe took pictures of Comet Wild 2 (fig. 11.18A) and collected samples of dust from near the comet. The samples were successfully returned to Earth by parachute in early 2006. The dust particles included small crystals of silicate rock as well as a wide range of organic compounds, and some chemicals that require liquid water in order to form. The *EPOXI* spacecraft made even more detailed pictures of Comet Hartley 2 (fig. 11.18B), even observing chunks of ice ejected from the nucleus by jets of evaporating carbon dioxide.

A more intrusive comet sampling was made by NASA's *Deep Impact* mission, which in 2005 smashed a 370-kilogram (~800-lb) probe into Comet Tempel 1 at a relative speed of just over 10 kilometers per second (about 23,000 mph). The impact was designed to break through the comet's outer crust and stir up and release dust and gas. The impact event is shown in figure 11.19. The first two images are from the point of view of the impact probe, which sent pictures up until a few seconds before it smashed into the surface. The third image is from the main spacecraft, showing a cloud of very

FIGURE 11.18

(A) The nucleus of Comet Wild 2 (*Wild* is pronounced "vilt"). A longer-exposure image is superimposed, showing jets of gas. (B) The nucleus of Comet Hartley 2.

FIGURE 11.19
Comet Tempel 1 before and after NASA's *Deep Impact* mission. (A) The comet nucleus minutes before impact. (B) A composite image made from the impactor probe, showing the collision point. (C) An image made by the main spacecraft about a minute after impact shows a spray of fine particles blasted out by the impact and brightly lit by sunlight. (D) The *Stardust* probe imaged the impact crater about 6 years later.

fine dust blasted out by the impact, which created a crater about 150 meters across. The spectra of the material blasted out by the impact showed the presence of water and silicates as well as clays and other water-based crystals.

In 2014 the European Space Agency put the *Rosetta* spacecraft in orbit around Comet Churyumov-Gerasimenko, and placed a probe on the surface. The comet has a peculiar shape suggesting it might have formed from two objects or been reshaped as ices evaporated from around its narrow "neck" connecting the two halves (fig. 11.20). *Rosetta* is observing the comet's increasing activity as it approaches the Sun.

Spectroscopic studies of the water vapor flowing off Comet Hartley indicated that its isotopic composition closely matches that of Earth's oceans, but the water from Churyumov-Gerasimenko has a very different composition. This has renewed the debate about whether comets could have been the major source of Earth's water.

These various space missions have been able to estimate the masses and densities of comet nuclei by measuring their gravitational pull and detailed shape. The results indicate that the nuclei have densities in the neighborhood of 0.5 gm/cm^3. This implies an icy composition that is loosely packed, more like a snowball than a chunk of ice.

This loose structure eventually erodes away if a comet passes by the Sun too many times. Also, some comets literally fall into the Sun. For example, the *SOHO* satellite (which observes the Sun's outer atmosphere) has imaged dozens per year falling into the Sun. New comets show up frequently, so there must be a source to replace those devoured by the Sun, and it is to their origin that we now turn.

FIGURE 11.20
Views of Comet Churyumov-Gerasimenko from the *Rosetta* spacecraft. (A) The nucleus has a peculiar dumbbell shape. (B) As the comet moved closer to the Sun, gas and dust seemed to flow mostly from the "neck" region. (C) A close-up image shows surface variations.

Oort cloud and Kuiper belt

FIGURE 11.21
Schematic drawing of the Oort cloud, a swarm of icy comet nuclei orbiting the Sun out to about 100,000 AU. Also shown is the Kuiper belt, another source of comet nuclei, with a size exaggerated for clarity.

Origin of Comets

Astronomers think that most comets come from the **Oort cloud,** a swarm of trillions of icy bodies thought to lie far beyond the orbit of Neptune, as we discussed in chapter 8. Astronomers think the Oort cloud formed from planetesimals that originally orbited near the giant planets and were tossed into the outer parts of the Solar System by the gravitational force of those planets. There, they form a spherical shell that completely surrounds the Solar System and extends to perhaps as much as 150,000 AU from the Sun, as illustrated in figure 11.21. Astronomers deduce this shape for the Oort cloud from the many comet orbits that are highly tilted with respect to the main plane of the Solar System. However, as we will discuss later, some comets also seem to come from a flatter, less remote region—the so-called **Kuiper belt,** also shown schematically in figure 11.21. The Kuiper belt begins at about the orbit of Neptune and extends to approximately 50 AU.

Each comet nucleus moves along its own path, and those in the Oort cloud take millions of years to complete an orbit. With orbits so far from the Sun, these icy bodies receive essentially no heat from the Sun, and calculations indicate that their temperature is a mere 3 K, or about –454°F. Thus, the gases and ices remain deeply frozen. Such cold and distant objects are invisible to us on Earth; so if we are to see a comet, its orbit must somehow have been altered to carry it closer to us and the Sun.

Astronomers think that orbital changes may arise from the chance passage of a star far beyond the outskirts of the Solar System. For example, astronomers recently identified a faint star that passed only about 50,000 AU from the Sun about 70,000 years ago. The gravitational effects of its passage would disturb the orbits of the comet nuclei in the Oort cloud, altering their paths and making some drop in toward the inner Solar System, as illustrated in figure 11.22. A single disturbance may shift enough orbits to supply comets to the inner Solar System for tens of thousands of years.

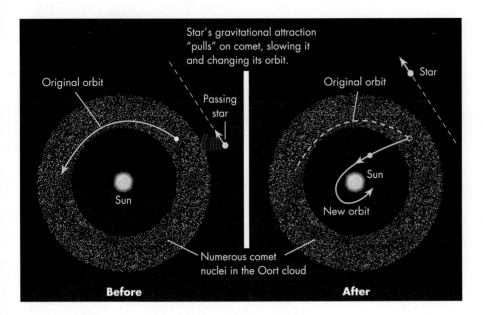

Before

After

FIGURE 11.22
Sketch of how a passing star alters the orbit of a comet nucleus. On its new orbit, the comet will pass by the Sun and be visible from Earth. Although only one comet's orbit is shown, such encounters typically will affect many comets. Note: The distance and size scales are greatly exaggerated. No star presently gets anywhere near this close to the Oort cloud.

Short-Period Comets and the Kuiper Belt

Although most comets that we see from Earth swing by the Sun on orbits that will bring them back to the inner Solar System only after millions of years, a small number of **short-period comets** reappear at time intervals less than 200 years. The origin of short-period comets is still under study. At one time it was thought that they came from the Oort cloud but as they moved through the region of the Solar System containing the giant planets, their orbits were shifted by a close encounter with one of the planets into smaller orbits with periods of centuries rather than millennia. Many astronomers think this is the history of Comet Halley, which has a period of 76 years.

However, astronomers now think that the majority of short-period comets come from the icy nuclei orbiting in the Kuiper belt. Support for this origin comes from the detection of hundreds of small, presumably icy, bodies orbiting near and somewhat beyond Pluto. Astronomers estimate that the Kuiper belt contains well over 30,000 icy objects bigger than 100 kilometers in diameter, and its total mass may be hundreds of times larger than that of the asteroid belt between Mars and Jupiter. These frozen objects are probably survivors of the Solar System's birth—icy planetesimals still orbiting in the disk—but they are too far apart to form additional planets.

The outer radius of the Kuiper belt is uncertain. There appear to be few comets or larger bodies such as plutoids orbiting beyond 50 AU, but even the largest of these bodies is extremely difficult to detect. Much still remains to be learned about this remote part of our Solar System.

Fate of Short-Period Comets

A short-period comet's repeated orbits past the Sun gradually whittle it away: all the ices and gases evaporate, and only the small amount of solid matter, dust and grit, remains. This fate is like that of a snowball made from snow scooped up alongside the road, where small amounts of gravel have been packed into it. If such a snowball is brought inside, it melts and evaporates, leaving behind only the grit accidentally incorporated in it. So too, the evaporated comet leaves behind in its orbit grit that continues to circle the Sun.

As a comet orbits the Sun and its icy material evaporates, it leaves in its path a trail of dust and small bits of solid material ejected from its nucleus. When we cross through or closely approach such a trail, our planet is blasted by this microscopic debris, which rains into our atmosphere, burning up and creating a meteor shower.

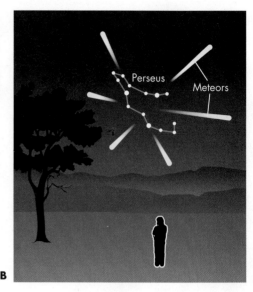

FIGURE 11.23

Sketch showing how (A) in mid-August, at the time of the Perseid meteor shower, the Earth is moving along its orbit. When the Earth crosses the debris strewn along a comet's orbit, the scattered material plunges into our atmosphere, producing (B) the diverging pattern of meteors characteristic of a meteor shower. (Bodies and orbits are not to scale.)

Meteor Showers

ANIMATION ▶

A meteor shower

If you go outside on a clear night and have an unobscured view of the sky, you will see, on average, one meteor every 15 or so minutes. Most of these meteors are stray fragments of asteroids that arrive at the Earth randomly. At some times of year, however, you may observe one every few minutes. Furthermore, if you watch such meteors carefully, you will see that they appear to come from the same general direction in the sky. Meteors of this type are part of what is called a **meteor shower,** which is made of the debris from a comet.

One of the best-known meteor showers occurs each year in mid-August. Peaking in intensity around August 13, meteors shoot through our atmosphere from a direction that lies toward the constellation Perseus. The meteors themselves have no association with Perseus. Rather, they are following an orbit around the Sun that happens to move toward us from roughly that direction (fig. 11.23A), and the Earth happens to cross their orbits in mid-August. Thus, at that time we encounter far more meteoroids than usual.

This encounter creates an effect similar to what you observe when you drive at night through falling snow: the flakes seem to radiate from a point in front of you, the location of which depends on a combination of the direction and speed of both the wind and your car. Thus, during the time that the Earth crosses the path they follow, the meteoroids seem to diverge from a common point (fig. 11.23B), called the **radiant.** Meteor showers are generally named for the constellation from which they appear to diverge, and appendix table 6 lists several of the brighter and more impressive showers and their dates. Each shower therefore marks when the Earth crosses the path of a comet.

On rare occasions the Earth will pass through a particularly dense clump of material left by the comet. If that happens, thousands of meteors per hour may spangle the sky. Such a display happened in November 1966, when dawn observers on the West Coast of the United States and Canada saw literally dozens of meteors per second! Likewise, in November 2001 and 2002, observers were treated to a similar cosmic fireworks display, as shown in the chapter's opening "What Is This?" figure.

Spectacles of this kind are one of the delights of astronomy, even viewed from orbit (fig. 11.24). However, meteor showers can pose a risk for orbiting satellites, and on even rarer occasions, far more sinister meteoritic events may occur.

FIGURE 11.24

Astronauts aboard the International Space Station caught this glimpse of a Perseid meteor seen from above the atmosphere.

11.5 GIANT IMPACTS

Every few thousand years the Earth is hit by a large meteoroid, a body tens of meters or more in size. Such bodies will produce not only a spectacular glare as they pass through the atmosphere but also an enormous blast on impact. Such large meteoroids have a very large kinetic energy—that is, energy of motion. If the meteoroid does not burn up on passage through the atmosphere, its remaining kinetic energy is released when it hits the ground.

Meteor Impacts on Earth

About 100 tons of meteoroids strike the Earth every day! Fortunately, the Earth's atmosphere significantly slows down or burns up the great majority of these, which then drift down as dust and ash (see Astronomy by the Numbers: "The Energy of Impacts"). However, our atmosphere can do little to slow meteoroids that are more than a few feet across, and so although rare, they can potentially cause serious damage. An impact by a meteoroid the size of a football stadium would have the explosive power of the largest thermonuclear bombs ever built. When a body this size strikes the Earth at orbital speeds, the energy released travels away from the site of impact in a shock wave that blasts out a deep crater, as illustrated in figure 11.25. Were such a body to hit a heavily populated area, the results would be catastrophic. Fortunately, we have been spared such disasters recently, but there have been some close calls and some truly horrific impacts in the distant past.

Many giant meteor craters scar our planet. One of the most famous is in northern Arizona (fig. 11.26). About 50,000 years ago, a meteoroid estimated to have been

FIGURE 11.25
A computer simulation of a crater's formation.

FIGURE 11.26
Photograph of the Arizona meteor crater, which is more than 1 kilometer across. The inset image shows one of the iron meteorites found at the site.

ASTRONOMY *by the numbers*   THE ENERGY OF IMPACTS

The energy released when a meteoroid strikes the Earth can be huge, as we can easily show from the formula for a body's kinetic energy, E, given by the expression

$$E = \frac{mV^2}{2},$$

where m is the body's mass and V is its velocity.

An iron meteoroid 3 meters (10 feet) in diameter, or a rocky meteoroid 4 meters (13 feet) in diameter, has a mass of about 100,000 kilograms (about 100 tons). If it were traveling at 30 kilometers per second ($= 3 \times 10^4$ meters per second) relative to the Earth, the kinetic energy of impact would be

$$E = 10^5 \times (3 \times 10^4)^2/2 = 4.5 \times 10^{13} \text{ joules},$$

which is about the same as that released when 10,000 tons of TNT are exploded. This is comparable to the nuclear bomb that destroyed Hiroshima.

Such an object hitting the Moon would make a crater roughly 100 meters (300 feet) in diameter. Striking the Earth, it would be just big enough that some fragments might reach the ground.

Q: What might explain why Manicouagan Crater is so much less sharply defined than the Arizona meteor crater?

|← Approx. 70 km →|
(about 43 miles)

FIGURE 11.27
Picture (from Earth orbit) of the Manicouagan Crater in Quebec. This winter view shows the lake that fills the crater covered with snow.

some 50 meters in diameter hit the Earth about 40 miles east of what is now Flagstaff. Its impact vaporized tons of rock, which expanded and peeled back the ground, creating a crater about 1.2 kilometers across and 200 meters deep.

More ancient impact scars occur in many places on our planet. The huge, ring-shaped Manicouagan Lake, about 70 kilometers (43 miles) in diameter and pictured in figure 11.27, is a meteor crater, as is Wolf Creek Crater in northwestern Australia. Astronomers have found even larger craters but are not sure they are impact features. Two such craters are the vast arc (nearly 500 kilometers across) on the east edge of Hudson Bay and a basin (about 300 kilometers across) in central Europe. Seismic studies have found still other craters, hidden under layers of sediment.

More recently, in 1908 an asteroid broke up in our atmosphere over a largely uninhabited part of north-central Siberia. This so-called Tunguska event, named for the region where it hit, leveled trees radially outward from the blast point to a distance of some 30 kilometers. The blast was preceded by a brilliant fireball in the sky and was followed by clouds of dust that rose to the upper atmosphere. Sunlight reflected off this dust gave an eerie glow to the night sky for several days. According to some accounts, the blast killed two people. Casualties were few because the area was so remote. Unfortunately, scientists didn't visit the site until about two decades later, because of the political turmoil in Russia at that time. When they did reach the site, they found no crater, just the felled trees. Interestingly, trees at the center of the damaged area were left standing vertically but with their branches stripped off. Thus, the explosion must have occurred in the air (see Science at Work: "Ghost Craters, or No Telltale Fragments").

Tunguska-like events may not be as rare as once thought. Hundreds of cameras caught a large meteor explosion over Russia in 2013 (fig. 11.28). The shock wave from the blast blew in windows, injuring more than a thousand people in an explosion estimated to have been the equivalent of about 500 kilotons of TNT. Only relatively small amounts of meteorite material were recovered from this event, although it is estimated that the original asteroid was about 17 meters (55 feet) in diameter. Accounts of fireballs and widespread damage in two remote regions of South America in the 1930s also resemble this event. Another recent event that did not do major damage because of its remoteness was a brilliant flash detected in February 1994 over the South Pacific ocean by a spy satellite. Analysis of the flash suggests that it marked the breakup of a 3-meter (10-foot) diameter meteoroid. The resulting blast had an energy between 10 and 20 kilotons of TNT, comparable to the bomb that destroyed Hiroshima. Collisions of this size occur about once a decade, but usually over unpopulated regions.

Although we have no well-documented cases of people being killed by meteorites, such deaths have almost certainly occurred. Careful searches through newspapers

FIGURE 11.28

Pictures from the meteor explosion over Russia in February 2013. From left to right: a "dash-cam" image of the asteroid exploding in the atmosphere, a factory roof and wall collapsed by the shock wave, a hole broken through the ice of a frozen lake by a fragment of the asteroid.

have turned up many near misses. Moreover, reports of meteorites hitting buildings and even cars are surprisingly common. Some tsunamis (tidal waves) may have been caused by impacts in the ocean. One suspected event about 500 years ago left deposits of beach sand more than 200 meters above sea level in New Zealand. This catastrophic event may explain why the Maori people abandoned many coastal settlements at that time, although this is debatable. Searching old records from Europe and China, astronomers have found instances of fatalities attributed to meteorites. One of the grimmer of these reports is from a fifteenth-century Chinese chronicle that describes "stones that fell like rain" that killed 10,000 people. Even allowing for exaggeration or misinterpretation of these old records, meteoroids are certainly far from harmless.

SCIENCE *at work* GHOST CRATERS, OR NO TELLTALE FRAGMENTS

Hypotheses to explain the Tunguska event offer a good example of how science tests and retests ideas. The lack of a big crater and the absence of meteorites at the site at first seemed to be evidence against the idea that it was caused by a small asteroid. This led some astronomers to propose that a comet might have caused the blast. Being icy, a comet would break up more easily in the atmosphere, so there would be no crater. Moreover, any fragments that survived would rapidly disappear, explaining the absence of large rock or iron meteorites. However, computer simulations of how an asteroid travels at high speed through our atmosphere showed what at first seemed a surprising result: stony or iron asteroids can also create devastation without leaving a crater or telltale fragments. A typical asteroid approaches Earth at a speed of 10 to 30 kilometers per second. When it hits our atmosphere, it compresses and heats the air ahead of it. The hot compressed air obeys Newton's law of action–reaction and exerts a tremendous force back on the asteroid that may shatter it. The resulting fragments plunge deeper into our atmosphere, where air resistance heats them further until they vaporize, creating a fireball at a height of 20 kilometers (12 miles) or so above the ground. No trace of the asteroid survives to form a crater or fragments at the ground. All that remains is the

FIGURE 11.29

The site of the Tunguska event photographed by an expedition in 1927, almost 20 years after the event occurred.

6000 K incandescent air of the fireball that blasts downward and out, stripping branches from trees and blasting them outward as was seen (fig. 11.29).

Such calculations do not prove that an asteroid caused the Tunguska event, but they do show that the lack of meteorites does not prove that a comet was to blame. The results also help to explain the relatively small amount of meteoritic material recovered in Chelyabinsk given the size of the meteoroid that we know must have given rise to the blast.

Mass Extinction and Asteroid/Comet Impacts

At the end of the Cretaceous period, about 65 million years ago, an asteroid or comet hit the Earth. Its impact and the subsequent disruption of the atmosphere are blamed for exterminating the dinosaurs and many less conspicuous but widespread creatures and plants. In fact, the sudden disappearance of large numbers of life forms at the end of the Cretaceous period defines the end of the Mesozoic era.

The evidence that an extraterrestrial body caused this devastation comes from the relatively high abundance of the otherwise rare element iridium found in sediments from that time. Iridium, a heavy element similar to platinum, is very rare in terrestrial surface rocks because it dissolves easily in molten iron and is one of the densest elements, so most of it sank to the Earth's deep interior at the time our planet formed its core. On the other hand, samples of meteoritic material contain moderate amounts of iridium because most of these bodies have not formed iron cores. Thus, the presence of so much iridium in a layer of clay laid down 65 million years ago is suggestive of a link to meteoritic material. The amount of iridium in the earth sediments at that layer (fig. 11.30) is the quantity that would be dispersed from a piece of meteoritic material 10 kilometers in diameter. Therefore, many astronomers and paleontologists think the Earth was hit by an asteroid of that size.

A 10-kilometer diameter asteroid hitting the Earth would produce an explosion on impact equivalent to that of several billion nuclear weapons. The impact would not only make an immense crater, it would also blast huge amounts of dust and molten rock into the air. The molten rock raining down would raise the surface temperature as high as that under an electric broiler and ignite global wildfires. The hot fragments and blast would also create nitrogen oxides, which would combine with water to form a rain of highly concentrated nitric acid. This devastating combination of heat, acid rain, and blast would then be followed by months of darkness and intense cold caused by the dust shroud blotting out the Sun. It seems likely that the biosphere would be devastated, leading to mass extinctions, just as the fossil record shows.

This frightening picture is supported not only by the iridium layer and the sudden disappearance of dinosaurs but also by a layer of soot, as well as a layer of tiny quartz pellets believed to have been created by the melting and blast of a violent impact. Moreover, geologists have found a large circular impact feature (figure 11.31) about 65 million years old near Chicxulub (*cheek-shoo-loob*, meaning "flea of the devil" in Mayan) in the Yucatán region of Mexico.

The Cretaceous mass extinction may have played an especially important role in our own evolution. Before that event, reptiles were the largest animals on Earth.

FIGURE 11.30
A layer of dark, iridium-rich clay marks the end of the Cretaceous period on Earth. The layer was probably formed when an asteroid struck the Earth. Fossils of many species are found below the layer (running diagonally across the image) but are not present in the more recent layers above it.

FIGURE 11.31
(A) The Chicxulub impact structure from 65 million years ago is buried under a kilometer of sedimentary rock in the Yucatan, so it is difficult to trace on the surface. (B) However, geological maps of subsurface features clearly trace the edge of a 180-km (110-mile) diameter impact crater. (C) A depiction of the moment of impact by NASA artist Don Davis.

Subsequently, mammals have assumed that niche. Small mammals may have escaped the fury of the heat and acid rain by remaining in burrows, and they may have survived the subsequent cold by virtue of their fur. You may be running your fingers through your hair, rather than your claws across your scales, because of that impact.

Other mass extinctions have occurred earlier and later than the Cretaceous event, but many scientists believe that massive volcanic eruptions or major changes in sea level might have caused them. Thus, like so many of the most interesting issues in science, no single accepted explanation has yet emerged.

SUMMARY

Our Solar System contains numerous small bodies: dwarf planets, asteroids, comets, and meteoroids. They are important astronomically because they give us information about the time of formation, composition, and physical conditions in the solar nebula.

Asteroids are rocky, metallic, or carbon-rich objects found mainly in the asteroid belt between the orbits of Jupiter and Mars. Comets are icy bodies found mainly in the Oort cloud, far beyond the orbit of Pluto.

A comet becomes visible if a passing star or some other event disturbs the comet's orbit so that it drops in toward the inner Solar System. There, solar heating thaws the frozen nucleus and evaporates gases. Radiation pressure and the solar wind then sweep the liberated gas and dust into a tail.

Some comets are caught in short-period orbits where they may be melted away to a collection of dust and grit. If the Earth crosses or nears the path of such a skeletal comet, the debris falling into the atmosphere causes a meteor shower.

Some asteroids and comets are large enough that their gravity has pulled them into a spherical shape. These bodies, which include Pluto, are now called dwarf planets. These are not considered planets because a large relative mass of other objects orbit at the same distance.

The Earth is hit frequently by small pieces of asteroidal material that may reach the ground and are called meteorites. Occasionally a large asteroidal or cometary body strikes the Earth, producing craters or, in very rare instances, mass extinctions.

QUESTIONS FOR REVIEW

1. (11.1) What makes a "shooting star"?
2. (11.1) What is the difference between a meteor, a meteoroid, and a meteorite?
3. (11.1) How is a meteor heated?
4. (11.1) What kinds of meteorites are there?
5. (11.2) Where are most asteroids found?
6. (11.2) What shape are typical asteroids and how do we know? Why does Ceres not have this shape?
7. (11.2) How do we know that asteroids have a composition similar to that of some meteorites?
8. (11.2) What do asteroids tell us about the formation of the Solar System?
9. (11.2) What are near-Earth objects?
10. (11.3) What evidence makes astronomers believe that Pluto is strongly influenced by Neptune?
11. (11.3) How did the discovery of a moon orbiting Pluto help astronomers better understand this object?
12. (11.3) Where did Pluto and other TNOs form? How did they get to where they are today?
13. (11.4) What parts make up a comet? What are they made of? How do we know?
14. (11.4) Why are there two tails to some comets? What are they made of?
15. (11.4) What is the Oort cloud? What is the Kuiper belt?
16. (11.4) What is the life history of a comet from the Oort cloud that has become an object we see?
17. (11.1/11.4) What creates meteor showers? When do some occur?
18. (11.5) What evidence is there that the Earth has been hit by asteroids or comets?
19. (11.5) What was the Tunguska event?
20. (11.5) Why do some scientists believe that asteroids and comets play a role in mass extinctions?

THOUGHT QUESTIONS

1. (11.1/11.4) Explain the difference between the "tail" of a meteor and the "tail" of a comet. Are both of them hot gas and debris?
2. (10.1) Suppose you see a meteor about 45° above the horizon and that the meteor is 100 km above the ground when you see it (fig 11.2). What is the farthest distance from you at which another person could also see the meteor? You may find making a sketch helpful.
3. (11.2) The total mass of the asteroid belt is much less than the mass of any of the planets. If there were many, many more asteroids, do you think they could form a planet?

4. (11.2) Suppose the coast guard is monitoring satellite pictures of the ocean, looking for vessels traveling at night in faint moonlight. The images do not have enough detail to resolve the boats—they appear as a single dot in the image. If they see a dot, how can they estimate the size of the boat? How would this depend on the color of the boat? How would infrared images help determine the size of the boat? Compare this situation to the visual and infrared study of asteroids described in this chapter.

5. (11.2) Compare the compositions of asteroids with the compositions of Mars and of Jupiter. How does the composition of bodies across the asteroid belt support the solar nebular theory?

6. (11.3) Why do some astronomers think that Pluto should be considered a planet? Do you agree? What happens if you apply this logic to Sedna and Eris?

7. (11.4) Examining the images of the nuclei of comets shown in section 11.4, do you think any of the crater-like depressions were caused by impacts like the ones that cause craters on asteroids? Explain your reasoning.

8. (11.5) If an asteroid were heading toward a possible future impact with Earth, what might be some advantages and disadvantages of setting off a nuclear explosion on it to deflect it to a different path?

7. (11.4) Comet Swift-Tuttle has a period of about 133 years and leaves the debris that causes the Perseid meteor shower. Calculate the semimajor axis of the comet's orbit. If the orbit is highly elliptical, approximately how far from the Sun is the comet at its greatest distance from the Sun? What is found at this distance?

8. (11.4) Use Kepler's third law to determine the period of a comet whose orbit extends to 50,000 AU, within the inner Oort cloud.

9. (11.4) Given that the temperature of a body decreases as the square root of its distance from the Sun increases, estimate the temperature of a comet nucleus in the Oort cloud. Take the temperature at 1 AU to be 300 kelvin.

10. (11.1/11.5) Use the formula for kinetic energy of a moving body to estimate the energy of an SUV (mass about 2700 kilograms) traveling at 65 miles per hour (you'll need to do a few conversions on units). Compare this to the kinetic energy of a 0.010-kilogram meteoroid (about as heavy as two quarters) that collides with the Earth at 30 kilometers/second.

11. (11.5) Calculate the kinetic energy of impact of a 1000-kilogram (roughly 1-ton) object hitting the Earth at 30 kilometers per second. Express your answer in kilotons of TNT, using the conversion that 1 kiloton is about 4×10^{12} joules. Be sure to convert kilometers/second to meters/second.

PROBLEMS

1. (11.1) The speed with which a meteoroid hits the atmosphere is roughly the speed of the Earth in its orbit (you might say the Earth hits the meteoroid). Show that the speed of the Earth in its orbit is about 30 kilometers/second.

2. (11.2) Calculate the surface gravity and escape velocity for Ceres, assuming it has a radius of 487 kilometers and a mass of 9.43×10^{20} kilograms.

3. (11.2) The asteroid Icarus has an elliptical orbit that carries it between 0.19 and 1.97 AU from the Sun. What is its semimajor axis? How often does it cross the Earth's orbital radius?

4. (11.3) Show that Pluto's orbital period is very close to 1.5 times Neptune's. Use the data in the appendix.

5. (11.3) Use the modified form of Kepler's third law to calculate the sum of the masses of Pluto and Charon from Charon's orbital period of 6.387 days and its semimajor axis of 17,530 km. Compare your result to the sum of the masses given for Pluto and Charon found in the appendix. Be sure to convert the orbital period into appropriate units before putting those numbers into the formula. (Note that Charon's mass is a significant fraction of Pluto's so it is not a good approximation to ignore it.)

6. (11.3) Calculate the density of Charon, given that its radius is approximately 600 kilometers and its mass is about 1.52×10^{24} grams. (Be sure to convert kilometers to centimeters or meters.) Is it likely that Charon has a large iron core? Why?

TEST YOURSELF

1. (11.1) The bright streak of light we see, as a meteoroid enters our atmosphere, is caused by
   (a) sunlight reflected from the solid body of the meteoroid.
   (b) radioactive decay of material in the meteoroid.
   (c) a process similar to the aurora that is triggered by the meteoroid's disturbing the Earth's magnetic field.
   (d) frictional heating as the meteoroid speeds through the gases of our atmosphere.
   (e) the meteoroid's disturbing the atmosphere so that sunlight is refracted in unusual directions.

2. (11.1/11.4) Meteor showers such as the Perseids in August are caused by
   (a) the breakup of asteroids that hit our atmosphere at predictable times.
   (b) the Earth passing through the debris left behind by a comet as it moved through the inner Solar System.
   (c) passing asteroids triggering auroral displays.
   (d) nuclear reactions in the upper atmosphere triggered by an abnormally large meteoritic particle entering the upper atmosphere.
   (e) none of the above.

3. (11.1) What is(are) the source(s) of *most* meteorites? (More than one answer may apply.)
   (a) Comets   (d) Mars
   (b) The Moon   (e) Material from the solar wind
   (c) Asteroids

4. (11.2) The asteroid belt lies between the orbits of
   (a) Earth and Mars.   (d) Mars and Jupiter.
   (b) Saturn and Jupiter.   (e) Pluto and the Oort cloud.
   (c) Venus and Earth.

5. (11.2) How do astronomers estimate the sizes of most asteroids?
   (a) By observing angular diameters from space telescopes
   (b) By timing how long they take to pass in front of stars
   (c) By determining their gravitational influence on other asteroids
   (d) By measuring the amount of light they reflect

6. (11.2) Asteroids in the asteroid belt are made up of
   (a) iron.   (b) silicates (rock).   (c) organic compounds.
   (d) a and b only, in amounts that vary depending on where they orbit (and were formed).
   (e) a, b and c, in amounts that vary depending on where they orbit (and were formed).

7. (11.3) How was the mass of Pluto determined?
   (a) By measuring the effect of its gravity on the terrestrial planets
   (b) By observing its effect on the motion of an unmanned flyby of Pluto
   (c) By determining the orbit of its satellite, Charon
   (d) By measuring how it bends light rays passing near it

8. (11.4) The tail of a comet
   (a) is gas and dust pulled off the comet by the Sun's gravity.
   (b) always points away from the Sun.
   (c) trails behind the comet, pointing away from the Sun as the comet approaches it and toward the Sun as the comet moves out of the inner Solar System.
   (d) is gas and dust expelled from the comet's nucleus by the Sun's heat and radiation pressure.
   (e) is both (b) and (d).

9. (11.4) Short-period comets have a period of around _____ and mostly come from the _____.
   (a) decades to a few hundred years; Kuiper belt
   (b) a few hundred to a thousand years; Kuiper belt
   (c) decades to a few hundred years; Oort cloud
   (d) a few hundred to a thousand years; Oort cloud
   (e) a thousand to a million years; Oort cloud

10. (11.5) A moving object's kinetic energy depends on its _____ and _____.
   (a) size, density   (d) magnetic field, velocity
   (b) density, velocity   (e) velocity only
   (c) velocity, mass

11. (11.5) Strong evidence that the dinosaurs were killed by a meteor impact is provided by (there may be more than one correct answer, so select all that apply):
   (a) mass extinctions 65 million years ago.
   (b) a large crater in Arizona.
   (c) pieces of the asteroid that have been recovered.
   (d) an unusually rich layer of a rare element in the rock record at 65 million years ago.
   (e) a layer of 65-million-year-old soot in the rock record.

KEY TERMS

achondrite, 279
asteroid, 280
asteroid belt, 280
carbonaceous chondrite, 278
chondrite, 278
chondrule, 278
coma, 287
dust tail, 289
fluorescence, 290
ion tail, 288
Kirkwood gaps, 282
Kuiper belt, 292
meteor, 277

meteorite, 278
meteoroid, 277
meteor shower, 294
near-Earth object, 283
nucleus, 287
Oort cloud, 292
radiant, 294
radiation pressure, 288
short-period comet, 293
solar wind, 288
tail, 287
trans-Neptunian object (TNO), 284

: FIGURE QUESTION ANSWERS

WHAT IS THIS? (chapter opening): This image was made by combining a number of separate images of the Leonid meteor shower taken just before dawn on November 19, 2002. You can see that the meteors appear to radiate from a small area on the sky. That area, it turns out, lies in the constellation Leo, hence the name of the shower.

FIGURE 11.1: The star trails show that this was a fairly long time exposure, during which the Earth rotated, smearing the stars' images into long arcs. The meteor passed by very quickly; the Earth didn't rotate very far during its passage, and as a result the meteor's track is straight.

FIGURE 11.5: The Trojan asteroids are about equidistant from the Sun and Jupiter, forming a nearly equilateral triangle. This position is known as a Lagrange point, and is a stable location for orbits. There are similar Lagrange points relative to the Moon orbiting the Earth or the Earth orbiting the Sun.

FIGURE 11.9: The gaps in the distribution of the asteroids are very like the gaps in Saturn's rings, such as Cassini's division.

FIGURE 11.11: Pluto's orbit is relatively steeply tilted with respect to Neptune's. As a result, when Pluto "crosses" Neptune's orbit, Pluto is actually well above or below Neptune's path.

FIGURE 11.27: Manicouagan Crater is far enough north that it was buried by glaciers during the last ice age. Moreover, being in a wet climate, it erodes faster and more extensively than the Arizona meteor crater, which is in a dry climate.

12

A *Solar Dynamics Observatory* image of the Sun at 30 nanometer wavelength showing a huge flare erupting from the surface in 2012.

THE SUN, OUR STAR

LEARNING OBJECTIVES

Upon completing this chapter you should be able to:

- Describe the structure of the Sun from its core to the solar wind, and indicate the approximate temperatures and densities of the various layers.
- Define pressure and the ideal gas law, and explain what astronomers mean by hydrostatic balance.
- Explain why the Sun must continuously generate heat and what becomes of the energy it generates.
- Describe the process of nuclear fusion and what conditions are necessary for it to take place.
- Calculate the energy released by conversion of mass to energy according to Einstein's equation $E = mc^2$.

- Indicate the relative amount of mass the Sun loses in fusion and through the solar wind.
- Discuss the solar neutrino problem and its resolution.
- Explain how magnetic fields and gas in the Sun interact, and describe features on the Sun that arise as a result.
- Discuss how magnetic fields grow and remain strong in sunspots, and how they cause sunspots to be dark.
- Explain how magnetic fields can make the corona so hot.
- Describe the solar cycle, its cause, and how it has varied in the past.
- Describe how solar activity affects the Earth through both individual events and long-term changes in the Sun.

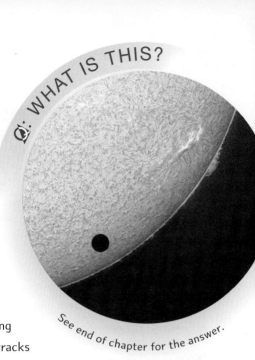
The Sun is a star, a dazzling, luminous ball of gas more than 100 times bigger in diameter than the Earth. The Sun is the source of light that heats the planets and maintains the comfortable temperatures that allow life to flourish on Earth. In fact, the Sun is the source of most of the energy that we use in our everyday lives. For example, the hydropower from a dam relies on solar heating to evaporate water so it can rise into the atmosphere, where it is carried by winds (also driven by the Sun) to precipitate over high elevations and then run downhill again. Even fossil fuels represent solar energy stored up by plants in chemical bonds millions of years ago.

Although the Sun gleams peacefully in our daytime sky, specially equipped telescopes and space observatories reveal that its surface is violently agitated, with rising fountains of incandescent gas and a twisted magnetic field. Even greater violence wracks its core. There, a nuclear furnace burns 600 million tons of hydrogen into helium every second, producing in one heartbeat the energy of 100 billion nuclear bombs.

How the Sun releases vast quantities of energy while managing to hold itself together is one of the main themes of this chapter. We will begin by describing the Sun: its radius, mass, and other properties. Then we will discuss how the crushing force of its gravity balances the explosive power in its core. The tension between these enormous forces is remarkably stable, but the energy escaping from the core interacts with magnetic fields, stirring up the atmosphere in complicated ways. This results in small changes in the Sun's energy output—small for the Sun, but with significant consequences for us on Earth.

CONCEPTS AND SKILLS TO REVIEW

- Relation between temperature of a hot object and the color of the light it emits (Wien's law) (4.3)
- Doppler shift (4.6)
- Structure of atoms (4.4)

12.1　SIZE AND STRUCTURE

The Sun is immense, dwarfing the Earth and even Jupiter, and its immensity is what makes it shine. With a radius over 100 times that of the Earth and a mass 300,000 times the Earth's, the Sun has an enormous gravity that crushes the material in its interior. To offset that crushing force and prevent its own collapse, the Sun must be extremely hot. But hot objects always lose energy, and the Sun is no exception. We see that lost energy as sunshine and welcome it as the source of our life. But sunshine is a death warrant for the Sun because the energy it carries off must be replenished or the Sun will collapse. Fortunately for us, the Sun does replace its lost energy, but only at the cost of consuming itself—a dilemma that is not unique to the Sun but is shared by most stars.

Measuring the Sun's Properties

Before we discuss how the Sun replaces its lost energy, we need to understand better some of its overall properties. How much mass does it contain? How rapidly does it lose energy? What resources are available to supply its energy needs?

The Sun is about 150 million kilometers (1 AU, or 93 million miles) from Earth. Astronomers originally measured its distance by triangulation, but they now use radar, bounced either directly from the Sun or from other bodies whose distance is known in AU from Kepler's third law. Knowing the Sun's distance, we can find its radius from its angular size, as we showed in chapter 2. We also need to know the Sun's distance

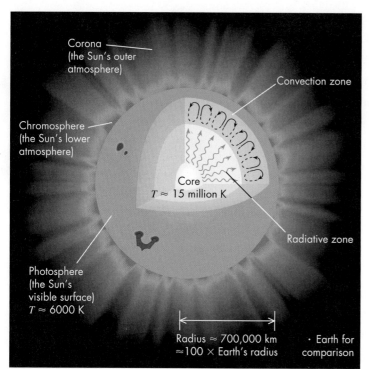

Table 12.1	Properties of the Sun
Radius ($R_\odot$*)	7×10^8 m = 7×10^5 km, or about 109 R_{Earth}
Mass ($M_\odot$*)	2×10^{30} kg, or about 333,000 M_{Earth}
Distance from Earth	1.5×10^8 km, or 1 AU
Temperature of surface	5780 kelvin (~9900 °F)
Temperature of core	15 million kelvin (~27 million °F)
Composition by mass	71% hydrogen 27% helium 2% heavier elements
Power output	4×10^{26} watts

* Astronomers use the symbol $\odot$ to stand for the Sun. Thus, $R_\odot$ is the Sun's radius and $M_\odot$ is its mass. The symbol is the ancient Egyptian hieroglyph for "Sun."

FIGURE 12.1
A cutaway sketch of the Sun.

from Earth if we are to measure its mass with the help of Newton's laws, as we showed in chapter 3. From its mass and radius, we can calculate that its surface gravity is about 30 times the Earth's. The Sun's internal heat gives the gas a pressure that counterbalances this crushing force of gravity.

The Sun's surface temperature can be found from its color and Wien's law, using the methods discussed in chapter 4. Astronomers cannot measure its interior temperature directly; rather, from calculations and indirect measurements discussed below, they deduce that its core temperature is about 15 million K (about 27 million °F).

From measurements of the amount of solar energy that reaches the Earth, astronomers calculate that the Sun radiates a total of 4×10^{26} watts of power into space from its surface, energy that must be replenished by the fusion of hydrogen in its core, as we will see in section 12.2. Fortunately, the Sun has a plentiful supply of hydrogen: its spectrum shows it is about 71% hydrogen, 27% helium, and 2% heavier elements such as carbon and oxygen, similar in composition to Jupiter and Saturn. But unlike these mostly liquid bodies, the Sun is gaseous throughout because its high temperature breaks almost all molecular bonds, vaporizing even iron, and allowing the atoms to move freely.

Table 12.1 lists some of these vital statistics, and the portrait of the Sun in figure 12.1 shows its main features. It will be helpful many times in this chapter to refer to this picture.

The Solar Interior

When we observe the Sun, we see through the low-density, tenuous gases of its outer atmosphere. Our vision is ultimately blocked, however, as we peer deeper into the Sun. There, the material is compressed to high density by the weight of the gas above it. In this dense material, the atoms are sufficiently close together that they strongly absorb the light from deeper layers, blocking our view of them much like frosted glass obscures what lies behind it. Above these layers, however, the absorption of light is weaker and the light emitted by hot material there escapes freely into space as the sunlight we see. These layers where the Sun's gases change from transparent to opaque form what is called the **photosphere,** the visible surface of the Sun. Although the blocking of radiation by denser gas below the photosphere limits our view into the Sun, it helps the Sun retain heat and, like a well-insulated house, thereby reduces the amount of fuel it must consume.

If we could probe inside the Sun, we would find that its density and temperature rise steadily as we neared the core. In the photosphere, the density is comparable to that of the air around us; deeper, the material above pushes down on that below, compressing the gas like a pile of pillows. A similar compression occurs in the atmosphere of the Earth and other planets, as we discussed in chapters 6 and 10. But the greater mass of the Sun leads to a vastly greater compression of its gas, and so, near its core, the density is more than 100 times that of water. Despite this great density, the Sun is gaseous throughout because its high temperature gives the atoms so much energy of motion that they are unable to bond with one another to form a liquid or solid substance.

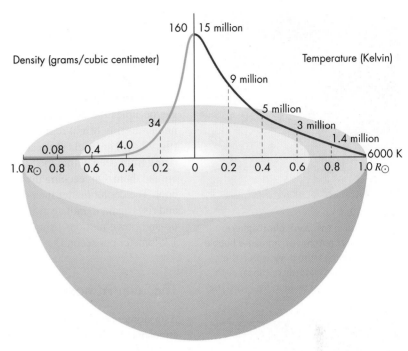

FIGURE 12.2

Plots of how density and temperature change within the Sun.

The temperature also rises as we plunge into the Sun's interior. The photosphere is almost 6000 K. Below it, heat is partially trapped and the temperature soars to about 15 million K at the core. Figure 12.2, based upon theoretical calculations, illustrates how the temperature and density change through the Sun. No spacecraft has made, or is ever likely to make, the measurements directly, but we are confident that they are correct because the Sun needs such high temperatures and densities to keep it from collapsing under its own gravity.

Q: About how deep in the Sun is the density the same as that of water?

Energy Flow in the Sun

Astronomers can measure some of the Sun's properties, such as its diameter and surface temperature. However, they can only deduce most of its other properties, such as its internal temperature and density. Such deductions are based on computer models that use the laws of physics to calculate the Sun's properties. The accuracy of the models' predictions is then judged by whether the predictions agree with measured properties. For example, if a model predicted the Sun's surface temperature to be 10,000 K, it would be rejected as incorrect. Thus, our understanding of the Sun's properties comes from a combination of theory and measurement.

Your own experience and experiments in the laboratory show that heat always flows from hot to cold. Applying this principle to the Sun, we can infer that because its core is hotter than its surface, heat will flow outward from its center, as illustrated in figure 12.3. Near the core, the energy moves by radiation carried by photons through what is called the **radiative zone.** Because the gas there is so dense, a photon travels less than an inch before it is absorbed by an atom and stopped. The photon is eventually re-emitted, but it will be almost immediately reabsorbed. The constant absorption and re-emission slows photons like cars caught in stop-and-go traffic. The question of how long it takes a photon to travel from the Sun's center to its surface turns out to be tricky. One reason is that a single photon never makes such a journey. A typical photon leaving the core is a gamma ray, which will be converted to many lower-energy photons before its energy reaches the surface. Therefore, it is perhaps better to ask how long it takes energy to travel from the Sun's core to the Sun's surface. Calculations

FIGURE 12.3

Sketch of how energy flows from the Sun's core to its surface. In the deep interior, radiation carries the energy. Near the surface, convection carries the energy.

FIGURE 12.4
Granulation in the Sun's photosphere. Each brighter patch is an immense bubble of hot gas rising from deep within the Sun. The darker material outlining each bubble is cooler gas sinking back into the Sun's interior.

Q: How big across is one of the rising bubbles that forms the granulation? What has a similar size on Earth?

suggest that it takes about 16 million years for the energy to emerge from the surface. Thus, today's sunshine was born in the Sun's core before we existed as a species!

The flow of energy from the core toward the Sun's surface is slowed in the region just below the photosphere, where the gas is cooler and less transparent. Here, photons are even less effective in moving energy, and convection currents like those in the Earth and giant planets carry the energy to the surface. The rising and sinking gas occupies the **convection zone,** and we can infer the gas's motion there from the numerous tiny bright regions surrounded by narrow darker zones, called **granulation** (fig. 12.4). The bright areas are bubbles of hot gas many hundreds of kilometers across, rising from deep within the Sun. They are brighter because they are hotter than the gas around them. On reaching the surface, these hot bubbles radiate their heat to space and cool. The cooler matter then sinks back toward the hotter interior, where it is reheated and rises again to radiate away more heat. Astronomers can measure the speed of these up-and-down motions using the Doppler effect and find that the bubbles rise at about 1 kilometer per second.

The Solar Atmosphere

Astronomers refer to the low-density gases that lie above the photosphere as the Sun's atmosphere. This region marks a gradual change from the dense gas of the photosphere to the extremely low-density gas of interplanetary space. A similar transition occurs in our own atmosphere, where the gas density decreases steadily with altitude and eventually merges with the near-vacuum of space.

Although the gas density in the Sun's atmosphere decreases above the photosphere, the gas temperature behaves very differently. Immediately above the photosphere the temperature decreases, but at higher altitudes the gas grows hotter, reaching temperatures of several million kelvin. Why the Sun's atmosphere is so hot remains a mystery, though astronomers believe that the Sun's magnetic field plays a role in somehow heating these low-density gases, as we will discuss in section 12.4.

The Sun's atmosphere consists of two main regions. Immediately above the photosphere lies the **chromosphere,** the Sun's lower atmosphere. It is usually invisible against the glare of the photosphere but can be seen at a total eclipse of the Sun as a thin red zone around the Sun (fig. 12.5). The red color is what gives the chromosphere its name—the prefix *chromo* comes from the Greek for color. With a telescope equipped with an appropriate filter, you can see that the chromosphere contains millions of thin columns called **spicules** (fig. 12.6), each a jet of hot gas thousands of kilometers long.

The chromosphere's color comes from the strong red emission line of hydrogen, Hα. We saw in chapter 4 that emission lines arise in hot, low-density gas. From these lines astronomers can infer the gas's temperature. Just above the photosphere, the temperature is about 4500 K, but 2000 kilometers higher, it reaches 50,000 K. Here, the chromosphere ends and the temperature shoots up to about 1 million kelvin as we enter the **corona,** the Sun's outer atmosphere.

The corona's extremely hot gas has such low density that under most conditions we look right through it. But like the chromosphere, it can be seen during a total solar eclipse when the Moon covers the Sun's brilliant disk. Then the pale glow of the corona can be seen to extend

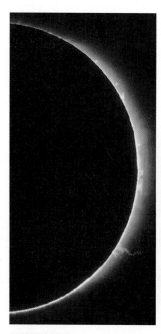

FIGURE 12.5
Photograph of a portion of the solar chromosphere at a total solar eclipse.

FIGURE 12.6
Photograph of spicules in the chromosphere. The spicules are the thin, stringy features that look like tufts of grass.

FIGURE 12.7
Photograph of the corona during a total eclipse of the Sun.

FIGURE 12.8
Image of the Sun in ultraviolet light made with the *SOHO* satellite. The dark region is a coronal hole where gas is streaming out from the Sun.

beyond the Sun's edge to several solar radii (fig. 12.7). Pictures of the Sun made at ultraviolet and X ray wavelengths show that the corona is not uniform but has extremely hot streamers along the Sun's magnetic field. As shown in figure 12.8, the corona also contains huge low-density regions called **coronal holes** through which gas escapes from the Sun into space, as we will discuss in section 12.4.

Because the corona is so tenuous, it contains very little energy despite its high temperature. It is like the sparks from a Fourth of July sparkler: despite their high temperature, you hardly feel them if they land on your hand because they are so tiny and carry very little total heat.

12.2   HOW THE SUN WORKS

Our discussion so far has centered on the structure of the Sun. Our task now is to understand *why* it has that structure and how it works. For example, why is the Sun hot? What makes it shine? How does it generate energy?

Internal Balance (Hydrostatic Equilibrium)

The structure of the Sun depends on a balance between its internal forces. One force holds the Sun together. A second force prevents the Sun from collapsing. This balance is technically called **hydrostatic equilibrium.** Our atmosphere is in hydrostatic equilibrium, its gases pulled downward by the Earth's gravitational force but supported by collisions of the air molecules, creating pressure. Likewise, pressure in an automobile tire supports the weight of the car.

The Sun's inward force arises from its own gravity. The outward force arises from the rapid motion of its atoms, a motion that gives rise to a **pressure.** Thus, in the Sun, as in virtually all stars and planets, the balance of hydrostatic equilibrium requires that the *outward* force created by pressure exactly balance the *inward* force of the Sun's gravity (fig. 12.9). Without such a balance, the Sun would rapidly change. For example, if its pressure were too weak, the Sun's own gravity would rapidly crush it. Therefore, to understand the Sun, we need to discuss in more detail how its pressure arises.

Surface of Sun

Gravity
Pressure

Hydrostatic equilibrium
The outward pressure force must balance the inward gravitational force at every distance from the center of the Sun.

Center of Sun

FIGURE 12.9
A sketch illustrating the condition of hydrostatic equilibrium, the balance of pressure (blue arrows) and gravitational force (purple arrows) in the Sun.

ANIMATION ▶

How heating raises pressure

Gas cool
Atoms move slowly
↓
Pressure small

Gas hot
Atoms move rapidly
↓
Pressure large

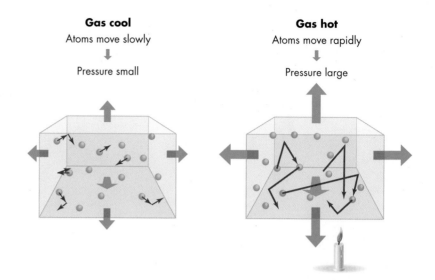

FIGURE 12.10

Sketch illustrating the perfect, or ideal, gas law. Gas atoms move faster at the higher temperature, so they collide both more forcefully and more often than atoms in a cooler gas. These effects combine to create a higher pressure. Thus, other things being equal, a hotter gas exerts a greater pressure.

Pressure in a gas comes from collisions among its atoms and molecules. If the gas is squeezed, atoms are pushed toward each other. As they collide, they rebound, resisting the compression, as you can feel by squeezing a balloon. The strength of the pressure depends on how often and how hard the collisions occur. Raising the density increases the frequency of collisions by moving atoms closer together. Raising the temperature speeds atoms up, making them collide harder and more often (fig. 12.10). Thus, the strength of the pressure is proportional to the density times the temperature of the gas; that is:

$$\text{Pressure} = \text{Density} \times \text{Temperature} \times \text{a constant,}$$

a result known as the **perfect, or ideal, gas law.** The value of the constant in the equation depends on the units used to measure the pressure, density, and temperature.

Although the perfect gas law shows us that the pressure inside the Sun depends on temperature, it does not by itself show us how hot that temperature needs to be. We might reasonably guess that because the Sun has a huge mass (and therefore a huge gravity crushing it), it needs a huge temperature to offset that crushing force.

Indeed, the Sun's core is very hot. To determine exactly how hot, astronomers solve a computer model of the Sun, and the answer turns out to be about 15 million K for the temperature and about 160 grams per cubic centimeter for the density, as we discussed in section 12.1 (see fig. 12.2). But because heat flows out of the hot core, the Sun needs a mechanism to keep replacing its core heat.

Powering the Sun

Energy that flows outward from the Sun's core eventually escapes into space as the sunshine that warms the Earth and other planets. The energy lost by the Sun must be replaced, or the Sun's core would begin to cool, the pressure would begin to drop, and the Sun would shrink under the force of its own gravity. The Sun is therefore like an inflatable chair with tiny leaks through which the air escapes. If you sit in the chair, it will gradually collapse under your weight unless you pump air in to replace that which escapes. What acts as the energy pump for the Sun?

Early astronomers thought that the Sun might burn ordinary fuel such as coal. But even if the Sun were pure coal, it would have only enough energy to produce its current light output for a few thousand years.

At the end of the nineteenth century, the English physicist Lord Kelvin and the German scientist Hermann Helmholtz independently proposed that the Sun is not in hydrostatic equilibrium but that gravity slowly compresses it, making it shrink. They thought that the Sun's weight squeezed gas in the interior, increasing its density and heating it, similar to why the gas giant planets have such hot interiors. However, gravity could power the Sun by this mechanism only for 10 million years, and we know the Sun has been shining for billions of years because we have fossils of life on Earth that old. Furthermore, if gravity were the power source, the Sun would be steadily shrinking, and it isn't. Therefore, something else must supply energy.

In 1899 the American geologist Thomas Chamberlin suggested that subatomic energy—energy from the reactions of atomic nuclei—might power stars, but he could offer no mechanism for how the energy was liberated. In 1905, Einstein proposed that energy might come from a body's mass. His formula,

$$E = mc^2$$

c = Speed of light: 3×10^8 m/sec
m = An amount of mass
E = An amount of energy

states that a mass, m, can become an amount of energy, E, equal to the mass multiplied by the square of the speed of light, c, where $c = 3 \times 10^8$ meters per second. The amount of energy available from reactions involving atomic nuclei is vastly larger than that from chemical processes such as burning. For example, if 1 gram of mass—less than the amount in an aspirin tablet—is converted to energy, it releases 9×10^{13} joules, the equivalent of about 20 kilotons of TNT* or a small nuclear weapon. If the Sun could convert even a tiny amount of its mass into energy, it would have an enormous source of power.

In 1919 the English astrophysicist Arthur Eddington, a pioneer in the study of the physics of stars, showed that the conversion of hydrogen into helium would release enough energy to power the Sun, but even his theory lacked the necessary details, which were supplied independently in the late 1930s by the physicists Hans Bethe and Carl von Weizsäcker. Their work showed that the Sun could generate its energy by converting hydrogen into helium by a process called **nuclear fusion,** a process that bonds two or more nuclei into a single, heavier one.

Nuclear Fusion

Fusion is possible in the Sun because its interior is so hot. Under normal conditions hydrogen nuclei repel each other, pushed apart by the electrical charge of the protons. However, at high temperatures, nuclei move so fast that when they collide, they are driven extremely close together. At such a collision, the nuclei may get close enough to each other that the electrical repulsion between their protons is overwhelmed by the **strong force** (sometimes called the strong nuclear force), which holds the protons and neutrons together in a normal atom. Thus, the two separate nuclei can merge, or fuse, into a single new nucleus. Because this fusing process requires such a high temperature, the only place in the Sun hot enough for fusion to occur is its core. The core therefore is where the Sun makes its energy. However, before we can understand how fusion creates energy, we need to look at the structures of hydrogen and helium.

In their usual form, hydrogen consists of one proton and an orbiting electron, and helium consists of two protons, two neutrons, and two orbiting electrons. In the Sun's hot interior, however, atoms collide so violently that the electrons are generally removed. Atoms that are missing one or more electrons are said to be ionized. A gas in which the atoms are ionized is technically called a "plasma." The electrons are removed from most atoms and move independently in the gas, and therefore we can ignore them in much of what follows.

Hydrogen and helium always have one and two protons respectively, but they can have other forms (called "isotopes") with different numbers of neutrons (fig. 12.11). To identify the isotopes, we write their chemical symbol with a superscript that shows the total number of protons and neutrons. The usual form of hydrogen with one proton

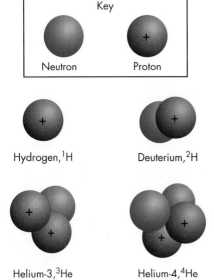

FIGURE 12.11
Schematic diagrams of the nuclei of hydrogen, its isotope deuterium, and two isotopes of helium.

---

* One kiloton is about 4.2×10^{12} joules of energy.

The Proton-Proton Chain

FIGURE 12.12

Diagram of the proton–proton chain. (A) Hydrogen (H) nuclei first combine to make deuterium (^{2}H), and (B) the positron created by the process immediately annihilates an electron. (C) Deuterium and hydrogen combine to make ^{3}He. (D) ^{3}He nuclei combine to form ^{4}He, returning two hydrogen nuclei into the gas.

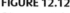

ANIMATION ▶

The reactions of the proton–proton chain

Although the Sun's power comes from the conversion of its mass to energy, the change in the Sun's mass is extremely small compared to the overall mass of the Sun. The decrease in the Sun's mass over its entire lifetime will be less than 1%.

and no neutrons is ^{1}H, while the form of hydrogen containing a proton and a neutron is ^{2}H, which is also known as deuterium or "heavy hydrogen." Likewise, the normal form of helium with two protons and two neutrons is ^{4}He, while helium with two protons but only one neutron is ^{3}He. These are often called "helium-4" and "helium-3," respectively. As we will see, these isotopes play a critical role in the Sun's energy supply.

The Proton–Proton Chain

Hydrogen fusion in the Sun occurs in three steps, called the **proton–proton chain.** In the first step, two ^{1}H nuclei collide and fuse to form the isotope of hydrogen ^{2}H. In the collision, one proton becomes a neutron by ejecting two particles: a positron (denoted as e$^+$) and a **neutrino** (denoted by ν). The neutrinos so generated play no further role in the Sun's energy generation, but we will encounter them again because they can be detected when they leave the Sun and may help astronomers learn more about conditions in the Sun's interior.

This first step in converting mass into the energy in the Sun is depicted in figure 12.12A and can be written symbolically as

$$^1\text{H} + {}^1\text{H} \rightarrow {}^2\text{H} + e^+ + \nu$$

The terms to the left of the arrow are the normal hydrogen nuclei that start the process. The terms to the right are the heavy isotope of hydrogen that results (^{2}H), plus the positron and the neutrino. Energy is released by this reaction because the mass of ^{2}H is slightly less than the mass of the two ^{1}H's that were used to make it. And the positron and a normal electron interact, annihilating each other (fig. 12.12B). The lost mass is converted to energy, as described in Einstein's formula $E = mc^2$, and ultimately that energy becomes the electromagnetic radiation of sunlight.

In the second step, the ^{2}H nucleus collides with a third ^{1}H to make the isotope of helium containing a single neutron, ^{3}He. This releases a high-energy photon (gamma ray), which we denote by γ. Figure 12.12C shows this step, which can be written as:

$$^1\text{H} + {}^2\text{H} \rightarrow {}^3\text{He} + \gamma$$

Here again, the resulting particle, ^{3}He, has a smaller mass than the particles from which it was made, and again energy is released.

The third and final step is the collision and fusion of two ^{3}He nuclei. Here, the fusion results not in a single particle but rather in one ^{4}He and two ^{1}H nuclei. You can

ASTRONOMY
by the numbers

THE MASS LOST IN HYDROGEN-TO-HELIUM FUSION

We can find the quantity of energy released during hydrogen fusion by comparing the initial and final masses when four ^{1}H's are converted into one ^{4}He. The mass of a single hydrogen nucleus is 1.673×10^{-27} kilograms. The mass of a helium nucleus is 6.645×10^{-27} kilograms. Thus,

$$4 \text{ hydrogen} = 6.693 \times 10^{-27} \text{ kg}$$
$$-1 \text{ helium} = 6.645 \times 10^{-27} \text{ kg}$$
$$\overline{\text{Mass lost} = 0.048 \times 10^{-27} \text{ kg}}$$

Multiplying this by c^2 gives the energy yield per helium atom made, E. That is,

$$E = mc^2 = 0.048 \times 10^{-27} \times (3 \times 10^8)^2$$
$$= 0.048 \times 10^{-27} \times 9 \times 10^{16} = 4.3 \times 10^{-12} \text{ joule}$$

To supply the power of the Sun's luminosity of 4×10^{26} watts (a watt is an energy consumption of 1 joule per second), 4 million tons (4×10^9 kg) of mass are converted to energy each second.

think about this reaction as the attempt to form a nucleus with 4 protons and 2 neutrons, except that 2 protons are ejected by their electric repulsion, as shown in figure 12.12D. Symbolically, we can write it as

$$^3\text{He} + {}^3\text{He} \rightarrow {}^4\text{He} + {}^1\text{H} + {}^1\text{H},$$

where again, the final mass is less than the initial mass. Therefore, altogether six ^{1}H's are used, but two are returned in the last step, and so a total of four ^{1}H's are consumed in making each ^{4}He. The Astronomy by the Numbers box above shows that this means the Sun must annihilate 4 million tons of matter each second, the equivalent of exploding a billion 100-megaton H-bombs per second! Our sunshine has a violent birth.

12.3 PROBING THE SUN'S CORE

Although the immense bulk of the Sun hides its core from view, two types of experiments allow us an indirect look. Solar neutrinos escaping from the interior carry information about conditions there, and waves on the Sun's surface reveal the structure of the gas far below as waves in a stream reveal the presence of submerged rocks.

Solar Neutrinos

We saw in section 12.2 that the Sun makes neutrinos as it converts hydrogen into helium. The number of released neutrinos therefore tells us how rapidly hydrogen is being converted into helium, and from that we can deduce the temperature and density in the Sun's core.

But counting neutrinos is extremely difficult. Neutrinos have no electric charge, a mass so tiny it has not yet been measured, and a phenomenal penetrating power. They escape from the Sun's core through its outer 700,000 kilometers and into space like bullets through wet Kleenex. They pass straight through the Earth and anything on the Earth, such as you, and keep going. In fact, roughly a trillion neutrinos from the Sun pass harmlessly through your body every second whether it is day or night.

The elusiveness that allows neutrinos to slip so easily through the Sun makes them slip with equal ease through detectors on Earth. Nevertheless, because so many neutrinos are produced, we need trap only a tiny fraction of them to get useful information. This trapping is done with neutrino detectors.

Neutrino "telescopes" look nothing like optical telescopes. They are generally built deep underground, made of a chamber filled with water or other material. They hunt for the signals produced on the rare occasion when a neutrino interacts with the

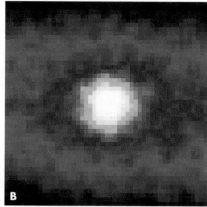

FIGURE 12.13
(A) An inside view of the Sudbury Neutrino Observatory. Almost 10,000 photon detectors cover the inside surface of this tank, which is about 12 meters (about 40 feet) in diameter. Neutrinos that interact with water in the tank produce a tiny flash of light, which the detectors can use to identify neutrinos. (B) An image of neutrinos from the Sun's core made with the Super-Kamiokande neutrino detector.

nucleus of an atom. For example, the Sudbury Neutrino Observatory (fig. 12.13) was built more than a mile underground in a nickel mine in northern Ontario, Canada. It is filled with water, and when a neutrino strikes one of the neutrons, the neutron breaks down into a proton and electron. As the electron streaks off, it produces a tiny flash of light. Ten thousand detectors (which form the grid visible around the inside of the sphere) record the emitted light, thereby allowing scientists to detect the neutrino's passage. The Super-Kamiokande detector located deep in a zinc mine west of Tokyo has a similar design, and it was able to produce a neutrino image of the Sun, shown in figure 12.13B. To make even larger neutrino observatories, scientists are now building them at the bottom of the ocean and under the Antarctic ice.

Neutrino detectors are constructed deep underground or underwater to shield them from the many other kinds of particles besides neutrinos that constantly bombard the Earth. For example, protons, electrons, and a variety of other subatomic particles constantly shower our planet. These particles, traveling at nearly the speed of light, are called **cosmic rays** and are thought to be particles blasted across space when a massive star explodes. Most cosmic rays penetrate only a short distance into the Earth; so if a detector is located deep underground, nearly all the cosmic rays are filtered out. Neutrinos, however, are unfazed by a mere mile of solid ground: on the average they could travel through a light-year of lead!

When astronomers first tried to detect solar neutrinos in the 1960s, they found only one-third the expected number coming from the Sun. At first it was assumed that the detectors were simply not working as well as expected, but over decades of efforts to improve the detectors the discrepancy remained. What was wrong? Did the Sun not fuse hydrogen into helium, as predicted? Did neutrinos somehow escape detection? It turned out that neutrinos had a property that physicists had not expected. They can change form as they travel through matter. Thus, the solar neutrino experiments provided us with new information not only about the Sun but about neutrinos themselves. These intriguing experiments provide us with a chance to see the scientific process at work, as explained in Science at Work: "Solving the Solar Neutrino Puzzle."

SCIENCE *at work*   SOLVING THE SOLAR NEUTRINO PUZZLE

When the discrepancy in solar neutrino counts was discovered, astronomers first checked their calculations for the predicted number of neutrinos. Might the Sun's interior be much different from what we thought? Was our understanding of the nuclear reactions taking place in the Sun complete? All these checks led to roughly the same result, implying that there was no obvious flaw in astronomers' understanding of how the Sun works. Was it possible, then, that neutrinos had undiscovered properties that affected their detectability on Earth or their production in the Sun?

As scientists looked more closely at the properties of neutrinos, they found that one theory suggested the existence of three kinds of neutrinos, not just one. This seemed promising because the first neutrino detectors could detect only one of the three kinds, thus perhaps explaining why only one-third the expected number were observed. Moreover, the revised theory of neutrinos implied that these elusive particles were not massless and that one kind could change into another. With this new hypothesis about the nature of neutrinos, scientists built a second generation of neutrino detectors that could detect all three varieties. The results from these new detectors agreed perfectly with the predictions of the solar theory. Thus, astronomers have confirmed their understanding of the Sun and have established new properties of neutrinos, including the fact that they have a mass.

In 2002 the American scientist Raymond Davis and the Japanese scientist Masatoshi Koshiba received a Nobel Prize for their work on neutrinos from astronomical sources. Davis led the first experiment to detect solar neutrinos and revealed the solar neutrino discrepancy. Koshiba headed the Kamiokande group whose work led to a resolution of the discrepancy.

Solar Seismology

Solar seismologists study the Sun's interior by analyzing waves in the Sun's atmosphere. We learned in chapter 6 that scientists can study the Earth's interior by analyzing earthquake waves. Astronomers can also learn about the Sun's interior by analyzing waves in its gases.

Waves similar to those of earthquakes travel through the Sun and make its surface heave like the ocean or a bubbling pan of hot oatmeal. The rising and falling surface gas makes a regular pattern (fig. 12.14), which can be detected as a Doppler shift of the moving material. Astronomers next use computer models of the Sun to predict how the observed surface waves are affected by conditions in the Sun's interior. With this technique, astronomers can measure the density and rotations deep within the Sun from the pattern and speed of the waves in its atmosphere.

Solar seismology studies have been able to determine the fraction of helium in the Sun's core. The fraction of helium there confirms the idea that the Sun has consumed about half of the hydrogen there, fusing it into helium. The results from solar seismology provide a strong independent confirmation of current models of the Sun, indicating that we have a good understanding of the Sun's structure.

FIGURE 12.14
Computer diagram of solar surface waves.

12.4   SOLAR MAGNETIC ACTIVITY

The surface waves described above are but one of many kinds of disturbances in the Sun's outer regions. A wide class of spectacular phenomena on the Sun are caused by the interaction of its magnetic field with the hot gas that makes up the Sun. The Sun has an overall magnetic field similar to the Earth's, with a magnetic north pole and south pole. At the surface the average magnetic field strength on the Sun is only about twice as strong as on the Earth's surface, but the effects are far more dramatic. To understand why, we need to consider how magnetic fields interact with electrically charged particles. This magnetic activity is also of interest because it affects the Earth, where it triggers auroral displays and affects our climate.

Magnetic field

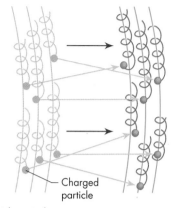

Charged
particle

Particles spiral
around the field lines.

FIGURE 12.15
Charged particles are tied to magnetic field lines—spiraling around a line as they move along it. If forces move the gas containing the particles, it can drag the field with it, and vice versa.

ANIMATION ▶

Prominences on the Sun's limb

Solar Magnetic Fields

Like the Earth's, the Sun's magnetic field is thought to arise from the rotation of electrically conductive material deep in its interior, although this is hot ionized gas in the Sun instead of the molten iron in the Earth. The electromagnetic force, as its name implies, is a force not just between electrically charged particles, but with magnetic fields as well. When electrons and other charged particles encounter a magnetic field, they spiral around the field lines, as illustrated in figure 12.15.

The stronger the magnetic field is, the more tightly the electrically charged particles spiral around the lines, although they can move freely parallel to the field lines. The particles are essentially "frozen" to travel along a magnetic field line when the field is strong enough. The Earth's magnetic field deflects energetic charged particles from space along the field lines toward the magnetic poles, where they create auroras. Charged particles in the Sun experience similar deflections, but the Sun is entirely made of charged particles, so the interactions shape motions throughout the Sun.

By Newton's law of action and reaction, charged particles in the Sun not only feel a force from the magnetic field, they exert a force that "freezes" the magnetic field in the gas. Thus, if external forces push the gas to the side (fig. 12.15), the magnetic field is dragged along with the gas. This interaction between magnetic fields and the gas causes a variety of "storms" on the Sun that are unlike anything we see on the Earth.

Sunspots, Prominences, and Flares

One of the most readily seen effects of the Sun's magnetic field is a **sunspot.** This is a large, dark-appearing region (fig. 12.16) that may have a size up to many thousands of kilometers across. Sunspots last from a few days to over a month. They are darker than the surrounding gas because they are cooler—about 4500 K, as opposed to the 6000 K of the normal photosphere. Spattering a few drops of water onto a hot electric stove burner will show a similar effect. Each drop momentarily cools the burner, making a dark spot. If they were seen in isolation, sunspots would look bright; they look dark only in contrast to the brighter surrounding regions.

What makes sunspots grow cooler? One clue is that they are regions where the Sun's magnetic field is concentrated to more than a thousand times its normal strength. The strong magnetic field blocks the normal convection beneath the Sun's surface, preventing hot gas from rising in the region of the sunspot (fig. 12.17). As a result, the

FIGURE 12.16
Visible-light image of a large group of sunspots. The darker areas are cooler than the surrounding regions of the photosphere.

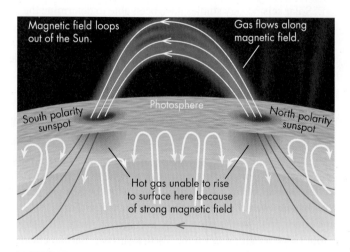

Magnetic field loops
out of the Sun.

Gas flows along
magnetic field.

South polarity
sunspot

Photosphere

North polarity
sunspot

Hot gas unable to rise
to surface here because
of strong magnetic field

FIGURE 12.17
Sunspots occur where an intense magnetic field emerges through the surface of the Sun and blocks the flow of hot gas upward.

EXTENDING *our reach*   DETECTING MAGNETIC FIELDS: THE ZEEMAN EFFECT

Astronomers can detect magnetic fields in sunspots and other astronomical bodies by the **Zeeman effect,** a physical process in which the magnetic field splits some of the spectral lines of the gas into two, three, or more components. The splitting occurs because the magnetic field alters the atom's electron orbits, which in turn alter the wavelength of its emitted light.

Figure 12.18A and B show the Zeeman effect splitting spectral lines in a sunspot. The line has a single wavelength outside the spot but splits into three within. By mapping the splitting across the Sun's face, astronomers make a magnetogram (fig.12.18C) showing the strength and polarity of the magnetic field. Note that opposite polarity regions occur in pairs.

Spectrum of photosphere collected along this line

Zeeman effect splits spectral line where magnetic field is strong

Portion of spectrum in sunspot

Wavelength ⟶

FIGURE 12.18
(A) Sunspot that was examined spectroscopically. (B) In the spectrum the line splits over the sunspot because the magnetic field is strong. (C) Magnetogram of the Sun. Yellow indicates regions with north polarity, whereas blue indicates regions with south polarity.

surface cools and darkens. To understand how astronomers can measure the strength of a sunspot's magnetic field, see the Extending Our Reach: "Detecting Magnetic Fields: The Zeeman Effect."

Left to themselves, magnetic field lines push away from each other. This is like the repulsion you have felt between a pair of magnets oriented in the same direction. So the final piece of the puzzle is understanding how magnetic fields grow so concentrated in sunspots. The tight connection between the magnetic fields and the gas provides the answer. A solar storm is a little like a low-pressure storm system on Earth. It draws surrounding gas in toward it, and because the magnetic field is frozen into the gas, the magnetic field is dragged inward and grows more concentrated. This in turn cools the region, intensifying the storm.

Prominences are another manifestation of the intense magnetic disturbances in the low-density, virtually transparent hot gases above the Sun's visible surface. Prominences are huge plumes of glowing gas that jut from the lower chromosphere into the corona. Figure 12.19 shows an example. You can get some sense of their immensity from the pale blue dot in the image, which shows the size of the Earth. Prominences form where the Sun's magnetic field reduces heat flow to a region. They are *cooler* than the gas around them, which means, according to the perfect gas law, that the pressure inside is less than the pressure outside. Thus, the hot external gas "bottles up" the

Q: Approximately how high does the prominence reach above the Sun's surface?

Earth for comparison

FIGURE 12.19
Image of a solar prominence made from the *SOHO* satellite.

FIGURE 12.20
(A) A solar flare—the bright spot on the right side of the image. This image, made in high-energy ultraviolet light with *SOHO* in 2003, is one of the strongest ever recorded. (B) *SOHO* visible-light image of a coronal mass ejection event with an image of the Sun from about the same time superimposed over the "occulting disk" used to block the light of the photosphere.

cooler gas of the prominence. Under favorable conditions, this cooling gas, trapped in its magnetic prison, may glow for weeks.

Time-lapse movies show gas streaming through prominences, sometimes rising into the corona, sometimes raining down onto the photosphere. The flow is channeled by and supported by the magnetic field, which often arcs between sunspots (fig. 12.17). Thus, prominences also are related to sunspots.

Sunspots also give birth to **solar flares,** brief but bright eruptions of hot gas in the chromosphere (fig. 12.20A). Over a few minutes or hours, gas near a sunspot may dramatically brighten. Such eruptions, though violent, are so localized that they hardly affect the total light output of the Sun at all. Generally, you need a telescope to see their visible light, though they can increase the Sun's radio and X ray emission by factors of a thousand in a few seconds.

Flares are not fully understood, but magnetic fields appear to play an important role. One hypothesis suggests that the field near a spot gets twisted by gas motions, a bit like winding up a rubber-band-powered toy. But such twisting can go only so far before the rubber band breaks. So, too, the magnetic field can be twisted only so far before it suddenly readjusts, whipping the gas in its vicinity into a new configuration. The sudden motion heats the gas, and it expands explosively. Large solar flares are sometimes followed by coronal mass ejections. These ejections are enormous bubbles of hot gas and trapped magnetic fields that burst from the corona out into space as shown in figure 12.20B. If directed toward the Earth, the gas can reach the Earth in a few days, producing auroral events. Such a burst created the spectacular auroral displays seen in March 1989, shown in figure 12.21.

An aurora is a beautiful sight, but the stream of charged particles that causes them can have more serious consequences. Communications satellites have been disabled by some major flares, and even electrical power grids can be destabilized as the magnetic field carried by the stream of gas causes electric currents to surge in electrical transmission wires. The 1989 solar outburst caused blackouts in several locations across North America. An even more remarkable event occurred in 1859, when the aurora was seen even in the tropics, and telegraph equipment was set afire by huge magnetic surges. Today an array of satellites monitors "space weather," but we can only guess how an event like that in 1859 might affect modern electronics.

: Aurora was the goddess of dawn in Roman mythology. Why might that name have been chosen for this phenomenon?

FIGURE 12.21
Photograph of the great aurora of March 1989, which knocked out the power grid in parts of Canada.

Heating of the Chromosphere and Corona

Although the Sun's magnetic field cools sunspots and prominences, it heats the chromosphere and corona. Other stars also have chromospheres and coronas, although we cannot detect their light separately, so it is additionally important to understand how the Sun's atmosphere behaves to help us interpret the light from other stars. To begin, we need first to recall that the temperature of a gas is a measure of how fast its particles are moving. Anything that speeds atoms up increases their temperature.

An analogy may help you understand how magnetic waves can heat a gas. When you crack a whip, a slow motion of its handle travels as a wave along the whip. As the whip tapers, the wave's energy of motion is given to an ever smaller piece of material. With the same amount of energy and less mass to move, the tip accelerates and eventually breaks the sound barrier. The whip's "crack" is a tiny sonic boom.

A similar speedup occurs in the Sun's atmosphere when magnetic waves formed in the photosphere move into the corona along the Sun's field lines (fig. 12.22). As the atmospheric gas thins, the wave energy is imparted to an ever smaller number of atoms, making them move faster in random directions. But "faster" in this case means hotter. Thus, the upper atmosphere heats up as the waves travel into it.

How are the magnetic waves generated? They probably start in the convection zone where rising bubbles of gas shake the magnetic field and create magnetic waves, like shaking the loose end of a rope makes it wiggle. Thus, the high temperature of the chromosphere and corona is another example of the importance of the Sun's magnetic field and its convection zone. This theory is supported by observations of other stars with active convection zones—they also have active chromospheres and coronas.

The Solar Wind

In addition to the mass it loses in the outbursts of flares, the Sun undergoes a steady, less dramatic loss of mass. The corona's high temperature gives its atoms enough energy to escape the Sun's gravity. As these atoms stream into space, they form the **solar wind,** a tenuous flow of mainly hydrogen and helium that sweeps across the Solar System. The amount of material lost from the Sun is small: less than 1 ten-trillionth of its mass each year. Nevertheless, the solar wind sweeps the gas boiling off the surface of a comet into a tail that may extend tens of millions of miles from the nucleus, as we saw in chapter 11.

The solar wind arises because, unlike the rest of the Sun, the corona is *not* in hydrostatic equilibrium. Recall that the temperature in the Sun's atmosphere increases with altitude, making the corona much hotter than the photosphere. The corona's high temperature, according to the perfect gas law, creates a pressure within it larger than we might otherwise expect for its distance above the photosphere. The pressure is sufficient to overcome the Sun's gravitational force on gas in its upper atmosphere. As a result, it pushes that material outward into space. The expanding gas has a very low density, only a few hundred atoms in a cubic centimeter—the volume of a thimble. For comparison, a thimbleful of the air we breathe contains about 10^{19} molecules!

The gas atoms begin their outward motion slowly but accelerate with increasing distance as the Sun's gravitational attraction on them weakens. On average, the wind speed is about 500 kilometers per second at the Earth's orbit, but it speeds up and slows down in response to changes in the Sun's magnetic field. From the Earth outward, the wind coasts at a relatively steady speed that carries it at least to the orbit of Neptune. At some point it impinges on the interstellar gas surrounding the Solar System. The two *Voyager* spacecraft launched in 1977 are now several times farther from the Sun than Neptune, where they are probing the solar wind and magnetic field far from the Sun. In August 2012, *Voyager 1* crossed through a turbulent region that appears to mark the boundary between the "bubble" of solar wind surrounding the Sun, entering interstellar space about 120 AU from the Sun (fig. 12.23).

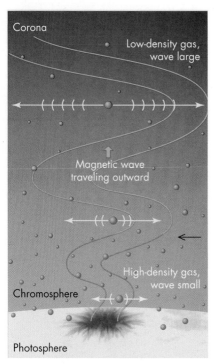

FIGURE 12.22

Diagram illustrating how magnetic waves (blue) heat the Sun's upper atmosphere. As the waves move outward through the Sun's atmosphere, they grow larger, imparting ever more energy to the gas (red dots) through which they move, accelerating and thereby heating it.

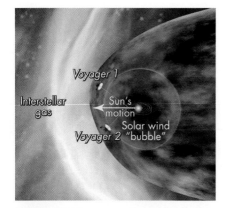

FIGURE 12.23

Diagram illustrating where the solar wind impinges on the interstellar gas and the trajectories of the two *Voyager* spacecraft. The Sun is surrounded by a "bubble" of solar wind gas that pushes into the interstellar gas, producing a bow wave in front of the Sun's direction of motion.

12.5 THE SOLAR CYCLE

The number of sunspots changes from year to year in what is called the **solar cycle.** This variability can be seen in figure 12.24, which shows the number of sunspots detected over the last century and a half. The number of sunspots rises and falls approximately every 11 years. For example, the cycle had peaks in 1958, 1969, 1980, 1990, 2001, and 2014. This cyclical change in the number of sunspots was discovered by Samuel Heinrich Schwabe, a pharmacist and amateur astronomer. He studied the Sun almost every day, hunting for a hypothesized planet orbiting closer to the Sun than Mercury, but after 17 years his detailed notes revealed this entirely unexpected phenomenon.

Flares, prominences, and occurrences of the aurora also follow the solar cycle, and climate patterns on Earth may, too. For this reason, astronomers have sought to understand not only the cause of the solar cycle but also how it influences terrestrial climate.

Cause of the Solar Cycle

As the Sun rotates, gas near its equator circles the Sun faster than gas near its poles; that is, it spins differentially, a property common in gaseous bodies (recall from chapter 10 that Jupiter and the other giant planets rotate differentially). The Sun's differential rotation is such that its equator rotates in about 25 days and its poles in about 35 days. Thus, a set of points arranged from pole to pole in a straight line would move over the course of time into a curve, as shown in figure 12.25.

Differential rotation should similarly distort the Sun's magnetic field, "winding up" the field below the Sun's surface. Astronomers think such winding of the Sun's magnetic field may cause the solar cycle. The exact mechanism is still not well understood, but spots and cycles of magnetic activity occur on other stars as well. According to one hypothesis, the Sun's rotation wraps the solar magnetic field into coils below the surface, making the field stronger and increasing solar activity: spots, prominences, and flares. The wrapping occurs because the Sun's magnetic field is "frozen" into the

FIGURE 12.24
Plot of sunspot numbers showing solar cycle.

FIGURE 12.25
Sketch showing solar differential rotation. Points near the Sun's equator rotate faster than points near the poles.

| Magnetic field at start of cycle | Differential rotation begins to twist magnetic field. | Subsurface magnetic field beginning to coil up on itself | Winding continues —only one field line is shown for clarity. | Subsurface magnetic field now in coils | Kinks in coils break through surface, making sunspots. |

FIGURE 12.26

Sketch showing how differential rotation can wind up the Sun's subsurface magnetic field. The field becomes twisted into coils that can break through the surface, making sunspots.

gas, as discussed in section 12.4. Thus, if the gas moves, so does the field, and vice versa.

Because the field and gas are tightly connected, differential rotation causes gas at the equator, which is moving faster than the gas at the poles, to drag the magnetic field with it, so that a field, initially straight north to south, is wound into two subsurface loops, as shown in figure 12.26. As the loops are wound tighter, they develop kinks, as when you twist a rubber band too tight. The cycle ends when the field twists too "tightly" and collapses, and the process repeats.

Sunspots form when kinks in the magnetic field rise to the Sun's surface and break through the photosphere (last panel of fig. 12.26). Here, the field slows the outward flow of heat, making the surface cooler and darker than in surrounding areas and thereby creating sunspots, as we discussed in section 12.4. Each kink breaks the photosphere in *two* places—one where it leaves and one where it enters. We therefore expect that spots will occur in pairs or paired groupings.

In such pairs, one grouping has a north polarity and the other a south polarity.* That is, in one the field emerges from the surface, while in the other it descends so that the field direction is reversed. The false-color map of the Sun's field in figure 12.18C shows this effect clearly. Yellow areas have a north polarity and dark blue areas a south polarity. Notice that the pattern of the polarity differs between the two hemispheres of the Sun. In one, the yellow areas are on the right, while in the other they are on the left. This reversal arises because the subsurface field is coiled in opposite directions in the two hemispheres, additional evidence that sunspots and the solar cycle are caused by winding of the Sun's magnetic field.

Changes in the Solar Cycle

The solar cycle is not always 11 years: it may be as short as 7 or as long as 16 years. Moreover, if we consider the polarity of the spot groups, the cycle averages 22 years, rather than 11, because the polarity of the Sun's field reverses at the end of each 11-year cycle. It therefore takes *two* 11-year cycles for the field to return to its original configuration.

Figure 12.27 illustrates this effect. In the first frame, we see spot pairs as the cycle begins. A right-hand spot in the top hemisphere (technically, the "leading spot" because it leads in the direction the Sun rotates) has a south polarity, while a left-hand spot (technically called a "trailing spot") has a north polarity.

Wrapping of the Sun's subsurface magnetic field

---

* Polarity of magnetic fields was described briefly in chapter 6, when we discussed the Earth's magnetic field.

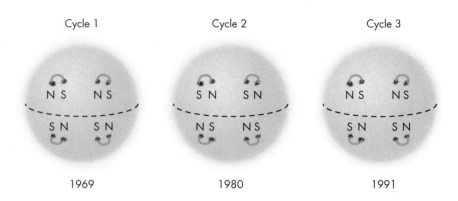

FIGURE 12.27
Diagram illustrating the approximate 22-year periodicity of solar magnetic activity.

You can also see an additional feature of the cycle: all leading spots in one hemisphere have the same polarity. In the other hemisphere, the leading spots have the opposite polarity. Eleven years later, the polarity of the pairs will be reversed, as you can see by looking at Cycle 2 in figure 12.27. Only in the next cycle, another 11 years later, will the spot fields have returned to their original directions. Thus, the full cycle of magnetic activity takes 22 years on the average.

Links Between the Solar Cycle and Terrestrial Climate

Climatologists find this 22-year period interesting because of a cycle in which droughts occur approximately every 22 years in the midwestern United States and Canada. Does the Sun's cycle affect the Earth's climate cycle? If so, how?

One possibility is as follows. The Sun's magnetic field heats the corona. The corona drives the solar wind. The solar wind alters the Earth's upper atmosphere; in particular, it changes the way the temperature varies with altitude. This in turn alters the atmosphere's circulation and may shift the jet stream to a new location. The jet stream steers storms and hence rainfall.

Although this hypothesis has not been verified yet, many scientists think that solar activity affects our climate. The evidence to support this hypothesis is based in part on the work of E. W. Maunder, a British astronomer who studied sunspots. Maunder noted in 1893 that, according to historical records, very few sunspots were seen between 1645 and 1715 (fig. 12.28). He concluded that the solar cycle essentially turned off during that period. The period is now called the **Maunder minimum** in honor of his discovery.

The Maunder minimum coincides with an approximately 70-year spell of abnormally cold winters in Europe and North America. Glaciers in the Alps advanced; rivers froze early and remained frozen late; the North Sea froze. The cold was so

FIGURE 12.28
Plot illustrating that the number of sunspots changes with time, showing the Maunder minimum and the solar cycle.

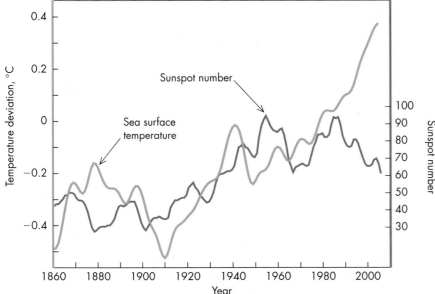

FIGURE 12.29
Curves showing the change in ocean temperatures on Earth and the change in sunspot numbers over more than a century. Notice that the curves were approximately in step until the 1980s. Over the last two decades the sea surface temperature has continued to rise while sunspot counts have declined. Astronomers deduce that solar activity affects our climate, but other factors must explain the recent temperature rise.

abnormal that meteorologists call the epoch part of the "little ice age." If only one such episode were known, we might dismiss the sunspot–climate connection as a coincidence, but three other cold periods have also occurred during times of low solar activity. This strengthens our conclusion that somehow the Sun's magnetic activity affects our climate.

Scientists have concluded that a link exists between solar activity and Earth's climate, but there are clearly other factors affecting our climate as well. Figure 12.29 shows changes in ocean temperature (expressed as deviations from the normal average) from 1860 to 2000. The figure also shows how the number of sunspots changed over the same time span. Notice that from 1860 to 1980, when the number of spots is high, the ocean is warmer than average, and when the number of spots is low, the ocean is colder than average. But from 1990 on, the Sun's activity has been low, with a weak maximum in 2014, yet ocean temperatures continue to increase. Because the curves no longer are in step, most scientists conclude that rising temperatures are due, not to the Sun, but instead to increasing levels of greenhouse gases in Earth's atmosphere.

SUMMARY

The Sun is composed mostly of hydrogen and helium atoms. It is held together by gravity and supported against gravitational collapse by the pressure created by its high internal temperature. This creates a stable structure in hydrostatic equilibrium.

Temperatures in the core of the Sun are so high that hydrogen can fuse to form helium, as protons combine through a sequence of steps called the proton–proton chain. Each step of the chain converts mass to energy according to Einstein's formula $E = mc^2$. That energy then slowly works its way to the surface by radiation and convection.

Nuclear fusion in the Sun's core also produces neutrinos that can travel rapidly out of the Sun. Neutrino detectors on Earth have seen these neutrinos, and discovered that neutrinos have unexpected properties.

Above the Sun's visible surface (the photosphere) lie the chromosphere and corona—low-density, hot atmospheric layers. They are heated by magnetic waves traveling outward from the photosphere. In the corona, the temperature reaches a million kelvin and drives the corona's gases into space, creating the solar wind.

The Sun's magnetic field changes in structure and strength over time. As the subsurface field strengthens, sunspots form where the field breaks through the surface. Near the sunspots, prominences and flares may occur that send streams of particles into space and produce auroras on Earth. Spot numbers rise and fall with a roughly 11-year period. For reasons that are not yet well understood, the strength of the solar magnetic activity also creates climatic variation on the Earth.

QUESTIONS FOR REVIEW

1. (12.1) How big is the Sun compared to the Earth?
2. (12.1) How can we measure the Sun's size, mass, and temperature?
3. (12.1) What is the Sun made of? How do we know this?
4. (12.1) What holds the Sun together?
5. (12.1–12.2) Why doesn't the Sun collapse?
6. (12.1) Why must the interior of the Sun be so hot?
7. (12.1) How does energy get to the Sun's surface from its core?
8. (12.1) What visible evidence do we have that the Sun has a convection zone?
9. (12.1) What are the photosphere, chromosphere, and corona? Which of these layers is hottest? How do we know this?
10. (12.2) How is solar energy generated? In what form(s) does it leave the core?
11. (12.3) Explain the "solar neutrino problem," and how its resolution was a good example of the scientific process.
12. (12.3) What is solar seismology? What does it tell us about the Sun?
13. (12.4) What is meant by solar activity?
14. (12.4) What role does magnetic activity play in solar activity?
15. (12.4) Why do sunspots appear dark?
16. (12.4) How do a prominence and a flare differ?
17. (12.4) How do we know there are magnetic fields in the Sun?
18. (12.5) What is the solar cycle?
19. (12.5) What is the period between maximum sunspot numbers? How does this differ from the full solar cycle?
20. (12.5) What is the Maunder minimum? Why is it of interest?

THOUGHT QUESTIONS

1. (12.1) The Sun's corona has a temperature of 1 million K. Why does it not incinerate us?
2. (12.1–12.2) Why is hydroelectric power generation an indirect form of solar power?
3. (12.1–12.2) Why is sailing an indirect form of solar power?
4. (12.1–12.2) Why is it impossible that the Sun might be powered by a chemical process such as the combustion of hydrogen and oxygen to form water?
5. (12.1–12.2) The Sun clearly has a hot interior, and must contain the same kinds of materials as the rest of the Solar System. Why doesn't differentiation produce an iron core at the center as it did in other solar system bodies?
6. (12.1–12.2) What kinds of physics do you think would need to be included in the kind of model used to determine the interior structure of the Sun?
7. (12.1/12.4) Explain why Wien's law can't be used to determine the temperature of a prominence or the corona.

8. (12.1–12.2) What would you expect the solar spectrum to look like if the photons produced by the proton–proton chain were able to exit the Sun without interacting with matter in the rest of the Sun? Would it still be continuous?
9. (12.4) Because sunspots are a little cooler than the average temperature of the photosphere, they prevent some energy from being released from the part of the surface they occupy, but this energy is usually released from nearby areas that are hotter and brighter than average. What would happen to the Sun if this did not happen?

PROBLEMS

1. (12.1) Using the values in the appendix, how many times larger in diameter is the Sun than the Earth? And in volume?
2. (12.1) Use the observation that the angular diameter of the Sun is $0.5°$ and that its distance is 1.5×10^8 kilometers to determine the Sun's diameter.
3. (12.1) Suppose you were an astronomy student on Neptune. Use the orbital data for Neptune (distance from Sun = 30.05 AU; period = 164.8 years) to measure the Sun's mass using the modified form of Kepler's third law.
4. (12.1) Show that the Sun's surface gravity is about 30 times Earth's.
5. (12.1) Calculate the escape velocity from the Sun. Compare this to the speed of rising material in the granulation given in section 12.1. If the escape velocity were less, what would happen at the center of the granulation cells?
6. (12.1) In this problem, you will calculate an estimate of the temperature of the Sun's core. You can do it either step by step or by writing out all the algebra to obtain a final result. You will need the following ideas: The pressure P in a gas is given by $P = \text{constant} \times \rho T$. If we measure the density ρ in kilograms/meter^3, and the temperature T in kelvin, then the constant has the value of about 8300 m^2 sec^{-2} K^{-1}. The volume of a sphere is $\left(\frac{4}{3}\right)\pi R^3$. The pressure force from the interior is $P \times A$, where A is the area over which the pressure acts. You can take that as the Sun's cross section, πR^2, so the pressure force is $\pi R^2 P$. Finally, you need to invoke hydrostatic equilibrium: pressure forces must balance gravitational forces. Approximate the gravitational force holding the Sun together by assuming it is split into two equal halves and apply Newton's law of gravity to calculate the force between the halves. Assume they are separated by 1 solar radius.
7. (12.1) Calculate the Sun's density in grams per cubic centimeter. The Sun's mass is approximately 2×10^{33} grams. Its radius is approximately 7×10^{10} centimeters. How does the density you find compare with the density of Jupiter?
8. (12.2) Calculate the energy (in joules) that would be released if a 60-kg person were converted entirely into energy.
9. (12.2) The Sun's total energy output is 4×10^{26} watts, and 1 watt is 1 joule/second. Use the energy yield from the

proton–proton chain to determine how many proton–proton chain fusion cycles must be happening each second in the solar core.

10. (12.2) Estimate the lifetime of the Sun. Assume that 10% of the Sun's total mass is processed through the proton–proton chain (there is only so much hydrogen available to the core), and that for each p–p chain reaction 6.6793×10^{-27} kg of mass is processed and 4.3×10^{-12} J of energy is liberated as per the values at the end of section 12.2. Assuming the Sun's power output given in table 1 is constant, how many years will it shine?

11. (12.5) Examining figure 12.29, what temperature would you expect for the sea surface at the end of the graph if the sunspot numbers had continued to rise after 1980?

TEST YOURSELF

1. (12.1) The Sun is supported against the crushing force of its own gravity by
   (a) magnetic forces.
   (b) its rapid rotation.
   (c) the force exerted by escaping neutrinos.
   (d) gas pressure.
   (e) the antigravity of its positrons.

2. (12.2) The Sun produces its energy from
   (a) fusion of neutrinos into helium.
   (b) fusion of positrons into hydrogen.
   (c) disintegration of helium into hydrogen.
   (d) fusion of hydrogen into helium.
   (e) electric currents generated in its core.

3. (12.2) According to the ideal gas law, if the temperature of a gas is made 4 times higher, which of the following is a possible result? (More than one answer may be correct.)
   (a) Its pressure increases by 4 times and its density remains the same.
   (b) Its density increases by 4 times and its pressure remains the same.
   (c) Its pressure and density both double.
   (d) Its pressure increases by 4 times while its density decreases by 4 times.
   (e) Its pressure and density both decrease by 2 times.

4. (12.3) During the daytime, about a trillion solar neutrinos per second pass through you. At night, the number is
   (a) zero.   (c) about half as much.
   (b) about the same.   (d) much, much smaller.

5. (12.3) The primary method astronomers use to measure oscillations on the surface of the Sun is by
   (a) comparing telescopic images.
   (b) magnetograms from measuring Zeeman splitting of spectral lines.
   (c) measuring the Doppler shift of absorption lines from the surface.
   (d) X ray and ultraviolet imaging by satellites.
   (e) sonic detection.

6. (12.4) Sunspots are dark because
   (a) they are cool relative to the gas around them.
   (b) they contain 10 times as much iron as surrounding regions.
   (c) nuclear reactions occur in them more slowly than in the surrounding gas.
   (d) clouds in the cool corona block our view of the hot photosphere.
   (e) the gas within them is too hot to emit any light.

7. Differential rotation results in
   (a) the solar wind.
   (b) a wound-up magnetic field.
   (c) the Maunder minimum.
   (d) the Sun's generation of energy.
   (e) all of the above.

8. (12.5) About how many years elapse between times of maximum solar activity?
   (a) 3   (b) 5   (c) 11   (d) 33   (e) 105

KEY TERMS

chromosphere, 306
convection zone, 306
corona, 306
coronal hole, 307
cosmic rays, 312
granulation, 306
hydrostatic equilibrium, 307
Maunder minimum, 320
neutrino, 310
nuclear fusion, 309
perfect gas law, or ideal gas law, 308

photosphere, 305
pressure, 307
prominences, 315
proton–proton chain, 310
radiative zone, 305
solar cycle, 318
solar flare, 316
solar wind, 317
spicule, 306
strong force, 309
sunspot, 314
Zeeman effect, 315

Q: FIGURE QUESTION ANSWERS

WHAT IS THIS? (chapter opening): This image shows Venus (the round dark circle) passing directly between us and the Sun on the June 8, 2004, transit of Venus. You can also see red columns of hot gas in the Sun's chromosphere and corona projecting beyond the edge of the Sun at right.

FIGURE 12.2: Water has a density of 1 gram per cubic centimeter. From the numbers at 0.4 and 0.6 of the Sun's radius, it reaches this density about halfway in.

FIGURE 12.4: Approximately 1000 km (600 miles). This is about the size of Texas or Egypt.

FIGURE 12.19: The prominence is ~200,000 km high.

FIGURE 12.21: The aurora's glow and sometimes rosy tints resemble the pale colors of dawn.

13

MEASURING THE PROPERTIES OF STARS

The "Jewel Box," a cluster of stars in the southern constellation Crux.

LEARNING OBJECTIVES

Upon completing this chapter you should be able to:

- Explain how astronomers measure stars' distances using parallax and describe the size of the parallactic shifts.
- Calculate a distance using the parallax formula.
- Describe the difference between the luminosity and the brightness of a star, and calculate one from the other using the inverse-square law.
- Estimate a star's distance based on the method of standard candles.
- Explain why a star's color indicates its surface temperature, and use Wien's law to determine this temperature.
- Recall the temperature dependence of the Stefan-Boltzmann law and describe how surface temperature and radius determine the luminosity of a star.

- Show how the radius of a star can be determined from its surface temperature and luminosity, as well as by interferometer or eclipse measurements.
- Describe the magnitude system for star brightness.
- List the characteristics of stellar spectra, and explain how the classification scheme indicates surface temperature.
- Describe luminosity classes and how they are determined.
- Explain how the masses of stars are estimated, and apply the modified version of Kepler's third law to solve for mass.
- Describe the layout of the H-R diagram, and infer a star's size and evolutionary state based on its location in the diagram.
- Describe how the mass–luminosity relationship was derived and what class of stars it can be applied to.

t is hard to believe that the stars we see in the night sky as tiny glints of light are in reality huge, dazzling balls of gas and that many are vastly larger and brighter than our Sun. They look dim to us only because they are so far away—several light-years (trillions of miles) to even the nearest. Such remoteness creates tremendous difficulties for astronomers trying to understand the nature of stars. We cannot physically travel to the stars, but in this chapter we will see how astronomers overcome the distance barrier that separates us from stars and how they learn many of the physical properties of these distant objects. How far away are stars? How big? What are they made of?

The answers to these questions show us that most stars are remarkably like the Sun. For example, like the Sun, they are composed mostly of hydrogen and helium, and most have fairly similar masses. A small percentage, however, are more than 30 times the Sun's mass ($30\ M_\odot$) and are much hotter than the Sun and blue in color. Others are much less massive than the Sun, only one-tenth its mass, and are cool, red, and dim. Moreover, even stars similar to the Sun in composition and mass may differ enormously from it in their radii and density. For example, some giant stars have a radius hundreds of times larger than the Sun's—so big that were the Sun their size, it would extend beyond the Earth's orbit. On the other hand, some stars are white dwarfs, with as much material as the Sun packed into a volume the size of the Earth.

Astronomers can learn all these properties of stars by using physical laws and theories to interpret measurements made from the Earth. For example, theories of light yield the surface temperature, distance, and motion of a star; theories of atoms yield the composition of a star; and a modified form of Kepler's third law yields the mass of a star. In using such laws, astronomers may sometimes employ more than one method to determine a desired property of a star. For example, a star's temperature may be measured from either its color or its spectrum. Such alternative methods serve as checks on the correctness of the procedures astronomers use to determine the properties of stars.

See end of chapter for the answer.

CONCEPTS AND SKILLS TO REVIEW

- Relation between temperature of a hot object and the color of the light it emits (Wien's law) (4.3)
- Doppler shift (4.6)
- Modified form of Kepler's third law (3.6)

13.1   MEASURING A STAR'S DISTANCE

Astronomers need to know an object's distance in order to determine its size and structure. For example, without knowledge of a star's distance, astronomers cannot learn many of the star's other properties, such as its mass, radius, or energy output. However, measuring the vast distances to stars at first seems an impossibility. Even *Voyager 1*, the fastest-moving spacecraft currently heading out of the Solar System, would take 10,000 years to reach the nearest stars at its speed. On its present trajectory, it will not encounter another star for another 40,000 years. How then can we determine the distance to something trillions of miles away?

The solution is to apply geometric techniques just as ancient Greek astronomers did thousands of years ago (chapter 2). By applying these methods, astronomers first determined the size of the Earth, then the distance to the Moon and Sun and eventually to other planets. Using these methods to find the distances of stars is no more difficult conceptually, but it had to await the invention of telescopes to make measurements precise enough to learn how far away stars are.

Scale drawing of measured triangle

$d = b \tan A$

Distance to be found

90°

Known

A

Baseline (known)

Scale: 1 cm = 1 meter

Distance (*d*)

90°

Baseline (*b*)

A

FIGURE 13.1
Sketch illustrating the principle of triangulation.

Measuring Distance by Triangulation and Parallax

Astronomers have several methods for measuring a star's distance, but for nearby stars the fundamental technique is **triangulation,** a method that early astronomers used to find the distance to the Moon. In triangulation, we lay out a triangle in which one side is the distance we seek but cannot measure directly and another side is a distance we can measure—a baseline, as shown in figure 13.1. For example, to measure the distance across a deep gorge, we construct an imaginary triangle with one side spanning the gorge and another side at right angles to it and running along the edge we are on, as shown in figure 13.1. By measuring the length of the side along the gorge edge and the angle *A,* we can determine the distance across the gorge either by a trigonometric calculation or from a scale drawing of the triangle.

Astronomers use a method of triangulation called **parallax** to measure the distance to stars. Parallax is a change in an object's apparent position caused by a change in the observer's position. An easy demonstration of this is to hold your hand motionless at arm's length and shift your head from side to side. Your hand seems to move against the background even though it is your head that has changed position, not your hand. This simple demonstration illustrates how parallax gives a clue to an object's distance. If you hold your hand at different distances from your face, you will notice that the apparent shift in your hand's position—its parallax—is larger if your hand is close to your face than if it is at arm's length. That is, *nearby objects exhibit more parallax than more remote ones,* for a given motion of the observer, a result true for your hand and for stars.

Using parallax to find distance may seem unfamiliar, but parallax creates our stereovision, the ability to see things three-dimensionally. When we look at something, each eye sends a slightly different image to the brain, which then processes the pictures to determine the object's distance. You can demonstrate the importance of the two images by covering one eye and trying to reach across a desk to a keyboard quickly and tap a particular key. Compare your success to hitting the right key with both eyes open.

To observe stellar parallax, astronomers take advantage of the Earth's motion around the Sun, as shown in figure 13.2A. They observe a star and carefully measure its position against background stars. They then wait 6 months until the Earth has moved to the other side of its orbit, a known distance of 2 AU (about 300 million kilometers), and make a second measurement. As figure 13.2B shows, the star will have a slightly different position compared to the background of stars as seen from the two points. The amount by which the star's apparent position changes depends on its distance from us. The change is larger for nearby stars than for remote stars, but for all

ANIMATION

Parallax

INTERACTIVE

Stellar Parallax

FIGURE 13.2

(A) Triangulation to measure a star's distance. The radius of the Earth's orbit is the baseline. (B) As the Earth moves around the Sun, the star's position changes as seen against background stars. (C) Parallax is defined as one-half the angle by which the star's position shifts. Sizes of bodies and their separation are exaggerated for clarity.

stars it is extremely small—so small that it is measured not in degrees but in fractions of a degree called "arc seconds," which are just 1/3600 of a degree (or 1/60 of an "arc minute").

For convenience, astronomers define a star's parallax, p, not by the angle by which its position appears to shift, but by half that angle (see fig. 13.2C). With that definition for parallax, the star's distance, d, is simply $1/p$ if we measure p in arc seconds and d, not in kilometers or light-years, but in a new unit called "parsecs" (abbreviated pc). That is,

$$d_{pc} = \frac{1}{p_{\text{arc seconds}}}.$$

> 1 arc second is extremely small, equivalent to the apparent size of a U.S. penny 4 km (2.5 miles) away.

> d_{pc} = distance measured in parsecs
> $p_{\text{arc seconds}}$ = parallax measured in arc seconds

With this choice of units, one **parsec** equals 3.26 light-years (3.09×10^{13} kilometers). The word *parsec* comes from a combination of *par*allax and "arc *sec*ond." The parallax formula is derived in Astronomy by the Numbers: "Deriving the Parallax Formula."

To determine a star's distance, we measure its parallax, p, and use $d_{pc} = 1/p_{\text{arc seconds}}$. For example, suppose that from the shift in position of a nearby star we find that its parallax is 0.25 arc seconds. Its distance is then $d = \frac{1}{0.25} = 4$ parsecs. Similarly, a star whose parallax is 0.1 arc second is $\frac{1}{0.1} = 10$ parsecs from the Sun. From this technique, astronomers have discovered that at present the nearest star is Proxima Centauri, which lies 1.3 parsecs (4.3 light-years) from the Sun. This spacing (about 1 parsec) is typical for stars near the Sun.

Although the parallax–distance relation is a simple formula mathematically, measuring a star's parallax to use in the formula is very difficult because the angle by which the star shifts is extremely small. It was not until the 1830s that the first parallax

Proxima Centauri is a companion to the much brighter star Alpha Centauri, which is slightly farther away. Both are shown in Looking Up #8 at the front of the book.

was measured by the German astronomer Friedrich Bessel at Königsberg Observatory (now in Kaliningrad). Except for the nearest stars, parallax measurements need to be made from space to avoid the blurring effects of the atmosphere, and an orbiting satellite, *Hipparcos,* launched by the Eurpean Space Agency, has done just that. With its data, astronomers have accurately measured distances to stars as far away as 250 parsecs. In 2014 the ESA launched another satellite, *Gaia,* which is currently collecting parallax data on more than a billion stars.

ASTRONOMY– *by the numbers*   DERIVING THE PARALLAX FORMULA

As the Earth circles the Sun, a nearby star will shift back and forth against background stars, and the parallax angle is defined as half of the overall shift, as shown in figure 13.3 below. To determine the star's distance from the parallax angle, p, we construct the triangle ACS. The line AC is the radius of the Earth's orbit, with point A chosen so that an imaginary line from the star (S) to the Sun, C, will be perpendicular to AC. This makes ACS a right triangle, with angle C equal to 90° and the side AC—the distance from the Earth to the Sun—1 astronomical unit.

To calculate the star's distance from the Sun, d, we draw a circle centered on the star and passing through the Sun. We call the radius of the circle d, to stand for the star's distance from the Sun. We next form the following proportion: AC (1 astronomical unit) is to the circumference of the circle as p is to the total number of degrees around the circle, which we know is 360°. That is,

$$\frac{1\,\text{AU}}{2\pi d} = \frac{p}{360°}.$$

We can invert this equation to solve for d, obtaining

$$d = \frac{360°}{2\pi p}\,\text{AU}.$$

Astronomers invented the new unit of a parsec as the distance when p is 1 arc second (1/3600 of a degree). A parsec equals $(360 \times 3600/2\pi)$ AU, so when the parallax is measured in arc seconds, all of the other factors in the equation combine to give parsecs, leaving $d = 1/p$.

For example, measurements show that Sirius has a parallax angle $p = 0.377$ arc seconds. Therefore, its distance is $1/0.377 = 2.65$ parsecs. To express this in light-years, we multiply by 3.26 (the number of light-years in a parsec) to find that Sirius is about 8.6 light-years away from us.

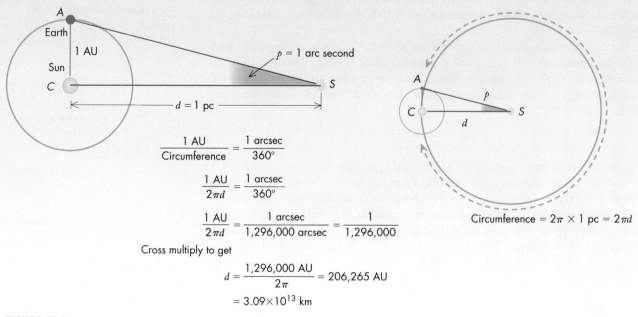

FIGURE 13.3
How to determine the relation between a star's distance and its parallax.

Astronomers study many properties of stars in order to understand how they work. The brightness of stars in particular is essential to unraveling their life story. Stars in the night sky appear quite peaceful, but once we have an idea about how far away they are, we gain quite a different picture of them. The amount of energy required to keep stars shining this bright demands that they must be consuming energy at a furious rate.

Luminosity

Astronomers call the amount of energy an object radiates each second its **luminosity** (abbreviated as L). An everyday example of luminosity is the wattage of a lightbulb: a typical table lamp has a luminosity of 100 watts, whereas a bulb for an outdoor parking lot light may have a luminosity of 1500 watts. Stars are enormously more luminous. For example, the Sun has a luminosity of about 4×10^{26} watts, which it obtains by "burning" its hydrogen into helium. Thus, a star's luminosity measures how fast it consumes its fuel, a vital quantity for determining its lifetime. Knowing a star's luminosity is also important because from it astronomers can measure a star's radius and distance. But to understand how such measurements are made, we must first discuss a relation between how bright an object appears, how bright it really is (its luminosity), and its distance—a relation known as the *inverse-square law*.

The Inverse-Square Law and Measuring a Star's Luminosity

The **inverse-square law** relates an object's luminosity to its distance and its apparent brightness—that is, how bright it looks to us. We all know that a light looks brighter when we are close to it than when we are farther from it. The inverse-square law puts that everyday experience into a mathematical form, describing how light energy spreads out from a source such as a star. As the light travels outward, it moves in straight lines, spreading its energy uniformly in all directions, as shown in figure 13.4A. Near the source, the light will have spread only a little, and so more enters an observer's eye, making it look brighter than if the observer were far away. That is, the farther away from us a source of light is, the less of its light enters our eyes. As a result, more-distant objects look dimmer (fig. 13.4B).

This decrease in brightness with increasing distance can also be understood if you think of light as photons. Photons leaving a light source such as a star spread out evenly in all directions. Now imagine a series of progressively larger spheres drawn around the source. If nothing absorbs the light, the same number of photons pass through each sphere. But because more-distant spheres are larger, the number of photons passing through any given *area* on any one sphere grows smaller as the spheres become more

FIGURE 13.4
Decrease of brightness with distance. (A) Light spreads out from a point source in all directions, so the rays become less concentrated. (B) You can see the inverse-square law at work in this picture of street lights.

A   **B**

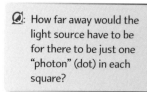

Q: How far away would the light source have to be for there to be just one "photon" (dot) in each square?

FIGURE 13.5
The inverse-square law. (A) As photons move out from a source, they are spread over a progressively larger area as the distance from the source increases. Thus, a *given* area intercepts fewer photons the farther it is from the source. (B) The area over which light at a distance *d* from the source is spread is $4\pi d^2$.

ANIMATION ▶

The inverse-square law

L = Luminosity of star
B = Brightness of star
d = Distance to star

distant and larger, as shown in figure 13.5A. Similarly, as you move away from a light source, the photons reaching you are spread more widely—therefore fewer enter your eyes and the source appears dimmer. Thus, a more distant source appears dimmer because more of its light has spread along lines that never reach you.

We can use this argument to show the following: at a distance *d* from a light source, the source's luminosity, *L*, has spread over a sphere whose radius is *d*, as shown in figure 13.5B. (We use *d* to stand for radius here to emphasize that we are finding a distance.) Any sphere has a surface area given by $4\pi d^2$, and so the brightness we observe, *B*, is just

$$B = \frac{L}{4\pi d^2}.$$

This relationship is called the inverse-square law because distance appears in the denominator as a square.

The inverse-square law is one of the most useful mathematical tools available to astronomers for measuring not only a star's luminosity but also its distance. To find a star's luminosity, astronomers measure its distance, *d*, by parallax, as described in section 13.1. Next, with a photometer, a device similar to the electric exposure meter in a camera, they measure how bright the star appears from Earth, *B*. Finally, with *B* and *d* known, they calculate the star's luminosity, *L*, using the inverse-square law. Such measurements reveal that the average star has a luminosity similar to that of our Sun. However, astronomers also find some stars that are millions of times more luminous than the Sun and other stars that are thousands of times less luminous than the Sun. Much of this range in luminosity comes from the great range in stellar radii. That is, some stars are vastly larger than others, as we will now show.

Finding a Star's Distance by the Method of Standard Candles

If you look at an object of known light output, you can estimate its distance based on how bright it appears. For example, if you look at two 100-watt lightbulbs, one close and one far away, you can tell fairly accurately how much farther away the dim one is. In fact, if you drive at night, your life depends on making such distance estimates when you see traffic lights or the headlights of oncoming cars. Knowing how bright they are, you are able to make a quick judgment about how distant they must be, allowing you to decide when to slow down or pass another car in the dark.

Astronomers call distance measurements based on a known luminosity the method of **standard candles** and use a more refined version of it to find the distance to stars

ASTRONOMY— *by the numbers*

FINDING THE DISTANCE OF A DISTANT STAR FROM A NEARBY STAR

Suppose we know two stars are nearly identical based on their spectra. If one of the stars, the nearer one, is close enough that we can find its distance by parallax to be 10 parsecs, we can find the distance to the farther one by comparing their brightness.

As an example, suppose the star whose distance we seek looks 25 times dimmer. We find its distance by writing out the inverse-square law for both stars:

$$B_{near} = \frac{L_{near}}{4\pi d_{near}^2} \quad \text{and} \quad B_{far} = \frac{L_{far}}{4\pi d_{far}^2}.$$

Next, we divide the expression on the right into the other to obtain after cancellation

$$\frac{B_{near}}{B_{far}} = \frac{4\pi L_{near} d_{far}^2}{4\pi L_{far} d_{near}^2} = \frac{L_{near} d_{far}^2}{L_{far} d_{near}^2}.$$

Since we believe the two stars have the same luminosity, we can also cancel L_{near} and L_{far}, leaving us with

$$\frac{B_{near}}{B_{far}} = \left(\frac{d_{far}}{d_{near}}\right)^2.$$

That is, for two stars of equal luminosity, the brightness ratio is the inverse of the distance ratio squared. We can therefore find the distance ratio by taking the square root of both sides, to get

$$d_{far}/d_{near} = \sqrt{B_{near}/B_{far}}.$$

For the problem we are solving, the brightness ratio is 25. Its square root is 5. Therefore, the farther star is 5 times more distant than the nearer one. Because we know that the nearer star is 10 parsecs away, the farther star is 50 parsecs from us.

---

and galaxies. To use this method, astronomers must have some independent way of learning a star's luminosity. This method then provides a powerful tool for finding distances.

Astronomers may be able to deduce a star's luminosity, *L,* by a variety of methods. For example, they may be able to use the star's spectrum, its position in the H-R diagram within a group of stars, or, for some pulsating variable stars, its period (as we will discuss in chapter 14). To determine the star's distance, it is then only necessary to measure the star's apparent brightness, *B,* with a photometer. With *B* and *L* known, the astronomers can then calculate the distance to the star, *d,* using an inverted version of the inverse-square law:

$$d = \sqrt{\frac{L}{4\pi B}}.$$

An example of applying this method is worked out in the Astronomy by the Numbers box above to obtain the distance of a star if we have a nearby example of the same kind of star (the standard candle) with a known distance. This method works best if the luminosity is known very accurately. Often the classification of a star is too uncertain to give a very accurate luminosity, but for some types of stars the luminosity is very well determined by certain features of the star, and these are the best suited for the method of standard candles.

The Magnitude System

About 150 B.C., the ancient Greek astronomer Hipparchus measured the apparent brightness of stars in the night sky using units he called **magnitudes.** He designated the stars that looked brightest as magnitude 1 and the dimmest ones (the ones he could just barely see) as magnitude 6. For example, Betelgeuse, a bright red star in the constellation Orion, is magnitude 1, while the somewhat dimmer stars in the Big Dipper's handle are approximately magnitude 2. The Little Dipper (fig. 13.6) offers a good way to see how faint the different magnitudes are. The four stars around the bowl of the Little Dipper are magnitudes 2, 3, 4, and 5. Depending on how clear and dark the sky is, you may not be able to see all of them, so this provides a way of judging the conditions.

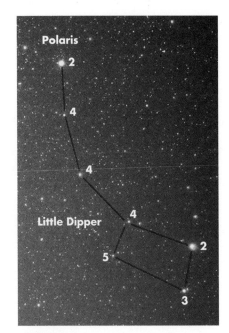

FIGURE 13.6
Apparent magnitudes of the stars in the Little Dipper.

You can use the Big Dipper to help you find the Little Dipper. See Looking Up #2 at the front of the book.

Astronomers still use this scheme to measure the brightness of astronomical objects today, but they now use the term *apparent magnitude* to emphasize that they are measuring how bright a star *looks* to an observer. A star's apparent magnitude depends on its luminosity and its distance; it may look dim either because it has a small luminosity—it does not emit much energy—or because it is very far away.

Astronomers use the magnitude system for many purposes (for example, to indicate the brightness of stars on star charts), but it has several confusing properties. First, the scale is "backward" in the sense that bright stars have small magnitudes, while dim stars have large magnitudes. Moreover, modern measurements show that Hipparchus underestimated the magnitudes of the brightest stars, and so the magnitudes now assigned them are negative numbers.

Second, magnitude *differences* correspond to brightness *ratios*. That is, if we measure the brightness of a first-magnitude star and a sixth-magnitude star, the former is 100 times brighter than the latter. Thus, a *difference* of 5 magnitudes corresponds to a brightness *ratio* of 100; so, when we say a star is 5 magnitudes brighter than another, we mean it is a *factor* of 100 brighter. Each magnitude difference corresponds to a factor of 2.512... (the fifth root of 100) in brightness. Thus, a first-magnitude star is 2.512 times brighter than a second-magnitude star and is 2.512 × 2.512, or 6.310, times brighter than a third-magnitude star. Table 13.1 lists the ratios that correspond to various differences in magnitude. For example, let us compare the apparent brightness of the planet Venus with the star Aldebaran. At its brightest, Venus has an apparent magnitude of −4.2; Aldebaran's apparent magnitude is 0.8, and so the difference in their magnitudes is 0.8 − (−4.2) = 5.0. Therefore we see from table 13.1 that Venus is 100 times brighter to our eye than Aldebaran.

Third, astronomers often use a quantity called "absolute magnitude" to measure a star's luminosity. Recall that a star's apparent magnitude is how bright it *looks* to an observer. But the apparent magnitude of a star depends on the distance from which we observe it. The same star will have one apparent magnitude if it is near to us but a different apparent magnitude if it is far from us. Astronomers therefore find it useful to have a way to describe a star's brightness that does not depend on the star's distance. One way to do that is to imagine how bright a star would look if we were to observe it from some *standard* distance. Astronomers have chosen 10 parsecs as that standard distance, and they call the apparent magnitude of a star seen from 10 pc its *absolute magnitude*. Because the absolute magnitude does not depend on distance, it is a measure of a star's luminosity. Table 13.2 illustrates how absolute magnitude is related to a star's luminosity.

Table 13.1	Relating Magnitude Differences to Brightness Ratios
Magnitude Difference	**Ratio of Brightness**
1	2.512:1
2	$2.512^2 = 6.31$:1
3	$2.512^3 = 15.85$:1
4	$2.512^4 = 39.8$:1
5	$2.512^5 = 100$:1
10	$2.512^{10} = 10^4$:1
15	$2.512^{15} = 10^6$:1

Table 13.2	Relating Absolute Magnitude to Luminosity
Absolute Magnitude	**Approximate Luminosity in Solar Units**
−15	100,000,000
−10	1,000,000
−5	10,000
0	100
5	1
10	0.01
15	0.0001

With stars' distances and brightnesses we have gained insight into their power output. With just one more piece of information that we can obtain from stars' light output—their colors—we can determine two more important features of stars: their temperatures and radii.

Temperature

Stars are extremely hot by everyday standards. The surface temperature of even cool stars is far above the temperatures at which most substances, even metals, vaporize. To learn how hot a star is, astronomers use an indirect method. Yet the method used is quite familiar. You use it yourself in judging the temperature of an electric stove burner.

An object's temperature can often be deduced from the color of its emitted light. As we saw in chapter 4, hotter objects emit more blue light than red. Thus, hot objects tend to glow blue and cooler ones red. You can see such color differences if you look carefully at stars in the night sky. Some, such as Rigel in Orion, have a blue tint. Others, such as Betelgeuse, are reddish (fig. 13.7). Thus, even our naked eye can tell us that stars differ in temperature.

We can use color in a more precise way to measure a star's temperature with **Wien's law,** which we introduced in chapter 4. It states that an object's temperature, T, in kelvin is given by the following:

$$T = \frac{2.9 \times 10^6 \text{ K nm}}{\lambda_{max}}$$

T = Temperature of star in kelvin
λ_{max} = Strongest emitted wavelength of starlight in nanometers

where λ_{max} is the wavelength in nanometers (nm) at which it radiates most strongly (the "max" subscript is a reminder of this). Thus, the longer the wavelength of the maximum emitted energy, the lower the temperature of the radiating object.

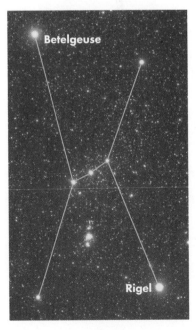

FIGURE 13.7

From the photograph at right you can see that Betelgeuse is red compared to Rigel. A plot of each star's relative brightness at different wavelengths shows that they peak at different wavelengths. Wien's law lets us determine their surface temperatures from the wavelengths at which their light output reaches a maximum.

A S T R O N O M Y
by the numbers

THE SURFACE TEMPERATURES OF RIGEL AND BETELGEUSE

For an illustration of how we can use Wien's law to measure a star's temperature, suppose we pick Rigel. To determine the wavelength at which Rigel radiates most strongly, we need to measure its brightness at many different wavelengths, as illustrated in figure 13.7. We find $\lambda_{max} = 240$ nanometers. Inserting this value in Wien's law, we find

$$T = 2.9 \times 10^6 \text{ K nm}/240 \text{ nm} \approx 12{,}000 \text{ K.}$$

For another example, we pick the red star Betelgeuse, which radiates most strongly at about 830 nanometers. Its temperature is therefore about

$$\frac{2.9 \times 10^6 \text{ K nm}}{830 \text{ nm}} = \frac{2.9 \times 10^6}{0.83 \times 10^3} \text{K} \approx 3.5 \times 10^3 \text{ K} = 3500 \text{ K.}$$

Examples of using Wien's law are given in Astronomy by the Numbers: "The Surface Temperatures of Rigel and Betelgeuse," and we will explore another way of finding a star's temperature in section 13.4. We can also find other properties from a star's light such as the star's radius, but before we discuss how, we need to discuss briefly another general property of stars and other hot objects—the amount of energy they radiate.

Radius

Common sense tells us that if we have two objects of the same temperature but of different sizes, the larger one will emit more energy than the smaller one. For example, three glowing charcoal briquettes in a barbecue emit more energy than just one briquette at the same temperature. Similarly,

if two stars have the same temperature but one is more luminous than the other, the more-luminous star must have a larger surface area, and therefore a larger radius than the dimmer star.

Therefore, we can infer a star's size from its temperature and the amount of energy it radiates. To calculate the star's radius we need a mathematical relation between luminosity, temperature, and radius—a relation known as the Stefan-Boltzmann law.

The Stefan-Boltzmann Law

Imagine watching an electric stove burner heat up. When the burner is on low and is relatively cool, it glows dimly red and gives off only a slight amount of heat. When the burner is on high and is very hot, it glows bright yellow-orange and gives off far more heat. Thus, you can both see and feel that raising a body's temperature increases the amount of radiation it emits per second—that is, its luminosity. Similarly,

if two stars have the same size but the surface of one is hotter than the other, the hotter star will have a greater luminosity.

A mathematical formulation of this law was deduced in the late 1800s by two German scientists, Josef Stefan and Ludwig Boltzmann.

The Stefan-Boltzmann law, as their discovery is now called, states that an object of temperature T radiates an amount of energy each second equal to σT^4 per square meter, as shown in figure 13.8. The quantity σ is called the "Stefan-Boltzmann constant," and its value is 5.67×10^{-8} watts m^{-2} K^{-4}. The Stefan-Boltzmann law affords a mathematical explanation of what we noticed for the electric stove burner. That is, a hotter burner has a larger T, and therefore, according to the law, the burner radiates

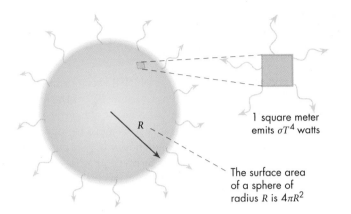

$$L \quad = \quad \sigma T^4 \quad \times \quad 4\pi R^2$$

Luminosity—total
energy radiated per
second by the star

Energy emitted
per second by
1 square meter

Number of square
meters in surface
area of the star

1 square meter
emits σT^4 watts

The surface area
of a sphere of
radius R is $4\pi R^2$

FIGURE 13.8
The Stefan-Boltzmann law relates a star's radius to its luminosity
and temperature. Each part of the star's surface radiates σT^4.
Multiplying σT^4 by the star's surface area ($4\pi R^2$) gives its total
power output—its luminosity, $L = 4\pi R^2\, \sigma T^4$.

more strongly. Similarly, the Stefan-Boltzmann law is the basis for dimmer switches
on lights: changing the amount of electricity going to the lamp changes its temperature and thereby its brightness. However, the Stefan-Boltzmann law does not apply
to all hot objects. For example, it does not accurately describe the radiation from hot,
low-density gas, such as that in fluorescent lightbulbs or interstellar clouds.

We can apply the Stefan-Boltzmann law to determine a star's luminosity as
follows. According to the law, if a star has a temperature T, each square meter of
its surface radiates an amount of energy per second given by σT^4. We can find the
total energy the star radiates per second—its luminosity, L—by multiplying the energy
radiated from 1 square meter (σT^4) by the number of square meters of its surface area
(fig. 13.8). If we assume that the star is a sphere, its area is $4\pi R^2$, where R is its radius.
Its luminosity, L, is therefore

$$L = 4\pi R^2\, \sigma T^4.$$

L = Luminosity of star
R = Radius of star
T = Surface temperature of star

That is, a star's luminosity equals its surface area times σT^4. This relation between L,
R, and T may at first appear complex, but its content is very simple: increasing either
the temperature or the radius of a star makes it more luminous. Making T larger makes
each square meter of the star brighter. Making R larger increases the number of square
meters.

In the above expression, L is measured in watts, but the "wattage" of stars is so
enormous that it is more convenient to use the Sun's luminosity as a standard unit.
Likewise, it is easier to use the Sun's radius as a standard size unit rather than meters
or even kilometers. If we need to convert to watts or meters, we simply remember or
look up that one solar luminosity, $L_\odot$, is 4×10^{26} watts and one solar radius, $R_\odot$, is $7\times
10^5$ kilometers.

Because a star's luminosity depends on its radius and temperature, if we know its
luminosity and temperature, we can find its radius, R. The method works because if we
have a mathematical relation between three quantities (in this case L, T, and R), we can
find any one of these given the other two. Thus, if we know a star's luminosity and its
temperature, we can use the Stefan-Boltzmann law to solve for its radius. Astronomy
by the Numbers: "Finding the Radius of the Star Sirius" explains how.

Direct Measurements of Radius

Astronomers can also measure a star's radius from its angular size, the technique we
used in chapter 12 to measure the radius of the Sun. Unfortunately, the angular size of
all stars except the Sun is extremely tiny because they are so far from the Earth, and
so even in powerful telescopes, stars generally look like smeary spots of light. That
smearing is caused by the blurring effects of our atmosphere and by a physical limitation of telescopes called diffraction, which we discussed in chapter 5.

ASTRONOMY *by the numbers*

FINDING THE RADIUS OF THE STAR SIRIUS

From its color (the wavelength at which it radiates most strongly), we found that the temperature ($T_\star$) of the star Sirius is about 10,000 K. From the amount of energy we receive from Sirius and its distance, which can be found by parallax, we can find that its luminosity ($L_\star$) is about 25 $L_\odot$. We solve for the radius of Sirius ($R_\star$) as follows. First write down the Stefan-Boltzmann law for Sirius to get $L_\star = 4\pi R_\star^2 \sigma T_\star^4$. Then write down the same relation for the Sun: $L_\odot = 4\pi R_\odot^2 \sigma T_\odot^4$.

Next, we divide the expression for Sirius by the expression for the Sun to get

$$\frac{L_\star}{L_\odot} = \frac{4\pi R_\star^2 \times T_\star^4}{4\pi R_\odot^2 \times T_\odot^4}.$$

We can simplify this expression by canceling the identical 4π and σ factors to get

$$\frac{L_\star}{L_\odot} = \frac{R_\star^2 T_\star^4}{R_\odot^2 T_\odot^4}.$$

We next collect the R's and T's as separate factors,

giving us

$$L_\star/L_\odot = (R_\star/R_\odot)^2 (T_\star/T_\odot)^4.$$

We now solve this expression for $(R_\star/R_\odot)^2$ to get

$$(R_\star/R_\odot)^2 = (L_\star/L_\odot)(T_\odot/T_\star)^4.$$

Finally, we take the square root of both sides to get

$$R_\star/R_\odot = (L_\star/L_\odot)^{1/2} (T_\odot/T_\star)^2.$$

We can now evaluate $R_\star/R_\odot$ by inserting the values for the luminosity and temperature of Sirius and the Sun. Notice that $T_\odot = 5800$ K, and by definition of our units for luminosity, $L_\odot = 1$. This gives us

$$\frac{R_\star}{R_\odot} = \left(\frac{25}{1}\right)^{1/2} \left(\frac{5800\text{ K}}{10,000\text{ K}}\right)^2 = 5(0.58)^2 = 1.7.$$

That is, the radius of Sirius is 1.7 $R_\odot$, a little less than twice the Sun's radius.

Diffraction limits a telescope's ability to measure tiny angular size, hopelessly blurring the light by an amount that depends on the diameter of the telescope's lens or mirror. This can be seen in the Hubble Space Telescope image in figure 13.9A, which is as sharp an image of the giant star Betelgeuse as can be made with a telescope this small. Larger ground-based telescopes can potentially record greater detail, but for the blurring effects of Earth's atmosphere. Those effects can be partially offset by a technique called "speckle interferometry" in which a high-speed camera takes many very short exposures of the star's light. A computer then combines the separate short exposures into a "deblurred" image, such as that in figure 13.9B.

To measure angles at the level of detail needed for most stars, a telescope with truly immense diameter is needed. For example, to measure the angular size of a star like the Sun if it were 50 light-years away would require a telescope 300 meters in diameter, about three times the length of a football field. To avoid the need for such enormous (and expensive) telescopes, astronomers have devised an alternative way to measure the angular size of stars by using not one huge telescope but two (or more)

FIGURE 13.9

The star Betelgeuse. (A) Hubble Space Telescope image made at ultraviolet wavelengths. (B) Image made by speckle imaging using a 4-meter ground-based telescope. (C) An image made at infrared wavelengths with the IOTA interferometer (see chapter 5). The size scales on the left are the same for all three images.

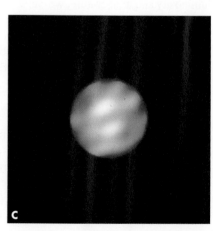

10 AU

0.05 arc seconds

A B C

smaller ones separated by large distances. Such a device is called an "interferometer" (discussed more fully in chapter 5), and its ability to measure angular sizes is equivalent to that of a single telescope whose diameter is equal to the distance that separates the two smaller ones. A computer then combines the information from the two telescopes to provide a reconstruction of the star's appearance (fig. 13.9C). Such interferometric observations are still hampered by blurring caused by our atmosphere, but with interferometers, astronomers have measured the radii of hundreds of stars.

The Stefan-Boltzmann law and interferometer observations show that stars differ enormously in radius. Although most stars have approximately the same radius as the Sun, some, like Betelgeuse, are hundreds of times larger, and astronomers call them **giants.** Smaller stars (our Sun included) are called **dwarfs.** At present, the angular sizes of only the nearest dwarfs can be measured with interferometers.

In using the Stefan-Boltzmann law to measure a star's radius, we have seen that L can be measured either in watts or in solar units. Astronomers sometimes use a different set of units, however, to measure stellar brightness.

13.4 SPECTRA OF STARS

A star's spectrum depicts the energy it emits at each wavelength and is perhaps the single most important thing we can know about the star. From the spectrum we can find the star's composition, temperature, luminosity, velocity in space, rotation speed, and some other properties as well. For example, under some circumstances we may also deduce the star's mass and radius.

Figure 13.10 shows the spectra of three stars. You can easily see differences between them. The middle spectrum is from a star similar to the Sun; the top spectrum is from a star hotter than the Sun; and the bottom spectrum is from a star cooler than

Stellar spectroscopy

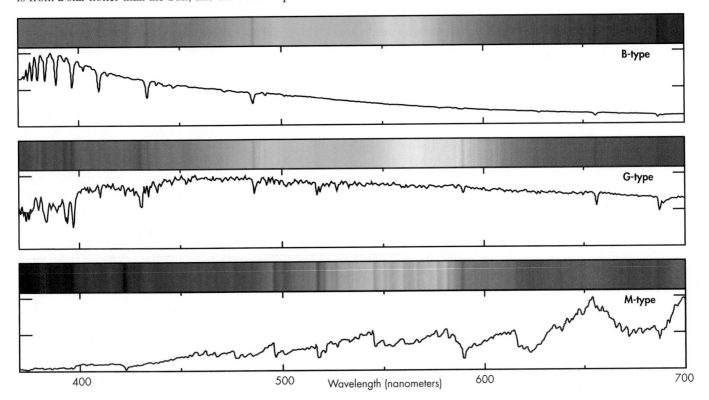

FIGURE 13.10

Spectra of three stars. The middle spectrum is similar to the Sun's. The one above it is hotter than the Sun. The one below is cooler than the Sun. A plot of the intensity of light at each wavelength is shown in the graph below each spectrum. Dark lines show up as dips in the graph.

the Sun. The hot star's spectrum has only a few dark lines, but they are strong and their spacing follows a regular pattern. The cool star's spectrum, however, shows a welter of lines with no apparent regularity. Understanding such differences and what they tell us about a star is one of the goals of studying stellar spectra.

Measuring a Star's Composition

As light moves from a star's core through the gas in its surface layers, atoms there absorb the radiation at some wavelengths, creating the dark absorption lines in the star's spectrum that we saw in figure 13.10. Each type of atom—hydrogen, helium, calcium, and so on—absorbs at a unique set of wavelengths. For example, hydrogen absorbs at 656, 486, and 434 nanometers, in the red, blue, and violet part of the spectrum, respectively. Gaseous calcium, on the other hand, absorbs strongly at 393.3 and 396.8 nanometers, producing a strong double line in the violet. Because each atom absorbs a unique combination of wavelengths of light, each has a unique set of absorption lines. From such absorption lines, we can determine what a star is made of.

To measure a star's composition, we make a spectrum of its light and then compare the absorption lines we see with tables that list the lines made by each atom. When we find a match between an absorption line in the star and a line in the table, we can deduce that the element listed exists in the star. To find the quantity of each atom in the star—each element's abundance—we use the darkness of the absorption line, as discussed in chapter 4. Such determinations of the identity and quantity of elements present are difficult, however. Even though an element may be present in the star, temperature effects might make the element's atoms unable to absorb light and make a spectral line.

How Temperature Affects a Star's Spectrum

To see why temperature affects a star's spectrum, recall that light is absorbed when its energy matches the energy difference between two electron orbits, as we discussed in chapter 4. For an atom to absorb light, its electrons must be in the proper orbit, or, more technically, energy level. An atom may be abundant in a star's atmosphere and create only very weak lines at a particular wavelength simply because the gas is either too hot or too cold, so that its electrons are in the "wrong" level to absorb light at that wavelength. Hydrogen illustrates this situation dramatically, as shown in figure 13.11.

The absorption lines of hydrogen that we see at visible wavelengths are made by electrons orbiting in its second level. These are called "Balmer lines" after the scientist who first studied their pattern and to distinguish them from other hydrogen lines in the ultraviolet and infrared wavelengths. Balmer lines occur at wavelengths where light has exactly the amount of energy needed to lift an electron from hydrogen's second energy level to the third level, or a higher level. These lines are especially important because their wavelengths are in the visible spectrum and are therefore easily observable. If the hydrogen atoms in a star have no electrons orbiting in level 2, no Balmer absorption lines will appear, even though hydrogen may be the most abundant element in the star.

Hydrogen Balmer lines appear weakly or not at all if a star is either very cold or very hot. In a cool star, most hydrogen atoms have their electrons in level 1 and so they cannot absorb Balmer line radiation, which can be absorbed only if the electron is in level 2. In a hot star the atoms move faster, and when they collide, electrons may be excited ("knocked") into higher orbits: the hotter the star, the higher the orbit. Moreover, in very hot stars the radiation is so energetic that the photons can knock electrons out of the atom, in which case the atom is said to be ionized. As a result of this excitation, proportionally more of the electrons in a very hot star will be in level 3 or higher. Only if the hydrogen has a temperature between about 8000 K and 15,000 K will

FIGURE 13.11

Formation of stellar absorption lines. Atoms in the cooler atmospheric gas absorb radiation at wavelengths corresponding to jumps between electron orbits. Absorption lines of hydrogen are illustrated.

enough atoms have their electrons in level 2 to make strong hydrogen Balmer lines. Therefore, if we are to deduce correctly the abundance of elements in a star, we must correct for such temperature effects. With these corrections, we discover that virtually all stars are composed mainly of hydrogen: their composition is similar to that of our Sun—about 71% H, 27% He, and 2% a mix of heavier elements. But despite their uniform composition, stars exhibit a wide range in the appearance of their spectra, as astronomers noted when they began to study stellar spectra.

Classification of Stellar Spectra

Stellar spectroscopy, the study and classification of spectra, was born early in the nineteenth century when the German scientist Joseph Fraunhofer discovered dark lines in the spectrum of the Sun. He later observed similar lines in the spectra of stars and noted that different stars had different patterns of lines. In 1866, Pietro Angelo Secchi, an Italian priest and scientist, noticed that the line patterns depended on the star's color. He assigned stars to four color types—white, yellow, red, and deep red—which he also classified according to the types and strengths of spectral lines present.

Technological improvements allowed astronomers to refine Secchi's system further. The first such improvement came in 1872, when Henry Draper,* a physician and amateur astronomer, recorded spectra on photographs. Upon Draper's death, his widow endowed a project at Harvard to create a compilation of stellar spectra, known as the Henry Draper Catalog. By using photography to compile the Draper Catalog, it was possible to obtain many more spectra and with better details than could be observed visually. In fact, details were so good that E. C. Pickering, the astronomer in charge of the project, began to use letters to subdivide the spectral types, assigning the letters *A* to *D* to Secchi's white type, *E* to *L* to the yellow type, and *M* and *N* to the red types, and adding more letters for types that did not fit Secchi's scheme.

About 1901, Annie Jump Cannon (fig. 13.12), the astronomer who was actually doing the classification for the Draper Catalog, discovered that the types fell in a more orderly sequence of appearance if rearranged by temperature as indicated by their color. She therefore reordered Secchi's types to obtain the sequence *O, B, A, F, G, K,*

FIGURE 13.12

Annie Jump Cannon.

---

*Draper's father was a professor of chemistry and physiology, and the first person known to have photographed an astronomical object—the Moon.

SCIENCE *at work*

NEW SPECTRAL TYPES

Since the establishment of the spectral types *O* through *M*, astronomers, aided by new technology, have made further progress in classifying stellar spectra. New infrared detectors in particular have allowed astronomers to detect dim stars far cooler than any previously known.

Some of these stars are so cool that not only do molecules form in their atmospheres but solid dust particles condense there as well. With atmospheres so different from those of ordinary stars, these extremely

cool objects have spectra that are also radically different. Therefore, astronomers have devised new spectral types—*L, T,* and *Y*—to describe these cool, dim objects.

The *L* types show strong molecular lines of iron hydride and chromium hydride, while *T* types show strong absorption lines of methane, and *Y* types have possible absorption lines of ammonia. Most of these objects appear to be "failed stars" in that they do not generate heat by fusion like the Sun and other stars. They are often called brown dwarfs after their dull reddish color.

FIGURE 13.13
Cecilia Payne.

and *M,* with *O* stars being the hottest and *M* stars being the coolest. Her work is the basis for the stellar **spectral types** we use today. Three additional types that are even cooler than *M* stars have been added in recent years, labeled *L, T,* and *Y,* as discussed in the Science at Work box above.

Stellar spectral types are based on the appearance of the spectrum. For example, *A*-type stars show extremely strong hydrogen lines, while *B*-type stars show helium and weaker hydrogen lines. This scheme, despite its wide adoption, lacked a physical basis in that it employed the appearance of the spectra rather than the physical properties of the stars that produced them. For example, astronomers at that time did not know what made the spectra of *A* and *F* stars differ. That understanding came in the 1920s from the work of the American astronomer Cecilia Payne (later Payne-Gaposchkin, fig. 13.13), who explained why the strength of the hydrogen lines depended on the star's temperature.

Payne's discovery was based on work by the Indian astronomer Meghnad Saha, who showed how to calculate the level in which an atom's electrons were most likely to be found. The levels occupied by electrons—a quantity crucial for interpreting the strength of spectral lines—depend on the star's temperature and density. From Saha's equation, Payne was able to show that Cannon's order for the spectral types was indeed determined mainly by stars' temperatures. She also proved that stars are composed mainly of hydrogen, surprising astronomers, who at the time believed the Sun was probably made mostly of heavier elements.

Payne's findings unfortunately left astronomers with Cannon's odd nonalphabetical progression for the spectral types, which—from hot to cold—ran *O, B, A, F, G, K,* and *M.* So much effort, however, had been invested in classifying stars using this system that it was easier to keep the types as assigned with their odd order than to reclassify them (Cannon, in her life, classified some quarter million stars). As a help to remember the peculiar order, astronomers have devised mnemonics. One of the first was "Oh, be a fine girl/guy. Kiss me." Another version is "Only brilliant, artistic females generate killer mnemonics." Choose whichever appeals most to you, or make up your own version, to help you learn this important sequence.

Definition of the Spectral Types

A star's spectral type is determined by the lines in its spectrum. Knowing which element makes which line is essential for understanding the different line patterns in different types of stars. As we noted earlier, almost all stars are made of a similar mix of elements, so the different elements and molecules that show up in the stars' spectra

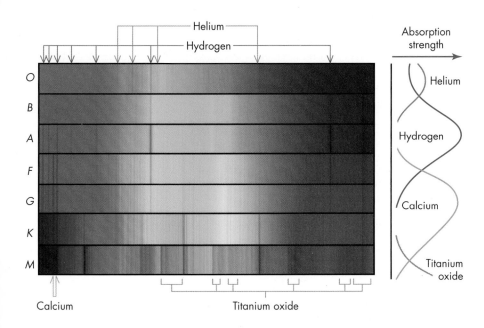

generally do not indicate different compositions. Rather, the spectra indicate differing conditions in the stars' atmospheres, especially their surface temperatures.

For example, because they are so hot, *O* stars have weak absorption lines of hydrogen but strong absorption lines of helium, the second most abundant element. At an *O* star's high surface temperature, hydrogen atoms collide so violently and are excited so much by the star's intense radiation that the electrons are stripped from most of the hydrogen, ionizing it. With its electron missing, a hydrogen atom cannot absorb light. Because most of the *O* star's hydrogen is ionized, such stars have extremely weak hydrogen absorption lines. Helium atoms are more tightly bound, however, and most retain at least one of their electrons, allowing them to absorb light. Thus, *O* stars have absorption lines of helium but only weak lines of hydrogen.

The differences in the line patterns of the seven main types—*O* through *M*—are shown in figure 13.14. The curves on the right-hand side of figure 13.14 show how the strength of the absorption lines of helium, hydrogen, calcium, and titanium oxide change with spectral type.

Stars of spectral type *A* have very strong hydrogen lines. Their temperature is just right to put lots of electrons into orbit 2 of hydrogen, which makes for strong Balmer lines. Balmer lines also appear in *F* stars but are weaker. *F* stars are distinguished by the multitude of lines from metals such as calcium and iron, elements that also appear strongly in *G* and *K* stars. Such elements are present in hotter stars but are ionized and generally create only very weak spectral lines under those conditions. In cooler stars, however, spectral lines of metals, particularly ionized calcium, are moderately strong, and hydrogen lines become weak.

In the very cool *K* and *M* stars, hydrogen is almost invisible because its electrons are mostly in level 1 and therefore cannot make Balmer lines. These stars have such cool atmospheres that molecules form, and as we saw in chapter 4, they produce very complex spectra. As a result, *K* and *M* stars have numerous lines from carbon compounds, as well as gaseous titanium oxide (TiO).

We have seen that *O* stars are hot and *M* stars are cool, but what, in fact, are their temperatures? Application of Wien's law and theoretical calculations based on Saha's law of how electrons are distributed in atomic orbits show that temperatures range from more than 30,000 K for *O* stars down to about 3500 K for *M* stars, with *A* stars being 7500 to 10,000 K and *G* stars, such as our Sun, being between 5000 and 6000 K. Because a star's spectral type is set by its temperature, its type also indicates its color.

To distinguish finer gradations in temperature, astronomers subdivide each type by adding a numerical suffix—for example, *B*1, *B*2, . . *B*9—with the smaller numbers indicating higher temperatures. With this system, the temperature of a *B*1 star is about 20,000 K, whereas that of a *B*5 star is about 13,500 K. Similarly, our Sun, rather than being just a *G* star, is a *G*2 star.

Spectral Type	Temperature Range (K)	Features	Representative Star
O	Hotter than 30,000	Ionized helium, weak hydrogen	
B	10,000–30,000	Neutral helium, hydrogen stronger	Rigel
A	7500–10,000	Hydrogen very strong	Sirius
F	6000–7500	Hydrogen weaker, metals (especially ionized Ca) moderate	Canopus
G	5000–6000	Ionized Ca strong, hydrogen even weaker	The Sun
K	3500–5000	Metals strong, CH and CN molecules appearing	Aldebaran
M	2000–3500	Molecules strong, especially TiO and water	Betelgeuse
L	1300–2000	Metal hydrides, water, and reactive metals strong, but no TiO	
T	700–1300	Strong lines of water and methane	
Y	<700	Absorption line at 1.55 μm, possibly of ammonia	

Table 13.3 Summary of Spectral Types

We know that hot objects are blue and cool objects are red (recall Wien's law), and so naturally *O* and *B* stars (hot types) are blue, while *K* and *M* stars (cool types) are red. Table 13.3 lists some of the main lines astronomers use to identify the different types, and the range of temperatures for each type.

Measuring a Star's Motion

All stars move through space, and these motions turn out to be important clues to a star's history and are vital in determining star masses and the structure of galaxies.

We can find a star's motion from its spectrum using the Doppler shift, which, as we saw in chapter 4, results in the alteration of the wavelength of light from a moving source. If a source moves toward us, its wavelengths are shortened, and if it moves away, they are lengthened, as illustrated in figure 13.15A. The amount of the wavelength shift depends on the source's speed along the line of sight, what astronomers call **radial velocity.** We find the shift by subtracting the wavelength the star would

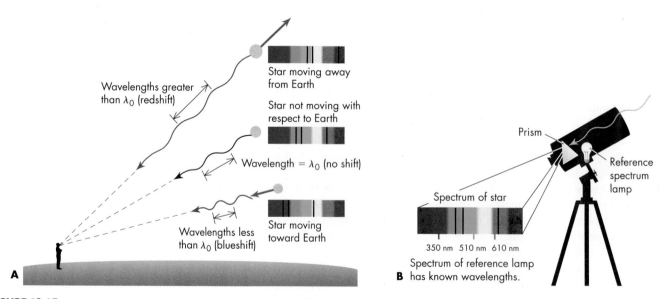

FIGURE 13.15
Measuring a star's radial velocity from its Doppler shift. (A) Spectrum lines from a star moving away from Earth are shifted to longer wavelengths (a redshift). Spectrum lines from a star approaching Earth are shifted to shorter wavelengths (a blueshift). (B) A standard lamp attached inside the telescope serves as a comparison spectrum to allow the wavelength shift of the star's light to be found.

ASTRONOMY
by the numbers
CALCULATING A STAR'S RADIAL VELOCITY

Suppose we observe a star and see a strong absorption line in the red part of the spectrum with a wavelength of 656.380 nanometers. From laboratory measurements on nonmoving sources, we know that hydrogen has a strong red spectral line at a wavelength of 656.280 nanometers. We infer that the wavelength is shifted by a $\Delta\lambda$ of 0.100 nanometers. The Doppler shift law tells us that

$$\frac{V}{c} = \frac{\Delta\lambda}{\lambda_0} = \frac{0.100}{656.28}.$$

To find the radial velocity, V, we multiply both sides by c and obtain

$$V = \frac{0.100c}{656.28} = 45.7 \text{ kilometers per second.}$$

Because $\Delta\lambda$ is positive, the star is moving away from us.

---

absorb (or emit) if it were not moving relative to us, λ_0, from the new wavelength, λ. We will call that difference $\Delta\lambda$. With these definitions, the Doppler shift law states that

$$\frac{\Delta\lambda}{\lambda_0} = \frac{V}{c}$$

where V is the source's radial velocity and c is the speed of light. That is, the change in wavelength created by motion of the star (or observer) divided by the wavelength in the absence of motion is just the radial velocity of the star divided by the speed of light. Thus, if we can measure $\Delta\lambda$, we can solve for V.

The quantity $\Delta\lambda$ may turn out to be either positive or negative, depending on whether wavelengths are increased or decreased. If the wavelengths are increased, the distance between the source and observer is increasing, meaning that from the observer's point of view the source is moving away. If the wavelengths are decreased, the distance between the source and observer is decreasing, and so from the observer's point of view the source is approaching. Thus, from the observer's point of view, a positive $\Delta\lambda$ implies the source is receding, while a negative $\Delta\lambda$ implies it is approaching.

Astronomers measure the Doppler shift of a star by recording its spectrum directly beside the spectrum of a standard, nonmoving source of light attached to the telescope, as shown in figure 13.15B. They then measure the wavelength shift by comparing the spectrum of the star to the standard. From the Doppler shift, they calculate the star's radial velocity as illustrated in the Astronomy by the Numbers box above.

Notice that a star will have a radial velocity only if its motion has some component toward or away from us. Stars moving across our line of sight, maintaining a constant distance from us, have no Doppler shift, although they do display a shift in position that astronomers call **proper motion.** Stars are so far away that their motion is imperceptible to the eye, much as a distant airplane seems to hang motionless in the sky. In fact, stellar motions across the sky are difficult to detect even in pictures of the sky taken years apart. Only after many thousands of years do proper motions make a clear change in the patterns of the stars (fig. 13.16).

Such measurements reveal that all stars are moving and that those near the Sun share approximately its direction and speed of revolution (about 200 kilometers per second) around the center of our Galaxy, the Milky Way. Superimposed on this orbital motion, however, are small random motions of about 20 kilometers per second. Stars therefore move rather like cars on a freeway—some move a little faster and overtake slower-moving cars, which gradually fall behind.

$\Delta\lambda =$ Change in wavelength due to motion of light source

$\lambda_0 =$ Wavelength of unmoving (zero-velocity) light source

$V =$ Radial velocity of light source*

$c =$ Speed of light (300,000 km/sec)

---

*We assume here that $V \ll c$.

50,000 years ago   Today   50,000 years from now

FIGURE 13.16
Proper motion of stars gradually alters their positions on the sky. The three figures show the pattern of the Big Dipper 50,000 years ago (left), today (middle), and 50,000 years in the future (right).

13.5 BINARY STARS

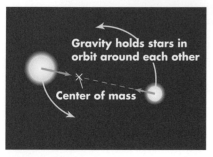

FIGURE 13.17
Two stars orbiting in a binary system, held together as a pair by their mutual gravitational attraction.

Binary stars

Mizar and Alcor are pictured in Looking Up #2 at the front of the book.

Many stars have a special motion: they orbit around each other, held together as stellar companions by their mutual gravitational attraction, as illustrated in figure 13.17. Astronomers call such stellar pairs **binary stars** and value them greatly because they offer one of the few ways to measure stellar masses. To understand how, recall that the gravitational force between two bodies depends on their masses. The gravitational force, in turn, determines the stars' orbital motion. If that motion can be measured, we can work backward to find the mass.

A star's mass, as we will discover in chapter 14, critically controls its existence, determining both how long it lasts and how its structure changes as it ages. Thus, if we are to understand the nature of stars, we must know their mass. Fortunately, this is not difficult to determine for stars belonging to a binary system, and such stars are relatively common. Among the hottest stars (*O* and *B* type), roughly 80% have orbiting companions. Of the Sun-like stars, a little over half have companions. But for the coolest stars (*M* type, in particular), the majority lack companions. Because such cool, dim stars vastly outnumber more luminous stars, single stars are the rule, although binary stars predominate among the brighter stars. We have described binaries as star pairs, but in many cases more than two stars are involved. Some stars are triples, others quadruples, and we know of at least one six-member system.

Most binary stars are only a few astronomical units apart. A few, however, are so close that they "touch," orbiting in contact with each other in a common envelope of gas. These contact binaries and other close pairs probably formed at the same time from a common parent gas cloud; that is, rather than becoming a single star surrounded by planets, as our Sun did, the cloud formed a close star pair. Widely spaced binaries, on the other hand, perhaps formed when one star captured another that happened to pass nearby, but astronomers still have many unanswered questions about how binary stars form.

A few binary stars are easy to see with even a small telescope. Mizar, the middle star in the handle of the Big Dipper, is a good example. When you look at this star with the naked eye, you will first see a dim star, Alcor, which is not a true binary companion but simply a star lying in the same direction as Mizar. With a telescope, however, Mizar can be seen to be two very close stars: a true binary.

Visual and Spectroscopic Binaries

For some stars (such as Mizar) we can directly see orbital motion of one star about the other* by comparing images made several years apart. Such star pairs are called **visual binaries** because we can see two separate stars and their individual motion.

Some binary stars may be so close together, however, that their light blends into a single blob that defies separation with even the most powerful telescopes. In such cases, the orbital motion cannot be seen directly, but it may nevertheless be inferred from their combined spectra. Figure 13.18A shows a typical **spectroscopic binary,** as such stars are called. As the stars orbit each other about their shared center of mass, each star alternately moves toward and then away from Earth. This motion creates a Doppler shift, and so the spectrum of the star pair shows lines that shift—figure 13.18B—first to the red as the star swings away from us and then toward the blue as it approaches. From a series of spectra, astronomers can measure the orbital speed from the Doppler shift of the star, and by observing a full cycle of the motion,

---

*More technically, the two stars orbit a common center of mass located on the imaginary line joining them. The exact location of the center of mass depends on the star masses—if the masses are the same, the center of mass lies exactly halfway between them.

FIGURE 13.18

Spectroscopic binary star. (A) The two stars are generally too close to be separated by even powerful telescopes. (B) Their orbital motion creates a different Doppler shift for the light from each star. Thus, the spectrum of the stellar pair contains two sets of lines, one from each star.

they can find the orbital period. From the orbital period and speed, they can find the size of the orbit, and with the size of the orbit and the period, they can find the stars' masses using a modified form of Kepler's third law, as we will now show.

> **ANIMATION** ▶
>
> Orbital motion is reflected in the shifting spectral lines of a spectroscopic binary.

Measuring Stellar Masses with Binary Stars

In the early seventeenth century, Kepler showed that the time required for a planet to orbit the Sun is related to its distance from the Sun. If P is the orbital period in years and a is the semimajor axis (half the long dimension) of the planet's orbit in AU, then $P^2 = a^3$, a relation called Kepler's third law, as we discussed in chapter 2.

Newton discovered that Kepler's third law in a generalized form applies to any two bodies in orbit around each other. If their masses are m and M and they follow an elliptical path of semimajor axis, a, relative to each other with an orbital period, P, then

$$(m + M)P^2 = a^3$$

> $(m + M) =$ Sum of orbiting stars' masses
> $P =$ Period of orbit
> $a =$ Semimajor axis of orbit

where P is expressed in years, a in astronomical units, and m and M in solar masses. This relation is our basic tool for measuring stellar masses, as we will now describe.

To find the mass of the stars in a visual binary, astronomers first plot the stars' orbital motion, as depicted in figure 13.19. It may take many years to observe the entire orbit, but eventually the time required for the stars to complete an orbit, P, can be determined. From the plot of the orbit and with knowledge of the star's distance from the Sun, astronomers next measure the semimajor axis, a, of the orbit of one star about

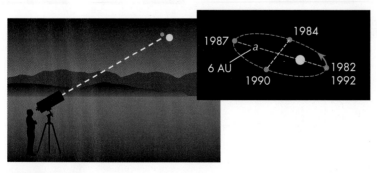

Plot of star positions ⟶ Period of 10 years

Measure semi-major axis $= a = 6$ AU

Use modified form of Kepler's third law

$$M_A + M_B = \frac{a_{AU}^3}{P_{yr}^2}$$

$$= \frac{6^3}{10^2}$$

$$= 2.16 \; M_\odot$$

FIGURE 13.19

Measuring the combined mass of two stars in a binary system using the modified form of Kepler's third law.

Alpha Centauri can be seen in Looking Up #8 at the front of the book.

the other. For an example, see Astronomy by the Numbers: "The Combined Mass of Alpha Centauri."

Additional observations of the stars' orbits allow us to find their individual masses, but for simplicity we will omit the details here. From analyzing many such star pairs, astronomers have discovered that star masses fall within a fairly narrow range from about 30 to 0.1 $M_\odot$. We will discover in chapter 14 why stars are limited to this range, but once again, we find our Sun is about midway between the extremes.

In the preceding discussion, we assumed that we can actually see the two stars as separate objects—that is, that they are a visual binary. If the stars are spectroscopic binaries and cannot be distinguished as separate stars, their period and separation can be determined by observing the shifting spectral lines in a series of spectra.

Eclipsing Binary Stars

An eclipsing binary and its light curve

On rare occasions the orbit of a binary star will be almost exactly edge-on as seen from the Earth. Then, as the stars orbit, one will eclipse the other as it passes between its companion and the Earth. Such systems are called **eclipsing binary** stars, and if we watch such a system, its light will periodically dim. During most of the orbit we see the combined light of both stars, but at the times of eclipse, the brightness of the system decreases as one star covers the other, producing a cycle of variation in light intensity called a "light curve." Figure 13.20 contains a graph of such change in brightness over time.

Eclipsing binary stars are useful to astronomers because the duration of the eclipses depends on the stars' diameters. Figure 13.20 shows why. The eclipse begins as the edge of one star first lines up with the edge of the other, as shown in the figure. The eclipse ends when the opposite edges of each star line up. During the eclipse, the covering star has moved relative to the other a distance equal to its diameter plus that

If we imagine star B at rest, star A moves a distance equal to its own diameter between the times 2 and 3. That time interval, when multiplied by star A's orbital speed, gives its diameter. Similarly, the time between points 5 and 6 multiplied by star A's orbital speed will give star B's diameter.

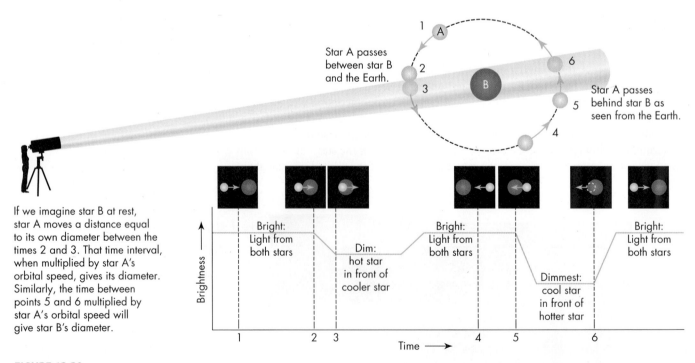

FIGURE 13.20
An eclipsing binary and its light curve. From the duration of the eclipse, the diameter of each star may be found. Note: In this illustration, the yellow star is hotter than the red star.

ASTRONOMY
by the numbers   THE COMBINED MASS OF ALPHA CENTAURI

The star Alpha Centauri has two stars orbiting with a period $P = 68$ years and the semimajor axis $a = 20.6$ AU. We can then find their combined mass, $m + M$, by solving the modified form of Kepler's law, and we get (after dividing both sides by P^2)

$$m + M = a^3/P^2.$$

Inserting the measured values for P and a, we see that

$$m + M = 20.6^3 / 68^2$$
$$= 8740 / 4620$$
$$= 1.9\ M_\odot.$$

That is, the combined mass of the stars is 1.9 times the Sun's mass.

---

of the star it has covered. From such observations and knowledge of the orbital velocities of the eclipsing stars, astronomers can calculate the diameter of each star.

Light curves of eclipsing stars can also give information about the stars' shapes and the distribution of brightness across their disks. For example, "star spots"—analogous to sunspots—can be detected on some stars by this method. If, in addition, astronomers can obtain spectra of the eclipsing system, the stars' masses may be found using the method we described for spectroscopic binaries. Thus, astronomers can obtain especially detailed information about eclipsing binary stars.

13.6   THE H-R DIAGRAM

By the early 1900s, astronomers had learned to *measure* stellar temperatures, masses, radii, and motions, but they *understood* little of how stars worked. What made an *A* star different from a *K* star? Why were some *M* stars more luminous and others much less luminous? What supplied the power to make stars shine? Our examination of stars in this chapter has reached a similar stage. We know that there are many varieties of stars. But what creates that variety and what does it mean? A fundamental step in the discovery of the nature of stars was examining the relationship of stellar properties on the Hertzsprung-Russell diagram.

In 1912 the Danish astronomer Ejnar Hertzsprung and the American astronomer Henry Norris Russell independently found that if stars are plotted on a diagram according to their luminosity and their temperature (or, equivalently, spectral type), then most of them fall in a few specific areas of the diagram. The nearly simultaneous discovery of this relation by astronomers working on opposite sides of the Atlantic Ocean is an interesting example of how conditions can be "ripe" for scientific advances.

An example of a Hertzsprung-Russell diagram, now generally called an **H-R diagram** for short, is shown in figure 13.21. Most stars in the diagram lie along a curving diagonal line, with hot luminous stars at the upper left end and cool dim stars at the lower right. The Sun lies almost in the middle. A fraction of stars fall in the upper right or lower left portion of the diagram. Note that the H-R diagram does not indicate anything about the position of stars' locations in space. It shows merely a correlation between stellar properties, much as we might plot the heights and weights of people in a graph and see a correlation between these properties.

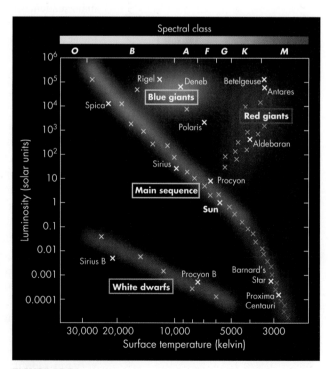

FIGURE 13.21
The H-R diagram. Bright stars are at the top of the diagram and dim stars at the bottom, but notice that hot (blue) stars are on the left and cool (red) stars are on the right. Several different regions and well-known stars are identified in the diagram.

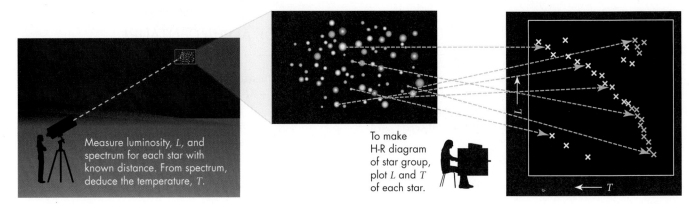

FIGURE 13.22
Constructing an H-R diagram from the spectra and luminosities of stars in a cluster.

Constructing the H-R Diagram

To construct an H-R diagram of a group of stars, astronomers plot each star according to its temperature (or spectral type) and luminosity, as shown in figure 13.22. By tradition, they put bright stars at the top and dim stars at the bottom—hot stars on the left and cool stars on the right. Equivalently, blue stars lie to the left and red stars to the right. Notice that temperature therefore increases to the left, rather than to the right, as is conventional in graphs. The approximately straight line along which the majority of stars lies is called the **main sequence.**

When plotting stars on an H-R diagram, generally about 90% will lie along the main sequence, but some will lie off it. Of these, a few will be in the upper right, where stars are cool but very luminous, and some will be below the main sequence to the lower center, where stars are hot but dim. What makes these stars different?

Interpreting the H-R Diagram

The Stefan-Boltzmann law, described in section 13.3, explains why these stars are so different. A star's luminosity depends on its surface area and temperature. If two stars have the same temperature, any difference in luminosity must reflect a difference in area. But a star's area depends on its radius—a large area means a large radius. Stars that lie in the upper part of the H-R diagram have the same temperature as stars directly below them on the main sequence, so the more-luminous stars must be larger. In fact they must be immensely larger, because some of these bright cool stars emit thousands of times more energy than main-sequence stars of the same temperature, implying that their surface areas are thousands of times larger. Astronomers therefore call these bright, cool stars "giants." The region in the H-R diagram populated by giants is sometimes called the "giant branch." Because many of the stars are cool and therefore red (recall that color is related to temperature), these huge cool stars are called **red giants.** For example, the bright star Aldebaran in the constellation Taurus is a red giant. Its temperature is about 4000 kelvin and its radius is about 30 times larger than the Sun's.

A similar analysis shows that the stars lying below the main sequence must have very tiny radii if they are both hot and dim. In fact, the radius of a typical hot, dim star is roughly 100 times smaller than the Sun's, making it about the same size as the Earth. Because these stars are so hot, they glow with a white heat and are therefore called **white dwarfs.** Sirius B, a dim companion of the bright star Sirius, is a white dwarf. Its radius is about 0.008 the Sun's—a little smaller than the Earth's radius—and its surface temperature is about 27,000 kelvin.

We can illustrate such extremes in size by drawing lines in the H-R diagram that represent a star's radius. That is, we use the Stefan-Boltzmann law, which relates the star's luminosity L, radius R, and temperature T. From this law, $L = 4\pi R^2 \sigma T^4$, we plot the values of L that result from a fixed choice of R as we vary T. Any stars lying along the plotted line must have equal radii. Figure 13.23 shows that such lines run diagonally from upper left to lower right. Notice that as we move to the upper right, we cross lines of progressively larger radius, implying that stars there are very large. Not only are they giants in luminosity, they are also giants in radius. Comparing the radii of stars across the H-R diagram, we see three main types: main-sequence stars, red giants, and white dwarfs.

Giants and Dwarfs

Giants, white dwarfs, and main-sequence stars differ in more than just diameter. They also differ dramatically in their overall density. Recall that density is a body's mass divided by its volume. A larger body of the same mass has a lower density, and so a giant star is much less dense than a main-sequence star if their masses are similar. For example, the density of the Sun, a main-sequence star, is about 1 gm/cm^3, while the density of a typical giant star is about 10^{-6} gm/cm^3—1 million times less.

Why do stars have such disparate densities? Russell suggested about 1918 that as a star evolves, its density changes. He believed that stars began their lives as diffuse, low-density red giants. Gravity then drew the stars' atoms toward one another, compressing their matter and making stars smaller, denser, and hotter, so that they turned into hot, blue, main-sequence stars. He was partly right about the idea of contraction during formation, but most of his ideas proved incorrect.

As we will discuss in chapter 14, today we know that stars evolve in a different way. After they live their lives as main-sequence stars, they turn into giants. But although Russell was wrong about how stars evolve, his ideas led to many other important discoveries about stars, such as the mass–luminosity relation.

Luminosity Classes

Because astronomers use a star's luminosity to find its distance, radius, and life span, they have sought other ways to measure it. In the late 1800s, Antonia Maury, an early stellar spectroscopist at Harvard, noticed that absorption lines were extremely narrow in some stars compared with other stars of the same temperature. In the early 1900s, Hertzsprung, too, noticed that narrowness and, even before developing the H-R diagram, recognized that luminous stars had narrower lines than less luminous stars, as illustrated in figure 13.24.

The width of the absorption lines in a star's spectrum turns out to depend on the star's density: the lines are wide in high-density stars and narrow in low-density ones.

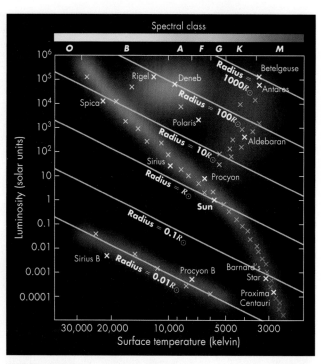

FIGURE 13.23
Lines showing where in the H-R diagram stars of a given radius will lie.

Q: On the basis of these lines, what is the approximate radius of Betelgeuse? Does this agree with the size you estimate from figure 13.9?

INTERACTIVE

Stellar evolution and the H-R diagram

Q: All three of these spectra are of stars with the same spectral type. Comparing these spectra to the ones in figure 13.14, what spectral type is it?

FIGURE 13.24
Spectral lines are narrow in class I bright supergiants, and progressively wider in lower-luminosity classes of stars.

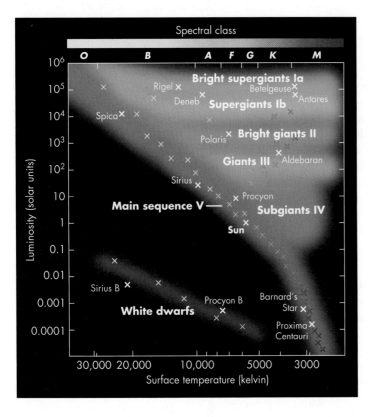

Table 13.4 **Stellar Luminosity Classes**

Class	Description	Example
Ia	Bright supergiants	Betelgeuse, Rigel
Ib	Supergiants	Antares (brightest star in Scorpius)
II	Bright giants	Polaris (the North Star)
III	Ordinary giants	Aldebaran (brightest star in Taurus)
IV	Subgiants	Procyon (brightest star in Canis Minor)
V	Main sequence	The Sun, Sirius

FIGURE 13.25
Stellar luminosity classes.

The density of a star's gas is related in turn to its luminosity, because a large-diameter star has a large surface area, causing it to emit more light—other things being equal—and a large volume, giving it a lower density. A small-diameter star, on the other hand, generally has a high density because its gas is compressed into a small volume. Such a star also tends to be less luminous because its surface area is small and thus emits less light.

Using this relationship between spectral line width and luminosity, astronomers divide stars into five luminosity classes (fig. 13.25), denoted by the Roman numerals I to V. Class V stars are the dimmest, and class I stars are the brightest. In fact, class I stars are split into two classes, Ia and Ib. Table 13.4 shows the correspondence between class and luminosity, and although the scheme is not very precise, it allows astronomers to get an indication of a star's luminosity from its spectrum.

A star's luminosity class is often added to its spectral type to give a more complete description of its light. For example, our Sun is a G2V star, while the blue giant Rigel is a B8Ia star. The luminosity class is especially useful for indicating the difference in luminosity between main-sequence and giant stars of the same spectral type, such as the low-luminosity, nearby, main-sequence star 40 Eridani (K1V) and the high-luminosity giant star Arcturus (K1III).

The Mass–Luminosity Relation

In 1924 the English astrophysicist Arthur Eddington discovered that the luminosity of main-sequence stars is determined by their masses. That is, main-sequence stars obey a **mass–luminosity relation** such that the larger a star's mass, the larger its luminosity is, as illustrated in figure 13.26. A consequence of this relation is that stars near the top of the main sequence, brighter stars, are more massive than stars lower down.

Eddington made his discovery by measuring the masses of stars in binary systems and determining their luminosity from their distance and apparent brightness.

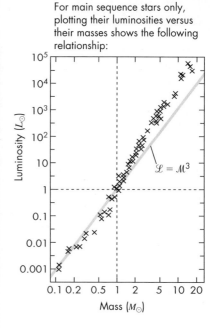

For main sequence stars only, plotting their luminosities versus their masses shows the following relationship:

FIGURE 13.26

(A) The mass–luminosity relation shows that along the main sequence, more-massive stars are more luminous. The law does not work for red giants or white dwarfs. (B) On the H-R diagram, high-mass stars lie higher on the main sequence than low-mass stars.

Moreover, he discovered that the mass–luminosity relation could be expressed by a simple formula. If $\mathscr{M}$ and $\mathscr{L}$ are given in solar units, then for main-sequence stars

$$\mathscr{L} \approx \mathscr{M}^3.$$

$\mathscr{L}$ = Luminosity of stars in solar units
$\mathscr{M}$ = Mass of stars in solar units

If we apply the relation to the Sun, its mass is 1 and its luminosity is 1 in these units, and the relation is clearly obeyed because $1 = 1^3$. If we consider instead a 2-$M_\odot$ star (that is, $\mathscr{M} = 2$), then according to the relation its luminosity is

$$\mathscr{L} \approx 2^3 = 8, \quad \text{or} \quad L \approx 8 L_\odot.$$

Eddington went on to show from the physics of stellar structure that just such a mass–luminosity relationship is required if a star's energy flows through its interior by radiation and it is to avoid collapsing under its own gravitational forces.

The mathematical relationships examined in this chapter have provided us with the means for uncovering stellar properties that we will probably never get the opportunity to measure directly. It is worth taking a moment to think about what an amazing feat this is. We are like the ancient Greek astronomers who used geometry and mathematics to determine the size and distances of the Earth, Moon, and Sun. They had no way to send space probes to observe the round Earth or to test their theories, but today we know they were correct. The calculations we make today, which are reviewed in Astronomy by the Numbers: "Calculating Stellar Properties," are confirmed in many ways, allowing us to speak with great confidence about stars with an amazing array of properties, from enormous giant stars to incredibly dense dwarfs. In chapter 14, we will see how a star's need to balance energy generation with its internal gravity is what gives rise to many of the properties of stars, ultimately determining how long a star can live.

CALCULATING STELLAR PROPERTIES

Using the methods we have described in this chapter, astronomers can determine many properties of stars: their distance, luminosity, temperature, radius, mass, composition, and radial velocity. The methods used are summarized in table 13.5 and illustrated in figure 13.27.

From stellar parallaxes we can find distances, and once we have distances, we can determine the luminosity of stars from their apparent brightness. By measuring the color of a star and the wavelength of maximum emission,

we can find the star's temperature using Wien's law. And if we know both luminosity and temperature, we can also calculate a star's radius using the Stefan-Boltzmann law.

Many stars are in orbit around another star, which allows us to measure their masses using a modified form of Kepler's third law. Examining spectral lines allows us to measure the compositions of stars, and it also allows us to measure the motion of stars toward or away from us and analyze orbital motions more fully.

Table 13.5	**Methods for Determining Stellar Quantities**
Quantity	**Method**
Distance (d)	1. Parallax (triangulation)—for nearby stars (distance less than 250 pc) 2. Standard-candle method for stars of known luminosity
Luminosity (L)	1. Measure star's apparent brightness and distance and then calculate with inverse-square law. 2. Luminosity class of spectrum
Temperature (T)	1. Wien's law (color–temperature relation) 2. Spectral type (O hot; M cool)
Radius (R)	1. Stefan-Boltzmann law (measure L and T, solve for R) 2. Interferometer (gives angular size of star; from distance and angular size, calculate radius) 3. Eclipsing binary light curve (duration of eclipse phases)
Mass (M)	Modified form of Kepler's third law applied to binary stars
Composition	Strength of spectral lines observed in star, adjusted for surface temperature
Radial velocity (V)	Doppler shift of spectrum lines

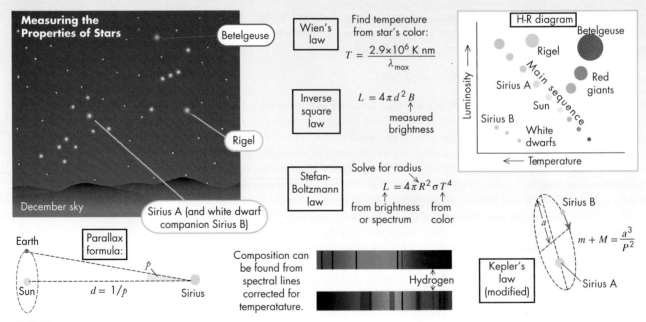

FIGURE 13.27

Summary of how astronomers measure the distance, temperature, mass, composition, and radius of a star.

SUMMARY

Astronomers use many techniques to measure the properties of a star. Parallax—triangulation of the star from opposite sides of the Earth's orbit—gives a distance for nearby stars, with the nearest being about 1.3 parsecs (about 4 light-years) away from the Sun. The parallax method fails beyond several hundred parsecs because the angle becomes too small to measure. Astronomers must therefore use the inverse-square law, comparing apparent brightness and luminosity to find the distance of more remote stars.

Stars have a wide range of surface temperature as measured by their color. According to Wien's law, hot stars will be blue and cool stars red. Most stars have surface temperatures between about 3000 and 30,000 K. Once the temperature and luminosity of a star are known, we can calculate its radius with the Stefan-Boltzmann law.

We can find a star's composition by looking at the lines in its spectrum. These measurements show that all stars have nearly the same composition of about 71% hydrogen and 27% helium, with a trace of the heavier elements. The spectrum can also be used to assign stars to the spectral types *O, B, A, F, G, K, M,* where *O* stars are the hottest and *M* stars are the coolest.

We can find a star's mass if it is in a binary system—two (or more) stars bound together gravitationally. For such stars, the orbital period and size, when combined in the modified form of Kepler's third law, give the masses of the orbiting stars. Most stars have masses between about 0.1 and 30 $M_{\odot}$.

Once a star's luminosity and temperature are known, the star can be plotted on an H-R diagram. In this plot, most stars lie along a diagonal line called the main sequence, which runs from hot, blue, luminous stars to cool, red, dim ones. Main-sequence stars obey a mass–luminosity relation such that high-mass stars are more luminous and hotter than low-mass stars. The Sun is a *G*2 main-sequence star.

A small percentage of stars fall above the main sequence, being very luminous but cool. Their high luminosity implies they must have a large diameter, and so they are called red giants. Although they are large, they do not have higher masses than main-sequence stars. Some stars fall below the main sequence and are hot but dim. Their low luminosity implies a very small diameter, comparable to that of the Earth. These hot, small stars are called white dwarfs. Their masses are always found to be less than 1.4 $M_{\odot}$.

QUESTIONS FOR REVIEW

1. (13.1) Describe the method of distance measurement by triangulation.
2. (13.1) How do astronomers triangulate a star's distance?
3. (13.1) How is the parsec defined? How big is a parsec compared with a light-year?
4. (13.2) What do astronomers mean by an inverse-square law?
5. (13.2) What is luminosity? What two characteristics of a star determine its luminosity?
6. (13.2) What does a star's magnitude measure? Which is brighter, a star of magnitude 1 or a star of magnitude 3?
7. (13.2) What is a standard candle and how is it used?
8. (13.3) How do astronomers measure a star's temperature?
9. (13.3) How do astronomers measure a star's radius?
10. (13.4) What are the stellar spectral types? Which are hot and which are cool?
11. (13.4) Why do stars have dark lines in their spectra?
12. (13.4) What is different about the spectra of different types of stars?
13. (13.5) What are binary stars? How are they useful to astronomers?
14. (13.5) How do visual and spectroscopic binaries differ?
15. (13.5) What is an eclipsing binary? What can be learned from eclipsing binaries?
16. (13.6) What is the H-R diagram? What are its axes?
17. (13.6) What is the main sequence?
18. (13.6) How do we know that giant stars are big and dwarf stars are small?
19. (13.6) How does mass vary along the main sequence?
20. (13.6) What is the mass–luminosity relation?

THOUGHT QUESTIONS

1. (13.1) Using the method of parallax, what observational result would disprove that Proxima Centauri is the closest star to Earth?
2. (13.1) Would it be easier to measure a star's parallax from Pluto? Why?
3. (13.2) Astronomers can directly measure the diameters of a few nearby, large stars using interferometry (see figure 13.9). We might wonder why interferometry is necessary. Looking at many of the images in this book, the stars do not look like points. Why can't you use a simple photograph of Betelgeuse (like fig. 13.7) to determine its size?
4. (13.2) Explain the difference between luminosity and brightness. In what way is this similar to the difference between force and pressure?
5. (13.4) The spectra of *B*-type and *M*-type stars feature different absorption lines from different elements and molecules (see fig. 13.14). Is this a result of a difference in the composition or the temperature of the star's surface? Explain.
6. (13.5) When we observe spectroscopic binaries, we often do not know how the plane of the orbit is oriented to our

line of sight. How will the Doppler shift change if the same orbit is oriented edge-on, face-on, or somewhere in between?

7. (13.5) Redraw the brightness vs. time plot for the stars in figure 13.20 as it would appear if the smaller star were a red *M*-type star and the larger star were a yellow giant.

8. (13.1/13.5) How could you tell the difference between a visual binary with a one-year orbit and a star moving as a result of parallax?

9. (13.6) If instead of plotting luminosity vs. temperature to make an H-R diagram, we made a plot of stars' distances vs. temperatures, what sort of pattern would we find?

10. (13.2/13.6) Suppose you are placing a star on an H-R diagram. Using a telescope you note a red star of apparent magnitude 3. What else would you need to know?

PROBLEMS

1. (13.1) Sirius has a parallax of 0.377 arc seconds. How far away is it?

2. (13.1) The parallax of Proxima Centauri is about 0.763 arc seconds. How far away is it?

3. (13.1) The parallax of the red giant Betelgeuse is just barely measurable and has a value of about 0.005 arc seconds. What is its distance? Suppose the measurement is in error by + or −0.003 arc seconds. What limits can you set on its distance?

4. (13.1) Estimate the distance from where you live to a neighboring building or object using triangulation (as shown in fig. 13.1 for a tree). Mark the location of the 90° angle with a stick. Pace off the distance along the baseline or measure it with a meterstick or tape measure. Now from the other end of the baseline, measure the angle A between the stick and the building with a protractor. Make a scale drawing (1/4 inch on the drawing to 1 foot in real life works well) to determine the distance.

5. (13.2) The two stars τ Ceti and α Centauri B have nearly the same luminosity (and absolute magnitudes) but very different apparent magnitudes: τ Ceti is 3.5, and α Centauri B is 1.34. (a) Determine the ratio of brightness of α Centauri B to τ Ceti. (b) Use the method from section 13.2 to determine how much farther away τ Ceti is than α Centauri B.

6. (13.3) The star Rigel radiates most strongly at about 200 nm. What is its temperature? How does this compare to the Sun?

7. (13.3) The bright southern star Alpha Centauri radiates most strongly at about 500 nm. What is its temperature? How does this compare to the Sun's?

8. (13.3) Arcturus is about half as hot as the Sun but is about 100 times more luminous. What is its radius compared to the Sun's?

9. (13.3) A stellar companion of Sirius has a temperature of about 27,000 K and a luminosity of about 10^{-2} $L_\odot$. What is its radius compared to the Sun's? What is its radius compared to the Earth's?

10. (13.4) If a star has very faint Balmer lines, what are the next steps involved in determining its spectral type?

11. (13.4) A line in a star's spectrum lies at 400.0 nanometers. In the laboratory, that same line lies at 400.2 nanometers. How fast is the star moving along the line of sight? Is it moving toward or away from us?

12. (13.5) Two stars in a binary system have an orbital period, *P*, of 5 years and an orbital separation, *a*, of 10 AU. What is their combined mass?

13. (13.5) Two stars in a binary system have an orbital period, *P*, of 2 years and an orbital separation, *a*, of 4 AU. What is their combined mass?

14. (13.5) Two stars in a binary system are determined from their position on the H-R diagram and the mass–luminosity relation to have a combined mass of 8 $M_\odot$. Their orbital period, *P*, is 1 year. What is their orbital separation, *a*?

15. (13.6) Estimate the luminosity and temperature of both Rigel and Barnard's Star based on the position of their plotted points as marked in the H-R diagram in figure 13.23. Calculate the radii of both stars compared to a solar radius using the Stefan-Boltzmann equation, and show that the values agree with the radii indicated by the diagonal lines in figure 13.23.

16. (13.6) If a main-sequence star has a luminosity of 5000 $L_\odot$, what is its mass in relation to the Sun's?

17. (13.6) Use the data in table 13.6 to plot an H-R diagram. Which stars are red giants? Which are white dwarfs? Note: Plotting will be much easier if you plot the logarithm of the luminosity; that is, express it in powers of 10 and use the power. For example, if the luminosity is 100, plot it as 2 for 10^2. You can use a calculator to find the power: enter the luminosity in solar units and hit the "log" key. (If the luminosity is 300,000, the answer you get should be approximately 5.5.)

Table 13.6	Data for Plotting an H-R Diagram*		
Star Name	Temperature	Spectral Type	Luminosity (Solar Units)
Sun	6000	G2	1
Sirius	10,000	A1	25
Deneb	9000	A2	100,000
Betelgeuse	3500	M2	100,000
40 Eridani B	15,000	DA†	0.01
Proxima Centauri	3000	M5	0.001
Spica	20,000	B1	2000

* These data have been rounded off to make plotting easier.

† Given the low luminosity of this star, what do you think the *D* in its spectral type stands for?

TEST YOURSELF

1. (13.1) A star has a parallax of 0.04 arc seconds. What is its distance?
   (a) 4 light-years   (c) 40 parsecs   (e) 250 parsecs
   (b) 4 parsecs   (d) 25 parsecs
2. (13.2) Which statement about stellar luminosities is FALSE?
   (a) A small drop in temperature has a much larger impact on the star's brightness than does a small change in size.
   (b) Most naked eye stars are more luminous than our own Sun.
   (c) Red stars with higher luminosity than our Sun must be much larger than our Sun.
   (d) Luminosities have a smaller range than masses.
3. (13.2) If two stars have the same luminosity, but one appears only one-quarter as bright as the other, you can conclude that the dimmer star is _____ times farther away.
   (a) 2   (b) 4   (c) 8   (d) 16   (e) 64
4. (13.3) A star radiates most strongly at 400 nanometers. What is its surface temperature?
   (a) 400 K   (c) 40,000 K   (e) 7500 K
   (b) 4000 K   (d) 75,000 K
5. (13.3) A star that is cool and very luminous must have
   (a) a very large radius.
   (b) a very small radius.
   (c) a very small mass.
   (d) a very great distance.
   (e) a very low velocity.
6. (13.4) Which of the following stars is hottest?
   (a) An *M* star   (c) A *G* star   (e) An *O* star
   (b) An *F* star   (d) A *B* star
7. (13.5) A binary system has a period of about 100 years and an average separation of 30 AU. Its combined mass is about _____ solar masses.
   (a) 27   (b) 4   (c) 9   (d) 3   (e) 1000
8. (13.6) In what part of the H-R diagram do white dwarfs lie?
   (a) Upper left
   (b) Lower center
   (c) Upper right
   (d) Lower right
   (e) Just above the Sun on the main sequence
9. (13.6) In a sample of nearby stars, about what percentage will lie on the main sequence?
   (a) 99%   (b) 90%   (c) 50%   (d) 35%   (e) 9%

KEY TERMS

binary star, 344
dwarf, 337
eclipsing binary, 346
giant, 337
H-R diagram, 347
inverse-square law, 329
luminosity, 329
magnitude, 331
main sequence, 348
mass–luminosity relation, 350
parallax, 326

parsec, 327
proper motion, 343
radial velocity, 342
red giant, 348
spectral type, 340
spectroscopic binary, 344
standard candle, 330
triangulation, 326
visual binary, 344
white dwarf, 348
Wien's law, 333

Ⓠ: FIGURE QUESTION ANSWERS

WHAT IS THIS? (chapter opening): This is a photograph of the Hyades star cluster made through a thin prism. The prism spreads light from each star into a spectrum. The brightest spectrum toward the top is that of Aldebaran, a red giant.

FIGURE 13.5: The 36 dots would have to be spread out over 36 square meters. This would happen at a distance of 6 meters.

FIGURE 13.23: Betelgeuse lies close to the line for a radius of 1000 $R_\odot$, or about 7×10^8 km. In figure 13.8, Betelgeuse's *diameter* is somewhat less than 10 AU $= 1.5 \times 10^9$ km, which corresponds to a radius half that number, 7.5×10^8 km, so the value read from the graph is pretty close to the direct measurement.

FIGURE 13.24: The spectra shown here are of spectral type *A*.

14

STELLAR EVOLUTION

Hubble Space Telescope image of the Carina Nebula. Radiation and outflowing gas from young stars sculpt the dusty gas clouds where the stars are forming.

LEARNING OBJECTIVES

Upon completing this chapter you should be able to:
- Explain the basic principles of stellar evolution models.
- Discuss how the mass of a star plays a critical role in determining the evolution and fate of the star.
- Compare the lifetimes of high- and low-mass stars, and calculate a star's lifetime given its mass and luminosity.
- Describe how and where astronomers detect forming stars, and list the stages a forming star evolves through.
- Compare the CNO cycle with the proton–proton chain.
- Compare convection in high-mass stars to that in the Sun.
- Describe the changes that occur as a star consumes hydrogen in its core and when it is finally exhausted.

- Explain why a star's core shrinks while its atmosphere expands when it leaves the main sequence.
- Describe and explain the processes that will change low-mass stars like the Sun from a red giant to a Mira variable to a planetary nebula and white dwarf at the end of its life.
- List different kinds of variable stars and discuss the physical conditions that give rise to pulsating variables.
- Describe the period–luminosity relationship.
- Describe the sequence of fusion reactions that take place in a high-mass star as it nears the end of its life.
- Explain the conditions that result in a supernova.
- Discuss how observations of star clusters can be used to test stellar evolution models.

To us, stars appear permanent and unchanging, but like many other things, they are born, grow old, and die. Stars are not alive, of course; they are immense balls of gas undergoing gradual changes in structure and appearance. Astronomers refer to these changes as stellar evolution, which is driven by the physical laws that govern a star's structure.

For example, gravity holds a star together, and the pressure of the star's gas supports it against gravity's inward pull. A star generates its supporting pressure from heat energy created by gravity compressing the material in its core. That heat energy steadily escapes from the star as the starlight we see, and it must be replenished or the star will collapse.

A star replenishes its energy by nuclear reactions in its core. In the early stages of its evolution, the nuclear reactions fuse hydrogen into helium. But eventually the star consumes all of the hydrogen in its core and then its structure must change, and it swells into a red giant. Stars like the Sun consume their core's hydrogen in about 10 billion years, but more-massive stars use up their hydrogen much faster. They therefore evolve much faster than the Sun and burn out in only a few million years. To understand why massive stars evolve faster, recall that according to the mass–luminosity relation, massive stars are far more luminous than stars like the Sun. Thus, to supply their greater luminosity, massive stars must burn their fuel faster than less-massive stars.

When a star runs out of fuel, its evolution is nearly finished. A star like the Sun dies relatively quietly and becomes a white dwarf. A more massive star explodes and leaves a neutron star or black hole as a remnant. These exotic objects are studied further in chapter 15. The stellar explosions that mark the death of one star may trigger the birth of another. Moreover, the explosion of a dying star blasts into space elements vital to human life, such as carbon, oxygen, and iron, elements that eventually become incorporated into new stars, planets, and, ultimately, people. We therefore owe our very existence to stellar evolution. The goal of this chapter is to illustrate how such evolution occurs.

CONCEPTS AND SKILLS TO REVIEW

- Relation between the temperature of a hot object and the color of the light it emits (Wien's law) (4.3)
- Structure of atoms (4.4)

14.1   OVERVIEW OF STELLAR EVOLUTION

Stars change so slowly that very little change can be detected in our lifetimes, or indeed in the entire history of astronomical observations. Indeed, most stars present today would have looked identical to the first humans to gaze at the sky. The great majority would even have looked the same in the age of the dinosaurs. Astronomers can, however, deduce from observations some of the basic features of how a star evolves. For example, the existence of main-sequence stars, red giants, and white dwarfs suggests to astronomers a picture of how stars age, much the way a snapshot of a baby with its parents and grandparents suggests to a viewer a picture of human aging.

Astronomers have been able to piece together this evidence to deduce how stars are born, how long they live, and the spectacular ways they die. We can predict how much longer the Sun will live, and how the nature of our own Solar System depended on generations of stars before the Sun. In this first section of the chapter we will explore some of these general ideas about stellar evolution and summarize the lives of stars. We then will examine each phase in more detail in subsequent sections.

The Importance of Gravity

Gravity drives stellar evolution. From the moment of a star's birth, when an interstellar gas cloud collapses to become a star, to the moment of its death, gravity is the unrelenting force that drives the star toward its final collapse. What begins as a gentle tug grows into a crushing force that heats up a star's interior to temperatures hotter than any furnace on Earth. The steady contraction caused by gravity pauses only if a star can generate internal pressures sufficient to counterbalance gravity's pull.

Despite the titanic forces involved, the evolutionary changes at work in stars require millions to billions of years. While we cannot directly observe most of the changes that occur in a star's lifetime, astronomers have a powerful tool for examining the lives of stars. We can "watch" a star's aging with computer calculations that solve the equations that govern the star's structure. Stellar models, as such calculations are called, allow us to trace a star's life, revealing billions of years of a star's history. Figure 14.1 outlines a few of the most important physical processes governing a star's structure and evolution that go into such a calculation.

Gravity squeezes a star until it drives the gas temperature up so high that nuclear fusion occurs. No other source of energy is powerful enough to bring the star's gravitational contraction to a standstill. The energy from this fusion replenishes the heat that is constantly flowing out of the star, establishing a balance with gravity for as long as the nuclear fuel lasts. When the fuel is exhausted, a star then must readjust and enter a new phase of its lifetime.

All stars need fuel to keep their cores hot, and all begin their lives by fusing hydrogen into helium. When they use up their hydrogen fuel, their structure must change. For example, when a star runs out of hydrogen fuel in its core, gravity's force is no longer counterbalanced, and the core of the star resumes its contraction, heating up to even higher temperatures until it reaches a temperature high enough to fuse helium into carbon. The outer parts of the star are driven outward in response to the heat pouring out from the core, and the star begins its red giant phase.

With these ideas in mind, let us follow the life story of a star from its birth to its death. In telling that story, we will discover that its ending depends critically on how much material the star contains—its mass. A star's mass determines not only how strong its gravity is, but how much fuel it has. Thus, stars that begin their lives with the same mass as the Sun live out lives very similar to the Sun's.

There can be many minor differences between stars as we move up to incrementally larger masses, but there is a major change in the fates of stars that weigh more than about eight times the Sun's mass. These high-mass stars have such a powerful gravitational pull that they undergo a cataclysmic collapse after they have exhausted all possible nuclear fuels. We will therefore divide stars into two groups—low-mass stars, such as the Sun, and high-mass stars. We will briefly describe the stages that each of these kinds of stars passes through, beginning with one such low-mass star, our Sun.

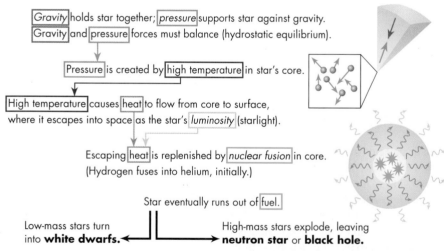

Principles Governing the Structure of a Star

Gravity holds star together; pressure supports star against gravity.
Gravity and pressure forces must balance (hydrostatic equilibrium).

Pressure is created by high temperature in star's core.

High temperature causes heat to flow from core to surface, where it escapes into space as the star's luminosity (starlight).

Escaping heat is replenished by nuclear fusion in core.
(Hydrogen fuses into helium, initially.)

Star eventually runs out of fuel.

Low-mass stars turn into **white dwarfs.**

High-mass stars explode, leaving **neutron star** or **black hole.**

Note: High-mass stars require more pressure to support their greater mass.
Greater pressure is produced by higher temperature.
Higher temperature produces higher luminosity.
Higher luminosity leads to faster fuel usage.
Faster fuel usage means a high-mass star burns out sooner than a low-mass star.

FIGURE 14.1
Outline of the processes that govern the structure of stars.

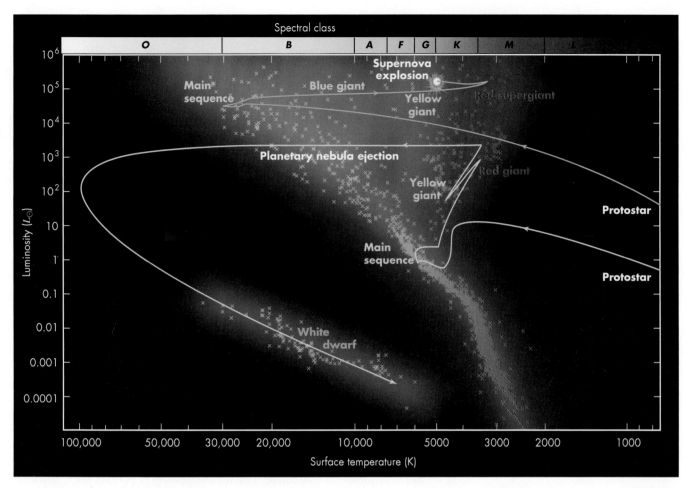

FIGURE 14.2
Schematic diagram of the evolutionary tracks of a low-mass star like the Sun (yellow) and a high-mass star (blue) through the H-R diagram.

The Life Story of the Sun—A Low-Mass Star

The Sun began its life as an **interstellar cloud,** a tenuous, cold, dark mass of gas drifting around the Milky Way Galaxy. The cloud—perhaps because of a collision with a neighboring cloud or the explosion of a nearby star—began to collapse. As it shrank under the influence of its increasing gravity, it grew smaller and hotter. Within a few million years—an instant, to the Sun—hydrogen began to fuse into helium in its core. The energy released by the fusion sustained the pressure inside the Sun, stopping its collapse and leaving it looking much as it does today: a small, yellow star. If we plotted it on the H-R diagram, it would lie on the main sequence as illustrated in figure 14.2.

The Sun will remain a main-sequence star until it consumes about 90% of the hydrogen in its core, a process that will take a total of 10 billion years, about half of which time the Sun has already used up. When the hydrogen in the Sun's core is nearly spent, its core will shrink and grow hotter. The rising temperature in the core will make the remaining hydrogen there burn faster, generating more energy. As that energy flows outward through the Sun's outer layers, it will lift them and they will cool as they expand farther from the source of heat. As the surface of this bloated Sun cools, it will turn red, and its light will reveal to our distant descendants a Sun swollen into a red giant. The Earth will have become uninhabitable, however, the intense luminosity of the Sun having boiled away our oceans and perhaps even melted Earth's crust.

The Sun will shine as a red giant for perhaps another billion years but will then shrink and grow hotter, transforming—when its core finally becomes hot enough to fuse helium—into a yellow giant. During this stage, it will pulsate as if taking slow, deep breaths, but these are dying gasps. Eventually the Sun will consume most of

An H-R diagram showing the evolutionary track of a 1-$M_\odot$ star

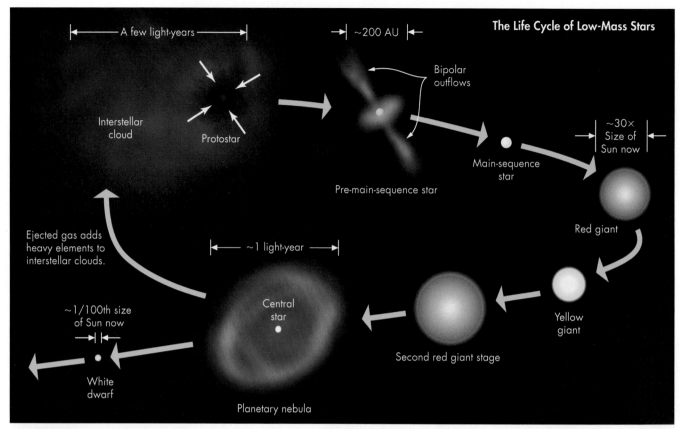

FIGURE 14.3

Evolution of a low-mass star such as the Sun. Stars with masses below 8 $M_\odot$ share a similar fate.

the helium in its core and will once again change into a red giant, but a giant even larger and more luminous than before. That high luminosity is its death knell: radiation streaming outward through its atmosphere will drive its gas into space, stripping the Sun to its bare core. The ejected gas will form a shell that will gradually disperse. The tiny core, fiercely hot but with no further energy supply, will cool and dwindle to a white dwarf. These changes, summarized in figure 14.3, are the fate of the Sun.

The Life Story of a High-Mass Star

A very different fate awaits stars significantly more massive than the Sun: they explode when they run out of fuel. Astronomers are uncertain precisely what mass is needed to make a star explode but believe it is about 8 $M_\odot$. For that reason, we take 8 $M_\odot$ as the dividing line between high-mass and low-mass stars.

We chart the history of a high-mass star in figure 14.4, where we see that its early life is similar to the Sun's: it originates from the collapse of an interstellar cloud but from a larger and more massive clump than the one that formed the Sun. The clump's greater mass "squeezes" it more, thereby heating it more than the young Sun was heated. Thus, when the clump becomes a main-sequence star, it is much hotter, bluer, and more luminous than the Sun. Its greater luminosity indicates that it burns its fuel faster, and so rather than taking 10 billion years to expend its hydrogen, it consumes its fuel in 100 million years or less, aging far more rapidly than the Sun. As it runs out of hydrogen, a massive star behaves much like a low-mass star, swelling steadily in size and growing cooler. However, a massive star passes through a pulsating yellow-giant stage before it turns into a red giant.

The greater mass of such a star now becomes a primary influence on its evolution, creating an intense gravitational compression of its core. As the core is compressed, its temperature rises, and it burns fuel ever more furiously, thereby maintaining the

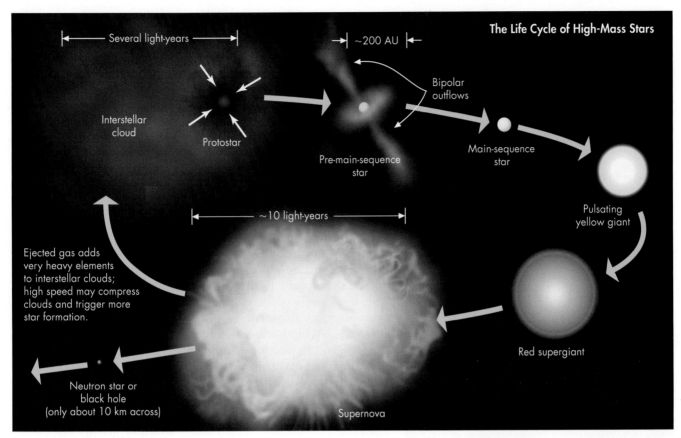

The Life Cycle of High-Mass Stars

Several light-years

Interstellar cloud

Protostar

~200 AU

Bipolar outflows

Pre-main-sequence star

Main-sequence star

Pulsating yellow giant

~10 light-years

Ejected gas adds very heavy elements to interstellar clouds; high speed may compress clouds and trigger more star formation.

Red supergiant

Neutron star or black hole (only about 10 km across)

Supernova

FIGURE 14.4

Evolution of a high-mass star, more than about 8 $M_\odot$.

pressure that supports the star. The higher temperature also permits the star to fuse progressively heavier elements to supply its energy—helium is fused into carbon, carbon into oxygen, and, ultimately, silicon into iron. Iron, however, does *not* release energy when it is fused, and thus a star with a massive iron core cannot support itself against gravity's inward pull. Lacking support, its core collapses in less than a second, triggering a cataclysmic explosion that showers into space the heavy elements it has made. Such a star does not leave a white dwarf; rather, its core becomes either a tiny, incredibly compressed ball of neutrons (a neutron star) or an even denser body—a black hole, gravity's ultimate triumph.

Stellar Recycling

Stars form from the matter in interstellar clouds, but they also return much of that matter—in some cases as much as 90% of their original mass—to interstellar space. The great energy they release during their lifetimes drives a steady flow of matter during their lifetimes, and as they die, they may blast huge quantities out at high speeds. Much of this material is the original gas that the star formed from, but added to this are heavy elements formed by fusion during the star's lifetime. This material may even trigger more stars to form as the gas collides with and compresses interstellar clouds.

This gas–star–gas cycle has steadily increased the amount of heavy elements in interstellar clouds. In the cloud from which the Sun formed, about 2% of its mass was composed of elements heavier than hydrogen and helium, created by several previous generations of stars. Indeed, the terrestrial planets would not exist without that material. This process has been going on for more than ten billion years and has produced more than 100 billion stars in our own Galaxy.

In the remainder of this chapter, we examine each stage of stellar evolution in detail, beginning with the formation of stars from interstellar clouds.

| 14.2 | STAR FORMATION |

Interstellar Gas Clouds

FIGURE 14.5
Radio map of a star-forming interstellar cloud in Taurus. The image shows a region about 60 light-years across.

Stars form from interstellar matter—huge clouds of gas and dust that orbit with the stars inside our Galaxy. The gas is mostly hydrogen (71% by mass) and helium (27%); the dust is composed of solid, microscopic particles made primarily of silicates, carbon, and iron compounds. Interstellar gas is generally cold, perhaps only 10 K (hundreds of degrees below zero on the Fahrenheit scale). At such low temperatures, atoms and molecules in a gas move too slowly to generate much pressure. As a result, the pressure in such cold gas may not be large enough to support the cloud against its own gravity; if the cloud is disturbed, it may then collapse. The collapse may be triggered by collision with a neighboring cloud, the explosion of a nearby star, or other processes.

Maps made with radio telescopes show that gas clouds are not uniform: they contain smaller clumps of gas where the density is higher than the average (fig. 14.5). When a cloud with such clumps collapses, each clump is also compressed, growing denser. Thus, a single cloud may break into many smaller pieces, each of which may form a star. The result is that stars generally form in groups, not in isolation, and all the stars within a group form at approximately the same time.

The transformation from gas cloud to star proceeds through several stages. In the first stage, a dense clump, called a **Bok globule,** within a cloud begins to collapse, its gas drawn inward by gravity, which compresses and heats it. In the second stage, any rotation of the gas clump makes it flatten into a disk, as we described in chapter 8 on the origin of the Solar System. In about a million years, the disk forms a small, hot, dense core at its center called a **protostar,** which marks the third stage. These stages are illustrated in figure 14.6.

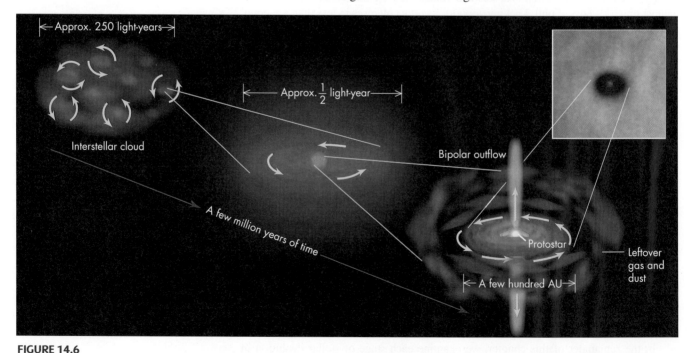

FIGURE 14.6
An artist's sketches depicting the birth of a star. (Insert, far right) A Hubble Space Telescope picture of the disk around a star forming in Orion.

FIGURE 14.7
The Eagle Nebula (M16), a star-forming region estimated to be about 2 million years old. (A) Visible-light image made by the Hubble Space Telescope of dense, dusty clouds, which contain many protostars. (B) A false-color infrared image of the same region made with the Hubble Space Telescope shows infrared-bright protostars hidden in the visible-light image. The images are about 7 light-years across.

Protostars

A protostar is hotter than the gas from which it condensed—perhaps 1500 K—but it is still much cooler than ordinary stars. Its relatively low temperature makes a protostar "shine" mainly at infrared wavelengths rather than at visible ones. Thus, to observe a protostar, astronomers use telescopes that detect infrared wavelengths. Visible and infrared images of an interstellar cloud are shown in figure 14.7. Protostars show up as bright objects in areas of the infrared image both because they emit strongly at infrared wavelengths and because the dusty clouds block visible wavelengths.

Protostars do not remain cool for long. As their gravity continues to draw surrounding material inward, they grow hotter from the gravitational energy released by the infalling material. Gravitational energy is released whenever something falls. For example, when a cinder block falls onto a box of tennis balls, the impact scatters the balls in all directions, giving them kinetic energy—energy of motion. In a gas, energy of motion appears as heat, raising the temperature; as material falls onto a protostar, its gas atoms speed up and it grows hotter.

When the temperature in the protostar's core reaches about 1 million kelvin, some nuclear reactions begin there. These reactions fuse the trace amounts of a hydrogen isotope known as **deuterium** (^{2}H) into helium and release energy. Deuterium, sometimes called "heavy hydrogen," is the isotope of hydrogen that has one neutron in addition to a proton in the nucleus. It is about 100,000 times less abundant than the common form of hydrogen, which has no neutrons. The energy supplied by deuterium fusion replenishes the energy that is radiating away from the star, so the star stops contracting while this fuel is consumed. Surrounding material, especially from the disk, continues to fall onto the star for millions of years. The infalling matter releases gravitational energy, which supplies additional heat to the protostar. For young stars, the gravitational energy of infalling matter can greatly exceed that released by fusion in the core. Once the deuterium is used up, the protostar resumes its contraction. A protostar passes through these stages in a time that is brief by astronomical standards: a few million years for a star like the Sun, and even less time for a more massive star.

Bipolar Flows from Young Stars

As stars form, the infalling material causes violent changes in brightness and, more unexpectedly, a strong *outflow* of gas. The outflow of gas does not take place in all directions. Instead, it generally is "focused" into narrow jets of gas pointed in opposite directions called **bipolar flows.** Bipolar flows (which get their name because there are two jets of gas ejected parallel to the star's polar axis) can be seen in figure 14.8, which shows a radio map of the area around a protostar. Here, a computer has been used to color red the gas moving away from us (that is, red-shifted material) and to color blue the gas moving toward us. What powers the jets and what makes them so narrow are not yet fully understood. Computer simulations suggest that the jets' power somehow comes from energy released as matter falls from the inner parts of the disk onto the star. The focusing of the jets probably results from magnetic fields in the star and disk.

Jets from young stars play an important role in "cleaning up" after a star has formed. They may also affect the formation of other stars in their vicinity. Figure 14.9 shows jets emerging from a dark cloud. You can see long, thin jets of gas squirting out from the dark cloud in which a young star lies. The jets have blown "bubbles" of hot gas whose edges are visible in green in the false-color infrared image. The jets of gas from the star spread when they hit surrounding gas, and the impact heats both the jets and the gas around them, making them glow, as shown in the picture.

Bipolar flows clear away gas and dust from around protostars and thereby allow astronomers to see them directly with visible light. However, even powerful bipolar flows leave much of the surrounding material behind. Thus, many young stars are partially immersed in interstellar matter, as illustrated in figure 14.10A. Such stars often vary erratically in brightness and have gas streaming away from their surfaces, as can be detected from the Doppler shift of their spectral lines. This flow occurs in the star's upper atmosphere and is distinct from the bipolar flow we described above. It is more similar to the solar wind but far more intense. Young stars that exhibit variable light and outflowing gas are called **T Tauri stars,** and some of their variability is probably caused by magnetic activity much like that observed on our Sun. For example, T Tauri stars show evidence of giant "starspots" and magnetic flares thousands of times more intense than the activity we observe on our Sun. Astronomers cannot see the spots directly with optical telescopes, but by combining data on variations of the star's brightness obtained with X ray, radio, and infrared telescopes, they can deduce the presence of spots on these peculiar young stars.

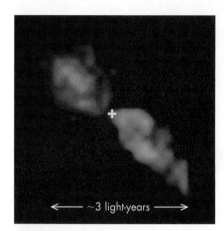

FIGURE 14.8
Bipolar flow of gas ejected from a young star. This is a false-color picture of the radio emissions from the gas. The gas receding from us is shown in red, and the gas approaching us is shown in blue. The cross marks the location of the young star.

~3 light-years

A

B

FIGURE 14.9
(A) Visible-light image of a dense cloud in which star formation is occurring. The cloud is seen in silhouette against distant background stars. (B) A false-color infrared image of the same region made with the Spitzer Space Telescope showing a bipolar outflow from the forming star. The images cover a region 3 × 2 light-years across.

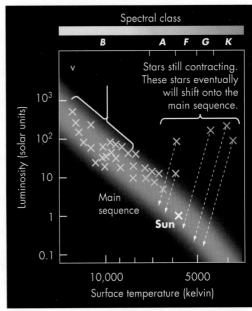

FIGURE 14.10
(A) The young star cluster NGC 2264 and (B) its H-R diagram. Notice the large amounts of leftover gas heated by the brilliant young stars. Massive blue stars have already reached their main-sequence phase even while lower-mass stars are still in their pre-main-sequence phase.

Figure 14.10B shows the H-R diagram for the young star group shown in figure 14.10A. The T Tauri stars lie a little above the lower main sequence in the H-R diagram because they are still shrinking and are thus larger and brighter than main-sequence stars. The position of protostars in the H-R diagram for this and other young clusters shows that lower-mass stars take longer to contract than high-mass stars, which reach the main sequence more rapidly.

Q: What creates the red glow of the gas near these stars?

Stellar Mass Limits

We saw in chapter 13 that the mass of most stars is between about 0.1 and 30 $M_\odot$. These limits are not absolute—higher- and lower-mass stars exist—but our theory of star formation can help us understand why very high- or low-mass stars are uncommon.

Stars much smaller than about 0.1 $M_\odot$ are rarely seen because their mass is too small to compress them enough to begin steady nuclear burning. Such low-mass stars are very dim, and are called "brown dwarfs" because of the dim red light they emit. These objects, midway between stars and planets in mass, are massive enough to reach a temperature to fuse deuterium, but they are too low in mass to sustain hydrogen fusion. They might be very numerous, but their dimness makes them extremely hard to detect. Searches with sensitive detectors have begun only in recent years to detect these tiny, dim, failed stars.

Very high-mass stars are rare for a different reason. As their great gravity compresses them, they rapidly become extremely hot and luminous, emitting such intense radiation that they heat the gas around them, raising its pressure and preventing additional material from falling onto them. Thus, their high temperature limits the amount of material they can accumulate. Furthermore, once formed, high-mass stars are so luminous (recall the mass–luminosity relation in section 13.6) that their radiation strips gas from their outer layers, driving it into space. Thus, a 200-$M_\odot$ star would quickly "shine" itself apart.

We have followed the stages of a protostar's early development. Now we will proceed to the next stage of its life.

One of the most massive stars known, Eta Carinae, is shown in Looking Up #8 at the front of the book.

A star's life on the main sequence begins when its core temperature rises above about 5 million kelvin. At this temperature, normal hydrogen fuses into helium, and the surrounding envelope of gas transports the energy to the star's surface. The properties of the core and envelope, however, depend greatly on the star's mass.

Structure of High-Mass and Low-Mass Stars

We saw in chapter 13 that high-mass stars are both hotter and more luminous than low-mass stars. This relation between mass, temperature, and luminosity results from the fact that the gravity and pressure inside a star must be in balance, a condition we called "hydrostatic equilibrium." This equilibrium develops as gas squeezes inward on the core, raising its temperature by compression until the gas pressure there is high enough to support the weight of all the gas above the core. The amount of heating is set by the forming star's mass, such that the larger the star's mass, the greater its heating. High-mass stars end up hotter (and thereby more luminous) in their cores than low-mass stars, although not as dense (fig. 14.11).

The high temperature at the core of a massive star allows it to fuse its hydrogen into helium by a mechanism different from that of the Sun. In the Sun, hydrogen is converted to helium by the proton–proton chain: two protons fuse to form deuterium, to which a third proton is added to make ^{3}He. Two ^{3}He then fuses to form ^{4}He (fig. 14.12). In a massive star, the conversion of hydrogen to helium takes place by means of the CNO cycle, whereby carbon, nitrogen, and oxygen atoms already present in the star's core act as catalysts to aid the reaction (fig. 14.13). The CNO cycle, however, requires a very high temperature to operate and thus can generate energy only at the very center of the star. Photons are unable to carry this huge energy flow from so small a volume, and so the gas in the core begins to rise in irregular clumps and carries the heat outward by convection.

As we saw in chapter 12, convective motions occur in the outer portion of the Sun just below its surface, where the radiant energy flow is impeded by the strong absorption of the relatively cool subsurface gas. In a high-mass star, however, the convection currents are confined to the core. They do not extend close enough to the star's surface to allow hydrogen-rich gas there to mix into its core and replenish the core's depleted fuel. Neither high-mass nor low-mass stars such as the Sun can use the fuel in their outer layers. Only stars less than about $0.4\,M_{\odot}$ mix completely. Most stars' cores eventually become starved for fuel, even though the outer layers are still fuel-rich.

Main-Sequence Lifetime of a Star

The period of time when a star fuses hydrogen in its core is called its **main-sequence lifetime.** A star's main-sequence lifetime depends on its mass and luminosity. A simple analogy shows why. Suppose we want to know how long a car can run on a tank of gasoline. That time clearly depends on how much fuel it has and how rapidly the fuel is consumed. If the car has a 16-gallon tank of gasoline and burns fuel at the rate of 2 gallons per hour, it can cruise for 8 hours ($= \frac{16}{2}$) before running out of gas. When the same formula is applied to a star, its mass determines how much fuel it has, and its luminosity determines how rapidly the fuel is burned. Thus, to calculate the star's main-sequence lifetime, we divide the amount of fuel by the rate at which it is consumed.

Calculations show that the Sun should consume the hydrogen in its core in about 10^{10} years, as shown in Astronomy by the Numbers: "The Lifetime of the Sun." By comparison, a star whose mass is $\mathscr{M}$ times the Sun's has $\mathscr{M}$ times more fuel. However, if its luminosity is $\mathscr{L}$ times the Sun's, it will burn up its fuel $\mathscr{L}$ times faster. Therefore,

FIGURE 14.11
Schematic view of the interiors of low- and high-mass stars, illustrating the higher density but lower temperature of a low-mass star's core.

Low-mass stars must compress their cores to a higher density than high-mass stars for the core to reach fusion temperature.

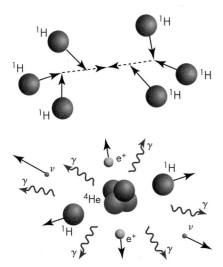

FIGURE 14.12
In the proton–proton chain, 6 protons react in sequence to produce a helium nucleus, 2 neutrinos, 2 positrons, and radiation, returning two protons at the end.

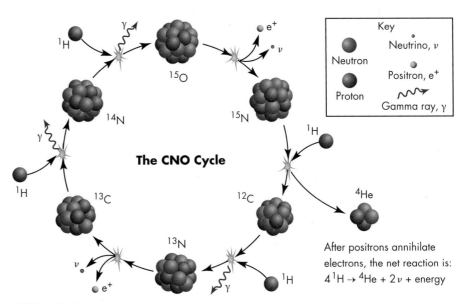

The CNO Cycle

Key

Neutrino, ν

Neutron

Positron, e^+

Proton

Gamma ray, γ

After positrons annihilate electrons, the net reaction is:
$4\,^1H \rightarrow \,^4He + 2\nu + energy$

FIGURE 14.13
In the CNO cycle, carbon, nitrogen, and oxygen act as catalysts to help hydrogen fuse into helium, but the net result is the same as the proton–proton chain.

a star's total main-sequence lifetime can be expressed as

$$t_\star = 10^{10}\,\frac{\mathscr{M}}{\mathscr{L}}\ years,$$

where $\mathscr{M}$ is in solar masses and $\mathscr{L}$ is in solar luminosities.

$t_\star$ = Lifetime of star
$\mathscr{L}$ = Luminosity of stars in solar units
$\mathscr{M}$ = Mass of stars in solar units

For the Sun ($\mathscr{M} = 1$ and $\mathscr{L} = 1$), this is 10^{10} (or 10 billion) years. The star Sirius ($\mathscr{M} = 2$ and $\mathscr{L} = 20$) has a main-sequence lifetime of $10^{10} \times \frac{2}{20}$ years $= 10^9$ years. For a star whose mass is $30\ M_\odot$ and whose luminosity is $10^5\ L_\odot$, the main-sequence lifetime is a mere 3×10^6 years. Notice that *massive stars have much shorter lifetimes* than the Sun, despite having more fuel. This is because they burn their fuel so much faster to supply their greater luminosity. They are like a gas-guzzler car that has a big gas tank, but runs out of fuel sooner than an economy-sized car.

The shortness of massive stars' lives implies that any we see must have formed recently. Their youth is also indicated by the fact that they are frequently found in the presence of the gas left over from their formation. Because massive stars are blue and very luminous, we can find regions of recent star formation by looking for blue stars.

ASTRONOMY
by the numbers   THE LIFETIME OF THE SUN

The Sun must generate an amount of power in its core equal to the luminosity that escapes to space (chapter 12). Based on the amount of power that comes from hydrogen fusion, we can determine that the Sun fuses about 2×10^{19} kilograms of hydrogen into a slightly lower mass of helium each year.

The lifetime of the Sun, $t_\odot$, should then be the mass of hydrogen it has available divided by this rate, or

$$t_\odot = \frac{mass\ of\ available\ hydrogen}{2 \times 10^{19}\ kg}\ years.$$

The Sun cannot use its entire mass of hydrogen for fuel, because only its core is hot enough for nuclear reactions to occur, and the core is only about one-tenth of its total mass. Therefore, the value of the mass we use in this equation should be about 0.1 times the Sun's mass, $2 \times 10^{30} \times 0.1 = 2 \times 10^{29}$ kilograms. Using this value for the amount of fuel available, we obtain

$$t_\odot = \frac{2 \times 10^{29}\ kg}{2 \times 10^{19}\ kg}\ years = 10^{10}\ years.$$

14.4   GIANT STARS

Stars are remarkably stable throughout their main-sequence lifetimes. A star like the Sun may increase in luminosity by a factor of 3 over about 10 billion years as it consumes the hydrogen in its core, but huge changes begin once it runs out of this core hydrogen. Once this happens a star does not have much longer to live, and the changes start occurring much faster.

Leaving the Main Sequence

The structure of a star when hydrogen burning in the shell begins

When a main-sequence star has consumed most of the hydrogen in its core, its structure begins to change. With too little fuel available, energy cannot be generated fast enough. That in turn allows the weight of the star's outer layers to compress the core, heating it and making it denser, as shown in figure 14.14. As a result, the hydrogen just *outside* the core now reaches a high enough temperature that it begins to fuse, creating what is called a **shell source,** as illustrated in the middle part of figure 14.14.

Heat from the shell source and contracting core raises the pressure around the core. That stronger pressure pushes the surrounding gas outward, making the star expand and its radius grow by a factor of anywhere from 5 to several hundred, depending on the star's mass. But this expansion cools the outer layers, and so the star grows red (recall that cooler bodies radiate more strongly at longer, redder wavelengths). Because of its size and color, we call such a star a **red giant.** Thus, a main-sequence star becomes a red giant when it uses up the hydrogen in its core.

A giant star has a structure very different from that of a main-sequence star. As its name implies, its radius is much larger: hundreds of times larger than the Sun's in some cases. But this enormous size is deceptive. Although most of the star's *volume* is taken up by its very low-density, tenuous atmosphere, most of its *mass* lies in a tiny, hot, compressed core not much larger than the Earth. The core, or its surrounding shell source, supplies the star's energy. The huge envelope is relatively cool and the gas in it readily absorbs the photons flowing from the core toward the star's surface. This slows the photons, and so convection rather than radiation carries energy to the star's surface.

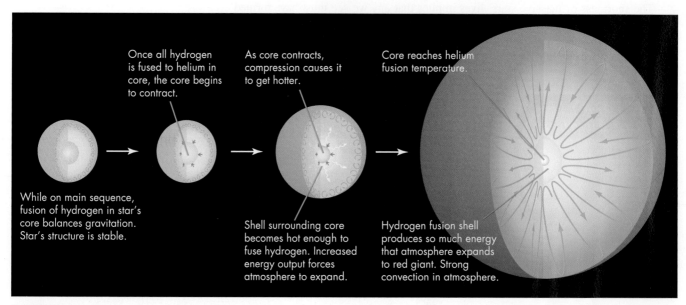

FIGURE 14.14
The end of main-sequence fusion and the start of the red giant phase.

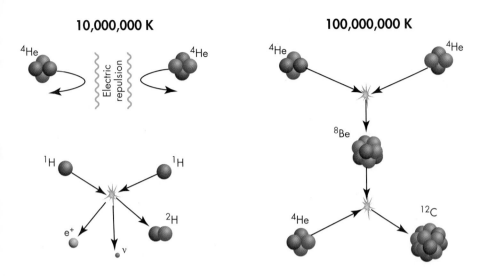

10,000,000 K

⁴He   Electric repulsion   ⁴He

¹H   ¹H

e⁺   ²H

ν

100,000,000 K

⁴He   ⁴He

⁸Be

⁴He   ¹²C

FIGURE 14.15
How a high temperature acts to overcome the electrical repulsion of the nuclei and brings them close enough for the nuclear force to fuse them.

Some of that energy continues to come from burning hydrogen, but many stars switch to a new energy supply: helium burning.

Nuclear Fuels Heavier than Hydrogen

Helium burning begins very differently in high-mass and low-mass stars and creates significant differences in their structures. To see why, we need to look in more detail at how nuclear fusion occurs and why it requires a very high temperature.

Two nuclei fuse when they are brought close enough together for the strong nuclear force to bind them to one another. That force, which is also what holds the protons and neutrons together in a nucleus, operates only over very short distances. If nuclei are more than a few diameters apart, the nuclear force is too weak to bond them. In fact, they will be repelled by the electrical force between the similarly charged protons, as illustrated in figure 14.15. That electric repulsion grows as the number of protons increases. It is a significant barrier for fusion of hydrogen, with its single proton, but even stronger for heavier elements, with their greater number of protons. It is much more difficult, therefore, for two helium atoms, with their two protons, to fuse than for two hydrogen atoms to do so.

Nuclei can overcome this electrical repulsion and fuse if they collide at a high enough speed. That speed is achieved in hot gases: the hotter a gas, the faster the nuclear particles move, and so the more easily the nuclei can fuse. Hydrogen can fuse at about 5 million kelvin, but helium, with its two protons, must be heated to about 100 million kelvin. Carbon, with its six protons, must be still hotter to fuse.

If a star's core reaches 100 million kelvin, helium nuclei there will fuse into carbon. The nuclear reaction combines 3 ⁴He into a ¹²C. Helium nuclei are also called "alpha particles," so this reaction is sometimes called the **triple alpha process,** or simply "helium burning." The reaction releases energy but only about one-tenth the amount per kilogram of fuel that hydrogen burning yields. Nevertheless, once a star has consumed all its core hydrogen, it must use its helium or die. How does a star make the switch from hydrogen to helium fusion? It must heat its core to 100 million K, and any star more massive than about 0.5 $M_\odot$ can do this by gravitational compression.

A high-mass star needs to compress its core only a little before helium begins to fuse, because its core is already extremely hot. Thus, helium burning begins easily for high-mass stars. A low-mass star such as the Sun, however, must compress its core enormously to make it hot enough for helium fusion. The compression of such a star packs its gas atoms so closely that they no longer behave like an ordinary gas—in technical terms, the gas becomes a **degenerate gas,** which responds to changes in temperature and pressure in unfamiliar ways.

Degeneracy in Low-Mass Stars

In a degenerate gas, the matter is so densely packed that the electrons interact not as ordinary charged particles but according to laws of subatomic physics. One consequence of these laws is that no more than two electrons of the same energy can be packed into the same submicroscopic volume, the size of which is determined by quantum physics. This packing limit makes a degenerate gas "stiff," so it behaves almost like a solid.

When a nuclear fuel begins to fuse in a normal gas, the energy released heats the gas, raising its pressure so that the gas expands. The expansion cools the gas, making fusion reactions slow down, acting like a thermostat that switches off the furnace when the house gets too hot. In normal stars this mechanism keeps them from collapsing or exploding. In degenerate matter, this thermostat mechanism fails.

For a low-mass star with a degenerate core, when a fuel ignites, the energy released does not raise the pressure because the pressure does not depend on the temperature. Thus, the gas does not expand and cool. Instead, it simply gets hotter and releases more energy, which makes it hotter still. Under these circumstances, the release of energy accelerates explosively in what is called the **helium flash.** In only a few minutes, the star's energy production increases several thousand times, yet this outburst is totally hidden from our view by the star's outer layers, just as a firecracker set off under a mattress creates little visible disturbance. However, the energy released heats the core enough to change it back to a normal gas, ending its degeneracy. With its degeneracy gone, the gas can now expand. That expansion in turn makes the star's outer layers adjust—shrinking, compressing, and heating. The star's reheated surface changes color from red to yellow, and the star becomes a yellow giant. The helium flash therefore ends the red-giant stage of low-mass stars.

Stars more than about twice as massive as the Sun do not have a helium flash. They are hot enough initially that their cores ignite helium with less compression before they become degenerate. They continue to expand and change gradually from yellow to red giants, but for both high-mass and low-mass stars, the onset of helium burning may trigger instabilities in the star and make it pulsate.

LOOKING UP

Aldebaran, a red giant shown in Looking Up #5 at the front of the book, has a mass estimated to be less than $2\,M_\odot$. It is probably in the stage of compressing its core toward helium-fusion temperature.

14.5 YELLOW GIANTS AND PULSATING STARS

Variable Stars

Not all stars have a constant luminosity. Stars that change in brightness are called **variable stars,** and many of these are associated with stages in the evolution of a star. For example, the red giant star Mira changes brightness over a regular period of 332 days (fig. 14.16). The T Tauri stars we discussed in section 14.2 are irregular variables.

When stars are plotted on the H-R diagram, some lie between the hot, luminous, blue, main-sequence stars and the red giants. Such stars are called "yellow giants." The most luminous yellow giants are aging high-mass stars. Less luminous yellow giants are old low-mass stars that have completed their first red-giant stage and are burning helium in their cores. Regardless of their mass, many yellow giants have the unusual property of swelling and shrinking rhythmically: they pulsate. As they change size, their luminosity changes, and so astronomers refer to them as "pulsating stars."

Two particularly important types of pulsating stars are the RR Lyrae (pronounced *lie-ree*) and Cepheid (pronounced *sef-ee-id*) variables. RR Lyrae stars have a mass comparable to the Sun's and are yellow-white giants with about 40 times the Sun's luminosity. The time it takes them to complete a pulsation cycle from bright to dim and back to bright is called their "period," as illustrated in figure 14.17. RR Lyrae variables have periods of about half a day and are named for RR Lyrae, a star in the constellation Lyra, the harp, which was the first star of this type to be identified.

FIGURE 14.16
Two images of the star Mira in the constellation Cetus. Mira varies from an easily visible star to about 100 times fainter.

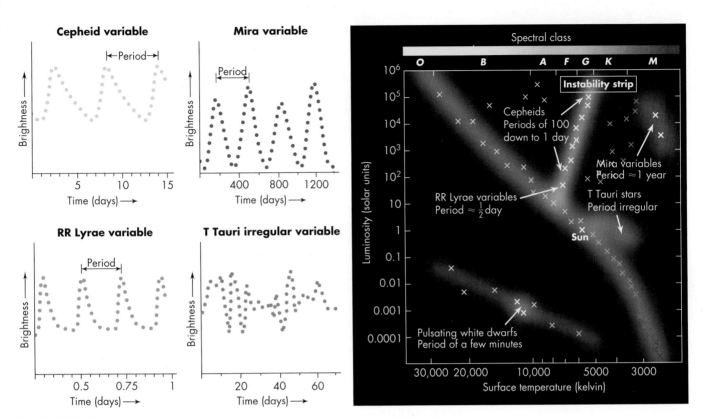

FIGURE 14.17

Brightness variations of various kinds of variable stars and their location in the H-R diagram. RR Lyrae and Cepheid variables are usually yellow in color and lie along a narrow "instability strip" in the diagram.

Cepheid variables are yellow supergiants that are more massive than the Sun and range in luminosity from several hundred times the Sun's luminosity to several tens of thousands times its luminosity. They are named for the star Delta Cephei, and their periods can be as short as about 1 day or as long as about 70 days.

Giant stars pulsate because their atmospheres trap some of their radiated energy. This heats their outer layers, raising the pressure and making the layers expand. The expanded gas cools, and the pressure drops, so gravity pulls the layers downward and recompresses them. The recompressed gas begins once more to absorb energy, leading to a new expansion. These stars continue alternately to trap and release the energy, and so they continue to swell and shrink, as shown in figure 14.18.

LOOKING UP 🔺

Delta Cephei, the prototype of Cepheid variables, is visible in Looking Up #1 at the front of the book.

FIGURE 14.18

Schematic view of a pulsating star.

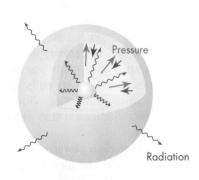

Radiation (purple wiggles) partially trapped. Pressure (blue arrows) increases, overcomes star's gravity (red arrows), and makes it expand.

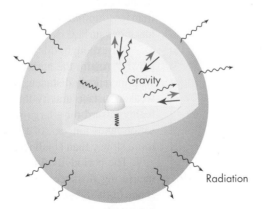

Expansion allows trapped radiation to escape. Star cools and pressure decreases. Gravity now exceeds pressure, so star contracts.

Pressure rises inside star as gas is compressed. After contraction halts, the cycle begins again.

A covered pan of water boiling on a stove behaves similarly. The lid will trap the steam so that pressure inside rises. Eventually the pressure becomes strong enough to tip the lid, and steam escapes. The pressure decreases, and the lid falls back. It again traps the steam, the pressure again builds up, and the cycle is repeated.

A similar process occurs in pulsating stars, with the role of steam played by the star's radiation and the role of the lid played by the star's atmosphere. For a star to trap radiation this way, its atmosphere must have special absorbing properties—technically called "opacity"—that occur only if its surface temperature and radius fall in a narrow range. That range, called the **instability strip,** is shown in the H-R diagram in figure 14.17. In this region the star's opacity "puts a lid on" light coming out of the star's core. The atmosphere then expands and cools until its opacity declines enough to let the light out. The atmosphere compresses and heats, and the cycle repeats. A star that assumes these characteristics as it evolves through the H-R diagram will begin to pulsate and will continue to pulsate until its temperature or radius changes enough to remove it from the instability strip.

When a low-mass star crosses the region of the instability strip during its evolution as a red giant, it becomes an RR Lyrae variable star. When a high-mass star crosses the instability strip in its red giant phase, being more luminous it instead becomes a Cepheid variable. The amount of time a given star spends in the instability strip depends on its mass. Massive stars such as the highly luminous Cepheids evolve across the strip in less than 1 million years, but some cross the region several times as their interior structures alter. Low-mass stars such as the RR Lyrae variables spend more time in the strip, perhaps a few million years, but cross it less often. In either case, stars pulsate for only a brief portion of their lives.

Astronomers have identified many other types of pulsating variables besides Cepheids and RR Lyrae stars. For example, ZZ Ceti stars, a kind of pulsating white dwarf with periods as short as a few minutes, are found in the lower portion of the H-R diagram. There are many stars like Mira (known as Mira variables) that have pulsation periods of about 1 year. These lie in the upper right of the H-R diagram. The Sun will probably become a Mira variable near the end of its lifetime, as we will discuss in section 14.6.

Q: What is the approximate luminosity of a Cepheid that has a 3-day period?

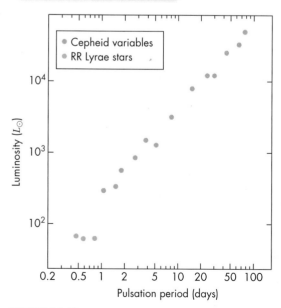

FIGURE 14.19

The period–luminosity relation. More-luminous stars tend to pulsate more slowly.

The Period–Luminosity Relation

Many pulsating variable stars obey a law that relates their luminosity to their period—the time it takes them to complete a pulsation. Observations (and theoretical calculations) show that the more slowly a star pulsates, the more luminous it is (fig. 14.19). This so-called **period–luminosity relation** arises because, other things being equal, more-luminous stars have a larger radius than less-luminous ones. This follows simply from the fact that a bigger star has more surface, so it can emit more light (see section 13.3).

Why, though, does the larger radius make the larger star pulsate more slowly? Again, the answer is simple. Because an object's gravity weakens with increasing distance, other things being equal, a large-radius star has a weaker surface gravity than a small-radius star. Hence, gravity pulls more feebly inward for a large-radius star than for a small-radius one, so its pulsation takes longer. As a result, big (and therefore luminous) stars pulsate more slowly than small (and therefore dim) stars.

The period–luminosity relation gives astronomers a powerful tool for measuring distances. By measuring the star's period and identifying the shape of its light curve with stars of the same class, astronomers can find its luminosity. From its luminosity and apparent brightness, they find the star's distance, as discussed in chapter 13. We will see in chapters 16 and 17 some specific applications of this method to measuring the distances of galaxies.

14.6   DEATH OF STARS LIKE THE SUN

Once the Sun becomes a yellow giant, it will be nearing the end of its life, and as death approaches, its evolution will speed up. The evolution will be faster because helium yields less energy than hydrogen when it is burned, and the Sun will be more luminous, so it will burn it faster. During the time it consumes the helium in its core, the star will lie between the main sequence and the giant region in the H-R diagram, a part of the diagram known as the horizontal branch.

Ejection of a Low-Mass Star's Outer Layers

As the core burns its helium into carbon, its radius will once again shrink, compressing and heating the gas there. That compression is unable to make the core hot enough to burn carbon, but it does make it hot enough to increase significantly the rate at which the star burns its fuel. The more rapid fuel consumption makes the star more luminous, inflating it to an even larger radius than before, making it a bright red giant. As the star swells in size, its outer layers cool to about 2500 K, becoming a Mira variable.

The atmosphere is so cool that carbon and silicon atoms condense and form "flakes," just as water in our atmosphere forms snowflakes when it cools. These carbon and "rock" flakes (technically, what are called grains) do not fall into the star, however. Rather, they are pushed *outward* by the flood of photons pouring from the star's luminous core, just as dust in a comet is blown into a tail by the Sun's radiation. The rising grains drag gas with them and drive it into space, forming a huge shell around the star (fig. 14.20A).

The star Mira, prototype of the Mira variable class, is in this phase, and is shedding gas that streams out behind it as it travels through interstellar space (fig. 14.20B). Each pulsation of Mira probably sends an outflow of gas and dust, and the buildup of opaque dust grains in the star's atmosphere during each pulsation cycle is part of the reason this class of stars shows such a dramatic change in brightness (section 14.5). The carbon grains (as well as other solid particles that condense in the star's cool atmosphere) survive their trip out of the star and drift through space. Some have been found in meteoritic material within our Solar System—"fossils" from a bygone star.

As the former atmosphere expands, it thins and becomes transparent, allowing us to see through to the star's furiously hot core. Because it is so hot, the core's radiation is rich in ultraviolet light, which heats and ionizes the inner edge of the shell of gas it has expelled, making it glow. These shells of glowing gas can be quite beautiful, but they mark the death of a star.

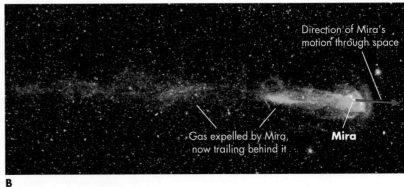

Carbon "flakes" form in red giant's outer atmosphere.

Radiation pushes on flakes. Flakes push on gas, pushing away outer atmosphere.

Direction of Mira's motion through space

Gas expelled by Mira, now trailing behind it

Mira

A   **B**

FIGURE 14.20

Artist's conception of how radiation pressure in a luminous red giant pushes on carbon flakes in its atmosphere and drives off the gas. Magnetic forces probably play a role, too. The ejected gas forms a shell around the star—a planetary nebula.

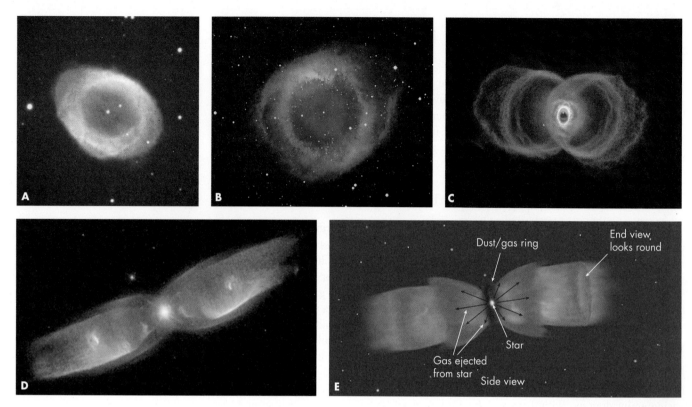

FIGURE 14.21
Pictures of several planetary nebulas. (A) Helix Nebula. (B) Ring Nebula. (C) Hourglass Nebula. (D) Butterfly Nebula. Notice the central star in each. Other stars that look as if they are inside the shell are foreground or background stars. (E) Sketch of how such a curious object as (D) might form.

Q: What will eventually happen to the central star? What will eventually happen to the nebula itself?

The Planetary Nebula Stage

Astronomers call the glowing shell of gas around a dying star a **planetary nebula.** This term is unfortunate because a planetary nebula has nothing to do with planets. The name survives from times when observers had only poor telescopes through which a planetary nebula looked like a small disk that might be mistaken for a planet, but the name can be confused with the solar nebula out of which the Solar System formed.

For many years, astronomers thought that the typical planetary nebula was ejected uniformly in all directions so that it formed a huge "bubble" around the star's core, as you might infer from figure 14.21A. Astronomers knew of oddly shaped planetary nebulas but thought them to be the exception.

This view has changed markedly over the last decade, especially now that more detailed pictures (such as fig. 14.21C, made with the Hubble Space Telescope) are available. Many such pictures show that the shell is not spherical and that the gas has been ejected mainly along two directions, presumably the star's rotation axis. Such ejection creates not a bubble but two oppositely directed cones, as sketched in figure 14.21E. Moreover, many planetary nebulas have narrow jets of hot gas that are directed more or less along the cones, as you can see in figure 14.21D. Why this happens is unclear, but it has forced astronomers to rethink their hypotheses about how planetary nebulas form.

For example, some astronomers think that magnetic fields create the narrow jets and that the fields also create a pressure that helps expel the gas. Such magnetic effects occur on our Sun and help expel gas from it, although on a much smaller scale. Another hypothesis is that a ring of dust and gas in orbit around the star's equator blocks

LOOKING UP

Two bright planetary nebulae in the Summer Triangle, easily seen with a small telescope, are shown in Looking Up #4 at the front of the book.

the ejection in those directions. Such a ring might form if a companion star's gravity draws some of the ejected material into a disk around both stars.

Whether such disks or rings give shape to all planetary nebulas is still unclear, but some of those that look circular merely have a rotation axis that happens to point toward us so we see their cones end-on, not from the side. Thus, we see how better data may require new ideas for its interpretation.

A planetary nebula shell contains about one-fourth of a solar mass of glowing gas but may have as much as several solar masses of cooler, nonluminous gas around it. Shells are typically about one-fourth of a light-year in diameter and expand at about 20 kilometers per second. We can see this expansion by comparing photographs taken several decades apart. The shells eventually grow so big and diffuse that they mingle with the interstellar medium. Thus, the cycle of a star's life is complete: gas to star and back to gas. The Sun will spend about 10 billion years consuming its hydrogen and becoming a red giant. It will spend only about 100 million years (1% of its main-sequence lifetime) consuming its helium. Its lifetime as a planetary nebula is measured in thousands of years, but the core of the star remains behind as a tiny, glowing ball that astronomers call a "white dwarf." Figure 14.22 shows these changes in an H-R diagram, affording a shorthand view of the life of a star like the Sun.

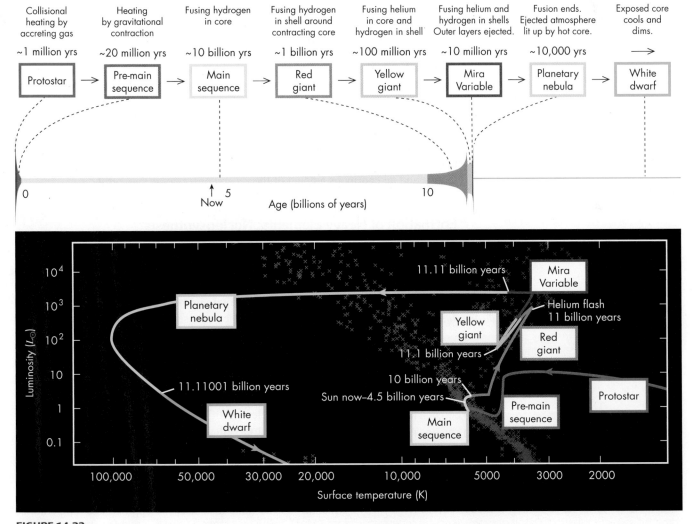

FIGURE 14.22

The evolution of the Sun plotted in an H-R diagram. The time line along the top illustrates the relative times spent in each stage and gives a general idea of the changing size of the star (not to scale). The fuel powering the star at each stage is described above the time line.

The Fates of Other Low-Mass Stars

Calculations indicate that all stars that begin their lives with less than 8 $M_\odot$ end up as white dwarfs, but their life stories may differ a little from the Sun's. The exact evolution depends on how rapidly the star loses mass in its red giant phase, and this is difficult to predict precisely. However, it appears that low-mass stars that begin with more than a few solar masses may even grow hot enough in their cores to fuse carbon into heavier elements toward the end of their lives. On the other hand, stars that have less than half the Sun's mass may never ignite helium, and their mass loss may be too little and too slow to form a planetary nebula.

The total lifetimes of low-mass stars vary greatly, from several *trillion* years for the lowest-mass stars to under 100 million years at the upper end of the mass range. The white dwarfs they leave behind all have masses less than about 1.4 $M_\odot$, and although they initially glow white hot from the leftover heat from fusion, they gradually darken and fade away over billions of years. A very different fate, however, awaits a more massive star.

14.7   OLD AGE OF MASSIVE STARS

Stars that begin their lives with more than 8 solar masses do not become planetary nebulas or white dwarfs. Their great mass compresses and heats their cores enough to ignite carbon and allows them to keep burning when their helium is gone. A variety of other nuclear reactions may also occur that create heavy elements and at the same time supply energy for the star. Astronomers call the formation of heavy elements by such nuclear burning processes **nucleosynthesis.** Nucleosynthesis in stars was discussed in detail by E. Margaret Burbidge, Geoffrey Burbidge, William Fowler, and Fred Hoyle, who proposed in 1957 that *all* the chemical elements in our Universe heavier than helium were made in this way. Their idea is one of the triumphs of stellar evolution theory, and an enormous amount of theory and observation support it.

Formation of Heavy Elements: Nucleosynthesis

Heavy elements form when two or more light nuclei combine and fuse into a single heavier one. The new nucleus must have a number of protons + neutrons equal to the sum of the numbers found in its parent nuclei. Usually the total number of protons and neutrons each remain the same separately, but in some reactions, such as the proton–proton chain, a proton may turn into a neutron or a neutron into a proton.

For example, suppose ^{12}C and ^{16}O fuse. Carbon has 6 protons and 6 neutrons, and oxygen has 8 protons and 8 neutrons. Their fusion makes a nucleus with $6 + 8 = 14$ protons and $6 + 8 = 14$ neutrons (that is, an atom with 14 protons and 28 particles total in the nucleus). Reference to a table of the elements shows that such an atom is silicon, ^{28}Si. Table 14.1 shows some of the primary reactions that make heavy elements in high-mass stars. The table also lists the approximate temperatures needed for the reactions and the length of time the fusion lasts in a 25-$M_\odot$ star before that fuel is exhausted.

Table 14.1	Sequence of Major Fusion Reactions in Massive Stars	
Fusion Reaction	**Minimum Temperature**	**Duration in 25-$M_\odot$ Star**
$4\,^1\text{H} \rightarrow\ ^4\text{He}$	5,000,000 K	7,000,000 years
$3\,^4\text{He} \rightarrow\ ^{12}\text{C}$	100,000,000 K	700,000 years
$^{12}\text{C} +\ ^4\text{He} \rightarrow\ ^{16}\text{O}$	200,000,000 K	
$^{12}\text{C} +\ ^{12}\text{C} \rightarrow\ ^{20}\text{Ne} +\ ^4\text{He}$	600,000,000 K	300 years
$^{20}\text{Ne} +\ ^{20}\text{Ne} \rightarrow\ ^{24}\text{Mg} +\ ^{16}\text{O}$	1,500,000,000 K	8 months
$^{16}\text{O} +\ ^{16}\text{O} \rightarrow\ ^{28}\text{Si} +\ ^4\text{He}$	2,000,000,000 K	3 months
$^{28}\text{Si} +\ ^{28}\text{Si} \rightarrow\ ^{56}\text{Fe}$	2,500,000,000 K	1 day

KEY: H, hydrogen; He, helium; C, carbon; O, oxygen; Ne, neon; Mg, magnesium; Si, silicon; Fe, iron.

In massive stars, these reactions provide enough energy to support the stellar mass against gravity's unceasing pull. As each fuel—hydrogen, helium, carbon, and so forth—is exhausted, the star's core contracts and heats by compression. The higher temperature then allows the star to burn still-heavier elements, the "ashes" of one set of nuclear reactions becoming the fuel for the next set. Oxygen, neon, magnesium, and, eventually, silicon are formed, but because progressively higher temperatures are needed for each new burning process, each new fuel is confined to a smaller and hotter region around the star's core. Thus, the star develops a layered structure, as illustrated in figure 14.23.

The star's surface, where the temperature is too low for nuclear reactions to occur, remains hydrogen, but beneath the surface lies a series of nested shells, each made of a heavier element than the one surrounding it. Helium surrounds a shell of carbon, which in turn surrounds a shell of neon, and so on to the core. By the time the star is burning silicon into iron, its core has shrunk to a diameter smaller than the Earth's, and the core's temperature is about 2.5 billion K. A very massive star ($20\,M_\odot$ or more) may take less than 10 million years from its birth to form an iron core. This brief lifetime results because (1) the star burns its fuel rapidly to offset the energy lost because of its high luminosity, and (2) the structure of atomic nuclei is such that fusion of elements heavier than hydrogen yields much less energy than the fusion of a similar amount of hydrogen, and so more fuel must be burned to supply the same amount of energy.

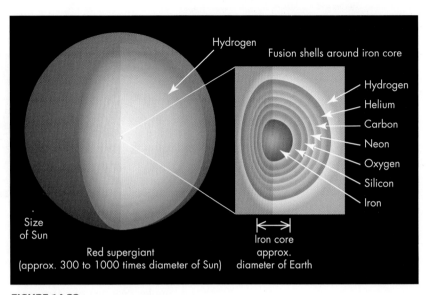

FIGURE 14.23

The layered structure of a massive star as it burns progressively heavier fuels in its core. (These inner zones are enormously magnified for clarity.)

Betelgeuse, shown in Looking Up #6 at the front of the book, is a red supergiant. Astronomers are uncertain whether it has completed its helium fusion or not.

Core Collapse of Massive Stars

The formation of an iron core signals the end of a massive star's life. Iron cannot fuse and release energy: the iron nucleus turns out to be the most tightly bound of all nuclei. As a result, attempting to fuse additional protons or neutrons to it *weakens* the bonds and *absorbs* energy, rather than releasing it. Thus, nuclear fusion stops with iron, and a star with an iron core is out of fuel.

Without fuel, the star's core begins to shrink and heat. But in high-mass stars, the shrinkage presses the iron nuclei so tightly together that a new reaction can occur: protons and electrons may themselves "merge," neutralizing their charge and becoming neutrons. The shrinking core is thus transformed from a sphere of iron into a sphere of neutrons, with catastrophic results for the star. Most of the pressure that supported the core was supplied by electrons, but they have been absorbed by protons, so the star's core pressure suddenly drops. Nothing remains to support the star, and so its interior begins to collapse. Because the matter is so dense, its force of gravity is immense, and it crushes the core. In less than a second, the core drops from an iron ball the size of the Earth to a ball of neutrons about 10 kilometers (about 6 miles) in radius. The star's outer layers, with nothing to support them, plummet inward like a tall building whose first floor is suddenly blown away. The plunging outer layers of the star strike the neutron core, crushing it still more, while the impact heats the infalling gas to billions of kelvin. The pressure surges and lifts the outer layers away from the star in a titanic explosion in which a star can briefly shine as brightly as all of the stars in a galaxy combined (fig. 14.24). Astronomers call this remarkable event a **supernova** explosion.

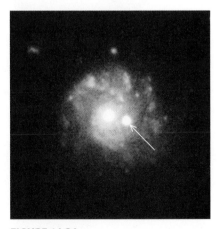

FIGURE 14.24

A supernova explosion in a distant galaxy observed in 2002 shines brightly compared to the billions of stars in the galaxy.

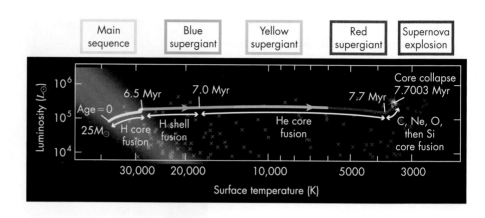

Main sequence		Blue supergiant		Yellow supergiant		Red supergiant		Supernova explosion

FIGURE 14.25

A massive star's evolution through the H-R diagram. Massive stars usually begin fusing helium in their cores while they are still moving across the H-R diagram, but depending on their mass and heavy-element composition, this may occur when the star is in its blue, yellow, or red supergiant phase.

Supernova Explosions

A supernova explosion marks the death of a massive star,* and we chart its complete life in the H-R diagram shown in figure 14.25. We can see that its path through time is far simpler than that of a low-mass star. The massive star begins its life on the upper main sequence as a blue star. As its core hydrogen is consumed, it leaves the main sequence, swelling and cooling to become a yellow supergiant while it begins to burn helium in its core. When the helium there is consumed, the star switches to other fuels, all the while contracting and heating its core and expanding and cooling its outer layers to become a red supergiant. Overall, its luminosity stays approximately constant as the star ages, but some massive stars, depending on how many heavy elements are incorporated in them at their birth, may return partway to the main sequence as they switch to new fuels. The resultant heating may turn them into blue supergiants before they explode as supernovas. Such was the behavior of Supernova 1987A, whose explosion is documented in figure 14.26.

For the star, the supernova explosion is a quick and glorious death. In a few minutes, it releases more energy than it has generated by nuclear burning during its entire existence. In a matter of hours, the dying star brightens to several billion times the luminosity of the Sun, radiating as much energy in its death throes as the Sun emits over its entire lifetime.

A supernova emits more than just visible light: most of the energy of its blast is carried by a burst of neutrinos, the same tiny, highly penetrating particles generated in the Sun as it converts hydrogen into helium. Just as the Sun's neutrinos escape freely into space, the supernova's neutrinos escape as well. A pulse of such neutrinos was detected in February 1987 when Supernova 1987A blew up in the Large Magellanic Cloud. Despite its enormous distance, more than a trillion neutrinos from that explosion passed through each person on Earth.

LOOKING UP

The Large Magellanic Cloud is visible in Looking Up #9 at the front of the book.

* We will see in section 15.1 that a low-mass star can also undergo a supernova explosion if it is a member of a binary star system.

FIGURE 14.26

Photograph of a portion of the Large Magellanic Cloud (a small, nearby galaxy) before Supernova 1987A exploded, and a second photograph taken during the event.

FIGURE 14.27
Several supernova remnants arranged according to their age: (A) *Chandra* X ray image of Cassiopeia A supernova remnant, which is about 300 years old. In this image, the red on the left is radiation from hot iron atoms and the bright green is radiation from silicon and sulfur atoms. (B) Hubble Space Telescope image of the Crab Nebula, which is about 1000 years old. (C) Hubble Space Telescope image of a supernova remnant that is several thousand years old.

Supernova Remnants

The explosion of a massive star mixes the elements synthesized by nuclear burning during its evolution with the star's outer layers and blasts them into space. This incandescent spray expands away from the star's collapsed core at more than 10,000 kilometers per second. Depending on the star's mass, 10 or so solar masses of hot gas may be flung outward. In addition to being enriched by heavy elements produced in the star's core and surrounding fusion shells during its lifetime, the supernova outburst itself generates even more heavy elements. The core collapse creates neutrons, many of which are blasted outward, where they rapidly combine with other atoms in the star to build up heavy and rare elements such as gold, platinum, and uranium.

Gas ejected by the supernova blast plows through the surrounding interstellar space, sweeping up and compressing other gas that it encounters. The huge, glowing cloud of stellar debris—a **supernova remnant**—steadily expands. At the end of one year, it is 300 billion kilometers (about 0.03 light-years) across. At the end of one century, it is several light-years in diameter, but its expansion slows as it runs into more surrounding gas. Figure 14.27 shows pictures of several supernova remnants of different ages. Notice how ragged the remnants in figure 14.27 look compared with the smoothly ejected bubbles of planetary nebulas—evidence that massive stars die more violently than low-mass stars.

One such violent outburst was seen about 1000 years ago by astronomers in China and elsewhere in Asia. On a date that according to our calendar was July 4, 1054, the records describe a "guest star" that appeared in the evening sky and that was visible even in broad daylight for several weeks. Today, with even a small telescope, you can still see the glowing gases ejected from that dying star. Figure 14.27B is a picture of this perhaps most famous of all supernova remnants, now known as the Crab Nebula. In chapter 15 we will learn that the star's core survived the blast and lies in the expanding debris.

Supernova remnants eventually slow down, cool, and mingle with the interstellar clouds around them. The compression of interstellar clouds caused by the supernova remnant running into them can trigger their gravitational collapse, leading to a new round of star formation. When the remnant's gas, rich in heavy elements, mixes with the interstellar gas, it also creates interstellar clouds enriched in heavy elements. These new stars will, therefore, contain more heavy elements than earlier generations of stars. In this way, the amount of heavy elements in stars increases from generation to generation.

What remains when the supernova remnant dissipates? For stars that began their lives with less than about 20 $M_{\odot}$, the core of the star—a few solar masses—survives as a ball of neutrons (a neutron star). Astronomers think that more-massive stars may end up as even more compressed bodies—black holes—as we will discover in chapter 15.

LOOKING UP

The location of the Crab Nebula is shown in Looking Up #5 at the front of the book.

14.8 HISTORY OF STELLAR EVOLUTION THEORIES

Aristotle speculated more than 2000 years ago that stars are heated by their passage through the heavens, with no thought of their evolution. In the eighteenth century, the German philosopher Immanuel Kant described stars slightly more scientifically, envisioning the Sun as a fiery sphere, formed when gases gravitated to the center of the solar nebula. Developing modern theories about how stars work and evolve required the development of far more complex physical models than ever before.

The Development of Astrophysical Models of Stars

The English physicist Lord Kelvin and the German physicist Hermann von Helmholtz were the first scientists to use the equations of basic physics to arrive at a mathematical description of what we would today call "model stars." In the 1850s and 1860s they independently solved the mathematical equations for a gas sphere in which pressure is balanced by gravity and showed that a star's core is compressed by the weight of the overlying gas and is much denser than the outer parts of the star. They further showed that as gravity compresses a star, that compression converts gravitational energy into heat, raising the star's temperature. This proves to be the main power source during a star's pre-main-sequence phase, but far too little to power a star for billions of years.

Astronomers had to await discoveries about the nature of atoms in the early years of the twentieth century before they could solve the mystery of stellar energy. The British astrophysicist Sir Arthur Stanley Eddington, who discovered the mass–luminosity relation, was one of the first scientists to recognize the immense power available in the form of nuclear energy, but at the time too little was known about how atoms interacted to make models of stars powered by hydrogen burning. Moreover, Eddington discovered that to make a model star, more was required than simply balancing pressure and gravitational forces. He demonstrated that a correct model must also account for the outward leakage of energy from the star's core to its surface, a process that made solving the equations much more difficult.

The equations that govern a star's structure express a complicated set of relations between the star's pressure, density, temperature, gravitational force, energy generation, composition, and luminosity. The equations must balance at each point in the star and yield answers that agree with observations. Solving such complex equations became easier in the 1940s when astronomers began using mechanical calculators. These machines—incredibly primitive compared with a modern computer—were operated by turning a hand crank. Nevertheless, they were faster than pencil-and-paper solutions, and the models so produced revealed more clearly the physical processes in stars. Then, in the 1950s, when astronomers began to use electronic computers to calculate model stars, only a few minutes were needed to solve the equations of a stellar model. By combining the results of dozens of such models, each of which represented a star with successively more of its precious fuel consumed, astronomers could see how stars evolved. Such work continues to this day, with astronomers still trying to understand the late stages of a star's life, when a star is losing mass and burning elements heavier than helium in its core.

Testing Stellar Evolution Theory

Although we now understand that main-sequence stars turn into red giants, as recently as the 1940s some astronomers thought the reverse. What evidence supports our current theories of stellar evolution? The best demonstration that the modern theory is correct comes from comparing the H-R diagrams of real groups of stars with theoretically determined H-R diagrams (see the Extending Our Reach box).

EXTENDING
our reach

MEASURING THE AGE OF A STAR CLUSTER

Astronomers can deduce the age of a star cluster from the shape of its H-R diagram. To understand why its H-R diagram reveals the cluster's age, recall that in a star cluster, all the stars form at approximately the same time. Moreover, at their birth, all the stars are burning hydrogen in their core. Because the main sequence is determined by the location of stars burning hydrogen in their core, all the stars in a newly formed cluster lie on or near the main sequence. Not all the cluster's stars burn their fuel at the same rate, however. To maintain their high luminosity, massive stars burn their fuel more rapidly than low-mass stars do. With their hydrogen used up, high-mass stars leave the main sequence and turn into red giants. The low-mass stars, on the other hand, still have hydrogen to burn, and so they remain on the main sequence longer.

Because the high-mass stars in an old cluster turn into red giants (and thus lie off the main sequence), a line in the H-R diagram connecting the position of the cluster's stars bends away to the right, as shown in figure 14.28. The point where that line bends away from the main sequence is called the **turnoff point**. At the turnoff point, a star is not quite old enough to have used up the hydrogen in its core. That is, its age is just a tiny bit less than its main-sequence lifetime. But all stars in the cluster are the same age—namely, the age of the cluster. To get the cluster's age, we therefore measure the age of the star at the turnoff point by calculating its main-sequence lifetime

FIGURE 14.28
The turnoff point in the H-R diagram of a star cluster indicates its age.

from its mass and luminosity, as described in section 14.3. The answer we get is the age of the star cluster.

Figure 14.29 shows a series of H-R diagrams for clusters ranging in age from very young to very old (a few million years for the youngest to more than 10 billion years for the oldest). Notice that old clusters have few, if any, stars on the upper part of the main sequence. Young clusters, on the other hand, do have stars on the upper main sequence.

FIGURE 14.29
Schematic H-R diagrams of clusters of different ages, illustrating the turnoff point and how it shifts down and to the right for older clusters.

We learned in section 14.2 that stars probably form in groups—star clusters. The stars in a cluster are bound together by their mutual gravitational attraction, although each star moves along its own orbit within the cluster. All the stars within a given cluster form at approximately the same time and are therefore nearly the same age. If we look at the stars in a cluster shortly after the cluster's birth, when they have all completed their contraction stage from protostars, they will all lie on the main sequence.

If we look at some later time, some of the stars will have consumed their core hydrogen and evolved off the main sequence. To check our theory of stellar evolution, we calculate evolutionary tracks, such as those in figure 14.2, for every star on the main sequence to show where each star will be at, say, 10 million years, 100 million years, and so on. The resulting curves show us what the H-R diagram of the entire cluster should look like 10 million years, 100 million years, and so on, after its birth. When we compare such curves with H-R diagrams of actual star clusters, the match is excellent. If our theory of stellar evolution were wrong, the shapes would be unlikely to agree so well. In addition, the models help astronomers understand why so many kinds of stars exist. For example, the models not only offer a natural explanation of main-sequence, red giant, and white dwarf stars but also help us see how such totally different objects, such as pulsating variables, planetary nebulas, and supernovas, fit into the scheme of stellar evolution. This success in interpreting such stellar diversity is evidence that the theory is essentially correct. Moreover, that success offers astronomers a way to measure the age of a star cluster.

SUMMARY

A star forms from interstellar gas drawn together by gravity, which compresses and heats the gas to form a protostar. Further heating causes the core of the protostar to fuse hydrogen into helium. The energy released keeps the core hot and thereby maintains an outward pressure that stops the protostar's collapse and changes it into a main-sequence star. When the hydrogen is exhausted in a star's core, the core shrinks and heats, making the star swell into a red giant.

A low-mass star such as the Sun continues to burn hydrogen in a shell during its first red giant stage but eventually compresses the helium in its core enough to ignite the gas in a helium flash. The energy released expands the core, and the outer layers shrink, turning the star into a yellow giant. As the star burns helium, its core again shrinks, and it once more becomes a red giant. It is so luminous now that its radiation drives off its outer layers to form a planetary nebula shell. The core remains as a small, dense star: a white dwarf.

Because of the tremendous heating from gravitational compression of their cores, high-mass stars more easily ignite fuels heavier than hydrogen and undergo less dramatic changes initially than low-mass stars do. They build in their cores layers of heavy elements: carbon, neon, silicon, and, eventually, iron. The iron core cannot burn and will collapse, triggering a supernova explosion. The heavy elements produced during the star's life and in the supernova that destroys it are blasted into space, enriching the interstellar medium.

The lifetime of a star depends on its mass. High-mass stars burn their fuel quickly to supply the energy they need to support their great weight. These stars are blue because they are so hot. Thus, blue stars burn out quickly and must be young.

Both high- and low-mass stars pass through a stage late in their lives when they pulsate as yellow giants. Low-mass stars become RR Lyrae variables, whereas high-mass stars become Cepheids. Table 14.2 summarizes these stages. Some stars vary in brightness and are called variable stars. Many of these variables lie in a narrow region of the H-R diagram called the "instability strip."

Table 14.2 Evolution of Low-Mass and High-Mass Stars	
Low-Mass Star (Mass Less Than 8 $M_\odot$)	**High-Mass Star (Mass More Than 8 $M_\odot$)**
Protostar (gravity supplies energy)	Protostar (gravity supplies energy)

Main sequence (hydrogen fuses in core)
    Core hydrogen is consumed.
	Main sequence (hydrogen fuses in core)
	    Core hydrogen is consumed.

Red giant (hydrogen fuses in shell)
    Core contracts, heats, and ignites helium (helium flash).
	Yellow supergiant (helium fuses in core)
	    Star pulsates.

Yellow giant (star pulsates)
Red giant (helium fuses in shell)
    Outer layers are ejected.
	Red supergiant (heavier elements fuse in a series of shells in the star's core)
	    Iron core builds up and collapses.

Planetary nebula
    Core cools to become white dwarf.
	Supernova explosion when core collapses;
	    remnant may be a neutron star or black hole.

QUESTIONS FOR REVIEW

1. (14.1) What processes and forces determine the structure of stars?
2. (14.1) Through what stages will the Sun evolve? Through what stages will a high-mass star evolve?
3. (14.2) What heats a protostar? How can we observe protostars? Why are they surrounded by dust and gas?
4. (14.2) What is a bipolar flow?
5. (14.2) What is a T Tauri star?
6. (14.2) What is a Bok globule?
7. (14.2–14.3) What determines when a star becomes a main-sequence star?
8. (14.3) How long do stars stay on the main sequence?
9. (14.4) What makes a star move off the main sequence?
10. (14.4–14.5) Where do main-sequence stars end up as they evolve?
11. (14.4) Why is it easier for a high-mass star than for a low-mass star to burn helium?
12. (14.4) Why do high- and low-mass stars evolve differently as they become red giants?
13. (14.5) What is a variable star? What is meant by the period of a variable star?
14. (14.5) Where in the H-R diagram are variable stars found?
15. (14.5) What is meant by a pulsating star? Why do stars pulsate?
16. (14.4/14.6) What happens to a solar-mass star when it starts to burn helium in its core? What does it turn into?
17. (14.6) What is a planetary nebula?
18. (14.6) What is one explanation for how a low-mass star expels its outer layers to make a planetary nebula?
19. (14.6) What is left when a planetary nebula dissipates?
20. (14.7) What makes a high-mass star's core collapse?
21. (14.7) Why do neutrons form in a massive star's iron core?
22. (14.7) What is a supernova explosion?
23. (14.7) What kind of subatomic particles have been observed when a supernova explodes?
24. (14.8) How are clusters of stars used to test theories of stellar evolution?

THOUGHT QUESTIONS

1. (13.6 /14.1) Stars of all colors are visible to the naked eye, from Betelgeuse's red to Sirius's white to Spica's blue-white. If, late in their lives, stars did not evolve into giants but just became dimmer and dimmer, what would you expect for the relative numbers of blue, white, yellow, and red stars visible in the night sky? Would any kind of yellow stars be visible to the naked eye?

2. (14.2) Inflate a balloon and carefully measure its size. Put it in the freezer for a few hours. Does it look the same when it is cold? How does this relate to how stars form in cold regions of space?
3. (14.2) Describe how a protostar is believed to form.
4. (14.2–14.4) Take a plastic bottle and put a little soapy water in it. Run your finger across the mouth to make a soap film. Now, without breaking the film, run hot water over the bottle. What happens to the soap film? How does this relate to what happens to a star when it is heated?
5. (14.3–14.5) Suppose that stars could mix unburned fuel from their outer layers into their cores. Would that alter the way they evolve? Can you suggest some possible differences?
6. (14.6) Hold a clear cylindrical water glass below a bright light so that it casts a shadow on a piece of paper. If you hold the glass at different angles, you can create different shadow patterns. Can you produce shapes similar to the planetary nebula images in figure 14.21? Is there another shape that would create patterns more similar to some images?
7. (14.7) Hold a small rubber ball on top of a basketball, and drop them together toward the floor. What happens to the small ball? Does that help you understand what happens to the outer layers of a supernova as they collapse on the core?
8. (14.7) Thinking about why massive stars start fusing heavy elements, and examining the periodic table in the back of the book, explain why there should be more neon, magnesium, sulfur, argon, and calcium in a supernova remnant than fluorine, sodium, phosphorus, and chlorine.
9. (14.7) Explain why gold and platinum are much scarcer than iron and silicon (found in most rocks), thinking about how massive stars live and die.
10. (14.7/14.8) In some very old, dense star clusters, a few blue stars, known as "blue stragglers," are seen. Why are blue stars unexpected in such clusters? If stars that stick together mix their material thoroughly when they collide, how might you explain such blue stars?

PROBLEMS

1. (14.1) Sketch an H-R diagram of what you expect the evolutionary paths of a 2-$M_\odot$ and a 15-$M_\odot$ star to look like.
2. (14.2) The density of a cold interstellar cloud prior to collapsing might be about 1.67×10^{-20} kg/m^3. How many cubic meters of this cloud would contain enough mass to build the Sun? What is the ratio of this volume to the volume of the Sun? What is the volume of cloud required if written in cubic light-years?

3. (14.2) The jets from a T Tauri star are measured to have speeds of about 300 km/sec. At this speed, how long would it take the ejected material to travel 2 light-years (similar to the jets shown in figure 14.8)?

4. (14.3) Calculate the main-sequence lifetime of the Sun by first determining the rate at which it burns hydrogen (refer to chapter 12) and then dividing that rate into its core mass. If your answer is in seconds, convert to years, given that there are about 3.2×10^7 seconds in a year.

5. (14.4/14.6) Calculate the escape velocity from a red giant's atmosphere (use the formula for escape velocity from chapter 3). Assume that the star's mass is $1\,M_\odot$ and its radius is $100\,R_\odot$. How does this compare with the speed at which a planetary nebula shell is ejected?

6. (14.6) Calculate the average density of a red giant star in grams per cubic centimeter. Take the star's mass to be 2×10^{33} grams and its radius to be 3×10^{12} centimeters. How does this density compare with that of the air we breathe (about 10^{-3} gm/cm^3)?

7. (14.6) How long will it take a planetary nebula shell moving at 20 km/sec to expand to a radius of one-fourth of a light-year?

8. (14.7) Calculate the Doppler shift of X rays emitted from shocked silicon in a supernova remnant if the ejected material is moving at 5000 km/sec and the X ray line is at 0.688 nanometers.

9. (14.8) Figure 14.30 shows the H-R diagram of a star cluster. How old is it? (The dashed line is the position of the stars. The solid line is the main sequence.)

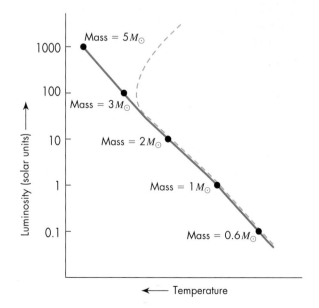

FIGURE 14.30
H-R diagram of a star cluster.

TEST YOURSELF

1. (14.1) Which of the following sequences correctly describes the evolution of the Sun from young to old?
   (a) White dwarf, red giant, main-sequence, protostar
   (b) Red giant, main-sequence, white dwarf, protostar
   (c) Protostar, red giant, main-sequence, white dwarf
   (d) Protostar, main-sequence, white dwarf, red giant
   (e) Protostar, main-sequence, red giant, white dwarf

2. (14.1) A star's _____ is the most critical factor determining what happens in every phase of a star's life.
   (a) composition
   (b) mass
   (c) temperature
   (d) angular momentum
   (e) radiation pressure

3. (14.2) All of the following are terms related to star formation except
   (a) T Tauri stars.
   (b) Bok globules.
   (c) interstellar cloud.
   (d) bipolar outflow.
   (e) planetary nebula.

4. (14.3) A star whose mass is 5 times larger than the Sun's and whose luminosity is 100 times larger than the Sun's has a main-sequence lifetime about
   (a) 5 times longer than the Sun's.
   (b) 500 times longer than the Sun's.
   (c) 5 times shorter than the Sun's.
   (d) 100 times shorter than the Sun's.
   (e) 20 times shorter than the Sun's.

5. (14.4) The helium flash occurs in stars with a degenerate core because
   (a) degenerate gas expands faster than ideal gases.
   (b) stars with degenerate cores are larger.
   (c) a degenerate gas can get very hot over a large region all at once.
   (d) stars with degenerate cores have much lower densities in the core.

6. (14.5) Variable stars are useful because (choose all that apply)
   (a) they can be used as standard candles.
   (b) they provide ways to test stellar structure theories.
   (c) their luminosity can be determined from the period.
   (d) many variable stars have large luminosities.

7. (14.6) A planetary nebula is
   (a) another term for the disk of gas around a young star.
   (b) the cloud from which protostars form.
   (c) a shell of gas ejected from a star late in its life.
   (d) what is left when a white dwarf star explodes as a supernova.
   (e) the remnants of the explosion created by the collapse of the iron core in a massive star.

8. (14.6) As a star like the Sun evolves into a red giant, its core
   (a) expands and cools.
   (b) contracts and heats.
   (c) expands and heats.
   (d) turns into iron.
   (e) turns into uranium.

9. (14.6–14.7) Stars like the Sun do not form iron cores during their evolution because
   (a) all the iron is ejected when they become planetary nebulas.
   (b) their cores never get hot enough for them to make iron by nucleosynthesis.
   (c) the iron they make by nucleosynthesis is all fused into uranium.
   (d) their strong magnetic fields keep their iron in their atmospheres.
   (e) None of the above

10. (14.7) A supernova (select all that are correct)
    (a) creates elements heavier than iron.
    (b) is the final state of all stars.
    (c) disperses metals into the interstellar medium.
    (d) can outshine the rest of a galaxy.
    (e) can trigger star formation in a nearby cloud.

11. (14.8) An H-R diagram of a cluster created from data that included the spectral types of the stars could be used to determine (select all that are correct)
    (a) the age of the cluster.
    (b) the angular momentum of the stars in the cluster.
    (c) the gravitational force in the cluster.
    (d) the distance of the cluster.
    (e) None of the answers are correct.

KEY TERMS

bipolar flows, 364
Bok globule, 362
degenerate gas, 369
deuterium, 363
helium flash, 370
instability strip, 372
interstellar cloud, 359
main-sequence lifetime, 366
nucleosynthesis, 376
period–luminosity relation, 372

planetary nebulas, 374
protostar, 362
red giant, 368
shell source, 368
supernova, 377
supernova remnant, 379
triple alpha process, 369
T Tauri stars, 364
turnoff point, 381
variable star, 370

ⓠ: FIGURE QUESTION ANSWERS

WHAT IS THIS? (chapter opening): This is a Hubble Space Telescope image of the Eskimo nebula, so-named because it resembles a head surrounded by a parka hood. This planetary nebula is composed of gas that was the outer layers of a star, expelled beginning about 10,000 years ago.

FIGURE 14.10: The red glow is caused by hydrogen emission lines.

FIGURE 14.19: About 1000 times the Sun's luminosity.

FIGURE 14.21: The central star will eventually cool off and become a white dwarf. The nebula will expand, becoming ever larger, and eventually will mingle with the surrounding interstellar gas.

Composite X ray, infrared, and optical image of Tycho's supernova remnant.

STELLAR REMNANTS: White Dwarfs, Neutron Stars, and Black Holes

LEARNING OBJECTIVES

Upon completing this chapter you should be able to:

- Describe the kinds of objects that are left behind by stars of different masses after they have consumed their nuclear fuel.
- Explain the role of gravity and other forces in determining the characteristics of these stellar remnants.
- List the characteristics of white dwarfs and explain why there is an upper limit to their possible mass.
- Explain the mechanisms that cause a nova or a type Ia supernova to occur.
- List the characteristics of neutron stars and recall how they were discovered.
- Explain why pulsars rotate so rapidly, and calculate the effect a size change has on the rotation speed of an object.

- Explain the mechanism that generates pulses from pulsars and contrast this with how X ray pulses may be generated.
- Describe how pulsars may slow down or speed up.
- Explain how gravity can prevent an object from emitting any light, creating a "black hole."
- Explain what is meant by the Schwarzschild radius of an object and calculate its value.
- Describe the nature of the "curvature" at the edge of a black hole.
- Discuss possible differences in a supernova that result in a black hole instead of a neutron star.
- Describe methods of detecting black holes.

During its life, every star supports itself against gravity by fusing nuclear fuel in its core. When its fuel is spent, the star's core collapses. This fate awaits the Sun 5 billion years from now. Gravity will crush what remains into a white dwarf—a star roughly the size of the Earth. More-massive stars surrender sooner and are more drastically squeezed to even smaller dimensions, becoming either neutron stars or black holes.

Compact stars, as these three kinds of stellar remnants are known, are the end points of stellar evolution. Because nearly all stars must eventually reach this stage, the Galaxy is littered with them. With no nuclear fuel remaining, they shine— if at all—with heat inherited from their previous state. Their matter, too, is unusual. In crushing compact stars to their tiny dimensions, gravity squeezes them into exotic materials. For example, a piece of white dwarf material the size of an ice cube would weigh about 16 tons. Matter in a neutron star is so compressed that electrons merge with protons, making the star resemble a giant atomic nucleus. So unmercifully has gravity squeezed the most massive of compact stars that they have collapsed completely, their immense gravity warping the space around them so that no light escapes, making them black holes in space.

Compact, however, does not mean inconspicuous: some of these crushed stars radiate intensely despite having no fuel of their own. Some compact stars "parasitize" a companion star, capturing matter from it that may burn explosively and allow the crushed star to "rise from the dead" as a nova—a "new" star—or a much brighter supernova. And because of a compact star's intense gravity, any material that falls on it releases such immense amounts of gravitational energy that this dead star may be thousands of times brighter than a "live" star like the Sun. But this brightness is bought at a dreadful price. The star's mass may increase to the point that gravity causes it to collapse even further, or the energy released may blast the star apart into atoms.

Q: WHAT IS THIS?

See end of chapter for the answer.

CONCEPTS AND SKILLS TO REVIEW

- Density (6.1)
- Escape velocity (3.8)

15.1 WHITE DWARFS

General Properties, Origin, and Fate

White dwarfs are hot, compact stars whose mass is comparable to the Sun's but whose diameter is about the same as the Earth's (about 1% that of the Sun). Their tiny size gives them so small a surface area that they are very dim despite being hot. Even the brightest white dwarfs have only about 1% the brightness of the Sun. Their pale light is not supplied by burning fuel, for they have exhausted all fuel supplies. Instead it comes from residual heat that leaks from the star's interior. Because they shine by residual heat and are therefore steadily cooling, white dwarfs show a wide range of surface temperatures. Some are as hot as 25,000 K; others are as cool as about 4000 K. A typical surface temperature might be about 10,000 K, so most have a bluish-white color.

A white dwarf forms when a low-mass star such as the Sun nears the end of its life. When the star swells into a red giant, its intense radiation strips off its outer layers to form a planetary nebula shell. This process exposes the star's core, which becomes the white dwarf. The composition of the core (and thus of the white dwarf) is mainly carbon and oxygen, the end products of the parent star's nuclear burning. The white dwarf's surface is composed of a thin layer of hydrogen and helium, but its core has

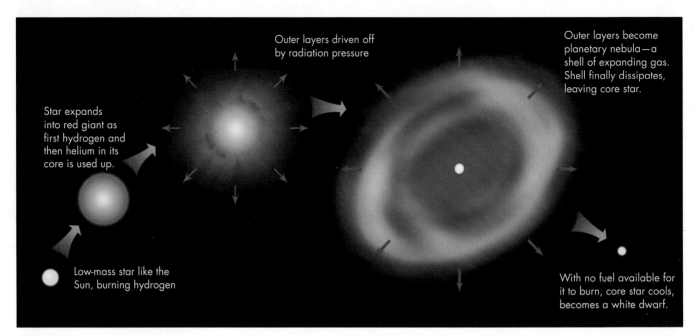

Outer layers driven off by radiation pressure

Outer layers become planetary nebula—a shell of expanding gas. Shell finally dissipates, leaving core star.

Star expands into red giant as first hydrogen and then helium in its core is used up.

Low-mass star like the Sun, burning hydrogen

With no fuel available for it to burn, core star cools, becomes a white dwarf.

FIGURE 15.1

Origin of a white dwarf. White dwarfs form when a low-mass star such as the Sun reaches the end of its life. Having expanded into a red giant for the second time (see fig. 14.2), the star loses its outer atmosphere and becomes a planetary nebula. Its core is exposed, and eventually it turns into a white dwarf, a star with no source of nuclear fuel. (*Note:* Not all stages in a star's evolution are shown.)

too little mass—and therefore insufficient gravity—to contract and heat itself to the ignition temperature of carbon. So, unable to burn further, the dying core simply cools and grows dim. This evolutionary sequence is illustrated in figure 15.1.

White dwarfs form when a low-mass star (up to about 8 $M_\odot$) ejects a planetary nebula shell. In theory, some could form from very low-mass stars (less than about 0.5 $M_\odot$) without such ejection. These stars are relatively cool throughout, and so their gas, like the relatively cool matter just below the Sun's surface, must carry energy by convection. The convection currents in these small stars extend throughout their entire volume, mixing their gas and preventing a core from forming. Moreover, their weak gravity cannot compress the spent fuel to make them red giants. Because low-mass stars evolve so slowly, however, they require longer than the current age of the Universe to reach this point in their evolution, and therefore none of these very low-mass stars has turned into white dwarfs yet.

What happens to a white dwarf as time passes? Although it is born very hot (recall that it was the core of a star), it has no fuel to burn to keep it hot. Like a dying ember, therefore, it cools rapidly (by stellar standards). Astronomers calculate that it takes only about 10 million years (roughly analogous to a month in the human life span) for the temperature of a white dwarf to drop to 20,000 K, and the current age of the Universe for it to drop to 4000 K. In trillions of years it would be so dark as to be undetectable, a dead, dark star that some refer to as a black dwarf. Cooling white dwarfs are probably very abundant in our Galaxy. How abundant is still widely debated.

Structure of White Dwarfs

Because white dwarfs are very compact and have no fuel supply, their structure differs significantly from that of ordinary stars. Although they are in hydrostatic equilibrium, with pressure balancing gravity, their pressure arises from a peculiar interaction among the electrons that limits the number that can occupy a given volume. This gives white dwarfs an odd property: increasing their mass makes them shrink! Even more crucial, however, white dwarfs must have a mass below a critical limit, or they collapse. To see why, we must consider conditions in a white dwarf's interior.

White dwarfs are very dense, having formed from the shrunken cores of their parent stars. If we divide the mass of a white dwarf by its volume, we find that its density is about 10^5 grams per cubic centimeter (roughly 1 ton per cubic inch). This high density packs the star's atoms so closely that they are separated by less than the normal radius of an electron orbit. If electrons could remain attached to nuclei under these conditions, their orbits would overlap those of neighboring atoms. But electrons can easily escape from their nuclei at such high density and move freely between nuclei, allowing the electrons to interact in a peculiar way.

Degeneracy and the Chandrasekhar Limit

A law of subatomic physics called the **exclusion principle** limits the number of electrons that can be put into a given volume. This, in turn, limits how closely electrons can be squeezed, thereby creating a pressure that, to distinguish it from that of an ordinary gas, is called **degeneracy pressure.** Degeneracy pressure depends only on the gas density, not on its temperature. In an ordinary gas, pressure depends on temperature and density. When such a gas is compressed, both the heat and the higher density contribute to raising its pressure. When a degenerate gas is compressed, it heats up, but the heat does not raise its pressure as it does for an ordinary gas. This makes degenerate gas less "springy" than ordinary gas.

If mass is added to a white dwarf, its radius *shrinks*. The extra mass increases the star's gravity, compressing the star. Its pressure rises, but because the star is degenerate, the increase is small and is unable to prevent the star from being squashed into a smaller volume. Squashing the star decreases the distance between particles and makes its gravity larger still, squashing the star even more. In fact, gravity may ultimately overwhelm the star's pressure forces and crush the star completely. Thus, too much mass makes a white dwarf collapse, as was demonstrated by theoretical calculations published in 1931 by the Indian astrophysicist Subrahmanyan Chandrasekhar (pronounced *chan-dra-say-car*), who was just 21 years old at the time.

The limiting mass of a white dwarf is now called the **Chandrasekhar limit** in his honor. For a typical white dwarf, the limiting mass is about 1.4 $M_\odot$, but this depends somewhat on the white dwarf's composition. Measurements of white dwarf masses after the completion of Chandrasekhar's work showed that no white dwarfs exceed this limit. For those in binary star systems, the masses can be measured using the modified form of Kepler's third law. For isolated white dwarfs, calculations are based on the shift of their spectral lines caused by their immense gravity. Chandrasekhar won the Nobel Prize for Physics in 1983 for these calculations and related work.

When light escapes from a body, it must work against the force of gravity. The escaping light acts like a ball rolled up a slope: as the ball moves higher, it slows down as it loses energy. Light cannot slow down—it must always travel at the constant speed, c—but it can lose energy, and it does. Light's energy, as we saw in chapter 4, determines its wavelength: lower energy corresponds to longer wavelengths. Longer wavelengths, in turn, correspond to redder colors; thus, as light "climbs" away from a body, the body's gravity stretches the wavelengths and creates a **gravitational redshift,** as illustrated in figure 15.2. The amount of gravitational redshift depends on the star's escape velocity, which depends in turn on its mass and radius. If the white dwarf's radius can be estimated from its luminosity and temperature (see section 13.3), the gravitational redshift can be used to determine the white dwarf's mass.*

Spectral lines also permit astronomers to measure the magnetic fields of white dwarfs. We discussed in chapter 12 how the Sun's magnetic field can be measured from the splitting of its spectral lines caused by the Zeeman effect. Similar observations show that some white dwarfs have magnetic fields 10,000 times stronger than sunspots—10^8 times the strength of the Earth's magnetic fields. These immense field strengths occur because as the white dwarf shrinks, it draws its magnetic field lines closer together, amplifying the field.

---

* To use this method, astronomers need to determine the kinds of atoms producing the spectral lines, and the wavelengths need to be adjusted for any Doppler-shift resulting from the white dwarf's motion.

A   Ball rolling up a hill loses energy and slows down.

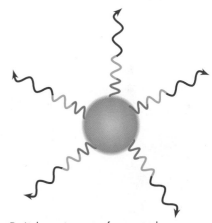

B   Light moving away from a star loses energy.

FIGURE 15.2
The gravitational redshift. (A) Just as a ball rolling up a hill loses energy, light loses energy as it moves away from a star. (B) Light does not slow down, however. Instead, because its energy, E, determines its wavelength, λ, according to the relation $E = \text{constant}/\lambda$, as the light's energy decreases, its wavelength increases. Longer wavelengths correspond to redder colors, and so we refer to the lengthening of the light's wavelength caused by the star's gravity as the gravitational redshift.

Stages in the evolution of a low-mass binary

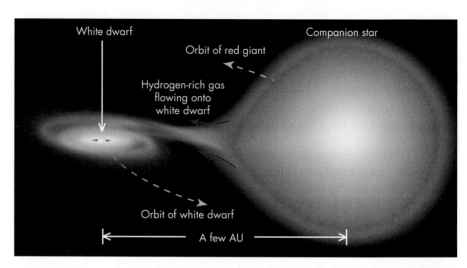

FIGURE 15.3

A white dwarf accreting (pulling in) mass from a binary star companion, which in this case is shown as a red giant. The typical separation of the stars is a few AU.

White Dwarfs in Binary Systems: Novas and Type Ia Supernovas

Isolated white dwarfs cool off and disappear from sight, but in binary systems they may heat up and explode. If a white dwarf has a nearby companion, gas expelled from the companion may fall onto the dwarf, as illustrated in figure 15.3. Coming from the companion's outer layers, such gas is generally rich in hydrogen, and it may briefly replenish the white dwarf's fuel supply. The new fuel forms a layer on the dwarf's surface, where gravity compresses and heats it. The gas layer eventually reaches the ignition temperature for hydrogen, but as we saw in chapter 14, nuclear burning in a degenerate gas can be explosive. The detonating hydrogen is blasted into space and forms an expanding shell of hot gas, as figure 15.4 shows. This shell radiates far more energy than the white dwarf itself does. Sometimes these stellar explosions are visible to the naked eye. Early astronomers called this type of event a *nova stella,* from the Latin for "new star," because the explosion would make a bright point of light appear in the sky where no star was previously visible. Today, we shorten *nova stella* to **nova.**

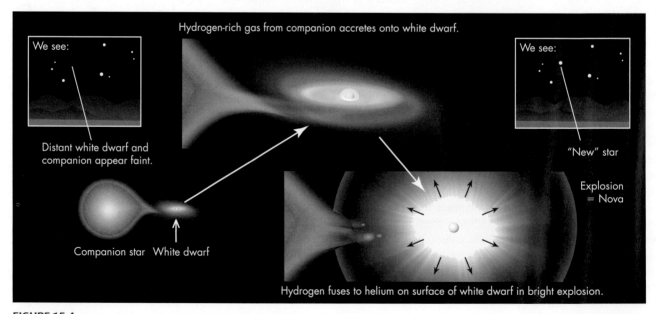

FIGURE 15.4

A white dwarf exploding as a nova. The hydrogen that falls onto the surface of a white dwarf from its companion may suddenly fuse into helium, creating an explosion that makes the star brighten.

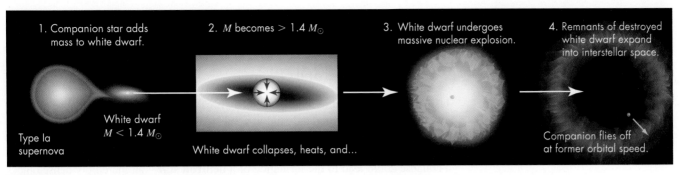

FIGURE 15.5

A white dwarf exploding as a type Ia supernova. If gas from a companion accumulates on the white dwarf and raises the mass above the Chandrasekhar limit (about 1.4 $M_\odot$), the white dwarf collapses. Heat from the resulting compression causes the white dwarf to explode.

A nova may erupt repeatedly if it does not accumulate too much mass. If enough gas is added to push it over the Chandrasekhar limit, however, the white dwarf will collapse. Carbon and oxygen in the collapsing star will be compressed and heated and, although initially not hot enough to fuse, they quickly reach their ignition temperature and begin nuclear fusion. Carbon and oxygen fuse to form silicon (^{12}C + ^{16}O becomes ^{28}Si), which in turn fuses into nickel (two ^{28}Si nuclei fuse to ^{56}Ni). The energy released in this process is enough to blow the entire star apart, resulting in a **type Ia supernova** (fig 15.5). The nickel, sprayed into space by the explosion, is highly radioactive and rapidly decays into cobalt (^{56}Co), which finally decays to iron (^{56}Fe), each decay adding additional energy to the outburst.

A type Ia supernova leaves no remnant star, except for the white dwarf's binary companion. This star sails off at its former orbital speed because the gravitational pull from its companion has disappeared. All that is left of the white dwarf is a cloud of rapidly expanding gas rich in the leftover carbon and oxygen, plus the silicon, iron, and other elements made during the explosive nuclear burning during the supernova, as illustrated by several examples in figure 15.6. Much of the iron in your blood was made when a white dwarf met its doom.

While astronomers have been able to identify a former companion star in a number of type Ia supernova remnants, others show no evidence of the former companion. This has caused astronomers to hypothesize that some type Ia explosions occur when two white dwarfs in a binary system merge and exceed the Chandrasekhar limit.

FIGURE 15.6

False-color images of three type Ia supernova remnants, with *Chandra* X ray data (green through violet) superimposed on optical images. (A) Kepler's supernova exploded in 1604 and is about 20 ly across. (B) A supernova about 30 ly across that exploded several hundred years ago in the Large Magellanic Cloud. No former companion has been identified. (C) A supernova seen in 1006, now about 60 ly across.

15.2 NEUTRON STARS

General Properties and Origin

Neutron stars

Just a year after Chandrasekhar published his calculations about the collapse of white dwarfs, physicists in England studying subatomic particles discovered the neutron. In 1934 this led astronomers Walter Baade (pronounced *Bah-deh*), at Mount Wilson Observatory, and Fritz Zwicky, at the California Institute of Technology, to predict that a collapse could lead to the formation of a **neutron star**—an incredibly dense object made of neutrons, like a giant atomic nucleus. They were also the first to suggest that this collapse would lead to a massive explosion and coined the term *supernova*.

As described in chapter 14, this type of supernova occurs when a star more massive than about 8 $M_\odot$ forms a large iron core. Just as Chandrasekhar found for white dwarfs, an iron core cannot support its own weight when its mass grows larger than about 1.4 $M_\odot$ (or somewhat higher, since high temperatures are supplying part of the pressure against gravity). Because iron does not release energy if it fuses, when the core collapses there is not a sudden ignition of nuclear burning as in a type Ia supernova. This is instead called a **type II supernova,** or **core-collapse supernova,*** as illustrated in figure 15.7.

Astronomers readily accepted Baade and Zwicky's idea that a massive star dies as a supernova—such violent explosions had been observed—but they paid little attention to their suggestion that a neutron star might be born in the blast. Such stars would be tiny compared even with white dwarfs and were thought to be unobservable. Nevertheless, several physicists calculated the properties of neutron stars and found that their radii should be about 10 kilometers (about 6 miles) and their masses should typically be between one and several times that of the Sun. The calculations also showed that, like white dwarfs, neutron stars have a maximum possible mass, which for them is about 2 to 3 solar masses.

Pulsars and the Discovery of Neutron Stars

Despite this theoretical model of neutron stars, there was no evidence that they existed, and so Baade and Zwicky's concept of tiny, collapsed stars lay dormant for over 30 years. Then in 1967 a group of English astronomers led by Anthony Hewish

---

* Before the exact mechanisms were understood, astronomers classified several different kinds of supernovas based on their spectra and pattern of brightening. Less common types known as Ib and Ic also come from massive stars whose iron cores collapse and only superficially resemble type Ia. Some astronomers therefore prefer the names nuclear detonation (type Ia) and core-collapse (types II, Ib, Ic) supernova.

1. Iron core collapses in star with mass between about 8 and 20 $M_\odot$.

2. Outer layers fall onto core, heat, "bounce" off of nuclear density material forming there.

3. Core generates huge neutrino burst, outer layers fuse and explode, leaving...

4. Neutron star (which may be pulsar)

Type II supernova

FIGURE 15.7
When a star that begins its life with more than about 8 $M_\odot$, it builds up an iron core that collapses to a neutron star. Surrounding materials are blasted outward by a type II supernova.

was observing fluctuating radio signals from distant, peculiar galaxies. Jocelyn Bell, a graduate student working with the group, noticed an odd radio signal with a rapid and astonishingly precise pulse rate of one burst every 1.33 seconds. The precision of the pulses led some astronomers to wonder whimsically if they had perhaps stumbled onto signals from another civilization, and informally the signal was known as LGM-1, for "little green men #1." Over the next few months, Hewish's group found several more pulsating radio sources, which came to be called **pulsars** for their rapid and precisely spaced bursts of radiation (fig. 15.8).

The discovery of other pulsars convinced astronomers that they were observing a natural phenomenon, and so their explanations focused on peculiar varieties of pulsating stars. Ordinary pulsating stars, such as Cepheid variables, obey a period–density relation such that each star's pulsation period depends inversely on the square root of its density. That law applies in a generalized form to all pulsating stars. Thus, the extreme shortness of pulsar periods indicates that they must be extremely dense objects, and the first pulsars discovered had periods that were consistent with their being white dwarfs. But soon other pulsars were discovered whose periods were much shorter, implying densities far larger than those of white dwarfs, and so astronomers turned back to Baade and Zwicky's ideas about neutron stars.

Neutron stars, according to their theory, not only would be extremely dense (and thus have very short periods) but also would be associated with supernova explosions. Thus, when a very-short-period pulsar was found in the Crab Nebula supernova remnant, it seemed to confirm Baade and Zwicky's model. But calculations of the pulsation period of neutron stars indicated that they should pulsate much faster than any of the pulsars so far observed. The solution to this puzzle came from the independent work of Franco Pacini and Thomas Gold, astrophysicists at Cornell University.

Even before the discovery of pulsars, Pacini had been studying the link between neutron stars and supernova remnants that had been proposed by Baade and Zwicky some 40 years earlier. He found that a neutron star remaining in the debris of a supernova explosion could supply large amounts of energy to the gas in its vicinity if it were rapidly spinning and if it had a very strong magnetic field. Rapid rotation and a strong magnetic field were also the keys to Gold's hypothesis about the rapidity of pulsar pulses, rather than the energizing of supernova debris. Gold suggested that a pulsar is a rapidly *spinning,* not pulsing, neutron star. Furthermore, its magnetic field causes it to emit radiation in two narrow beams so that, as the star spins, its beams sweep across space, as illustrated in figure 15.9. We see a burst of radiation when the beam points at Earth. Thus, a pulsar shines like a cosmic lighthouse whose beams swing around as its lamp rotates. But the model worried some astronomers because at least one of the pulsars then known—that in the Crab Nebula supernova remnant—rotates 30 times per second. What could make a star spin that fast? For comparison, consider that the Sun spins once a month.

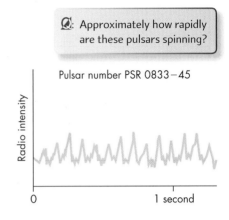

Q: Approximately how rapidly are these pulsars spinning?

Pulsar number PSR 0833−45

Radio intensity

0   1 second

Pulsar number PSR 0329+54

Radio intensity

0   2 seconds   4 seconds

FIGURE 15.8
Pulsar signals recorded from a radio telescope.

ANIMATION ▶

The oblique rotator model of a pulsar

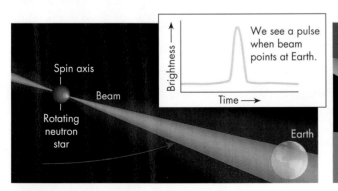

Spin axis

Beam

Rotating neutron star

Brightness →

We see a pulse when beam points at Earth.

Time →

Earth

FIGURE 15.9
A pulsar's pulses are like the flashes of a lighthouse as its lamp rotates.

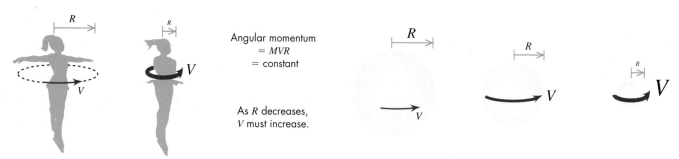

FIGURE 15.10

Conservation of angular momentum spins up a skater and a spinning star. Angular momentum is proportional to a body's mass, M, times its rotation speed, V, times its radius, R. Conservation of angular momentum requires that mathematically MVR = constant. Thus, if a given mass changes its R, its V must also change. In particular, if R decreases, V must increase, as the ice skater demonstrates by spinning faster as she draws her arms in. Similarly, a rotating star spins faster if its radius shrinks.

If pulsars are neutron stars, their rapid spin has a natural explanation. When a massive star collapses, forming a neutron star and exploding as a supernova, its core shrinks to a tiny radius—perhaps only 10 kilometers or so—and any rotation it has will be vastly increased. This rapid rotation results from the same effect that causes ice skaters to spin faster. As ice skaters draw their arms inward, an initially slow pirouette turns into a blur of rotation.

Any rotating body spins faster when it shrinks—a principle called the **conservation of angular momentum,** illustrated in figure 15.10. Angular momentum measures rotational inertia. For a spherical mass, the angular momentum is approximately given by the body's mass, M, times its equatorial rotation velocity, V, times its radius, R; or

$$\text{Angular momentum} = MVR.$$

M = Mass of an object

V = Rotation velocity at the object's equator

R = Radius of the object

The principle of the conservation of angular momentum states that this product must be constant unless some force acts to brake or accelerate the spin. Therefore, when the mass, M, does not change, *if we make R smaller, V must increase* to keep the angular momentum constant. Thus, a spinning body that shrinks in size (decreasing its radius) must spin faster, and a simple calculation using this principle shows that if even a slowly spinning star such as the Sun were to shrink to the size of a neutron star, it would rotate several thousand times per second! Such a shrinkage is exactly what happens when the iron core of a massive star collapses. Thus, the rapid spin of neutron stars becomes entirely explainable. For an example calculation, see the Astronomy by the Numbers box below.

Collapse to such a small radius has another effect: it squeezes the star's magnetic field into a denser state, amplifying the field's strength to about 1 trillion times that of the Earth's. Pulsars are thus extremely powerful magnets, and the combination of such a strong magnetic field and the rapid rotation generates the pulsar's radiation beams.

ASTRONOMY *by the numbers*   ROTATION RATE OF A SHRUNKEN STAR

Suppose a star that rotates once every 10 days shrinks to 1/10th its current diameter. How long will it take to complete a rotation when it reaches its new smaller size?

This problem can be solved using the conservation of angular momentum formula, and the answer is different than you might first guess. Because the mass of the star does not change when it shrinks, its speed of rotation V must be 10 times larger to keep the product of MVR constant (when R is 10 times smaller).

This does not mean that the star takes 1 day to rotate. Its circumference is 1/10th as big after it shrinks, so if it rotated in 1 day (1/10th the original time), the speed V would be the same as before. It must rotate in just 1/10th of a day for the speed to be 10 times faster.

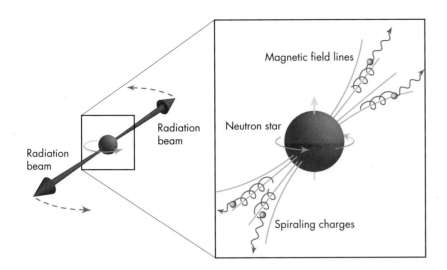

FIGURE 15.11
The strong magnetic field of a spinning pulsar generates a powerful electric field, which strips charges from its surface. The charges spiral in the star's magnetic field and emit radiation along their direction of motion. Because the charges stream in a narrow beam confined by the magnetic field, their radiation is also in a narrow beam at each magnetic pole of the star.

Emission from Neutron Stars

A varying magnetic field creates an electric field—a principle we use here on Earth to generate electricity by spinning magnets in dynamos. Similarly, the spin of a neutron star and its magnetic field generates powerful electric fields. These fields rip positively and negatively charged particles off the star's surface and accelerate them to nearly the speed of light. No metal wires confine the charges; rather, they are channeled by the pulsar's intense magnetic field. We saw in chapter 6 how the Earth's magnetic field directs solar particles toward the poles to create the aurora. The magnetic field of a spinning pulsar, in a like manner, generates two narrow beams of charged particles at the star's magnetic poles.

To explain how the pulsar's particle beams generate radiation, we invoke the physical law that accelerating charges radiate electromagnetic energy. A radio transmitter works according to this principle: electric currents are pulsed through the broadcast antenna, accelerating electrons in it; the accelerated electrons in turn produce the radio waves we detect. In a pulsar, too, the charges radiate as they accelerate, in this case along the magnetic field, spiraling as they go (fig. 15.11). The spiraling charges radiate electromagnetic energy over a range of wavelengths that depends on their energy. Higher-energy charges generate more energetic—and therefore shorter-wavelength—photons. The radiation is beamed because the charges are traveling along the field lines that, near the star's surface, form a tight cone. Because of this beaming, we see only those neutron stars whose sweeping beams point to Earth; those radiating in other directions are invisible to us. Thus, many more neutron stars must exist than the few hundred we have detected so far. Figure 15.12 shows the very young pulsar in the Crab Nebula, which emits visible light flashes 30 times per second in addition to its radio waves. A high-speed video camera was used to capture the flashes caused by its rapid spin.

The emission created by the accelerating charges is called **nonthermal radiation** because its properties depend on the charges' acceleration and on the strength of the magnetic field, rather than on

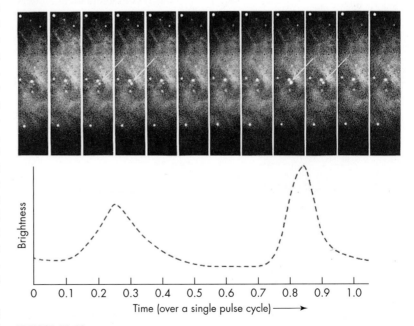

FIGURE 15.12
The rotation of the pulsar in the Crab Nebula makes the star appear to turn on and off 30 times per second as its beam of radiation sweeps across the Earth.

the temperature of a heated gas. It is also sometimes called **synchrotron radiation,** after a type of atomic particle accelerator in which such radiation was first observed on Earth. For most pulsars, the charges generate mostly low-energy radiation at radio wavelengths. However, in some very young pulsars the charges are so energetic that they produce radiation across the entire electromagnetic spectrum, including visible light and gamma rays. Pulsars presumably emit some thermal radiation—ordinary visible light and X rays—from their hot surfaces, but because they are so tiny, that radiation is very dim and difficult to detect.

In everyday life, spinning objects slow down. Pulsars are no exception. As a pulsar spins, it drags its magnetic field through the particles that boil off its surface into the surrounding space. The magnetic field exerts a force on those particles and speeds them up. But as the particles speed up, they exert a reaction force back on the magnetic field that slows it and the pulsar down. This process is a little like what happens if you pour the ingredients for batter onto the blades of a rapidly spinning electric blender: the batter begins to spin, but the dropping pitch of the motor tells you the blades slow down. Astronomers measure this "spindown" of a pulsar by precisely timing the interval between pulses. Such measurements indicate that the time interval—the period of the spinning pulsar—is lengthening. The slower rotation also limits the energy of the radiation the pulsar emits. Young, rapidly spinning pulsars emit visible light and radio waves; old, slowly spinning pulsars generate only radio waves.

As the pulsar spins ever more slowly, its emission of electromagnetic radiation weakens, and ultimately the star becomes undetectable. Thus, there could be many neutron stars scattered throughout space that are invisible to us because they are no longer emitting pulses.

Structure of Neutron Stars

Although in general pulsars are slowing down, some occasionally speed up suddenly. Such jumps in their rotation speed are called **glitches** (fig. 15.13A), and they can tell us about the interior structure of the pulsar or neutron star. Astronomers study the structure of neutron stars with the same methods used to study ordinary stars; that is, they construct mathematical models of the neutron star. Such models are based on physical laws—for example, the requirement that, in a stable star, gravitational forces must balance pressure forces. From these models, astronomers can deduce the structure of the neutron star's interior—its density and pressure. Such calculations show that neutron stars contain at least three distinct regions: a thin, outer gaseous atmosphere about 1 millimeter thick; a thin, solid (perhaps iron) crust a few hundred meters thick, below the atmosphere; and the bulk of the star, a liquid "sea" of neutrons lying below the crust (fig. 15.13B). As the neutron star's spin slows, the bulge at the neutron star's equator decreases. Astronomers think that glitches occur when the rigid crust suddenly adjusts to the changing shape of the neutron star.

Neutron Stars in Binary Systems

Like other types of stars, some neutron stars belong to binary star systems. But such binary pairs are rare, for several reasons. First, the supernova explosion that creates the neutron star might expel so much mass that gravity is no longer strong enough to hold the pair of stars together—they rapidly escape from the debris of the explosion that spawned them. This helps explain why so few are seen in supernova remnants. Second, astronomers have calculated that the supernova blast or other processes "kick up" the neutron star to speeds of hundreds of kilometers per second, again making it difficult for its companion to hold on to it. In fact, a number of such speeding pulsars have been detected. Despite such difficulties, some neutron stars do remain bound to a companion, and to such odd couples we now turn.

FIGURE 15.13
(A) A pulsar's rotation gradually slows down, but occasionally it will speed up briefly during a "glitch," which probably occurs when a rigid crust adjusts its shape. (B) Schematic structure of the interior of a neutron star.

X Ray Binary Stars

Neutron stars in binary systems sometimes emit intense X ray radiation. Such stars are called **X ray binaries.** Astronomers have identified several types of X ray binaries. In one type, the X ray emission occurs in intense bursts at irregular time intervals, and so these stars are called X ray bursters. According to one hypothesis, the X rays come from gas that falls onto the neutron star's surface, where the star's intense gravity compresses and heats it to millions of degrees. This heating triggers a thermonuclear explosion like those that cause nova outbursts. The exploding gas is so hot that most of its radiation is at X ray wavelengths, in accordance with Wien's law (the hotter a gas, the shorter the wavelength at which it radiates). Such explosions on the surface of a neutron star emit thousands of times more energy than the Sun, an impressive achievement for a "dead" star.

In another type of X ray binary, the X ray emission comes in rapid, regular pulses every few seconds or less. The X ray pulses from these stars are reminiscent of the radio pulses from ordinary pulsars, and so astronomers refer to such objects as "X ray pulsars." An X ray pulsar generates its emission from small "hot spots" near its magnetic poles. Gas from a companion star flows in along the field lines, much as gas particles from the Sun stream in toward Earth's magnetic poles, creating the aurora. As the gas falls toward the neutron star, the star's intense gravity accelerates the gas. As it crashes onto the neutron star at the magnetic poles (fig. 15.14), the gas is compressed and heated intensely, causing it to emit X rays. As the neutron star spins, the hot regions near each pole move into and then out of our field of view, creating the X ray pulses that we observe.

Because pulsars generally slow down as they age, astronomers were initially puzzled when they discovered some extremely rapidly spinning—and therefore presumably very young—pulsars in several very old star clusters. Some of these seemingly young pulsars rotate approximately 1000 times per second, so that the pulses we receive from them are separated by little more than a thousandth of a second. Astronomers accordingly call them "millisecond pulsars." Millisecond pulsars are distinguished not only by their rapid rotation but also by having companion stars in most cases. In fact, astronomers think they probably all had companions at one time, and the companion is the cause of their rapid spin.

If a neutron star has a companion that ejects mass, the neutron star may capture some of the mass by its gravitational attraction. However, the captured mass generally does not fall directly onto the neutron star. Instead, it temporarily goes into orbit around the neutron star, accumulating in a flat, spinning disk that astronomers call an **accretion disk.** The matter in the accretion disk eventually falls onto the neutron star, and when it does, the material carries some of its spinning motion—technically, angular momentum—with it. That spinning motion is imparted to the neutron star when the gas strikes its surface, making the neutron star spin faster. Thus, the usual slowdown of an aging pulsar may be reversed if it has a companion star. But

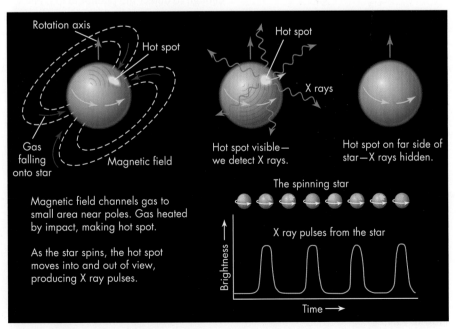

FIGURE 15.14

Gas falling onto a neutron star follows the magnetic field lines and makes a hot spot on the star's surface, creating X rays. As the star rotates, we observe X ray pulses.

what of those millisecond pulsars without companions? What made their rotations speed up?

One possibility is that such pulsars originally had companions that have since disappeared. In some cases, the companion may have been lost if the pulsar passed too closely to a neighboring star. Gravitational forces from that neighbor may have pulled the pulsar's companion away. A more sinister possibility is that the pulsar destroyed its companion, a process that astronomers believe they may be witnessing in the case of the "black-widow pulsar." This peculiar pulsar has a low-mass companion star that is evaporating under the intense heating created by the pulsar. Given enough time, the companion star may totally evaporate, and future astronomers may see the millisecond pulsar without its original companion.

Gravitational Waves from Binary Neutron Stars

Astronomers have discovered a few neutron stars in tight orbits around each other. This unusual situation occurs when two massive stars that formed together both collapse without disrupting the binary system. The neutron stars probably spiraled close to each other during their red giant phases because of the frictional drag as they orbited within their companion's atmosphere. The intense gravity and rapid motion of these stars has allowed astronomers to carry out a critical test of Albert Einstein's theory of gravity developed in 1916, known as general relativity (see essay 2). General relativity is based on the idea that gravity causes a **curvature of space.** According to general relativity, masses generate a curvature or distortion of space and time that causes objects to accelerate toward the masses.

Where gravity is weak, Einstein's theory matches the predictions of Newton's law of gravity, but intense gravitational fields produce some unexpected effects. The orbits of objects such as neutron stars can generate strong **gravitational waves.** Just as ripples spread away from a stone tossed into a pond, gravitational waves spread across space, sending waves of distortion in space and time as suggested in figure 15.15. Gravitational wave detectors are currently being built, but they are not yet sensitive enough to detect the phenomenon. Astronomical "detectors" offer a shortcut.

Early in the 1900s, Einstein predicted that rapidly orbiting stars should generate gravitational waves. These waves carry energy away from the orbiting stars, making them gradually move together. In 1974, Joseph Taylor and Russell Hulse, then at the University of Massachusetts, discovered a binary pulsar: two neutron stars in orbit around each other. From observations of the pulses, Taylor and Hulse deduced the orbital speeds and how the separation of the neutron stars was steadily decreasing.

To their delight the speed and separation changed exactly in the manner predicted by Einstein's general theory of relativity. For this discovery, which showed that Einstein's theory of gravity was right and that gravitational waves do exist, Taylor and Hulse were awarded the 1993 Nobel Prize in Physics. These results provide strong support for the theory of general relativity, a theory that predicts even stranger objects—the existence of "rips" in the fabric of space and time called black holes.

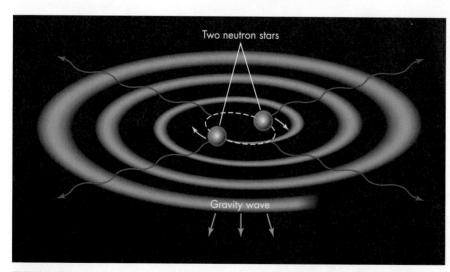

Two neutron stars

Gravity wave

FIGURE 15.15

Gravitational waves generated by a rapidly orbiting pair of neutron stars (purple). Such waves are "ripples" in space and time—and, like ripples in water, they make objects bob (move up and down) slightly as they pass.

15.3 | BLACK HOLES

When a star that was initially more massive than about 20 $M_\odot$ reaches the end of its life and collapses, it may create a compact star whose properties differ dramatically from those of white dwarfs or neutron stars. Its greater mass can compress its core so much that pressure is unable to support it, and it totally collapses to form what astronomers call a **black hole.**

Black holes have very strange properties, and to understand them we need to review the concept of **escape velocity,** which we discussed in chapter 3. Escape velocity is the speed a mass needs to avoid being drawn back by another object's gravity. For a body of mass M and at a radius R from the center of that object, the escape velocity, V_{esc}, for an object to travel away from the body is

$$V_{esc} = \sqrt{\frac{2GM}{R}}$$

where G is the gravitational constant and has a value of 6.67×10^{-11} if V_{esc} is in meters per second, R is in meters, and M is in kilograms (fig. 15.16).

For the Earth, the escape velocity is about 11 kilometers per second; for the Sun, it is about 600 kilometers per second. You can see from the formula that the escape velocity for a body of a given mass will be larger at a smaller radius. The same mass squeezed to a smaller radius results in a larger surface gravity, making it more difficult to escape its surface. Thus, a white dwarf whose mass is the same as the Sun's has an escape velocity from its surface much larger than the Sun's because its radius is so small. In fact, because its radius is about 1% of the Sun's and because the escape velocity depends on the square root of the radius, its escape velocity is larger by a factor of $\sqrt{100}$ ($=10$) and equals about 6000 kilometers per second. An object leaving the surface of a white dwarf at a velocity of 6000 kilometers per second would therefore overcome the star's gravity and escape into space.

The escape velocity for a neutron star, whose radius is about 10^5 times smaller than the Sun's but whose mass is comparable, is about 180,000 kilometers per second, more than half the speed of light. Thus, if a neutron star were a little smaller or a little more massive, its escape velocity would exceed the speed of light. Such an object would become a black hole.

As long ago as 1783, the English cleric John Michell discussed the possibility that there might exist objects with escape velocities exceeding the speed of light. A little more than a decade later, the French mathematician and physicist Pierre Simon Laplace entertained the same idea. Following their logic, we can calculate the radius at which a body of a given mass would become a black hole by equating its escape velocity to the speed of light. That is, $V_{esc} = c = \sqrt{GM/R}$. Notice that the calculation involves only two properties of the body itself: its mass and its radius.

The mathematically correct treatment of a black hole's structure using general relativity is much more difficult, but the German astrophysicist Karl Schwarzschild showed that relativity theory gives the same formula for a black hole as we derived based on Newton's law of gravity. Thus, we can solve for the **Schwarzschild radius,** R_S, at which a body of mass M becomes a black hole and find that

$$R_S = \frac{2GM}{c^2}.$$

For a body of 1 solar mass, this turns out to be about 3 kilometers, or 1.9 miles (see the Astronomy by the Numbers box on the next page). In other words, if the Sun could be compressed that small, it would become a black hole. Because the radius of a black hole is proportional to the mass, we find that the radius of any black hole is approximately $3 \times \mathcal{M}$ kilometers, where $\mathcal{M}$ is the mass of the body in solar units.

V_{esc} = Velocity necessary to escape an object's surface
M = Object's mass
R = Object's radius

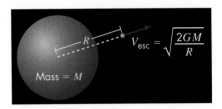

FIGURE 15.16
Diagram illustrating the principle of escape velocity. An object traveling at the escape velocity or greater can escape from the mass M, never to return.

R_S = Schwarzschild radius—radius at which an object of a particular mass becomes a black hole
M = Object's mass
c = Speed of light

ASTRONOMY *by the numbers*

THE SCHWARZSCHILD RADIUS OF A 1-SOLAR-MASS BLACK HOLE

How small would the Sun have to become for its escape velocity to reach the speed of light? In other words, what is the Schwarzschild radius of an object with a mass equal to the Sun's?

We need the Sun's mass in kilograms along with the value of G and c in standard MKS units. These are: $M_\odot = 1.99 \times 10^{30}$ kg; $G = 6.67 \times 10^{-11}$ m^3 sec^{-2} kg^{-1}; $c = 3.00 \times 10^8$ m sec^{-1}. Putting these values into the equation for the Schwarzschild radius, we find:

$$R_S = \frac{2GM}{c^2}$$

$$= \frac{2 \times 1.99 \times 10^{30} \text{ kg} \times 6.67 \times 10^{-11} \text{ m}^3 \text{ sec}^{-2} \text{ kg}^{-1}}{(3.00 \times 10^8 \text{ m sec}^{-1})^2}$$

$$= 2.97 \times 10^3 \text{ m}.$$

Thus, the Sun would become a black hole if it were compressed to about a 3-km radius.

The Nature of Space Around Black Holes

It was not until the advent of Einstein's general theory of relativity that astronomers obtained the mathematical and physical understanding to properly study black holes. As discussed at the end of section 15.2, relativity theory showed that gravity is related to the curvature of space and time. A black hole, then, is a place where the curvature of space and time become so extreme that motion away from the black hole becomes impossible.

An analogy may help illustrate how gravity and curvature of space are related. Imagine a waterbed with a plastic surface on which you have placed a baseball (fig. 15.17). The baseball makes a small depression in the otherwise flat surface of the bed. If a marble is now placed near the baseball, it will roll along the curved surface into the depression. The bending of its environment made by the baseball therefore creates an "attraction" between the baseball and the marble. Now suppose we replace the baseball with a bowling ball. It will make a bigger depression and the marble will roll in farther and be moving faster as it hits the bottom. We therefore infer from the analogy that the strength of the attraction between the bodies depends on the amount by which the surface is curved. Gravity also behaves this way, according to the general theory of relativity. According to that theory, mass creates a curvature of space, and gravitational motion occurs as bodies move along the curvature.

We now extend our analogy to black holes by supposing that we remove the bowling ball and place a large boulder on the waterbed. This depresses the plastic even

Baseball:
Marble rolls into depression

Bowling ball:
Marble rolls in faster

Big rock:
Marble disappears into hole

FIGURE 15.17
Objects on a waterbed make depressions analogous to the curvature of space created by a mass. According to the theory of general relativity, that curvature produces the effect of gravity. We can see the similarity by placing a marble at the edge of the depression and watching it roll inward, as if attracted to the body that makes the depression. Bigger bodies make bigger depressions, and so a marble rolls in faster. However, a very big body may tear the waterbed, creating an analogue of a black hole.

more and rips a hole in it. A marble placed on the torn surface will fall toward the boulder, drop into the hole, and disappear. Analogously, a black hole produces a "rip" in space where the curvature has become so great that the structure of space is disrupted.

All masses, including the Sun, curve space to some degree. Far from the Sun, space is "flat," nearly unaffected by the Sun's gravity. As we approach the Sun, however, the curvature increases. The path of light or radio waves is bent by gravity, so astronomers can *see* the curvature of space. In essay 2 we described how this curvature can also be thought of as an intrinsic flow of space toward massive objects. When traveling through space, then, the flow of space changes the direction in which light travels, much as the direction of a boat is changed by currents in the water.

The extreme curvature of a black hole corresponds to an inward flow of space that reaches the speed of light at the Schwarzschild radius, as illustrated in figure 15.18. Because nothing can travel through space faster than the speed of light, objects crossing this boundary cannot travel outward. Because nothing can go out, this boundary is sometimes called the **event horizon.** Just as the Earth's horizon blocks our view of what lies beyond it, so too we cannot see past the event horizon. All that happens within it is hidden forever from our view. No radiation of any sort, nor any material body—rocket, spacecraft, and so on—can break free of its gravity. Because we cannot observe the interior of a black hole even in principle, there is only a limited number of physical properties we can ascribe to it. For example, it is meaningless to ask what a black hole is made of. When it warps space inside its event horizon, we can no longer observe its interior—it could be made of neutrons, corn flakes, or gerbils. Only the amount of mass is important, not what composes it.

We can measure the mass of a black hole because the mass generates a gravitational field. In fact, at large distances the gravitational field generated by a black hole is no different from that generated by any other body of the same mass. For example, if the Sun were suddenly to collapse and become a black hole with the same mass it has now, the Earth would continue to orbit it just as it does now: it would not be pulled in. Nor would Mercury. For something to fall into a black hole, it must approach the black hole to nearly its Schwarzschild radius, and recall that this is about 3 kilometers for a solar-mass black hole. Because astronomical objects are generally so immensely far apart, the likelihood that any object will approach another to within a few kilometers is extremely small. In fact, at a given distance, a black hole exerts exactly the same gravitational force as *any* object of the same mass. That force is given (essentially) by Newton's law of gravity, just as it is for the Sun. So, we are no more likely to be pulled into a black hole than we are to be pulled into the Sun.

Black holes can make themselves felt beyond their Schwarzschild radius by more than just their gravity. They may possess electric charge and angular momentum. For example, a black hole may become charged if an excess of one kind of charge, say, positive, falls into it. Angular momentum is much more likely to be a significant property of a black hole. Material nearly always has some angular momentum, and if it becomes a black hole, its angular momentum is retained. As a result, a spinning black hole will not be a sphere. The distance from its center to the event horizon is smaller along the rotation axis than it is perpendicular to it.

The model of inward flowing space provides another way of interpreting the gravitational redshift discussed in section 15.1. Light leaving the surface of a white dwarf is redshifted by an amount that is the same as if the surface were moving away from us at the escape velocity of the white dwarf.

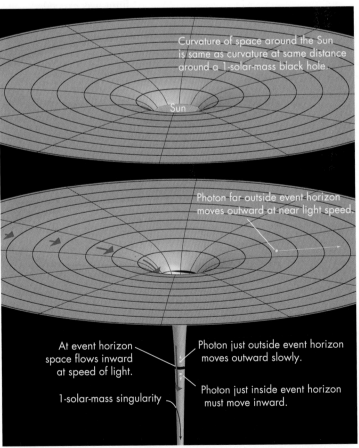

FIGURE 15.18
Space "flows" toward masses like water down a drain. At the Schwarzschild radius, the inward flow reaches the speed of light, so nothing, not even light, can travel out from inside this radius.

1. Iron core collapses in star more massive than about 20 $M_\odot$.

2. Inner layers disappear into black hole forming at center. Outer layers flatten into disk as they shrink and spin faster.

3. Collapsing material heats, explodes in jets, leaving...

4. Black hole

Hypernova

FIGURE 15.19

Very massive stars may collapse directly into a black hole. When this happens, infalling matter may collapse to a thin disk and produce intense jets blasted out perpendicular to the disk.

The Formation and Observation of Black Holes

Astronomers have excellent evidence that some supernova explosions form neutron stars, as we saw in section 15.2. The step from a neutron star to a black hole does not require a breakdown of the degeneracy pressure between neutrons comparable to the way the degeneracy pressure between electrons fails in a white dwarf or a massive star's iron core. If we simply have more than about 3 $M_\odot$ of material of nuclear density, its Schwarzschild radius will exceed its size. We might imagine gently dropping matter onto a neutron star until its mass exceeded this limit. Light from the surface of the neutron star would become steadily more gravitationally redshifted, and then when the limit was exceeded, the star would disappear from sight.

In recent years astronomers have attempted to model the collapse of very massive stars. A star that begins its life with more than about 20 $M_\odot$ will probably have a collapsing core that exceeds 3 $M_\odot$ at the end of its life. The models suggest some differences from a type II supernova for the death of these very massive stars, as illustrated in figure 15.19. The main difference is caused by the disappearance of matter straight into a black hole that forms during the collapse. In a type II supernova, much of the energy that blasts material outward comes from the impact of the collapsing material with the surface of the neutron star as well as the flood of neutrinos produced when protons and electrons combine to make neutrons. Without this energy source, the matter falls even farther in, crashing together to form a disk around the black hole, somewhat like the disks of material that collect around young stars (chapter 14). Like these young stars, extremely hot gas is ejected in jets perpendicular to the disk. These jets could explain intense bursts of gamma rays that are seen coming from distant galaxies (see Extending Our Reach: "Exploring New Wavelengths: Gamma Rays" in chapter 5, p. 137) when the beams happen to point toward Earth. Some astronomers have termed this type of collapse a "hypernova."

While these theoretical ideas are intriguing, it is a much more difficult matter to show by *observation* that black holes exist. An object that emits no light or other electromagnetic radiation is not easy to observe! But just as you may "see" the wind by its effect on leaves and dust, astronomers can see black holes by their effects on their surroundings.

Suppose a massive star in a binary system explodes, and its core mass is 5 $M_\odot$, rather than 2 $M_\odot$, so that the core collapses to become a black hole rather than a neutron star. Gas from the companion star may be drawn toward the hole by its gravity, as we know happens for neutron stars that are X ray bursters. The infalling matter swirls around the black hole and forms an accretion disk whose inner edge lies just outside the hole's Schwarzschild radius, as depicted in figure 15.20A. Here, where the disk orbits at nearly the speed of light, turbulence and friction heat the swirling gas to a furious 10 million K, making it emit X rays and gamma rays.

As the black hole orbits its neighbor, the X ray-emitting gas may disappear from our view as it is eclipsed by the companion star. An X ray telescope trained on such a

FIGURE 15.20
(A) Black holes may reveal themselves by the X rays emitted by gas orbiting them in an accretion disk. (B) The X ray emission from gas around a black hole is sometimes eclipsed by the black hole's companion star. Sizes and separations are not to scale.

star system will show a steady X ray signal that disappears at each eclipse, as shown in figure 15.20B. Such a signal might be the sign of a black hole, and astronomers have detected several X ray sources that bear that signature. But how do we know an X ray source is not just a neutron star?

Because X ray sources are binary stars, we can measure their masses. If X ray-emitting gas surrounds a body we cannot see but whose mass exceeds about 3 $M_\odot$, we can be reasonably confident that the invisible object is a black hole, because neutron stars cannot be that massive. For example, Cygnus X-1, the first X ray source detected in the constellation Cygnus, the swan, consists of a *B* supergiant star and an invisible companion whose mass—based on an application of the modified form of Kepler's third law—is at least 15 $M_\odot$. Nothing we know of but a black hole can be so massive and yet invisible. An even better candidate is A0620–00, an X ray-emitting object in the constellation Monoceros, the Unicorn. For A0620–00, the invisible star's mass is probably 16 $M_\odot$.

The evidence suggests that when the most massive stars die, they leave behind black holes with masses sometimes more than 10 times that of the Sun. Curiously, in recent decades astronomers have found convincing evidence that enormously more massive black holes—with masses millions or even billions of times that of the Sun—exist in the centers of many galaxies, including our own. We will discuss the evidence for these supermassive black holes in chapters 16 and 17.

X rays from a black hole binary

Hawking Radiation

Up to this point we have described black holes as intrinsically black. In 1974, however, the English mathematical physicist and cosmologist Stephen Hawking predicted that black holes can radiate! As so often happens in science, others were working on similar ideas. For example, in the previous year the physicist Jacob Bekenstein, who was studying the relation between gravity and thermodynamics, noted that black holes can be assigned a temperature. But Hawking went further and showed that—astonishingly—a black hole emits blackbody radiation, such as we discussed in chapter 4. In fact, we can use Wien's law, which relates a body's temperature to the wavelength at which it radiates most strongly, to estimate the temperature of a black hole.

Hawking calculated that the wavelength of maximum emission is approximately 16 times the black hole's Schwarzschild radius (R_S), a result also found by Bekenstein. Wien's law states that an object's temperature is inversely proportional to the wavelength of the radiation it emits, and so the temperature of a black hole is equal to a constant divided by R_S. For a solar-mass black hole, this turns out to be about 6×10^{-8} K. This is very cold, but it is not absolute zero. Accordingly, a black hole, like any other body whose temperature is not absolute zero, emits energy in the form of electromagnetic waves, energy now known as **Hawking radiation.** However, for black holes of stellar mass, the amount of Hawking radiation emitted is far too small to be detected at this time. Nevertheless, the existence of such radiation means that black holes are not truly black.

Hawking radiation is created by quantum physical processes near the event horizon of a black hole. The quantum processes allow tiny fluctuations in energy, everywhere in space, frequently in the form of pairs of photons that rapidly re-merge. If one of these fluctuations occurs near the event horizon, however, one of the pair may cross the event horizon, leaving the other free to travel away, carrying away energy (fig. 15.21). Because the only source of energy available to a black hole is its mass, as a black hole "shines" (albeit very dimly), its mass must decrease. In other words, black holes must eventually "evaporate." However, the time it takes for a solar-mass black hole to disappear by "shining itself away" is *very* long: approximately 10^{67} years! This is immensely long—vastly longer than the current age of the Universe—but the implications are important: even black holes evolve and "die."

Event horizon of black hole

Quantum fluctuation produces pair of photons. Normally they recombine an instant later, but if one of the pair enters the black hole, the other may travel away, subtracting energy from the black hole.

FIGURE 15.21
Quantum fluctuations near the event horizon of a black hole can gradually subtract energy (and therefore mass) from a black hole.

SUMMARY

White dwarfs, neutron stars, and black holes are the remnants of dead stars. A white dwarf forms when a low-mass star expels its outer layers to form a planetary nebula shell and leaves its hot core exposed. The radius of a white dwarf is about the same as the radius of the Earth. Its matter is degenerate, and its mass must be less than the Chandrasekhar limit or it collapses. Having no fuel supply, a white dwarf gradually cools and dims. If a white dwarf is in a binary system, it may accrete mass from its neighbor and explode either as a nova or as a type Ia supernova.

A neutron star forms when a massive star's iron core collapses and triggers a supernova explosion. The collapse compresses the core's protons and electrons together to make neutrons, forming a ball of neutrons with a radius of a mere 10 kilometers that may contain up to about 2 to 3 solar masses. Conservation of angular momentum during its collapse accelerates its rotation rate to about 1000 times per second. This spin and the star's magnetic field generate beams of radiation that sweep across space, making the spinning neutron star a pulsar.

If the massive star's iron core contains more than about $3\,M_\odot$, a black hole, rather than a neutron star, is born. Gravity overwhelms the pressure forces of the core, crushing it and bending space around it so that light cannot escape. The black hole lies invisibly inside its Schwarzschild radius. If it has a companion star, however, we may be able to see X rays from an accretion disk as gas from the neighbor falls toward the hole.

QUESTIONS FOR REVIEW

1. (15.1) What are the approximate mass and radius of a white dwarf compared with those of the Sun?
2. (15.1) How does a white dwarf form?
3. (15.1) What keeps a white dwarf hot?
4. (15.1) Can a white dwarf have a mass of 10 solar masses? Why? What happens if a white dwarf increases in mass?
5. (15.1–15.2) What is meant by degeneracy pressure? How is it related to white dwarfs and neutron stars?
6. (15.1) Explain what makes a nova occur.
7. (15.2) What is a neutron star?
8. (15.2) What are the mass and radius of a typical neutron star compared with those of the Sun? Can a neutron star have a mass of 10 solar masses?
9. (15.2) How does a neutron star form?
10. (15.2) How do we observe neutron stars?
11. (15.2) What is a pulsar? Does it pulsate?
12. (15.2) Are all neutron stars pulsars? Are all pulsars neutron stars?
13. (15.2) What creates the beams of radiation seen in pulsars?
14. (15.2) What is nonthermal radiation?
15. (15.2) What happens when a gravitational wave moves? What does it affect? Compare this to how light waves move.
16. (15.3) What is a black hole? Are they truly "black"? What properties can they have?
17. (15.3) What is the Schwarzschild radius?
18. (15.3) Why might the distance to the event horizon of a black hole vary depending on which direction you measure from the center?
19. (15.3) Can astronomers see black holes? Explain.
20. (15.3) What is Hawking radiation?

THOUGHT QUESTIONS

1. (15.1) Some astronomers have suggested that cooled white dwarfs are made of diamond. Why might it be impractical to mine them?
2. (15.1) Is a white dwarf significantly different from the core of a low-mass main-sequence star? Thinking about how the speed of stellar evolution relates to the mass of the stars, and about what happens when novas occur, can you explain how you might have a binary system with an 8-solar-mass blue main-sequence star and a 1-solar-mass white dwarf?
3. (15.2) Is it surprising that a pulsar is not seen in every supernova remnant? Why?
4. (15.2) The period of a star is equal to its circumference divided by its rotational velocity. If a star collapses to 1/X of its previous radius, how many times shorter is the period?
5. (14.5/15.2) Explain how neutron star "pulses" are different from the "pulsing" of variable stars. Are neutron stars variable stars?
6. (15.1–15.3) White dwarfs, neutron stars, and black holes can be found in binary star systems with normal or giant companion stars. What happens in each of these systems?

7. (2.8/15.1–15.3) A 1-solar-mass black hole would be smaller than a 1-solar-mass neutron star, which in turn would be smaller than a 1-solar-mass white dwarf. If the Sun were replaced by each of these objects, what would be the impact on the Earth's orbit? Compare the conditions close to the surface or event horizon of these objects.
8. (15.3) How would you explain to an 8-year-old child that black holes may exist?
9. (15.3) Suppose you jumped into a black hole feet first. What would happen to you as your feet approached its Schwarzschild radius? *Hint:* Think about tides on the Earth created by the Moon.
10. (15.3) What differences would you observe if you were orbiting a 16-solar-mass black hole versus a 16-solar-mass star at a distance of 5 AU?

PROBLEMS

1. (15.1) Calculate the density of a white dwarf star of 1 solar mass that has a radius of 10^4 kilometers.
2. (15.1–15.2) Calculate the escape velocity from a white dwarf and a neutron star. Assume that each is 1 solar mass. Let the white dwarf's radius be 10^4 kilometers and the neutron star's radius be 10 kilometers.
3. (15.2) A neutron star has a radius of about 10 km. What is the circumference of the equator of a neutron star? If you could stand on it, it would be a very visibly curved surface. Imagine walking along the neutron star's equator. How many kilometers along the circumference would correspond to moving 7° around the star? How far is 7° along the surface of the Earth (radius of 6400 km)? Seven degrees was the difference in angle Eratosthenes measured when he determined the circumference of the Earth in ancient times, a task that would be a little faster to do on a neutron star (if you could survive the high temperature and gravity!).
4. (15.2) The mass of a neutron is about 1.7×10^{-27} kg. Suppose the mass of a neutron star is about 3.4×10^{30} kg. How many neutrons does such a star contain?
5. (15.2) The volume of a neutron is about 10^{-45} cubic meters. Suppose you packed the number of neutrons you found for problem 4 (above) into a cube so that the neutrons touched edge to edge. How big would the volume of the cube be? How big across would the cube be? *Hint:* The volume of a cube with sides of length X is X^3. How does the cube's size compare to the size of a neutron star? What can you conclude about the spacing of neutrons in a neutron star?
6. (15.2) Very approximately, the core of a high-mass star about to collapse to form a neutron star in a type II supernova might be as large as 10^4 km and have a rotational period of around a half a day. If the core collapsed to a radius of 10 km, what would be the period of the resulting neutron star? (Use the answer to Thought Question 4.) How does this compare to the periods of the pulsars discussed in the chapter?

7. (15.3) Calculate the Schwarzschild radius of the Sun.

8. (15.3) Calculate your Schwarzschild radius. How does that compare to the size of an atom? How does it compare to the size of a proton?

9. (4.2/13.2/15.3) Use Wien's law to determine the wavelength of light generated by 10-million-kelvin gas found in an accretion disk. What are photons of this wavelength called?

10. (15.1–15.3) You observe a main sequence $K0$-type star ($M = 0.79\ M_\odot$) that moves as if it is in a binary system, but no companion is visible. If the period of the system is 34 days and the semimajor axis is 0.5 AU, what is the mass of the system? (Remember to convert 34 days to years to use Kepler's law as discussed in section 13.5.) What is the mass of the companion? What kind of compact star do you think the companion is? What other observational evidence would you look for to confirm this hypothesis?

11. (15.1–15.3) You observe a main sequence $B5$-type star ($M = 5.9\ M_\odot$) that moves as if it is in a binary system, but no companion is visible. If the period of the system is 8.4 years and the semimajor axis is 8 AU, what is the mass of the system? What is the mass of the companion? What kind of compact star do you think the companion is? What other observational evidence could you look for to confirm this hypothesis?

12. (15.1–15.3) You observe a main-sequence $A0$-type star ($M = 2.9\ M_\odot$) that moves as if it is in a binary system, but no companion is visible. If the period of the system is 4 years and the semimajor axis is 4 AU, what is the mass of the system? What is the mass of the companion? What kind of compact star do you think the companion is? What other observational evidence could you look for to confirm this hypothesis?

TEST YOURSELF

1. (15.1) If a quantity of hydrogen is added to a white dwarf, then (select all that apply)
   (a) its radius increases.
   (b) its radius decreases.
   (c) its density increases.
   (d) it may exceed the Chandrasekhar limit and collapse.
   (e) the hydrogen may explosively fuse to helium.

2. (15.1–15.3) For a white dwarf or a neutron star to shrink,
   (a) it must emit Hawking radiation.
   (b) some of the electrons or neutrons must gain higher energies.
   (c) material must be removed from the star through a burst event.
   (d) a black hole must develop inside the white dwarf or neutron star.
   (e) it must be losing material to a companion star.

3. (15.1–15.2) The spectrum of a supernova shows lots of iron and nickel, but no hydrogen. This is probably because
   (a) the explosion was of a white dwarf pushed over the Chandrasekhar limit.
   (b) the explosion was of an old star that had used up all its hydrogen.
   (c) the hydrogen was fused into heavier elements in the explosion.
   (d) the explosion was the result of core-collapse of a massive star.
   (e) the shock of the explosion pushed all the hydrogen far away from the exploding stellar remnant.

4. (15.2) Which of the following has a radius (linear size) closest to that of a neutron star?
   (a) The Sun
   (b) The Earth
   (c) A basketball
   (d) A small city
   (e) A gymnasium

5. (15.2) What causes the radio pulses of a pulsar?
   (a) The star vibrates.
   (b) As the star spins, beams of radio radiation from it sweep through space. If one of these beams points toward the Earth, we observe a pulse.
   (c) The star undergoes nuclear explosions that generate radio emission.
   (d) The star's dark orbiting companion periodically eclipses the radio waves emitted by the main star.
   (e) A black hole near the star absorbs energy from it and re-emits it as radio pulses.

6. (15.2) In binary systems, accreting material that falls on a neutron star produces an X ray burst, whereas material that falls on a white dwarf produces a visible-light nova burst because
   (a) the acceleration of gravity is stronger near the surface of a neutron star.
   (b) neutron stars are larger than white dwarfs.
   (c) the accreting material comes from a hotter source in an X ray binary.
   (d) in a nova, the accreting matter is hydrogen, but in an X ray burst it is iron.
   (e) neutron stars rotate much faster than white dwarfs.

7. (15.3) What evidence leads astronomers to believe that they have detected black holes?
   (a) They have seen tiny dark spots drift across the face of some distant stars.
   (b) They have detected pulses of ultraviolet radiation coming from within black holes.
   (c) They have seen X rays, perhaps from gas around a black hole, suddenly disappear as a companion star eclipses the hole.
   (d) They have seen a star suddenly disappear as it was swallowed by a black hole.
   (e) They have looked inside a black hole with X ray telescopes.

8. (15.3) The Schwarzschild radius of a body is
   (a) the distance from its center at which nuclear fusion ceases.
   (b) the distance from its surface at which an orbiting companion will be broken apart.
   (c) the maximum radius a white dwarf can have before it collapses.
   (d) the maximum radius a neutron star can have before it collapses.
   (e) the radius of a body at which its escape velocity equals the speed of light.
9. (15.1/15.3) Spectral lines emitted by material close to the event horizon of a black hole appear redshifted to us because
   (a) blue light cannot escape from a black hole.
   (b) the photons lose energy in the strong gravitational field.
   (c) there is always a strong Doppler effect near the black hole.
   (d) accreting material is heated and glows red.
10. (15.1–15.3) The Sun will ultimately
   (a) suffer core collapse and explode in a type II supernova and become a neutron star.
   (b) collapse without exploding and become a black hole.
   (c) eject its outer layers and leave behind its core as a white dwarf.
   (d) suffer core collapse and explode in a type II supernova and become a black hole.
   (e) explode as a type Ia supernova.

KEY TERMS

accretion disk, 397
black hole, 399
Chandrasekhar limit, 389
compact star, 387
conservation of angular momentum, 394
core-collapse supernova, 392
curvature of space, 398
degeneracy pressure, 389
escape velocity, 399
event horizon, 401
exclusion principle, 389
glitches, 396

gravitational redshift, 389
gravitational waves, 398
Hawking radiation, 404
neutron star, 392
nonthermal radiation, 395
nova, 390
pulsar, 393
Schwarzschild radius, 399
synchrotron radiation, 396
type Ia supernova, 391
type II supernova, 392
white dwarf, 387
X ray binary, 397

ⓠ: FIGURE QUESTION ANSWERS

WHAT IS THIS? (chapter opening): This image shows a Hubble Space Telescope image of an unusual nova, an exploding star. Most of what is visible in the image is from the "light echo" of the flash lighting up material expelled in the late stages of the star's life.

FIGURE 15.8: Count the peaks in a time interval of, say, 1 or 2 seconds. Then divide the number of peaks by the time interval. The one in the upper graph spins 9 times per second. The one in the lower graph spins 3 times in 2 seconds for a spin rate of 1.5 times per second. If pulses were visible from both poles, the number of peaks should be first divided by 2.

16

THE MILKY WAY GALAXY

THE MILKY WAY GALAXY

Infrared image of the center of the Milky Way made by the *Spitzer Space Telescope*

LEARNING OBJECTIVES

Upon completing this chapter you should be able to:

- Explain how the shape and size of the Milky Way Galaxy were deduced from observations.
- Discuss the reasons astronomers disagreed about the size and shape of the Milky Way until the last century.
- Describe the main components of the Galaxy.
- Compare the characteristics of the different populations of stars in the Milky Way.
- Explain how to apply gravitational formulas to an extended system like the Milky Way, and use them to derive masses.
- Discuss the two-stage collapse and merger hypotheses and how they explain features of the Milky Way.

- Classify the different kinds of clouds found in interstellar space and explain how each kind is detected.
- Explain how the composition of interstellar clouds can be determined by observations at a variety of wavelengths.
- Describe the properties of interstellar dust and how it interacts with light.
- Discuss hypotheses about how spiral arms form in a galaxy.
- Describe the evidence for dark matter in the Milky Way as well as the evidence for a massive central black hole.
- Describe how the Milky Way is thought to have evolved over time, and discuss how it might evolve in the future.

On a clear, moonless night, far from city lights, you can see a pale band of light spangled with stars stretching across the sky. The ancient Hindus thought this shimmering river of light in the heavens was the source of the sacred river Ganges. From their temple observatories in the Andes, ancient Incan astronomers noted that dark blotches interrupt the glowing river. They gave these dark areas names, just as peoples of the classical world named the constellations. To the ancient Greeks, this glowing celestial band looked like milk spilled across the night sky, and so they called it the *galaxias kyklos*, literally the "milky circle." This gave us both our terms **Milky Way** and Galaxy. The first recorded use of these terms in English are found in Geoffrey Chaucer's 1380 poem "House of Fame":

> See yonder, lo, the Galaxyë,
>
> Which men clepeth the Milky Wey,
>
> For hit is whyt.

With some of his first telescopic observations in the seventeenth century, Galileo showed that the Milky Way contains uncounted stars, most too dim to be seen with the naked eye as individual points of light. Now, in the twenty-first century, we know that there are hundreds of billions of stars along with our Sun forming a huge, slowly spinning disk, and the dark regions named by the Incans are vast clouds of dust and gas that give birth to new stars. The Milky Way is not just a band of light across the night sky, but our home Galaxy.

CONCEPTS AND SKILLS TO REVIEW

- Inverse-square law (13.2)
- Standard-candle method (13.2)
- Newton's law of gravity (3.5)
- Modified Kepler's third law (3.7)
- Wavelengths of light (4.1)
- Evolution and colors of stars (13.6)

16.1 OVERVIEW OF THE MILKY WAY

A view of the Milky Way on a clear, dark night is one of nature's finest spectacles. Superposed on the dim background glow are most of the bright stars and star clusters that we can see, which all belong to our Galaxy. There was a great deal of mostly philosophical speculation about the nature of the Milky Way for many centuries. It has required modern telescopes operating at many wavelengths to unravel the mystery of this faint band of light in the sky. Understanding the immensity and structure of the Milky Way has been a puzzle only finally solved in the 20th century when it was also discovered that the Milky Way is but one among hundreds of billions of other galaxies in the Universe.

Shape of the Milky Way

Our understanding of the Milky Way as a star system dates to the eighteenth century, when Thomas Wright, an English astronomer, and Immanuel Kant, the great German philosopher, independently suggested that the Milky Way is a flattened swarm of stars. They argued that if the Solar System were near the center of a spherical cloud of stars, we would see roughly the same number of stars in all directions. Instead, we see vastly more stars in the direction of the Milky Way than in directions perpendicular to it, as shown in the "fish-eye" photograph in figure 16.1.

FIGURE 16.1
Wide-angle photo of the Milky Way. Trees and structures are visible on the horizon around the outer edge of the image.

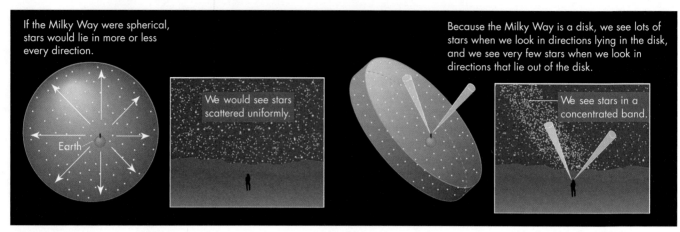

If the Milky Way were spherical, stars would lie in more or less every direction.

We would see stars scattered uniformly.

Earth

Because the Milky Way is a disk, we see lots of stars when we look in directions lying in the disk, and we see very few stars when we look in directions that lie out of the disk.

We see stars in a concentrated band.

FIGURE 16.2

(A) If the Milky Way had a spherical shape, we would see about the same number of stars in every direction. (B) However, because it is a disk, we see stars concentrated into a band around the sky.

In this and the remaining chapters, we will be dealing with huge objects at immense distances. Here is a quick review of some units and powers-of-ten notation used for this purpose.

Astronomers measure large distances in either light-years (ly) or parsecs (pc). One parsec is 3.26 light-years and is about 3×10^{13} (30 trillion) kilometers.

Galaxy dimensions are so huge that even parsecs are inappropriate for measuring their size, so astronomers use kiloparsecs (kpc) for that purpose. 1 kpc = 1000 parsecs.

Astronomers use an even larger unit, the megaparsec (Mpc) for measuring the distances to galaxies.

This is what we expect to see if the Milky Way is a disk rather than a ball of stars, as illustrated in figure 16.2. A simple analogy may help you better understand why we see the Milky Way as a band on the sky. When you are sitting in a classroom, if you look at the heads of other students throughout the room, they all lie roughly in a band that encircles you. You see no heads when you look up toward the ceiling or down toward the floor, but you see lots of heads lying in a plane roughly parallel to the floor of the classroom. So, too, it is with us and stars in the disk of the Milky Way.

These arguments were refined toward the end of the eighteenth century by the English astronomer Sir William Herschel, who counted the number of stars visible in different directions, and estimated that the disk was about five times wider than it was thick. Moreover, because he saw roughly the same number of stars in all directions around the disk, he concluded that the Sun must lie near its middle, as shown in figure 16.3. Based on the brightness of the dimmest stars he could see, Herschel estimated that the Milky Way was about 8000 light-years, or 2500 parsecs, across. However, this was very uncertain because stellar parallaxes were not detected until the next century.

Size of the Milky Way

The size of the Milky Way was remeasured with much more precision in the early 1900s by the Dutch astronomer Jacobus Kapteyn (pronounced *cap-tine*) and the American astronomer Harlow Shapley. Like Herschel, Kapteyn concluded that the Milky Way is a disklike system with the Sun located near the center, but about 18,000 parsecs in diameter.

Kapteyn estimated the distance to remote stars based on their motion. Stars move relative to each other, and astronomers can often detect this shift in photographs taken several years apart. Kapteyn realized that nearby stars change their positions faster than more distant stars, just as a ball tossed past your head moves across your field of view more rapidly than one thrown at the far end of a playing field. By comparing the

FIGURE 16.3

Herschel's 1784 sketch of the Milky Way. He based this cross-sectional view on the number of stars he could see in different directions. He correctly deduced that the Milky Way was wider than it was thick, but *incorrectly* concluded that the Sun (the orange dot) is near the Milky Way's center.

motions of remote stars with those of nearby stars whose distances were known, he estimated the distance to the remote stars and thus the size of the Milky Way.

Almost immediately after Kapteyn's publication of his model of the Milky Way, Shapley published a model in strong disagreement with it. Shapley argued that the Milky Way is larger than Kapteyn believed and that the Sun is not near the center but about two-thirds of the way out in the disk. Shapley drew his conclusions from a study of globular star clusters—dense groupings of up to a million stars, such as the one shown in figure 16.4. Because these clusters contain so many stars, they are very luminous and can be seen at large distances—across the Galaxy and beyond. Moreover, many of them lie above the disk, and so we have a clear view of them.

Shapley noticed that the globular clusters are not scattered uniformly across the whole sky but are concentrated in the same place that the stars are concentrated, that is, toward the constellation Sagittarius. He hoped, therefore, that a detailed map of the distribution of the globular clusters would show him the shape of the Milky Way.

To make his map of the clusters, Shapley needed to know their distance from the Sun. He found the distances to the clusters by observing the variable stars in them. These stars obey a law that relates their luminosity to how long it takes them to brighten, dim, and brighten again (their period) that had just recently been discovered by Henrietta Leavitt. From the luminosity of the stars, their apparent brightness, and the inverse-square law, Shapley calculated the distances to the clusters they were in. Then, having found the distance to each cluster, Shapley plotted where they lay in the Milky Way. Figure 16.4 shows his results.

The clusters fill a roughly elliptical region about 100,000 parsecs (roughly 300,000 light-years) across. Moreover, the Sun lies *not* at the middle of the region, but about two-thirds of the way from its center. Shapley concluded that our Galaxy is ~100,000 parsecs in diameter and that the Sun is nearer to its edge than to its center (fig. 16.4).

ANIMATION ▶

The Milky Way

LOOKING UP ⬆

The Sagittarius region is shown in Looking Up #7 at the front of the book.

FIGURE 16.4
Schematic version of Shapley's plot of globular clusters from which he inferred the size of the Milky Way and the Sun's location in it. Notice that the clusters fill a roughly elliptical region and that the Sun is *not* at the center of their distribution. An image of a typical globular cluster is shown at top, and Kapteyn's and Herschel's models are shown to scale.

The great difference between Shapley's value for the diameter of the Milky Way and the Sun's location and Kapteyn's much smaller size created a major controversy among astronomers. Like so many other controversies, both sides were partially right and partially wrong. For example, Kapteyn deduced a size for the Milky Way that was far too small because he assumed that space is transparent: he did not recognize the dimming effect of interstellar dust.

On the other hand, Shapley was correct about the Sun not being at the center, but he greatly overestimated the diameter of the Milky Way, finding a value about three times larger than the one currently accepted. Part of the reason for his error was that Shapley also, like Kapteyn, did not recognize the dimming effect of interstellar dust. Thus, Shapley mistook the dimness of the globular clusters to mean that they were very far away. Moreover, he did not realize that he was dealing with two types of pulsating variables: RR Lyrae stars and Cepheids. Leavitt had determined the period–luminosity law for Cepheids, but Shapley was measuring the Galaxy's size with RR Lyrae stars. Because RR Lyrae stars are much dimmer than Cepheids, Shapley thought they were much farther away than they really are. Thus, he overestimated the size of the Milky Way. Kapteyn came closer to the presently accepted size of our Galaxy, but he was wrong about where the Sun lies in it.

Structure of the Milky Way

Today astronomers have revised and improved Shapley's model of the Milky Way. Much of that improvement comes from studies of other galaxies. For example, many external galaxies are flat disks with conspicuous spiral arms, as illustrated in figure 16.5. Such arms are hard to detect in our own Galaxy because from our location within the Milky Way we cannot get an overview of its disk. Nevertheless, given that the Milky Way is a flat disk and that other disk galaxies have spiral arms, astronomers have searched for, and found, spiral arms in our Galaxy. Thus, by combining

FIGURE 16.5

Image of the spiral galaxy NGC 4303 (also known as M61), thought to be quite similar to the Milky Way. A position corresponding to the Sun's distance from the center is also indicated.

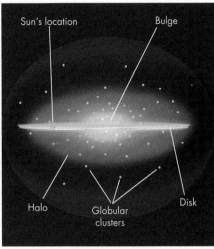

Top view

Side view

FIGURE 16.6
Artist's sketch of the Milky Way showing top and side views, illustrating the disk, bulge, and halo. Notice how thin the disk is and how the halo surrounds the disk much as a bun surrounds a hamburger. Because the Milky Way's stars gradually thin out and do not just stop at some distance, its size is labeled only approximately. The size of the halo in which the Milky Way is embedded is very uncertain, so it is shown only schematically. It is omitted from the top view for clarity.

observations of our own Galaxy with the general features known to occur in other galaxies, astronomers can assemble a detailed model for the Milky Way.

The Milky Way's structure has three main parts, as illustrated in figure 16.6: a **disk** about 30 kiloparsecs (100,000 light-years) in diameter, a **bulge** at its center, and a **halo** that extends out to large distances. Most of the gas and dust clouds lie within the disk, along with numerous bright young stars that gather into **spiral arms** that wind outward from near the center. Our Solar System lies on the inner edge of a spiral arm about 8 kiloparsecs (roughly 26,000 light-years) from the center. The orbital plane of the planets in the Solar System is tilted by about 60° with respect to the Galaxy's disk. This tilt is the reason the band of the Milky Way on the sky is tipped with respect to the ecliptic.

The bulge at the center of the Milky Way is a dense concentration of stars primarily. The bulge is somewhat flattened and elongated, like the shape of a flattened football. The halo of the Galaxy contains the globular clusters, scattered stars, and hot gas, and is more spherically distributed. It extends from the center of the Galaxy outward, reaching out beyond where the disk can be detected. You may find the following analogy helpful in visualizing the scale of our Galaxy: if the Milky Way were the size of the Earth, the Solar System would be the size of a cookie.

The gravity of all the components holds our Galaxy together—but that gravity would draw everything inward and collapse the Milky Way, were it not for the orbital motion of the stars and gas. Astronomers have measured that motion for our Sun and neighboring stars: the Sun and its neighbors move around the center of the Milky Way at a speed of about 220 kilometers per second. That motion around the center is what originally gave the disk its flattened shape, much as the rotation of the presolar nebula made our Solar System flat. Moreover, just as planets near the Sun complete their orbits faster than planets farther out, the stars near the center of the Galaxy complete their orbits faster than stars at the edge of the Galaxy. Thus, the inner parts of the Galaxy spin faster than the outer parts, a phenomenon called "differential rotation." In section 16.4 we will see that this plays an important role in making the Galaxy's spiral arms.

The Sun's large orbital speed (roughly 10 times the speed of the Earth around the Sun) might make you think the Milky Way spins rapidly, but because our Galaxy is so huge, the Sun takes approximately 220 million years to complete one trip around it. Since the death of the dinosaurs about 65 million years ago, our Galaxy has traveled only about one-quarter of its orbit.

Within this swirling mass of stars, the Sun is nearly lost—like a speck of sand on a beach. But unlike the close packing of grains of sand, stars are widely separated. For

> Because the central bulge of our Galaxy is elongated, the Milky Way should be described as a "barred" spiral (see chapter 17).

example, near the Sun and throughout most of the disk, stars are typically several light-years apart: the Sun's nearest neighbor is 4.2 light-years away, a separation relative to its size like that of grains of sand 15 miles apart. Near the core of the Milky Way, stars are packed far more densely, with a separation roughly 1500 times less than that near the Sun, a separation like grains of sand at opposite ends of a football field. The density of stars, as measured by the number of stars per cubic light-year, is probably more than 10 million stars per cubic light-year near the core but only about 0.003 stars per cubic light-year near the Sun. Near the edge of the disk, stars are spread even more thinly, much as our atmosphere thins out as it merges into space. Thus, the Milky Way has no sharply defined outer edge.

Composition and Mass of the Milky Way

Many of the interesting objects visible in figure 16.7 are identified in Looking Up #8 at the front of the book

Mingled with the stars of the Milky Way are huge clouds of gas and dust that amount to about 15% of the mass of stars. We can see some of these clouds by the visible light they emit, whereas others reveal themselves by blocking the light of background stars, as can be seen in figure 16.7. In fact, dust scattered throughout the Galaxy prevents us from seeing farther than about 3000 parsecs within the disk at visible wavelengths. This interstellar material is examined in section 16.3.

Interstellar dust prevents us from seeing our **Galaxy's nucleus**—its core—when we look at the night sky. However, radio, infrared, and X ray telescopes can "see" through the dust and reveal that the core of our Galaxy contains a dense swarm of stars and gas as well as a massive black hole. These features are discussed in section 16.6.

Astronomers can calculate the Milky Way's mass from the gravitational attraction needed to hold its stars in orbit. Such calculations show that the Milky Way must contain at least 10^{11} $M_\odot$ of matter inside the Sun's orbit, and even more matter farther out, where its gravitational effect on nearby galaxies implies an overall mass as large as 2×10^{12} $M_\odot$. These calculations are puzzling, because when astronomers add up the mass of the Milky Way's *visible* stars and gas, they find only about 10 to 20% of the predicted masses. The other 80 to 90% is undetectable except by its gravitational effect. Thus, astronomers conclude that the major part of our Galaxy is composed of some unseen material, which they therefore call **dark matter.** This is discussed further is section 16.5

Because dark matter emits no radiation of any sort, it is very difficult to study. Its composition is one of the major mysteries in astronomy today, and astronomers know few of its properties. They do know that dark matter is spread throughout our Galaxy within the halo. They also know that it does not interact with ordinary stars and gas except through its gravitational pull. Finally, as we will see in chapter 17, astronomers know that the Milky Way is not alone in having a massive dark-matter halo: other galaxies are also embedded in dark-matter haloes of their own.

Our uncertainty about the nature of dark matter and the value of our Galaxy's mass makes it difficult to determine the number of stars our Galaxy contains. However, if we assume that the average star in our Galaxy has a mass comparable to the Sun's (an assumption based on censuses of stars, as we will see in section 16.2), then from the best estimates of the Galaxy's total mass and light output, we deduce that it contains about 10^{11} (100 billion) stars, each moving along its own orbit, held in its path by the collective gravity of the other stars, gas, and dark matter.

Age of the Milky Way

The immensity of the Milky Way's size and number of stars is matched by the Milky Way's great age. Astronomers measure this age from the age of our Galaxy's oldest stars, using the techniques for measuring stellar ages that we discussed in chapter 14. Such measurements (and others) show that the oldest stars in our Galaxy have an age

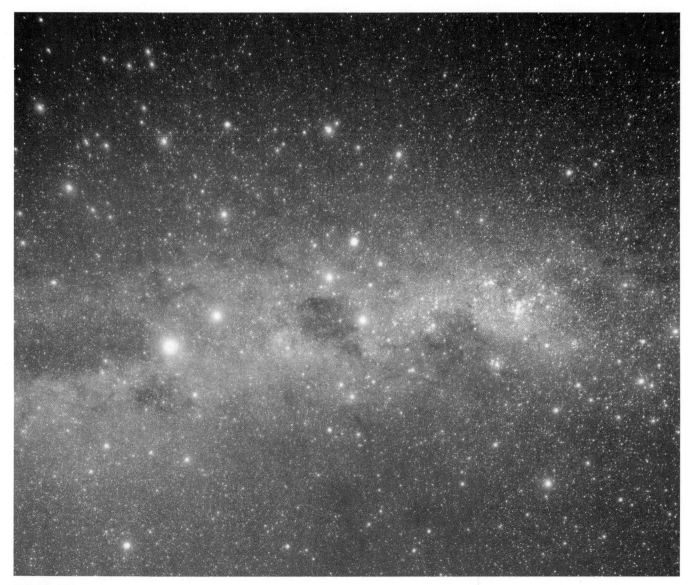

FIGURE 16.7
Clouds of gas and dust in the Milky Way. Some hot clouds glow pink with hydrogen emission. The dust in some cold clouds can be seen silhouetted against background stars.

of approximately 13 billion years, which appears to be when the Milky Way began to assemble itself.

Our Galaxy's stars were not all created at a single time in that distant past. As we discussed in chapter 14, stars are forming even today. If we divide the total number of stars in the Galaxy by the Galaxy's age, we can find an average rate of star formation over the Galaxy's lifetime. This gives a birthrate of approximately 7 to 8 stars born in the Milky Way each year, although it appears that the birthrate was much higher when our Galaxy was young, and it has slowed down substantially today.

Given the declining rate of star formation and the amount of interstellar gas remaining in our Galaxy, the Galaxy can continue to form new stars for billions of years into the future. However, interactions with other galaxies may have played a big role in the Milky Way's past as well as its future, as we will explore in section 16.7. Some of our best evidence of how the Milky Way formed and interacted comes from identifying stars of different ages and determining their distribution within our Galaxy, which we turn to next.

Stellar Censuses

The Milky Way contains many types of stars: giants and dwarfs, hot and cold, young and old. Stellar censuses, which list all known stars of a given type in a given region of space, show that all the types described in chapters 13 to 15 occur in our Galaxy. By analyzing the types of stars and their relative numbers, astronomers can deduce some of the ancient history of our Galaxy, much as archeologists can learn about life in an ancient city from the kinds of buildings it contained. Such studies of the stellar contents of the Milky Way show that stars similar to or smaller than the Sun vastly outnumber more-massive stars.

The most numerous stars in our Galaxy turn out to be dim, cool, red dwarfs—stars that lie on the main sequence but whose mass is about 0.1 to 0.5 $M_\odot$. Astronomers do not understand yet what determines the distribution of stellar masses, though it probably depends on the conditions in the cloud where the stars form. The average mass for Milky Way stars is a little smaller than 1 $M_\odot$. Some are significantly more massive (up to 100 or so $M_\odot$), whereas others are much less massive (perhaps as small as 0.08 $M_\odot$). Stars much more massive than about 30 $M_\odot$ form only very rarely, as we discussed in chapter 14. Objects less massive than about 0.08 $M_\odot$ were once also believed to be rare, but astronomers now think they may be common and were simply missed in previous stellar surveys.

How could astronomers miss these objects if they are so abundant? Recall that low-mass stars tend to be cooler than the Sun. In fact, if a star is less massive than about 0.08 $M_\odot$, it is so cool that it is unable to fuse hydrogen into helium. Thus, these "brown dwarfs" (as we called them) are more like huge planets than tiny stars, and their extreme dimness and small size make them very difficult to detect. However, searching the sky at the wavelengths at which these objects are brightest greatly increases the chance of finding them. What wavelengths are best? Because they are so cool, brown dwarfs are brightest in the infrared, so it is with infrared surveys such as 2MASS (2 Micron All Sky Survey) that astronomers are now finding large numbers of these "failed" stars. In fact, astronomers now think brown dwarfs may outnumber ordinary stars, making them one of the most common kinds of astronomical objects in our Galaxy.

Despite their great numbers, low-mass stars are hard to see because they are so intrinsically dim. Thus, when we look out at the night sky, the stars we see tend to be giants. Such stars are rare, but because we can see them even at great distance, we see greater numbers of them. Astronomers refer to this difference between what is seen and what is truly present as a **selection effect.**

> Selection effects can occur whenever one samples a population. For example, when the U.S. Census Bureau counts our population, it must make corrections for people who are homeless or who otherwise might be overlooked. Similarly, astronomers must guard against drawing false conclusions from data that are biased by selection effects.

Two Stellar Populations: Population I and Population II

Hidden in the great diversity of star types is an underlying simplicity astronomers first noted in the 1940s. At that time, the 100-inch reflector at Mount Wilson Observatory, located outside Los Angeles, California, was the largest telescope in the world and the best for observing galaxies. The glow from city lights, however, made it hard to see faint galaxies. But blackouts during World War II darkened the night sky, and Walter Baade, an astronomer at Mount Wilson, took advantage of the darkness to make a series of photographs of neighboring galaxies. He noticed that stars in these nearby galaxies were segregated by color. Red stars were concentrated in the bulges and halos of the galaxies, whereas blue stars were concentrated in their disks and especially in the spiral arms. To distinguish these groups, Baade called the blue stars in the disk **population I** and the red stars of the bulge and halo **population II** (Pop I and Pop II

ANIMATION ▶

The Pop I and II orbits

Side view

A

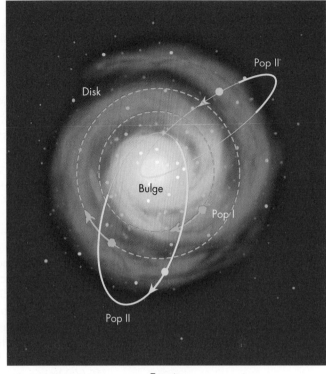

Top view

B

FIGURE 16.8
Stellar orbits in the Milky Way. Pop I stars orbit in the disk
(blue lines). Pop II stars orbit in the halo (yellow lines); notice
the elongation and inclination of their orbits.

Table 16.1	Properties of Population I and II Stars	
	Pop I	**Pop II**
Age	Young to old (10^6 to a few times 10^9 years)	Old (about 10^{10} years)
Color	Blue (overall)	Red
Location	Disk and concentrated in arms	Halo and bulge
Orbit	Approximately circular in disk	Plunging through disk
Heavy-element content	High (similar to Sun's)	Low (1% to 0.1% of Sun's)

for short). As he pursued his studies of the stellar populations of other galaxies, he saw
that Milky Way stars showed the same division.

Population I and population II stars generally differ in five ways: location in Galaxy, motion, composition, color, and age. Population I stars lie in the disk and follow
approximately circular orbits, as shown in blue in figure 16.8. Their composition is
like the Sun's, mostly hydrogen and helium, but about 3% of their mass is heavy elements such as carbon and iron. They are generally young (10^6 to a few times 10^9 years
old) compared to the Galaxy, and many of them are blue.

Population II stars, on the other hand, lie in the bulge and halo of the Galaxy and
move along highly elliptical orbits that may be tilted strongly with respect to the disk,
as shown in yellow in figure 16.8. They too are mostly hydrogen and helium, but only
a few hundredths percent of their mass is composed of heavy elements, roughly a hundred times less than in stars like the Sun. Unlike population I stars, population II stars
are very old (about 10^{10} years old) and generally red. Table 16.1 summarizes some of
these properties.

Some astronomers now prefer the
simple labels *disk, bulge,* and *halo* stars
to describe the stellar populations of
the Milky Way and other galaxies.

FIGURE 16.9

The positions of *O* and *B* stars give some indication of the spiral arms near the Sun. The stars are plotted in the plane of the Milky Way, with 0° indicating the direction toward the Galaxy's center.

Division of all stars into two broad categories is an oversimplification. For example, the Sun fits in neither category precisely, but it is considered a Pop I star because of its relatively high heavy-element content and approximately circular orbit in the disk. Although not all stars fit within the precise division of these two populations, the division still provides a fairly good description of the majority of stars. Many astronomers think that the presence of these two different populations show that star formation has not occurred continuously in the Milky Way. Population II stars probably formed in a major burst at the time of the Galaxy's birth during its initial collapse, whereas population I stars formed later and continue forming even today. This hypothesis explains the differences between the two populations, as we will discuss in greater detail in section 16.7.

One of the first uses astronomers made of stellar population differences was to map the Milky Way's spiral arms. As Baade noted, the spiral arms of other galaxies are demarcated by their Pop I stars, especially the hot blue ones of spectral classes *O* and *B*. It is therefore logical to assume that the same is true for the Milky Way. By measuring the location of *O* and *B* stars near the Sun as shown in figure 16.9, astronomers can make a crude picture of the spiral arms near us, even though they cannot view the arms from a distance. This method is limited, however, because dust in space prevents us from seeing even the brilliant *O* and *B* stars if they are farther from the Sun than about 3000 parsecs (10,000 light-years). Nevertheless, such maps were the first evidence that we live in a spiral Galaxy.

Star Clusters

Some of the Milky Way's stars are bound together gravitationally into groups called **star clusters.** Within a cluster, each star moves along its own orbit about the center of mass of the cluster, held to it by the group's gravity. You can see several such clusters with your naked eye. The most obvious one is the Pleiades, or the Seven Sisters (shown in fig. 16.10), named for the daughters of the giant Atlas, who in Greek mythology carried the world on his shoulders. The Pleiades (pronounced *plee-a-dees*) are visible from late August through March as a tiny group of stars north of the V in the constellation of Taurus. With binoculars, you can see that the Pleiades contain many more than the half-dozen stars visible to the naked eye. Likewise, by scanning the Milky Way with binoculars, you will see many additional clusters dimmer than the Pleiades.

The Milky Way has two main types of star clusters: open clusters and globular clusters. **Open clusters**—the Pleiades, for example—contain up to a few hundred members in a volume with a radius of typically 7 to 20 light-years. They are called "open" because their stars are scattered loosely, as you can see in figure 16.10. Sometimes they are also called "galactic clusters" because most of them lie in the disk of the Galaxy.

Astronomers think that open clusters form when giant, cold, interstellar gas clouds move into the Galaxy's spiral arms. The clouds are compressed by the stronger gravitational field within the arm and collapse, breaking up into hundreds of stars whose mutual gravity binds them into the cluster. Once formed, an open cluster continues to orbit the Galaxy in the disk, but over hundreds of millions of years, stars gradually escape from it and so the cluster eventually dissolves. Assuming that open clusters are as abundant elsewhere in the Galaxy as within the regions visible from Earth, astronomers estimate that our Galaxy contains about 20,000 open clusters, though dust hides most of them from our view. Our own Sun probably once was a member of such a star group, its companion stars having long since scattered across space.

The Pleiades are shown in Looking Up #5 at the front of the book

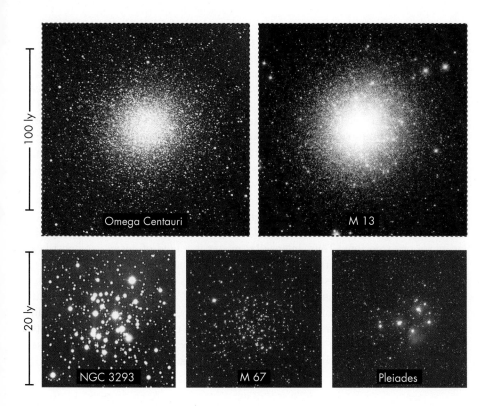

FIGURE 16.10
Star clusters. Two globular clusters are shown on the top at the same scale, and three open clusters are shown along the bottom, at a larger scale.

Within our Galaxy's spiral arms, extremely young stars sometimes occur in loose groups called **associations** that are a few hundred light-years across. Associations typically spread out from a single large open cluster near their center and may contain other, smaller star groupings. Moreover, the stars in associations are usually still mingled with the massive clouds of dust and gas from which they formed.

Globular clusters contain far more stars than found in open clusters, from a few hundred thousand to several million per cluster. We described them briefly in our discussion of Shapley's study of the shape of the Milky Way. They have larger radii than open clusters have—40 to 160 light-years—but the larger number of stars in them creates a stronger gravity than that found in open clusters. This stronger gravity pulls their stars into a denser ball, as figure 16.10 shows. Astronomers estimate that about 150 to 200 globular clusters orbit the Milky Way. Table 16.2 summarizes some properties of open clusters, globular clusters, and associations.

Open clusters and globular clusters reflect the properties of Pop I and II stars that we discussed earlier. Open clusters are generally Pop I, whereas globular clusters are always Pop II. This difference in their stellar population makes clusters especially useful to astronomers for studying the structure of the Milky Way. For example, globular clusters (among the oldest stars in the Galaxy, many of them roughly 12 billion years old) outline the halo. On the other hand, the younger open clusters (some a mere million years old) trace the Galaxy's arms.

Table 16.2	Properties of Clusters and Associations		
Type	**Number of Stars**	**Radius***	**Location**
Open cluster	Tens to a few thousand	7 to 20 ly (2 to 6 pc)	Along spiral arms and in disk
Globular cluster	10^5 to 10^6	40 to 160 ly (12 to 50 pc)	Halo and bulge of galaxy
Associations†	5 to 70 O or B stars	65 to 325 ly (20 to 100 pc)	Spread along a spiral arm

*Because star clusters do not have sharp edges but instead gradually thin out, quoted dimensions differ substantially.
†Astronomers identify several other types of associations as well. For example, T associations are regions with above-average numbers of T Tauri stars.

16.3   GAS AND DUST IN THE MILKY WAY

The space between stars is not empty: it contains **interstellar matter** composed of gas and dust particles. By terrestrial standards, this space is almost a perfect vacuum. On the average, each cubic centimeter contains only a few atoms of gas. For comparison, the air we breathe contains about 10^{19} atoms per cubic centimeter. The density of gas in interstellar space compared to air is like one marble in a box 5 miles on a side compared to the same box filled completely with marbles.

Despite interstellar matter's very low density, it plays several important roles in the Milky Way. First, it is the material from which new stars form, as we discussed in chapter 14. With no interstellar matter, star birth would stop—though star deaths would, for a while, replenish the material with gas and dust blown out of stars by stellar winds or explosions. Thus, interstellar matter is a kind of cosmic compost heap with matter from dying stars being recycled back into new stars.

A second reason interstellar matter is important is that the dust mixed in with it strongly limits what we can observe within the Milky Way. Because interstellar dust dims and blocks our view of distant parts of our Galaxy, it seriously affects our ability to measure distances to stars and even to detect stars in the more distant regions of the Milky Way. But before we turn to how astronomers overcome the difficulties created by interstellar dust, it will be helpful to look more closely at some general properties of interstellar matter.

Distribution and Composition of Interstellar Matter

Interstellar matter is scattered throughout the Milky Way, but gravity has pulled most of it into a thin layer in the disk, as can be seen by the dark band of dust in other galaxies (fig. 16.11). Within this layer, gravity and gas pressure have further clumped the dust and gas into clouds whose sizes range from a few light-years to a few hundred light-years. We can see such clouds directly (see fig. 16.6), but we can also detect them by their effect on light from distant stars.

As light from a background star passes through an interstellar cloud, the cloud's gas and dust imprint narrow absorption lines in the star's spectrum (fig. 16.12). By analyzing these imprinted lines, astronomers can deduce the cloud's composition, temperature, density, and speed. Such observations show that the gas in interstellar

Q: Look carefully above and below the disk of the Galaxy. Do you see some small, fuzzy, round spots? On the basis of their location, what do you think they are?

FIGURE 16.11
Interstellar matter, outlined by dark dust clouds, is concentrated in the plane of a galaxy, forming a dark band across it. This galaxy is sometimes called the "Sombrero galaxy." Technically, it is called either NGC 4594 or M104.

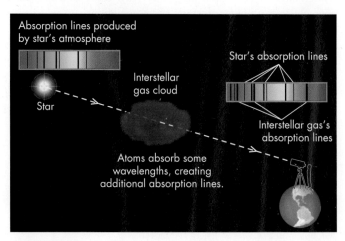

FIGURE 16.12
Gas atoms in an interstellar cloud between us and a distant star absorb some of the star's light, adding their own absorption lines to those in the star's spectrum.

clouds has a composition similar to that of the Sun and other stars—namely, about 71% hydrogen and 27% helium, with trace amounts of heavier elements such as carbon and neon. Analysis of the light absorbed by the dust particles shows that they are made of silicates and carbon compounds, coated with a thin layer of ices made of molecules such as water, carbon monoxide, and methyl alcohol. These particles are extremely tiny, ranging from about a micrometer to a nanometer or so in diameter (roughly the size of a single large molecule).

Interstellar Dust: Dimming and Reddening

We know there is dust in interstellar space because it dims the brightness and alters the color of the stars we see through it, much as haze in our atmosphere dims and reddens the setting Sun. Dimming and reddening result when light strikes a tiny particle such as a dust grain. The light is reflected from it in random directions in a process called **scattering.** A particle's ability to scatter light depends on its size compared to the light's wavelength. Light is scattered strongly if its wavelength is smaller than the particles, which for interstellar dust and molecules in our atmosphere means that short-wavelength blue and ultraviolet light is scattered extremely well. Scattering of sunlight creates the blue color of our sky. Scattering of starlight in space also creates visible effects.

Scattering dims a star's light because radiation that would have reached us in the absence of dust is sent off in random directions and is lost to our sight. Because red light scatters less than blue, more of the star's red light reaches us than its blue. As a result the star looks redder than it really is, a phenomenon called **reddening** (fig. 16.13A). The same effect occurs in our atmosphere and makes the setting Sun look red.

If a dust cloud is extremely thick or dense, it may dim the light of background stars to the point where they become invisible, an effect called "obscuration." This creates a dark region, seemingly empty of stars, which astronomers call a **dark nebula.** The images in figure 16.13B and C show regions of the Milky Way in which a dark nebula is silhouetted against faint background stars, blotting out their light.

Only about 1% of the interstellar matter is dust, but this dust scatters visible light so well that it strongly limits our view of the Milky Way. We can see no more than a few thousand light-years away from the Sun into the disk because the dust is

Q: How is reddening similar to the effect on the Sun's color when it sets?

Star emits photons with wide range of wavelengths.

Interstellar cloud of gas and dust

Dust particles scatter blue light. Red light passes through cloud.

Star looks red.

A

B

Canada-France-Hawaii Telescope/J. C. Cuillandre/Coelum

C

FIGURE 16.13

(A) Dust in an interstellar cloud scatters the blue light from a distant star, removing it from the radiation reaching Earth and making the star look redder than it really is. (B) Barnard 86 and (C) the "snake nebula" are dust clouds visible in silhouette against background stars. Note the reddening of some stars near the edges of the dark clouds.

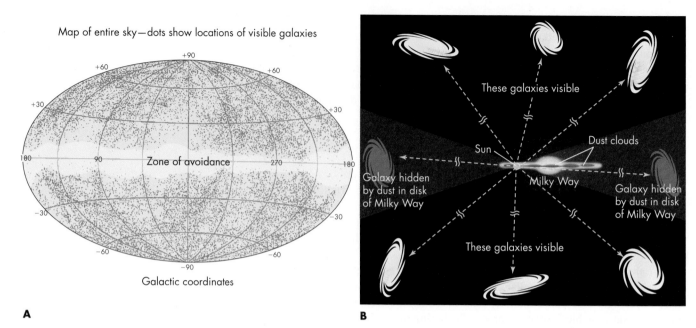

A

B

FIGURE 16.14

Dust limits our view in the disk of the Milky Way, creating the "zone of avoidance" where we see no galaxies. (A) A plot showing how galaxies are distributed on the sky. Notice how few lie along the central line (the Galactic equator). This blank region is the zone of avoidance. (B) A sketch illustrating why we see so few galaxies along the plane of the Milky Way. Dust clouds in the disk of our Galaxy block our view in directions that lie in the Milky Way's plane. The other galaxies depicted in the sketch are in reality about the same size as the Milky Way and are, of course, much farther away than can properly be shown in the figure.

concentrated there. Only by looking above or below the disk can we see more-distant objects. Dust in the disk also hides galaxies that lie behind it, creating a band around the sky called the **zone of avoidance,** in which essentially no galaxies can be seen. Figure 16.14A shows a chart of over 34,000 galaxies illustrating this effect. The narrow horizontal region where no galaxies appear is the zone of avoidance and coincides with the plane of the Milky Way. We know today that galaxies do occur in this region because they can be seen with radio and infrared telescopes sensitive to radiation that can penetrate the dust.

Infrared and radio waves pass more readily through dusty clouds than visible light does because their wavelengths are much larger than the dust particles. This allows infrared pictures to reveal the structure of the Milky Way far better than visible light pictures can, as illustrated in figure 16.15. The disk and central bulge of our Galaxy are

FIGURE 16.15

An all-sky view of our Galaxy, as we see it from Earth. The image was made by plotting on a map of the whole sky the millions of stars observed in the infrared by the 2-Micron All Sky Survey (2MASS). It is therefore like a map of the Earth, flattened out so as to show all continents. The map clearly reveals the flatness of our Galaxy and the bulge of its central regions.

easy to see in this wide-angle picture, which was made by plotting the positions of millions of stars observed as part of the 2MASS infrared survey of the sky.

Dust is not all bad. Most of the atoms that compose the Earth and our bodies came from such dust grains. Dust grains may also scatter enough light to become visible, thereby creating beautiful **reflection nebulas** (fig. 16.16). Scattered starlight makes the interstellar dust shine, just as scattered sunlight makes dust in the air visible. Reflection nebulas are not only lovely but additional proof that dust exists in space.

Interstellar Gas

Interstellar gas plays several crucial roles in our Galaxy. It is the material from which stars form, and it is the repository of matter blown off dying stars. Without interstellar matter to make new stars to replace those that die, the Milky Way would be far dimmer and filled only with stellar corpses and low-mass stars. Interstellar gas is also a help to astronomers seeking to map the Milky Way. Dust blocks our view of the Milky Way at visible wavelengths, making it difficult for us to observe its structure, but interstellar gas emits radiation that penetrates the dust and allows us to "see" regions of the Galaxy that would otherwise be invisible to us. Finally, interstellar gas creates some of the most spectacular structures in our Galaxy: huge glowing clouds heated by stars embedded in them, as shown in figure 16.17. Another example of such a gas cloud is the Great Nebula in Orion, which is easy to see with a small telescope or binoculars. If you look at the middle star in Orion's sword you will see a pale, fuzzy glow created by luminous interstellar gas surrounding a dense star cluster.

Interstellar gas clouds that emit visible light are called **emission nebulas.** Emission nebulas generate their own light in visible and other wavelengths, unlike the dust in reflection nebulas, which simply reflects and scatters light from neighboring stars. However, a cloud must be hot to emit visible light, and so such clouds need a source of heat (generally a nearby star), as shown in figure 16.17C. Hot, blue stars are especially

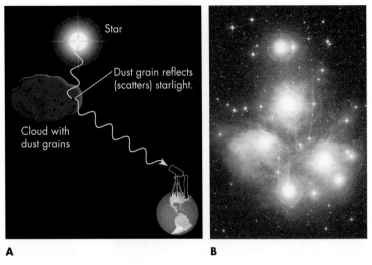

A   **B**

FIGURE 16.16

Reflection nebulas. (A) Light from a star near a dust cloud is reflected from the dust, making it visible. (B) The Pleiades star cluster. Note the bluish reflection nebulas around many of the brighter stars.

Q: Why is red light not scattered?

LOOKING UP ⬆

The Orion Nebula is shown in Looking Up #6 at the front of the book.

A   **B**   **C**

FIGURE 16.17

(A), (B) Emission nebulas. Notice the bright blue stars near the glowing gas. (C) These stars emit ultraviolet light, which ionizes hydrogen gas. Pink-colored visible light is emitted when electrons recombine with the hydrogen and drop down to its second energy level.

FIGURE 16.18

Map of the Milky Way made by combining radio observations (since radio can penetrate dust and thereby allow one to see all the way across the Milky Way's disk) and optical observations of the position of H II regions, outlining our Galaxy's spiral arms.

effective at heating because they emit large amounts of ultraviolet radiation whose short wavelength is much more energetic than that of visible light. Ultraviolet light whose wavelength is less than 91.2 nm is so energetic that when it is absorbed by hydrogen, it tears the electron free of the nucleus, ionizing the gas. In an ionized gas (one in which atoms are missing one or more electrons), the free electrons collide with atoms of oxygen, nitrogen, and other elements and excite them, making them emit light of their own. Eventually the electrons recombine with the ionized hydrogen atoms and emit more radiation, especially the red light of hydrogen alpha, which gives these nebulas their lovely reddish color. Because hydrogen in these hot gas clouds is ionized, they are sometimes called **H II regions,** the *H* denoting the hydrogen and the Roman numeral *II* indicating that it is ionized.

H II (pronounced *H-two*) regions have a temperature of approximately 10,000 K and are heated by the electrons freed by the ionizing ultraviolet radiation. The liberated electrons collide with gas atoms around them and speed them up, increasing their energy and thus their temperature. The freed electrons are themselves a powerful source of radio emission, emission that helps astronomers see H II regions that might otherwise be hidden from us by dust. Because these H II regions are generally located in spiral arms, radio maps of their location can reveal the spiral structure of our Galaxy. In fact, maps made by combining radio observations and optical observations of H II regions are the most direct evidence we have that the Milky Way has spiral arms (fig. 16.18).

Cold Interstellar Gas

Although we can easily see the glow from gas clouds that happen to be near hot stars, much interstellar gas is too cold to emit visible light, and we can detect it only with radio telescopes. Clouds more than a few tens of light-years from hot stars may be only a few kelvin (more than 400° below zero on the Fahrenheit scale). Such cold material emits no visible light because it has too little energy to generate visible-wavelength photons. However, it can emit radio waves of very low energy because radio photons carry much less energy than visible photons do.

Neutral hydrogen gas in H I (*H-one*) clouds emits 21-centimeter radiation that is particularly useful in mapping the Milky Way, as described in Extending Our Reach: "Mapping the Milky Way with Radio Waves." Because hydrogen is the most common element in the Universe, the 21-centimeter radiation traces much of the cold interstellar gas; however, at extremely low temperatures (15 K or so), hydrogen atoms bond to one another to make hydrogen molecules. In its molecular form hydrogen cannot emit 21-centimeter radiation, and it emits very little radiation at other wavelengths either. Although hydrogen is difficult to detect in these very cold clouds, other molecules form that emit detectable radio waves.

More than 100 interstellar molecules have been identified to date, including common compounds such as carbon monoxide (CO), formaldehyde (HCHO), and ethyl alcohol (CH_3CH_2OH), and also more exotic substances. Table 16.3 lists a few of the many interstellar molecules now known. Carbon monoxide is particularly common and widely used to trace out the molecular gas in interstellar regions.

Collisions among neighboring molecules make them spin, giving them energy, as shown in figure 16.19, but they quickly radiate that energy away and reduce their spin by emitting radio waves. Each molecule has a distinctive set of wavelengths at which it can emit its rotational energy, just as each atom has a distinctive set of wavelengths (its spectral lines) at which it can radiate energy from electron jumps. For example, carbon monoxide emits at wavelengths of 2.6 and 1.3 millimeters, and ethyl alcohol emits at 3.5 millimeters. Both these molecules emit at many other wavelengths as well.

Table 16.3	Some Interstellar Molecules
H_2	Molecular hydrogen
OH	Hydroxyl radical
CN	Cyanogen radical
CO	Carbon monoxide
H_2O	Water
HCN	Hydrogen cyanide
NH_3	Ammonia
HCOH	Formaldehyde
HCCH	Acetylene
HCOOH	Formic acid
CH_3OH	Methanol
CH_3CH_2OH	Ethanol

Hydrogen   Oxygen

Carbon

Hydrogen molecule   CO molecule   CO molecule   2.6-mm wavelength
collides with   set spinning. radiation

Spin slows as
radiation is emitted.

2.6 mm

FIGURE 16.19
Emission of radio waves by spinning molecules in cold interstellar clouds. Collisions set the molecules spinning. When their rotation speed changes, they emit radio waves. The wavelength emitted depends on the kind of molecule and by how much its rotation speed changes.

The strength of the molecular radio emission from a gas cloud depends on the number of molecules it contains, and the speed and frequency of collisions. This gives astronomers information on the cloud's density and temperature, and if the gas is moving, the wavelength of the emission will be slightly altered by the Doppler shift, displacing and broadening the spectral lines. From such shifts and broadening, astronomers can deduce the speed of the cloud and its rotation. With that information, and by mapping out where this cold, dense molecular gas is located, astronomers are able to study how gas and stars interact, as well as patterns of star formation in our Galaxy.

EXTENDING *our reach*   MAPPING THE MILKY WAY WITH RADIO WAVES

One of the most important sources of radio emission is cold hydrogen atoms,* which radiate at a wavelength of 21 centimeters. This radiation, called **21-centimeter radiation,** arises because subatomic particles such as protons and electrons possess a property, known as "spin," that makes them behave like tiny magnets. Magnets have north and south poles with the property that like poles repel and unlike poles attract. If an electron has the same alignment as the proton it is orbiting—both with north poles "up," for example—the electron will flip over, as shown in figure 16.20. As it flips, the changed magnetic interaction between the electron and the nucleus makes the electron shift to a slightly smaller orbit with a slightly lower energy. However, energy must be conserved, and so when the electron shifts to the new orbit, it emits the excess energy as electromagnetic radiation. We discussed in chapter 3 that this process can produce visible light. In cold hydrogen, however, the flip of the electron liberates too little energy to make visible radiation and can create only radio emission at a wavelength of 21 centimeters.

The 21-centimeter radiation has proved extremely valuable for studying the Milky Way and other galaxies. Hydrogen is abundant in space, and so the signal of the 21-centimeter radiation is strong. From the strength of the signal, astronomers can deduce the amount of hydrogen in an area. If the gas is moving, the wavelength will be

FIGURE 16.20
Radio radiation at 21 centimeters is emitted by cold hydrogen as an electron "flips" and spins in the opposite direction.

Electron   Proton

21 cm

Electron "flips" over, emitting
21-cm wavelength radio waves.

Doppler-shifted, from which the gas's speed along our line of sight can be found. Radio observations thereby allow astronomers to map not only where the gas is concentrated in the Galaxy but also how it is moving. Although details in such 21-centimeter maps are uncertain because of assumptions that must be made about how the gas moves, they nevertheless are extremely useful because the radiation is not absorbed by interstellar dust, thereby allowing astronomers to "see" the entire Milky Way in its radiation. Such maps show that the gas is confined to a thin disk and its distribution is suggestive of a spiral pattern, confirming the picture deduced by other means.

*Cold hydrogen atoms with their electrons attached are called "H I" (pronounced *H-one*) to distinguish them from ionized hydrogen atoms, H II.

16.4 MOTION OF STARS AND GAS IN THE MILKY WAY

The motion of stars and gas gives important clues to the structure of the Milky Way. In accordance with Newton's first law (the law of inertia, see chapter 3), if no force acted on the stars, each would move in a straight line. However, each star is acted on by a force: the gravitational force created collectively by all the other stars, gas, and other matter in the Galaxy. That force acts to draw each star inward toward the center of the Galaxy. The result is that each star follows its own orbit around the center of the Galaxy, just as each planet moves in orbit about the Sun.

Although all stars within the Milky Way move around its center, the paths followed by stars in the disk and halo are very different, as we discussed in section 16.2. We saw there that all the stars in the disk orbit in the same direction and in nearly the same plane, giving the disk its flat shape, just as the planets' motion in approximately the same plane gives the Solar System its flattened shape. On the other hand, stars in the halo and bulge follow orbits that are steeply inclined to the disk's plane (see fig. 16.8).

Motion within the disk also creates the Milky Way's spiral arms. We know that gravity holds a galaxy together, and so we might infer that gravity is also what makes a spiral arm. Gravity does in fact help hold a spiral arm together, but a spiral arm is not a fixed mass of material. Rather, the stars and gas in an arm slowly drift out and are replaced by new stars and gas, according to the density-wave theory of spiral structure. In the **density-wave model,** waves of stars and gas sweep around the Galactic disk. The wave is not primarily an up-and-down motion like an ocean wave; instead, it is a place where the density of stars and gas is large compared to the surroundings. This denser region is the wave crest, and we see it as a spiral arm. Spiral waves are not unusual in rotating matter. Batter stirred in a bowl with an electric mixer forms spiral waves, but to better understand how arms form in the Milky Way, consider the following analogy.

As cars move along a freeway, most move at nearly the same speed and keep approximately the same separation. If one car moves slightly slower than the others, however, traffic will begin to bunch up behind it. Cars can pass the slower vehicle but must change lanes to do so. The result is a clump of cars behind the more slowly moving one. But that clump is not composed permanently of the same cars. Cars join the clump from behind, pass the slower moving car, and then leave the clump at its front.

A similar phenomenon happens to gas and stars orbiting in the Milky Way, as illustrated in figure 16.21. As stars circle the Galaxy, a few closely spaced stars at a given location will make the local gravity slightly stronger than elsewhere in the disk. That excess gravity draws stars into the region so that the clump grows a little. But stars do not remain in the clump. They continue along their orbits and eventually leave the

FIGURE 16.21
Sketch of a density wave in a galaxy, showing the progression from dust and gas to stars across the arm.

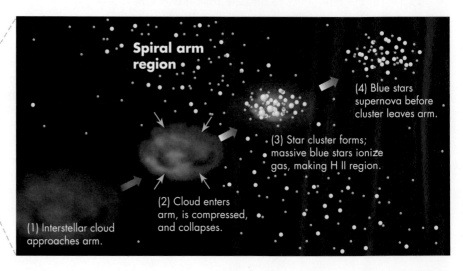

Spiral arm region

(4) Blue stars supernova before cluster leaves arm.

(3) Star cluster forms; massive blue stars ionize gas, making H II region.

(2) Cloud enters arm, is compressed, and collapses.

(1) Interstellar cloud approaches arm.

Star and gas orbit

clump. Likewise, gas clouds drift through the clump, initially accelerated inward by the clump's gravity and then slightly slowed as they move out the other side. The result is a wave in the Galaxy's disk where stars and gas accumulate before moving through. The accumulation of gas compresses clouds and makes them collide with one another, which in turn makes some of the clouds collapse and form stars. This is the first stage of the star formation process described in chapter 14. The brilliant young *O* and *B* stars so born illuminate the arms and make them stand out against the more ordinary stars of the disk. In fact, most of a galaxy's stars form in its spiral arms, and as we will discover in chapter 17, galaxies without arms generally contain very few young stars.

Although the density-wave theory has difficulty explaining the longevity of spiral arms, many observations support the theory. For example, by combining radio and visual pictures of a galaxy, astronomers can see a progression through a spiral arm as gas enters on one side, is compressed, and forms stars, taking a few million years to sweep through the arm. On one side of the arm, you can see gas and young stars; on the other side, older, more evolved stars, as indicated in figure 16.21. Doppler-shift studies of the gas motion also clearly show the gas moving into the arm on one side and leaving on the other.

Although many spiral galaxies probably make their arms this way, another hypothesis has been proposed to explain the very ragged-appearing arms in some galaxies, such as M63 in figure 16.22. This second hypothesis is called **self-propagating star formation.**

The idea of this second hypothesis is that star formation starts at random points in the disk of a galaxy when gas clouds collapse and turn into stars. As the stars heat the gas around them and explode as supernovas, they generate a disturbance that makes the surrounding gas clouds collapse and turn into stars. The original stars burn out, but the new stars—formed as the old ones evolve and explode—trigger more gas clouds to collapse and form additional stars. In this fashion, the region of star formation spreads across the galaxy's disk, much as a forest fire burns outward in a circle through a forest. But unlike a forest where the trees stand still, stars orbit in a galaxy, and those near the center orbit in less time than those farther out. Thus, the zone in which star formation occurs is drawn out into a spiral by the difference in rotation rate between the inner and outer parts of the disk. The result is spiral structure, as illustrated in figure 16.23.

Astronomers are not certain which process—density-wave or self-propagating star formation—is most important in forming spiral arms. There is probably a continuum, with both processes contributing different amounts in different galaxies.

FIGURE 16.22
The galaxy M81 (top) has smooth spiral arms, but M63 (bottom) has ragged arms whose spiral pattern is difficult to trace. M63's arms may be the result of self-propagating star formation.

FIGURE 16.23
Computer model of self-propagating star formation.

We described earlier in this chapter how astronomers measure the basic properties of the Milky Way, such as its radius and mass. In this section, we look in greater detail at those methods.

Diameter of the Milky Way

FIGURE 16.24
Finding the distance to the center of the Milky Way using globular clusters. Make a scale plot of where the clusters lie with respect to the Sun. Outline the region filled by the clusters and note where they are most concentrated. The center of their concentration is presumably the center of the Galaxy. Measure the distance from the center to the Sun, and determine the actual distance from the known scale of the plot.

Astronomers have several ways to measure the Milky Way's diameter, but all such methods depend on first knowing how far from the center the Sun is. In section 16.1 we saw how Shapley used globular clusters to map out the structure of the Galaxy and thereby find its size and where the Sun lies in it. This method remains one of the best techniques for measuring our Galaxy's scale. Astronomers measure the distance and direction from the Sun to a large number of globular clusters or some type of star that can be seen for great distances and then plot their positions, as shown in figure 16.24. They next locate the point around which the clusters seem to be most symmetrically located and assume that this point is the center of the Galaxy. Finally, they measure the distance from that point to the Sun, as shown in the figure. This method is somewhat uncertain and gives distance estimates of about 6 to 10 kiloparsecs to the center.

Newer methods give a more precise distance estimate by using radio observations of cool giant stars. Some such stars emit powerful beams of energy at microwave wavelengths analogous to the light emitted by lasers. These are called "maser sources." The word *maser* is an acronym for *m*icrowave-*a*mplification by *s*timulated *e*mission of *r*adiation. The radiation comes from molecules in an expanding shell of gas around the star. Maser sources are especially common in the inner bulge of the Milky Way because of the many red giants there, and with the masers, astronomers can measure the distance from the Sun to the Milky Way's center.

The maser sources expand away from the star they surround. Some move along our line of sight and therefore show a Doppler shift, which gives their velocity. Others move perpendicular to our line of sight, and we can detect that sideways motion by comparing observations a few years apart. To obtain the distance to the star at the center of the maser sources, astronomers assume that the maser sources moving toward and away from us have the same average speed as sources moving across our line of sight. From the change of position and speed, astronomers can deduce the stars' distances and therefore the distance to the center of the Milky Way (the point about which the stars orbit).

Recently astronomers have also been able to detect stars in orbit near the center of the Galaxy by observing at infrared wavelengths. The estimates of the distances for these stars and the maser sources give a distance of about 8 kiloparsecs (approximately 26,000 light-years). Once the Sun's distance from the center is known, the overall size of the Milky Way follows when the distance to objects lying near the Galaxy's edge is measured and the two distances are added. From such measures, astronomers estimate that the Milky Way has an overall diameter spanning 40 kiloparsecs or more, but much of the vast outer part beyond the Sun's orbit is so sparsely populated, and the numbers taper off so gradually, that no precise "edge" can be specified.

Mass of the Milky Way

Astronomers can deduce the Milky Way's mass from its gravitational attraction on matter in and near it. The procedure they use is very similar to the one described in chapter 13 to determine the mass of stars. You may recall that if one star is in orbit around another, we can calculate their combined mass using the modified form of

The mass interior to Sun's orbit can be treated as a single mass.

$$M = \frac{R^3}{P^2}$$

FIGURE 16.25

Finding the mass of the Milky Way. Represent the entire Milky Way within the Sun's orbit by a single mass, M (the white ball in the figure). The Sun (mass m) orbits the mass M at a distance R and speed V, and with an orbital period P. Apply the modified form of Kepler's third law to calculate $M + m$. Because the Sun's mass is so tiny, it can be ignored and we can set $M = R^3/P^2$.

Kepler's third law. Similarly, astronomers can find the mass of the Milky Way by treating it as a scaled-up version of a binary star in which one star, the Sun, for example, orbits under the gravitational influence of the Galaxy (fig. 16.25). The details are given in Astronomy by the Numbers: "Measuring the Mass of the Milky Way" on the next page.

More-refined methods build on this same technique but account for the mass of the entire Galaxy and use the rotation speed of stars at a variety of distances (the so-called **rotation curve**). The rotation curve, shown in figure 16.26, shows how fast the Galaxy spins at each distance from its core. That speed must be exactly sufficient to offset the gravity of the Milky Way that pulls orbiting stars in toward the center. The gravity in turn depends on the Milky Way's mass, and so from the rotation curve and Newton's laws of motion and gravity, astronomers can calculate the mass of our Galaxy. The first task, therefore, is to measure the rotation curve.

To determine the rotation curve, we must remember that we are observing from the Earth, which is traveling with the Sun in the Sun's orbit around the center of the Galaxy. Astronomers thus have the same problem that the police in a moving patrol cruiser have in measuring the speed of a traffic violator. That is, to find the speed of the star around the Milky Way, they must measure how fast it moves with respect to the Sun and then add the Sun's motion about the Galaxy. Astronomers have several methods for making this measurement. For example, we can measure the Doppler shifts of globular clusters. These Doppler shifts reflect not only their orbital motions but the Sun's as well. Because these clusters fill an approximately spherical volume around the Galaxy, they, as a group, presumably revolve only slowly compared with

FIGURE 16.26
Rotation curve of the Milky Way.

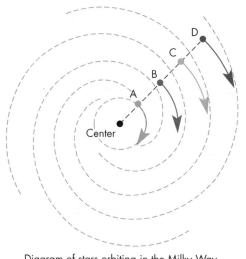

Diagram of stars orbiting in the Milky Way

Rotation curve of the Milky Way

ASTRONOMY *by the numbers* — MEASURING THE MASS OF THE MILKY WAY

To estimate the Milky Way's mass, we proceed as follows. We represent the mass of the Milky Way inside the Sun's orbit as a single mass, M, and assume the Sun (with mass m) moves in a circular orbit of radius R about M, as shown in figure 16.25. We now use the modified form of Kepler's third law described in chapter 13 for measuring the masses of binary stars. The modified form of Kepler's third law states that

$$m + M = R^3/P^2$$

m = mass of smaller object
M = mass of larger object
R = semimajor axis (radius) of orbit
P = years to complete one orbit

where the masses m and M are measured in solar masses, the orbital period P is measured in years, and the radius of the orbit R is measured in astronomical units.

The orbital period of the Sun around the Milky Way is about 220 million years (as determined in the text). Its orbital radius is about 8 kiloparsecs, or about 1.65×10^9 astronomical units. If we insert these values into the equation, we find that the mass of the Sun (m) plus that of the Milky Way (M) is

$$m + M = \frac{(1.65 \times 10^9)^3}{(2.2 \times 10^8)^2} M_\odot = \frac{1.65^3 \times 10^{27}}{2.2^2 \times 10^{16}} M_\odot$$

$$= 9.3 \times 10^{10} M_\odot \approx 10^{11} M_\odot$$

Because m, the mass of the Sun, is so tiny compared with the mass of the Galaxy, we can ignore the Sun's mass and conclude that the Milky Way's mass is about $10^{11} M_\odot$.

In our calculation, we made use of a result from physics that, if the Milky Way were spherical, the gravity of all the mass interior to the Sun's orbit acts as if it were packed into a point at the orbit's center. Note that only that part of it inside the Sun's orbit affects the Sun's motion. And although the Milky Way is not spherical, detailed calculations using its actual shape differ only slightly from what we find with our simple model.

A more serious limitation of our calculation is that it omits all the mass in the Milky Way that lies outside the Sun's orbit. That is, the Milky Way contains *at least* the amount of mass given by our calculation, and possibly much more beyond the Sun's orbit.

---

the flat disk. Thus, the Sun's average speed with respect to the globular clusters is approximately its own rotation speed around the Milky Way.

From many such measurements, astronomers conclude that the Sun is orbiting the Galaxy with a velocity of about 220 kilometers per second (about 130 miles per second!). From the Sun's orbital speed, we can calculate its orbital period, P, by dividing the speed, V, into the circumference of its orbit. That is,

P = Sun's orbital period
R = Radius of Sun's orbit
V = Sun's orbital speed

$$P = 2\pi R/V,$$

where R is the radius of the Sun's orbit, which we showed earlier is about 8 kiloparsecs. Inserting the values of R and V and converting to appropriate units, we find that the orbital period P is about 2.2×10^8 years, or 220 million years.

There are uncertainties in measuring of the speed of the Sun and other stars as they orbit within the Milky Way, as well as their exact distances from the Galaxy's center. Because these influence the mass that we calculate for the Milky Way, our Galaxy's mass is also somewhat uncertain. A far more serious problem is accounting for the large mass of our Galaxy that such calculations imply. The calculated mass greatly exceeds the mass of detectable stars and gas. As we saw in section 16.1, this discrepancy has led astronomers to hypothesize that our Galaxy is embedded in a huge, massive halo of dark matter. The dark matter exerts a gravitational force on the detectable stars and gas, but it emits no as-yet observed radiation of its own. Moreover, observations of its gravitational effects suggest that this immense halo has a radius of at least 100 kiloparsecs, so that it dwarfs the visible Milky Way. What can this dark matter be made of? This is one of the greatest mysteries in astronomy today, which we will explore further in chapter 17.

16.6   THE GALACTIC CENTER

The Sun lies at the center of the Solar System. What lies at the center of the Milky Way Galaxy? Even though the core of our Galaxy is difficult to observe because interstellar dust clouds totally block the visible light emitted by objects there, radio, infrared, X ray, and even gamma-ray telescopes allow astronomers to see through the dust. Infrared observations reveal a dense swarm of stars orbiting around the center of our Galaxy (fig. 16.27A). Radio observations have revealed an intense source of emission emanating from the very core of our Galaxy known as Sagittarius A*. Measurements of this source show that it is less than 10 AU in diameter—about the size of Jupiter's orbit—and that hot gas and stars are in rapid orbit around it. The position of this mysterious object at the center of figure 16.27A looks unremarkable, but something quite amazing is occurring there.

> The asterisk in Sagittarius A* is part of its name, said as "Sagittarius A star." Astronomers often shorten this further to Sgr A*, which they pronounce *sadj ay star*.

Over the past two decades, astronomers have tracked the motion of a number of the stars near Sagittarius A*. Some of these stars are moving at hundreds of kilometers per second (fig. 16.27B). From these motions, and using the modified form of Kepler's third law, astronomers calculate that the stars are orbiting an object containing about 4 million solar masses (see Astronomy by the Numbers: "The Mass of the Milky Way's Central Black Hole" on the next page). Only such an enormous mass could keep such high-speed stars from escaping from our Galaxy's core. Yet no object is visible there with any kind of telescope. Thus, the almost inescapable conclusion is that a massive black hole lies at the center of the Milky Way and holds those stars in its grip. In fact, these observations are perhaps the best evidence yet that black holes exist.

How might such a huge black hole form? We saw in chapter 14 that when a massive star reaches the end of its life, it may explode as a supernova, leaving a black hole as a remnant. Such black holes contain only 10 or so solar masses, but a black hole that

FIGURE 16.27
(A) Infrared image shows the inner 2 light-years of the core of the Milky Way. (B) The position of Sagittarius A* has been determined by tracking the orbits of stars as they orbit the massive black hole at the center of the Galaxy. The positions of eight stars tracked by the Keck/UCLA Galactic Center Group since 1995 are shown by dots of different colors, which are given a stronger color in each successive year along with an ellipse that extrapolates the rest of the orbit.

ASTRONOMY
by the numbers

THE MASS OF THE MILKY WAY'S CENTRAL BLACK HOLE

The mass of Sagittarius A* can be found using the orbit of any of the stars that orbits it. Referring to figure 16.28, the star SO-2 has completed an orbit in a period P of about 15.5 years. The orbit's semimajor axis a is about 1000 astronomical units.

With this information, we can use Newton's modified version of Kepler's third law to find the mass of the body that SO-2 is orbiting. And just as for planets orbiting a star, the mass of SO-2 is minor compared to the object it

orbits, so:

$$M(\text{Sgr A*}) = a^3/P^2 = 1000^3/15.5^2 = 4.2 \times 10^6 \, M_\odot.$$

This implies that the mass of the object SO-2 is orbiting is over 4 million solar masses.

Other stars' orbits imply that this massive object is no bigger than the Solar System, and it emits very little visible or infrared light. All evidence points to it being a massive black hole.

formed near the core of the Milky Way during our Galaxy's youth had opportunities for growth. At that long-ago time, gas naturally tended to fall inward toward the core, where any black hole lurking there might capture it. As such a black hole swallowed material, its mass and hence its gravity would increase, allowing it to capture material ever more easily. In addition, gas lost from stars by explosion, stellar wind, or tidal force stripping from neighbors in that crowded environment might also be drawn into the black hole. With so many potential sources of matter, once a black hole formed in the center of the Milky Way, it would not stop growing until it ran out of available matter. In that way, a small black hole might have grown to $10^6 \, M_\odot$ in less than a billion years.

The huge accelerations caused by the gravity of this object lead to high-energy collisions outside the black hole's event horizon, and this can explain the intense X ray radiation seen coming from around Sagittarius A* (fig. 16.28A). A recent gamma-ray study also reveals immense "bubbles" of energetic radiation that originate from the Galaxy's center and span almost half the sky (fig. 16.28B). The Milky Way is not alone in exhibiting these phenomena. Other galaxies also contain massive black holes in their central regions and emit energetic radiation, as we will discuss in chapter 17. For now we will simply say that, according to some hypotheses, such black holes may be a nearly inevitable by-product of galaxy formation.

A   **B**

FIGURE 16.28

(A) The central region of the Milky Way as observed at X ray wavelengths by the orbiting *Chandra* X ray telescope. In this false-color image, the colors represent X rays of different energies. The small bright spots are white dwarfs, neutron stars, and probably gas surrounding a few black holes. The diffuse glow is from extremely hot gas surrounding the Galaxy's core. (B) Illustration of giant "bubbles" of faint gamma-ray radiation discovered with NASA's *Fermi* telescope. The gamma-ray bubbles appear to be the result of energetic particles emitted from the Galaxy's center.

16.7   EVOLUTION OF THE MILKY WAY

One of the major unsolved problems in astronomy is how galaxies form. For many years astronomers thought that the process was simply a large-scale version of star formation; that is, instead of a small gas cloud collapsing under its own gravity to form a single star, a huge cloud collapses to form billions of stars. Although this model can explain many properties of the Milky Way, it does not explain all of them. Thus, astronomers have explored how to modify this model so that it better explains the Milky Way's properties. Astronomers now think that the Milky Way evolved through a number of stages as it built up to its present form.

Birth of Population I and II Stars

In trying to explain the origin of the Milky Way, astronomers draw on a variety of observations. In developing the theory of how the Solar System formed (chapter 8), we found that its disklike shape, the orbital regularities and common age of the planets, and the existence of two families of planets all provided important clues. Similar clues guide astronomers in studying the formation of the Milky Way. For example, our Galaxy has a flattened shape and contains two main groups of stars—populations I and II—that differ greatly in their properties. In particular, Pop II stars are old, contain few heavy elements, and follow orbits that are steeply inclined to the disk. Pop I stars, on the other hand, are younger, contain more heavy elements, and follow roughly circular orbits that lie in the disk.

Using such information, British astronomer Donald Lynden-Bell and American astronomers Olin J. Eggen and Allan R. Sandage proposed in 1962 a two-stage collapse model to explain the birth of the Milky Way. They proposed that our Galaxy probably began as a vast, slowly rotating gas cloud, perhaps a million light-years in diameter and containing about 10^{11} $M_\odot$ of gas. The cloud was composed of almost pure hydrogen and helium because no stars yet existed to make the heavier elements. As gravity began shrinking this immense cloud, clumps of gas within it grew more dense and gradually turned into stars (fig. 16.29A). These stars, composed of essentially pure hydrogen and helium, may have been exceptionally massive compared to stars forming today. Because of their greater mass, they evolved quickly and blew up

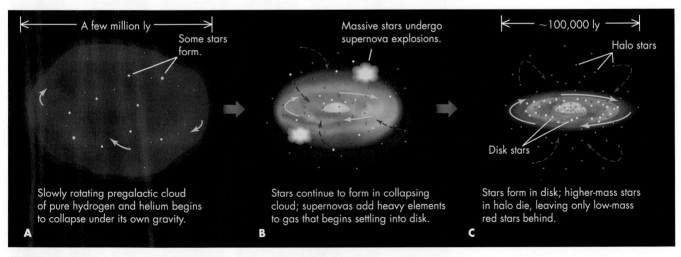

FIGURE 16.29

(A) Birth of the Milky Way as a gas cloud collapses. A first generation of stars forms. (B) Collapse continues with more population II stars forming and some exploding as supernovas. Heavy elements created by the stars exploding as supernovas are mixed into the gas. Gas not used in making stars settles into a rotating disk. (C) Population I stars form in the disk.

as supernovas, which added the first heavy elements to the gas cloud even before it had finished collapsing.

It may seem odd that stars could form and die before the cloud's collapse was complete. But collapse of such an immense cloud takes hundreds of millions of years, whereas massive stars evolve in only a few tens of millions of years—10 or so times faster. As the collapse continued, more stars formed, now with at least some heavy elements. Because these stars formed from infalling gas, they follow elongated "plunging" orbits—becoming the population II stars that we see today.

Star formation during this initial stage was not efficient enough to consume all the collapsing gas, and some gas survived to form a rotating disk, as sketched in figure 16.29B. That disk gas, now rich in heavy elements from the death of Pop II stars, then began to form the first population I stars. These stars followed the motion of the gas they formed from. Pop I stars, therefore, contain more heavy elements than Pop II stars and move along approximately circular orbits lying in the disk, as illustrated in figure 16.29C.

Computer simulations support this collapse model for the birth of the Milky Way. For example, figure 16.30 shows one of the early computer simulations that explored how an unstructured region of gas, of the appropriate size and rotation speed, collapses. You can see that it forms a thin disk with spiral arms, surrounded by a halo. Computer simulations have become steadily more realistic by including more of the physical processes that take place in addition to gravity, such as the hydrodynamics of the colliding gas clouds and the effects of magnetic fields. This allows astronomers to test hypotheses about galaxy evolution that would be impossible to observe otherwise. Such simulations also show that the unseen dark matter plays an important role in making galaxies, because computer models that include dark matter much more closely resemble real galaxies than models that do not include dark matter.

FIGURE 16.30

An early computer simulation of the birth of a galaxy similar to the Milky Way. Frames (A), (B), and (C) show the initial state of the gas and the development of a clump that becomes the galaxy, illustrated in frames (D) and (E). Notice the thin disk and spiral arms. Mergers are not part of this simulation.

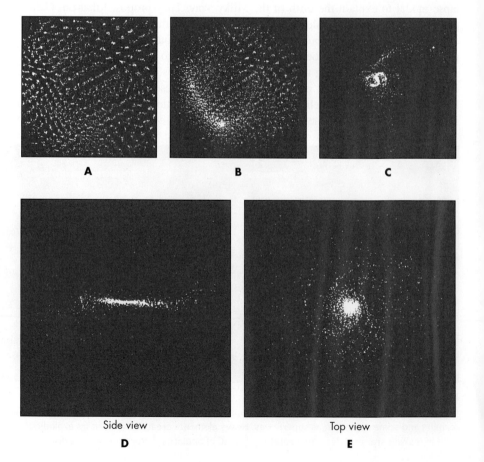

A   B   C

Side view   Top view
D   E

Evolution by Mergers

Although the two-stage model explains many of the features of our Galaxy, it fails to explain some details of stellar populations. In the collapse model, population II stars should have all formed during the relatively brief period of the Galaxy's initial collapse. But observations of Pop II stars show that they formed over a much longer time span than the model predicts. There is also evidence that Pop I stars did not form steadily, but had some periods of rapid formation and others where the Galaxy was relatively quiescent. Astronomers now believe an additional process plays a major role in shaping galaxies: collisions with other galaxies.

As galaxies move through space, they occasionally collide with one another. These collisions may cause bursts of star formation when interstellar gas is compressed by the gravitational disturbance. Sometimes these collisions lead to a merger of the two galaxies into a single larger system. Astronomers today can see collisions occurring in a number of systems (fig. 16.31), and with the Hubble Space Telescope they can see that such collisions and mergers were far more common in the past. How can the Hubble Space Telescope see into the past? Recall that light takes 5 billion years to reach us from a galaxy 5 billion light-years away. Thus, when we see distant galaxies, we observe them as they were long ago, when they were young. With this "window" back in time, astronomers see that many young galaxies are smaller than galaxies we see today, such as the Milky Way. In addition, many more of these young galaxies are colliding with neighbors than galaxies that we see today. Astronomers conclude that large galaxies like the Milky Way may have begun their life with the collapse of a single gas cloud, but that they have grown to their present size by swallowing smaller galaxies.

This hypothesis is supported by recent discoveries of a number of dim, dwarf galaxies orbiting very close to the Milky Way. Several of these are so close that the Milky Way's gravity is tearing them apart,* leaving long faint streams of stars within the Milky Way's halo. The streams are stars that appear to have been ripped out of small galaxies as they were falling into the Milky Way, and these stars will eventually be widely distributed in the Milky Way's halo. For example, figure 16.31B illustrates one such stream of stars detected in an infrared survey of the sky. Thus, galaxy evolution is a dog-eat-dog event.

FIGURE 16.31
(A) Two colliding galaxies known as The Mice or more formally as NGC4676. The gravitational interaction between galaxies pulls out long streams of stars. (B) Diagram illustrating a stream of stars pulled out of a dwarf galaxy orbiting the Milky Way. Based on estimated distances to the stars in the stream, the figure shows what the stream (shown in red) might look like from outside our Galaxy (shown as a blue spiral).

Population III

The two-stage collapse model also predicts that the low-mass stars born in that first generation that have not burned out should have no heavy elements, but no such stars have ever been observed. We should expect such a composition if our Galaxy formed from a cloud in which no stars had yet made heavier elements. Up to now, all stars found in the Milky Way contain at least a small percentage of heavy elements (typically at least 1/1000 the amount seen in the Sun), although one star has recently been observed with a mere 5×10^{-6} times the heavy element abundance of the Sun.

Astronomers have devised different explanations for why they have so far seen no **population III stars,** as these hypothetical ancient pure hydrogen and helium stars are

---

* This is similar to how gravitational tidal forces can pull apart satellites orbiting within the Roche limit. See chapter 10.

called. First, conditions in the early Galaxy may have allowed only short-lived massive stars to form. For example, the gas may have been too hot or turbulent to make small stars that would have lived long enough for us to see them today. Alternatively, because massive stars can form and die before very low-mass stars even complete their contraction, the surface layers of Pop III stars may have been contaminated by gas ejected during the deaths of more-massive neighboring stars. Thus, it is possible that there exist Pop III stars that are masquerading as Pop II stars because they contain heavy elements that were not part of their original makeup.

The Future of the Milky Way

We know that stars change dramatically as they age. Will the Milky Way change as well? Astronomers think the answer is yes. As our Galaxy ages, it steadily uses up its interstellar gas to make stars. Although on reaching the end of their lives, stars return some of their material to space as planetary nebula shells, supernova remnants, and stellar winds, eventually that gas will also end up in stellar remnants such as white dwarfs, neutron stars, and black holes, leaving the Milky Way with less and less gas from which to make new stars.

This gradual decline in star formation will likely be affected by some major mergers that lie in the Milky Way's future. Two relatively large satellite galaxies of the Milky Way—the Large and Small Magellanic Clouds—are spiraling into our Galaxy and will eventually merge with it. A far more dramatic merger lies further in our future. From the speed of our motion toward the galaxy M31 in Andromeda, it appears that the Milky Way will merge with this larger galaxy in about 4 to 5 billion years. What this might look like to our distant descendants is illustrated in figure 16.32. Eventually, tens of billion years from now our Galaxy will contain only dead stars and dim, low-mass red dwarf stars. Gradually even these dwarfs will burn out, and our Galaxy will fade, slowly spinning in space, a dark disk of stellar cinders.

The Magellanic Clouds are shown in Looking Up #9, and M31 is shown in Looking Up #3 at the front of the book.

FIGURE 16.32

This series of images illustrates how the merging of the Milky Way with the galaxy M31 might appear over the next 7 billion years. The top two images show the approach of M31. The middle row shows the collision in about 4 billion years. Many stars form as interstellar clouds collide and consume most of the gas. The last two images show how the two galaxies merge, with little interstellar gas or dust remaining.

2 billion years

3.75 billion years

3.85 billion years

4 billion years

5.1 billion years

7 billion years

SUMMARY

Our Galaxy, the Milky Way, consists of a disk about 100,000 light-years across embedded in a halo that extends out several times farther. Together, the disk and halo contain about 10^{11} stars. Near the Sun, stars are a few light-years apart. Because the disk rotates, it is flat, and it contains star groups arranged in a pattern of spiral arms. Near its center the Galaxy is thicker (several kiloparsecs), forming the Galactic bulge, which we can see toward the constellation Sagittarius. The Sun lies about two-thirds of the way out in the disk. From Earth, we see the Milky Way's disk as a band of dim light stretching across the night sky.

The Milky Way's stars form two main groups—population I and population II—which differ in many ways. Population I stars are found mainly in the disk. They are relatively young, tend to be blue, move in circular orbits, and are rich in heavy elements. Population II stars are found mainly in the halo and bulge. They are old, tend to be red, move in elongated and tilted orbits, and are poor in heavy elements.

Gas and dust clouds orbit with our Galaxy's stars and contribute about 15% as much mass as the stars. The dust dims the light of distant stars, making it difficult for astronomers to see beyond 10,000 light-years in the disk. The gas, mostly hydrogen and helium, plays a crucial role in the evolution of the Milky Way because it is the building material for new stars. Radio emission from the gas, both cold hydrogen emitting 21-centimeter radiation and interstellar molecules, is not blocked by dust, allowing us to map the Galaxy's structure.

At the very center of the Milky Way lies the Galactic nucleus, a dense collection of stars and gas. Astronomers think a black hole with a mass of about 4 million solar masses lies at the center because stars orbiting there move so rapidly.

Astronomers can measure the diameter of the Milky Way and the Sun's location in it by plotting the position of globular star clusters. They measure its mass by applying the modified form of Kepler's third law to the motion of the Sun and other stars, but such measures give values of the mass that disagree with the value arrived at when stars are simply counted and their masses are added. Astronomers, therefore, infer that the Milky Way is embedded in a halo of dark matter that makes up most of its mass, but whose nature is unknown.

The existence of two different stellar populations, I and II, leads astronomers to think that the Milky Way formed by the collapse of a vast gas cloud about 13 billion years ago. The first stars to form became the population II stars. Some gas did not turn into stars during collapse but instead collected into a rotating disk. From supernova explosions of population II stars, heavy elements became part of the disk of gas. Stars eventually formed from this material to become the population I stars such as our Sun. Subsequently the Milky Way has captured smaller galaxies and grown larger. Eventually the Milky Way will convert all its available gas to stars, and as the surviving stars die, our Galaxy will grow dark.

QUESTIONS FOR REVIEW

1. (16.1) What does the Milky Way look like in the night sky?
2. (16.1) How do we know our Galaxy is a flat disk?
3. (16.1) How did Shapley deduce the Milky Way's size and the Sun's position in it?
4. (16.1) Draw a sketch of the Milky Way Galaxy and label its major components. Where is the Sun in the Milky Way?
5. (16.1) Roughly how big in diameter is our Galaxy, and how much mass does it contain?
6. (16.2) What are some differences between Pop I and Pop II stars?
7. (16.2) Summarize the differences between open and globular clusters. Are both still forming today in the Milky Way?
8. (16.3) How do we know interstellar matter exists? What different kinds and forms are there?
9. (16.3) How do we know that some interstellar matter is dust? What is the other main category of interstellar matter?
10. (16.3) How does interstellar dust affect our observations of stars, the Milky Way, and other galaxies? What is the zone of avoidance?
11. (16.3) What is the difference between an emission and a reflection nebula?
12. (16.3) What are some ways that we know cold clouds of gas exist in space?
13. (16.3–16.4) What evidence makes astronomers conclude the Milky Way has spiral arms?
14. (16.4) How is star formation related to spiral arms?
15. (16.5) How can we determine the Milky Way's diameter?
16. (16.5) How can we determine the Milky Way's mass?
17. (16.5) What is meant by dark matter? Why do astronomers conclude that the Milky Way may contain such as-yet unobserved material?
18. (16.6) What is the evidence for a black hole at the center of the Milky Way? How does it compare to the black holes discussed in chapter 15?
19. (16.7) Describe one model for the origin of the Milky Way. How does this model explain the difference between Pop I and Pop II stars?
20. (16.7) How do mergers affect the evolution of the Milky Way?
21. (16.7) What is meant by Pop III stars?

THOUGHT QUESTIONS

1. (16.1–16.6) What would the Milky Way look like in the night sky to an observer on a planet at the very edge of the Galaxy? What would it look like to an observer in the Galactic bulge?

2. (16.3) Compare what happens in interstellar reddening to what occurs in our atmosphere at sunset. Does this also occur earlier in the day? What is different?

3. (16.3) Knowing whether a selection effect is affecting your data is very important in astronomy. Imagine you are an alien visitor who only has an hour to study how long humans are babies, children, young adults, middle-aged, and elderly. Describe how your results would be affected by selection effects if you visited a shopping mall for an hour on a Wednesday afternoon in October, if you visited an elementary school, if you visited the mall at 9 P.M., or if you visited the mall on the weekend. What if instead you were studying human clothing, and you landed in Chicago or Florida in December?

4. (16.3) Understanding possible selection effects is very important in polling to make sure results are accurate. If you were conducting a telephone poll about two presidential candidates, one of whom was more likely to appeal to older people and one more likely to appeal to younger people, what effects might you have to think about if you (1) could only call people with land lines (no cell phones), (2) could only call people with a cell phone, (3) could call both groups?

5. (16.4) If you were explaining the density-wave theory to another person, what additional examples can you think of besides the slow car described in section 16.4 where a density-wave causes clumps of people, cars, or something else?

6. (16.4) Cut four circles of cardboard whose diameters are 2, 4, 6, and 8 inches. Fasten them together through their centers with a brass paper fastener. Draw a circle about 2.5 inches in radius, centered near the edge of the 4-inch disk that marks all the disks. Now turn the circles in a way that reflects differential rotation. Use what you see to explain how self-propagating star formation might make spiral arms.

7. (16.4–16.5) Are spiral arms permanent structures? Explain, using the idea of the density-wave model in the book. What would happen to a spiral arm if the same stars stayed in it? (Consider how long it would take a star on the end of an arm to orbit once around compared to a star halfway in on the arm, if both have a velocity of about 225 km/sec).

8. (16.5) How could we measure the mass of the Milky Way using a distant orbiting globular cluster?

9. (15.2–15.3/16.6) How might you argue that there is a black hole at the center of the Milky Way, instead of, for example, a region filled with many neutron stars? What observations would help you rule out such a case?

10. (16.1/16.3/16.7) What would the night sky look like if there were no dust or gas in the Milky Way, but the Galaxy were otherwise the same as it is today? What will the night sky look like when the Milky Way is very old and has used up all its star-forming materials?

PROBLEMS

1. (16.1) Make a scale diagram of the Galaxy looking down on it from above. Let 1 inch = 4 kpc (about 17,000 ly), and draw a 7.5-inch diameter circle to represent the 30-kpc wide disk of the Galaxy (see fig. 16.6). The Sun is about 8 kpc out from the center, so place a small dot 2 inches out from the center. The most distant stars visible to the naked eye would be no more than about 3000 pc (3 kpc) from the Earth, and most bright naked-eye stars are less than 1200 pc away. Draw circles around the Sun at distances corresponding to 1200 pc and 3000 pc. Are any of the stars in the night sky outside the Milky Way Galaxy?

2. (16.1/16.6) How long does it take light from the center of the Milky Way, 8 kpc away, to reach the Solar System?

3. (16.3) On average, a hydrogen atom emits a 21-cm photon once every 15 million years. How many 21-cm photons does a 100-$M_\odot$ gas cloud emit per second?

4. (16.3) The dust in an interstellar cloud blocks blue light in the following way: For every 1 pc light travels through the cloud, only 90% of the light continues. Thus, after 2 pc, 90% of the remaining 90%, or 81% ($0.9 \times 0.9 = 0.81$), remains.
   (a) You might initially expect that after 10 pc, the ten 10% reductions would have removed all of the blue light. How much blue light actually remains?
   (b) The same cloud removes about 7% of the red light every parsec, so after 1 pc, the ratio of blue to red has dropped to 90/93 ≈ 97% of its unreddened value. What is the ratio of blue to red after 5 pc? After 10 pc?

5. (16.5) Given that the Sun moves in a circular orbit of radius 8 kiloparsecs around the center of the Milky Way and that its orbital speed is 220 kilometers per second, work out how long it takes the Sun to complete one orbit of the Galaxy. (*Hint:* Remember to convert kiloparsecs to kilometers and convert your final answer to years, not seconds.)

6. (16.5) According to figure 16.26, the Milky Way's rotation speed is about 230 km/sec at a distance of about 16 kpc from its center. Using the equation from figure 16.25, estimate the mass of the Milky Way. Is this an upper or a lower limit? Compare with the mass of the Andromeda Galaxy, where the rotation speed is 230 km/sec at a radius of 35 kpc.

7. (16.5) How long is the period for a star orbiting the black hole at the center of our Galaxy that is about 2800 AU from the black hole at the farthest part of its orbit, but that gets as close as 1200 AU at the nearest? Take the mass of the black hole to be 4×10^6 solar masses. Compare your answer to the path of star S0-20 in figure 16.27B.

8. (16.6) Calculate the Schwarzschild radius of the black hole at the center of the Galaxy using 4×10^6 solar masses for its mass. How many AU is the Schwarzschild radius?

9. (16.1/16.6) The 4×10^6-solar-mass black hole at the center of the Galaxy may have gained its first million solar masses in the first billion years of the Milky Way's existence. How much mass must it have absorbed on average each year since then to reach 4 million solar masses today?

TEST YOURSELF

1. (16.1) One way astronomers deduce that the Milky Way has a disklike shape is that they see
   (a) stars forming a circle around the north celestial pole.
   (b) the great majority of stars in a band that encircles us.
   (c) a large dark circular shadow of the southern Milky Way.
   (d) about the same number of stars in all directions in the sky.
   (e) dark clouds of spinning gas and dust that must be flat.

2. (16.2) Select all that are correct. Most of the stars in the Milky Way
   (a) are not visible to the naked eye.
   (b) have plunging orbits around the Galactic center.
   (c) have the same or less mass than the Sun.
   (d) are surrounded by cold gas clouds.
   (e) burn out before they leave a spiral arm.

3. (16.2) A young blue star moving along a circular orbit in the disk is a
   (a) Pop I star   (b) Pop II star   (c) Pop III star.

4. (16.2) Globular clusters have mainly _____ stars.
   (a) Pop I   (b) Pop II   (c) Pop III

5. (16.3) Astronomers know that interstellar matter exists because (select all that apply)
   (a) they can see it in dark clouds and clouds that absorb light.
   (b) the matter creates narrow absorption lines in the spectra of some stars.
   (c) X rays cannot penetrate the clouds toward the center of the Galaxy.
   (d) they can detect radio waves coming from atoms and molecules in the cold gas.
   (e) spacecraft have sampled clouds near Orion.

6. (16.3) What kind of interstellar cloud generally has a bluish color?
   (a) H I cloud   (c) Emission nebula   (e) Dark nebula
   (b) H II region   (d) Reflection nebula

7. (16.4) Astronomers think the Milky Way has spiral arms because (select all that are correct)
   (a) they can see them unwinding along the celestial equator.
   (b) radio maps show that gas clouds are distributed in the disk with a spiral pattern.
   (c) young star clusters, H II regions, and associations outline spiral arms.
   (d) globular clusters outline spiral arms.
   (e) other galaxies that also have gas and dust have spiral arms.

8. (16.5) The modified form of Kepler's third law allows astronomers to determine the Milky Way's
   (a) mass.   (c) composition.   (e) number of spiral arms.
   (b) age.   (d) shape.

9. (16.6) Astronomers think there is a massive black hole at the center of the Galaxy because (select all that are correct)
   (a) without it, there does not seem to be enough mass to hold the Galaxy together.
   (b) the center of the Galaxy is dark in visible light images.
   (c) other galaxies have massive black holes at their centers.
   (d) they have plotted the orbits of stars and calculated a large, dark mass is present there.
   (e) an enormous supernova explosion was observed in the direction of the center.

10. (16.7) Faint streams of stars in the Milky Way's halo are interpreted as evidence that
    (a) parts of the Milky Way are still collapsing.
    (b) the Milky Way is losing gas and will eventually have none.
    (c) supernova explosions are pushing gas out of the disk.
    (d) the Milky Way is stripping material from smaller dwarf galaxies.
    (e) the Milky Way does not have enough mass to hold all the gas and dust in the disk.

KEY TERMS

association, 419	population I stars, 416
bulge, 413	population II stars, 416
dark matter, 414	population III stars, 435
dark nebula, 421	reddening, 421
density-wave model, 426	reflection nebula, 423
disk, 413	rotation curve, 429
emission nebula, 423	scattering, 421
globular cluster, 419	selection effect, 416
halo, 413	self-propagating star
H II region, 424	formation, 427
interstellar matter, 420	spiral arm, 413
Milky Way, 409	star clusters, 418
nucleus, Galaxy's, 414	21-centimeter radiation, 425
open cluster, 418	zone of avoidance, 422

🔍: FIGURE QUESTION ANSWERS

WHAT IS THIS? (chapter opening): This infrared image shows the inner 2 light-years of the core of the Milky Way. The exact center of the Galaxy is indicated by the two small arrows. Observations of the motion of a star near this spot are strong evidence that a black hole lies there.

FIGURE 16.11: They are globular clusters.

FIGURE 16.13: The Sun looks red when it sets because molecules and dust in our atmosphere interact more strongly with the blue light than with the red light and scatter the blue light. As a result, less blue light reaches us, so the Sun looks redder. Interstellar dust does the same thing to starlight—it preferentially removes the blue light, leaving the star looking redder than it really is.

FIGURE 16.16: Red light has a longer wavelength than blue and so it does not interact as strongly with the dust. It is therefore scattered less.

17

GALAXIES

A cluster of galaxies (Abell S0740) that contains a diverse assortment of galaxy types.

LEARNING OBJECTIVES

Upon completing this chapter you should be able to:

- Classify galaxies according to their appearance, and compare how stars, gas, and dust contribute to differences.
- Explain how galaxies' distances are determined, and discuss the limitations of different methods.
- Describe Hubble's law and apply it to find the distance of a galaxy.
- Explain the limitations of Hubble's law.
- Explain how galaxies can be studied at different stages of their evolution by observing distant galaxies.
- Describe the range of galaxy sizes, and explain how galaxies may evolve through interactions and mergers.
- Explain how galaxies' colors may change as they evolve and grow in size.

- Recall how active galaxies were discovered, and describe the various phenomena associated with each type.
- Explain how supermassive black holes can produce the phenomena associated with galaxy activity.
- Explain how astronomers detect intergalactic matter.
- Describe the Local Group and Virgo Cluster, and the large-scale structure of the Universe, and discuss how these affect galaxy evolution.
- Explain how galaxy rotation curves imply the existence of dark matter in galaxies.
- Discuss the MACHO and WIMP hypotheses and the evidence for dark matter in galaxy clusters.
- Describe the effects of gravitational lensing on the light from distant objects.

Beyond the edge of the Milky Way and filling the depths of space are billions of other star systems. These remote, immense star clouds are called galaxies, named after our own Galaxy.

While some galaxies resemble the Milky Way, many others differ in shape, content, and size. Some are "egg-shaped" and contain mostly old stars; others are irregular and contain predominantly young stars. Some are rich in interstellar matter; others have hardly any at all. The largest galaxies contain millions of times more stars than the smallest galaxies. Still other galaxies have tremendously powerful energy sources at their cores that emit as much energy as the entire Milky Way but from regions less than a few light-years in diameter. Many of these so-called active galactic nuclei eject gas in narrow jets at nearly the speed of light.

Galaxies are not spread uniformly throughout space. Just as stars often lie in clusters, so too galaxies often lie in galaxy clusters. These immense collections of galaxies are not stationary. They are separating from one another, caught up in the expansion of the Universe, like snowflakes carried on the wind.

Astronomers can see all this diversity in galaxies, but as yet they only partially understand how it arises. Unlike stars, whose structure and evolution are well understood, galaxies remain mysterious in many ways. For instance, astronomers do not yet agree about why some galaxies are disks while others are egg-shaped. Although this lack of certainty can be frustrating, it offers us a chance to see how scientists work, and, in particular, how astronomers begin with basic observations and go on to construct hypotheses to explain what they observe. For example, many galaxies appear to be colliding with their neighbors. Could past interactions of this kind account for the different galaxy shapes?

CONCEPTS AND SKILLS TO REVIEW

- Inverse-square law (13.2)
- Standard-candle method (13.6)
- Red light has a longer wavelength than blue (4.1)
- Newton's law of gravity (3.5)
- Modified form of Kepler's third law (3.6)

17.1 DISCOVERING GALAXIES

A **galaxy** is an immense swarm of hundreds of millions to hundreds of billions of stars and vast clouds of interstellar gas. Each star moves along its own orbit, held within the galaxy by the combined gravitational force of the rest of the matter within the galaxy. Each galaxy (apart from those undergoing collisions) is therefore an independent and isolated star system. Until the early 20th century, astronomers remained divided about whether these objects were actually vast collections of stars or if they were instead something much smaller and nearer to us. The discovery that they were actually objects like the Milky Way gave us new insights into the nature of our own Galaxy and the entire Universe.

Early Observations of Galaxies

All galaxies (except our own) are extremely distant from Earth, and few are visible without a telescope. Of galaxies similar in size to the Milky Way, the nearest—M31, in the constellation Andromeda (fig. 17.1)—is about 2.5 million light-years away and about 25% more luminous. You can see M31 on clear, dark nights in autumn and winter in the Northern Hemisphere. It is the most distant object visible with the unaided

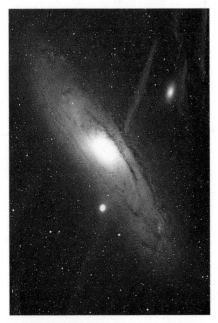

FIGURE 17.1
M31, the Andromeda galaxy,

FIGURE 17.2
Photograph of the Large and Small Magellanic Clouds. These are two small galaxies orbiting the Milky Way.

The Magellanic Clouds are shown in Looking Up #9, and M31 is shown in Looking Up #3 at the front of the book.

eye, looking like a pale elliptical smudge on the sky. Early observers such as the tenth-century Persian astronomer Al-Sufi noted this object, and it appeared on early star charts as the "Little Cloud." In the Southern Hemisphere, you can easily see the Large and Small Magellanic Clouds, two satellite galaxies of the Milky Way (fig. 17.2). Today, astronomers have cataloged millions of galaxies besides the Magellanic Clouds and M31, and they estimate that the visible Universe contains tens of billions more.

The study of other galaxies began in the eighteenth century, when the French astronomer Charles Messier (pronounced *mess-yay*) accidentally discovered many galaxies during his searches for new comets. He noticed a large number of faint, diffuse patches of light, and to avoid confusing them with comets, he assigned them numbers and made a catalog of their positions. Although many of Messier's objects have since been identified as star clusters or glowing gas clouds in the Milky Way, several dozen are galaxies. These and the other objects in Messier's catalog are still known by their Messier, or M number, such as M31, the Andromeda galaxy.

In the nineteenth century, other astronomers, such as Sir William Herschel, began systematically to map the heavens, finding and cataloging numerous astronomical objects, including galaxies. Herschel's work, revised and added to by his son and by John Dreyer, is now known as the New General Catalog (or NGC, for short), and it gives names to many galaxies, such as NGC 1275, a peculiar, active galaxy. But many galaxies appear in more than one catalog and so bear several names. For example, Messier's M82 is the same galaxy as NGC 3034.

We know today that galaxies are immense star systems like our Milky Way, but this idea was not widely accepted by astronomers until the 1920s. Before that time, astronomical opinion was strongly divided as to whether galaxies were huge star systems like the Milky Way or merely small satellites of our own Galaxy. This controversy was aired in detail during a debate organized by the National Academy of Sciences, in which two American astronomers, Harlow Shapley and Heber Curtis, took opposing sides. Shapley argued that the Milky Way is huge and that other galaxies (at that time called nebulas) are merely small, nearby companions. Curtis, on the other hand, argued that the Milky Way is smaller than Shapley claimed, and that the nebulas are star systems like our own and are immensely distant from it.

Historians differ on who "won" the debate. Although Shapley was more nearly correct about our Galaxy's size, Curtis was right that the nebulas are distant systems like the Milky Way. In any case, once the distance and true nature of galaxies were understood, astronomers could more successfully study galaxies' properties.

FIGURE 17.3
Pictures of typical spiral galaxies. (A) NGC 4565; (B) NGC 4414; and (C) NGC 1288. These examples show spiral galaxies with their disks oriented from nearly edge-on (A) to nearly face-on (C).

Types of Galaxies

Even early nineteenth-century observers noticed that not all galaxies look the same. However, it was Edwin Hubble, an American astronomer working first as a graduate student at the University of Chicago and later at Mount Wilson in California in the 1920s, who demonstrated that galaxies could conveniently be divided on the basis of their shape into three main types: spirals, ellipticals, and irregulars. **Spiral galaxies** have two or more arms winding out from the center. We saw in chapter 16 that the Milky Way is of this type, and figure 17.3 shows photographs of several others seen with their disks oriented at different angles to our line of sight.

The second type shows no signs of spiral structure. These galaxies have a smooth and featureless appearance and a generally elliptical shape, as can be seen in figure 17.4. Accordingly, astronomers call them **elliptical galaxies.** These galaxies have no disk or arms—instead the stars are distributed smoothly around the center in all three dimensions, gradually declining in density at larger distances from the center, and they lack the massive young blue stars found in the disk of galaxies like the Milky Way. Some elliptical galaxies appear elongated (fig. 17.4A) while others look nearly round (fig. 17.4B). Unlike the cause of the shapes of spiral galaxies, this does not appear to be just an effect of orientation. Elliptical galaxies span a range from having their stars distributed in a shape like an American football to being completely spherical.

FIGURE 17.4
Hubble Space Telescope images of the elliptical galaxies (A) NGC 4660 and (B) NGC 4458. The images show how elliptical galaxies can sometimes be very elongated and other times almost spherical.

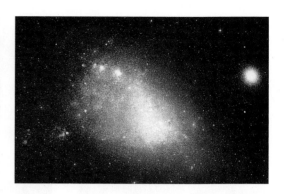

FIGURE 17.5

Photographs of two irregular galaxies, NGC 6822 and the Small Magellanic Cloud.

Galaxies of the third major type show neither arms nor a smooth uniform appearance. In fact, they generally have stars and gas clouds scattered in random patches. For this reason, they are called **irregular galaxies** (fig. 17.5). These galaxies often appear bluish in color because of the large number of massive blue stars, and part of their irregular appearance results from the light of these bright young stars coming from individual star associations scattered around the galaxy. It turns out that most of the irregular galaxies are quite small, so individual star-forming regions represent a large percentage of their total light output. On the other hand, some of the galaxies classified as irregulars by Hubble are actually colliding galaxies, and their irregular appearance results from the gravitational distortions as the pair merges.

Spiral, elliptical, and irregular galaxies are sometimes denoted by *S*, *E*, and *Irr*, respectively. In addition to these three main types, astronomers recognize two additional galaxy types closely related to spiral systems. The first of these are **barred spiral galaxies,** which have arms that emerge from the ends of an elongated central region, or bar, rather than from the core of the galaxy, as shown in figure 17.6. It is this bar that gives them their name, and they are denoted *SB* galaxies to distinguish them from normal spirals. The outer parts of barred spiral galaxies resemble other spiral galaxies, and astronomers today suspect that the bars may be temporary features that come and go in a spiral galaxy depending on gravitational interactions with other galaxies.

The other additional type of galaxy that Hubble identified is a kind of "spiral without a spiral." These are galaxies with disks with no evidence of arms (fig. 17.7), which he called *S0* (pronounced "S"-zero) galaxies. These resemble elliptical galaxies in their overall yellow color, but the degree to which stars are concentrated to the center and the presence of a disk are more similar to a spiral, so they appear transitional between those two major types.

Studies of the stars in the central region of the Milky Way show that it is a barred spiral.

FIGURE 17.6

Barred spirals. NGC 1365 (left) has a prominent bar, while NGC 5236 or M83 (right) has a weaker bar—more like the one thought to be at the center of the Milky Way.

FIGURE 17.7

Two S0 galaxies, NGC 3384 and NGC 4552. These galaxies look similar to ellipticals, but some of the stars are distributed in a disk, as occurs in spiral galaxies.

FIGURE 17.8

The Hubble tuning-fork diagram. Elliptical galaxies are subdivided by how "flattened" they look. Spirals are subdivided according to how large their bulges are and how tightly wound their arms are. (Images are drawn from the Sloan Digital Sky Survey, with foreground stars removed to make the galaxies' structures clearer.)

Hubble refined his classification system one step further, noticing that both spiral and elliptical galaxies can be subdivided into additional classes that, when properly ordered, show a smooth transition from spherical E type of galaxies to flatter E galaxies, to spirals, and to irregulars. This sequence of types was laid out by Hubble in his "tuning fork" diagram, illustrated in figure 17.8. Particularly notable in this diagram is the sequence of spiral types from S0, which have no arms, to Sa galaxies, which have smooth arms and a bright nuclear region, through Sd galaxies, which have loosely wrapped arms and a dim nucleus. Hubble believed this might be an evolutionary sequence, so he called the ellipticals "early" types, transitioning through the spirals to the irregulars, which he called "late" types. The galaxies also become bluer along this sequence, showing that the "late" types must have had star formation more recently, because the luminous blue stars are massive and have short lifetimes (chapter 14).

Interpreting Hubble's tuning fork diagram in some ways resembles the early attempts by astronomers to interpret the H-R diagram of stars. Early hypotheses proposed that stars traveled along the main sequence, but astronomers discovered later that stars were born at a particular position on the main sequence, and remained near that position throughout the hydrogen-fusion lifetime (chapter 14).

So which way is it for galaxies? Are they born at some point along Hubble's tuning fork diagram and then remain there, or do they evolve from one end of the sequence to the other? We will find that the answer is not as simple as either of these choices, but first we need more information. Just as for stars, we need to learn more about other properties of galaxies, such as their sizes, masses, and luminosities, in order to understand them. And to find any of those properties, we need to find galaxies' distances.

| 17.2 | THE DISTANCES OF GALAXIES AND HUBBLE'S LAW |

Our ability to study galaxies is hindered by their great distance from us. In fact, the extent of their distance from us was a major controversy among astronomers for centuries. For example, in 1755 the German philosopher Immanuel Kant, whose views on the origin of the Solar System were so ahead of their time, suggested that galaxies were remote star systems—Island Universes—similar to the Milky Way. On the other hand, as we saw in section 17.1, the American astronomer Harlow Shapley argued as recently as the 1920s that galaxies were merely small satellites of the Milky Way. The reason for such disagreement comes from the difficulty of *measuring* how far away from us galaxies are.

Galaxy Distances

Galaxies are separated from each other by enormous distances. Even the Large Magellanic Cloud, a small companion galaxy of the Milky Way, is more than 150,000 light-years distant. Astronomers cannot use parallax to measure such vast distances because the angle by which the galaxy's position changes as we move around the Sun is too tiny to be detected. Astronomers must therefore use other ways to find distances, such as the method of standard candles.

We discussed the method of standard candles in detail in chapter 13. We saw there that if we know the luminosity of a light source, we can use the inverse-square law to deduce its distance from how bright it looks to us. This was the technique we used in chapter 16 to find the size of the Milky Way, and we can apply it here to find the distance to galaxies.

Astronomers use many different astronomical objects as standard candles, but one of the most reliable is the class of luminous, pulsating variable stars known as the Cepheids (described in chapter 14). Cepheids have a luminosity about a million times the luminosity of the Sun, and because of their brilliance, they are easy to see in nearby galaxies. Furthermore, because their brightness changes as they pulsate, they are relatively easy to identify. The time it takes these variable stars to go from bright to dim to bright again (their period) is related to their luminosity. (Recall our discussion of the period–luminosity relation in chapter 14.) This then allows astronomers to determine a Cepheid's luminosity by simply measuring how long it takes the star to change its luminosity over one cycle. Once the luminosity of a variable star is known, astronomers can find its distance from a simple measurement of how bright the star appears as seen from Earth and an application of the inverse-square law. (See Astronomy by the Numbers: "Measuring the Distance of a Galaxy Using Cepheid Variables.")

Cepheids are luminous enough that astronomers can use them to measure the distance to nearby galaxies, but despite their luminosity, Cepheids are not observable in very distant galaxies. Therefore, astronomers use even more luminous events such as supernova explosions when determining the distance to more remote systems. Not all such events are ideal standard candles, however, because they are rare and their luminosity is not always accurately known. Astronomers therefore have sought other ways to measure the distance to galaxies.

One method is similar to the simple observation you have made if you look closely at a picture printed in a newspaper or comic book. Holding such a picture near your eyes, you can see that it consists of many tiny separate dots. As you move the picture farther from your eyes, the dots disappear and merge together to form the picture. Similarly, in a galaxy near us, we can easily see the separate spots of its more luminous stars. However, in more distant galaxies, the stars blend together into a uniform blur of light. Careful measurement of the graininess of a galaxy's picture can therefore give an indication of its distance.

ASTRONOMY
by the numbers

MEASURING THE DISTANCE OF A GALAXY USING CEPHEID VARIABLES

Suppose we detect a Cepheid in a galaxy and find that the Cepheid appears 100 million (10^8) times dimmer than a Cepheid of the same pulsation period in the Milky Way. Because the stars have the same period, we assume that they have the same luminosity. Let us denote the brightness and distance of the Cepheid in the Milky Way as $B_\star$ and $d_\star$, respectively. Similarly, we will denote the brightness and distance of the Cepheid in the distant galaxy by B_g and d_g. According to the inverse-square law, if two objects have the same luminosity, their apparent brightnesses $B_\star$ and B_g, are related to their respective distances, $d_\star$ and d_g, by

$$\frac{B_\star}{B_g} = \left(\frac{d_g}{d_\star}\right)^2.$$

$B_\star$ = Brightness of Cepheid in Milky Way
B_g = Brightness of Cepheid in distant galaxy
$d_\star$ = Distance to Cepheid in Milky Way
d_g = Distance to Cepheid in distant galaxy

To find their distances, we take the square root of both sides, getting

$$\sqrt{\frac{B_\star}{B_g}} = \frac{d_g}{d_\star}.$$

In other words, the square root of the ratio of their brightness equals the inverse of the ratio of their distances. Because the distant Cepheid in our problem looks 10^8 times dimmer from Earth than the Cepheid in the Milky Way, $B_\star/B_g = 10^8$. Thus, according to the inverse-square law, the ratio of the distance of the Cepheid in the galaxy to the distance of the Cepheid in the Milky Way must be

$$\frac{d_g}{d_\star} = \sqrt{\frac{B_\star}{B_g}} = \sqrt{10^8} = 10^4.$$

Thus, $d_g = 10^4 d_\star$. That is, the Cepheid in the remote galaxy is 10^4 times farther away than the one in our galaxy.

Suppose we know that the Cepheid in our own galaxy is 1000 light-years away. Then the Cepheid in the other galaxy must be 10^4 times 1000 light-years away. The distant Cepheid, and hence the galaxy in which it is located, is therefore 10^7 light-years distant (fig. 17.9).

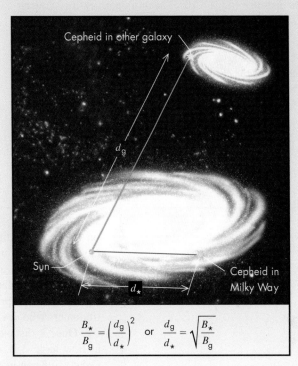

$$\frac{B_\star}{B_g} = \left(\frac{d_g}{d_\star}\right)^2 \quad \text{or} \quad \frac{d_g}{d_\star} = \sqrt{\frac{B_\star}{B_g}}$$

FIGURE 17.9
Finding the distance to a galaxy using Cepheid variables as standard candles. $B_\star$ and B_g are the apparent brightness of the Cepheid in the Milky Way and in the Cepheid distant galaxy, respectively. Their distances are $d_\star$ and d_g, respectively. From a measurement of the brightness ratio of the Cepheids, astronomers can calculate the ratio of their distances and, knowing the distance of the Cepheid in the Milky Way, they can therefore find the distance to the other galaxy.

Another method widely used for finding galaxy distances is based on the standard-candle method but uses the galaxies themselves as the standard candle. This method relies on the observation that the rotation velocity of a spiral galaxy is related to its luminosity. In particular, the more luminous a galaxy, the more rapidly it rotates. We might anticipate this result because if a galaxy rotates very rapidly, it must be very massive to hold itself together gravitationally. Moreover, a massive galaxy contains more stars and thus emits more light than a small galaxy. Thus, the faster a galaxy rotates, the more luminous it is.

Astronomers can find a galaxy's rotation speed by obtaining a spectrum of the galaxy at optical or radio wavelengths. The spectral lines will shift to shorter wavelengths

ASTRONOMY *by the numbers*   MEASURING THE DIAMETER OF A GALAXY

Astronomers measure the diameter of a galaxy using the same method we discussed in chapter 2 for measuring the diameter of the Moon or the Sun. There, we showed that an object's true diameter, D, is given by

$$D = \frac{2\pi A d}{360°}$$

where A is its angular size on the sky in degrees and d is its distance (fig. 17.10). The units of d are the same as

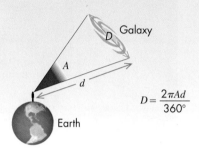

FIGURE 17.10
Measuring the diameter of a galaxy from its angular diameter.

for D; for example, in measuring galaxy sizes, we might express both in light-years or both in parsecs.

To illustrate the method, we will measure the diameter of M31, a close neighbor of the Milky Way. M31 is a spiral galaxy, so when we speak of its diameter, we are referring to the diameter of its disk. Even that is somewhat ambiguous, because galaxies do not have sharp edges—the stars just thin out gradually—and it is hard to say exactly where the galaxy ends and intergalactic space begins.

The angular diameter, A, of M31 is about 5° and its distance, d, is about 750,000 parsecs. Using the expression for D above and inserting these quantities, we find that the diameter, D, of M31 is

$$D = \frac{2\pi \times 5° \times 750{,}000 \text{ pc}}{360°}$$
$$\approx 65{,}000 \text{ pc or 65 kpc.}$$

Based on similar detailed measurements, astronomers conclude that M31 is larger than the Milky Way.

on the side moving toward us, and to longer wavelengths on the other side, by an amount that tells us the rotation speed according to the Doppler formula. From the speed of rotation, astronomers can estimate the galaxy's luminosity, and the distance can then be calculated from the luminosity and apparent brightness using the inverse-square law. This procedure is known as the "Tully-Fisher method," after the two astronomers who devised it.

Once a distance is found, we can determine other properties of a galaxy such as how large it is. (See Astronomy by the Numbers: "Measuring the Diameter of a Galaxy.") Keep in mind that most methods for finding the distance to a galaxy have many underlying assumptions. For example, many methods require knowing the distance to remote stars or other objects in our own galaxy, and those distances depend in turn on the distance to nearby stars. Astronomers refer to such multistep distance determinations as the distance ladder, and they must account for the inevitable errors and uncertainties at each step of the ladder. As a result, the final answer to how far away a particular galaxy is can be quite uncertain.

The Redshift and Hubble's Law

We saw in chapter 13 that we can learn much about a star from its spectrum. Likewise, we can learn much about a galaxy from its spectrum.* One of the most important such discoveries was made in the early 1900s by the American astronomer Vesto Slipher. Slipher found that of the galaxies he observed, most of their spectral lines were strongly shifted toward longer wavelengths—the red end of the spectrum. This so-called **redshift** is seen in nearly all galaxies. (The exceptions are a few nearby galaxies, whose motion is influenced by the gravitational attraction of the Milky Way and its neighbors.)

---

* We need to keep in mind, however, that a galaxy's spectrum is the spectrum of all of its stars, gas clouds, and other contents added together.

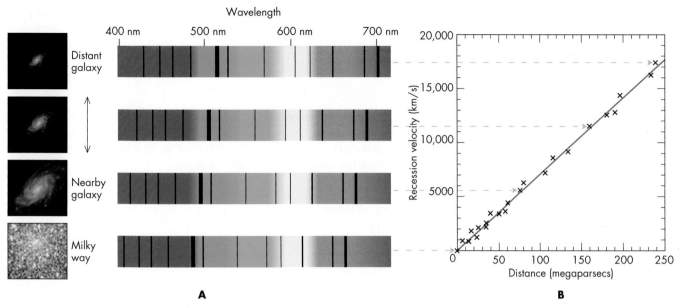

FIGURE 17.11

Redshift and recession velocity. (A) The more distant a galaxy, the larger its redshift. The spectrum of the total light of the Milky Way (which has zero redshift) is shown for comparison. (B) Hubble's law: Galaxies farther away recede faster.

Galaxy redshifts are *not* simply Doppler shifts. Rather, as astronomers know today, they result from the expansion of the Universe, as we will discuss in chapter 18. In Slipher's time, however, the cause of galaxy redshifts was a major puzzle. This puzzle became even more mysterious when Edwin Hubble and others noticed that the redshift was larger for dimmer (and therefore presumably more distant) galaxies. This implied that the speed with which a galaxy moves away from us—its **recession velocity**—increases with distance, as figure 17.11 shows. That is, *the faster a galaxy recedes, the farther away it is.* This, then, becomes a powerful tool for measuring galaxy distances. In 1929, Hubble discovered that a simple formula related the recession velocity, *V,* and the distance, *d.* He found that

$$V = Hd$$

where *H* is a constant. Because of his discovery, this relation is called **Hubble's law,** and *H* is called the **Hubble constant.**

The value of the Hubble constant depends on the units used to measure *V* and *d.* Astronomers generally measure *V* in kilometers per second and *d* in **megaparsecs,** where 1 megaparsec (abbreviated Mpc) is a million parsecs (= 3.26 million light-years). In these units, *H* is about 70 kilometers per second per megaparsec. For many decades astronomers disagreed about the value of *H.* The reason for this controversy was that, to determine *H,* one had to know *both* the distance and the velocity of at least a few galaxies in order to calibrate the law. *H* was then simply the average value of each calibration galaxy's velocity divided by its distance. However, as we have seen, it is difficult to accurately measure the distance to even a nearby galaxy. A new technique for determining *H* has essentially resolved this problem—establishing that $H \approx$ 70 kilometers per second per megaparsec. This new method uses information about irregularities in the brightness of radiation emitted by the young Universe, which we will discuss in chapter 18. For now, we will simply say that the method appears to give the most precise value yet for *H.*

Once a value for *H* is known, we can turn the method around and use Hubble's law to find the distance to a galaxy. The method is as follows: Take a spectrum of the galaxy whose distance you want to know. From the spectrum, we measure the shift of the spectral lines. From the Doppler-shift formula, we calculate the recession velocity.

V = Recession velocity of galaxy
d = Distance of galaxy
H = Hubble constant

Finally, we use that velocity in Hubble's law to find the distance by rearranging the equation.

Starting from Hubble's law $V = Hd$, to find d we divide both sides by H. This gives us an alternative form of Hubble's law:

$$d = V/H.$$

Thus, if we can measure the redshift of a galaxy, we can use the Hubble constant to find the galaxy's distance. An example is worked out in Astronomy by the Numbers: "Finding a Galaxy's Distance from Its Redshift," but we also want to sound a word of caution. Hubble's law does not work in all situations.

Limitations of Hubble's Law

Hubble's law describes a basic property of the Universe, that the more distant an object is, the faster it moves away from us. Indeed, we will discuss in chapter 18 the profound implications this has about the expansion of space. The law is so simple and so far-reaching that it is tempting to apply it everywhere, but it turns out this can lead to serious errors.

The problem is that galaxies and other objects may have motions that have nothing to do with the expansion of space, and there is no way to distinguish a redshift due to an orbit from a redshift due to the expansion of space. The problem is illustrated in figure 17.12, which plots recession velocities against distance. When galaxies orbit each other, sometimes their orbital speed is much larger than the redshift caused by expansion. As a result, some nearby galaxies have negative redshifts (blueshifts), or their distances might be greatly under- or overestimated if we used Hubble's law to try to find their distances from their recession velocities.

If a galaxy is isolated, it is usually safe to assume that its recession velocity is within about 100 km/sec of what Hubble's law would predict, but otherwise the recession velocity may be off by as much as about 1000 km/sec. As a result, we cannot use Hubble's law for nearby galaxies, and certainly not for any objects orbiting inside our own Galaxy.

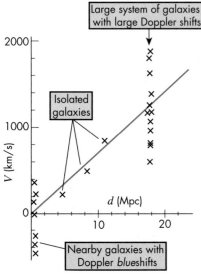

FIGURE 17.12
Hubble's diagram for galaxies that orbit one another. Orbital motions can make it difficult to find a distance from a redshift.

ASTRONOMY
by the numbers   FINDING A GALAXY'S DISTANCE FROM ITS REDSHIFT

Suppose we observe a galaxy and discover that a spectral line normally seen at $\lambda_0 = 500$ nanometers is instead detected at 570 nanometers, so its wavelength is redshifted by $\Delta\lambda = 70$ nm. How distant is the galaxy?

The first step to solving this problem is to find the galaxy's recession velocity. We use the Doppler formula from chapter 4 to find:

$$V = c \times \frac{\Delta\lambda}{\lambda_0} = 300{,}000 \text{ km/sec} \times \frac{70 \text{ nm}}{500 \text{ nm}}$$

$$= 42{,}000 \text{ km/sec}.$$

Next, we insert this value for V into Hubble's law:

$$d = (42{,}000 \text{ km/sec})/H.$$

For $H = 70$ kilometers/second per megaparsec, the value of d is then

$$d = \frac{42{,}000 \text{ km/sec}}{70 \text{ km/sec/Mpc}} = 600 \text{ Mpc}.$$

Thus, the galaxy is 600 megaparsecs away from us.

In applying this method, care must be taken with the units. When we divide V in kilometers per second by H in kilometers per second per megaparsec, the answer we get will be in megaparsecs. Thus, the 600 found in this example is not in parsecs or light-years but in megaparsecs. After carrying out a conversion, this distance corresponds to about 2 billion light-years.

17.3   GALAXY INTERACTIONS AND EVOLUTION

Astronomers continue to use Hubble's galaxy classifications today as a useful way to organize galaxy studies. However, many ideas have changed about what features are significant. For example, the categorization into barred and unbarred spirals is now seen as a short-term difference. Computer models of galaxies show that most disk-shaped systems will form a bar if they undergo a gravitational disturbance, perhaps as the result of a close encounter with a neighboring galaxy. Our own Milky Way has a weak bar. The bar forms because a gravitational disturbance alters the orbits of the stars, causing them to bunch up in an elliptical pattern that we see as the bar. Some models also suggest that over hundreds of millions of years, the orbits spread out again and the bar dissolves. Thus, bars may appear and disappear, and so an ordinary spiral galaxy may change into a barred spiral and then back again.

Since Hubble's time, it has also been recognized that the irregular class is composed of two very different types. The majority are smaller galaxies, often called dwarf galaxies, and they lack a coherent structure. Examples of these were shown in figure 17.5. However, some large spirals and ellipticals can also become classified as irregulars when they undergo interactions with neighboring galaxies that distort their shapes.

When galaxies collide, stars do not actually crash into each other, but the gravitational force of the galaxies alters the orbits of their stars and thereby changes the shape of each galaxy. The top row of figure 17.13 shows a computer simulation of such a close passage, and you can see the stars torn from each galaxy by their mutual gravitational force and flung outward in long arcs. The bottom two pictures in the figure show an images of two real pairs of galaxies with just such plumes of stars.

Galaxies with tails

Although interactions of this sort are disruptive to galaxies, they rarely harm the individual stars. Stars are so far apart within a galaxy that most stars move harmlessly past each other. Thus, galaxy collisions are something like tossing two handfuls of sand at each other: most of the grains simply pass by one another without hitting. Clouds of gas and dust in the galaxies are not so lucky. Being much larger than stars, clouds do collide, and their impact may compress them enough to trigger a burst of star formation within them. Such "starburst galaxies," as these collisionally stimulated galaxies are called, are some of the most luminous galaxies known.

Simulations suggest that even when galaxies pass close by each other, disturbances to the orbits will cause gas clouds to collide near the galaxy's center, leading to

FIGURE 17.13
Colliding galaxies. The top row shows a computer simulation of two galaxies colliding. The bottom row of the figure shows images of two such collisions: "The Antennae" (bottom left) and "The Mice" (bottom right).

FIGURE 17.14

(A) A starburst in the central portion of the galaxy M82 may be the result of a gravitational interaction with a neighboring galaxy. (B) The "Cartwheel" galaxy's ring shape was probably caused when another galaxy passed through it a few hundred million years ago. The likely trajectory is shown, and several candidate galaxies can be found in the direction of the arrow.

M81 and M82 are shown in Looking Up #1 at the front of the book.

a starburst within a few hundred million years. A burst of star formation in the galaxy Messier 82 (fig. 17.14A) may have been triggered by neighboring galaxy M81.

Figure 17.14B shows what can happen when one galaxy plows directly into another. The head-on collision created what astronomers call a ring galaxy. The colliding galaxy sent a compression wave traveling out through the disk, causing interstellar clouds to collapse and form stars. This is somewhat like the compression that occurs in the spiral arms of the Milky Way (chapter 16), but more extreme. We might imagine that over time, after the brightest young stars have died and the galaxy's rotation has smeared out the stars formed here, the galaxy will again show a spiral pattern.

Although the odd shapes of colliding galaxies might cause them to be temporarily labeled as "irregulars," they will probably settle back into spirals or ellipticals after the collisions. This is a possible way for galaxies to change shape over time.

Differences in the Stellar and Gas Content of Galaxies

Since Hubble's time, astronomers have discovered that spiral, elliptical, and irregular galaxies differ not only in their shape but also in the types of stars they contain. For example, spiral galaxies contain a mix of young and old stars (Pop I and II, see chapter 16), but elliptical galaxies contain mostly old (Pop II) stars. This difference in the kinds of stars composing S and E galaxies is understandable in terms of their different gas content. To make young (Pop I) stars, a galaxy must contain dense clouds of gas and dust. Spiral galaxies like the Milky Way have about 15% of the mass of their disk in the form of such interstellar clouds, and so they can easily make the young stars of their spiral arms. Radio images of hydrogen emission (fig. 17.15) show the presence of gas in spirals and irregulars. On the other hand, ellipticals contain much less interstellar matter. In fact, astronomers used to think that the E type of galaxies contained no interstellar gas or dust at all. However, more recent observations made with X ray telescopes show that many E galaxies contain very low density but very hot (10^7 K) gas. Such hot gas does not readily form stars, and so it is no surprise that elliptical galaxies rarely contain young, blue stars like the ones we see in the disks of spiral systems.

Although young blue stars are rare in elliptical systems, such stars are common in most irregular galaxies. Many of these Irr galaxies also contain large amounts of interstellar gas. In fact, some have more mass in gas than in stars, and they are very actively forming new stars.

Apart from their different star and interstellar matter content, S, E, and Irr galaxies have few other differences to generalize about. For example, if we look at the mass

FIGURE 17.15

The spiral galaxy M81. (A) Visible-light image of M81. Note also its small companion to its lower left. (B) False color map of the 21-centimeter radio emission from hydrogen. Hydrogen is also visible in a faint irregular galaxy left of M81. Note the concentration of hydrogen in M81's spiral arms, and the lack in the bulge, where the stars are yellow.

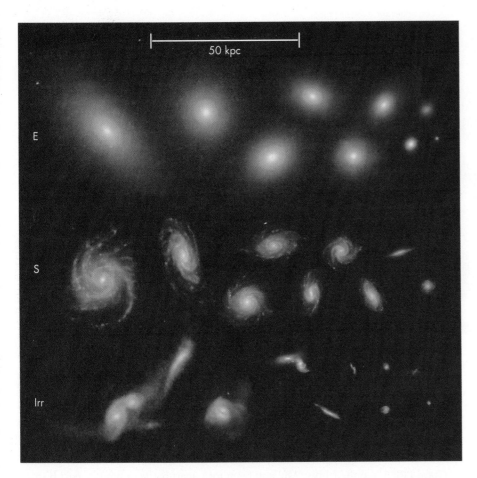

FIGURE 17.16

Images of galaxies shown with their correct relative sizes. The spirals are all in Hubble's Sc class (see fig. 17.8). The irregulars are a mix of interacting galaxies and dwarf galaxies. Blue colors indicate current star formation. (Galaxy images were drawn from the Sloan Digital Sky Survey. Foreground stars have been digitally removed to make the galaxy structure clearer.)

and radius of galaxies, we do not find that all ellipticals are huge and all spirals are small. Instead, we discover that elliptical galaxies range enormously in size. Some are not much larger than globular clusters and contain only 10^7 solar masses of material; others are monster galaxies, 10 to 100 times the mass of the Milky Way. This spread is shown in figure 17.16, which shows sets of ellipticals, spirals, and irregulars all reproduced to their correct relative sizes.

What fraction of galaxies makes up each type? Even that question has no simple answer. If we make a census of the galaxies near the Milky Way, we find that most galaxies are dim dwarf E and Irr systems, sparsely populated with stars. These dim galaxies are undetectable at greater distances, where we can see only the more luminous galaxies. If we make a census of those luminous objects, we find that the majority of galaxies are spirals, but the percentage depends on whether we look in crowded or in more sparsely populated regions. Where galaxies are crowded together, only about 10% of them are spirals, but in less crowded regions that proportion is about 80%.

The Evolution of Galaxies: Collisions and Mergers

The diversity of galaxy form and content requires an explanation, and a major area of astronomical research today is exploring why galaxies come in such distinct forms. In recent years, with the aid of the Hubble Space Telescope, astronomers have begun to develop a picture of how galaxies form and evolve. That instrument's ability to see fine details in distant, ancient galaxies has allowed astronomers to observe galaxies at various ages all the way back to when they were less than a billion years old.

The most distant galaxies are so faint that they require hundreds of hours to detect. Because it would take millions of years to study the whole sky at this depth, astronomers selected two tiny regions on the sky to concentrate their efforts. They

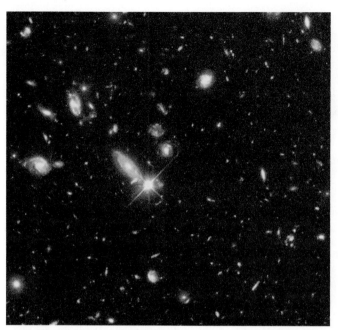

FIGURE 17.17

The Hubble Space Telescope observed two "deep fields," small regions of the sky, for more than 100 hours to detect extremely distant galaxies. Nearly every dot in this image is a galaxy. The galaxies are not close together in space, but are spread out over more than 10 billion light-years along our line of sight.

The position of the first Hubble Deep Field is shown in Looking Up #2 at the front of the book.

produced "deep fields" by pointing the Hubble Space Telescope at these regions for hundreds of hours, allowing them to examine galaxies all the way back to near the beginning of time. A portion of one of these deep fields is shown in figure 17.17. This region is tiny—if you hold a penny at arm's length, the eye of Abraham Lincoln could completely cover this area—yet it contains thousands of galaxies. The galaxies appear crowded together, but they are actually very spread out—distributed from nearby to almost 13 billion light-years' distance.

When we observe the most distant galaxies we can detect, the light we see left them when they were still very young. At this early time, galaxies are generally much smaller than the Milky Way. Some examples of distant galaxies detected with the Hubble Space Telescope are shown in figure 17.18. Not only are these young galaxies small, they are more numerous than galaxies today. Thus, these young galaxies must disappear, probably by colliding and merging into bigger systems. This would account both for the formation of larger systems like the Milky Way and for the smaller number of galaxies we see today.

The importance of interactions is strikingly confirmed in the images of distant systems made with the Hubble Space Telescope (fig. 17.18). Remote galaxies show signs of disturbance by neighboring galaxies—the odd shapes, bent arms, and twisted disks caused by collisions—much more frequently than nearby galaxies.

We saw in chapter 16 that our own Galaxy, the Milky Way, shows signs of past mergers. For example, astronomers can see streams of stars in our Galaxy's halo, which appear to be debris from small galaxies that have fallen into the Milky Way and been torn apart by its gravity. A past history of mergers also explains a puzzling feature of stars in the bulge of the Milky Way. Astronomers at one time thought that the stars in our Galaxy's bulge must have formed in one single, massive event at the Galaxy's birth. However, stars in the bulge have a wide range of ages, suggesting that the bulge grew gradually over time. Such gradual growth is easy to explain by mergers. Each time a large galaxy (such as the Milky Way) swallows a smaller companion, some of the gas from the companion sinks to the center of the larger system, where it forms new stars. Successive mergers lead to successive generations of stars, so that the bulge stars have the wide age range that is observed.

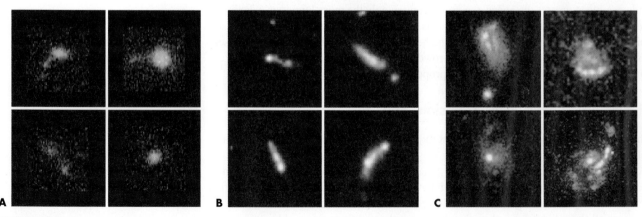

FIGURE 17.18

Hubble Space Telescope images of distant galaxies. Each small square shows an area approximately 25 kiloparsecs on a side, about the size of the Milky Way's disk today. The panels show galaxies (A) less than 1 billion years old, (B) a few billion years old, and (C) about 6 billion years old—half the current age of the Milky Way. (A) and (B) are from the Hubble Ultra Deep Field; (C) is from the Medium Deep Survey.

Streams of stars from small galaxies that have been torn apart are seen around other galaxies as well, as shown in figure 17.19. Astronomers call this process of a large galaxy capturing and absorbing a small galaxy **galactic cannibalism.** Astronomers have found evidence that the Andromeda galaxy, M31, also has cannibalized at least one small galaxy, based on the detection of a small bright clump of stars near the center of M31 that appears to be the remains of a galaxy it absorbed.

Cannibalism is one of the main mechanisms by which galaxies grow, but what makes one series of mergers result in a spiral galaxy, while another becomes an elliptical galaxy? Astronomers have proposed a model illustrated in figure 17.20. Most galaxies start out as small, gas-rich systems, which actively form stars and are therefore blue in color. Subsequently, when these galaxies collide and grow into larger systems, they undergo bursts of star formation, and perhaps consume much of their gas. Computer models suggest that if enough stars form at once, the supernova explosions of many massive stars may drive gas out of the galaxy. If the collision/merger leaves little gas remaining in the system, an elliptical galaxy results. Without gas to form new stars, the elliptical galaxy grows redder with time. If the collision/merger leaves moderate to large amounts of gas in the system, a large spiral results. The spiral continues to form stars and remains relatively blue in color, but still consumes its gas and grows redder more gradually. Such interactions could potentially allow some ellipticals to turn back into spirals as well. This might happen if an elliptical captures a gas-rich system and the gas then settles into a disk around the elliptical galaxy's nucleus.

This hypothesis of the origin of spiral and elliptical systems may explain why galaxies exhibit a division into red and blue "sequences" when their colors are plotted against their masses as shown in the graph in figure 17.21. Galaxies grow more massive through mergers, moving to the right in the diagram. They also steadily consume their gas, so the more massive galaxies have fewer young blue stars and become redder overall. Once the gas is exhausted, perhaps as a result of a collision, the galaxies shift to the red sequence after the short-lived blue stars die. Through repeated cannibalism, a galaxy may grow into a giant, which perhaps explains why abnormally large elliptical galaxies are found at the centers of groups and clusters of galaxies (see section 17.5). Cannibalism may also account for much of the activity observed in some galaxies (see section 17.4).

FIGURE 17.19

Streams of stars around the galaxy NGC 5907 are the remnants of a small galaxy that has been pulled apart by the larger galaxy's gravity.

The first galaxies were mostly small, gas-rich, and blue from star formation.

Galaxies merge and some use up or lose their gas, so star formation ceases.

Today we have more large and gas-poor galaxies, but some small ones remain.

FIGURE 17.20

A possible scenario of how galaxies evolve. The first galaxies were mostly small and gas-rich (left panel). Larger galaxies formed through mergers, as illustrated by the next two panels. Some of these galaxies consumed their gas or lost it during collisions, and ceased forming stars, so they grew redder in color. Others merged to form large spiral galaxies that continue to form stars and remain blue in color.

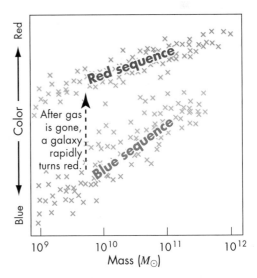

FIGURE 17.21

When we plot the colors and magnitudes of galaxies today, some galaxies are distributed along a "blue sequence," which are probably galaxies that grew through mergers without losing their gas. Gas-poor galaxies lie along the "red sequence."

FIGURE 17.22

Galaxies changing character. (A) A merger in which two spirals are colliding and forming an elliptical galaxy. (B) As the spiral galaxy above and to the right of the center of this image travels through intergalactic gas, clouds of gas are being stripped from the galaxy. The blue blobs below and to the left of the galaxy are regions where stars are forming in the stripped gas. The cluster, Abell 2667, is a little over 3 billion light-years distant.

Astronomers have identified some galaxies that appear to be undergoing the conversion from spirals to ellipticals. For example, the galaxies in figure 17.22A are colliding. Both galaxies show evidence of clouds of dust and gas, resembling spirals, but the collision has disrupted any clear disklike pattern of circular rotation in the larger galaxy. Stars have been sent into new orbits oriented in various directions, and the resulting distribution of stars resembles an elliptical galaxy.

S0 galaxies are particularly interesting because they appear to be transitional between ellipticals, which have almost no gas, and spirals, which have a substantial gas content. The S0 galaxies still have stars orbiting in a flat disk, so it is as if the gas in a spiral had simply been removed without major disturbance to the orbits that would turn them into ellipticals. According to one hypothesis, S0 galaxies are spirals whose gas has been stripped by their motion through the hot, tenuous gas in intergalactic space. The impact of such gas from outside might "sweep" gas and dust out of the moving spiral, much like a leaf-blower cleans debris from between well-rooted plants. Figure 17.22B shows what may be an example of this. The blue blobs down and to the left of the spiral galaxy appear to be gas swept out of the galaxy which is forming stars. An alternative idea is that S0 galaxies form when the original gas in a disk galaxy is all turned into stars when the galaxy was relatively young, so the stars remained in their disklike orbits. In either case, without gas the disk galaxy can make no new stars and hence it shows no spiral structure and grows steadily redder.

In this discussion of the factors that might shape a galaxy, we have said nothing about how spiral arms form. In chapter 16, we discussed some hypotheses about how the Milky Way may have formed its spiral arms. For example, we saw that spiral arms might form as a density wave travels around a galaxy's disk, compressing gas clouds and triggering star formation in a spiral pattern. Astronomers assume that this mechanism operates in other galaxies and forms their arms, too. We also discussed in chapter 16 a different model in which spiral arms form as the galaxy's rotation stretches regions of star formation into elongated arcs. This mechanism probably also forms arms in other galaxies.

The mergers of galaxies, starbursts, and the depletion of interstellar gas can explain many features of galaxies and their evolution. However, some galaxies exhibit much more extreme phenomena, generating far more light and high-energy radiation than can be explained by any starburst. The production of this high-energy radiation appears to be connected to galaxies' interaction and evolution, as we examine next.

17.4   ACTIVE GALAXIES

Active galaxies are galaxies whose centers (nuclei) emit abnormally large amounts of energy from a tiny region in their core. Approximately 10% of galaxies are found to be active, but astronomers suspect that all large galaxies may go through active phases during their lives, especially when they are young and still actively merging. Not only is the emitted radiation intense, but it also usually fluctuates in intensity. The centers of these active systems are sometimes called AGNs (active galactic nuclei), and they are now suspected to harbor enormous black holes.

The Discovery of Nuclear Activity

Even before astronomers had determined what galaxies were, they noticed that some had peculiar features at their centers. For example, in the early 1900s, astronomers noticed that the bright elliptical galaxy M87 (fig. 17.23) has a "jet" of gas coming out of its nucleus. By measuring the shift in position of the individual blobs of gas in the jet, astronomers in recent decades have shown that the gas in this jet is being ejected from the nucleus at nearly the speed of light.

In the 1940s, the American astronomer Carl Seyfert (pronounced *see-firt*) identified a class of unusual spiral galaxies. These spirals looked normal except that their nuclei were remarkably luminous. An example of such a **Seyfert galaxy** is shown in figure 17.24. The luminosity of the nucleus of a typical Seyfert galaxy is immense, amounting to the entire radiation output of the Milky Way, but coming from a region far less than a light-year across. The nuclear regions of Seyfert galaxies also contain gas clouds moving at speeds of about 10,000 kilometers per second or more. In some instances, the nucleus is also ejecting gas in jets, as in M87. The intense radiation from a Seyfert galaxy's core can also fluctuate rapidly in intensity, sometimes changing appreciably in a few minutes.

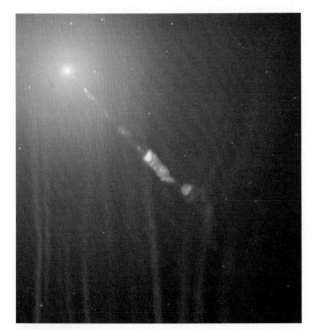

FIGURE 17.23
Hubble Space Telescope image of the center of the elliptical galaxy M87. A jet of hot gas can be seen coming from the bright nuclear source at the center of the galaxy.

FIGURE 17.24
The Seyfert galaxy NGC 7742, imaged with the Hubble Space Telescope. The nucleus of this face-on spiral galaxy is exceptionally luminous.

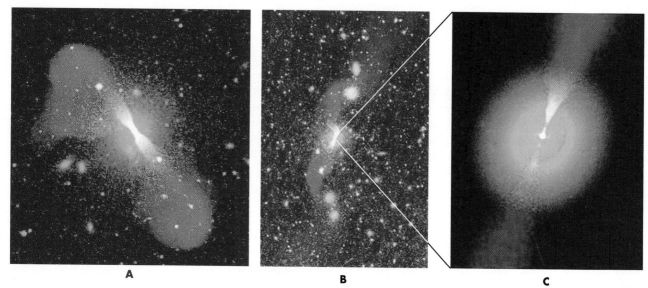

A   B   C

FIGURE 17.25
Radio galaxies (A) NGC 5532, and (B & C) NGC 383. Radio emission is shown in red, superimposed on visible emission in blue. The image in (C) is a close-up view of the nucleus of NGC 383.

By the 1960s, astronomers began to identify another kind of active galaxy, which they labeled **radio galaxies.** As their name suggests, these galaxies emit large amounts of energy in the radio part of the spectrum, typically millions of times more than the radio energy emitted by a normal galaxy. Most are elliptical galaxies, and the radio emission comes primarily from regions outside the galaxy on either side of it, as you can see in figure 17.25. This figure shows superimposed visible and radio wavelength images of the galaxy, with red representing the radio emission.

The radio-emitting regions visible outside the radio galaxies in figure 17.25 are called "radio lobes." These lobes, often located symmetrically on opposite sides of the galaxy, may span hundreds to millions of light-years. Although most of the radio emission comes from the lobes, some comes from within the visible body of the galaxy, from narrow jets and a tiny nuclear region much less than a light-year across.

What causes this intense radio emission? Astronomers can tell from its spectrum that the emission is synchrotron radiation, created by electrons traveling at nearly the speed of light and spiraling around magnetic field lines. We discussed this mechanism in chapter 15, where we described radio emission from pulsars. Pulsars generate high-speed electrons at their surface, but radio galaxies generate them in their core. The electrons are then shot out of the core in narrow **jets** (such as the ones shown in figs. 17.23 and 17.25) that generally point at the radio lobes. Streaming away from the end of the jet, the electrons eventually collide with the gas around the galaxy. There, they spread out to form the lobes, all the while emitting synchrotron radiation.

Q: What evidence in this picture suggests this might be an active galaxy?

Quasars

Quasars get their name from a contraction of the words *quasi-stellar radio source,* where *quasi-stellar* means "almost starlike." This name refers to their appearance in photographs, where they look like dim stars (fig. 17.26). However, quasars are some of the brightest radio sources in the sky despite their dimness at visual wavelengths.

Quasars were initially thought to be some kind of exotic star in our Galaxy, but today they are known to be extremely luminous, extremely distant, active galaxies. We deduce their huge distance from the huge redshift of their spectra (recall that a galaxy's distance can be found from its redshift using Hubble's law). On the basis of their redshifts, the most distant ones are more than 10 billion light-years from Earth. To be so bright at such an immense distance, an object must be immensely luminous, and a typical quasar turns out to be about 1000 times more luminous than the Milky Way.

Quasar   Star in Milky Way

FIGURE 17.26
Photograph of the quasar 3C273. This is the brightest quasar in the sky, but it looks like little more than a dim star. ("3C" stands for *Third Cambridge Catalog of Radio Sources.*)

Despite their huge power output, quasars, like Seyfert galaxies, fluctuate in brightness. Large changes in brightness can occur within hours. Rapid changes of this sort give clues to the size of the emitting object, as discussed in Extending Our Reach: "Estimating the Diameter of Astronomical Objects by Using Their Light Variability." However, astronomers need not rely solely on light variations to determine the size of an active galaxy's core. Observations with radio interferometer telescopes show that the active cores are extremely small (smaller than a fraction of a light-year, in some cases).

These findings tell us that some quasars have the power output of trillions of stars packed into a volume smaller than the Solar System. This is such an extraordinary claim that astronomers searched for alternative possibilities for quasars for a number of years. For example, the extraordinary luminosities derived for quasars are based on their huge distances, so some astronomers worked on the hypothesis that quasars' large redshifts were perhaps caused by gravitational redshifts instead (see chapter 15). If quasars were nearby, the implied luminosity would be smaller, but this model led to a host of other problems and self-contradictions.

EXTENDING
our reach

ESTIMATING THE DIAMETER OF ASTRONOMICAL OBJECTS BY USING THEIR LIGHT VARIABILITY

Changes in brightness give an important clue to the size of a quasar. Suppose we have a ball of glowing material that we can turn on and off. Suppose further that when we turn it on, it begins to glow everywhere at the same time, and when we turn it off, it stops glowing everywhere at the same time. If we now look at this ball as it begins to glow, light from its nearest part, A, will reach us before light from its farthest part, B (fig. 17.27). An exactly similar effect occurs with sound produced from a lightning stroke. The lightning stroke lasts only fractions of a second, but the sound it generates lasts many seconds. The reason the sound is drawn out for such a long time is that sound travels relatively slowly (only about 1000 feet/second). Thus, if the stroke is 5000 feet long (a typical size for a bolt of lightning), sound from the near part reaches us 5 seconds before sound from the far part does. So, too, even though the ball begins to glow everywhere on its surface at exactly the same moment, the light we see does not turn on instantly. Instead, it gradually increases in brightness over a time interval equal to the time required for light to travel across the ball. That is, if the ball has a diameter of 1 light-second (the distance light travels in 1 second = 3×10^5 kilometers), the light reaching an observer does not reach full brilliance in less than 1 second. Likewise, if it is a light-month across (about 10^{12} kilometers) and it is turned off instantly, it does not dim away to darkness in less than 1 month.

From this, we can conclude that if a quasar varies in brightness in a time as short as a few days (and assuming simultaneous illumination and dimming), then the emitting region can be no more than a few light-days across. A core a few light-days across is tiny compared to the size of a typical galaxy (100,000 light-years across). In fact, such a core is about 10 million times smaller than a galaxy, or about 200 AU in diameter, about 5 times the size of Pluto's orbit around the Sun.

Distance = velocity × time
$$D = ct$$
$$\frac{D}{c} = t$$

FIGURE 17.27
Sketch illustrating that when a light source turns on, we see the light from its near side before we see the light from its far side, and so we do not see its light turn on instantly. For the objects we encounter in everyday life, the time delay is so short that we do not notice it. However, for astronomical-sized bodies, the effect can create delays ranging from days to months, from which we can deduce that the emitting body is light-days or light-months across.

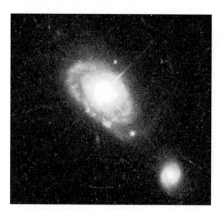

FIGURE 17.28
PG 0052+251 as viewed by the Hubble Space Telescope. The image shows the galaxy surrounding the quasar, demonstrating that it is an active galactic nucleus. ("PG" stands for *Palomar Green* survey.)

That quasars actually *are* distant and almost unbelievably luminous was ultimately settled by images such as the one in figure 17.28, which clearly reveal that quasars lie in the centers of distant galaxies. The host galaxy of a quasar is difficult to see except from space because of the intense brightness of the quasar and the small angular size of the galaxy at its huge distance. It is now clear that quasars are the most powerful extreme of a broad range of active galaxies. Astronomers also now recognize that some other kinds of peculiar objects are related to quasars. For example, BL Lacertae objects (BL Lac, for short) were originally believed to be peculiar variable stars within the Milky Way. They are now known to be distant radio galaxies in which a jet of ejected matter is pointing almost straight at the Solar System.

Today, astronomers have cataloged hundreds of thousands of quasars and active galaxies. Apparently, something goes on at the very center of galaxies that can generate enormous power. What could possibly produce such vast amounts of energy within such a small volume?

A Unified Model of Active Galaxies

Because the different classes of active galaxies share many features in common, astronomers have sought a model for nuclear activity that could explain this huge power output. At the same time, a successful model must also explain the differences between the types of active galaxies. For example, why are some powerful radio emitters and others not? Why are the most powerful quasars seen at large distances but not nearby?

Any model for active galaxies must explain how such a small central region can emit so much energy over such a broad range of wavelengths. For example, some quasars are powerful sources of both radio waves and gamma rays. Astronomers therefore hypothesize that the core must contain something very unusual. No ordinary single star could be so luminous. No ordinary group of stars that might produce so much radiation could be packed into so small a region. One kind of object that is very small but that can also emit intense radiation is the accretion disk around a black hole. We saw in chapter 15 that such accretion disks can emit intense X ray radiation. Could the center of an active galaxy contain an immense black hole? Could the radiation we observe from the core be hot gas in an accretion disk around such a black hole? This is precisely the model for active galaxies most widely favored by astronomers today. According to this model, an active galaxy has in its core a **supermassive black hole** that contains anywhere from millions to up to several billions of solar masses of matter. Around the black hole an accretion disk as large as the entire Solar System swirls furiously under the intense gravitational pull of the black hole. Orbiting under the black hole's fatal attraction, the gas in the accretion disk is heated to incandescence by frictional forces as it collides with other gas in the disk at a significant fraction of the speed of light.

Figure 17.29, an image made with the Hubble Space Telescope, supports this model. The image shows just such a disk in the central regions of the radio galaxy NGC 4261. Gas falling into the disk releases gravitational energy, which heats the material to temperatures as hot as several million K. Although

FIGURE 17.29
The radio galaxy NGC 4261. The image on the left shows this elliptical galaxy and its radio jets in a composite optical and radio picture. The jets extend outward about 50,000 light-years. The image on the right, made with the Hubble Space Telescope, shows the central 1000 light-years, in which a dusty disk-shaped cloud surrounds and probably feeds the central black hole that powers the radio jets.

Looking into jet:
Blazar

If jets present:
Radio galaxy

Looking at hot
accretion disk:
Quasar or
broad-line Seyfert

View blocked by
dusty outer disk:
Narrow-line Seyfert

A

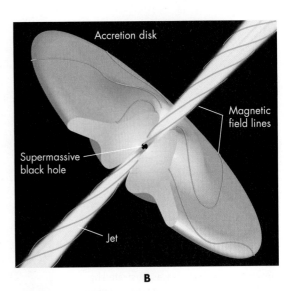

B

FIGURE 17.30

Sketches of (A) the disk surrounding a (B) accretion disk, and the jets in the nucleus of an active galaxy. Spiralling magnetic fields may help confine the jets. Depending on the angle of view and the presence of jets, an active galaxy may be classified in different ways.

An Explanation of Superluminal Motion

most of the mass ultimately falls into the black hole, some material from the accretion disk "boils off" at enormously high temperatures and escapes. This outflow is channeled into two oppositely directed jets by the black hole's magnetic field, as illustrated in figure 17.30. Depending on the angle of view and the strength of the jets. we arrive at different classifications for the active galaxy. Computer models show that the intense magnetic fields become tightly twisted as the black hole spins. Magnetic fields trap the charged particles found in a hot gas (as seen in pulsars in chapter 15). The spinning and twisting of the magnetic field can, therefore, capture hot gas from the accretion disk and pump it outward along the field at very high speed (see Science at Work: "Superluminal Jets"). This model may help explain the narrowness of the jets and their kinks and turns. Moreover, it taps into a powerful supply of energy for the jet—the rotational energy of the spinning black hole.

How might such a massive black hole form? We saw in chapter 16 that astronomers have strong evidence that our own galaxy has a black hole at its center. That black hole has a mass of a few million solar masses. This is much smaller than the black holes thought to be in some active galaxies, but it is still enormously larger than the black holes produced when the most massive stars die. Astronomers hypothesize

SCIENCE *at work* — SUPERLUMINAL JETS

The jets observed in some quasars and other active galaxies exhibit a very curious property: they contain clumps that appear to move faster than the speed of light. Astronomers refer to such seeming faster-than-light motion as superluminal motion. According to Einstein's theory of relativity, no object can travel faster than light, but under certain circumstances it is possible for an object to *appear* to move faster than light.

This appearance of superluminal motion can occur when an object is moving almost straight toward us at close to the speed of light. Because the object is traveling toward us at nearly the same speed as the light it emits, we observe its motion in a compressed way. For example,

light emitted by a clump a thousand years ago and light emitted now may arrive at Earth just a year apart if the object has moved 999 light-years closer to Earth in the meantime. From Earth, if the motion of the clump causes it to shift by more than a light-year relative to our line of sight during those thousand years, it will appear to us that it is moving faster than the speed of light.

While superluminal motion is a kind of optical illusion, it tells something important about the objects that power active galaxies: they must be able to accelerate matter to near the speed of light. This is something that the intense gravity of black holes is capable of doing.

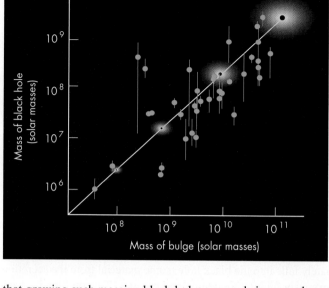

FIGURE 17.31

The correlation between central black-hole mass and the masses of the galaxies' central bulge. The green dots indicate the masses of central black holes and bulges in more than 30 galaxies studied with the Hubble Space Telescope. The vertical green bars indicate the uncertainty in the measurements. The mass of a central black hole grows with galaxy mass, as illustrated by the background diagrams of galaxy bulges.

that growing such massive black holes proceeds in several steps. First, a massive star in the central region of a galaxy reaches the end of its life, explodes as a supernova, and forms a black hole perhaps as small as 5 $M_\odot$. Next, this small black hole grows as interstellar matter falls into it. Recall that this matter is most concentrated at the centers of galaxies, so a black hole there has greatest opportunity to grow by this means—especially when the galaxy is young and less of its gas is converted into stars. Because the Schwarzschild radius of a black hole is only a few kilometers for each solar mass it contains (chapter 15), the likelihood of an encounter close enough to capture a star is extremely unlikely. On the other hand, gas in the accretion disk loses energy through friction, so it spirals into the black hole.

As gas falls in, the hole's mass and radius increase, making it easier for the hole to attract and swallow yet more material. Eventually the hole becomes large enough to swallow entire stars, at which point it grows rapidly to a million or more solar masses. Some galaxies may have had more mass at their cores to begin with, or they may have gained mass by collisions with small, neighboring galaxies. In some galaxies this extra mass thereby allowed the galaxy's central black hole to grow far larger than the black hole in the Milky Way.

Astronomers today have evidence that many (perhaps even all) large galaxies have a massive black hole at their core. They come to this conclusion from the observation that stars and gas at the core of many galaxies are orbiting the core at extremely high speed. To prevent such rapidly moving material from flying out of the galaxy's core requires the huge masses of millions or even billions of times the Sun's mass. Yet this huge mass seems to emit no light or other radiation, and astronomers know of no other objects that are both so massive and dark. Thus, by elimination, they are led to think that many galaxies have a massive black hole at their core.

Another recent discovery has led astronomers to think that the growth of massive black holes must be an integral part of the growth of galaxies. For galaxies in which the mass of the central black hole has been measured, that mass closely correlates with the mass of the galaxy's central stellar bulge (fig. 17.31). If growth of supermassive black holes were a chance event, this correlation would be unlikely. Instead, the central black hole must grow at the same rate as the bulge grows, from the initial collapse of gas when a galaxy first forms its bulge to the mergers and cannibalism that cause the bulge to grow. Part of this correlation may result from a "feedback" process in which the energy output of the central black hole affects star formation within a galaxy. It probably also indicates that central black holes merge when galaxies merge, as is suggested by images of double nuclei in some distant galaxies (fig. 17.32).

FIGURE 17.32

Hubble Space Telescope image of a galaxy with two active nuclei. The galaxy is probably the result of a merger in which the central black holes have not yet merged.

If black holes are so common in galaxies, why are none of the galaxies close to us as active as quasars? The youth of quasars may be the key. Because quasars are so distant, we see them the way they were in the distant past, not the way they are "now." For example, when we look at a quasar that is 10 billion light-years away, we are seeing light that has taken 10 billion years to reach us. Thus, we are seeing the quasar as it was 10 billion years ago. The fact that most quasars are billions of light-years away implies that the phenomenon causing them occurred billions of years ago, when the Universe was younger. Galaxies whose cores were once active quasars probably surround us, but their activity has long since died away. They still harbor huge black holes, but they have used up the gas that powered their activity.

Left to itself, every active galaxy's accretion disk would gradually drain into the central black hole and activity would cease. But the accretion disk may be replenished with matter from at least two sources. Sometimes stars within the galaxy may approach too closely to the black hole and be torn apart by its tidal forces. Their gaseous debris then falls into the accretion disk as a new source of material. Alternatively, from time to time, collisions with neighboring galaxies may add new material as the active galaxy cannibalizes smaller neighbors. Each such event gives the accretion disk a new burst of energy, causing the active galaxy to revive and increase in brightness.

Probing Intergalactic Space with Quasar Absorption Lines

Quasars are also useful to astronomers as probes of intergalactic space. Because quasars are so distant and bright, their light passes through other galaxies and intergalactic gas clouds on its way to Earth. Even if the intervening object is too dim to see, its matter may imprint its signature on the quasar's light, allowing us to detect these otherwise invisible objects.

For example, many quasars show absorption lines in their spectra that have very different Doppler shifts from those of the emission lines in the quasars themselves (fig. 17.33). These absorption lines are created by cool gas clouds lying between us and the quasar, much as similar lines in stars are created by cool interstellar clouds (chapter 16). Some of the cool gas responsible for the absorption lines is located within galaxies, but most of it is intergalactic: gas that has been ejected from galaxies by active galactic nuclei and supernova explosions, or gas in clouds that have not yet formed into galaxies. This intergalactic gas can also interact directly with galaxies, affecting their evolution, particularly in regions where galaxies cluster together.

FIGURE 17.33
The light from a distant quasar produces a uniform light source against which we can see hundreds of absorption lines produced by hydrogen clouds along the line of sight to the quasar.

17.5 GALAXY CLUSTERS

The Local Group

Galaxies are often found in groupings called **galaxy clusters.** A galaxy cluster is a group of galaxies held together by their mutual gravity. Each galaxy moves along its own orbit, just as within a galaxy each star moves on its own orbit. Our own Galaxy, the Milky Way, belongs to a very small cluster or "group" called the **Local Group.** The Local Group contains over forty members and is illustrated in figure 17.34. Its three largest members are spiral galaxies: the Milky Way, M31, and M33. The smaller members include the satellite galaxies orbiting M31 and the Milky Way, such as the Large and Small Magellanic Clouds.

The Magellanic Clouds are about one-tenth the size of the Milky Way. They appear to be dwarf irregular galaxies, although the large cloud has some spiral structure and a bar. Both galaxies are approaching the Milky Way along very elliptical orbits and are being pulled into a distorted shape by the Milky Way's gravity.

The Local Group is particularly interesting to astronomers, not just because it is our home galaxy cluster, but also because it allows us to study the "demographics" of galaxies. That is, we can see the true population of galaxies in this region of space "close" to home. For example, most of the galaxies in the Local Group are faint, small "dwarf" galaxies. The inset image of one of those dwarf galaxies in figure 17.34 shows a typical dwarf galaxy. It is a dim, small group of stars with little or no interstellar gas. Such galaxies are too faint to be visible except close by. Thus, without the Local Group to let us see the real abundance of these small systems, we might greatly underestimate their true numbers.

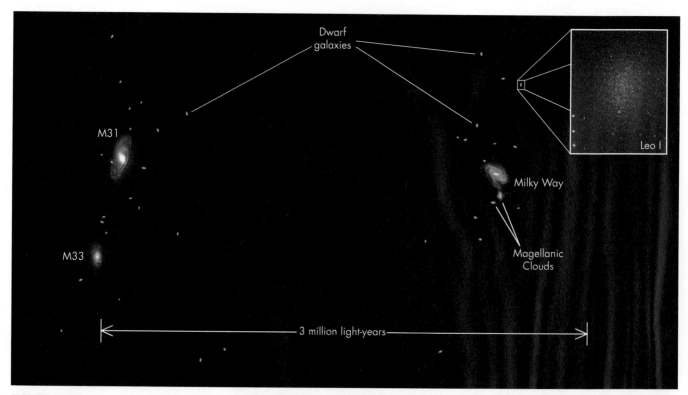

FIGURE 17.34

Illustration of what the Local Group would look like from outside. The Local Group's largest member is M31. The Milky Way is intermediate in size between it and M33. The positions of several dozen other dwarf galaxies in the group are also shown, and the inset image of one of these dwarf galaxies illustrates how faint they are.

The Relationship of Cluster Size and Galaxy Type

Small clusters, or groups, like the Local Group, contain only a few large galaxies. Large clusters of galaxies may contain hundreds to thousands of large galaxies of the size of the Milky Way or M31. For example, the Local Group lies on the outskirts of the Virgo Cluster, a moderately large cluster in the direction of the constellation Virgo (fig. 17.35).

Not only do clusters and groups differ in their number of galaxies, but they also tend to contain different types of galaxies. The more galaxies a cluster contains, the more elliptical and S0 galaxies it has. Moreover, the few spiral galaxies that they contain tend to be found in the outer parts of the cluster. On the other hand, groups tend to have a high proportion of spiral and irregular galaxies. Spirals are rare in the inner regions of large clusters, probably because of galaxy collisions. In the core of a cluster, the galaxies are close together, and so collisions between them are frequent. As we saw in section 17.3, collisions can change spiral galaxies into ellipticals. Thus, although clusters may have once contained many spirals, today they contain few, and those are the lucky ones that escaped running into a neighbor. The many collisions in the crowded inner part of these clusters also create giant elliptical galaxies by "cannibalism," as we described in section 17.3.

Observations with X ray telescopes show that large clusters usually contain large amounts (10^{12} to $10^{14}\ M_{\odot}$) of extremely hot gas that emits X rays (fig. 17.36). In fact, some clusters contain ten times more mass in hot gas than they do in the stars of their member galaxies. Some of this gas may be left over from the cluster's formation. Some may have been ejected from the galaxies within the clusters. For example, when a supernova explodes, it blasts material to such a high velocity that it may drive out much of a galaxy's interstellar gas into intergalactic space.

Gas also escapes from galaxies when they collide, as they move in the crowded cluster environment. Such collisions affect the stars hardly at all. Because they are so small compared to the distance between them, nearly all of stars in colliding galaxies simply shoot harmlessly by one another. The gas, on the other hand, is spread widely in clouds throughout the galaxies, so it cannot escape collisions. Thus, in a collision, the gas in one galaxy is driven at high speed into the gas of the other, compressing the gas and heating it to millions of degrees. In addition, as the stars of one galaxy move past the stars of the other, much of the gas may remain behind. Such gas, stripped from galaxies by collisions, generally contains heavy elements synthesized in stars within the galaxies. It then mixes into the otherwise pure hydrogen and helium gas from which the galaxies formed. But the essential point is that, regardless of its source, clusters contain huge amounts of hot gas.

Astronomers think that clusters grow in much the same way as galaxies, by merging of smaller clusters. Evidence for this hypothesis of how clusters evolve comes, once again, from our ability to "see back in time" by looking out into space. The galaxies in nearby clusters are generally fairly smoothly spread within the clusters. On the other hand, in some remote clusters the galaxies have a clumpy distribution, as if they have recently joined it and have not had time to spread themselves smoothly.

From such evidence, astronomers think that galaxy clusters form by galaxies attracting one another into small groups. Over time, these small groups attract other galaxies and even other entire clusters to make yet larger clusters, eventually growing into the major clusters we see today.

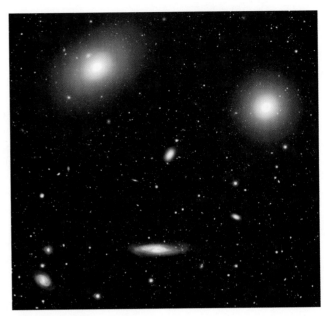

FIGURE 17.35
The Virgo galaxy cluster. Many of the small dots in the image are not stars, but small galaxies within this cluster.

Q: What types of galaxies can you identify in this photograph?

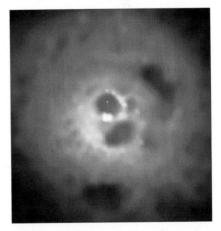

FIGURE 17.36
A false-color X ray image of the Perseus galaxy cluster. The purple glow is hot gas in the cluster. The picture shows a region about 600,000 light-years across. The dark spots near the center are "holes" in the gas blown out by jets from the active galaxy at the cluster's core.

Superclusters

The great mass of galaxy clusters creates gravitational interactions between them and draws them together into larger structures called "superclusters." A **supercluster** contains a half dozen to several dozen galaxy clusters spread throughout a region of space tens to hundreds of millions of light-years across. That is, a supercluster is a cluster of galaxy clusters. For example, the Local Group is part of the Local Supercluster, centered on the Virgo galaxy cluster and encompassing more than a dozen other smaller clusters spread over about 150 million light-years (about 50 megaparsecs). It is sketched in figure 17.37.

Superclusters have irregular shapes and, because of the enormous distances separating the galaxy clusters, are only loosely bound together. To map the distribution of galaxies on even larger scales, astronomers measure the redshifts of many thousands of galaxies, then use Hubble's law to estimate their distances. An example of such a redshift survey is shown in figure 17.38. In this survey the redshifts were measured in a thin "slice" across the sky to produce a cross section of the galaxy distribution.

One feature of this large-scale structure is clear: galaxy clusters line up in chains or shells surrounding regions nearly empty of galaxies. These empty spaces are called *voids,* and although they are not totally empty, the few galaxies they contain are generally small and dim. At the limits of our current observations, superclusters and voids do not appear evenly distributed but tend to be concentrated in some directions. The gravitational attraction from the densest parts of this large-scale structure pull on the Local Supercluster, altering its overall velocity from what is predicted by Hubble's law. Astronomers have found that the Local Supercluster is being pulled in the direction of the constellation Centaurus, where there is a concentration of galaxies dubbed the "Great Attractor." This assemblage of superclusters is about 160 million light-years (50 megaparsecs) away from us, containing the equivalent of tens of thousands of galaxies like the Milky Way.

The existence of such immense concentrations of galaxies and their effect on their neighbors is also revealed in computer simulations of the large-scale structure of the Universe. In such simulations, astronomers write out the equations that govern the motion of matter (basically Newton's laws) and then solve the equations assuming that the material is expanding in a way that mimics the expansion of the Universe. They then watch to see what happens as the gravitational force of the material acts within the

FIGURE 17.37
Sketch of the Local Supercluster, showing the relative location of some of the major groups and clusters.

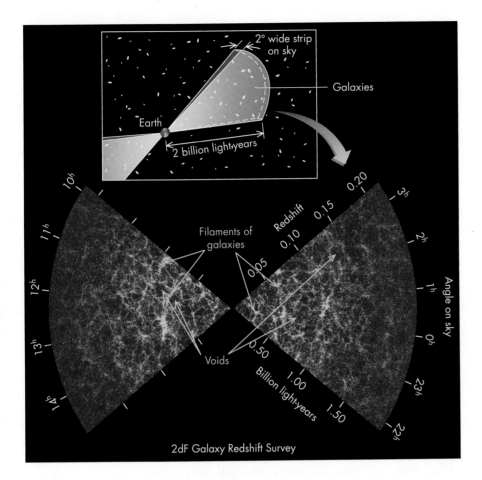

FIGURE 17.38

FIGURE 17.38
A map of how galaxies are distributed in space. The shape of the region shown in the figure is like two slices of pizza, tip-to-tip, as illustrated in the upper sketch. Each dot in the lower figure represents one of some 100,000 galaxies that lie in the wedge-shaped regions that extend out to about 2 billion light-years from Earth. Note the nearly empty regions (voids), the stringy structures, and the overall cobwebby appearance. The distances to the galaxies were estimated from their redshifts. The angles refer to a set of coordinates on the sky.

simulated volume. Figure 17.39 shows a series of images from such a simulation. You can see that matter that initially had a relatively smooth distribution is drawn by gravity into a network of filaments resembling the large-scale structure. The outcome of many such simulations by many different astronomers show that one additional ingredient is essential for the simulations to reproduce the kinds of structure we see—some kind of "dark matter" that is far more massive than the matter we see. There is simply not enough mass in the Universe for the structures we observe to have grown otherwise. What is this dark matter? We look at the clues about its nature in the next section.

FIGURE 17.39
A computer simulation of the growth of structure as the Universe expands. You can see how an initially smooth distribution of matter becomes stringy, much like the features in figure 17.37. Each box spans about 140 million light-years. The first box on the left represents a time when the Universe was only about half a billion years old. The last box on the right shows that same piece of the model of the Universe as it is today.

17.6 DARK MATTER

Measuring the Mass of a Galaxy

We can measure the mass of a galaxy the same way we measure the mass of a star or planet: by an application of the modified form of Kepler's third law. As you may recall, this method uses the principle that the orbital motion of one object around another is set by their mutual gravity. Their gravity is, in turn, set by their combined mass. Thus, from the orbital motion we can find the mass.

We went through the details of determining the mass of our own Galaxy, the Milky Way, in chapter 16. There we explained that we begin by measuring the orbital speed of a star far out in the disk of the Milky Way. With the star's speed determined, and knowing the radius of its orbit, we then found its orbital period (see section 16.5 for details). Finally, we inserted the orbital period and radius in the modified form of Kepler's third law to give us the Milky Way's mass.

To measure the mass of another galaxy, we follow the same procedure. We use the Doppler shift of the galaxy's spectrum lines to find the orbital speed of a stars near the galaxy's outer reaches. Then, from the size of those stars' orbit and the modified form of Kepler's third law, we calculate the galaxy's mass.

Within the Milky Way it is sometimes challenging to study the effects of dark matter because of our location deep amongst the stars, gas, and dust, but for other galaxies we have a much clearer view. It was studies of other galaxies that made clear the discrepancy between the mass we calculate and the mass of the visible material in the galaxy. Almost without exception, the mass calculated from the modified form of Kepler's law is *much larger* than the mass of stars and interstellar matter detectable by optical, infrared, and radio telescopes. Thus, galaxies seem to contain matter that cannot be seen, what astronomers therefore call "dark matter."

> Astronomers often find the rotation speed from the Doppler shift of radio lines produced by interstellar gas.

Dark Matter in Galaxy Haloes

Dark matter is the material that astronomers have concluded accounts for the discrepancy between the mass of a galaxy as found from the modified form of Kepler's third law and the mass observable in the form of stars and gas. The amount of dark matter required to resolve this discrepancy is very large, amounting in typical galaxies to about 10 times their visible mass. That is, when we look at a galaxy, we may in some cases be seeing only one-tenth of its matter.

Some of the strongest evidence that dark matter exists comes from galaxy-rotation curves. We discussed such curves in chapter 16 when we described the rotation of the Milky Way, but let us review them here briefly. A galaxy's rotation curve is a plot of the orbital velocity of the stars and gas moving around it at each distance from its center (curve A in fig. 17.40). The rotation curve is useful for studying dark matter because the orbital velocity depends on the mass, visible or not, that lies between the star and the galaxy's center. We can understand how that dependence arises by using the formula for orbital velocity that we developed in chapter 2 or by simply thinking about what keeps a star in orbit.

According to Newton's first law, if an object does not move in a straight line (as when, for example, a star moves in a circular orbit around a galaxy), a force must be acting on it. We know that for a star orbiting in a galaxy, that force is gravitational attraction. We also know that the strength of the gravitational force depends on the mass of the attracting body (in this case, the galaxy). All other things being equal, a greater mass therefore implies a greater force. It turns out, however, that to a good approximation, *only the material between the star's orbit and the galaxy's center contributes*

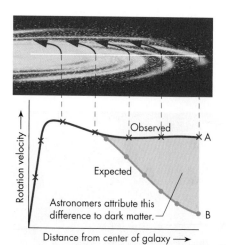

FIGURE 17.40

A schematic galaxy-rotation curve. The line with dots represents the curve expected if the galaxy's mass comes only from its luminous stars and gas. The line with crosses represents the observed curve, implying that the galaxy contains dark matter.

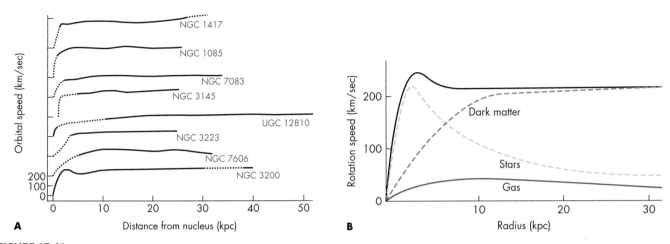

FIGURE 17.41

(A) Rotation curves for an assortment of spiral galaxies. Note that each curve starts at 0, but the curves are offset for clarity. (B) Example of a detailed analysis of a rotation curve. The expected rotation due to stars and gas alone can be determined from observations. The missing part that astronomers attribute to dark matter is shown by the dashed green curve. The combination of all three gives the overall rotation curve (in black).

to the gravitational force. Stars near the center of a galaxy will therefore have only a small force acting on them because there is little mass between them and the center. Because the force acting on them is small, such stars cannot be orbiting very rapidly, or they would have flown out of the galaxy. On the other hand, stars orbiting far from the center should also feel a small gravitational force acting on them from the galaxy. They feel the effect of the galaxy's full mass, but the gravitational force on them is nevertheless weak because they are so far from the main mass (recall that the force of gravity grows rapidly weaker at greater distances from a mass). These outermost stars must also orbit slowly if they are not to fly out of the galaxy. Thus, both stars near the center and stars far from the center should have low orbital velocities. Accordingly, the galaxy's rotation curve should drop steadily in its outer regions, as indicated in curve B of figure 17.40.

However, when astronomers measure the rotation curves of galaxies, they essentially never find this behavior, as illustrated by several representative examples in figure 17.41A. The rotation curves do rise near the center, but they then flatten out and virtually never drop off to low velocities again. What can explain such a major discrepancy between our hypothesis and observation?

A detailed analysis of one galaxy is shown in figure 17.41B. By observing the starlight and density of gas within the galaxy, the rotation speed expected based on each of those components is graphed. The stars and gas can account for the rising rotation speed near the center of the galaxy, but if our law of gravitational attraction is correct, a third component, dark matter, is needed to explain the rotation farther out. As we saw in chapter 16, astronomers also find evidence of such unseen mass in the Milky Way. The dark matter provides the gravitational force needed to hold stars in orbit in the outer parts of a galaxy despite their large velocity. The large gravitational force needed to bind such stars in a galaxy implies a huge mass of dark matter (many times the galaxy's luminous mass). Thus, we deduce that each galaxy is embedded in its own dark-matter halo, vastly more massive and larger than the "visible" galaxy.

What can this dark matter be? Astronomers have succeeded mostly in showing what it cannot be. They can rule out ordinary dim stars because such objects would emit enough infrared radiation to be detectable. They can rule out cold gas because that would be detectable by radio telescopes. They can rule out hot gas because that would be detectable by ultraviolet, radio, or X ray telescopes. Even black holes tend to draw in matter that collides violently outside the event horizon, producing X ray radiation. What then could make up the missing matter?

Dark matter

The MACHO Hypothesis

One candidate for dark matter is large numbers of black holes or other stellar remnants in the haloes of galaxies, an idea called the **MACHO (*m*assive *c*ompact *h*alo *o*bject)** hypothesis. In fact, there are some kinds of objects that emit little or no electromagnetic radiation and are therefore difficult to rule out directly. These include small, cold, planetesimal-size bodies, very old white dwarfs or neutron stars, isolated black holes, and lost socks.

It is difficult to develop a plausible model in which such objects would be common enough to explain the amount of dark matter indicated by the observations. To produce sufficient numbers of stellar remnants, we would normally expect the formation of even larger numbers of lower-mass, long-lived stars that would be easily seen, so we must postulate some type of formation process that avoided the production of these low-mass stars while producing vast numbers of high-mass stars that died very early in the history of the Universe. Even planetesimals (and socks) depend on the earlier production of carbon, oxygen, and other heavy elements, which also requires an early episode of extremely high rates of high-mass star formation, so again we would expect to find many MACHOs.

If the MACHO hypothesis is correct, when we look at distant galaxies, early in their lifetimes, we would expect to see a period of extremely high rates of massive star formation. Astronomers do not find such an episode, but we might speculate that the burst of massive star formation was so early that we have not yet been able to detect it. However, there is a more direct way of searching for such stellar remnants: the gravity of MACHOs within our Galaxy's halo would bend the light of background objects, according to Einstein's theory of general relativity (see essay 2), just as the Sun is observed to bend the light from background stars during a total eclipse. And if precisely aligned with a background object, MACHOs can focus its light, making it brighten by **gravitational lensing,** just as we saw for exoplanets in chapter 8.

Astronomers have searched for MACHOs within the Milky Way's halo using the gravitational lensing technique. The chance of a MACHO aligning closely enough with a background star to produce a measurable brightening has a low probability, so astronomers have set up telescopes to monitor the light from millions of stars within the Magellanic Clouds, as illustrated in figure 17.42. If a MACHO passes in front of one of those stars, it focuses the star's light for several days until it moves out of alignment. The brightening caused by lensing follows a known pattern, and several dozen objects have been detected in this way. Astronomers can estimate their mass from the lensing effect they create. Moreover, from a statistical analysis of the number of such events, astronomers can deduce the number of objects creating the lensing effect. From the evidence obtained so far, the objects creating the lensing effect are primarily dim, low-mass stars rather than a new population of dark objects. It does not appear that MACHOs can account for more than about 10% of the Milky Way's halo mass.

INTERACTIVE

Gravitational lensing

FIGURE 17.42
Gravitational lensing by a MACHO. If a MACHO passes in front of a star in the Magellanic Clouds, the star brightens as the star's light is focused by gravitational lensing.

Stars in the Milky Way

MACHOs orbiting in Milky Way's halo

Stars in the Magellanic Clouds

These light rays are bent toward the Earth by the MACHO as it passes directly in front of a star.

Earth

Dark Matter in Galaxy Clusters: The Case for WIMPs

While MACHOs do not appear to solve the dark matter problem, gravitational lensing observations have provided a clearer picture of dark matter on a larger scale. One of the first clues that gravitational lensing occurs on intergalactic scales was the discovery of quasars that seemed to have nearby companion quasars with essentially identical spectra but slightly different brightnesses and shapes. The existence of nearly identical quasars so close together is highly unlikely, and astronomers soon realized that the "companions" were actually multiple images of a single quasar created by a gravitational lens (fig. 17.43). A foreground galaxy or galaxy cluster has sufficient gravity that it curves the space around it so that light rays that would otherwise travel off in other directions are bent toward the Earth.

Gravitational lenses may result in forms other than multiple images of quasars. For example, figure 17.44 shows arcs created when light from distant galaxies is bent into elongated shapes as it passes through a galaxy cluster. Because the shape of these arcs depends on the mass of material within the cluster, astronomers can use the arcs to measure the cluster's mass. The masses found show that galaxy clusters contain large amounts of dark matter—nearly ten times more than all the stars and gas detected at all wavelengths. Galaxy clusters must therefore contain large amounts of dark matter.

In fact, some of the first evidence for dark matter came from observations of the motion of galaxies within clusters. Just as stars within galaxies orbit too rapidly for the gravitational force that can be attributed to their luminous mass, so too some galaxies in clusters orbit at more than a thousand kilometers per second, far too rapidly to be explained by their luminous mass. This was discovered in the 1930s, and confirmed decades later by observations of the hot gas in clusters. To hold such hot gas within the cluster and prevent its expansion, the cluster must exert a strong inward gravitational pull on the gas. The gravity needed to confine the gas depends on the cluster's total mass, and, once again, the calculated mass greatly exceeds the mass directly observable. Combining the evidence from these very different methods leads astronomers to deduce that the amount of dark matter in some galaxy clusters is so immense that it amounts to perhaps 100 times the amount of mass contained in stars.

Although dark matter is not directly visible, astronomers have now developed techniques to map out where dark matter is located by analyzing distortions in the shapes of distant galaxies. This has provided a powerful new tool for understanding the nature of dark matter. For example, two clusters that are colliding are shown in

FIGURE 17.43
The "Einstein cross," a complex image of a quasar created by a gravitational lens.

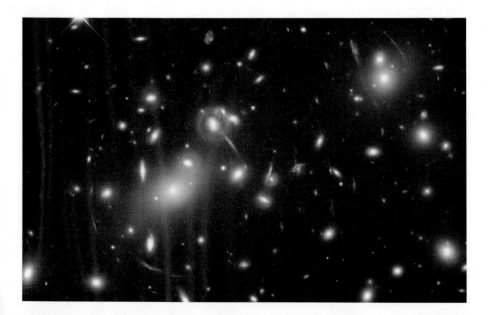

FIGURE 17.44
The arcs in this picture of the galaxy cluster Abell 2218 are the images of distant galaxies distorted by the gravitational lens effect created by the large mass of the galaxy cluster.

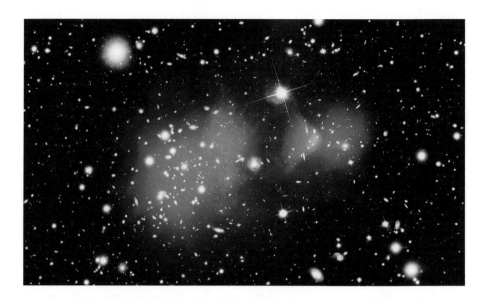

FIGURE 17.45

An image of two galaxy clusters that are colliding, made by superposing three images: a picture made at optical wavelengths (showing the galaxies in the two clusters); an image made at X ray wavelengths (showing in red the hot gas in each cluster); and a map of the dark matter in the two clusters (the blue blobs) as deduced from the gravitational lensing of the background galaxies. In the collision, the dark-matter blobs have passed through each other. The gas in the two clusters, on the other hand, collided and has piled up in between the two clusters.

figure 17.45. The location of the mass mapped by distortions is shown in blue. In this collision, the gas from the two clusters has collided and been compressed and heated to millions of degrees, so that it emits X rays, which are shown in red. The separate masses in the two clusters passed through each other as they collided, indicating that dark matter does not interact except through its gravity.

This kind of evidence suggests that dark matter is a kind of matter unlike the matter we know. This is supported by particle physicists who suspect that there may be other forms of matter not yet observed in the laboratory, such as "weakly interacting massive particles" or **WIMPs.** These are promising candidates because they do not interact with normal matter except through their gravitational pull. Other astronomers are reexamining the laws of gravity (see Science at Work: "An Alternative to Dark Matter?"), but the majority of astronomers now think the Universe is made mostly of a kind of matter unlike anything we know. It is beginning to look as though galaxies are actually large self-gravitating clouds of WIMPs in which a sprinkling of "normal" matter has formed the stars and clouds that we can see.

SCIENCE
at work

AN ALTERNATIVE TO DARK MATTER?

Although the great majority of astronomers think that the existence of dark matter is the best way to explain the observed high rotational speed of disk galaxies and the high speed of galaxies within clusters of galaxies, not all astronomers agree. A few have proposed that Newton's laws of motion need a correction for situations where the acceleration is very small, such as occurs with motions on the immense scale of galaxies.

One such proposal is referred to as the MOND hypothesis, for Modified Newtonian Dynamics. Proponents of the MOND hypothesis have had success in using the theory to explain motions within galaxies (including the Milky Way) and within galaxy clusters. Moreover, they claim that MOND models fit the observed motions as well as dark-matter models without requiring the existence of dark matter. On the other hand, observations using the theory of gravitational lenses to map the distribution of mass in galaxy clusters strongly favor the dark-matter hypothesis, and MOND does not appear to be consistent with general relativity.

For the time being, MOND provides an interesting example of how science works. Alternative ideas are proposed and explored by a subset of scientists. And perhaps if the prevailing theory reaches an impasse, scientists will find that a competing idea, such as MOND, offers a better explanation of what we observe.

SUMMARY

Astronomers classify galaxies into three main types: spirals, ellipticals, and irregulars. Spirals are disklike, with arms of young stars and gas winding out from the nucleus. Elliptical galaxies are spherical or egg-shaped and contain hardly any young stars because they have essentially no dense interstellar matter from which new stars could form. Irregular galaxies lack a symmetry to their shape, but many are extremely rich in interstellar matter and thus contain many young stars.

The first galaxies formed from the collapse of primordial gas clouds, probably as gas-rich irregular systems. Subsequently, mergers of these galaxies may produce spirals with gas orbiting in a circular disk, or may consume the gas and "scramble" the orbits of their stars and turn the merged galaxy into an elliptical.

Almost all galaxies are moving away from the Milky Way and each other as part of a universal expansion. The motion away from us causes galaxies' light to be redshifted. Hubble's law relates the speed of recession to the distance by a simple constant (the Hubble constant). Astronomers frequently use Hubble's law to estimate a galaxy's distance based on its observed redshift.

Many galaxies have extremely small but very energetic and luminous nuclei. These include radio and Seyfert galaxies as well as quasars. The amount of energy generated may dwarf the light output from the galaxy's stars and some may outshine the Milky Way by a factor of a thousand. According to current models, the activity results from matter falling in toward a supermassive black hole at the galactic center. These black holes have masses that are as large as several billion solar masses packed into a region not much larger than the Solar System. An accretion disk forms around the hole as matter falls in, and hot gas is channeled by the disk into jets and lobes of radio-emitting gas. Studies suggest that the black holes grow in size as galaxies grow and merge. Large galaxies today may have supermassive black holes at their centers that are quiet because no matter is falling into them.

Galaxies are usually grouped into clusters a few million light-years across. Small clusters or "groups" tend to contain spiral galaxies. The Milky Way belongs to a small cluster called the Local Group, which contains over forty member galaxies, mostly small "dwarf" galaxies. Large clusters contain thousands of galaxies, which are primarily ellipticals. Clusters themselves appear to be clustered into superclusters that are many tens of millions of light-years across and are arranged into long "filaments" surrounding voids.

Studies of the motions of stars in galaxies reveal that most of the mass of a galaxy is made up of dark matter, not stars or gas. Studies of galaxy clusters reveal a similar deficit of mass explained by dark matter. Astronomers do not yet know the composition of dark matter, but it may be some kind of weakly interacting massive particles (WIMPs) that do not interact with normal matter except gravitationally.

QUESTIONS FOR REVIEW

1. (17.1) What are the three main types of galaxies? How do the basic galaxy types differ in shape, stellar content, and interstellar matter?
2. (17.1) Why is it not surprising that elliptical galaxies contain mostly Pop II stars?
3. (17.2) Explain several ways astronomers measure the distance to nearby galaxies. Why is it important to know the distance?
4. (17.2) What is Hubble's law? In what situations does Hubble's law not apply?
5. (17.3) Why are galaxy collisions of interest?
6. (17.3) What happens to the stars when two galaxies collide?
7. (17.4) What are the three main types of active galaxies? List the main characteristics of each.
8. (17.4) Sketch of a radio galaxy illustrating its structure. Why do astronomers call these objects "radio" galaxies?
9. (17.4) What process generates the radio emission in radio galaxies?
10. (17.4) What is the evidence that quasars are very distant?
11. (17.4) How is it known that active galaxies have small core regions?
12. (17.4) What mechanism has been suggested as powering active galaxies?
13. (17.4) How might a large black hole form in a galaxy's core?
14. (17.5) What is the Local Group, and why is it a group and not a cluster?
15. (17.5) What are some differences between clusters that contain large and small numbers of galaxies?
16. (17.5) What is the Local Supercluster? How many clusters might be in a supercluster, and how big is a supercluster?
17. (17.5) What is a void?
18. (17.5) What does the large-scale structure of the Universe look like? How do astronomers study it?
19. (17.6) How do astronomers measure the mass of a galaxy?
20. (17.6) What is meant by dark matter? What observations support its existence? What types have been proposed?
21. (17.6) What is a gravitational lens? What can observations of lenses tell us about clusters?
22. (17.6) What is the apparent ratio of dark matter to normal matter in a typical galaxy? In a cluster of galaxies?

THOUGHT QUESTIONS

1. (17.1) Hubble proposed that his tuning-fork diagram might represent evolutionary changes with E galaxies gradually changing to S galaxies. What evidence based on the amount of interstellar matter and kind of stars in E and S galaxies indicates that this is very unlikely?

2. (17.1) Take a look at an egg from different angles. Do you think the classification of elliptical galaxies is always accurate? Examining figure 17.8, could an E0 be mistaken for an E7, or an E7 for an E0? Explain why there is or is not a similar problem for identifying spirals. What does this imply about the actual number of each type compared to what is observed?

3. (17.2) The value of the Hubble constant was only recently determined to high accuracy. If H were equal to 100 km/(sec Mpc) instead of $H = 70$ km/sec/Mpc, as some astronomers thought it would be, how many times larger or smaller would the distance to a distant galaxy be? How much larger or smaller would the diameter of the galaxy be?

4. (17.2–17.3) If you wanted to look at your mother when she was a baby, how far out in space would you need to be? (Assume you have a sufficiently powerful telescope and pretend that you can travel instantly to any distance.) What if you wanted to look in on astronomers in ancient Egypt around 3000 B.C.? How does this relate to astronomers studying young galaxies?

5. (16.3/17.3/17.5) Explain how selection effects might make it difficult to obtain an accurate census of galaxy types. What might happen at near and far distances? What about if you picked an area in the middle of a supercluster, or an area in a void? (You don't know what's there until you look!)

6. (14.2/17.4) Are the jets that come out of active galaxies similar to bipolar outflows? Explain your answer.

PROBLEMS

1. (17.2) A Cepheid variable in a nearby galaxy looks 10^6 times fainter than an identical Cepheid (in the Milky Way) 1000 parsecs away. How far away is the nearby galaxy?

2. (17.2) What would be the angular diameter of the Milky Way as seen by an alien astronomer in the Virgo Cluster at a distance of 18 Mpc?

3. (17.2) A galaxy has a recession velocity of 19,250 kilometers per second. What is its distance in megaparsecs?

4. (4.6/17.2) A quasar's spectrum shows a shift in spectral lines ($\Delta\lambda/\lambda$) of 0.158. Use the Doppler-shift formula to determine what recession velocity this shift indicates, and then use Hubble's law to determine the distance to the quasar in Mpc if $H = 70$ km/sec/Mpc. Be sure to calculate the recession velocity in kilometers per second.

5. (17.4) A radio galaxy shows lobes that extend 1.5 megaparsecs away from the galaxy on each side. If the accelerated electrons in the lobe travel at nearly the speed of light, how long will it take them to travel to the edge of the lobe?

6. (17.4) A nearby radio galaxy 50 megaparsecs from the Earth has a 1-billion-solar-mass black hole at its center. If the diameter of this black hole is twice its Schwarzschild radius, calculate the angular size of the black hole as seen from Earth. Convert the angular size from degrees to arc seconds (3600 arc seconds per degree). Can we directly image the black hole as a dark spot in the accretion disk? Compare your result with 1×10^{-5} arc seconds, which is about the best resolution currently possible with the radio astronomy technique of very long baseline interferometry.

7. (17.4) What is the minimum size of a quasar that shows variations in brightness over a period of 36 hours? Express your answer in kilometers.

8. (17.5) Estimate from figure 17.34 the ratio of the diameter of the local group to the diameter of the Milky Way (100,000 light-years across). Compare this with the ratio of the diameter of the Milky Way to the diameter of the Solar System (including the Oort cloud, the Solar System's diameter is about 200,000 AU).

9. (17.5–17.6) A galaxy cluster has a total mass of $10^{15}\ M_\odot$. If the cluster contains 100 galaxies, each with typical masses of $10^{11}\ M_\odot$, and 10 times as much mass in hot X ray gas as in galaxies, how much dark matter is contained in the cluster? What percentage of the total mass is this?

10. (17.6) NGC 3200 has a rotation speed of 200 kilometers per second at 5 kiloparsecs from its center, and 210 kilometers per second at 50 kiloparsecs. What is its mass within each distance, and what makes up most of the mass in each region?

TEST YOURSELF

1. (17.1) A large galaxy contains mostly old (Pop II) stars spread smoothly throughout its volume, but it has little dust or gas. What type of galaxy is it most likely to be? Select all that are possible.
   (a) Irr   (b) S   (c) SB   (d) E

2. (17.2) A galaxy's spectrum has a redshift of 14,000 kilometers per second. If the Hubble constant is 70 kilometers per second per megaparsec, how far away from Earth is the galaxy?
   (a) 2 megaparsecs   (d) 980 megaparsecs
   (b) 0.005 megaparsecs   (e) 5 megaparsecs
   (c) 200 megaparsecs

3. (17.2) Why might a galaxy such as M31 (Andromeda) have a blueshift instead of a redshift?
   (a) All nearby galaxies have a blue color.
   (b) The effect of its orbital or local motion is bigger than the effect of the universe's expansion.
   (c) It is undergoing a starburst, so it has many young, high-mass blue stars.
   (d) It contains no Cepheid variables.
   (e) It is "behind" us, expanding in the opposite direction.

4. (17.3) Which of the following would be a probable result of a collision between two galaxies?
   (a) Complete destruction of all the stars
   (b) Formation of a giant spiral from two elliptical ones
   (c) Formation of a single open star cluster
   (d) Formation of a ring galaxy
   (e) All of the above

5. (17.4) A spiral galaxy has a small bright central region, and its spectrum shows that it contains hot, rapidly moving gas. It is most likely a _____ galaxy.
   (a) barred spiral   (c) radio   (e) quasar
   (b) Seyfert   (d) BL Lac

6. (17.5) The Local Group (select all that are correct)
   (a) contains about 40 member galaxies.
   (b) is one of the smaller galaxy clusters.
   (c) is the galaxy cluster to which the Milky Way belongs.
   (d) is one of the larger galaxy clusters.
   (e) contains mostly medium-sized elliptical galaxies.

7. (17.5) The largest clusters probably contain many elliptical galaxies because
   (a) gas is stripped out of galaxies during collisions.
   (b) the dark matter in them preferentially attracts elliptical galaxies over spirals.
   (c) they are so old, all the galaxies have used up all their gas.
   (d) they were created at the same time in the history of the Universe as all the globular clusters.

8. (17.5) Galaxy clusters often contain _____ as much mass in hot gas as in stars in their member galaxies.
   (a) 100   (b) 10   (c) 1   (d) 1/10   (e) 1/100

9. (17.6) Most of the mass of a galaxy is contained in its
   (a) massive *O* and *B* stars.   (d) dark matter halo.
   (b) cold interstellar gas.   (e) stellar disk.
   (c) central black hole.

10. (17.6) The ratio of dark matter to visible matter in a typical elliptical or spiral galaxy is about
    (a) 100:1 (b) 10:1   (c) 1:1   (d) 1:10   (e) 1:100

KEY TERMS

active galaxy, 457
barred spiral galaxy, 444
dark matter, 468
elliptical galaxy, 443
galactic cannibalism, 455
galaxy, 441
galaxy cluster, 464
gravitational lensing, 470
Hubble constant, 449
Hubble's law, 449
irregular galaxy, 444
jets, 458
Local Group, 464

MACHO (MAssive Compact Halo Object), 470
megaparsec, 449
quasar, 458
radio galaxy, 458
recession velocity, 449
redshift, 448
Seyfert galaxy, 457
spiral galaxy, 443
supercluster, 466
supermassive black hole, 460
WIMP (Weakly Interacting Massive Particle), 472

⊘: FIGURE QUESTION ANSWERS

WHAT IS THIS? (chapter opening): This is not a comet, but a radio picture of gas being stripped out of the Large and Small Magellanic Clouds by the gravitational (tidal) pull of the Milky Way. The Magellanic Clouds are the two bright blobs on the left of the image. The main arc of hydrogen gas is about 300,000 ly long, three times the diameter of the Milky Way.

FIGURE 17.26 There is a jet pointing out of the blob labeled "Quasar" at an angle of about "5 o'clock."

FIGURE 17.35 A spiral (Sb) galaxy lies near the bottom center, and a barred spiral (SBa) at bottom left. An elliptical (E1) galaxy is near the upper right corner, an elliptical (E3) or S0 near the upper left corner and a small elliptical (E5) near the center of the image.

18

Structure forming in the Universe as seen in the Millennium Simulation. This supercomputer model depicts a region several hundred million light years across 4.7 billion years after the Big Bang.

COSMOLOGY

LEARNING OBJECTIVES

Upon completing this chapter you should be able to:

- Explain what it means that the Universe is expanding, and discuss how Hubble's law arises in this context.
- Describe why the Universe has no center, and how it is possible for it to have no edge in the usual sense of an edge.
- Estimate the Universe's age based on Hubble's constant.
- Define Olbers' paradox and explain its resolution.
- Discuss the evidence that the Universe began in a hot, dense state and has been expanding ever since.
- Delineate the stages that matter passed through as the Universe expanded, and describe what present-day properties of the Universe were results of these stages.

- Describe the different possible curvatures of the Universe, and explain how its geometry was shown to be flat.
- Describe the role of dark matter in galaxy formation and why CMB measurements indicate how much is present.
- Describe why cosmologists think the Universe inflated when it was young and what features this model explains.
- Discuss the possible fates for the Universe if its expansion is affected just by gravity alone.
- Explain how supernova observations have led astronomers to conclude that the Universe is accelerating.
- Discuss the properties of dark energy, why it produces an acceleration, and its consequences for the Universe's fate.

Q: **WHAT IS THIS?**

See end of chapter for the answer.

osmology is the study of the structure and evolution of the Universe as a whole. Cosmologists seek answers to questions such as: How big is the Universe? What shape is it? How old is it? How did it form? What will happen to it in the future?

Given our insignificant size in the cosmos, such questions may seem futile or even arrogant, but most cultures have tried to answer them. Many of those answers have become part of humanity's religious heritage. For example, Hindus believe that the Universe is the work of several gods who give the Universe form from nothingness and then destroy it in a cyclical pattern. Jews, Christians, and Muslims share a belief that there is a single god who began the Universe in a single act of creation. This is expressed in the King James version of the Bible, in the book of Genesis (1:1–3), in the following way:

> In the beginning God created the heaven and the earth; and the earth was without form and void; and darkness was upon the face of the deep and the spirit of God moved upon the face of the waters. And God said, "Let there be light," and there was light. And God saw the light, that it was good.

These creation stories give us powerful poetry and imagery of the birth of the Universe, but they offer few specifics of what our Universe was like at its birth.

In this final chapter, we will learn how contemporary scientists view the creation and evolution of our Universe. In the past few decades a wide array of new observations have given scientists surprising new insights into the origin and fate of the Universe. These discoveries provide us with scientifically based answers to questions that were once believed to be solely the province of religion and philosophy.

We begin with the implications of the distribution of galaxies in the Universe and of their motion away from us discovered by Edwin Hubble. Because light travels at a finite speed and modern telescopes allow us to observe objects billions of light years away, we can actually see what the Universe looked like when it was billions of years younger. We even see radiation emitted shortly after the Universe was born 13.8 billion years ago in a hot, dense, violent burst of matter and energy called the Big Bang. Remarkably, theory and observation imply that the entire Universe we see today expanded out of a volume originally smaller than a proton, and that the Universe is now entering a new stage of accelerating expansion that may pull the Universe apart in the remote future.

CONCEPTS AND SKILLS TO REVIEW

- Hubble's law (17.2)
- Density (6.1)
- Newton's law of gravity (3.5)
- Red light has a longer wavelength than blue (4.1)
- Forces (Preview, p. 8)

18.1   OBSERVATIONS OF THE UNIVERSE

Distribution of Galaxies

Immanuel Kant, the German natural philosopher, suggested in the eighteenth century that the Milky Way is merely one of a multitude of galaxies, but his views were not widely accepted. Up to the beginning of the twentieth century, most astronomers thought that the Universe consisted of just the Milky Way Galaxy and a number of small companions surrounded by a vast void. By the 1920s, astronomers were able to show that there were indeed great numbers of other galaxies, millions of light-years from the Milky Way, and today we know there are hundreds of billions of other galaxies. As

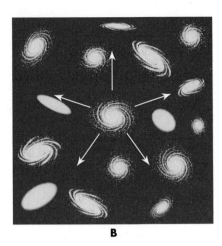

FIGURE 18.1

(A) An all-sky plot of the positions of over a million galaxies detected by 2MASS, an infrared survey of the sky. Each white dot represents a galaxy, or in clusters and superclusters of galaxies, they merge to form brighter areas. The blue-shaded portion of the diagram shows where the plane of the Milky Way crosses through this map. (B) The approximate constant number of galaxies in all directions implies that the Universe is uniform.

we scan the sky with ever more powerful telescopes, we see no evidence of an edge beyond which there are no more galaxies. What can we understand about the nature of the entire Universe from such basic observational facts?

Pictures of the night sky show that distant galaxies are scattered across it approximately uniformly, as shown in figure 18.1A. This is in sharp contrast to how stars are distributed in our Milky Way: its stars are concentrated in a narrow band, from which we deduce that our Galaxy is a disklike system. The observed distribution of galaxies leads astronomers to conclude that the Universe is filled approximately uniformly with galaxies everywhere, in somewhat the way raisins are distributed in raisin bread, perhaps with clumps here or there, but no favored position. Thus, our first deduction is that the Universe is approximately the same in all directions, as portrayed in figure 18.1B, but with no apparent center or edge.

Are We at the Center of the Universe?

In chapter 17 we learned that the spectral lines of nearly all galaxies are redshifted, and their shift to a longer wavelength implies that the galaxies are moving away from us. Moreover, the redshift increases with distance, d, in such a way that the recession velocity, V, of a galaxy obeys Hubble's law:

$$V = Hd$$

V = Recession velocity of galaxy
d = Distance of galaxy
H = Hubble constant

where $H \approx 70$ km/sec per megaparsec is the Hubble constant.

The recession of distant galaxies from the Milky Way often leads to the mistaken belief that the Milky Way is at the center of the Universe. A simple demonstration that this cannot be correct is illustrated in figure 18.2, which shows a small subset of galaxies distributed along a line. In this example, each galaxy is separated from its nearest neighbor by 10 megaparsecs.

Suppose the Milky Way is the third galaxy in the line and that galaxies A, B, and C obey Hubble's law, with each receding at a velocity given by $V = 70d$ (giving V in kilometers per second for d in megaparsecs). That is, A recedes from us at 700 km/sec, B at 1400 km/sec, and C at 2100 km/sec. Next, suppose we could communicate with an alien in galaxy C and ask what it sees. It sees us receding at 2100 km/sec, because from its perspective we are the ones moving away. It sees A receding at $2100 - 700 = 1400$ km/sec and B receding at $2100 - 1400 = 700$ km/sec, as shown along the bottom portion of figure 18.2.

	Milky Way	Galaxy A	Galaxy B	Galaxy C	Alien's galaxy	
As seen from Milky Way:						
$V =$	←700 $\frac{km}{sec}$	0 $\frac{km}{sec}$	700 $\frac{km}{sec}$ →	1400 $\frac{km}{sec}$ →	2100 $\frac{km}{sec}$ →	2800 $\frac{km}{sec}$ →
$d =$	10 Mpc		10 Mpc	20 Mpc	30 Mpc	40 Mpc
As seen from alien's galaxy:	$\frac{2100}{-(-700)}$	$\frac{2100}{-0}$	$\frac{2100}{-700}$	$\frac{2100}{-1400}$	$\frac{2100}{-2100}$	$\frac{2800}{-2100}$
$V =$	←2800 $\frac{km}{sec}$	←2100 $\frac{km}{sec}$	←1400 $\frac{km}{sec}$	←700 $\frac{km}{sec}$	0 $\frac{km}{sec}$	700 $\frac{km}{sec}$ →
$d =$	40 Mpc	30 Mpc	20 Mpc	10 Mpc		10 Mpc

FIGURE 18.2

A line of galaxies illustrating that in a Universe that obeys Hubble's law of expansion ($V = Hd$), an observer in any galaxy sees the same law.

In other words, the alien would see galaxies receding from it exactly the same way we see galaxies receding from us. Furthermore, the galaxies seen by the alien will obey the same formula that Hubble derived if the alien uses distances measured from its own galaxy. A similar argument can be made for an observer in any of the galaxies and looking in any direction.

Hubble's law that the velocity increases in proportion to the distance, implies that all observers, no matter where in the Universe they are, will see everything in the Universe expanding away from them in the same way. The Universe has no point that can be identified as the "center" of expansion. In fact, observers in every other galaxy will have the impression that they are at the center of the expansion. Astronomers describe this lack of a preferred location as the **cosmological principle.** It is a statement of cosmic modesty: there are no special locations. We are not at the center of the Universe, nor is anyone else.

Expansion of the Universe

When we picture the motion of galaxies like an explosion outward from a point in space, we run into problems. For example, beyond some distance we would have to be moving through space at greater than the speed of light, which violates special relativity. The cosmological principle only makes sense if space itself expanding.

A better analogy than an explosion is a rising loaf of raisin bread. Before the dough has risen, the raisins are all close together, but as the loaf expands, the raisins move apart, as illustrated in figure 18.3. If we measure the space between the raisins and their speed moving apart, they obey a velocity–distance relation just like Hubble's law. In this analogy the loaf of bread represents space, which as it expands carries along the raisins, representing the galaxies. The raisins remain in the same region of

Raisin bread dough before rising

Raisin bread dough after rising

FIGURE 18.3

Representation of the expansion of the Universe. Raisins in a rising loaf of bread are like galaxies in expanding spaces. They both separate from each other following a Hubble-like law in the relationship between distance and velocity.

FIGURE 18.4
Curved 3-dimensional space can curve back on itself like the curved 2-dimensional surface of a balloon. As the surface of the balloon expands, separations between buttons glued to its surface grow following a Hubble-like law. This illustrates how expanding curved space might be finite in extent yet have no edge.

space where they started, but space expands, carrying them farther apart from other raisins. In imagining this analogy, the loaf would have to be infinitely large to explain why none of the raisins is sitting on the edge of the loaf where it sees raisins to one side and not the other.

Einstein's general theory of relativity and its new understanding about the curvature of space (section 18.4 and essay 2) can help us see how space need not be infinite to satisfy the cosmological principle. Picturing the curvature of three-dimensional space is almost impossible, but we can picture the curvature of a two-dimensional surface like that of a balloon. Imagine buttons, representing galaxies, glued to the balloon (fig. 18.4). If the balloon expands, the buttons are carried farther apart by the expansion of the balloon's surface, and every button sees all the other buttons moving away from it across the surface of the balloon, again following a velocity–distance relationship just like Hubble's law.

Note that although space expands, the objects themselves—galaxies, stars, planets, and so forth—do not. Gravity and other forces maintain such material bodies at a relatively constant size. On the other hand, light waves *are* altered by the expansion of space, and that expansion creates the redshift of the galaxies that we observe. As a light wave from a galaxy moves through space, it is stretched as the space expands, just as the coils in a spring are stretched if you pull on the ends. If you draw a squiggle on a balloon, you can see a similar effect—the "wavelength" of the squiggle increases as you blow up the balloon (fig. 18.4).

You might wonder in the balloon analogy what lies inside of the balloon and what it is expanding into. Because we are describing spacetime and because the balloon's surface represents the three dimensions of space, the balloon is expanding in time. Inside it lies the past. Outside it lies the future. At some time in the past, all of space was concentrated in a single point at the beginning of time.

Age of the Universe

The model of the Universe as an expanding space filled uniformly with galaxies allows cosmologists to estimate its age and to calculate conditions near the time of its birth. Alexander Friedmann, a Russian scientist, made such calculations in the 1920s, but his work attracted little attention. Not until 1927, when Abbé Georges Lemaître, a Belgian cosmologist and priest, independently made similar calculations, did astronomers appreciate that conditions at the birth of the Universe could be deduced from what is observed today.

Lemaître reasoned that because galaxies are moving apart now, they must in the past have been closer together. He went on to propose that at one time in the distant past, all matter in the Universe was packed into a region approximately the size of the present Solar System. With so much mass in such a relatively small volume, the density of the Universe was enormous. In fact, at that remote time in the past, the density was so immense that the protons and neutrons that now compose the far-flung stars and galaxies were touching, just as today the protons and neutrons within a single atom touch. For that reason, Lemaître called this dense stage of the early Universe the Primeval Atom.

Lemaître argued that the Primeval Atom "decayed" much like an ordinary radioactive atom such as uranium decays. Moreover, he identified this decay with the birth of the Universe, and he calculated how long ago that occurred. Astronomers no longer accept Lemaître's theory of the birth of the Universe, but they do think that it was created from a highly compressed state in the distant past, and they measure its age from that time. To determine that age, we can carry out a simple calculation, but an analogy may help us better understand this calculation.

Suppose a car leaves a starting point in a long race and drives steadily at 60 miles per hour. Suppose the driver looks at the mileage and notes that the car has traveled

ASTRONOMY
by the numbers

ESTIMATING THE AGE OF THE UNIVERSE

To calculate the age of the Universe, t_U, we proceed as follows: let two representative galaxies be a distance, d, apart and separating with a velocity, V. For simplicity, we will assume that V has remained constant. Then in the time, t_U, the age of the Universe, they will have separated by a distance given by

$$d = Vt_U$$

d = Distance between two galaxies
V = Velocity at which galaxies are separating
t_U = Age of the Universe

We can find how long ago they began moving apart by dividing both sides by V. This gives

$$t_U = \frac{d}{V}.$$

Thus, in this simple model, the age of the Universe is d/V.

Hubble's law, however, relates a galaxy's velocity to its distance by $V = Hd$. Therefore, substituting this expression for V into the expression for the age of the Universe, we find

$$t_U = \frac{d}{V} = \frac{d}{Hd} = \frac{1}{H}$$

where H is the Hubble constant.

For $H = 70$ km/sec per Mpc, we need to remember that one megaparsec is 3.09×10^{19} km, so if the Hubble constant is 70 km/sec/Mpc, this gives

$$H = \frac{70 \text{ km/sec}}{3.09 \times 10^{19} \text{ km}} = \frac{70}{3.09} \times 10^{-19} \text{ sec}^{-1}$$
$$= 2.27 \times 10^{-18} \text{ sec}^{-1}.$$

Therefore, we find that the age of the Universe is

$$t_U = \frac{1}{H} = \frac{1}{2.27 \times 10^{-18} \text{ sec}^{-1}}$$
$$= 4.41 \times 10^{17} \text{ sec.}$$

Because there are about 3.16×10^7 seconds per year, we can convert this to years as follows:

$$t_U = \frac{4.41 \times 10^{17} \text{ sec}}{3.16 \times 10^7 \text{ sec/yr}}$$
$$= 1.40 \times 10^{10} \text{ yr.}$$

Thus, we find that the age of the Universe is about 14 billion years. This is only an approximate value because it assumes that the expansion rate has remained constant.

120 miles. How long ago did the car leave the starting point? You can figure that out by dividing the distance traveled (120) by the speed (60), to get 2 hours. So, too, with measuring the age of the Universe. Astronomers pick some galaxy at a known distance, d, from the Milky Way. If they know the galaxy's speed, V, then they can determine that the two galaxies began moving apart at a time $t_U = d/V$. We will take that time as the beginning of the Universe (fig. 18.5).

It may seem surprising at first, but because of Hubble's law, we will find the same answer no matter which galaxy we choose to carry out this calculation. A little algebra (worked out in Astronomy by the Numbers: "Estimating the Age of the Universe") shows that

$$t_U = 1/H.$$

To evaluate t_U in years, we must know H, the Hubble constant. However, we cannot simply insert the value of H in the expression for t_U, because H is expressed in units of kilometers/second per megaparsec. If we were to use H in these units, we would get an answer for the age of the Universe in equally odd units. We must therefore express H in units of 1/seconds by converting megaparsecs to kilometers so that the distance units in H cancel out. As shown in the Astronomy by the Numbers box, the result is that $1/H$ is approximately 14 billion years.

Astronomers find this value for the age of the Universe plausible because it is much older than the Solar System (4.6 billion years, as we saw in chapter 6) and is also older than the oldest known stars (about 13 billion years). Something would be seriously wrong with our model if the Universe turned out to be younger than its stars, a clearly impossible situation.

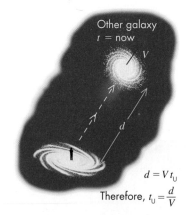

Other galaxy
t = now

$$d = Vt_U$$
Therefore, $t_U = \dfrac{d}{V}$

But according to Hubble's law,
$$V = dH$$

Therefore, $t_U = \dfrac{d}{dH} = \dfrac{1}{H}$

FIGURE 18.5

Estimating the age of the Universe from the recession of one galaxy from another galaxy.

18.2 LOOKING BACK TOWARD THE BEGINNING OF TIME

Olbers' Paradox

The Universe's finite age can also be demonstrated by the simple experiment of stepping outside at night and noting that the sky is dark. This relation between the darkness of the night sky and the structure of the Universe is known as **Olbers' paradox,** after the German astronomer Heinrich Olbers, who in 1823 presented an argument that if the Universe extends forever and has existed forever, the night sky should be bright. The paradox arises from following reasonable premises (or so it seemed at that time) to a clearly nonsensical conclusion. The argument is as follows:

Suppose you stand in a small grove of trees and look out between them to the surrounding landscape. If the grove is small, your line of sight will be blocked by trees in a few directions but will pass between the trees in most directions (fig. 18.6 left). If the grove is larger, more-distant trees in it will block your view in what were previously clear gaps (fig. 18.6 right). Then, no matter what direction you look, your line of sight is intercepted by a tree.

Now suppose that, rather than looking between trees, you look out into space through the galaxies and stars that compose the Universe. If space extends sufficiently far and is populated with galaxies and stars such as we see around us in the Local Group, then no matter in what direction you look, your line of sight will ultimately intercept a star just as in a sufficiently large forest your line of sight ultimately hits a tree. Therefore, the sky should be covered with stars, each glowing brilliantly, with no dark spaces between them. Even though such distant stars are very faint to us, there are so many of them that their collective brightness should be very large. In other words, the night sky should not be dark. This argument must be fallacious because the most obvious observation one can make about the sky is that it is dark at night. That is the paradox: the night sky should be bright, but it is not. Where, therefore, is the error in our reasoning?

The paradox is avoided if there are no stars beyond some distance. That is, just as we can see out of the woods if there are only a few trees, so too we will see a dark night sky if there is only a limited number of stars. The paradox is also avoided if we account for the time it takes light from the stars to reach us. Light from a star 1 billion light-years away takes a billion years to get here. To see such light requires that the Universe has existed long enough for the light to arrive. That is, if stars have not been shining for a long enough time, the night sky will be dark. A final way to avoid the paradox

Although the paradox bears Olbers' name, more than two centuries earlier Kepler pointed out that in an infinite Universe, the night sky should be bright, and he therefore concluded the Universe was finite in size.

In a small grove of trees, only a few block your view. Lots of space to view between trees.

In a larger forest, more distant trees block your view. No open space visible between trees.

FIGURE 18.6
(Left) An observer in a small grove of trees can see out through gaps between the trees to the surrounding countryside. (Right) In a larger forest, distant trees block the gaps so that no matter where you look, your line of sight ends on a tree.

is to note that light from remote stars is redshifted by the expansion of the Universe. The redshift reduces the light's energy (recall that longer-wavelength radiation carries less energy) and thereby makes the most distant stars and galaxies look very dim. An expanding Universe can shift visible light to wavelengths we cannot see with our eyes. Therefore, as we look out to the farthest distances we can see, we look back to the time when the first stars and galaxies formed, but to see what is there we must look at much longer wavelengths than those we use to observe nearby stars and galaxies.

The Cosmic Horizon

The age of the Universe limits the distance we can see. Because light requires time to move from one point to another, we can see no farther than the distance light travels in a time span equal to the age of the Universe. If the Universe did not expand, calculating the distance that light travels in that time span would be simple. For example, if the Universe were not expanding and were 14 billion years old, we could see nothing farther than 14 billion light-years away (fig. 18.7). The Universe may well extend beyond that distance, but the maximum distance that light can travel in the Universe's age therefore defines a **cosmic horizon.** Astronomers call the space within the horizon the visible Universe.

Because the Universe is expanding, it turns out that there is no simple expression for the distance to the cosmic horizon, or even a distant galaxy, for that matter. A simple example will show the problem. Recall that when we see a galaxy, we see it the way it was at the time when its light left it, perhaps billions of years ago. But when it emitted its light, the galaxy was closer to us than it is now because the Universe was smaller. Furthermore, since the time when the light we now receive was emitted, the galaxy has moved yet farther from us, carried by the expansion of the Universe. Thus, which distance do we apply to the galaxy? Its distance when the light was emitted? Its distance when the light was received? Some other distance? To avoid such confusion (and for other reasons), astronomers usually use the "*light travel time distance*" when speaking of the distance to remote galaxies.

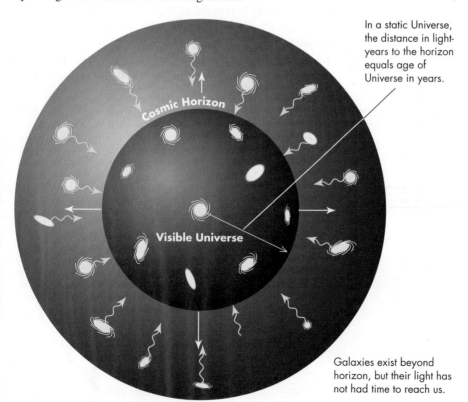

In a static Universe, the distance in light-years to the horizon equals age of Universe in years.

Cosmic Horizon

Visible Universe

Galaxies exist beyond horizon, but their light has not had time to reach us.

FIGURE 18.7
We cannot, even in theory, see beyond the cosmic horizon. Light takes more time than the age of the Universe to reach us from there. (The straight arrows indicate that the horizon moves outward as the Universe ages.)

The light travel time distance measures how far the light from the galaxy (or other object) would have traveled if the Universe were not expanding. This distance is simply a measure of how much older the Universe is when the light is received as opposed to when it was emitted. With that definition of distance, we can then say that a galaxy, which we observe today and whose light was emitted 10 billion years ago, is 10 billion light-years away from us. Defining distance in this way allows us to be more precise when we say that the cosmic horizon is approximately 14 billion light-years away. Thus, we will take the size (radius) of the visible Universe to be approximately 14 billion light-years.

The Cosmic Microwave Background

Because the Universe is expanding, as we look farther away we are also seeing the Universe when it was younger and more dense. Pursuing this idea, George Gamow, a Russian astrophysicist who fled to the United States before World War II, realized that it must also have been extremely hot. His argument was based on the observation that compression heats a gas, and so the enormous compression of the early Universe suggests that it too was as hot and dense as the interior of a star when the Universe was young. Gamow's ideas were developed further by two of his collaborators, the American physicists Ralph Alpher and Robert Herman. In 1948 they predicted that after the young Universe had expanded for several hundred thousand years,* it would have cooled and expanded enough that the Universe became transparent. Astronomers often call this the **recombination era** because the ionized atoms would have combined with their electrons to make a transparent neutral gas. From this time onward, radiation was able to travel large distances and reach us today, though in altered form. To understand how that alteration occurred, we need to look at the properties of the radiation in greater detail.

According to Wien's law, a dense gas radiates most intensely at a wavelength determined by its temperature, as discussed in chapter 4. In particular, hot matter radiates most strongly at short wavelengths. The Universe became transparent when the temperature dropped to about 3000 K (similar to the surface of a cool star). At this temperature, the wavelength of maximum emission is about 1000 nanometers, as we found in chapter 13. As space expanded, it stretched the wavelength of radiation traveling through it, increasing its wavelength as illustrated in figure 18.8. (Recall the analogy used earlier of a squiggle on an expanding balloon.) Calculations show that

The Expansion of the Universe

*Alpher and Herman first estimated this would take about 10 million years, but subsequent calculations have revised this downward to 380,000 years.

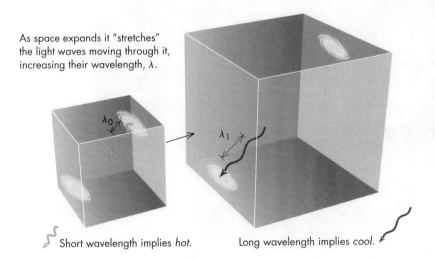

As space expands it "stretches" the light waves moving through it, increasing their wavelength, λ.

λ_0

λ_1

Short wavelength implies *hot*.

Long wavelength implies *cool*.

FIGURE 18.8

As space expands, it stretches radiation moving through it, making the wavelength longer. Because longer wavelengths are associated with cooler objects, the stretching of the radiation has the effect of cooling it.

as the Universe expanded to its present size, the wavelength of the radiation emitted at this early time was stretched by a factor of about 1000, so it is now about 1 millimeter (10^6 nanometers). This is in the microwave portion of the electromagnetic spectrum. Thus, Alpher and Herman showed that the Universe today should be filled with faint microwave radiation created when the Universe was very young. Their prediction was confirmed in 1965, when Arno Penzias and Robert Wilson of Bell Telephone Laboratories accidentally discovered cosmic microwave radiation having just that property.

Penzias and Wilson were trying to identify sources of background noise on telephone satellite links and noticed a weak radio signal with the unusual property that its strength was constant in every direction in the sky in which they pointed their detector. They found that the signal came not from isolated objects such as stars or galaxies but rather from all space. Initially mystified by the signal, they soon discovered that scientists at Princeton University had repeated Alpher, Herman, and Gamow's calculations and were searching for the predicted radiation. Once aware of that work, Penzias and Wilson realized that the background interference they had found was radiation created in the young Universe. Their discovery of the **cosmic microwave background (CMB),** as it is now called, won them the Nobel Prize in Physics in 1978.

Figure 18.9 shows measurements of the intensity of the CMB and demonstrates that it has a nearly perfect blackbody spectrum, most intense at about 1 millimeter as theory predicts. From this wavelength and using Wien's law ($\lambda_{max} = 2.9 \times 10^6$ nm K/T, where λ_{max} is the wavelength where the emission is strongest and T is the temperature), the temperature that describes the radiation turns out to be about 3 kelvin. A more precise measurement of λ_{max} shows that the temperature that describes the CMB is 2.725 kelvin, only a little warmer than absolute zero, the lowest temperature anything can have.

We see this radiation coming from every direction because as we look out in every direction, we look out into the past, eventually seeing nearly to the cosmic horizon when the Universe was a few hundred thousand years old. Note that when the Universe was young, this radiation in every direction would have appeared as hot as the surface of a star, and Olbers' prediction would have been true. However, the expansion of the Universe has stretched short wavelengths into long ones, so today the CMB radiation looks like the emission from an extremely cold body.

The discovery of this predicted radiation is one of the cornerstones of the **Big Bang** theory, that the Universe was born in a hot, dense state and expanded rapidly. No other theory has been successful at explaining the presence of this radiation. The high temperature and rapid expansion from a hot, dense state led the English astronomer Sir Fred Hoyle to jokingly refer to this theory of the birth of the Universe as the "Big Bang," a name that has stuck despite his sarcasm. Hoyle long championed an alternative "steady-state" model, which proposed that new matter forms continually to create new galaxies in the empty, expanding space between older galaxies. This could keep the Universe looking the same at all times, giving the steady-state model its name. However, modern observations, especially of the CMB, now show that the Universe has been changing throughout its history. The CMB radiation also provides us with a picture of how gas was distributed when the Universe was young, providing a starting point for understanding how galaxies formed.

FIGURE 18.9
Spectrum of the cosmic microwave background. The shape of the spectrum matches perfectly a blackbody with a temperature of 2.7 K. Notice the wavelength at which it peaks. The red curve shows the theoretical curve, while black dots show data points schematically.

The Formation of Galaxies

From the ages of the stars within galaxies, astronomers deduce that the Milky Way and other galaxies are about 13 billion years old. In fact, gravity began to pull together the regions of slightly higher density soon after protons and electrons recombined to make the first hydrogen atoms in the early Universe.

Astronomers think that galaxies formed by a process similar to the way stars form, but on a much larger scale. That is, gravity drew the gas expanding from the Big Bang

Small portion of cosmic microwave background
mapped by the Arcminute Cosmology
Bolometer Array Receiver (ACBAR)

$-300\mu K$ $0\mu K$ $+300\mu K$

FIGURE 18.10

A portion of the cosmic microwave background radiation (left) mapped by a microwave telescope at the South Pole (right). The red bumps are slightly hotter and denser regions where galaxies will likely form. At this redshift, 1° is about 800,000 light-years.

Small density fluctuations occur where
underlying dark matter is concentrated.

Since dark matter is unaffected by photons,
concentrations of mass grow at early times.

FIGURE 18.11

Schematic illustration of peaks in the density of dark and normal matter at the time the CMB was produced. Dark matter has pulled itself into dense clumps. Normal matter is attracted by the gravity of the dark-matter clumps, but cannot settle there as it is constantly struck by high-energy photons.

into gigantic clouds, which eventually collapsed under gravitational attraction and formed stars, as we discussed in describing the birth of the Milky Way in chapter 16. But to observe this process occurring in the first galaxies is extremely difficult because it happened so long ago. That is, to look back 13 billion years in time, we need to look out into space 13 billion light-years. At such enormous distances even an entire galaxy appears very dim. Nevertheless, with the Hubble Space Telescope astronomers have captured images of galaxies close to that remote time, as shown in the chapter-opening figure and discussed in chapter 17.

To understand galaxy formation, astronomers also rely heavily on computer simulations. Such simulations can model how the force of gravity pulls clumps of gas toward each other so that small irregularities in the Universe's matter as seen in the CMB (fig. 18.10) grow into galaxies. The simulations also show that dark matter plays a critical role in the process.

The fluctuations in the CMB are tiny, at most just a few hundred *micro*-kelvin (fig. 18.10) compared to the average temperature of 2.725 kelvin. This implies that the matter was almost uniformly spread out a few hundred thousand years into the Universe's history, smoother than the surface of a billiard ball. When astronomers simulate the growth of galaxies from such a uniform gas, they find these slight differences in density are not large enough for the slightly larger gravitational pull of a higher-density region to pull matter together as soon as it is observed.

In order for these models to build structures similar to the observed distribution of galaxies and voids, as described in chapter 17, dark matter is needed. It is not just that dark matter provides more mass; what is more important is its property of not interacting with light or normal matter except via gravity. During the first few hundred thousand years of the Universe, when matter and light were constantly interacting, normal matter could not begin to clump together. However, dark-matter clumps could begin to draw together, forming high-density regions. These dark-matter clumps provided a gravitational "seed" to attract normal matter as soon as light stopped interacting with matter after the epoch of CMB production (fig. 18.11).

Although we cannot see light coming from times before the production of the CMB, the even hotter, denser conditions during the first moments after the Big Bang left other traces of the Universe's violent beginning.

18.3 THE ORIGIN OF THE UNIVERSE

Where did the Universe come from, and what was it like in its earliest stages? Cosmologists study that remote time by extrapolating the present conditions in the Universe backward in time. That is, they use the laws of physics and present conditions to infer what the Universe was like at the time of its birth. We have already described one important conclusion of such extrapolations: the early Universe was hot and dense, as confirmed by the cosmic microwave background radiation. Figure 18.12 shows how its temperature has changed over the last 13.8 billion years.

You can see from the figure that the young Universe was extremely hot—trillions of degrees in its earliest moments. That very high temperature implies that in its youth the Universe was a mix of gas and radiation, much like the core of our Sun, although far hotter. Such high-temperature gases turn out to obey relatively simple laws. For example, the temperature and density depend in known ways on the Universe's size. This allows astronomers to deduce the Universe's past properties.

As figure 18.12 illustrates, there are several earlier stages we can identify. Each is associated with transitions known to take place at particular temperatures, and each leaves certain tell-tale evidence of what happened in that period. Thus, although we cannot see past the "wall" of CMB radiation, we have a number of clues to work with: the composition of the Universe, the prevalence of matter versus antimatter, and the distribution of matter on the largest scales. As we work back toward discovering the nature of the Universe at earlier and hotter times, it will be helpful to refer back to figure 18.12. Our first step back in time from the CMB epoch takes us to just a few minutes after the Big Bang. During the intervening several hundred thousand years the Universe had been expanding and cooling with little else changing. When the Universe was a few minutes old, however, it was as hot as the center of the Sun, and that had important consequences for the Universe.

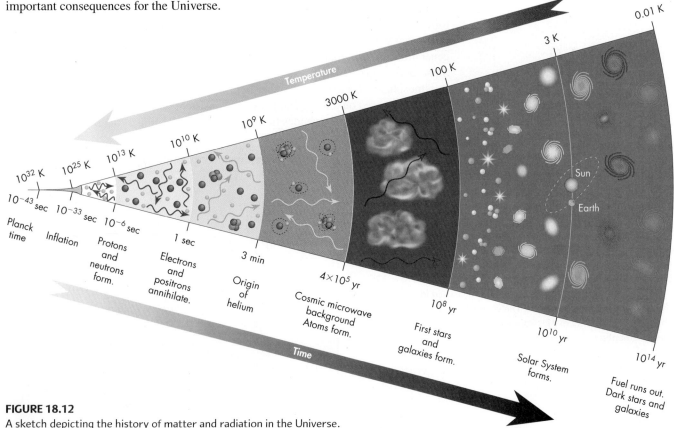

FIGURE 18.12
A sketch depicting the history of matter and radiation in the Universe.

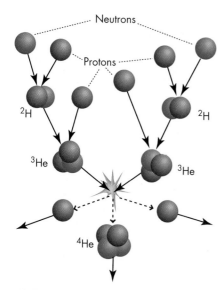

FIGURE 18.13

Nuclear fusion in the early universe.

The Origins of the Elements

As we extrapolate to times before the CMB radiation was produced, the Universe must have been hotter and more dense. George Gamow realized that at early enough times, conditions must have been similar to those in the core of a star, and so he predicted that helium should have been created by nuclear fusion as it is in stars.

Gamow's hypothesis that helium formed in the young Universe is supported by more-recent theory as well as by observations. According to modern theory, the young Universe consisted of protons, neutrons, and electrons (rather than just neutrons, as Gamow believed). Nevertheless, his basic idea is correct: in its youth, the Universe was hot and dense enough for nuclear reactions to fuse some of the protons and neutrons into helium (fig. 18.13). Moreover, current calculations show that the amount of helium and other light elements formed depends on how hot and dense the Universe was and how long those conditions lasted. The conditions at this early time can be related to the density of matter in the Universe today. If the Universe was lower in density, less ^{4}He would have been produced, because the particles would not have collided and fused as often. This also would leave higher amounts of some partial fusion products like deuterium (^{2}H) and ^{3}He (see fig. 18.13). By about 10 minutes after the expansion began, the Universe had cooled to about 1 billion K and spread apart so much that fusion stopped, leaving about 24% of the matter transformed into ^{4}He. The curves in figure 18.14 show theoretical predictions of how much of each element should form, depending on the Universe's density.

This prediction can be tested by measuring the amount of helium and other light elements in the oldest stars in the Milky Way and other galaxies. Those stars formed from the gas left over from the Big Bang, and therefore their composition reflects that produced in the Big Bang. Although additional fusion has occurred deep in their cores since they formed, the composition of the outer layers of these ancient stars is nearly the same as when they were born: about 24% helium, in excellent agreement with the Big Bang model. This is not the abundance of helium in stars like the Sun, which formed well after the Milky Way's birth. These later-generation stars have formed from interstellar gas that has been enriched in helium and other heavy elements from earlier generations of stars that have exploded or lost mass in other ways. Consequently, the Sun—born some 9 billion years after the Big Bang—contains about 27% helium in its atmosphere. The amounts of other light elements and isotopes also agree well with predictions based on the Big Bang theory, consistent with a specific density of matter today (fig. 18.14). This provides further strong confirmation of the Big Bang theory.

FIGURE 18.14

The amounts of different elements expected to be produced by nuclear fusion in the young Universe. The curves show how the amount of each element varies depending on the density of matter in the Universe. The values along the bottom axis show the density of normal matter in the Universe today, while the vertical axis indicates the amount of each element that would be produced. The red horizontal bars indicate the best estimates of the amount of each element measured in the oldest stars and gas clouds. The vertical blue bar shows the range of densities consistent with these measurements.

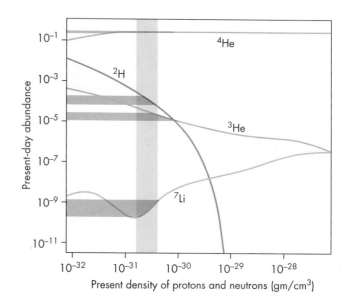

The Early Universe: Radiation, Matter, and Antimatter

Einstein's equation $E = mc^2$ equates an amount of energy, E, to an amount of mass, m, times the speed of light, c, squared. It tells us that stars can obtain their energy from mass lost when, for example, hydrogen is converted to helium, as discussed in chapters 12 and 14. In the early Universe, the implication of this energy–mass relation was reversed: a quantity of energy E may have turned into particles whose mass equaled E/c^2. Scientists know that matter can be created this way: in laboratory experiments they can observe high-energy radiation (gamma rays) forming particles. Given that energy can change into matter, it seems likely that high-energy radiation in the early Universe must have done the same. Such transformations of energy into mass, however, always create a *pair* of particles, as illustrated in figure 18.15A. Moreover, the particle pair has two special properties: the particles must have opposite charge (or no charge), and one of the pair must be made of ordinary matter, whereas the other must be made of what is called **antimatter.**

We briefly discussed antimatter in chapter 12, where we saw that the electron has an antimatter companion particle, the positron, with identical mass but opposite electric charge. Protons also have a corresponding antiparticle called the antiproton, and all other subatomic particles are similarly paired. Antimatter has the property that a particle and its antiparticle are annihilated on contact, leaving only high-energy radiation, as depicted in figure 18.15B. Thus, radiation (electromagnetic energy) can become matter, and matter can become radiation. In the early Universe, this shifting between mass and energy happened continually as particles formed and were annihilated.

Energetic radiation cannot turn into just any pair of particles: the pair's combined mass multiplied by c^2 must be less than the radiation's energy. For example, to become an electron–positron pair, the radiation needs an energy of about 1.6×10^{-13} joules, equivalent to a wavelength of about 0.0012 nanometers (a gamma ray). To become a proton–antiproton pair, the radiation needs roughly 1800 times more energy, making it a very high-energy gamma ray. Radiation with such large energies occurs only at very high temperatures. For example, to have enough energy to make an electron–positron pair, the temperature must be about 10^{10} kelvin. To make a proton–antiproton pair, the temperature must be about 10^{13} kelvin. Such high temperatures are not reached in ordinary stars, but they occurred in the Big Bang.

Astronomers estimate that approximately 1 microsecond after the Big Bang, the Universe had a temperature of about 10^{13} K, hot enough for its radiation to be converted into not only protons and antiprotons but even quarks and antiquarks, the particles that make up protons and neutrons. At that time, the Universe we now see was packed into a volume smaller than the Solar System. This volume, an inferno jammed with quarks and antiquarks created from the high-energy radiation, was expanding at a fantastic rate. As space expanded, this dazzlingly hot matter and radiation cooled. The cooler radiation was no longer energetic enough to create new quark–antiquark pairs. Thus, as quarks annihilated antiquarks, they were no longer re-created, and most of the quarks therefore disappeared from the Universe.

This self-annihilation was not total, however. While the process is not yet fully understood, there appears to have been a slight imbalance in interactions of subatomic particles that allowed a tiny excess of quarks, literally *one in a billion,* to survive with no antiquark to annihilate. These survivors then combined with one another to make the protons and neutrons of the Universe around us. Without that tiny "asymmetry" in the annihilation of quarks, our Universe would have been essentially empty. Every particle would have had an antiparticle with which it could combine and be annihilated into radiation, leaving the Universe filled only with the cosmic background radiation and containing no matter at all.

To understand our next step back toward the beginning of time, we must first understand how all this matter "curves" space and time in the Universe. We therefore next examine concepts of curvature and then investigate a process called inflation.

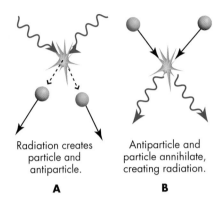

Radiation creates particle and antiparticle.

A

Antiparticle and particle annihilate, creating radiation.

B

FIGURE 18.15
(A) The energy of electromagnetic radiation can be converted into mass with the formation of a particle and its antiparticle. (B) The collision of a particle and its antiparticle leads to their annihilation and the conversion of their mass back into energy.

How matter goes to energy and vice versa

All of the mass and energy contained in the Universe causes space to be "curved," according to Einstein's general theory of relativity (essay 2). It is easier to picture how space is curved by the gravity of isolated objects such as exoplanets, black holes, and galaxy clusters. For example, astronomers can measure the bending of the path of light when one object lies in front of another.

How can we picture the **curvature of the Universe?** Our home planet, Earth, offers a simple example. Athletic fields and parking lots appear flat and, if we ignore hills, the surface of the Earth certainly *looks* flat. We know, however, that if we could walk in a "straight" line along this "flat" surface, we would return to our starting point. We are therefore like an ant that accidently crawls into a garden hose. The unbent hose represents a flat, uncurved space, and so, if the ant crawls in a straight line, it will reach the hose's end and emerge into the garden again. However, suppose we pick up the hose with the ant still in it and bend the hose so that its ends meet. Now the ant could crawl forever and would never reach the hose's end.

A similar situation exists for a Universe curved around on itself. Such a space looks as if it extends forever in all directions, but if we could travel in a spaceship along a "straight" line, we would eventually return to our starting position. Thus, we would have moved through space forever without coming to an end point. A universe with this property is said to have **positive curvature** and is "closed" because it has no "outside." We can infer that the Earth has a "positively curved surface" because if we travel in a straight line on it, we return to our starting point. Positively curved surfaces have other properties as well: lines that start out parallel eventually meet if they extend far enough, and the sum of the interior angles of a triangle drawn on such surfaces is greater than 180°, as illustrated in figure 18.16A.

The Universe might have other types of curvature, however. It might be a **flat Universe** (that is, have no curvature) or have **negative curvature.** Flat space is what most people picture when they think of space. In it, parallel lines do not meet, and a triangle's interior angles add up to exactly 180°, as illustrated in figure 18.16B. Negatively curved space is harder to visualize, but you can think of it as being bent into a saddle-like shape, as shown in figure 18.16C. Such curvature corresponds to bending space one way along one line and the opposite way along another line. In negatively curved space, the sum of the interior angles of a triangle is less than 180°.

Can we measure the curvature of space and determine whether it is positive, negative, or flat? Yes. After nearly half a century of attempts that gave only inconclusive results, astronomers at last have a method for measuring the curvature of the Universe that works with good accuracy. This method uses observations of the small temperature fluctuations of the cosmic background radiation, shown in figure 18.17. The size of these fluctuations gives information not only about

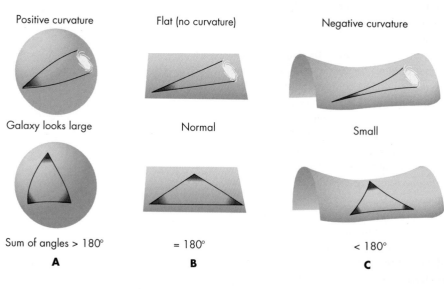

FIGURE 18.16
Figures illustrating how the shape of a surface determines the geometry of a triangle on it. (A) The interior angles of a triangle on the surface of the Earth add up to more than 180°. (B) The interior angles of a triangle on a flat surface add up to 180°. (C) The interior angles of a triangle on a saddle-shaped surface add up to less than 180°. The top row of figures illustrates that a distant object will therefore appear larger if the curvature is positive (A) and smaller if the curvature is negative (C).

FIGURE 18.17

Plot showing the strength of fluctuations of different angular sizes measured in the cosmic microwave background (CMB). The fluctuations in brightness are caused by the clumping of matter at the last scattering epoch. The position of the first, strongest peak at about an angular size of 1° implies that the geometry of the universe must be very nearly flat. Other peaks in the graph depend sensitively on the amount of dark matter and cosmological parameters. The green line shows a plot of the predicted size of the fluctuations for the best-matching model containing dark matter and dark energy. The map at right shows the pattern of bumps in the CMB as found by the *Planck* satellite in 2013.

the curvature of the Universe but also about the Universe's age and the amount of ordinary and dark matter.

To understand the method, it may help to think of a simple experiment. Imagine blowing across the mouth of an empty bottle. As you blow, you make the air in the bottle vibrate and generate sound waves. The pitch (tone) of sound waves is set by their wavelength, with shorter wavelengths having a higher pitch. You can demonstrate that effect by putting some water into the bottle and blowing across it again. The pitch of the sound will now be higher. This happens because the water gives the sound waves less space in the bottle and makes their wavelength shorter. Thus, the length (pitch) of the sound waves gives us a clue to the amount of air in the bottle.

Scientists use waves to probe a variety of objects. For example, seismic waves allow us to probe the interior of the Earth, and waves in the Sun's atmosphere allow us to determine properties of the Sun's interior. We can use similar techniques to probe the Universe. As the Universe expanded in earliest times, waves formed in its hot gas. Those waves took the form of fluctuations in its density and temperature. We observe those waves today as variations in the brightness of the CMB radiation across the sky, as seen in figure 18.17. Because the radiation comes from a time 380,000 years after the Big Bang, regions could have interacted with each other only if they were less than 380,000 light-years apart (the cosmic horizon). As a result, fluctuations in density most commonly extend out to the cosmic horizon. Combining this known size with the observation that these fluctuations are about 1° across allows us to determine the Universe's curvature, showing that it is flat to very high precision.

Additional analyses of the brightness variations in the CMB allow astronomers to deduce how much ordinary and dark matter is present. This is because waves in dark matter, which is frictionless, behave differently than waves in normal matter, which collides and heats. Astronomers show this with a graph of the strength of different-sized fluctuations (fig. 18.17), and matching the whole range of observed fluctuations requires about 5 times more dark matter than normal matter.

18.5 THE INFLATIONARY UNIVERSE

Physicists can study the behavior of matter at extremely high temperatures by smashing together particles in accelerators at nearly the speed of light (fig. 18.18). The highest temperature reached is several trillion kelvin. This recreates what conditions were like about 1 microsecond after the Universe's birth.

The Universe must have been even hotter, denser, and more energetic at even earlier times, and physicists have attempted to extrapolate what happened at these extraordinarily high temperatures. In principle, physical laws can be applied back to a time as early as about 10^{-43} second after the Big Bang, known as the Planck time, named for the German physicist who developed quantum theory. He recognized that at times smaller than this, the scale of quantum fluctuations becomes as large as the Schwarzschild radius of a black hole. This would result in random fluctuations in the curvature of the entire Universe that physicists do not know how to model. Some physicists speculate that before the Planck time all four forces of nature were merged into a single force.

The idea that forces merge at high enough energies has already been demonstrated in experiments studying the weak and electromagnetic forces. Although these forces look very different today, when physicists study interactions at higher and higher temperatures, the forces grow more similar. The weak force is only as weak as it is today because the particles that carry the force have a mass that requires a large amount of energy to generate. When the whole Universe was as hot as trillions of kelvin, these particles were as easily generated as massless photons, which carry the electromagnetic force, and the forces looked the same. These connections between the forces and how they split apart are sketched in figure 18.19. This has led to an intriguing hypothesis of how the Big Bang was set in motion that explains some of the Universe's unusual properties.

Inflation

When the Universe was about 10^{-35} second old, at a time when everything in our visible Universe today was packed into a volume smaller than a proton, the radiation and matter had a temperature that was trillions of times hotter than trillions of kelvin—about 10^{27} K. Physicists extrapolate that the strong and electroweak forces may have split apart when temperatures began to drop below this level. Under these extraordinary conditions, an unusual situation may arise that drives the Universe's expansion at an ever accelerating rate, which cosmologists call **inflation.**

FIGURE 18.18
Particle accelerator. This is a view of a small portion of the 27-km-long tunnel of the Large Hadron Collider. Subatomic particles are accelerated to near the speed of light in these evacuated tubes before they collide at speeds like those in the early Universe.

FIGURE 18.19
Schematic diagram of the relative strength of the forces through time. Physicists suspect that at the energies present at the time of inflation the electroweak and strong forces may merge in one of several possible "Grand Unified Theories" or GUTs. To understand even earlier times will require the development of a theory of quantum gravity.

Q: By what factor has the Universe increased *since* inflation?

FIGURE 18.20
The size of the Universe through time. The size of today's visible Universe is shown through time. Note the sudden inflation in size starting at 10^{-35} seconds.

The principles of inflation were worked out by Einstein and other cosmologists in the early 1900s for quite a different reason. Einstein was looking for a solution to the equations of general relativity that would allow for the Universe to be static, which is what was believed to be the case before Hubble's discoveries. In their most straightforward interpretation, the equations of general relativity seemed to imply that the Universe should be changing in size, but Einstein saw that there were solutions allowing for a constant level of energy throughout space. This term in the equations became known as the **cosmological constant.** If the constant was zero, the Universe would either expand or collapse under gravity's influence, but if its value was set to just the right level, it could create a repulsion that would counterbalance the pull of gravity.

The cosmological constant has some curious features. If it is even a little stronger than the gravitational pull, the Universe would expand at an accelerating rate. If it is a little too small, a static Universe would collapse in on itself. When Hubble discovered the expansion of the Universe a few years later, Einstein regretted having ever introduced the cosmological constant. He might have predicted the expansion of the Universe if he had left it out. However, astronomers now think that a cosmological constant might be what gave rise to the Universe's expansion.

As the strong and electroweak forces were splitting, space may have been filled everywhere with a constant energy until the split was completed. This would have caused an exponential expansion of the Universe beginning about 10^{-35} seconds after its birth and lasting until about 10^{-32} seconds. During this time, space expanded explosively in size, its radius swelling by a factor of 10^{25}, as shown in figure 18.20. If the period at the end of this sentence expanded by a similar factor, it would end up the size of the Local Group of galaxies. The inflationary stage "launched" the Big Bang.

Inflationary models of the first moments of the Universe mark the frontier of our understanding of the origins of the cosmos and offer tentative explanations of some of its features. For example, the models suggest that the Universe may have been created literally from nothing. Some models imply the existence of other universes forever separated from our own. Still others suggest that space has more dimensions than the four we are familiar with (three of space and one of time). In fact, according to some versions of these models, space has as many as 10 or 11 dimensions. Inflation may even have produced many "universes" isolated from our own, as discussed in Extending Our Reach: "Other Universes?"

Perhaps the most intriguing aspect of inflation is that it links the properties of the Universe at large with the properties of subatomic particles and the fundamental forces of nature. This link helps to explain some puzzles about the Universe, such as why the curvature of the Universe is "flat," and why the Universe (especially the cosmic microwave background) is so uniform. Astronomers refer to the first of these puzzles as the "flatness problem," and they refer to the second one as the "horizon problem."

EXTENDING *our reach* — OTHER UNIVERSES?

The inflationary model makes a very curious prediction: it suggests that our Universe may be just a tiny part of a vast "multiverse." Different regions within the multiverse can begin inflating at different times and at different rates. Such isolated regions might well be considered separate universes, independent of ours and completely unobservable by us. Furthermore, according to inflationary models, some of these other universes may be expanding much faster than others or recollapsing.

Inflation might create such sections of space much as bubbles form in a pot of boiling water. As water begins to boil, regions of the liquid suddenly change from liquid to gas (steam) and begin expanding. We see that region of steam as a bubble. Some bubbles expand, rise to the surface, and burst, whereas others

"Bubbles" grow.   Expansion cools Universe.   "Bubbles" become separate universes?

FIGURE 18.21
A sketch depicting bubbles that might become separate universes forming in an inflating universe.

collapse. So, too, in inflationary theories, entire universes may form and dissolve, entirely unknowable to us (fig. 18.21). It would be amazing indeed if all the wonderful intricacy and beauty of our Universe is merely a single bubble in an even vaster sea of space, a space that we can never see.

The Flatness Problem

We saw in section 18.4 that astronomers find the curvature of our Universe to be flat to within the measurement uncertainties. Because the curvature of the Universe depends on such things as the density of the matter it contains and how rapidly it is expanding, the Universe might have had any curvature, positive or negative. What is even more unusual about this is that a flat Universe is highly unstable. If the Universe began with a curvature that was even the slightest bit closed (or open), as it expanded it would grow rapidly more strongly closed (or open). A flat universe is balanced on a knife's edge and the least disturbance in either direction sends it falling over. For the Universe to be nearly flat today, it must have been almost perfectly flat ("fine-tuned") at the start.

Inflation explains in the following elegant way why the Universe is so flat. Recall that during the time of inflation, the observable Universe expanded by a colossal factor—from smaller than a proton to a few centimeters in size. Such an enormous expansion of an object flattens any curvature that it has. To see why, imagine inflating a balloon (fig. 18.22). While the balloon is small, the curvature of its surface is very obvious. However, if the balloon could be inflated to the size of the Earth, its surface, for all practical intents, appears flat, just as the surface of the Earth appears "flat" to us.

FIGURE 18.22
An initially small region with obvious curvature looks flatter and flatter if it is inflated by a large factor. At the same time, inflation carries regions of space far away from each other very rapidly.

The rapid inflation of the early Universe does the same to space, flattening any initial curvature of space.

The Horizon Problem

We saw in section 18.2 that the brightness of the cosmic microwave background is highly uniform across the sky. Although very small variations exist, they average only about 1 part in 100,000 across the sky. Such uniformity was difficult to explain prior to inflation theory. For example, if we look at the radiation coming from one spot in the sky and then look at radiation coming from exactly the opposite direction in the sky, the two locations are separated by the diameter of the observable Universe—over 27 billion light-years (fig. 18.23A). But the Universe is less than 14 billion years old, so no exchange can have occurred between these regions to average out any difference. Using cosmologists' wording, these regions lie outside each other's cosmic horizon.

The region encompassed by the cosmic horizon was only smaller in the past, in our simple expanding Universe model, so there seems to have been no time when these regions could have "communicated" to balance their temperatures. Inflation solves this in the period before inflation began. The entire visible Universe today was packed together in such a small volume that during the first 10^{-35} second, light could travel across it thousands of times, smoothing out any differences that may have existed. That in turn makes the cosmic microwave background almost exactly the same in every direction within the observable universe.

Careful study of the cosmic microwave background provides further evidence that the inflationary model is correct. The characteristics of the variations that are seen in the background radiation match another kind of variation: tiny quantum fluctuations. Figure 18.23B shows that there are regions where the radiation is slightly stronger or weaker than average, and they span distances much larger than $1°$, which was the size of the cosmic horizon at the time the CMB radiation was produced. Fluctuations spanning such large distances could not have grown by this early time, so they must have existed before inflation began. The number and sizes of these fluctuations match the characteristics of subatomic quantum fluctuations. Thus, the first steps toward the formation of the large-scale structure of the Universe may have been taken when the visible Universe was smaller than a proton.

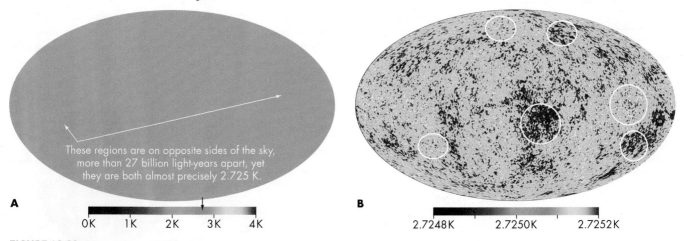

FIGURE 18.23

Maps of the cosmic microwave background, made by the *Wilkinson Microwave Anisotropy Probe* satellite. The maps show the brightness distribution of the CMB across the entire sky, much as a map of the Earth depicts its features on a flat chart. (A) This map depicts the uniformity of the radiation seen—equal to 2.725 K everywhere to within about one-hundredth of 1 percent. (B) The small deviations from uniformity are brought out in this map. Red regions are warmer (by about 200 micro-kelvin) and blue regions are cooler than average. They show irregularities in the temperature of the Universe at the time of recombination, when the universe was about 400,000 years old. Note that some warmer and cooler regions, such as those that are circled, are so large they must have been present before inflation.

18.6 DARK ENERGY AND THE FATE OF THE UNIVERSE

The Universe is currently expanding, and its curvature is approximately flat. We saw in section 18.4 that the nearly flat curvature of the Universe may tip toward a closed or open curvature as the Universe continues to expand. This can lead to different fates for the Universe—expansion continuing forever, or an eventual halt to the expansion and a subsequent collapse, as illustrated in figure 18.24. How can we determine what fate awaits the Universe in the distant future?

The Future Expansion of the Universe

We have seen that the Universe began with all matter flying away from all other matter at enormous speeds. In the first few minutes of this expansion, all the hydrogen in the Universe was created along with most of the helium. Then for several hundred thousand years this hot gas expanded and cooled, finally cooling to the point where it stopped glowing, and the matter began clumping into galaxies and larger-scale structures. Stars that formed have been gradually consuming the hydrogen from the Big Bang, generating heavier elements that ultimately became planets, molecules, and even us.

If the Universe expands forever, its stars will eventually consume all the hydrogen and burn out, leaving the Universe a black, empty space—a dismal prospect. However, recollapse of the Universe would mean that all the objects and atoms (and us) within it will be compressed into what some astronomers have called the "Big Crunch." The Big Crunch might lead to a rebirth of the Universe, perhaps leading to a new Big Bang, although it seems almost impossible that anything would survive being crushed to subatomic size.

To determine which fate awaits the Universe, we need to understand what factors affect its expansion or recollapse. But before we look at details, a simple analogy may help. Suppose you throw a stone into the air. As the stone moves upward, Earth's gravitational force slows its rise and eventually stops it so the stone falls back to Earth. On the other hand, if you could throw the stone faster than the escape velocity, it would rise forever. Thus, the stone's behavior depends on the strength of gravity and the upward force given to it. So, too, with the Universe.

Near the time of its birth, inflation or some other process began the Universe's expansion. Since that time, the gravitational force between its galaxies and other contents has generally acted to slow its expansion. If the gravitational force is weak or if

In his 1920 poem *Fire and Ice*, Robert Frost contemplates possible fates for the world similar to those we discuss for the Universe:

Some say the world will end in fire,
Some say in ice.
From what I've tasted of desire
I hold with those who favor fire.
But if it had to perish twice,
I think I know enough of hate
To say that for destruction ice
Is also great
And would suffice.

FIGURE 18.24
A sketch of how gravity and the energy of expansion determine the future expansion of the Universe.

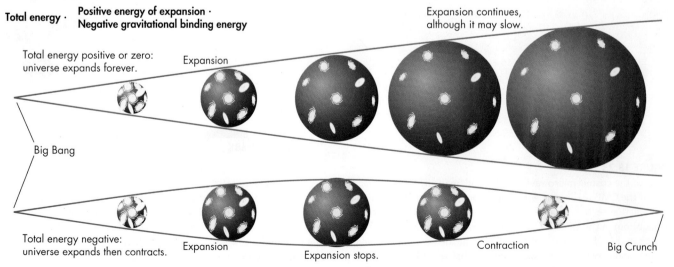

Total energy · **Positive energy of expansion · Negative gravitational binding energy**

Expansion continues, although it may slow.

Total energy positive or zero: universe expands forever.

Expansion

Big Bang

Total energy negative: universe expands then contracts.

Expansion

Expansion stops.

Contraction

Big Crunch

the impulse of expansion is strong, we might expect the Universe to expand forever. On the other hand, if the gravitational force is strong or the initial impulse is weak, we might expect that the expansion will stop and the Universe will collapse.

Is there some way we can measure the relative strength of these effects for the Universe? The answer is yes. We can compare the gravitational energy working to hold the Universe together with the energy of its expansion. If gravity dominates, the expansion will stop. If the expansion energy dominates, the expansion will continue forever. These scenarios are illustrated in figure 18.24 by the changing size of a representative portion of the Universe. Thus, we can learn (at least in theory) whether the Universe will expand forever or recollapse by seeing if the Universe has a gravity strong enough to stop its expansion.

We saw in chapter 3 that the gravitational force produced by an object depends on its mass—a result we can extend to the Universe itself. In particular, if astronomers know the Universe's mass, they can calculate the gravitational force it exerts to slow its expansion. However, to make such a calculation, it is easier to use the Universe's density (its mass divided by its volume) rather than its mass alone.

The Density of the Universe

To find the density of the Universe, astronomers choose a representative volume of space and add up the mass of all the material in it. In doing so, they must be sure to include the mass of dark matter as well as the visible matter—including dark matter is important because, as we have seen, it greatly outweighs the visible matter. Even though the dark matter is invisible, its mass can be determined within galaxies and galaxy clusters from its effect on the orbits of the stars and galaxies (chapter 17). Astronomers then divide the mass by the volume to obtain the density. When such a calculation is made, the density of the combined visible and dark matter in the Universe turns out to be about 3×10^{-30} gm/cm^3. This value is about ten times higher than the density of matter made of protons and neutrons based on the fusion of elements during the first few minutes of the Universe (section 18.3), which makes sense based on the percentage of dark matter found in galaxies and galaxy clusters.

To determine whether the Universe will expand forever or recollapse, astronomers compare the observed density with a theoretically calculated **critical density.** If the observed density is greater than the critical density, the Universe is expected to recollapse; if it is less, the Universe will expand forever. It turns out mathematically that the critical density is $\rho_c = 3H^2/(8\pi G)$, where H is the Hubble constant and G is the gravitational constant. Notice that the critical density depends on the Universe's rate of expansion (the Hubble constant), because if the expansion rate is very large, the critical density must also be large in order to have enough gravity to stop the expansion.

Astronomers use a shorthand notation for describing the Universe's density: they take the ratio of the actual density of matter ρ_M to the critical density ρ_c, which they call Ω_M. This is the Greek letter omega, and the subscript "M" denotes that this is the density of matter. Cosmologists predicted that if Ω_M ($= \rho_M/\rho_c$) is greater than 1, then the Universe would eventually recollapse. This is illustrated in figure 18.25, where different curves show how the size of the Universe would change, depending on the value of Ω_M.

Cosmology and the cosmos's fate

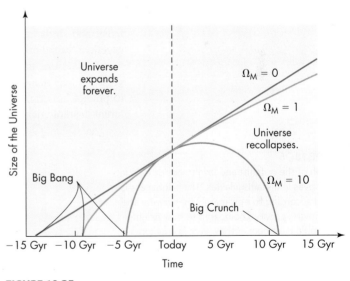

FIGURE 18.25

The size of the Universe in the past and future as calculated for different scenarios. Curves are shown for three cases in which ($\Omega_M = 0$) there is no matter, so the Universe expands without deceleration; ($\Omega_M = 1$) the pull of gravity and the expansion rate balance out, so the Universe constantly slows down but never recollapses; and ($\Omega_M = 10$) the pull of gravity greatly exceeds the energy of expansion, so the Universe recollapses in about 12 billion more years. All curves have the same slope today, which is determined by the Hubble constant. Curves lying in the blue region of the diagram have enough energy to expand forever, while those in the yellow region will eventually collapse back on themselves.

Estimates of the Hubble constant by the *Planck* satellite in 2015 give a value of $H = 67.3$ km/sec per Mpc.

When we compare the observed and critical densities, what do we find? If we use the value of H determined by observation (70 kilometers per second per megaparsec), we find that the critical density is about 10^{-29} grams per cubic centimeter, the equivalent of six hydrogen atoms per cubic meter. Thus, the actual density of matter is approximately one-third the critical density, which implies that the Universe should expand forever.

Although astronomers were pleased that the question of whether the Universe would expand forever or recollapse appeared to have been settled, they were troubled by the value found for the Universe's density. According to the inflation model for how the Big Bang began, the actual density should *exactly* equal the critical density, rather than be approximately one-third of that value, as was measured. Was this model, therefore, wrong? Or had some component of the Universe been overlooked? Perhaps, for example, there is even more dark matter than our estimate above suggests.

To confirm their models of expansion, astronomers sought to compare the present expansion speed of the Universe to its speed in the past. Note that each of the curves in figure 18.25 has a different history of sizes at different times. The curves all match the rate of expansion of the Universe today, which is determined by the Hubble constant, but each makes a different prediction of the distance to objects when the Universe was, for example, half its current size. If astronomers could determine the actual distances to extremely remote galaxies, they could compare those results to the different expansion curves. But how might such a comparison be made?

ANIMATION

The future of expansion

The Acceleration of the Universe

Astronomers are fortunate in that they have a remarkably precise way to determine the distances to extremely remote galaxies. By observing type Ia supernova blasts in remote galaxies, they can find both the distances and the redshifts of these systems. Recall that type Ia supernovas are produced when a white dwarf collapses (chapter 15). This process occurs when a white dwarf accretes enough material to exceed a specific mass, and so the explosions are always very similar to each other. This makes them excellent "standard candles."

At the same time, from the galaxy's redshift we know how much bigger the Universe is today than it was at the time the light left the galaxy. This allows astronomers to predict how faint the supernova should appear, depending on its redshift for different cosmological models, as shown in figure 18.26. Note that the nearby galaxies

FIGURE 18.26

A graph of the redshift and brightness of type Ia supernovas in distant galaxies. The brightness is labeled according to how many times fainter the supernovas appeared compared to how bright they would have been if they were in a galaxy 1 Mpc away. The curves show the predicted brightness if the Universe has expanded according to models of the Universe in which $\Omega_M = 0$, 1, or 10 (as in fig. 18.24). The horizontal lines through each data point show the uncertainty for each supernova distance measurement. The graph is based on observations by S. Perlmutter and collaborators. Note that the distant galaxies tend to lie below any of the models—indicating that the light is dimmer than any model in which the Universe has undergone only deceleration.

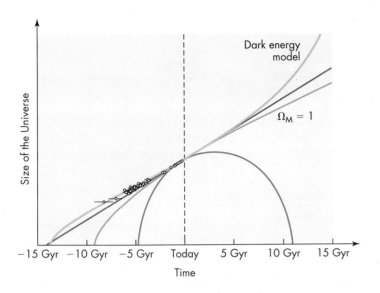

FIGURE 18.27
The data from type Ia supernovas is shown in a graph of the expansion of the Universe. At times in the past, the data lie above curves based on expansion with no dark energy. The best-fit curve, in which the Universe contains nearly three times as much dark energy as matter, is shown in blue. A Universe with dark energy will accelerate more and more rapidly in the future.

(in the lower left-hand part of the diagram) lie in a part of the diagram where the curves are so close together, they might be consistent with $\Omega_M = 1$. However, the more remote galaxies (upper right) clearly do not follow that law. Rather, the distant galaxies lie *below* even the line predicted for $\Omega_M = 0$, even less than what we would expect in a Universe containing no matter at all! This shows that expansion in the past was slower than it is now. That is, the expansion of the Universe seems to have *sped up!* Gravity's pull of one galaxy on another should only slow the expansion, so how is this possible?

The answer again goes back to Einstein's idea of a cosmological constant. Just as for the process of inflation, if the cosmological constant is not zero, it contributes an energy to the Universe that generates a repulsive force that acts on space itself. We cannot directly detect this energy, and so astronomers call it **dark energy,** in analogy to dark matter. Dark energy is quite different from other forms of energy because it is a property of space, not of objects. You cannot bottle dark energy or make an antigravity machine out of it. It acts only to make space expand faster and faster. This accelerating expansion is shown on a graph of the expanding size of the Universe in the past and extrapolated into the future in figure 18.27.

Although dark energy is invisible, its effect on the Universe is immense. In addition to creating the repulsive force that speeds up the recession of galaxies, dark energy contributes a colossal effective mass to the Universe. Using the relation $E = mc^2$ of relativity theory, astronomers find that dark energy has the equivalent of more than twice as much mass as that of all the other mass in the Universe, dark and observable, combined. Figure 18.28 shows the results from the *Planck* satellite's studies of the cosmic microwave background, which give the overall composition of the Universe. Ordinary matter is a very small portion of the Universe's composition, and even dark matter is outweighed by dark energy.

With this more complete understanding of how the Universe has been expanding, astronomers are able to predict the size of the Universe in the past, and from this we

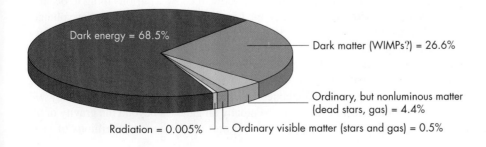

FIGURE 18.28
The makeup of the Universe as deduced from observations of the brightness variations in the cosmic microwave background by the *Planck* satellite in 2015.

Table 18.1		Relationships Between Redshift, Distance, and Time		
Redshift (z)	Length of Time Since Emitted (Light Travel Time)	Time After Big Bang	Size of Universe Compared to Now	Distance When Light Was Emitted
0	Today	13.8 billion yr	1	0
0.1	1.3 billion yr	12.5 billion yr	0.91	1.3 billion ly
0.5	5.2 billion yr	8.6 billion yr	0.67	4.2 billion ly
1.0	7.9 billion yr	5.9 billion yr	0.50	5.5 billion ly
2.0	10.5 billion yr	3.3 billion yr	0.33	5.8 billion ly
3.0	11.6 billion yr	2.2 billion yr	0.25	5.3 billion ly
5.0	12.6 billion yr	1.2 billion yr	0.17	4.3 billion ly
7.0	13.0 billion yr	800 million yr	0.13	3.6 billion ly
10	13.3 billion yr	500 million yr	0.09	2.9 billion ly
1000	13.8 billion yr	400,000 yr	0.001	45 billion ly

can determine what the redshift is at different times after the Big Bang. Table 18.1 shows how the redshift of a galaxy relates to the time when the light was emitted based on the current estimates of the amount of matter and energy in the Universe. The table also shows the interesting effect that, because of the Universe's expansion, some high-redshift objects were actually nearer us when their light was emitted than they are today. Cosmologists can also use our new understanding of the accelerating expansion to predict what the Universe will do in the future.

The Future of the Universe

For many years astronomers have used the terms *open* and *closed* to describe the shape of the Universe and whether its expansion will continue forever or not. The discovery that the cosmological constant (usually expressed by Λ, the capital Greek letter lambda) is probably not equal to zero greatly complicates the use of these two seemingly simple terms. If $\Lambda = 0$, then an open Universe expands forever and a closed Universe recollapses. If $\Lambda \neq 0$, that simple linkage does not hold. *Open* and *closed* still describe the curvature of the Universe, but not its fate. An open Universe may recollapse, and a closed Universe may expand forever.

Based on our current understanding of dark energy, we have a new picture of the expansion and fate of the Universe. When the Universe was young, the density of matter exceeded the density of dark energy, but the matter grew more dilute as the expansion proceeded. Dark energy remained a constant of space, and now exceeds the density of matter, so the Universe's expansion has begun to speed up (fig. 18.29), and it will speed up exponentially faster in the future. The density of dark energy and matter add up to make the overall density of the Universe equal to the critical density. Therefore $\Omega = 1$, as astronomers had predicted, but in a most unexpected way.

Dark energy at the level currently observed will cause a curious thing to occur in the remote future. Without dark energy, the Local Group would probably have plunged into the Virgo Cluster in a few hundred billion years. Instead, dark energy will cause the expansion of space to overcome the Local Group's gravitational attraction toward the Virgo Cluster, and then the separation will grow steadily faster. The accelerating expansion of space will move distant parts of space away from us at ever increasing speeds. The distances to these galaxies will grow so large so quickly that they will move outside of our cosmic horizon. Eventually all galaxies beyond the gravity of our Local Group will seem to disappear beyond our cosmic horizon. Trillions of years

Universe with dark energy

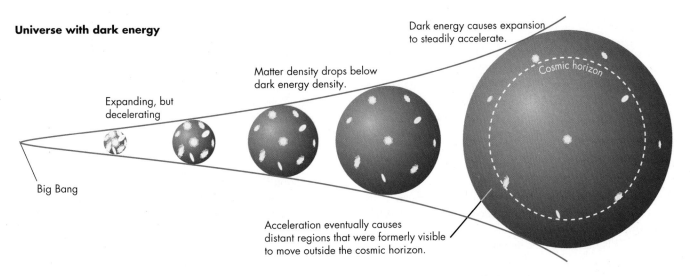

FIGURE 18.29

An expanding universe containing dark energy will initially slow down until the matter density drops below the level of the dark energy. Dark energy then causes accelerating expansion, which can eventually move remote regions of space away so fast that light can never reach us from them. This moves them beyond our cosmic horizon.

from now, our Galaxy will be essentially alone, an island of stars in a vast empty Universe.

Some cosmologists speculate that the Universe may have even stranger fates. Perhaps the cosmological constant is not constant but slowly grows with time. If this is the case, the Universe might start pushing itself apart on successively smaller scales, so even our Galaxy will tear itself apart. Others wonder whether we may eventually find pathways to other universes, perhaps ones that will not tear themselves apart.

SUMMARY

Because distant galaxies are receding from the Milky Way and each other, the Universe in its youth was much smaller, hotter, and denser than it is today. By extrapolating the present recessional motion of galaxies back in time, astronomers deduce that the Universe was created approximately 14 billion years ago in the Big Bang, a release of energy that initiated the expansion of space and the creation of matter. This age is consistent with that of other astronomical objects, in that it is greater than the age of the Earth and the age of the oldest stars.

Additional evidence for a hot, dense early Universe comes from the cosmic microwave background. This weak radiation was originally emitted as visible light long before galaxies formed and when the Universe was a glowing fireball of matter and energy. As the Universe expanded and cooled, so did the radiation; now its temperature is only 3 K, just slightly above absolute zero.

Analysis of tiny brightness variations in the cosmic microwave background allows astronomers to measure the shape of the Universe and its age to much higher accuracy

than was previously possible. They deduce from these measurements that the Universe is flat and 13.8 billion years old—both estimates accurate to a few percent.

The flatness of the Universe and the extreme uniformity of the cosmic microwave background lead astronomers to deduce that almost immediately after its formation, the Universe underwent a very brief but very rapid increase in volume—a process called inflation.

Whether the Universe will expand forever or collapse depends on its density and rate of expansion. If the density is less than a critical value, not enough mass is present for gravity to stop the cosmic expansion, and the Universe will expand forever. Observations indicate that the mass of matter is too small to ever halt the Universe's expansion.

Astronomers have discovered that the expansion of the Universe is actually speeding up. To explain this acceleration, astronomers hypothesize that the Universe contains dark energy that causes the speedup. Dark energy appears to make up nearly 70% of the Universe, but its nature is almost completely unknown.

QUESTIONS FOR REVIEW

1. (18.1) Why do astronomers think that the Universe is expanding?
2. (18.1) What is meant by the age of the Universe? How old is the Universe? How is its age found?
3. (18.1) How can there be no center to the Universe?
4. (18.2) What is Olbers' paradox?
5. (18.2) What is meant by a cosmic horizon?
6. (18.2) Is our cosmic horizon the same as that of beings in another galaxy? Why?
7. (18.2) What does "light travel time distance" mean?
8. (18.2) What happened at the recombination era? Why is this important today?
9. (18.2) What is the cosmic microwave background? What is its origin?
10. (18.2–18.3) Why do astronomers think that the early Universe was hot and dense?
11. (18.3) According to the Big Bang theory, what were the major eras in the history of the Universe? What was the Universe like in each of them? Approximately how long did each last?
12. (18.3) How was helium formed in the early Universe?
13. (18.3/18.6) Why is Einstein's equation $E = mc^2$ important when thinking about the early Universe?
14. (18.4) What do astronomers mean when they say that the Universe is flat? What are the other possibilities?
15. (18.4) What evidence do astronomers have that the Universe is flat?
16. (18.5) What is meant by inflation in cosmology?
17. (18.5) What problems of the Big Bang theory are resolved by inflationary models?
18. (18.5–18.6) What role does the cosmological constant play in cosmology? Is this the same role Einstein originally created it for? What evidence makes astronomers think it may exist?
19. (18.6) What determines whether the Universe will expand forever or recollapse? Explain what the critical density is.
20. (18.6) What is dark energy? What does its presence imply for the future of the Universe?

THOUGHT QUESTIONS

1. (18.1–18.6) Does it bother you that the Universe may expand forever?
2. (18.2) Explain why the cosmological redshift is not caused in the same way as a Doppler shift (chapter 4) even though galaxies are being moved away from us as the Universe expands.
3. (18.2) If the Universe had not expanded since the recombination era, what would the night sky look like? What would the daytime sky look like?
4. (18.2–18.3) Discuss three different kinds of observational evidence for the Big Bang theory you could present if you were challenged by a skeptical child.

5. (18.3) Why do most stars, including our own Sun, contain more helium than the 24% formed during the Big Bang?
6. (18.4) If you draw three lines that connect at their corners on a globe (like in fig. 18.16A), have you drawn one triangle or two? Explain your answer. If you draw three lines that connect on a flat or saddle-shaped surface, have you drawn one triangle or two?
7. (18.5–18.6) The recession velocities of distant galaxies suggest that the Universe is expanding faster today than in the early Universe. How is this similar to the process of inflation in the early Universe?
8. (18.6) Why are there points below the green line (instead of above it) in figure 18.26, if the Universe is expanding faster today than in the past?
9. (18.6) Dark energy accelerates the expansion of the Universe, but galaxies and solar systems hold themselves together while the distances between them increase. If the Universe continues to expand faster and faster, what will happen to our cosmic horizon? Will studying cosmology be easier or harder in the very distant future?

PROBLEMS

1. (18.1) Until recently, experimental results for the Hubble constant ranged from about 50 to 100 km/sec/Mpc. Calculate the age of the Universe for both of these extremes. Do you see any potential conflicts with other astronomical observations if H had been either of these numbers?
2. (18.2) The temperature of Universe at recombination was about 3000 K. If the ratio of the temperature of the background radiation at recombination to the temperature of the CMB radiation today is equal to the ratio of the wavelength of light today to the wavelength of the background radiation at recombination, how many times wider is the Universe today than it was at recombination? Remember that as the Universe expands, it stretches light's wavelength.
3. (18.2–18.3) Using the same logic as in problem 2, how many times wider is the Universe today than it was when the average temperature was 3×10^9 K, sometime between 1 and 3 minutes old? If you shrank the Milky Way's disk by that factor, would it fit inside the Solar System?
4. (18.2–18.3) If the factor you found in problem 3 is the change in diameter, how many times larger is the volume of the Universe today than when the temperature was 3×10^9 K?
5. (13.3/18.2–18.3) The temperature of the Universe at recombination was about 3000 K. Use Wien's law to calculate the peak wavelength of the radiation at that temperature. Which spectral type of star has this surface temperature?
6. (18.3) One second after the Big Bang, the density of the universe was about 0.1 gm/cm³. How big a volume of the universe contained as much mass as the Solar System?
7. (18.5) When the Universe was extremely young, it increased in size exponentially until it was 10^{25} times larger than it began. How many times would the Universe have to double in size to grow this much larger than it started?

(If you are unfamiliar with logarithms, you can estimate this by testing different powers of 2 on a calculator by trial and error.)

8. (4.2/18.6) If dark energy in the universe currently is about 68.5% of the critical density, then how many joules of energy are there in each liter of space? (Use the critical density given in the chapter to find a mass in kg, then use $E = mc^2$.) How many 1-mm-wavelength microwave photons would this correspond to in each liter? (The energy of a photon is discussed in section 4.2.)

TEST YOURSELF

1. (18.1) From what evidence do astronomers deduce that the Universe is expanding?
   (a) They can see the disks of galaxies getting smaller over time.
   (b) They see a redshift in the spectral lines of distant galaxies.
   (c) They detect cosmic background X ray radiation.
   (d) They can see distant galaxies dissolve, pulled apart by the expansion of space.
   (e) All of the above
2. (18.1) If Hubble's constant were half of the value we think it is, the Universe would be _____ as we think it is.
   (a) 4 times as old   (c) the same age   (e) 1/4 as old
   (b) 2 times as old   (d) 1/2 as old
3. (18.2) What is meant by the cosmic microwave background?
   (a) It is radiation from distant quasars.
   (b) It is radiation from hot gas in intergalactic space.
   (c) It is radiation from the first stars formed when the Universe was young.
   (d) It is radiation created in the early Universe.
   (e) It is the explanation of Olbers' paradox.
4. (18.3) Which of the following statements about the first few minutes of the Big Bang is true?
   (a) The universe was transparent.
   (b) The universe was very cool.
   (c) Hydrogen fused into helium.
   (d) Galaxies began to form.
   (e) Stellar cores began to form.
5. (18.3) For most of the time since the beginning of the Universe, conditions were
   (a) much hotter than today.
   (b) not very different from today.
   (c) similar to the inside of an atomic nucleus.
   (d) like the core of a star.
   (e) like the surface of a star.
6. (18.5) Which of the following are explained by a brief period of inflation? (Select all that apply.)
   (a) The high uniformity of the cosmic microwave background in multiple directions
   (b) The large number of black holes in the Universe
   (c) The fact that the Universe appears flat

   (d) The ratio of helium to hydrogen in very old stars
   (e) The existence of more matter than antimatter in the Universe
7. (18.6) If the density of the Universe were greater than the critical density and the Universe contained no dark energy, then
   (a) the Universe would expand forever.
   (b) the Universe would eventually collapse.
   (c) galaxies could never have formed.
   (d) the Universe would have negative curvature.
8. (18.6) Dark energy (select all that apply)
   (a) is represented in cosmological equations by the cosmological constant.
   (b) is the energy associated with dark matter.
   (c) acts against gravity.
   (d) causes space to expand ever faster.
   (e) is distributed evenly through space.
9. (18.6) Current measurements suggest the Universe is composed of about
   (a) 90% regular matter, 10% dark matter.
   (b) 27% dark matter, 5% dark energy, 68% regular matter.
   (c) 68% dark energy, 27% dark matter, 5% regular matter.
   (d) 33% dark matter, 33% regular matter, 33% dark energy.
   (e) 68% dark matter, 27% dark energy, 5% regular matter.

KEY TERMS

antimatter, 489
Big Bang, 485
cosmic horizon, 483
cosmic microwave background (CMB), 485
cosmological constant, 493
cosmological principle, 479
cosmology, 477
critical density, 497
curvature of the Universe, 490
dark energy, 499
flat Universe, 490
inflation, 492
negative curvature, 490
Olbers' paradox, 482
positive curvature, 490
recombination era, 484

Q: FIGURE QUESTION ANSWERS

WHAT IS THIS? (chapter opening): The dim red object at the center of this image is suspected to be one of the most distant galaxies yet discovered. The galaxy lies about 13.2 billion light-years from us, behind a massive cluster of galaxies whose gravity magnifies the galaxy's light. The galaxy has an estimated redshift of 9.6, so the wavelength of its light (and the Universe itself) have expanded by more than a factor of 10 since the light we see here left the galaxy.

FIGURE 18.20 By a factor of about 10^{25}, similar to the factor during inflation.

Life in the Universe

The evolution of the Universe from Big Bang to galaxies, stars, and planets has along the way created us. This evolutionary path from raw quarks to thinking beings who seek to understand how they fit into this grand scheme—what eighteenth-century scientists called "the great chain of being"—is one of many marvels of the Universe and is the theme of this essay.

In earlier chapters, we described astronomical evolution and how the Earth and Sun fit into the cosmos. Now we turn to life on Earth and how it might have formed and developed. We will see that the existence of life on our own planet is suggestive of the possibility of its existence elsewhere, and so we will also speculate about whether we are alone, leading to the question, Do other intelligent civilizations exist? Astronomers are strongly divided about such possibilities. Some argue that such life is highly probable, whereas others argue that it is highly improbable. One goal of this essay is to describe the evidence on both sides of this debate.

We will begin by examining the long history of life on Earth, looking at the factors thought to play a role in its development here. That history began quite soon after the Earth formed, as illustrated by the time line shown in figure E4.1. We will also discuss the likelihood of similar conditions elsewhere, discovering that planets suitable for life are probably numerous. In addition, we will describe how some astronomers have been motivated by such arguments to search for radio signals from other civilizations. Finally, we will speculate about whether our existence may even say something about the nature of the Universe itself.

LIFE ON EARTH

Life has existed on Earth for nearly the entire history of our planet. For example, fossil bacteria such as those illustrated in figure E4.2 occur in rocks perhaps as old as 3.5 billion years. The dates of these fossils have been measured by radioactivity in their associated rock, as described in chapter 6. For example, the ancient life forms "cyanobacteria" (cyan is a blue-green color associated with these photosynthesizing cells) form colonies called "stromatolites," which are found in fossils more than 3 billion years old and look identical to the stromatolites that exist today in shallow, warm, salty bays along the coast of western Australia (fig. E4.3).

There are possible signs of even earlier life in the isotopes of carbon found in deposits as much as 3.8 billion years old. Living organisms are selective in choosing lighter isotopes of carbon as they build various organic molecules, so deposits of carbon containing abnormally low amounts of the heavier

Atmosphere mostly carbon dioxide

Isotopic traces of possible life

Possible microfossils

Earliest stromatolites

Cyaonobacteria and other prokaryotes

| 5 | | 4 | | 3 |

Earth forms ---

Oldest surviving rocks ---

FIGURE E4.1
A time line illustrating the history of life on Earth.

FIGURE E4.2

Microscopic image of a probable microfossil of cells in rock 3.4 billion years old.

FIGURE E4.3

(A) Stromatolites in western Australia are formed by colonies of a cyanobacteria. They trap sediments, forming a layered structure in this characteristic dome shape. (B) 3.4-billion-year-old stromatolite fossil, sliced to show the layered growth pattern.

isotopes may indicate that life was present. Given that Earth formed less than 4.6 billion years ago and that its surface was probably molten rock for more than one hundred million years, and perhaps heavily bombarded until as late as 3.9 billion years ago, it appears that life may have begun quite quickly after the surface became hospitable.

Scientists conclude from the fossil evidence that only extremely simple life forms such as bacteria and algae existed for roughly three-fourths of Earth's history. Then, approximately 600 million years ago, more-complex forms such as invertebrates developed, and about 500 million years ago, only a little more than one-tenth of the Earth's lifetime, shells and

crustaceans appeared. Mammals and dinosaurs both appeared roughly 250 million years ago, but the latter were wiped out about 65 million years ago in the great Cretaceous extinction, probably caused by a totally chance event—an asteroidal collision, as described in chapter 11. Early hominids, our immediate ancestors, appeared roughly 5.5 million years ago, but our own species, *Homo sapiens,* evolved only about 500,000 years ago. Earth has therefore existed as a planet roughly 10,000 times longer than we have as a species. Thus, if all Earth's history were compressed into a year, humans would appear only in the last hour, civilization only in the last few minutes, and a human lifetime would be less than a second.

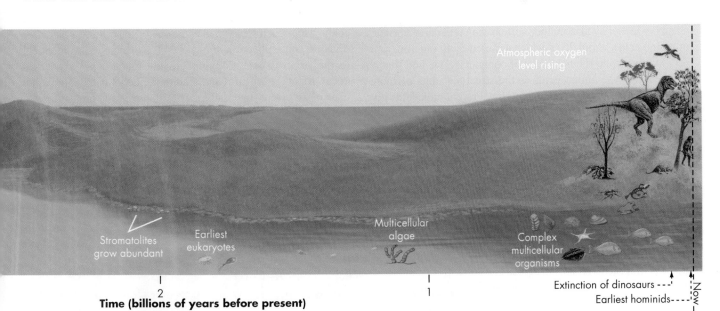

Atmospheric oxygen level rising

Stromatolites grow abundant

Earliest eukaryotes

Multicellular algae

Complex multicellular organisms

Extinction of dinosaurs ---

Earliest hominids----

Now

2

1

Time (billions of years before present)

FIGURE E4.4
A few of Earth's many complex life forms.

THE UNITY OF LIVING BEINGS

Life on Earth shows a dazzling variety of forms, from butterflies to birds, worms to whales, mushrooms to maples, as hinted at in figure E4.4. Despite that variety, all living beings on Earth have an underlying unity of structure, reproduction, and metabolism. All living organisms use the same kinds of atoms for their structure and function (mainly hydrogen, oxygen, carbon, nitrogen, phosphorus, and sulfur). These atoms are not only widely used, but they are also abundant throughout the Universe. Thus, life on Earth generally is very economical in its use of the same chemical substances in widely differing organisms. Moreover, life tends to draw on the materials with which our planet is most richly endowed: we are made primarily of the same substances that make up our oceans and atmosphere.

The chemical elements that compose living things are linked into long-chain molecules for their structure and function. The primary structural units of many of these chain molecules are amino acids. **Amino acids** are molecules built around the same basic building block of hydrogen, carbon, nitrogen, and oxygen atoms shown in figure E4.5. Several different side chains (which may also include sulfur) can yield many varieties of amino acids, about 20 of which are used by living things.

Moving up to the next level of complexity, we find that all living things use amino acids as the structural units of more-complicated molecules called **proteins.** Protein molecules give living things their structure. For example, the scales of reptiles and fish are composed of the protein keratin, and the stiff cartilage of our own bodies and other animals is composed of the protein collagen. Proteins not only give cells their structure but also supply their energy needs. That energy supply reveals yet another underlying unity among living things: all cells use the

same kind of molecule, adenosine triphosphate (ATP), to supply energy for action and growth.

This unity of structure and energy supply is echoed in the reproduction of life forms. All single-celled organisms divide to create new cells; all multicelled organisms produce egg cells, which divide to create new cells. Moreover, in all cases, "parents" pass on genetic information to their offspring by the same molecule, **DNA.**

DNA stands for deoxyribonucleic acid. It is a massive molecule consisting of two extremely long chains of smaller molecules wound around each other and linked by matched pairs of molecules called nucleo-bases or just bases. This gives DNA the appearance of a twisted ladder with the role of the connecting rungs played by the base pairs (fig. E4.6). There are just four different types of bases, and in groups of three they "spell out" codes corresponding to each of the amino acids. The order of the amino acids along the DNA molecule provides the assembly instructions for different kinds of proteins.

When a cell reproduces, the DNA separates into two individual chains. Each chain then adds the paired base at each point along its length to match the one already present on the other chain. Thus, each strand makes a copy of the other, so that after division and replication, a new DNA molecule exists identical to the original one. Copying errors—mutations—sometimes occur. Most are harmful and lead to the cell's death. Nonlethal ones make evolution possible. That all terrestrial organisms use DNA for their reproduction gives life a truly remarkable unity.

Unity can also be seen in the way higher organisms function. For example, plants derive their food from carbohydrates that they manufacture during photosynthesis. The chlorophyll they use in this process is structurally very similar to the hemoglobin that transports oxygen in the blood of animals. These two

FIGURE E4.5
Diagrams illustrating the structure of several amino acids. H, C, N, and O represent atoms of hydrogen, carbon, nitrogen, and oxygen, respectively. These amino acids all have the same basic building block (shown in yellow).

| Glycine | Serine | Threonine | Asparagine | Glutamine |

FIGURE E4.6
A diagram illustrating the structure of the DNA molecule. The blue horizontal structures in this model are the base pairs.

molecules differ primarily in that chlorophyll is built around a magnesium atom, whereas hemoglobin is built around an iron atom. Likewise, cellulose, the structural material of plants, has a molecular structure very similar to that of chitin, the structural material in insect bodies and crustacean shells.

Finally, living systems exhibit a unity in their functions, such as the chemical processes that move biologically important molecules into and out of cells.

DEDUCTIONS FROM THE UNITY OF LIFE AND THE TIME LINE

What conclusions can we draw from the chemical similarities of living beings and their ancient history, as deduced from the time line? First, the chemical similarity almost inescapably suggests that all life on Earth had a common origin. Second, the antiquity of life suggests that under suitable conditions, life develops rapidly from complex molecules. Thus, we might infer that wherever such conditions occur, so too will life.

We should treat these conclusions with great caution, however, for they are based on only a single case: life on Earth. For example, if the first time you hit a golf ball you make a lucky hole-in-one, it does not mean that you will do so again anytime soon. Thus, the fact that life exists on Earth may say next to nothing about its chances elsewhere. Supposing, however, that

we do choose to speculate, what can we say about the origin of life and its likelihood elsewhere?

THE ORIGIN OF LIFE

Most scientists today think that terrestrial life originated from chemical reactions among complex molecules present on the young Earth. This idea dates back to at least the time of Charles Darwin, but it was strongly bolstered in 1953 by an experiment performed by Stanley Miller—then a graduate student at the University of Chicago—and his professor, Harold Urey. Miller and Urey filled a sterile glass flask with water, hydrogen, methane, and ammonia—gases thought to be present in the Earth's early atmosphere (fig. E4.7). They then passed an electric spark through this mixture. The spark generated both visible and ultraviolet radiation, which triggered reactions in the gas-water mixture. At the end of a week, they analyzed the mixture, finding in it a variety of organic molecules as well as five of the amino acids used to make proteins. They therefore concluded that at least some of the ingredients necessary for life could have been generated spontaneously on the early Earth.

The success of Miller and Urey's experiment motivated other researchers to perform similar experiments. Subsequent experiments created conditions even more similar to the young Earth, and produced all the types of amino acids used by life, as well as the building blocks of DNA. Even the energy needs of living things can be produced in Miller-Urey type of experiments. For example, the Sri Lankan-American biochemist Cyril Ponnamperuma and the U.S. astronomer Carl Sagan showed

FIGURE E4.7
The Miller-Urey experiment simulated conditions on the early Earth and showed that complex organic molecules could form then.

FIGURE E4.8

Spheres about 1 to 2 microns in diameter form when protein-oids are heated and cooled. Similar pro-teinoids have been discovered around undersea vents of Hawaiian volcanoes.

FIGURE E4.9

(A) Undersea volcanic vents expel hot sulfur-rich minerals that provide the nutrients for very primitive types of bacteria. (B) These bacteria in turn are at the bottom of the food chain for colonies of unusual undersea creatures at such depths that sunlight never reaches them.

that ATP can be readily synthesized from compounds similar to those generated by Miller and Urey, demonstrating one additional step in creating living from nonliving material.

The U.S. chemist Sydney Fox demonstrated that another step in the process of forming life may occur by simple chemical processes. He took complex organic molecules such as those made by Miller and Urey and by repeatedly heating and cooling them in water was able to create short strands of proteins called "proteinoids." More interesting, however, Fox discovered that these proteinoids spontaneously formed small spheres reminiscent of cells (fig. E4.8). These small spheres have no power to reproduce, nor do they show any of the complex structure commonly found in living cells, but they do demonstrate that organic material can spontaneously build a structure that could wall off complex molecules from their environment, where they might take the next steps toward reproduction and life.

Why then do we not observe similar processes occurring today on Earth? One reason is that our atmosphere now contains oxygen, which quickly attacks organic molecules, breaking them down. Thus, the absence of oxygen earlier in the Earth's history—an absence shown by analysis of ancient rocks—was probably crucial for the development of life. Another reason was given by Charles Darwin in his speculations on life's origin: today's living things immediately consume such organic material.

The importance of low oxygen abundance to the origin of life has led some researchers to speculate that life may have originated deep in the ocean near sea-floor vents. These environments are rich in minerals and energy, and today they support a variety of life forms (fig. E4.9). In ancient days, these same environments might have been especially congenial to the assembly of complex molecules. Even now, bacteria and other life forms live in hot, oxygen-poor environments such as the boiling hot springs at places like Yellowstone National Park.

Even without the success of the Miller-Urey experiment in creating organic material, scientists studying the origin of life on Earth are confident that such material would be present on the young Earth. One source of organic material is interstellar matter. We saw in chapter 16 that many interstellar clouds contain a rich mix of organic molecules, including many that are precursors to amino acids. Because the Solar System and Earth formed from such clouds, complex organic molecules must have been present on or near the Earth from the time of its birth. Even if these molecules were destroyed during the Earth's

formation, fresh molecules may have been carried to its surface by planetesimals toward the end of their heavy bombardment of Earth's surface. In fact, carbonaceous chondritic meteors—probably surviving fragments of the ancient planetesimals—are rich in organic material, and some even contain amino acids, as we saw in chapter 11. Finally, some scientists have demonstrated experimentally that the compression and heating of gases resulting from planetesimal bombardment can be as fruitful as the electric discharges used by Miller and Urey in forming complex organic materials. Thus, astronomers are confident that organic molecules were common on the ancient Earth.

On the other hand, even if one assumes that organic matter existed on the early Earth (whether created by lightning, ultraviolet radiation, or planetesimal bombardment), many steps are needed to take it from amino-acid-filled droplets in a primordial ocean to living cells. For example, the structures created must be able to reproduce, a feature not yet seen in any origin-of-life experiments.

Cells replicate today using the double-chained molecule DNA. Some researchers have speculated, however, that the earliest life forms used the slightly simpler molecule, ribonucleic acid (RNA). Biochemists have shown that if the proper chemicals (especially certain enzymes) are present, RNA can replicate in the absence of living cells. Moreover, they have also discovered several other molecules far simpler in structure than RNA that can replicate, but so far only in a "mechanical" sense. That is, if such molecules are put in a solution containing the necessary raw materials, they will make copies of themselves. One particular mix of molecules not only replicates but also can "mutate" into a slightly better replicator. Even so, far more is needed to produce life as we know it. For example, simple life forms must be able to form cells that work cooperatively.

ORIGIN OF COMPLEX ORGANISMS

The original cellular life on Earth was probably far simpler than the cells we see today. The earliest cells, called **prokaryotes,** had no nucleus, a condition found now only in bacteria and certain algae. After billions of years, some prokaryotes may have merged accidentally, a merger that benefited both cells by allowing them to grow and reproduce more rapidly. A cell good at storing energy, for instance, might have combined with a cell good at reproducing, thereby creating a more complex cell. From such simple beginnings, **eukaryotes**—cells with nuclei—probably evolved. In addition to the nucleus, in which genetic information is stored, most eukaryotes also contain mitochondria, which are tiny bodies within the cell that convert food into energy that the cell utilizes. A clue that mitochondria may have begun as prokaryotes is that they contain their own DNA.

Multicellar organisms probably evolved as colonies of eukaryotes worked together to exploit resources more efficiently. With different sets of cells specialized to perform different tasks, much greater complexity became possible. For example, with some cells providing a skeletal structure, some working cooperatively to drawn in nutrients, and others specializing in digestion, a multicellular organism can become more successful and reproduce more prolifically than a single-celled creature. Such reproductive advantages and random genetic variations lead to a natural selection of organisms that are better adapted to their environments. The variety of environments on Earth and the changes they undergo have subsequently led to the evolution of the incredible diversity of life that we see around us today, as Charles Darwin first explained in 1859 in his monumental work *On the Origin of Species by Means of Natural Selection.*

LIFE ELSEWHERE IN THE UNIVERSE

Humans have speculated for thousands of years about life elsewhere in the Universe. Around 300 B.C., the ancient Greek philosopher Epicurus wrote in his "Letter to Herodotus" that "there are infinite worlds both like and unlike this world of ours." Epicurus went on to add that "in all worlds there are living creatures and plants." Likewise, the Roman scholar Lucretius wrote, around 50 B.C., that "it is in the highest degree unlikely that this Earth and sky is the only one to have been created."

Such views were not universal, however. For example, Plato and Aristotle argued that the Earth was the only abode of life, a view that prevailed through most of the Middle Ages. Today, however, influenced by science fiction movies, television, and books, many people find no difficulty at all in believing that extraterrestrial life exists. In fact, supermarket tabloids make silly claims almost weekly about encounters with aliens. If we look at the facts, however, there is not a shred of evidence that life exists elsewhere. This does not mean there is no extraterrestrial life; it simply means we do not *know* if there is. Thus, with no evidence, what can we say meaningfully about life elsewhere in the Universe?

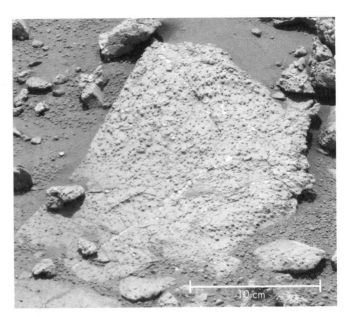

FIGURE E4.10
Geological and chemical analyses of this rock, found by the rover *Curiosity,* suggest that Mars had a habitable environment. The rock formed in water that was neutral in acidity, and the rock contains a variety of chemicals that scientists think would have been conducive to microbes.

SEARCHING FOR LIFE ELSEWHERE

Is there any evidence for life elsewhere in the Universe? Not really, although in the summer of 1996 two groups of scientists suggested that traces of organic chemicals that they detected in meteorites from Mars may indicate that life once existed there (see chapter 9 for more details). The absence of hard evidence may seem discouraging, but as British astronomer Sir Martin Rees has said, "Absence of evidence is not the same as evidence of absence."

Even if the Universe were teeming with life, we have so far examined only a tiny portion of it: some half dozen spots on the Moon and on Mars (by robot space-landers), and some asteroidal fragments picked up on Earth. The Moon's lack of atmosphere and water makes it so inhospitable that astronomers did not expect to find life there. Mars seemed more promising because its environment, though harsh, is more like the Earth's than that of any other planet in the Solar System.

Laboratory experiments have demonstrated that some terrestrial organisms (bacteria and lichens) could in fact survive Martian conditions, though they did not reproduce or grow. We know that the Martian environment was less harsh in the past. For example, Mars once had flowing water on its surface, an indication that a warmer climate and denser atmosphere existed long ago (fig. E4.10). Nevertheless, the two *Viking* spacecraft that landed there in 1976 found no trace of life, nor have any of the rovers or landers since, although these landers were outfitted primarily to study the planet's geology. The *Curiosity* rover

Ice covering

Liquid ocean
under ice

FIGURE E4.11
Beneath the icy crust of Jupiter's moon Europa, there may be an ocean of water harboring life. Astronomers are designing probes that could tunnel down to the water layer.

(see chapter 9) has experimental packages that will explore Mars's habitability in much greater detail, studying the climate record, and testing for the chemical building blocks of life.

Another intriguing possibility is that life may have formed in the oceans under the icy crust of moons like Europa, Ganymede, or Enceladus. Tidal and internal radioactive heating keep the interiors of these icy moons liquid (see chapter 10). Conditions there may be similar to the ocean on the young Earth, with underwater volcanic vents. Astronomers are considering how we might deliver a deep-sea vehicle to explore one of these moons. Just one of the challenges of such a mission will be to get through the kilometers of ice on the moon's surface. Perhaps the lander can heat the area beneath itself, melting the ice and gradually sinking down to the liquid interior (fig. E4.11).

PANSPERMIA

Most scientists think that life originated on Earth spontaneously, but is that necessarily true? The discovery of a meteorite from Mars with possible traces of life revived an older idea that life originated elsewhere and was later transported to Earth either accidentally or deliberately.

The hypothesis that terrestrial life descended from organisms created elsewhere in the Universe is called **panspermia.** According to this idea, simple life forms (perhaps bacteria) from some other location drifted from their place of origin across space to Earth. Bacteria spawned on some distant planet, for instance, may have hitched a ride on rocks or soil that were knocked into space by a meteorite impact. Astronomers had initially assumed that most of the debris of a meteorite impact would be superheated and sterilized, but the meteorite from Mars containing organic molecules made them reexamine that assumption. Although the immediate impact area of a collision is intensely heated, the shock wave that travels out from the impact site can launch substantial material at higher than the escape velocity (chapter 3). Because such collisions were even more common in the early Solar System, it is even possible that

we are all descended from microscopic Martians who arrived at the Earth when it was young.

It is not just Mars that might be a source for such material. Once in space, this material might travel almost anywhere. Although space is highly hostile to advanced life forms, simple organisms may be able to survive long passages through it if embedded in a rock. Eventually these rocks may have arrived in the Solar System and been captured by the Earth's gravity, initiating life on our planet.

Panspermia has a certain appeal, in that it suggests we may have distant cousins on remote planets elsewhere in our Galaxy. The theory was very popular in the 1920s, but critics point out that it does not really simplify the problem of the origin of life, it just shifts it elsewhere. It does, however, provide additional reasons for speculating that life may exist elsewhere.

ARE WE ALONE?

Do other life forms exist in the Universe? Is there somewhere, hundreds of light-years from Earth, a "student alien" reading a book discussing the possibility of life on planets such as Earth? Scientists do not know, and in fact they are strongly divided into two groups. One group of scientists (let us call them "many-worlders") believes that millions of planets with life exist in the Milky Way and that many of these have advanced civilizations. The other group (let us call them "loners") argues that we are the only intelligent life in the Galaxy. We discuss next how these two radically different points of view arise.

ARGUMENTS FOR MANY WORLDS

Many-worlders argue as follows: Earth-like planets are common. Life has formed on Earth. Therefore, it would be surprising if life did not exist elsewhere. In fact, even if only a small fraction of such planets have life, many advanced civilizations probably exist. Such arguments may actually vastly *underestimate* the number of habitable planets, because life forms elsewhere may be able to live in environments intolerable to terrestrial organisms. To see how such numbers are deduced, let us begin by estimating the number of Earth-like planets in the Milky Way.

We start by asking how many Sun-like stars are in the Milky Way. We limit ourselves to such stars because luminous blue ones burn their fuel so rapidly that they will burn out and explode as supernovas before life has much chance to evolve, and dim red stars are so cool that planets must be very close to them for conditions to be warm enough for a terrestrial type of life. Given those restrictions on star types, we turn to censuses of stars, which show that stars like the Sun make up about 10% of our Galaxy's stars, for a total of about 10^{10} such bodies.

Of these 10^{10} stars, we next ask, How many of them have Earth-like planets? One measure of this is whether a planet can sustain liquid water. While there may be liquid water heated by other processes, there is a range of distances from a star where

FIGURE E4.12

The habitable zone around different type stars. In the habitable zone, the star's luminosity is high enough to permit liquid water without vaporizing it.

temperatures are suitable. This depends on the type of star, as illustrated in figure E4.12. Our own system has two planets out of a total of eight that have had liquid water: Earth and Mars. Unfortunately, although astronomers have detected many planets around other stars, Earth-sized planets are still very hard to detect and so the probability of their occurrence is still unknown. If we are conservative, therefore, and say that other systems might not be quite so lucky as we are in the Solar System, we might choose the probability of a life-sustaining planet around a solar type of star to be 1 in 10, or 10%. Taking 10% of the 10^{10} Sun-like stars leaves us with 10^9 habitable planets.

How many of these actually have life on them? That will depend on how easy it is for life to form, a probability that the Miller-Urey experiments seem to indicate is quite high. The difficulty in assessing that probability is that we have only the single case of life here as our basis. If, however, we use Earth as the pattern, the rapidity with which life developed on Earth and the Miller-Urey experiment's success in making molecular precursors to life indicate that the chance of life starting is fairly high, say, 1 in 100 rather than 1 in a million. Given that probability, we end up with $10^9 \times 0.01 = 10^7 = 10$ million worlds with life in our Galaxy.

We now go to the next step and ask, On how many of these life-bearing planets do technical civilizations form? It is extremely difficult to assess the probability that higher life will evolve. Moreover, even if life succeeds in starting, perhaps it will be annihilated by asteroidal impact or, if it develops a high technology and is careless or thoughtless, life may destroy itself by pollution, nuclear war, or some other disaster. On the other hand, a disaster for one species provides an opportunity for another species to evolve—just as we might never have evolved if the dinosaurs had not been wiped out by a major impact. Some have speculated that major evolutionary changes may occur until a species arises that is capable of technology that can prevent cataclysmic events like an asteroid impact. Suppose we make a conservative guess and say that if life forms, it has a 1 in 1000 chance of developing a technical civilization. Thus, we find that the number of advanced civilizations in our Galaxy is $10^7 \times 0.001 = 10,000$. When we further consider that there are billions of other galaxies, it seems very likely that we are not alone.

An argument similar to this was first presented by the U.S. astronomer Frank Drake, who pointed out that one can estimate the number of civilizations by multiplying together the probability of each condition necessary for it, a result now called the **Drake equation.** Unfortunately, the accuracy of the probabilities is extremely uncertain, but they do help us to focus the argument. It is interesting that even using fairly conservative estimates, the equation suggests that *many* civilizations other than ours exist in our Galaxy. Assuming that they exist, we might then ask, How close is the nearest one? We can estimate the distance as follows.

Suppose the Milky Way contains N civilizations scattered randomly throughout its disk, as illustrated in figure E4.13. We can calculate their average separation, d, as follows. Imagine drawing a sphere of radius $d/2$ around each civilized world. Then N such spheres must cover the disk of the Milky Way. That is,

$$N\pi(d/2)^2 = \pi R^2$$

where we assume that the Milky Way's disk has radius R. We can now cancel π to obtain

$$Nd^2/4 = R^2.$$

Moving the N and 4 to the right-hand side of the equation gives

$$d^2 = 4R^2/N.$$

Finally, taking the square root of both sides gives

$$d = 2R/\sqrt{N}.$$

To evaluate d and thereby find how far away our neighbors are, we set R equal to the Milky Way's radius (taken to be 50,000 light-years) and $N = 10,000$, the value we deduced for the number of technical civilizations. These choices then give a value for d of 1000 light-years. Thus, even if such a large number of civilizations exist in our Galaxy, they are all likely to be extremely far from us. Such immense separations also make it very unlikely that we will be able to exchange information with other civilizations.

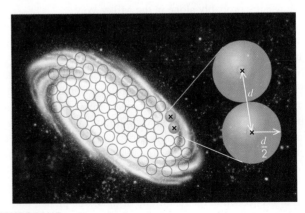

FIGURE E4.13

A sketch illustrating how to estimate the distance between civilizations in the Milky Way.

ARGUMENTS THAT WE ARE ALONE

Loners suggest that we are the only advanced life in the Milky Way. They point out that any of the probabilities going into the Drake equation could be much smaller than we have guessed. Perhaps so small that $N = 1$. Us.

For example, the British astrophysicist Fred Hoyle compared the likelihood of life arising spontaneously to the likelihood of a box containing several hundred tons of aluminum being shaken and by chance assembling itself into a 747 jumbo jet. While this comparison may seem silly, it illustrates the divergence of opinion on the likelihood of life forming. Perhaps Earth-like planets are far rarer than we have estimated. Perhaps evolution on other planets produces life forms that are so totally unlike those of Earth that we could not recognize them as being alive. And if life forms in oceans on worlds with an ice crust over the ocean, is there any reason to think that there is even a remote chance of a civilization arising? Furthermore, even if life forms similar to us do evolve elsewhere, the time period during which we could communicate with them might be so narrow that the opportunity passes. For example, if a civilization had existed on Mars even as recently as 200 years ago, we could not have communicated with them by radio, and if in the meantime they destroyed themselves somehow, the chance would have been lost forever.

The loners also base their argument on the rapidity with which our species has spread across Earth and the potential for its rapid spread across the Milky Way. This argument was first advanced by the Italian physicist Enrico Fermi, and is often called the **Fermi paradox.** Fermi argued that given the increasing rate of our technological advancement, it seems likely we will develop the ability for interstellar travel within a thousand years. And even if it takes us a thousand years to travel to a nearby star, hopping from star to star, it will take only a few million years to colonize our entire Galaxy. While this is long in human terms, it is extremely short compared to the long history of our Galaxy. The paradox Fermi pointed out is that if other civilizations exist, they should have already arrived at Earth long ago, and evidence of them would be plentiful. Thus, loners argue, we are probably alone in the Milky Way.

However, this argument, like that of the many-worlders for a multiplicity of civilizations, rests on several unproved assumptions. For example, it assumes that (1) most civilizations are driven to colonize, (2) they seek rather than avoid contact, and (3) they have mastered interstellar flight and communication.

The challenges for interstellar travel are daunting, and perhaps insurmountable except in very limited ways. We saw in essay 2 that it is possible, in principle, to travel so close to the speed of light that interstellar travel will take relatively short amounts of time to complete for the passengers. However, there are enormous costs in resources in energy and material, very long times needed in the rest frame of the home planet, and the dangers in traveling at such high speeds where encountering even an interstellar pebble could be disastrous. If anyone else is out there, we may have to learn it remotely.

FIGURE E4.14
When completed, the Allen Telescope Array, being constructed in California, will be able to simultaneously carry out astronomical surveys while it conducts a deep, targeted search for signals from other civilizations. The array will be made up of hundreds of small radio telescopes, using interferometry to simultaneously examine large areas of the sky over a wide part of the radio spectrum.

RADIO SEARCHES

Given the failure to find life near to us, and the difficulty of directly sampling distant locales, a few astronomers have searched for radio signals from other civilizations. Such signals might be either broadcast to us deliberately or communications directed elsewhere that we might simply "overhear." Astronomers began "listening" for extraterrestrial radio signals in the 1960s with Project Ozma, which monitored radio emission from several nearby star systems, hoping to detect emission from other than natural processes. More-recent projects, called **SETI** (*Search for Extra-Terrestrial Intelligence*), use receivers that automatically scan billions of radio wavelengths to search for signals from other civilizations (fig. E4.14).*

Searching for signals is not easy, because even if they exist, they are probably very weak and buried in cosmic static. Thus, astronomers use considerable care when choosing the wavelengths to monitor. For example, at very long wavelengths, interstellar gas in the Galaxy emits strongly and will overwhelm all but the most powerful signals. On the other hand, at very short wavelengths, molecules in our atmosphere block the signals. It turns out that the optimum wavelengths are near 21 centimeters, the wavelength emitted by neutral hydrogen in the cold gas clouds of our Galaxy. This is also near the wavelength of OH (hydroxyl) molecules, which emit strongly at 18 centimeters. These two wavelengths are very distinctive, and any other intelligent civilization would probably have discovered their

---

* Such searches for life elsewhere require huge amounts of computer time. The data in one project is e-mailed from the central search telescope to a network of millions of volunteers who allow their home computers to process the information when they are not using their machines.

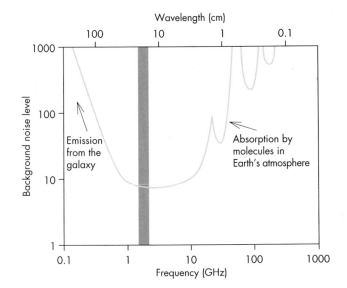

FIGURE E4.15

The waterhole is a range of wavelengths between prominent spectral lines of H and OH at 21 and 18 centimeters. It falls in a clear region of the radio spectrum that is free of interference from natural sources of radio emission and is not blocked by absorption in an Earth-like atmosphere.

significance and have equipment that could receive or broadcast them. Thus, several searches have been made for signals in this wavelength region, which is sometimes called the "waterhole" because it is bracketed on one side by emission from H and on the other by emission from OH (fig. E4.15). However, Ozma and all subsequent searches have so far proved fruitless.

One possible drawback to such searches is that other civilizations might not be transmitting signals. Like us, they may be listening. However, if other civilizations are like ours, they inadvertently broadcast a wide range of signals. For example, television stations and defense radar transmit extremely powerful signals that would be detectable many light-years from Earth. Moreover, on at least one occasion, a ceremony in 1974 celebrating the renovation of the huge Arecibo radio telescope in Puerto Rico, astronomers deliberately broadcast a signal toward the globular cluster M13. This cluster is more than 20,000 light-years away from us, so the signal has only begun its journey.

LIFE AND THE TRANSFORMATION OF PLANETS

The many difficulties associated with radio searches and the limited reach of our space probes have led scientists to consider other ways to search for signs of extraterrestrial life. One interesting method involves looking for unexpected gases in the atmosphere of a planet, gases that might be produced by living things.

In the late 1960s, James Lovelock, a British chemist, noted that several of the gases making up the Earth's atmosphere would disappear if Earth had no living beings to replenish them. For example, oxygen is very reactive and readily combines with surface rock. Lovelock went on to argue that the presence of such gases in a planet's atmosphere is an indication of life there. Thus, rather than searching for signals or sending robot spacecraft to explore a planet's surface, astronomers might find evidence of life by analyzing the light from a planet's atmosphere.

Astronomers have succeeded in detecting gases in the atmospheres of some exoplanets that pass in front of their stars, so that the gases in the atmosphere produce absorption lines in the star's spectrum. Indeed, oxygen was found in one gas giant, although it appears to be from the breakdown of water vapor as gas boils off this hot planet. Astronomers have also spectroscopically detected methane from Mars—this is another gas that might be associated with life, but it might also be produced by volcanic activity in Mars's interior.

While these results are inconclusive so far, these ideas have led to the realization that life may transform a planet in even more profound ways. In 1974, Lovelock and the U.S. microbiologist Lynn Margulis suggested that life creates a single "larger entity" with a planet, a "symbiosis" of life and planet, which they called the **Gaia hypothesis.** They chose this name from the Greek goddess of the Earth, Gaia (pronounced *Guy-uh* in this context but *Jee-uh* elsewhere).

According to the Gaia hypothesis, life does not merely respond to its environment but actually alters its planet's atmosphere and temperature to make it more hospitable. As we discussed in chapter 6, plant life has almost certainly done just that on Earth. For example, by photosynthesis, plants have created an oxygen-rich atmosphere on Earth that shields them from dangerous ultraviolet radiation. Similarly, plants may alter a planet's temperature through the greenhouse effect. If the planet is too cold, plant metabolism is slowed, and less CO_2 is converted to oxygen. The greater CO_2 abundance warms the planet and enhances plant growth. Conversely, if the planet gets too warm, plant growth, according to the hypothesis, runs amok and produces lots of oxygen, reducing the CO_2 abundance and therefore also reducing the greenhouse warming.

Lovelock and Margulis have also proposed that many other facets of the environment, such as humidity, salinity of the oceans, sea level, and even plate tectonics, may be controlled by living organisms so as to optimize their reproduction. Such an intimate linkage between the living and nonliving is a very appealing idea. It almost makes the planet "alive." But is the Gaia hypothesis correct? Some aspects certainly are, but many scientists are skeptical of its more extreme claims (for example, that life can control plate tectonics). Nevertheless, living beings do exert a remarkable control over their local environment, and many unsuspected links between life and the physical world surely remain to be found. One such possible connection has its roots in cosmology, a connection called the "anthropic principle."

THE ANTHROPIC PRINCIPLE

In 1961, Robert Dicke, a physicist at Princeton University, wrote a paper concerning some remarkable cosmological coincidences. Dicke noted, for example, that the age of the Universe is very close to the lifetime of stars like the Sun and went on to argue that this "coincidence" is in fact not remarkable at all. Rather, it is a *necessity* if cosmologists are to exist to note it. Dicke pointed out that life requires elements such as carbon, silicon, and iron, which are made in massive stars. Thus, for life to form, enough time must pass for massive stars to evolve and make the heavy elements and then eject them into space with a supernova explosion. Moreover, additional time is needed for the ejected material to be incorporated back into interstellar clouds and form new stars. Only with this second generation of stars does life become likely. Dicke therefore concluded that intelligent life is impossible in a Universe only a few thousand years old. He likewise argued that life requires stars, and so it can exist only in a Universe young enough that stars have not all burnt out. Thus, the very existence of an intelligent observer in the Universe requires that the age of the Universe fall within certain limits.

In 1974 the English physicist Brandon Carter took Dicke's idea further and proposed what he called the **anthropic principle.** According to the anthropic principle, "what we can expect to observe must be restricted by the conditions necessary for our presence as observers." Since Carter's work, many scientists have shown how "fine-tuned" the Universe must be for life to exist. Thus, no matter how unlikely some aspects of the Universe may appear to us, if those aspects are necessary for life, then that is what we will observe. For example, we might argue how truly marvelous it is that conditions on Earth are just right for life. According to the anthropic principle, the rude reply is, "Of course conditions are just right for life. If they were not, there would be no life here to do the marveling."

SUMMARY

Life has probably existed on Earth for at least 3.8 billion years, a large portion of our planet's existence. That long duration and the underlying similarities of all life forms in terms of their structure, energy supply, and methods of reproduction indicate that life had a common origin relatively soon after Earth's formation. Experiments such as Miller and Urey's show that the precursors to life can be easily created under natural conditions. Thus, many astronomers believe that life has formed in numerous other places in the Universe. All searches for life, however, have found nothing, but given the vast distances between stars, that negative result is not surprising.

Once life develops on a planet, it may subtly alter its environment to its own ends, a proposition known as the "Gaia hypothesis." Such alterations may be detectable more readily than signals from other civilizations and may therefore be signatures of extraterrestrial life.

QUESTIONS FOR REVIEW

1. What is the earliest evidence for life on Earth?
2. How old are the earliest life forms thought to be?
3. What features of life are suggestive of a common origin?
4. How might the building blocks of life have come to exist on the early Earth?
5. What is meant by panspermia?
6. What is the Drake equation?
7. What are the main arguments for whether life is common or uncommon in our Galaxy?
8. What is the Fermi paradox?
9. Why is 21 centimeters believed to be a good wavelength region to search for signals from other civilizations?
10. What is meant by the Gaia hypothesis?
11. What is meant by the anthropic principle?

THOUGHT QUESTIONS

1. Would you be upset if it turned out that we were the only intelligent life in the Universe? Why or why not?
2. If a long time ago the Sun formed along with many other stars from a single large collapsing cloud, how might that affect the likelihood of finding life on planets around nearby stars?
3. Do you think the "harder" step in the development of intelligent life is the initial development of life itself (self-replicating, very simple creatures) or the development of an intelligent civilization? Try to justify your argument with an appeal to—or argument about—how much time these steps took on Earth.
4. During the Apollo program, returning astronauts were briefly quarantined to ensure that any germs or other hazards they might have carried back from the Moon were not let loose on Earth. The danger may in fact be the opposite situation. How important do you think it is to ensure that spacecraft visiting Mars and the icy moons of Jupiter and Saturn have been sterilized? How might such planetary protection measures affect the search for primitive life on other planets in our Solar System?
5. Suppose astronomers did detect signals from another civilization. What effect do you think it would have on our society? What effect would it have on you?
6. Do you think it is a good idea to send out signals in the hope that aliens might detect them? Suppose, for example, they were more technologically advanced than us and could do to us what we have done to less developed societies on Earth.
7. Based on what you know about the composition of the Milky Way, are there are any places in it you think we would be more or less likely to find a civilization? If you were sending messages, to which parts of the Galaxy would you try sending them?
8. What message would you send to an alien civilization?

PROBLEMS

1. For what percentage of the Earth's lifetime has *Homo sapiens* existed?
2. From 1950 to 2000, the world population increased from 2.5 billion to 6 billion. If this trend continues, what will be the world population in 2050? In 2100? (Think carefully about the rate at which population increases.)
3. Look up the nearest stars in appendix table 8. How long would it take a meteoroid traveling at 10 kilometers per second to reach the nearest Sun-like star? If bacteria were harbored inside such meteoroids, do you think they might survive the trip? How long would it take a meteoroid to travel 50,000 light-years—effectively across the whole Galaxy? What would be the main challenges for bacteria surviving such a long trip?
4. After 20 years, the two *Voyager* spacecraft had reached about twice Pluto's distance from the Sun. How long will it take them to travel 4.2 ly—the distance to the nearest star?
5. Make your own estimate of the number of habitable planets within the Milky Way. What is the average distance between them?
6. Using the equation derived on page 511, how many current civilizations would you need in the Milky Way to expect one to be within 30 light-years of our own? (At this distance it would be possible to ask one or two questions and get one or two answers over a lifetime of conversation by radio transmissions.)

TEST YOURSELF

1. Evidence that life on Earth is very ancient comes from
   (a) fossil bacteria in rocks more than 3 billion years old.
   (b) fossils buried thousands of feet below the surface.
   (c) fossil algae in rocks a few million years old.
   (d) the discovery of silicon-based life forms in ancient rocks.
   (e) the discovery of silicon-based life forms in a Miller-Urey type of experiment.

2. The Miller-Urey experiment demonstrated that
   (a) very simple bacteria can be created from the chemicals present in the atmosphere of the ancient Earth.
   (b) Earth's early atmosphere was too harsh for life to have formed until recently.
   (c) conditions on the early Earth were suitable for the creation of many of the complex organic molecules found in living things.
   (d) if simple life forms landed on the early Earth, they could have survived.
   (e) early life forms were silicon-based.

3. High-mass stars probably do not harbor intelligent life because
   (a) they are so hot that no planet can have liquid water.
   (b) their ultraviolet radiation would destroy organic molecules.
   (c) their lives are too short for complex life to evolve.
   (d) their mass is so large that planets will be pulled in.
   (e) All of the above

4. The Drake equation attempts to determine
   (a) the conditions under which life originated on Earth.
   (b) the optimum wavelength for communicating with extraterrestrials.
   (c) the age of life on Earth.
   (d) the number of other technically advanced civilizations.
   (e) the lifetime of our own civilization.

5. Those who argue we are the only civilization in the Milky Way might support their argument by claiming that (select all that apply)
   (a) other civilizations might keep to themselves out of fear of being colonized.
   (b) otherwise we should have been colonized by another more advanced civilization.
   (c) primitive life may have been delivered to Earth in meteorites.
   (d) otherwise we should be inundated by radio signals from other civilizations.
   (e) just because life started early on Earth doesn't mean it is easy to start life—we might be an exception.

6. A planet on which life forms create an atmosphere that makes the planet more hospitable for life is an example of
   (a) the Drake equation.   (d) the anthropic principle.
   (b) the Gaia hypothesis.   (e) the Miller-Urey experiment.
   (c) SETI.

7. According to the anthropic principle,
   (a) we are the only civilization in the Universe.
   (b) there are 10^{12} civilizations more advanced than us in the Milky Way.
   (c) human beings are the highest form of life.
   (d) the Universe could not differ radically from its present properties, or we would not be here to observe it.
   (e) the greenhouse effect is beneficial to us.

KEY TERMS

amino acids, 506
anthropic principle, 514
DNA, 506
Drake equation, 511
eukaryotes, 509
Fermi paradox, 512

Gaia hypothesis, 513
panspermia, 510
prokaryotes, 509
proteins, 506
SETI, 512

Answers to
Test Yourself

Appendix

Scientific Notation

Scientific or powers-of-ten notation is a shorthand method for expressing and working with very large or very small numbers. With this method, we express numbers as a few digits times ten to a power or exponent. The power indicates the number of times that ten is multiplied by itself. For example, $100 = 10 \times 10 = 10^2$. Similarly $1{,}000{,}000 = 10 \times 10 \times 10 \times 10 \times 10 \times 10 = 10^6$. Note that we do not always need to write out $10 \times \ldots$. Instead, we can simply count the zeros. Thus, 10,000 is 1 followed by 4 zeros, so it is 10^4.

We can write the numbers 1 and 10 in powers-of-ten notation as well: $1 = 10^0$ and $10 = 10^1$.

To write a number like 300, we break it into two parts, $3 \times 100 = 3 \times 10^2$. Similarly, we can write $352 = 3.52 \times 100 = 3.52 \times 10^2$.

We can also write very small numbers (numbers less than 1) using powers of ten. For example, $0.01 = 1/100 = 1/10^2$. We can make this even more concise, however, by writing $1/10^2$ as 10^{-2}. Similarly, $0.0001 = 10^{-4}$. We can write a number like 0.00052 as $5.2 \times 0.0001 = 5.2 \times 10^{-4}$. Note that for numbers less than 1, the power is 1 more than the number of zeros after the decimal point.

Suppose we want to multiply numbers expressed in powers of ten. The rule is simple: we add the powers. Thus, $10^3 \times 10^2 = 10^{3+2} = 10^5$. Similarly $2 \times 10^8 \times 3 \times 10^7 = 2 \times 3 \times 10^8 \times 10^7 = 6 \times 10^{15}$. In general, $10^a \times 10^b = 10^{a+b}$.

Division works similarly, except that we subtract the exponents. Thus, $10^5/10^3 = 10^{5-3} = 10^2$. In general, $10^a/10^b = 10^{a-b}$.

The last operations we need to consider are raising a power-of-ten number to a power and taking a root. In raising a number to a power, we multiply the powers. Thus, $(10^3)^4 = 10^{3 \times 4} = 10^{12}$. Care must be used if we have a number like $(4 \times 10^2)^3$. Both the 4 and the 10^2 are raised to a power, so the result is $4^3 \times (10^2)^3 = 64 \times 10^{2 \times 3} = 64 \times 10^6 = 6.4 \times 10^7$.

Taking a root is equivalent to raising a number to a fractional power. Thus, the square root of a number is the number to the $\frac{1}{2}$ power. The cube root is the number to the $\frac{1}{3}$ power, and so forth. For example, $\sqrt{100} = (100)^{1/2} = (10^2)^{1/2} = 10^1 = 10$.

Metric Prefixes

You will find it easier to use scientific notation with the metric system if you remember the following prefixes (and their abbreviations), which denote powers of ten for very small to very large quantities:

milli (m)	$= 10^{-3}$	$= 1$ thousandth	kilo (k)	$= 10^3 = 1$ thousand
micro (μ)	$= 10^{-6}$	$= 1$ millionth	mega (M)	$= 10^6 = 1$ million
nano (n)	$= 10^{-9}$	$= 1$ billionth	giga (G)	$= 10^9 = 1$ billion

Solving Distance, Velocity, Time (d, V, t) Problems

A large number of problems in this book (and in science in general) involve the motion of something. In such problems, we often know two of the three quantities (d, V, t) and want to know the third. For example, we have something moving at a speed V and want to know how far it will travel in a time t. Or we know that something travels with a speed V and want to find out how long it takes for the object to travel a distance d. We all can solve such problems in our heads if the motion involves automobiles. For example, if it is 150 miles to the city and we travel at 50 miles per hour, how long does it take to get there? Or how far can we drive in 2 hours if we are traveling at 45 miles per hour? Because we all do such problems routinely, you might find it easier to think of astronomical problems in terms of cars. Regardless of your approach, the method of solution is simple.

Begin by making a simple sketch of what is happening. Draw some arrows to indicate the motion. Label the quantities known and put question marks beside the things you want to find. Then write out the basic relation $d = V \times t$. If you want to find d and know V and t, just multiply them for the answer. If you want to find the time and are given V and d, solve for t by dividing both sides by V to get $t = d / V$. If you want the velocity, divide d by t. In some problems, the motion may be in a circle of radius R. In that case, the distance traveled will be related to the circumference of the circle, $2\pi R$. For such cases, you may need to use the expression $d = 2\pi R$.

In most problems, you will find that it is extremely helpful to write the units of the quantities in the equation. For example, suppose you are asked how long it takes to travel 1500 km at a velocity of 30 kilometers per hour. First write out $d = V \times t$. Then rewrite it as

$$d/V = t.$$

Next insert the quantities so that

$$t = \frac{1500 \text{ km}}{30 \text{ km/hr}} = 50 \text{ hr.}$$

Note that the units of km cancel out and leave us with units of hours, as the problem requires.

Some Useful Formulas

Distance = velocity × time ($d = V \times t$)

Circumference of circle = $2\pi R$

Area of a circle = πR^2

Surface area of a sphere = $4\pi R^2$

Volume of a sphere = $\frac{4}{3}\pi R^3$

Table A.1   Physical and Astronomical Constants

Physical Constants	Luminosity (L$_\odot$)
Velocity of light (c)	$= 2.99792458 \times 10^8$ m/sec $\approx 3 \times 10^5$ km/sec
Gravitational constant (G)	$= 6.67259 \times 10^{-11}$ m$^3 \cdot$kg$^{-1} \cdot$sec^{-2} (or newton$\cdot$m$^2 \cdot$kg^{-2})
Planck's constant (h)	$= 6.62608 \times 10^{-34}$ joule$\cdot$sec
Mass of hydrogen atom (M_H)	$= 1.6735 \times 10^{-27}$ kg
Mass of electron (M_e)	$= 9.1094 \times 10^{-31}$ kg
Stefan-Boltzmann constant (σ)	$= 5.6705 \times 10^{-8}$ watts$\cdot$m$^{-2} \cdot$deg^{-4}
Constant in Wien's law ($T \times \lambda_{max}$)	$= 2.8978 \times 10^6$ K$\cdot$nm
Astronomical Constants	
Astronomical unit (AU)	$= 1.495978706 \times 10^{11}$ m $\approx 1.5 \times 10^8$ km
Light-year (ly)	$= 9.4605 \times 10^{15}$ m $= 9.4605 \times 10^{12}$ km $= 6.324 \times 10^4$ AU
Parsec (pc) $= 3.26$ ly	$= 3.085678 \times 10^{16}$ m $= 3.085678 \times 10^{13}$ km $= 206,265$ AU
Year (synodic or "tropical")	$= 365.2422$ days $= 3.1557 \times 10^6$ sec
Sidereal year	$= 365.256366$ days $= 3.155815 \times 10^7$ sec
Mass of Earth ($M_\oplus$)	$= 5.974 \times 10^{24}$ kg
Mass of Sun ($M_\odot$)	$= 1.989 \times 10^{30}$ kg
Equatorial radius of Earth ($R_\oplus$)	$= 6378.0$ km
Radius of Sun ($R_\odot$)	$= 6.96 \times 10^8$ m $= 6.96 \times 10^5$ km
Luminosity of Sun ($L_\odot$)	$= 3.83 \times 10^{26}$ watts
Hubble's constant (H)	≈ 70 km/sec per Mpc

Table A.2   Conversion Between American and Metric Units

Length				
	1 km	= 1 kilometer	= 1000 meters	= 0.6214 mile
	1 m	= 1 meter	= 1.094 yards	= 39.37 inches
	1 cm	= 1 centimeter	= 0.01 meter	= 0.3937 inch
	1 μm	= 1 micrometer	$= 10^{-6}$ meter	$= 10^{-4}$ cm $= 3.93 \times 10^{-5}$ inch
	1 nm	= 1 nanometer	$= 10^{-9}$ meter	$= 10^{-7}$ cm $= 3.93 \times 10^{-8}$ inch
	1 mile	= 1.6093 km		
	1 inch	= 2.5400 cm		
Mass	1 metric ton	$= 10^6$ grams	= 1000 kg	$= 2.2046 \times 10^3$ lb
	1 kg	= 1000 grams	= 2.2046 lb	
	1 g	= 1 gram	= 0.0022046 lb	= 0.0353 oz
	1 lb	= 0.4536 kg		
	1 oz	= 28.3495 gm		

Table A.3   Physical Properties of the Planets

Name	Equatorial Radius (Earth units)	(km)	Mass (Earth units)	(kg)	Average Density (gm/cm³)
Mercury	0.383	2440	0.055	3.30×10^{23}	5.43
Venus	0.949	6052	0.815	4.87×10^{24}	5.24
Earth	1.00	6378	1.00	5.97×10^{24}	5.52
Mars	0.532	3396	0.107	6.42×10^{23}	3.94
Ceres (dwarf planet)	0.076	487	0.00016	9.43×10^{20}	2.08
Jupiter	11.21	71,492	317.9	1.90×10^{27}	1.33
Saturn	9.45	60,268	95.16	5.68×10^{26}	0.69
Uranus	4.01	25,559	14.54	8.68×10^{25}	1.27
Neptune	3.88	24,764	17.15	1.02×10^{26}	1.64
Pluto (dwarf planet)	0.181	1186	0.0021	1.31×10^{22}	1.87
Haumea (dwarf planet)	0.11	~700*	0.00067	4.01×10^{21}	~2.8
Makemake (dwarf planet)	0.12	~750	0.0005	$\sim 3 \times 10^{21}$	~2
Eris (dwarf planet)	0.182	1163	0.0028	1.67×10^{22}	2.5

* Haumea is highly elongated, with its long axis about twice its shortest axis.

Table A.4   Orbital Properties of the Planets

Name	Distance from Sun* (AU)	(10⁶ km)	Period (Years)	(Days)	Inclination of Orbit†	Eccentricity of Orbit‡
Mercury	0.387	57.9	0.2409	87.97	7.00	0.206
Venus	0.723	108.2	0.6152	224.70	3.39	0.007
Earth	1.00	149.6	1.0	365.26	0.00	0.017
Mars	1.524	227.9	1.8809	686.98	1.85	0.093
Ceres (dwarf planet)	2.766	414.7	4.5990	1679.82	10.59	0.080
Jupiter	5.204	778.6	11.8622	4332.59	1.31	0.048
Saturn	9.582	1433.5	29.4577	10,759.22	2.49	0.056
Uranus	19.20	2872.5	84.011	30,685.4	0.77	0.046
Neptune	30.05	4495.1	164.79	60,189	1.77	0.010
Pluto (dwarf planet)	39.48	5906	247.68	90,465	17.15	0.248
Haumea (dwarf planet)	43.1	7710	283.3	103,500	28.2	0.195
Makemake (dwarf planet)	53.1	7940	309.9	113,200	29.0	0.159
Eris (dwarf planet)	67.7	10,120	557.4	203,600	44.2	0.442

* Semimajor axis of the orbit.   † With respect to the ecliptic.   ‡ Eccentricity ranges from 0 for a circular orbit to a maximum of 1 for the most extremely elongated orbit.

Table A.5 Larger Satellites of the Planets and Dwarf Planets*

Primary / Satellite	Radius[†] (km)	Distance from Planet (10^3 km)	Orbital Period[‡] (days)	Mass[§] (10^{15} kg)	Density (gm/cm³)
Earth					
Moon	1738	384.4	27.322	73,490,000	3.34
Mars					
Phobos	13 × 11 × 9	9.38	0.319	11	2.2
Deimos	8 × 6 × 5	23.5	1.263	1.8	1.7
Jupiter					
Metis	30 × 20 × 17	127.69	0.295	~55	
Adrastea	13 × 10 × 8	128.98	0.298	~6	
Amalthea	131 × 73 × 67	181	0.498	2080	0.86
Thebe	58 × 49 × 42	222	0.6745	~650	
Io	1822	422	1.769	89,320,000	3.53
Europa	1561	671	3.551	48,000,000	3.01
Ganymede	2631	1070	7.155	148,200,000	1.94
Callisto	2410	1883	16.689	107,600,000	1.83
Leda	5	11,188	241.8	~1	
Himalia	85	11,452	250.4	~3300	
Lysithea	12	11,740	259.9	~9	
Elara	40	11,778	261.1	~350	
Ananke	10	21,450	642R	~5	
Carme	15	23,200	692R	~18	
Pasiphae	18	23,600	742R	~32	
Sinope	14	24,100	762R	~15	
Saturn					
Pan	17 × 16 × 10	133.58	0.574	5	0.42
Atlas	20 × 18 × 9	137.67	0.602	7	0.46
Prometheus	68 × 40 × 30	139.35	0.613	160	0.48
Pandora	52 × 41 × 32	141.7	0.629	140	0.49
Epimetheus	65 × 57 × 53	151.42	0.694	530	0.64
Janus	102 × 93 × 76	151.47	0.695	1900	0.63
Mimas	199	185.52	0.942	37,900	1.15
Enceladus	252	238.02	1.37	108,000	1.61

(continued)

Table A.5	Larger Satellites of the Planets and Dwarf Planets (*Continued*)				
Primary / Satellite	Radius[†] (km)	Distance from Planet (10^3 km)	Orbital Period[‡] (days)	Mass[§] (10^{15} kg)	Density (gm/cm³)
Tethys	531	294.66	1.888	618,000	0.99
Calypso	15 × 12 × 7	294.66	1.888	4	1
Telesto	16 × 12 × 10	294.66	1.888	7	1
Dione	561	377.4	2.737	1,100,000	1.48
Helene	22 × 19 × 13	377.4	2.737	30	1.3
Rhea	764	527.04	4.518	2,306,500	1.24
Titan	2575	1221.8	15.945	134,550,000	1.88
Hyperion	180 × 133 × 102	1481.1	21.277	5600	0.55
Iapetus	734	3560.8	79.322	1,810,000	1.09
Kiviuq	~7	11,294	448.2	~2	
Ijiraq	~5	11,355	451.8	~1	
Phoebe	107	12,870	545.1R	8300	1.64
Paaliaq	~10	15,103	693	~5	
Albiorix	~13	16,227	774.6	~12	
Siarnaq	~16	17,777	884.9	~22	
Tarvos	~7	18,563	944.2	~2	
Ymir	~9	22,430	1254R	~4	
Uranus					
Cordelia	~20	49.75	0.336	~44	
Ophelia	~21	53.77	0.377	~51	
Bianca	~26	59.16	0.435	~96	
Cressida	40	61.77	0.465	~350	
Desdemona	~32	62.65	0.476	~180	
Juliet	~47	64.63	0.494	~570	
Portia	68	66.1	0.515	~1700	
Rosalind	~36	69.93	0.56	~250	
Cupid	~9	74.8	0.618	~4	
Belinda	40	75.25	0.624	~350	
Perdita	~10	76.42	0.638	~5	
Puck	81	86.01	0.764	~2900	

(continued)

Table A.5	Larger Satellites of the Planets and Dwarf Planets (*Continued*)				
Primary / Satellite	Radius† (km)	Distance from Planet (10³ km)	Orbital Period‡ (days)	Mass§ (10¹⁵ kg)	Density (gm/cm³)
Mab	~12	97.73	0.923	~9	
Miranda	236	129.4	1.413	66,000	1.2
Ariel	579	191.2	2.520	1,350,000	1.7
Umbriel	585	266.3	4.144	1,170,000	1.4
Titania	789	435.8	8.706	3,520,000	1.7
Oberon	761	582.6	13.463	3,010,000	1.6
Francisco	~11	48.2	0.296	~7	
Caliban	~36	7231	579.7R	~250	
Stephano	~16	20,901	2805.5R	~22	
Trinculo	~9	20,901	2805.5R	~4	
Sycorax	~75	12,179	1288.3R	~2300	
Margaret	~10	14,345	1687	~5	
Prospero	~25	16,256	1978.3R	~85	
Setebos	~24	17,418	2225.2R	~75	
Ferdinand	~10	20,901	2805.5R	~5	
Neptune					
Naiad	48 × 30 × 26	48.23	0.294	~200	
Thalassa	54 × 50 × 26	50.08	0.311	~380	
Despina	90 × 74 × 64	52.53	0.335	~2300	
Galatea	102 × 92 × 72	61.95	0.429	~3700	
Larissa	104 × 102 × 84	73.55	0.555	~4900	
S/2004N1	~9	105.283	0.94	~4	
Proteus	220 × 208 × 202	117.65	1.122	~50,000	
Triton	1353.4	354.76	5.877R	21,400,000	2.1
Nereid	170	5513.4	360.14	~27,000	
Halimede	~30	15,730	1879.7R	~150	
Sao	~20	22,420	2914.1	~40	
Laomedeia	~20	23,570	3167.9	~40	
Psamathe	~30	46,700	9115.9R	~150	
Neso	~20	48,390	9374.0R	~40	

(continued)

| Table A.5 | Larger Satellites of the Planets and Dwarf Planets (*Continued*) |

Primary / Satellite	Radius[†] (km)	Distance from Planet (10³ km)	Orbital Period[‡] (days)	Mass[§] (10¹⁵ kg)	Density (gm/cm³)
Pluto					
Charon	603	17.53	6.38718	1,520,000	1.65
Styx	~9	42	20.2	~5	
Nix	~45	48.71	24.9	~500	
Kerberos	~12	59	32.1	~10	
Hydra	~60	64.75	38	~1000	
Haumea					
Namaka	~80	25.657	18.278	~1800	
Hi'iaka	~150	49.88	49.462	17,900	
Eris					
Dysnomia	~150	37.4	15.77	18,400	

* Only satellites with a mass of at least 10¹⁵ kg or a radius of at least 5 km are included. Smaller objects are difficult to detect, but are steadily being discovered. A relatively up-to-date list is available at the website solarsystem.nasa.gov/planets. Click on the planet of interest and then click on the tab labeled "moons."

† Satellite radii marked with an approximate symbol (~) are estimated from their brightness. Some satellites have detailed measurements that show them to be irregular in shape. It the difference between their longest and smallest axes is larger than 10%, radial dimensions of the form $a \times b \times c$ are given.

‡ Satellites with retrograde orbits have an "R" after the orbital period.

§ Satellite masses marked with an approximate symbol (~) are estimated by assuming they have a density of 1.3 gm/cm³.

| Table A.6 | Meteor Showers |

Shower	Date of Peak	Hourly Rate	Radiant*			Associated Comet
			Constellation	R.A.	Dec.	
Quadrantids	Jan. 3–4	30–90	Boötes	15ʰ 24ᵐ	50°	2003 EH1
Lyrids	Apr. 22–23	10–15	Lyra	18ʰ 4ᵐ	33°	Thatcher
η Aquarids	May 7–8	10–20	Aquarius	22ʰ 24ᵐ	0°	Halley
δ Aquarids	July 28–29	15–20	Aquarius	22ʰ 36ᵐ	−10°	Marsden & Kracht
Perseids	Aug. 12–13	40–100	Perseus	3ʰ 4ᵐ	58°	Swift-Tuttle
Orionids	Oct. 21–22	15–20	Orion	6ʰ 20ᵐ	15°	Halley
Leonids	Nov. 17–19	10–100	Leo	10ʰ 12ᵐ	22°	Tempel-Tuttle
Geminids	Dec. 13–14	40–90	Gemini	7ʰ 28ᵐ	32°	Phaethon
Ursids	Dec. 22–23	10–15	Ursa Minor	14ʰ 40ᵐ	75°	Tuttle

* Meteors will appear to move away from the radiant, but they can be seen anywhere in the sky. Meteors from the shower will generally only be visible when the radiant is above the horizon, and they will be more numerous when the radiant is higher in the sky, but check for specific predictions at websites such as the American Meteor Society (www.amsmeteors.org). R.A. = right ascension; Dec. = declination.

Table A.7   The Brightest Stars*

Name	Apparent Visual Magnitude	Distance (ly)	Designation		Spectral Type	Luminosity Class	Luminosity ($L_\odot$)
Sun	−26.72	0.0000158			G2	V	1.00
Sirius	−1.46	8.58	α CMa	A	A1	V	26
Canopus	−0.74	310	α Car		F0	Ib–II	14,000
Rigel Kentaurus	−0.27	4.36	α Cen	A B	G2 K0	V V	1.5 0.5
Arcturus	−0.05	37	α Boo		K2	III	220
Vega	0.03	25	α Lyr		A0	V	59
Capella	0.08	42	α Aur	Aa Ab	G8 G1	III III	79 78
Rigel	0.10	860	β Ori		B8	Ia	110,000
Procyon	0.37	11.5	α CMi	A	F5	IV–V	7
Betelgeuse	0.42	640	α Ori		M1.5	Iab	120,000
Achernar	0.45	140	α Eri		B3	V	3200
Hadar	0.61	350	β Cen	A B	B1 B1	III III	40,000 40,000
Altair	0.77	16.7	α Aql		A7	V	11
Acrux	0.78	320	α Cru	Aa Ab B	B0.5 B3 B1	IV V V	25,000 7000 20,000
Aldebaran	0.85	65	α Tau	A	K5	III	400
Antares	0.96	550	α Sco	A	M1.5	Ib	60,000
Spica	1.04	250	α Vir	A B	B1 B2	III–IV V	12,000 1500
Pollux	1.14	34	β Gem		K0	III	46
Fomalhaut	1.16	25	α PsA		A3	V	17
Deneb	1.25	1400	α Cyg		A2	Ia	60,000
Mimosa	1.25	280	β Cru	A B	B0.5 B2	IV V	20,000 5000
Regulus	1.35	79	α Leo	A	B7	V	360
Adhara	1.50	405	ε CMa		B2	II	22,300
Castor	1.58	51	α Gem	Aa Ba	A1 A5	V V	37 13

* Many of these stars are actually multiple-star systems. Data is provided for companion stars that contribute a significant fraction of the visible light we see.

Table A.8   The Nearest Stars

Common Name	Distance (ly)	Spectral Type and Luminosity Class	Apparent Visual Magnitude	Absolute Visual Magnitude	Estimated Mass ($M_\odot$)
Sun	0.0000158	G2 V	−26.72	4.83	1.00
Proxima Centauri	4.24	M5.5 V	11.09	15.48	0.11
Alpha Centauri	4.36	G2 V K0 V	0.01 1.34	4.38 5.71	1.14 0.92
Barnard's Star	5.98	M3.5 V	9.57	13.25	0.16
Wolf 359	7.78	M5.5 V	13.53	16.64	0.09
Lalande 21185	8.29	M2 V	7.47	10.44	0.46
Sirius	8.58	A1 V DA2 (white dwarf)	−1.46 8.44	1.44 11.34	2.02 0.98
BL Ceti	8.73	M5.5 V M6 V	12.61 13.06	15.47 15.93	0.11 0.10
Ross 154	9.67	M3.5 V	10.44	13.08	0.17
Ross 248	10.31	M5.5V	12.29	14.79	0.12
Epsilon Eridani	10.48	K2 V	3.73	6.20	0.85
Lacaille 9352	10.69	M1 V	7.34	9.76	0.53
Ross 128	10.94	M4 V	11.16	13.53	0.16
EZ Aquarii	11.27	M5 V M V M V	13.03 13.27 15.07	15.33 15.58 17.37	0.16 0.11 0.11
61 Cygni	11.40	K5 V K7 V	5.20 6.03	7.48 8.31	0.7 0.63
Procyon	11.44	F5 IV–V DA (white dwarf)	0.37 10.70	2.65 12.98	1.42 0.60
Struve 2398	11.49	M3 V M3.5 V	8.90 9.69	11.17 11.96	0.35 0.26
Groombridge 34	11.65	M1.5 V M3.5 V	8.08 11.06	10.31 13.30	0.49 0.16
Epsilon Indi	11.81	K4 V T1 V (brown dwarf) T6 V (brown dwarf)	4.68 24.12 ~26.6	6.98 26.42 ~28.9	0.77 0.03 0.03
DX Cancri	11.83	M6 V	14.81	17.01	0.09
Tau Ceti	11.90	G8.5 V	3.49	5.68	0.92
Gliese-Jahreiss 1061	11.99	M5 V	13.09	15.26	0.11
YZ Ceti	12.12	M4 V	12.1	14.25	0.13
Luyten's Star	12.25	M3.5 V	9.85	11.98	0.26

Data based mainly on RECONS (Research Consortium on Nearby Stars) website.

Table A.9   Properties of Main-Sequence Stars*

Spectral Type	Luminosity ($L_\odot$)	Temperature (K)	Mass ($M_\odot$)	Radius ($R_\odot$)
O5	790,000	44,500	60	12.0
B0	52,000	30,000	17.5	7.4
B5	830	15,400	5.9	3.9
A0	54	9520	2.9	2.4
A5	14	8200	2.0	1.7
F0	6.5	7200	1.6	1.5
F5	3.2	6440	1.3	1.3
G0	1.5	6030	1.05	1.1
G5	0.79	5770	0.92	0.92
K0	0.42	5250	0.79	0.85
K5	0.15	4350	0.67	0.72
M0	0.08	3850	0.51	0.6
M5	0.01	3240	0.21	0.27
M8	0.001	2640	0.06	0.1

* Authorities differ substantially on many of the above values, especially at the upper- and lower-mass values. Note also that the values are generally not consistent with the Stefan-Boltzmann law.

Table A.10   Known and Suspected Members of the Local Group of Galaxies*

Name of Galaxy	Right Ascension (hours and minutes)	Declination (degrees and minutes)	Galaxy Type	Distance (Mpc)	Diameter (kpc)	Visual Apparent Magnitude	Approximate Luminosity ($10^6 L_\odot$)
WLM	0 02	−15 28	Irr	1.0	3	11	50
IC10	0 20	+59 18	Irr	0.7	1	11	300
Cetus dw	0 26	−11 02	E	0.8	—	14	1
NGC 147	0 33	+48 30	E5	0.7	2	10	100
Andromeda III	0 35	+36 30	E5	0.8	1	15	1
NGC 185	0 39	+48 20	E3	0.7	2	9	150
NGC 205 (M110)	0 40	+41 41	E5	0.8	5	8	300
Andromeda VIII	0 42	+40 37	E	0.8	14	9	150
M32	0 43	+40 52	E2	0.8	2	8	300
M31	0 43	+41 16	Sb	0.8	40	3	25,000
Andromeda I	0 46	+38 02	E3	0.8	0.5	14	4
SMC	0 53	−72 50	Irr	0.06	5	2	600
Sculptor dw	1 00	−33 42	E3	0.09	1	9	1
LGC 3	1 04	+21 53	Irr	0.8	0.5	18	1

(continued)

Table A.10   Known and Suspected Members of the Local Group of Galaxies (*Continued*)

Name of Galaxy	Right Ascension (hours and minutes)	Declination (degrees and minutes)	Galaxy Type	Distance (Mpc)	Diameter (kpc)	Visual Apparent Magnitude	Approximate Luminosity ($10^6 L_\odot$)
IC 1613	1 05	+02 07	Irr	0.7	3	9	100
Andromeda V	1 10	+47 38	E	0.8	—	15	1
Andromeda II	1 16	+33 25	E2	0.7	0.7	13	4
M33	1 34	+30 40	Sc	0.8	16	6	3000
Phoenix dw	1 51	−44 27	Irr	0.4	0.6	12	1
Fornax dw	2 40	−34 27	E3	0.14	0.7	9	15
UGCA 92	4 32	+63 36	Irr	1.44	0.6	14	5
LMC	5 23	−69 45	Irr	0.05	0.6	1	2000
Carina dw	6 42	−50 58	E4	0.10	0.7	16	1
Canis Major dw	7 35	−28 00	Irr	0.08	220	—	—
Leo T	9 35	+17 03	Irr	0.420	0.34	16	0.04
Leo A	9 59	+30 45	Irr	0.7	1	12	3
Sextans B	10 00	+05 20	Irr	1.41	1.6	12	30
NGC 3109	10 03	−26 09	Irr	1.38	5	10	100
Antlia dw	10 04	−27 20	E3	1.41	0.6	15	2
Leo I	10 08	+12 18	E3	0.2	0.7	10	4
Sextans A	10 11	−04 43	Irr	1.60	1.8	12	35
Sextans dw	10 13	−01 37	E	0.09	1	12	1
Leo II	11 13	+22 09	E0	0.2	0.7	12	1
GR 8	12 59	+14 13	Irr	2.4	0.4	14	8
Ursa Minor dw	15 09	+67 13	E5	0.06	0.5	11	1
Draco dw	17 20	+57 55	E3	0.08	0.8	10	1
Milky Way	17 46	−29 00	SBc	0.01	40	—	20,000
SagDEG	18 55	−30 29	Irr	0.03	—	10	30
SagDIG	19 30	−17 41	Irr	1.2	1.5	15	5
NGC 6822	19 45	−14 48	Irr	0.5	2	9	200
Aquarius dw	20 47	−12 51	Irr	1.0	0.6	14	2
Tucanae dw	22 42	−64 25	E4	0.9	0.5	15	1
UGCA 438	23 26	−32 33	Irr	1.4	0.5	14	5
Andromeda VII	23 27	+50 42	E	0.7	0.5	13	5
Pegasus dw	23 29	+14 45	Irr	0.8	1	13	7
Andromeda VI	23 52	+24 35	E	0.8	0.9	14	3

* Listing based largely on the SEDS Messier compilation of Local Group data (http://messier.seds.org/more/local.html).

Table A.11   The Brightest Galaxies Beyond the Local Group*

Name of Galaxy	Right Ascension (hours and minutes)	Declination (degrees and minutes)	Galaxy Type	Distance (Mpc)	Diameter (kpc)	Visual Apparent Magnitude	Approximate Luminosity ($10^6 L_\odot$)
NGC0253	00 48	−25 17	Sc	2.6	21	7.3	17,000
NGC0300	00 55	−37 41	Scd	2.2	13	8.2	2700
NGC1068 (M77)	02 43	−00 01	Sb	13.8	28	8.7	60,000
NGC1291	03 17	−41 06	S0-a	9.2	27	8.7	25,000
IC 342	03 47	+68 06	SBc	2.5	15	8.5	51,000
NGC2403	07 37	+65 36	Sc	3.4	23	8.4	6800
NGC3031 (M81)	09 56	+69 04	Sab	3.7	23	6.9	33,000
NGC3034 (M82)	09 56	+69 41	Irr	3.7	10	8.3	15,000
NGC4258 (M106)	12 19	+47 18	Sb	7.4	41	8.4	34,000
NGC4472 (M49)	12 30	+08 00	E	16.3	43	8.3	110,000
NGC4486 (M87)	12 31	+12 24	E	15.6	32	8.6	81,000
NGC4594 (M104)	12 40	−11 37	Sa	9.2	24	8.1	58,000
NGC4736 (M94)	12 51	+41 07	Sab	5.2	11	8.1	14,000
NGC4826 (M64)	12 57	+21 41	Sab	7.4	17	8.4	30,000
NGC4945	13 05	−49 28	SBc	4.6	29	8.6	37,000
NGC5055 (M63)	13 16	+42 02	Sbc	7.7	21	8.6	25,000
NGC5128 (Cen A)	13 25	−43 01	S0	3.7	32	6.7	32,000
NGC5194 (M51)	13 30	+47 12	Sbc	8.0	26	8.3	33,000
NGC5236 (M83)	13 37	−29 52	Sc	4.6	15	7.3	28,000
NGC5457 (M101)	14 03	+54 21	Sc	7.4	47	7.8	34,000
NGC6744	19 10	−63 51	Sb	10.7	65	8.4	62,000

* Galaxies in this table were selected from the HyperLeda database (http://leda.univ-lyon1.fr) based on their total visual apparent magnitude.

Glossary

A

absorption the process in which light or other electromagnetic radiation gives up its energy to an atom or molecule. For example, ozone in our atmosphere absorbs ultraviolet radiation.

absorption-line spectrum a spectrum showing dark lines at some narrow color regions (wavelengths). The lines are formed by atoms absorbing light, which lifts their electrons to higher orbits.

acceleration a change in an object's velocity (either its speed or its direction).

accretion the addition of matter to a body. Examples are gas falling onto a star and asteroids colliding and sticking together.

accretion disk a nearly flat disk of gas or other material held in orbit around a body by its gravity.

achondrite a meteorite lacking chondrules, associated with larger bodies whose gravity and internal heating has caused them to differentiate.

active galaxy a galaxy whose central region emits abnormally large amounts of electromagnetic radiation from a small volume. Examples are radio galaxies, Seyfert galaxies, and quasars.

adaptive optics a technique for adjusting a telescope's mirror or other optical parts to compensate for atmospheric distortions, such as seeing, thereby giving a sharper image.

æther a substance at one time proposed to be the substance that made it possible for light waves to travel through otherwise empty space.

AGN active galactic nucleus. The core of an active galaxy.

alpha particle a helium nucleus: two protons plus (usually) two neutrons.

altitude an object's angular distance above the horizon.

amino acid a carbon-based molecule used by living organisms to build protein molecules.

angstrom unit a unit of length used in describing wavelengths of radiation and the sizes of atoms and molecules. One angstrom = 10^{-10} meters.

angular momentum a measure of an object's tendency to keep rotating and to maintain its orientation. Mathematically, it depends on the object's mass, M, radius, R, and rotational velocity, V, and is proportional to MVR.

angular size a measure of how large an object *looks* to you. It is defined as the angle between lines drawn from the observer to opposite sides of an object. For example, the angular diameter of the Moon is about 0.5°.

annular eclipse an eclipse in which the body in front does not completely cover the other. In an annular eclipse of the Sun, a bright ring of the Sun's disk remains visible around the black disk of the Moon. We therefore see a ring (annulus) of light around the Moon.

anthropic principle the principle that the properties we observe the Universe to possess are limited to those that make our existence possible.

antimatter a type of matter that, if brought into contact with ordinary matter, annihilates it, leaving nothing but energy. The positron is the antimatter analog of the electron. The antiproton is the antimatter analog of the proton. Antimatter is observed in cosmic rays and can be created from energy in the laboratory.

aphelion the point in an orbit where a body is farthest from the Sun.

association a loose grouping of young stars and interstellar matter.

asterism an easily identified grouping of stars, often part of a larger constellation. For example, the Big Dipper.

asteroid a small, generally rocky, solid body orbiting the Sun and ranging in diameter from a few meters to hundreds of kilometers.

asteroid belt a region between the orbits of Mars and Jupiter in which most of the Solar System's asteroids are located.

astronomical unit (AU) a distance unit based on the average distance of the Earth from the Sun.

atmospheric window a wavelength band in which our atmosphere absorbs little radiation. For example, on Earth the visible window ranges from about 300 to 700 nanometers, allowing the light we can see with our eyes to pass through the atmosphere.

atom a submicroscopic particle consisting of a nucleus and orbiting electrons. The smallest unit of a chemical element.

aurora the light emitted by atoms and molecules in the upper atmosphere. This light is a result of magnetic disturbances caused by the solar wind. Often called the northern or southern lights.

autumnal equinox the autumn equinox in the Northern Hemisphere. Fall begins on the autumnal equinox, which is on or near September 22.

averted vision looking slightly to one side of a dim object so that you see it slightly away from the center of your field of view. This allows you to see a faint object better, although at a sacrifice of sharpness.

azimuth a coordinate for locating objects on the sky. Azimuth is the angle measured eastward from due north to the point on the horizon below the object.

B

barred spiral galaxy a galaxy in which the spiral arms wind out from the ends of a central bar rather than from the nucleus.

belt a dark band in a Jovian planet's atmosphere, encircling the planet. Gas in the belts appears to be descending and warming.

Big Bang the event that, according to many astronomical theories, created the Universe. It occurred about 13.7 billion years ago and generated the expanding motion that we observe today.

binary stars two or more stars in orbit around each other, held together by their mutual gravity.

bipolar flow narrow columns of high-speed gas ejected by a protostar in two opposite directions.

blackbody an object that is an ideal radiator when hot and a perfect absorber when cool. It absorbs all radiation that falls upon it, reflecting no light; hence, it appears black. Stars are approximately blackbodies. The radiation emitted by blackbodies obeys Wien's law and the Stefan-Boltzmann law.

black hole an object whose gravitational attraction is so strong that its escape velocity equals the speed of light, preventing light or any radiation or material body from leaving its "surface."

BL Lac object a type of active galaxy named for the peculiar galaxy BL Lac. These objects generally are strong radio sources, and their visible light varies rapidly and erratically.

blueshift a shift in the wavelength of electromagnetic radiation to a shorter wavelength. For visible light, this implies a shift toward the blue end of the spectrum. The shift can be caused by the motion of a source of radiation toward the observer or by the motion of an observer toward the source. For example, the spectrum lines of a star moving toward the Earth exhibit a blueshift. *See also* **Doppler shift.**

Bode's rule a numerical expression for the approximate distances of most of the planets from the Sun.

Bok globule small, dark, interstellar cloud, often approximately spherical. Many globules are the early stages of protostars.

brown dwarf a star that has a mass too low for it to begin nuclear fusion.

bulge the dense, central region of a spiral galaxy.

C

carbonaceous chondrite a type of meteorite containing many tiny spheres (chondrules) of rocky or metallic material stuck together by carbon-rich material.

CCD charge-coupled device. An electronic device that records the intensity of light falling on it. CCDs have replaced film in most astronomical applications.

celestial equator an imaginary line on the celestial sphere lying exactly above the Earth's equator. It divides the celestial sphere into northern and southern hemispheres.

celestial pole an imaginary point on the celestial sphere directly above the Earth's North or South Pole.

celestial sphere an imaginary sphere surrounding the Earth representing the sky. Ancient astronomers pictured celestial objects as attached to it.

Cepheid a class of yellow-giant pulsating stars. Their pulsation periods range from about 1 day to about 70 days. Cepheids can be used to determine distances. *See also* **standard candle.**

Chandrasekhar limit the maximum mass of a white dwarf above which it collapses. Approximately 1.4 solar masses. Named for the astronomer who first calculated that such a limit exists.

chondrite a meteorite containing small spherical grains called chondrules.

chondrule a small spherical grain embedded in a meteorite.

chromosphere the lower part of the Sun's outer atmosphere that lies directly above the Sun's visible surface (photosphere).

cluster a group of objects (stars, galaxies, and so forth) held together by their mutual gravitational attraction.

CNO cycle/process a reaction involving carbon, nitrogen, and oxygen (C, N, and O) that fuses hydrogen into helium and releases energy. The process begins with a hydrogen nucleus fusing with a carbon nucleus. Subsequent steps involve nitrogen and oxygen. The carbon, nitrogen, and oxygen act as catalysts and are released at the end of the process to start the cycle again. The CNO cycle is the dominant process for generating energy in main-sequence stars that are hotter and more massive than the Sun.

coma the gaseous atmosphere surrounding the head of a comet.

comet a small body in orbit around the Sun, consisting of a tiny, icy core and a tail of gas and dust. The tail forms only when the comet is near the Sun.

compact stars very dense stellar remnants or "dead" stars whose radii are much smaller than the Sun's. These stars include white dwarfs, neutron stars, and black holes.

condensation conversion of free gas atoms or molecules into a liquid or solid. A snowflake forms in our atmosphere when water vapor condenses into ice.

conjunction the appearance of two astronomical objects in approximately the same direction on the sky. For example, if Mars and Jupiter happen to appear near each other on the sky, they are said to be in conjunction. *Superior conjunction* refers to a planet that is approximately in line with the Sun but on the far side of the Sun from the Earth. *Inferior conjunction* refers to a planet that lies approximately between the Sun and the Earth.

conservation of angular momentum a principle of physics stating that the angular momentum of a rotating body remains constant unless forces act to speed it up or slow it down. Mathematically, conservation of angular momentum states that MVR is a constant, where M is the mass of a body moving with a velocity, V, in a circle of radius, R. One extremely important consequence of this principle is that if a rotating body shrinks, its rotational velocity must increase.

conservation of energy a principle of physics stating that energy is never created or destroyed, although it may change its form. For example, energy of motion may change into energy of heat.

constellation a grouping of stars in the night sky. Astronomers divide the sky into 88 constellations.

continuous spectrum a spectrum with neither dark absorption nor bright emission lines. The intensity of the radiation in such a spectrum changes smoothly from one wavelength to the next.

convection the rising and sinking motions in a liquid or gas that carry heat upward through the material. Convection is easily seen in a pan of heated soup on a stove.

convection zone the region immediately below the Sun's visible surface in which its heat is carried by convection.

Coriolis effect a deflection of a moving object caused by its motion across the surface of a rotating body. The Coriolis effect makes storms on Earth spin, generates large-scale wind systems, and creates cloud belts on many of the planets.

corona the outer, hottest part of the Sun's atmosphere.

coronal hole a low-density region in the Sun's corona. The solar wind may originate in these regions.

cosmic horizon the maximum distance one can see out into the Universe at a given time. The horizon lies at a distance in light-years approximately equal to the age of the Universe in years.

cosmic microwave background (CMB) radiation from the young Universe that began traveling through space after hot gas from the Big Bang expanded and cooled enough to become transparent. The radiation is visible in all directions and appears to have a temperature of only 2.73 K because of the redshift caused by the expansion of the Universe.

cosmic rays extremely energetic particles (protons, electrons, and so forth) traveling at nearly the speed of light. Some rays are emitted by the Sun, but most come from more-distant sources, perhaps exploding supernovas.

cosmological constant a term in the equations that Einstein developed to describe the expansion of the Universe. The cosmological constant has the effect of a repulsive "force" opposing gravity. *See also* **dark energy.**

cosmological principle the hypothesis that, on average, the Universe looks the same to every observer, no matter where he or she is located in it.

cosmology the study of the structure and evolution of the Universe.

crater a circular pit, generally with a raised rim and sometimes with a central peak. Crater diameters on the Moon range from centimeters to several hundred kilometers. Most craters on bodies such as the Moon are formed by the impact of solid bodies, such as asteroids.

critical density the minimum overall density of the Universe needed to provide enough gravity to make it eventually stop expanding and collapse. If the density is less than or equal to the critical density, the Universe will expand forever.

crust the rigid surface of a planet, moon, or other solid body.

curvature of space the bending of space by a mass, as described according to Einstein's general theory of relativity. Black holes bend the space around them, curving it so that the region within the black hole is cut off from the rest of the Universe. The Universe too may be curved in such a way as to make its volume finite.

curvature of the Universe the overall curvature of space in the Universe produced by a combination of all its matter and energy. The Universe appears to be very close to the critical density, at which there is no overall curvature, a condition astronomers describe as a "flat" Universe.

D

dark adaption the process by which the eye changes to become more sensitive to dim light.

dark energy a form of energy (or a cosmological constant) detected by its effect on the expansion of the Universe. It causes the expansion to speed up. The nature and properties of dark energy are unknown.

dark matter matter that emits no detectable radiation but whose presence can be deduced by its gravitational attraction on other bodies.

dark nebula a dense cloud of dust and gas in interstellar space that blocks the light from background stars.

daughter atoms the atoms produced by the decay of a radioactive element. For example, uranium decays into lead. These lead atoms are daughter atoms.

daylight saving time the time kept during summer months by setting the clock ahead one hour. This gives more hours of daylight after the workday.

declination one part of a coordinate system for locating objects in the sky north or south of the celestial equator. Declination is analogous to latitude on the Earth's surface.

degeneracy pressure the pressure created in a dense gas by the interaction of its electrons. Degeneracy pressure does not depend on temperature.

degenerate gas an extremely dense gas in which the electrons and nuclei are tightly packed. The pressure of a degenerate gas does not depend on its temperature.

density the mass of a body or region divided by its volume.

density-wave model a theory to account for the spiral arms of galaxies. According to the theory, waves of higher density traveling through the disk of a galaxy pull stars and interstellar gas into a spiral pattern.

deuterium a form of hydrogen in which the nucleus contains a neutron in addition to a proton. Sometimes called "heavy hydrogen."

differential gravitational force the difference between the gravitational forces exerted on an object at two different points. The effect of this force is to stretch the object. Such forces create tides and, if strong enough, may break up an astronomical object. *See also* **Roche limit.**

differentiation the separation of previously mixed materials inside a planet or other object. This is the same separation that occurs when a dense material, such as iron, settles to the planet's core while low-density material floats to the surface.

diffraction a bending of the path of light (or other electromagnetic waves) as it passes through an opening or around an obstacle. Diffraction limits the ability to distinguish fine details in images.

disk the flat, round portion of a galaxy. The Sun lies in the disk of the Milky Way.

dispersion the spreading of light or other electromagnetic radiation into a spectrum. A rainbow is an example of dispersion of light caused by raindrops.

DNA deoxyribonucleic acid. The complex molecule that encodes genetic information in all organisms here on Earth.

Doppler shift the change in the observed wavelength of radiation caused by the motion of the emitting body or the observer. The shift is an increase in the wavelength if the source and observer move apart and a decrease in the wavelength if the source and observer approach. *See also* **redshift** and **blueshift.**

Drake equation a formula, named after the astronomer who proposed it, used to estimate the number of civilizations that might be present in the Milky Way that are capable of communicating with us.

dust tail a comet tail containing dust that reflects sunlight. The dust in a comet tail is expelled from the nucleus of the comet.

dwarf a small, dim star.

dwarf planet an object that orbits the Sun and is massive enough that its gravity compresses it into an approximately spherical shape. However, it has not swept its orbital region clear of other objects of comparable mass.

dynamo model a proposed physical process for generating magnetic fields in astronomical bodies. The process involves the generation of electric currents from an interaction between rotation and convection in an electrically conductive liquid.

E

eclipse the blockage of light from one astronomical body caused by the passage of another between it and the observer. The shadow of one astronomical body falling on another. For example, the passage of the Moon between the Earth and Sun can block the Sun's light and cause a solar eclipse.

eclipse seasons the times of year, separated by about 6 months, when eclipses are possible. At any given eclipse season, both a solar eclipse and a lunar eclipse generally occur.

eclipsing binary a binary star pair in which one star periodically passes in front of the other, totally or partially blocking the background star from view as seen from Earth.

ecliptic the path that the Sun appears to make around the celestial sphere as the Earth moves along its orbit. The path gets its name because eclipses can occur only when the Moon crosses the ecliptic.

electric charge the electrical property of objects that causes them to attract or repel one another. A charge may be either positive or negative.

electric force the force generated by electric charges. It is attractive between unlike charges (+−) but repulsive between like charges (++ or −−).

electromagnetic force the force arising between electrically charged particles or between charges and magnetic fields. Forces between magnets are a special case of this force. This force holds electrons to the nucleus of atoms, makes moving charges spiral around magnetic field lines, and deflects a compass needle.

electromagnetic radiation a general term for any kind of electromagnetic wave.

electromagnetic spectrum the assemblage of all wavelengths of electromagnetic radiation. The spectrum includes the following wavelengths, from long to short: radio, microwave, infrared, visible light, ultraviolet, X rays, and gamma rays.

electromagnetic wave a wave consisting of alternating electric and magnetic energy. Ordinary visible light is an electromagnetic wave, and the wavelength determines the light's color.

electron a low-mass, negatively charged subatomic particle. Electrons orbit the atomic nucleus but may at times be torn free. *See also* **ionization.**

element a fundamental substance, such as hydrogen, carbon, or oxygen, that cannot be broken down into a simpler chemical substance. Approximately 100 elements occur in nature.

ellipse a geometric figure related to a circle but flattened into an oval shape.

elliptical galaxy a galaxy in which the stars smoothly fill an ellipsoidal volume. Abbreviated E galaxy. The stars in such systems are generally old (Pop II).

emission the production of light, or more generally, electromagnetic radiation by an atom or other object.

emission-line spectrum a spectrum consisting of bright lines at certain wavelengths separated by dark regions in which there is no light.

emission nebula a hot gas cloud in interstellar space that emits light.

energy a measure of the ability of a system to do work or cause motion.

energy level any of the numerous orbitals that an electron can occupy in an atom or molecule, roughly corresponding to an electron orbit.

epicycle a fictitious, small, and circular orbit superimposed on another circular orbit and proposed by early astronomers to explain the retrograde motion of the planets.

equator the imaginary line that divides the Earth (or other body) symmetrically into its northern and southern hemispheres. The equator is perpendicular to a body's rotation axis.

equinox the time of year when the Sun appears to cross the celestial equator. At this time, the number of hours of daylight and night are approximately equal. The vernal and autumnal equinoxes mark the beginning of the spring and fall seasons.

escape velocity the speed needed for an object to travel away from another body to overcome its gravitational attraction and not be pulled back. Mathematically, the escape velocity, V_{esc}, is defined as $\sqrt{2GM/R}$ where M is the body's mass, R is its radius, and G is the gravitational constant.

eukaryotes cells with nuclei. Most cells in current terrestrial organisms have nuclei and are thus eukaryotes.

Evening Star the planet Venus seen low in the western sky after sunset. (Sometimes used for other bright planets.)

event horizon the location of the "boundary" of a black hole. An outside observer cannot see in past the event horizon.

excited the condition in which the electrons of an atom are not in their lowest energy level (orbit).

exclusion principle the condition that no more than two electrons may occupy the same energy state in an atom. This limitation leads to degeneracy pressure.

exoplanet a planet not orbiting our Sun.

F

false-color picture/photograph a depiction of an astronomical object in which the colors are not the object's real colors. Instead, they are colors arbitrarily chosen to represent other properties of the body, such as the intensity of radiation, that we cannot see.

Fermi paradox an argument that we are probably the only technological civilization that has ever arisen in the Milky Way. Fermi suggested that if any such civilization had existed, it would have covered the whole Galaxy in a time much shorter than the evolution of humans.

fission the splitting of an atomic nucleus into two or more smaller nuclei.

flare an outburst of energy on the Sun. *See also* **solar flare.**

flat universe a universe that extends forever with no curvature. Its total energy is zero.

fluorescence the conversion of ultraviolet light (or other short-wavelength radiation) into visible light.

focus (1) one of two points within an ellipse used to generate the elliptical shape. Planets orbit along ellipses with the Sun at one focus of the ellipse. (2) a point in an optical system in which light rays are brought together. The location where an image forms in such systems.

frequency the number of times per second that a wave vibrates.

fundamental forces the four basic forces of nature: gravitation, electromagnetism, the weak force, and the strong force. Electromagnetism and the weak force are now recognized to both be part of an "electroweak" force. According to some modern theories, all of the forces are different forms of a single, more fundamental, unified force.

G

Gaia hypothesis the hypothesis that life does not merely respond to its environment but actually alters its planet's atmosphere and temperature to make the planet more hospitable. For example, by photosynthesis, plants have created an oxygen-rich atmosphere on Earth, which shields the plants from dangerous ultraviolet radiation.

galactic cannibalism the capture and merging of one galaxy into another.

galaxy a massive system of stars held together by their mutual gravity. Typical galaxies have a mass between about 10^7 and 10^{13} solar masses. Our Galaxy is the Milky Way.

galaxy cluster a group of galaxies held together by their mutual gravitational attraction. The Milky Way belongs to the Local Group galaxy cluster.

Galilean relativity a method for determining the relative speeds of motion seen by observers moving with respect to each other. This method works satisfactorily for motions at low speed, but it fails when the speed becomes an appreciable fraction of the speed of light.

Galilean satellites the four moons of Jupiter discovered by Galileo: Io, Europa, Ganymede, and Callisto.

gas giant a planet that grew largely by gravitationally attracting gas from the disk of gas out of which a planetary system forms. Jupiter and Saturn are thought to be gas giants.

general relativity Einstein's theory of gravity. The theory describes how mass and energy "curve" space and time so that objects move along the paths described by the gravitational "force."

geocentric models models of the Solar System centered on the Earth. Many of the earliest attempts to describe the Solar System were geocentric in that they supposed that the planets moved around the Earth rather than around the Sun.

giant a star of large radius and large luminosity.

glitches abrupt changes in the pulsation period of a pulsar, perhaps as the result of adjustments of its crust.

global warming a phenomenon in which the Earth's surface temperature has been observed to increase significantly over the last century. Most scientists attribute the change to increasing levels of gases that cause the greenhouse effect, released by human activities such as burning fossil fuels and deforestation.

globular cluster a dense grouping of old stars, containing generally about 10^5 to 10^6 members. They are often found in the halos of galaxies.

globule *See* **Bok globule.**

granulation texture seen in the Sun's photosphere. Granulation is created by clumps of hot gas that rise to the Sun's surface.

grating a piece of material that creates a spectrum by reflecting light from, or passing it through, many very fine and closely spaced parallel lines.

gravitational lens an object that bends space (and thereby the light passing through the space) by its gravitational attraction and focuses the light to create an image of a more distant object. *See also* **curvature of space.**

gravitational lensing the bending of light from a distant object to form an image, usually strongly distorted by the gravitation of a mass between the distant object and the observer. *See* **gravitational lens.**

gravitational redshift the shift in wavelength of electromagnetic radiation (light) created by a body's gravitational field as the radiation moves away from the body. Only extremely dense objects, such as white dwarfs, produce a significant redshift of their radiation.

gravitational waves a wavelike bending of space generated by the acceleration of massive bodies.

gravity the force of attraction that is between two bodies and is generated by their masses.

greatest elongation the position of an inner planet (Mercury or Venus) when it lies farthest from the Sun on the sky. Mercury and Venus are particularly easy to see when they are at greatest elongation. Objects may be at greatest eastern or western elongation according to whether they lie east or west of the Sun.

greenhouse effect the trapping of heat by a planet's atmosphere, making the planet warmer than would otherwise be expected. Generally the greenhouse effect operates if visible sunlight passes freely through a planet's atmosphere but the infrared radiation produced by the warm surface cannot escape readily into space, because of gases such as carbon dioxide or water vapor.

Great Red Spot a reddish colored region in Jupiter's atmosphere, larger than the Earth. It appears to be a stable atmospheric vortex, having been seen since some of the earliest telescopic observations in the 1600s.

Gregorian calendar the calendar devised at the request of Pope Gregory XIII and essentially the civil calendar used throughout the world today. It omits the leap year for century years not divisible evenly by 400.

H

H II region a region of ionized hydrogen. H II regions generally have a pink/red glow and often surround luminous, hot, young stars.

halo the approximately spherical region surrounding spiral galaxies that contains mainly old stars, such as the globular clusters. The halo also contains large amounts of dark matter.

Hawking radiation radiation that black holes are hypothesized to emit as a result of quantum effects. This radiation leads to the extremely slow evaporation of black holes.

heliocentric models models of the Solar System centered on the Sun. Compare to **geocentric models.**

helium flash the beginning of helium fusion in a low-mass star. The fusion begins explosively and causes a major readjustment of the star's structure.

highlands the old, heavily cratered regions on the Moon.

horizon the line separating the sky from the ground. *See also* **cosmic horizon.**

H-R diagram a graph on which stars are located according to their temperature and luminosity. Most stars on such a plot lie along a diagonal line, called the main sequence, which runs from cool, dim stars in the lower right, to hot, luminous stars in the upper left.

Hubble constant the multiplying constant H in Hubble's law, $V = Hd$. The reciprocal of the Hubble constant (in appropriate units) is approximately the age of the Universe.

Hubble's law a relation between a galaxy's distance, d, and its recession velocity, V, which states that more distant galaxies recede faster than nearby ones. Mathematically, $V = Hd$, where H is the Hubble constant.

hydrogen burning nuclear fusion of hydrogen into helium. It is not "burning" like ordinary fire but is instead the transformation of one kind of atom into another accompanied by the release of energy.

hydrostatic equilibrium the condition in which pressure and gravitational forces in a star or planet are in balance. Without such balance, bodies will either collapse or expand.

hypothesis an explanation proposed to account for some set of observations or facts.

I

ice giant a large, gas-rich planet that grew primarily through the accretion of ice-rich planetesimals. Uranus and Neptune are thought to be ice giants.

ideal gas law *See* **perfect gas law.**

inclination the tilt angle of an astronomical object's spin or its orbit.

inertia the tendency of an object at rest to remain at rest and of a body in motion to continue in motion in a straight line at a constant speed. *See also* **mass.**

inferior conjunction *See* **conjunction.**

inferior planet a planet whose orbit lies between the Earth's orbit and the Sun. Mercury and Venus are inferior planets.

inflation the rapid expansion of the early Universe by an enormous factor.

infrared a wavelength of electromagnetic radiation longer than visible light but shorter than radio waves. We cannot see these wavelengths with our eyes, but we can feel many of them as heat. The infrared wavelength region runs from about 700 nm to 1 mm.

inner core the innermost part of a planet, also called the *solid core*. The Earth's inner core is a mixture of solid iron and nickel.

inner planet a planet orbiting in the inner part of the Solar System. Sometimes taken to mean Mercury, Venus, Earth, and Mars.

instability strip a region in the H-R diagram indicating stars that pulsate.

interferometer a device consisting of two or more telescopes connected together to work as a single instrument. Used to obtain a high resolving power, the ability to see small-scale features. Interferometer telescopes have been contructed operating at radio, infrared, and visible wavelengths.

international date line an imaginary line from the Earth's North to South Pole, running approximately down the middle of the Pacific Ocean. It marks the location on Earth at which the date changes.

interstellar cloud a cloud of gas and dust in between the stars. Such clouds may be many light-years in diameter.

interstellar grains microscopic solid dust particles in interstellar space. These grains absorb starlight, making distant stars appear dimmer and redder than they truly are.

interstellar matter matter in the form of gas or dust in the space between stars.

inverse-square law (1) any law in which some property varies inversely as the square of the distance, d. Mathematically, as $1/d^2$. (2) the law stating that the apparent brightness of a body decreases inversely as the square of its distance.

ion tail a stream of ionized particles evaporated from a comet and then swept away from the Sun by the solar wind.

ionization the removal of one or more electrons from an atom, leaving the atom with a positive electric charge. Under some circumstances an extra electron may be attached to an atom, in which case the atom is described as negatively ionized.

ionized a condition in which the number of an atom's electrons does not equal the number of its protons. Typically, this means the atom is missing one or more electrons.

irregular galaxy a galaxy lacking a symmetric structure.

J

jets narrow streams of gas ejected from any of several types of astronomical objects. Jets are seen near protostars and in many active galaxies.

jet stream a narrow stream of high-speed wind that blows in the atmosphere of a planet. Such winds occur on Earth and many other planets.

joule a unit of energy. Expending one joule per second equals one watt of power.

Jovian planet one of the giant, gaseous planets: Jupiter, Saturn, Uranus, and Neptune, or most exoplanets. The name *Jovian* was chosen because the structure of Jupiter (or Jove) is representative of the others.

Julian calendar a 12-month calendar devised under the direction of Julius Caesar. It includes a leap year every four years.

K

Kepler's three laws laws that describe the motion of planets around the Sun. The first law states that planets move in elliptical orbits with the Sun off-center at a focus of the ellipse. The second law states that a line joining the planet and the Sun sweeps out equal areas in equal times. The third law relates a planet's orbital period, P, to the semi-major axis of its elliptical orbit, a. Mathematically, the law states that $P^2 = a^3$, if P is measured in years and a in astronomical units.

Kirkwood gaps regions in the asteroid belt with fewer than the average number of asteroids. The gaps result from the gravitational force of Jupiter removing asteroids whose orbits have periods that are a simple fraction of Jupiter's period.

Kuiper belt a region from which some comets come. The region appears to extend from the orbit of Neptune, past Pluto, out to approximately 55 AU.

L

late heavy bombardment a period about 600 million years after the Solar System formed when there appears to have been a high level of impact cratering events.

law of gravity a description of the gravitational force exerted by one body on another. The gravitational force is proportional to the product of the bodies' masses and the inverse square of their separation. If the masses are M and m and their separation is d, the force between them, F, is $F = GMm/d^2$, where G is a physical constant.

light electromagnetic energy.

light-gathering power a measure of the ability of a telescope (or other device) to collect light. It is generally proportional to the *area* of the telescope's mirror or lens. For example, a telescope with a 6-inch diameter lens has a four times larger light-gathering power than a telescope with a 3-inch diameter lens.

light-year a unit of distance equal to the distance light travels in one year. A light-year is roughly 10^{13} km, or about 6 trillion miles.

liquid core the molten interior of a planet, also called the *outer core.*

Local Group the small cluster or group of several dozen galaxies to which the Milky Way belongs.

Lorentz factor a term that designates how much an object's time, space, and mass are altered as a result of its motion. The factor is very close to 1 except at speeds approaching the speed of light.

luminosity the amount of energy radiated per second by a body. For example, the wattage of a lightbulb defines its luminosity. Stellar luminosity is usually measured in units of the Sun's luminosity (approximately 4×10^{26} watts).

lunar eclipse the passage of the Earth between the Sun and the Moon so that the Earth's shadow falls on the Moon.

M

MACHO (massive compact halo object) any body of planetary or stellar mass orbiting in the Milky Way's halo that produces very little detectable light. These objects have been hypothesized to be a significant part of the Milky Way's dark matter.

Magellanic Clouds two small companion galaxies of the Milky Way.

magnetic field a representation of the means by which magnetic forces are transmitted from one body to another. A compass needle points along the direction of the Earth's magnetic field.

magnetic lines of force fictitious lines used to visualize the orientation and strength of a magnetic field.

magnitude a unit for measuring stellar brightness. The smaller the magnitude, the brighter the star.

main sequence the region in the H-R diagram in which most stars, including the Sun, are located. The main sequence runs diagonally across the H-R diagram from cool, dim stars to hot, luminous ones. Stars on the main sequence fuse hydrogen into helium in their cores. *See also* **H-R diagram.**

main-sequence lifetime the time a star remains a main-sequence star, fusing hydrogen into helium in its core.

mantle the solid, outer part of a planet. This part is immediately below the crust and may flow very slowly due to heat rising from the core.

mare a vast, smooth, dark, and congealed lava flow filling a basin on the Moon and on some planets. Many maria have roughly circular shapes.

maria plural of *mare.*

maser an intense radio source created when excited gas amplifies some background radiation. *Maser* stands for *m*icrowave *a*mplification by *s*timulated *e*mission of *r*adiation.

mass a measure of the amount of material an object contains. A quantity measuring a body's inertia.

mass–luminosity relation a relation between the mass and luminosity of main-sequence stars. Higher-mass stars have higher luminosity.

Maunder minimum the time period, from about A.D. 1600 to 1740, during which the Sun was relatively inactive. Few sunspots were observed during this period.

mean solar day the standard 24-hour day. The mean solar day is based on the average day length over a year. (The time interval from solar noon to solar noon varies slightly throughout the year.)

megaparsec a distance unit equal to 1 million parsecs and abbreviated Mpc.

metal astronomically, any chemical element more massive than helium. Thus, carbon, oxygen, iron, and so forth are called metals.

meteor the bright trail of light created by small solid particles entering the Earth's atmosphere and burning up. A "shooting star."

meteor shower an event in which many meteors occur in a short space of time, all from the same general direction in the sky. The most famous shower is the Perseids in mid-August.

meteorite the solid remains of a meteor that falls to the Earth.

meteoroid the technical name for the small, solid bodies moving within the Solar System. When a meteoroid enters our atmosphere and heats up, the trail of luminous gas it leaves is called a meteor. When the body lands on the ground, it is called a meteorite. ("A meteoroid is in the void. A meteor above you soars. A meteorite is in your sight.")

method of standard candles *See* **standard candle.**

Milky Way Galaxy the Galaxy to which the Sun belongs. Seen from Earth, the Galaxy is a pale, milky-white band in the night sky.

Miller-Urey experiment an experimental attempt to simulate the conditions under which life might have developed on Earth. Miller and Urey discovered that amino acids and other complex organic compounds could form from the gases that are thought to have been present in the Earth's early atmosphere, if the gases are subjected to an electric spark or ultraviolet radiation.

millisecond pulsar a pulsar whose rotation period is about a millisecond.

minute of arc a measure of angle equal to one-sixtieth of a degree.

model a theoretical representation of some object or system.

molecule two or more atoms bonded into a single particle, such as water, H_2O (two hydrogen atoms bonded to one oxygen) or carbon dioxide, CO_2 (one carbon atom bonded to two oxygen atoms).

Moon illusion the illusion in which the Moon appears larger when near the horizon than when seen high in the sky.

Morning Star the planet Venus seen in the eastern sky before dawn. (Sometimes applied to other bright planets.)

N

nanometer a unit of length equal to 1 billionth of a meter (10^{-9} meters) and abbreviated nm. Wavelengths of visible light are several hundred nanometers. The diameter of a hydrogen atom is roughly 0.1 nm.

neap tide the abnormally small tide occurring when the Sun's and Moon's gravitational effects on the ocean partially offset each other.

near-Earth object an asteroid with an orbit that crosses Earth's orbit or comes close to it.

nebula a cloud in interstellar space.

negative curvature a form of curved space sometimes described as being "open" in that it has no boundary. Negative curvature is analogous to a saddle shape.

neutrinos tiny neutral particles with little or no mass and immense penetrating power. These particles are produced in great numbers by the Sun and other stars as they fuse hydrogen into helium, and also by supernova explosions.

neutron a subatomic particle of nearly the same mass as the proton but with no electric charge. Neutrons and protons compose the nuclei of atoms.

neutron star a very dense, compact star composed primarily of neutrons.

Newton's first law of motion the law that a body continues in a state of rest or uniform motion in a straight line unless made to change that state by forces acting on it. *See also* **inertia.**

Newton's second law of motion $F = ma$. In words, the amount of acceleration, *a,* that a force, *F,* produces depends on the mass, *m,* of the object being accelerated.

Newton's third law of motion the law that when two bodies interact, they exert equal and opposite forces on each other.

nonthermal radiation radiation emitted by charged particles moving at high speed in a magnetic field. The radio emission from pulsars and radio galaxies is nonthermal emission. More generally, nonthermal means "not due to high temperature."

north celestial pole the point on the celestial sphere directly above the Earth's North Pole. Objects on the sky appear to circle around this point.

North Star any star that happens to lie very close to the north celestial pole. Polaris has been the North Star for about 1000 years, and it will continue as such for about another 1000 years, at which time a star in Cepheus will be nearer the north celestial pole.

nova a process in which a surface layer of hydrogen builds up on a white dwarf and then fuses explosively into helium, making the white dwarf visible for a few weeks. Nova explosions do not destroy the white dwarf and may be recurrent.

nuclear fusion the binding of two light nuclei to form a heavier nucleus, with some nuclear mass converted to energy. For example, the fusion of hydrogen into helium. This process supplies the energy of most stars and is commonly called "burning" by astronomers.

nucleosynthesis the formation of elements, generally by the fusion of lighter elements into heavier ones. For example, the formation of carbon by the fusion of three helium nuclei.

nucleus the core of an atom around which the electrons orbit. The nucleus has a positive electric charge and constitutes most of an atom's mass.

nucleus of a comet the core, typically a few kilometers across, of frozen gases and dust that make up the solid part of a comet.

nucleus of a galaxy the central region of a galaxy, often containing a supermassive black hole.

O

Olbers' paradox an argument that in an infinite Universe, the sky should be bright at night, filled with the light from many distant stars and galaxies.

Oort cloud a vast region in which comet nuclei orbit. This cloud lies far beyond the orbit of Neptune.

opacity the blockage of light or other electromagnetic radiation by matter.

open cluster a loose cluster of stars, generally containing a few hundred members.

opposition the configuration of a planet when it is opposite the Sun in the sky. If a planet is in opposition, it rises when the Sun sets and sets when the Sun rises.

orbit the path in space followed by a celestial body.

orbital a description of an electron's possible location in an atom as it "orbits" the nucleus. At these tiny scales, the electron's position can be described only in terms of probabilities, due to the wave nature of matter.

outer core the molten interior of a planet; also called the *liquid core.*

outer planet a planet whose orbit lies in the outer part of the Solar System. Jupiter, Saturn, Uranus, and Neptune are outer planets.

ozone a form of oxygen consisting of three oxygen atoms bonded together. Its chemical symbol is O_3. The ozone in our atmosphere absorbs, and thus shields us from, the Sun's harmful ultraviolet radiation.

P

panspermia a theory that life originated elsewhere than on Earth and came here across interstellar space either accidently or deliberately.

parallax the shift in an object's position caused by the observer's motion. A method for finding distances based on that shift.

parsec a unit of distance equal to about 3.26 light-years (3.09×10^{13} km), defined as the distance at which an observer sees the maximum angle between the Sun and the Earth to be one arc second.

perfect gas law a law relating the pressure, density, and temperature of a gas. It states that the pressure is proportional to the density times the temperature. This is also called the *ideal gas law.*

perihelion the point in an orbit closest to the Sun.

period the time required for a repetitive process to repeat. For example, orbital period is the time it takes a planet or star to complete an orbit. Pulsation period is the time it takes a star to expand and then contract back to its original radius.

period–luminosity relation a law stating that the longer the period of a pulsating variable star, the more luminous it is.

phases the changing illumination of the Moon or other body that causes the body's apparent shape to change. The following is the cycle of lunar phases: new, crescent, first quarter, gibbous, full, gibbous, third quarter, crescent, new.

photo dissociation the breaking apart of a molecule by intense radiation.

photon a particle of visible light or other electromagnetic radiation.

photosphere the visible surface of the Sun. When we look at the Sun in the sky, we are seeing its photosphere.

Planck time the brief interval of time, about 10^{-43} second immediately after the Big Bang, when quantum fluctuations are so large that current theories of gravity can no longer describe space and time.

planet a body in orbit around a star that is large enough to have taken on a round shape, and that has cleared the path of its orbit of all bodies of comparable size.

planetary nebula a shell of gas ejected by a low-mass star late in its evolutionary lifetime. A planetary nebula typically appears as a glowing gas ring around a central white dwarf.

planetesimal one of the numerous small, solid bodies that, when accreted together, form a planet.

plate tectonics the idea that the crust of the Earth (or some other planet) is divided into large regions (plates) that move very slowly over the planet's surface. Interaction between plates at their boundaries creates mountains and activity such as earthquakes.

polarity the property of a magnet that causes it to have a north and south pole.

population (Pop) I the younger stars, some of which are blue, that populate a galaxy's disk, especially its spiral arms.

population (Pop) II the older, redder stars that populate a galaxy's halo and bulge.

population (Pop) III a hypothetical stellar population consisting of the first stars that formed in a galaxy, composed of only hydrogen and helium.

positive curvature bending of space leading to a finite volume. A space that is "closed." A universe with positive curvature is analogous to a spherical shape.

positron a subatomic antimatter particle with the same mass as the electron but a positive electric charge. An electron's antiparticle.

precession the slow change in direction of the pole (rotation axis) of a spinning body or of the orientation of an orbit.

pressure the force exerted by a substance such as a gas on an area divided by that area. That is, pressure × area = force.

principle of equivalence an idea developed by Einstein to understand that gravity is equivalent to an accelerating frame of reference.

prokaryotes cells without nuclei. The first lifeforms on Earth were prokaryotes.

prominence a cloud of hot gas in the Sun's outer atmosphere. This cloud is often shaped like an arch, supported by the Sun's magnetic field.

proper motion shift of a star's position on the celestial sphere, in other words, motion perpendicular to our line of sight. This motion is normally so gradual that thousands of years must pass before changes become noticeable to the unaided eye.

protein any of many complex organic molecules composed of a chain of amino acids. Proteins serve many functions in cells, including structure and metabolism.

proton a positively charged subatomic particle. One of the constituents of the nucleus of an atom along with neutrons.

proton–proton chain the nuclear fusion process that converts hydrogen into helium in stars like the Sun and thereby generates their energy. This is the dominant energy-generation mechanism in cool, low-mass stars.

protoplanetary disk a disk of gas and dust surrounding a young star out of which planetary systems form.

protostar a star still in its formation stage, heated by gravitational contraction.

pulsar a spinning neutron star that emits beams of radiation that happen to sweep across the Earth each time the star spins. We observe the radiation as regularly spaced pulses.

pulsate to expand and contract regularly. For example, pulsating variable stars swell and shrink in a predictable, regular fashion.

Q

quantized the property of a system that allows it to have only discrete values.

quark a fundamental particle of matter that interacts via the strong force; basic constituent of protons and neutrons.

quasar a peculiar galaxy characterized by a large redshift, high luminosity, and an extremely small, active core. Quasars are among the most luminous and most distant objects known to astronomers.

R

radial velocity the velocity of a body along the line of sight. That is, the part of its motion directly toward or away from the observer.

radiant the point in the sky from which meteors in showers appear to come. *See also* **meteor showers.**

radiation pressure the force exerted by radiation on matter.

radiative zone the region inside a star where its energy is carried outward by radiation (that is, by photons).

radioactive decay the breakdown of an atomic nucleus by the emission of subatomic particles.

radioactive element an element that undergoes radioactive decay and breaks down into a lighter element.

radio galaxy a galaxy, usually an elliptical, that emits abnormally large amounts of radio energy from huge lobes of hot gas ejected from the galaxy.

rays long, narrow, light-colored markings on the Moon or other bodies that radiate from young craters. Rays are debris "splashed" out of the crater by the impact that formed it.

recession velocity the velocity of an external galaxy (or other object) away from our Galaxy due to the expansion of the Universe.

recombination era a period of time about 400,000 years after the Big Bang when matter cooled enough for electrons to combine with nuclei to form neutral atoms.

reddening the alteration in a star's color as seen from Earth as the star's light passes through an intervening interstellar dust cloud. The dust preferentially scatters the blue light from the beam, leaving the remaining light redder.

red giant a cool, luminous star whose radius is much larger than the Sun's.

redshift a shift in the wavelength of electromagnetic radiation to a longer wavelength. For visible light, this implies a shift toward the red end of the spectrum. The shift can be caused by a source of radiation moving away from the observer or by the observer moving away from the source. For example, if a star is moving away from Earth, its spectrum lines exhibit a redshift. *See also* **Doppler shift.**

reflection nebula an interstellar cloud in which the dust particles reflect starlight, making the cloud visible.

reflector a telescope that uses a mirror to collect and focus light.

refraction the bending of light when it passes through one substance and enters another.

refractor a telescope that uses a lens to collect and focus light.

regolith the surface rubble of broken rock on the Moon or other solid body.

resolving power the ability of a telescope or instrument to discern fine details. Larger-diameter telescopes have greater (that is, better) resolving power.

resonance a condition in which the repetitive motion of one body interacts with the repetitive motion of another so as to reinforce the motion. Sliding back and forth in a bathtub to make a big splash is an example.

rest frame a system of coordinates that appear to be at rest with respect to the observer.

retrograde motion the drift of a planet westward against the background stars. Normally planets shift eastward because of their orbital motion. The planet does not actually reverse its motion. The change in its direction is caused by the change in the position from which we view the planet as the Earth overtakes and passes it.

rifting the breaking apart of a continental plate.

right ascension a coordinate for locating objects on the sky, analogous to longitude on the Earth's surface. Measured in hours and minutes of time.

rilles narrow canyons on the Moon or other body.

Roche limit the distance from an astronomical body at which its gravitational force can pull apart another astronomical body.

rotation axis an imaginary line through the center of a body about which the body spins.

rotation curve a plot of the rotation velocity of the stars or gas in a galaxy at different distances from its center.

RR Lyrae stars a type of white, giant, pulsating variable stars with a period of about one day or less. They are named for their prototype star, RR Lyrae.

runaway greenhouse effect greenhouse effect in which heat trapping grows as temperature rises, so the planet's atmosphere becomes extremely hot, as has occurred on Venus.

S

satellite a body orbiting a planet.

scattering the random redirection of a light wave or photon as it interacts with atoms or dust particles.

Schwarzschild radius the radius of a black hole. The distance from the center of a black hole to its event horizon.

scientific method the process of observing a phenomenon, proposing a hypothesis on the basis of the observations, and then testing the hypothesis.

scientific notation a shorthand way to write numbers using ten to a power. For example, $1,000,000 = 10^6$ and $2000 = 2 \times 10^3$. Also called "powers-of-ten notation."

scintillation the twinkling of stars caused by the atmosphere. *See* **seeing**.

seeing a measure of the steadiness of the atmosphere during astronomical observations. Under conditions of bad seeing, fine details are difficult to see. Bad seeing results from atmospheric irregularities moving between the telescope and the object being observed.

seismic waves waves generated in the Earth's interior by earthquakes. Similar waves occur in other bodies. Two of the more important varieties are S and P waves. The former can travel only through solid material; the latter can travel through either solid or liquid material.

selection effect an unintentional selection process that omits some set of the objects being studied and leads to invalid conclusions about the objects.

self-propagating star formation a model that explains spiral arms as arising from stars triggering the birth of other stars around them. The resulting pattern is then drawn out into a spiral by the galaxy's rotation.

semimajor axis half the long dimension of an ellipse.

SETI *S*earch for *E*xtra*t*errestrial *I*ntelligence. Some such searches involve automatic "listening" to millions of radio frequencies for signals that might be from other civilizations.

Seyfert galaxy a variety of active galaxy with a small, abnormally bright nucleus containing hot gas. Named for the astronomer who first drew attention to these objects.

shell source a region in a star where the nuclear energy generation occurs around the core rather than in it.

shepherding satellites satellites that by their gravitational attractions prevent particles in a planet's rings from spreading out and dispersing. Saturn's F-ring is held together by shepherding satellites.

short-period comet a comet whose orbital period is shorter than 200 years. For example, Halley's comet has a period of 76 years.

sidereal day the length of time from the rising of a star until it next rises. The length of the Earth's sidereal day is 23 hours 56 minutes.

sidereal period the time it takes a body to turn once on its rotation axis or to revolve once around a central body, as measured with respect to the stars.

sidereal time a system of time measurement based on the motion of stars across the sky rather than the Sun.

silicates materials composed of silicon and oxygen, and generally containing other substances as well. Most ordinary rocks are silicates. For example, quartz is silicon dioxide.

solar cycle the cyclic change in solar activity, such as sunspots and solar flares, rising and declining about every 11 years.

solar day the time interval from one sunrise to the next sunrise or from one noon to the next noon. That time interval is not always exactly 24 hours but varies throughout the year. For that reason, we use the mean solar day (which, by definition, is 24 hours) to keep time.

solar eclipse the passage of the Moon between the Earth and the Sun so that our view of the Sun is partially or totally blocked. *See also* **total eclipse.**

solar flare a sudden increase in brightness of a small region on the Sun. This flare is caused by a magnetic disturbance.

solar nebula the rotating disk of gas and dust from which the Sun and planets formed.

solar nebula theory the theory that the Solar System formed from a rotating cloud of gas and dust, the solar nebula.

Solar System the Sun, planets, their moons, and other bodies that orbit the Sun.

solar wind the outflow of low-density, hot gas from the Sun's upper atmosphere. It is partially this wind that creates the tail of a comet, by blowing gas away from the comet's immediate surroundings.

solid core the inner iron-nickel core of the Earth or another planet. Despite its high temperature, the core is solid because it is under great pressure. Also called the *inner core.*

solstice (winter and summer) the beginning of winter and summer. Astronomically the solstice occurs when the Sun is at its greatest distance north (June) or south (December) of the celestial equator.

south celestial pole the imaginary point on the celestial sphere directly over the Earth's South Pole.

special relativity a theory developed by Einstein to explain why the speed of light is always the same, regardless of the motion of the source of the light or the observer. The theory shows (and experiment has confirmed) that an object's mass, length, rate of time passage, and other quantities change depending on the speed of the observer relative to the object.

spectral type an indicator of a star's temperature. A star's spectral type is based on the appearance of its spectrum lines. The fundamental types are, from hot to cool: *O, B, A, F, G, K,* and *M.*

spectrograph a device for making a spectrum.

spectroscopic binary a type of binary star for which the spectrum lines exhibit a changing Doppler shift as a result of the orbital motion of one star around the other.

spectroscopy the study and analysis of spectra.

spectrum electromagnetic radiation (for example, visible light) spread into its component wavelengths or colors. The rainbow is a spectrum produced naturally by water droplets in our atmosphere.

spicule a hot, thin column of gas in the Sun's chromosphere.

spiral arm a long, narrow region containing young stars and interstellar matter that winds outward in the disk of spiral galaxies.

spiral galaxy a galaxy with a disk in which its bright stars form a spiral pattern.

spring tide the abnormally large tides that occur at new and full moon.

standard candle a type of star or other astronomical body in which the luminosity has a known value, allowing its distance to be determined by measuring its apparent brightness and applying the inverse-square law: Cepheid variable stars, supernovas, and so forth.

standard time a uniform time kept within a given region so that all clocks there agree.

star a massive, gaseous body held together by gravity and generally emitting light. Normal stars generate energy by nuclear reactions in their interiors.

star cluster a group of stars, numbering from hundreds to millions, held together by their mutual gravity.

Stefan-Boltzmann law the law that the amount of energy radiated from 1 square meter in 1 second by a blackbody of temperature T is proportional to T^4.

stratosphere a layer of Earth's atmosphere extending from about 12 to 50 km above the surface. A layer of ozone is found there.

strong force the force that holds protons and neutrons together in the atomic nucleus. Sometimes called the *nuclear force*.

subatomic particles particles making up an atom, such as electrons, neutrons, and protons, or other particles of similar submicroscopic size.

subduction the sinking of one crustal plate where it encounters another.

sunspot a dark, cooler region on the Sun's visible surface created by intense magnetic fields.

supercluster a cluster of galaxy clusters. One of the largest structures in the Universe.

super-earth a category of exoplanet with a radius about 1.25 to 2 times the Earth's radius.

supergiant a very large-diameter and luminous star, typically at least 10,000 times the Sun's luminosity.

superior conjunction *See* **conjunction.**

superior planet a planet orbiting farther from the Sun than the Earth. Mars, Jupiter, Saturn, Uranus, and Neptune are superior planets.

supermassive black hole a huge black hole, containing millions to billions of solar masses, at the center of most large galaxies.

supernova an explosion that occurs at the end of some stars' evolution. Astronomers identify two main kinds of supernovas: Type Ia and II. Type Ia occurs in a binary system in which one star is a white dwarf. The explosion is triggered when mass from a companion star falls onto the white dwarf, raising its mass above the Chandrasekhar limit and causing the star to collapse. Collapse heats the white dwarf so that its carbon and oxygen fuse explosively, destroying the star and leaving no remnant. Type II probably occurs when a massive star's iron core collapses. A type II supernova leaves either a neutron star or a black hole, depending on the mass of the collapsing core.

supernova remnant the debris ejected from a star when it explodes as a supernova. Typically, this material is hot gas, expanding away from the explosion at thousands of kilometers or more per second.

surface gravity the acceleration caused by gravity at the surface of a planet or other body.

synchronous rotation the condition where a body's rotation period is the same as its orbital period. The Moon rotates synchronously as it orbits the Earth.

synchrotron radiation a form of nonthermal radiation emitted by charged particles spiraling at nearly the speed of light in a magnetic field. Pulsars and radio galaxies emit synchrotron radiation. The radiation gets its name because it was first seen in synchrotrons, a type of atomic accelerator.

synodic period the time between repeated configurations of a planet or moon. For example, the time between oppositions of a planet or between full moons.

T

21-centimeter radiation a spectrum line at radio wavelengths produced by non-ionized (neutral) hydrogen.

tail the plume of gas and dust from a comet. The plume is produced by the solar wind and radiation pressure acting on the comet. The tail points away from the Sun and gets longer as the comet approaches perihelion.

terrestrial planet a rocky planet similar to Earth in size and structure. The terrestrial planets are Mercury, Venus, Earth, and Mars.

tidal braking the slowing of one body's rotation as a result of gravitational forces exerted on it by another body.

tidal bulge a bulge on one body created by another body's gravitational attraction on it. Two tidal bulges form, one on the side near the attracting body and one on the opposite side.

tides the rise and fall of the Earth's oceans created by the gravitational attraction of the Moon. Tides also occur in the solid crust of a body and its atmosphere.

time dilation the slowing of time due to the effects of special and general relativity.

time zone one of 24 divisions of the globe, centered at every 15 degrees of longitude. In each zone, a single standard time is kept. Most zones have irregular boundaries.

total eclipse an eclipse in which the eclipsing body totally covers the other body. Only at a total solar eclipse can we see the Sun's corona.

transit the passage of a planet directly between the observer and the Sun. At a transit, we see the planet as a dark spot against the Sun's bright disk. From Earth, only Mercury and Venus can transit the Sun.

trans-Neptunian objects (TNOs) numerous small, icy objects orbiting in the outer part of the Solar System beyond the orbit of Neptune.

triangulation a method for measuring distances. This method is based on constructing a triangle, one side of which is the distance to be determined. That side is then calculated by measuring another side (the base line) and the two angles at either end of the base line.

triple alpha process the fusion of three helium nuclei (alpha particles) into a carbon nucleus. This process is sometimes called helium burning, and it occurs in many old stars.

troposphere the lowest layer of the Earth's atmosphere, extending up to about 12 km, where convection leads to clouds and precipitation.

T Tauri star a type of extremely young star that varies erratically in its light output.

tuning-fork diagram a diagram devised by Hubble to classify the various forms of spiral, elliptical, and irregular galaxies. The diagram is named for its shape.

turnoff point the location on the main sequence where a star's evolution causes it to move away from the main sequence toward the red giant region. The location of the turnoff point can be used to deduce the age of a star cluster.

twin paradox a supposed paradox in special relativity arising from the difference in time measurement of two observers moving relative to each other. It is usually expressed in terms of a twin who travels to a distant star and returns to Earth to find that the other twin has aged much more than the first twin. The paradox is explained because the motion of the twins is not symmetric. The twin traveling away from Earth experiences accelerations that the twin remaining on Earth does not experience.

Type Ia supernova *See* **supernova.**

U

ultraviolet a portion of the electromagnetic spectrum with wavelengths shorter than those of visible light but longer than those of X rays. By convention, the ultraviolet region extends from about 10 nm to 400 nm.

Universal time the time kept at Greenwich, England. Universal time is the same as Greenwich mean time. Most local times (in the continental United States, these are Eastern, Central, Mountain, and Pacific) differ from it by a whole number of hours.

Universe the largest astronomical structure we know of. The Universe contains all matter and radiation and encompasses all space.

V

Van Allen radiation belts doughnut-shaped regions surrounding the Earth containing charged particles trapped by the Earth's magnetic field.

variable star a star whose luminosity changes over time.

vernal equinox the spring equinox in the Northern Hemisphere. Spring begins on the vernal equinox, which is on about March 20.

Virgo Supercluster the cluster of galaxy clusters in which the Milky Way is located. The Local Group is one of its members.

visible spectrum the part of the electromagnetic spectrum that we can see with our eyes. It consists of the familiar colors violet, blue, green, yellow, orange, and red, extending from about 400 nm to 700 nm.

visual binary star a pair of stars held together by their mutual gravity and in orbit about each other,

and that can be seen with a telescope as separate objects.

visual double star two stars that appear to lie very close together on the sky but in reality are at greatly different distances.

W

wavelength the distance between wave crests. It determines the color of visible light and is usually denoted by the Greek letter λ.

wave–particle duality the theory that electromagnetic radiation may be treated as either a particle (photon) or an electromagnetic wave.

weak force the force responsible for radioactive decay of atoms. Now known to be linked to the electromagnetic force at high energies and therefore called the electroweak force.

white dwarf a dense star whose radius is approximately the same as the Earth's but whose mass is comparable with the Sun's. White dwarfs burn no nuclear fuel and shine by residual heat. They are the end stage of stellar evolution for stars like the Sun.

white light visible light exhibiting no color of its own but composed of a mix of all colors. Sunlight and many artificial light sources are "white."

Wien's law a relation between a body's temperature and the wavelength at which it emits radiation most intensely. Hotter bodies radiate more intensely at shorter wavelengths. Mathematically, the law states that $\lambda_{max} = 2.9 \times 10^6 / T$, where λ_{max} is the wavelength of maximum emission in nanometers and T is the body's temperature on the Kelvin scale.

WIMP (weakly interacting massive particle) A hypothetical subatomic particle that interacts only

through the weak force and gravity. This could be a form of dark matter since it does not interact with electromagnetic radiation.

X

X rays a portion of the electromagnetic spectrum with wavelengths of about 0.01 nm to 10 nm.

X-ray binary a binary star system in which one of the stars, or the gas associated with a star, emits X rays intensely. Such systems generally contain a collapsed object such as a neutron star or a black hole.

Y

year the time it takes the Earth to complete its orbit around the Sun; that is, the period of the Earth's orbit.

Z

Zeeman effect the splitting of a single spectrum line into two or three lines by a magnetic field. A method for detecting magnetic fields in objects from their spectra.

zenith the point on the celestial sphere that lies directly overhead at your location.

zodiac a band running around the celestial sphere in which the planets move.

zone a white band in a Jovian planet's atmosphere, encircling the planet. It appears that in these zones gas is rising and cooling, forming cloud particles.

zone of avoidance a band running around the sky in which few galaxies are visible. It coincides with the Milky Way and is caused by dust that is within our Galaxy. This dust blocks the light from distant galaxies.

Credits

PHOTO CREDITS

Design Element

Faded Star: © Photodisc/RF.

Looking Up

Looking Up #1 Northern Circumpolar Constellations

Background (both): © Akira Fujii/DMI; M52: NOAO/AURA/NSF; M81 and M82: © Robert Gendler; M101: Adam Block/NOAO/AURA/NSF.

Looking Up #2 Ursa Major

Background: © Akira Fujii/DMI; M97: Gary White and Verlenne Monroe/Adam Block/NOAO/AURA/NSF; Mizar and Alcor: Courtesy of DSS/Processing by Coelum (www.coelum.com); M51: © Tony and Daphne Hallas.

Looking Up #3 M31 & Perseus

All photos: © Akira Fujii/DMI.

Looking Up #4 Summer Triangle

Background: © Akira Fujii/DMI; M57: Courtesy of H. Bond et al., Hubble Heritage Team (STScI/AURA), NASA; M27: © IAC/RGO/Malin; Alberio: Courtesy of Randy Brewer.

Looking Up #5 Taurus

Background: © Akira Fujii/DMI; M1: Courtesy of R. Wainscoat; M45: Courtesy of Anglo-Australian Observatory, photographs by David Malin; Hyades; © Akira Fujii/DMI.

Looking Up #6 Orion

Background: © Akira Fujii/DMI; Horsehead Nebula: Courtesy of Anglo-Australian Observatory, photograph by David Malin; Close-up of Horsehead Nebula: Courtesy NOAO/AURA/NSF; Betelgeuse: Courtesy of Dupree (CFA), NASA, ESA; Pink Orion Nebula: Courtesy of Carol B. Ivers and Gary Oleski; Orion Nebula with Dust and Gas: Courtesy Gary Bernstein (U. Pennsylvania), copyright U. Michigan, Lucent; Protoplanetary Disk: C.R. O'Dell/Rice University; NASA.

Looking Up #7 Sagittarius

Background: © William Attard McCarthy/Alamy Stock Photo; M16 Close-up: Courtesy of NASA, HST, J. Hester & P. Scowen (ASU); M16 with Gas Cloud: Courtesy of Bill Schoening/NOAO/AURA/NSF; M20: Courtesy of Jason Ware; M22: Courtesy of N.A. Sharp, REU program/NOAO/AURA/NSF.

Looking Up #8 Centaurus and Crux, The Southern Cross

Background: Anglo-Australian Observatory/David Malin Images; Centaurus A: Courtesy Peter Ward, 2004; Omega Centauri: Al Kelly; Eta Carinae: J. Hester/Arizona State University, NASA; The Jewel Box: Research School of Astronomy and Astrophysics, the Australian National University.

Looking Up #9 Southern Circumpolar Constellations

Background: Christopher J. Picking; Hourglass Nebula: Raghvendra Sahai and John Trauger (JPL), the WFPC2 science team, and NASA; Thumbprint Nebula: Courtesy of STScI; Tarantula Nebula: WFI/2.2-m/ESO.

Preview

Figure P.1: NASA; P.2a: © Vol. 74 PhotoDisc/Getty RF; P.2b: © Dr. F.A. Ringwald; P.2c: NASA; P.3(Mercury): NASA/John Hopkins University Applied Physics Laboratory/Carnegie Institution of Washington; (Venus): Courtesy of NASA/JPL; (Earth): © Lunar and Planetary Institute/NASA; (Mars): NASA/JPL/MSSS; (Earth): © Lunar and Planetary Institute/NASA; (Jupiter): NASA/JPL/Cassini; (Saturn, Uranus, Neptune): Courtesy of NASA/JPL; P.4: Courtesy of SOHO, NASA/ESA; P.7a: Courtesy of 2MASS/UMass/IPAC-Caltech/NASA/NSF; P.7b: NASA/JPL-Caltech; P.8: NASA, ESA, and M. Livio and the Hubble 20th Anniversary Team (STScI); P.10: S. Bechwith & the HUDF Working Group (STScI), HST, ESA, NASA; P.13(Milky Way): NASA/JPL-Caltech; (Earth): NASA.

Chapter 1

Opener: © Stocktrek Images/Getty Images RF; p. 15: © CNES; 1.3a(left): © Roger Ressmeyer/Corbis, digitally enhanced by Jon Alpert; (right): © Roger Ressmeyer/Corbis; 1.3b(left): © Roger Ressmeyer/Corbis, digitally enhanced by Jon Alpert; (right): © Roger Ressmeyer/Corbis; 1.14(both): © Stephen E. Schneider; 1.16a: www.stone-circles.org.uk; 1.17a: © Courtesy Ivan Ghezzi, PhD; 1.17b: © Photoimagerie/Alamy; 1.18: NASA/Goddard Space Flight Center Scientific Visualization Studio; 1.21: © Stephen E. Schneider; 1.22: © John Walker; 1.24a-e: © Stephen E. Schneider; 1.26: © Roger Ressmeyer/Corbis.

Chapter 2

Opener: © The Bridgeman Art Library/Getty Images; p. 37: © Laurent Laveder; 2.1a: © Copyright 2009 Anthony Ayiomamitis; 2.2: © Stephen E. Schneider; 2.5: NASA; 2.13: © Courtesy Tunç Tezel; 2.16: © Hulton Archive/Imagno/Getty Images; 2.18(both): Smithsonian Institution Library, Library of Congress, 46031926; 2.21: © Corbis; 2.22:

© Leemage/Getty Images; 2.25: © Courtesy of the Bridgeman Art Library/Getty; 2.26a-c: Galilei, Galileo, 1564-1642. Sidereus Nuncius. [1610] Library of Congress. Rare Book and Special Collections Division; 2.27: Statis Kalywas and ESO VT-2004 programme; 2.29: © Corbis; 2.30: © SSPL via Getty Images.

Essay 1

Figure E1.9: © Courtesy of Dominique Dierick; E1.11a,b: © Stephen E. Schneider; E1.12: © LazyPixel/Brunner Sebastien/Getty Images; E1.13a,b: © Stephen E. Schneider; E1.15a,b: © Roger Ressmeyer/Corbis.

Chapter 3

Opener: NASA/Rick Wetherington-Tim Powers; p. 71: Courtesy of NASA; 3.7: © Iconotec/Alamy RF; 3.12b: © Stone/Getty Images; 3.14: Courtesy of NASA.

Chapter 4

Opener: © Terje O. Nordvik; p. 87: Courtesy of NASA; 4.2a: © Phil Ashley/Getty Images; 4.4b: © Stephen E. Schneider; 4.5(Pulsar): © Courtesy of NASA/CXC/ISAO; (the Sun): Courtesy of SOHO, NASA/ESA; (other stars): NASA and The Hubble Heritage Team (STScI/AURA); (Interstellar cloud): A. Caulet (ST-ECF, ESA) and NASA; (Cosmic microwave background): Courtesy of NASA/WMAP Science Team; (Active galaxy): © NRAO/AUI/NSF; © Science Source; 4.7(Radio): Max-Planck-Institut & SPIRE Consortium, O. Krause, HSC, H. Linz; (Visible): Courtesy NOAO/AURA/NSF; (Ultraviolet): NASA/Swift/Stefan Immler (GSFC) and Erin Grant (UMCP); (X-ray): ROSAT, MPE, NASA; 4.12: Courtesy of SOHO/MDI consortium. SOHO is a project of international cooperation between ESA and NASA; 4.13a,b: © McGraw-Hill Education/Joe Franek, photographer; 4.23: © Courtesy of Mees Solar Observatory, University of Hawaii; 4.25a: © Courtesy of Richard Wainscoat; 4.25b: © NRAO/AUI/NSF; 4.25c: Courtesy of NASA/CXC/SAO; 4.28: NASA.

Essay 2

Figure E2.1: © Everett Collection, Inc.; E2.5: © Hulton-Deutsch/Corbis.

Chapter 5

Opener: © Getty Images; p. 123: Courtesy of MSSS, JPL, NASA; 5.1a,b: Courtesy of George Greaney; 5.4: © Jim Sugar/Corbis; 5.5: © Stephen E. Schneider; 5.7: Large Binocular Telescope Observatory; 5.9a: © Babak Tafreshi/Getty Images; 5.9b: © Santiago Ferrero/ X02264/Reuters/Corbis; 5.10: © ESO; 5.11a: 2001 Richard Megna, Fundamental Photographs, NYC; 5.11b: R. Thompson, (University of Arizona) and NASA; 5.12: Courtesy of Steve Criswell; 5.13a,b: Courtesy of Andrea Ghez; 5.15: California Institute of Technology; 5.16a: © NRAO/AUI/NSF; 5.16b: Courtesy of NASA/CXC/SAO/Rutgers/J. Hughes; 5.17: © ESO/F. Kamphues; 5.18(Chandra & EUVE): Courtesy of NASA, ESA, and Max Planck Institute for Extraterrestrial Physics, Germany; (Spitzer): Courtesy of NASA/JPL NASA; 5.19(Hubble): Courtesy of NASA/JPL; (Hourglass Nebula & Courtesy of NASA and The Hubble Heritage Team): Courtesy of NASA and The Hubble Heritage Team; 5.20: Courtesy of the NAIC-Arecibo Observatory, a facility of the NSF; 5.21: © James Lowenthal; 5.22: ALMA (ESO/NAOJ/NRAO)/W. Garnier (ALMA); 5.24a: Courtesy of USAF; 5.24b(both): Canada France Hawaii Telescope; 5.25c(all): © Pekka Parviainen/Polar Image; 5.26a: Courtesy of Kitt Peak National Observatory; 5.26b: Courtesy Astronomical Society of the Pacific; 5.26c: © 1993 W.T. Sullivan, III.

Chapter 6

Opener: NASA; p. 145(top): NASA/GSFC/METI/ERS-DAC/JAROS and U.S./Japan ASTER Science Team; 6.1: © 2005 Planetary Visions Limited; 6.3: © Stephen E. Schneider; 6.4: © Taxi/Getty Images RF; 6.13: NASA; 6.17: © Tom Arny; 6.20a: Courtesy Paul Schneider; 6.20b: Earth Science and Remote Sensing Unit, NASA Johnson Space Center/http://eol.jsc.nasa.gov/; p. 159: © Stephen E. Schneider; 6.22: © Stockbyte/Getty Images RF; 6.25: SVS/TOMS/NASA; 6.28: Courtesy of NOAA; 6.29: © Martin Hackworth; 6.30: Courtesy NASA/JPL.

Essay 3

Figure E3.5: © Arnulf Husmo/Stone/Getty Images; © Bettmann/Corbis.

Chapter 7

Opener & p. 179: NASA; 7.1: NASA/JPL/USGS; 7.2a,b: © UC Regents; UCO/Lick Observatory; 7.3a: NASA; 7.3b: © John Gillmoure/Corbis; 7.4a: © UC Regents; UCO/Lick Observatory; 7.4b: NASA; 7.5(both): NASA/GSFC/Arizona State University; 7.7a: Courtesy of NASA, KeithLaney.com; 7.7b: © Lunar and Planetary Institute/NASA; 7.8 & 7.10: NASA; 7.12: NASA/GSFC/Arizona State University; 7.13a: NASA/JPL/USGS; 7.15: NASA/GSFC/Arizona State University; 7.19(all): © Tom Arny.

Chapter 8

Opener: NASA/JPL-Caltech/T. Pyle (SSC); p. 197: Mark McCaughren (Max-Planck-Institute for Astronomy), C. Robert O'Dell (Rice University), and NASA; 8.1: Courtesy of SOHO, NASA/ESA; 8.3: Courtesy of SOHO, NASA/ESA; 8.4(Gaspra): NASA/JPL; (Vesta): NASA/JPL-Caltech/UCLA/MPS/DLR/IDA; (Earth): NASA; 8.5: Courtesy of Alan Dyer; 8.10: NASA/JPL; 8.11a,b: NASA, ESA, and R. Soummer (STScI); 8.11c: © ESO; 8.17: NASA/JPL-Caltech; 8.21: European Space Agency, Alfred Vidal-Madjar (Institut d'Astrophysique de Paris, CNRS), NASA; 8.22: NASA, ESA, and The Hubble Heritage Team (STScI/AURA); 8.24: NASA, ESA and L. Ricci (ESO); 8.25: © Comstock Images/PictureQuest RF; 8.27a: Smith (University

of Hawaii), G. Schneider (University of Arizona), E. Becklin and A. Weinberger (UCLA) and NASA; 8.27b: NASA, ESA, D. Golimowski (Johns Hopkins University), D. Ardila (IPAC), J. Krist (JPL), M. Clampin (GSFC), H. Ford (JHU), and G. Illingworth (UCO/Lick) and the ACS Science Team; 8.27c: NASA, ESA, P. Kalas, J. Graham, E. Chiange, E. Kite (University of California, Berkeley), M. Clampin (NASA Goddard Space Flight Center), M. Fitzgerald (Lawrence Livermore National Laboratory), and K. Stapelfeldt and J. Krist (NASA Jet Propulsion Laboratory); 8.29(Mercury): Courtesy NASA's Planetary Geology and Geophysics Program/Arizona State University; (Dione): NASA/JPL/Space Science Institute; 8.30: Erich Karkoschka (University of Arizona), and NASA.

Chapter 9

Opener: NASA/GSFC/Arizona State University; p. 223: Courtesy of NASA/JPL/Malin Space Science Systems; 9.1: NASA/John Hopkins University Applied Physics Laboratory/Carnegie Institution of Washington; 9.2a,b: NASA/John Hopkins University Applied Physics Laboratory/Carnegie Institution of Washington; 9.3a: NASA; 9.3b: NASA/John Hopkins University Applied Physics Laboratory/Carnegie Institution of Washington; 9.4a: NASA; 9.4b,c: NASA/John Hopkins University Applied Physics Laboratory/Carnegie Institution of Washington; 9.5: map created by Stephen Schneider based on data kindly provided by NASA/Johns Hopkins University Applied Physics Laboratory/Carnegie Institution of Washington/United States Geological Survey; 9.6: Courtesy of NASA; 9.7: NASA/JHU Applied Physics Lab/Carnegie Inst. Washington; 9.11: Courtesy of NASA; 9.12(all): ESA/VIRTIS/INAF-IASF/Obs. de Paris-LESIA/Univ. of Oxford; 9.13(both): © Ted Stryk; 9.15(all): NASA/Bert Ulrich bert.ulrich@nasa.gov; 9.16(both): NASA/JPL-Caltech/ESA; 9.19: NASA, J. Bell (Cornell U.) and M. Wolff (SSI); 9.20: NASA/JPL/Arizona State University; 9.22: Courtesy of NASA/JPL/USGS; 9.23a: Courtesy of NASA/JPL/USGS; 9.23b: Courtesy of A.S. McEwen, USGS; 9.23c: NASA/JPL-Caltech/ASU; 9.25a-c: Data from NASA/JPL, processed by the authors; 9.26: Courtesy of Mars Exploration Rover Mission, JPL, NASA; 9.27: NASA/JPL/Cornell University/Maas Digital; 9.28a: NASA/JPL/Cornell; 9.28b: Courtesy of NASA/JPL/Malin Space Science Systems; 9.29a: NASA/JPL/Malin Space Science Systems; 9.29b: NASA/JPL-Caltech/ASU; 9.29c: ESA/DLR/FU Berlin (G. Neukum); 9.30: NASA/JPL/JHUAPL/MSSS/Brown University.; 9.32: NASA/JPL-Caltech/University of Arizona/Texas A&M University; 9.33: NASA/JPL-Caltech/MSSS; 9.34a: NASA/JPL-Caltech/Cornell University; 9.34b: NASA/JPL-Caltech/MSSS; 9.35a: NASA/JPL-Caltech/Univ. of Arizona; 9.35b,c: Courtesy of NASA/JPL; 9.36(all): NASA/JPL-Caltech/Univ. of Arizona; 9.38(both): NASA/JPL-Caltech/University of Arizona; 9.40: NASA; 9.43(Mercury): NASA/John Hopkins University Applied Physics Laboratory/Carnegie Institution of Washington (Venus): Courtesy of NASA/JPL/USGS; (Earth, Moon): NASA; (Mars): NASA/JPL/MSSS; (swirling clouds): NASA Johnson Space Center-Earth Sciences and Image Analysis (NASA-JSC-ES&IA); (all others): Courtesy of NASA/JPL/USGS.

Chapter 10

Opener: NASA Jet Propulsion Laboratory (NASA-JPL); p. 253(top): NASA/JPL/Space Science Institute; 10.1: Courtesy of NASA/JPL; 10.2: Courtesy of NASA; 10.4: NASA/IRTF/JPL-Caltech/NAOJ/A. Wesley/A. Kazemoto/C. Go; 10.5c: Courtesy of NASA; 10.6: Courtesy of NASA/JPL; 10.7a: NASA/ESA, The Hubble Heritage Team. Acknowledgment: H. Weaver (JHU/APL) and A. Simon-Miller (NASA/GSFC); 10.7b: Courtesy of NASA/JPL; 10.8a: NASA/JPL; 10.8b: NASA/Johns Hopkins University Applied Physics Laboratory/Southwest Research Institute; 10.9: Courtesy Don Wheeler and Brian Dunn/Louisiana Delta Community College; 10.10(Moon): © Digital Vision RF/Punchstock; (Amaltheo): NASA/JPL/Cornell University; (Io): NASA Jet Propulsion Laboratory (NASA-JPL); (Europa): NASA; (Ganymede): NASA Jet Propulsion Laboratory (NASA-JPL); (Callisto): NASA Jet Propulsion Laboratory (NASA-JPL); 10.11a,b: NASA Jet Propulsion Laboratory (NASA-JPL); 10.12a: NASA; 10.12b: NASA Jet Propulsion Laboratory (NASA-JPL); 10.13a(top): NASA Jet Propulsion Laboratory (NASA-JPL); (bottom): NASA Goddard Space Flight Center; 10.13b&c(all) & 10.14: NASA Jet Propulsion Laboratory (NASA-JPL); 10.16a: Lawrence Sromovsky, University of Wisconsin-Madison/W.M. Keck Observatory; 10.16b: NASA/JPL-Caltech/Space Science Institute; 10.16c: NASA/JPL-Caltech/SSI; 10.17(left); NASA/JPL/Space Science Institute; (right): NASA and The Hubble Heritage Team (STScI/AURA) Acknowledgement: R.G. French (Wellesley College), J. Cuzzi (NASA/Ames), L. Dones (SwRI), and J. Lissauer (NASA/Ames); 10.18: Courtesy of NASA/JPL/Space Science Institute; 10.19: NASA/JPL/Space Science Institute; 10.20(Jupiter): NASA/JPL; (Uranus): NASA, ESA, and M. Showalter (SETI Institute); (Neptune, Saturn): NASA/JPL; 10.22: Courtesy of STScI/HST; 10.23-10.26c: NASA/JPL/Space Science Institute; 10.27a: NASA Jet Propulsion Laboratory (NASA-JPL); 10.27b: Courtesy of NASA/JPL/University of Arizona; 10.27c-d: NASA/JPL-Caltech/ASI; 10.28a-10.33: NASA/JPL; 10.34: NASA Jet Propulsion Laboratory (NASA-JPL); 10.35-10.39: Courtesy of NASA/JPL; Table 10.1(all): Courtesy of NASA/JPL.

Chapter 11

Opener: © Robert McNaught/Science Source; p. 277(top): Courtesy of NASA/JPL; 11.1: Courtesy of Ronald A. Oriti, Santa Rosa Junior College, Santa Rosa, Calif.; 11.3a,b: © Stephen E. Schneider; 11.4(Moon): NASA; (Ceres): NASA/JPL-Caltech/UCLA/MPS/DLR/IDA; 11.6a: NASA/JPL-Caltech/JAXA/ESA ; 11.6b: NASA Jet Propulsion Laboratory (NASA-JPL) ; 11.6c: ISAS/JAXA; 11.7: NASA/JPL-Caltech/UCLA/ASI/INAF; 11.10a: © UC Regents; UCO/Lick Observatory; 11.10b-d: NASA/Johns Hopkins University Applied Physics Laboratory/Southwest Research Institute; 11.12a(Triton): National Space Science Data Center; (Orcus): NASA Jet Propulsion Laboratory (NASA-JPL); (Makemaker): NASA; (Pluto): Images courtesy of Marc W. Buie/Lowell Observatory; (Haumea): NASA Jet Propulsion Laboratory (NASA-JPL); (Eris): NASA; (Charon): Images courtesy of Marc W. Buie/Lowell Observatory; (Quaoar & Sedna): NASA; 11.14a: National Space Science Data Center Photo Gallery, NASA; 11.14b: Copyright 1986, Max Planck Institute for Astronomy;

courtesy of Harold Reitsema, Ball Aerospace; 11.15: Courtesy of Mike Skrutskie, University of Virginia; 11.18a: Courtesy NASA/JPL-Caltech.; 11.18b: NASA/ JPL-Caltech/UMD; 11.19a: NASA/JPL-Caltech/UMD; 11.19b: NASA/JPL-Caltech/University of Maryland/ Cornel and NASA/JPL-Caltech/UMD; 11.19c: NASA/ JPL-Caltech/UMD; 11.19d: NASA/JPL-Caltech/ Cornell; 11.20a: ESA/Rosetta/MPS for OSIRIS Team MPS/UPD/LAM/IAA/SSO/INTA/UPM/DASP/IDA; 11.20b,c: ESA/Rosetta/NAVCAM – CC BY-SA IGO; 11.24: NASA/JSC; 11.26: Courtesy of David Roddy Meteor Crater, Northern Arizona, USA; (inset): © Tom Arny; 11.27: Courtesy of Landsat/ESO; 11.28(left): © Kyodo/Associated Press; (middle): © Associated Press; (right): © AFP/Getty Images; 11.30: NASA Jet Propulsion Laboratory (NASA-JPL); 11.31b,c: NASA (Artist: Don Davis).p. 303: Courtesy Stefan Seip/www.astromeeting.de;

Chapter 12

Opener: Courtesy of NASA/SDO and the AIA, EVE, and HMI science teams; p. 303: Courtesy Stefan Seip/www.astromeeting.de; 12.4: Hinode JAXA/ NASA/PPARC; 12.5: Courtesy of Jacques Guertin, Ph.D.; 12.6: Courtesy of NOAO/AURA/NSF; 12.7: 1988 eclipse image courtesy High Altitude Observatory (HAO), University Corporation for Atmospheric Research (UCAR), Boulder, Colorado. UCAR is sponsored by the National Science Foundation; 12.8: Courtesy of SOHO, NASA/ESA; 12.13a: Courtesy of Ernest Orlando, Lawrence Berkeley National Laboratory; 12.13b: R. Svoboda and K. Gordan (LSU); 12.14 & 12.16: Courtesy of Royal Swedish Academy of Sciences; 12.18a-c: Courtesy NOAO/AURA/NSF; 12.19: Courtesy of Eugene Lauria; 12.20a: Courtesy of SOHO-EIT Consortium, ESA, NASA, 12.20b: Courtesy NOAO/AURA/NSF; 12.21: Courtesy of Eugene Lauria; 12.23: NASA/Walt Feimer.

Chapter 13

Opener: © Johannes Schedler/Panther Observatory; p. 325: Courtesy of Dean Ketelsen; 13.4b: © Goodshoot/PunchStock RF; 13.6 & 13.7: © Akira Fujii/ DMI; 13.9a: Andrea Dupree (Harvard-Smithsonian CfA), Ronald Gilliland; 13.9b: © Stephen E. Schneider based on Richard Wilson's data; 13.9c: Haubois et al., A&A, 508, 2, 923,2009, reproduced with permission © ESO/Observatoire de Paris; 13.12: Courtesy of Harvard College Observatory; 13.13: Courtesy of Katherine Haramundanis; 13.24: Courtesy NOAO/ AURA/NSF.

Chapter 14

Opener: NASA, ESA, and M. Livio and the Hubble 20th Anniversary Team (STScI); p. 357: ESA/NASA; 14.5: Five College Radio Astronomy Observatory (FCRAO); 14.6(inset): © European Southern Observatory: 14.7a,b: NASA, ESA, and the Hubble Heritage Team (STScI/AURA); 14.8: Courtesy of Ronald Snell, FCRAO; 14.9a: NASA/JPL-Caltech/A. Noriega-Crespo (SSC/Caltech), Digital Sky Survey; 14.9b: NASA/JPL-Caltech/R. Hurt (SSC/ Caltech); 14.10a: © Subaru Telescope (NAOJ) & DSS; Assembly and Processing: Robert Gendler; 14.16(both): © Akira Fujii/DMI; 14.20b: NASA/JPL-Caltech; 14.21a: Courtesy of Hubble Heritage Team

(AURA,STScI/NASA); 14.21b: Courtesy of Australian Astronomical Observatory, photograph by David Malin; 14.21c: Raghvendra Sahai and John Trauger (JPL), the WFPC2 Science Team, and NASA; 14.21d: Bruce Balock, University of Washington; Vincent Icke, Leiden University, Netherlands; Garrett Mellema, Stockholm University/NASA; 14.24: Pablo Candia; 14.26(both): Courtesy of Australian Astronomical Observatory, photograph by David Malin; 14.27a: NASA/CXC/MIT/UMass Amherst/M.D. Stage et al; 14.27b: NASA, ESA, and J. Hester (Arizona State University); 14.27c: NASA and The Hubble Heritage Team (STScI/AURA).

Chapter 15

Opener: X-ray: NASA/CXC/SAO; Infrared: NASA/JPL-Caltech; Optical: MPIA, Calar Alto, O. Krause et al.; p. 387: NASA, ESA, and Z. Levay (STScI); 15.6a,b: NASA/CXC/SAO; 15.6c: X-ray: NASA/CXC/Rutgers/G.Cassam-Chenaï, J.Hughes et al.; Radio: NRAO/AUI/NSF/GBT/VLA/Dyer, Maddalena & Cornwell; Optical: Middlebury College/F.Winkler, NOAO/AURA/NSF/CTIO Schmidt & DSS; 15.12: Photos courtesy of NOAO.

Chapter 16

Opener: NASA/JPL-Caltech/S. Stolovy (Spitzer Science Center/Caltech); p. 409: Courtesy of Rainer Schodel (MPE) et al., NAOS-CONICA, ESO; 16.1: Courtesy of Steward Observatory and NOAO; 16.4: Courtesy of Hubble Heritage Team (AURA/STScI/ NASA); 16.5: Courtesy Sloan Digital Sky Survey; 16.7: Courtesy of Australian Astronomical Observatory, photograph by David Malin; 16.10(Omega Centauri): Max Kilmister; (M13); Robert Lupton and the Sloan Digital Sky Survey Consortium; (NGC3293); Courtesy of The Australian Astronomical Observatory; photograph by David Malin; (M67); Courtesy NOAO/ AURA/NSF; (Pleiades): H. Mikuz, Crni vrh Observatory; 16.11: Courtesy of NASA and The Hubble Heritage Team (STScI/AURA); 16.13b,c: Courtesy of The Australian Astronomical Observatory; photograph by David Malin; 16.15: Atlas Image courtesy of 2MASS/ UMASS/IPAC-Caltech/NASA/NSF; 16.16b-16.17b: Courtesy of The Australian Astronomical Observatory, photograph by David Malin; 16.22(top): Stefan Seip/ Adam Block/NOAO/AURA/NSF; (bottom): Bruce Hugo and Leslie Gaul/Adam Block/NOAO/AURA/ NSF; 16.27a: Courtesy of Rainer Scholdel (MPE) et al., NAOS-CONICA, ESO; 16.27b: Courtesy of Andrea Ghez; 16.28a: Courtesy of NASA, UMass, D. Wang et al.; 16.30a-16.31a: Courtesy of Neal Katz, University of Massachusetts, and James E. Gunn, Princeton University; 16.32(all): NASA, ESA, Z. Levay and R. van der Marel (STScI), T. Hallas, and A. Mellinger.

Chapter 17

Opener: NASA, ESA, and The Hubble Heritage Team (STScI/AURA); p. 441(top): Image created by M. Putman with data from the Parkes Radio Telescope published in Putman, M. Staveley-Smith, L., Freeman, K., Gibson, B. & Barnes, D., 2003, ApJ, 586, 170.; 17.1: Courtesy of George Greaney; 17.2: Courtesy of William C. Keel; 17.3a: Courtesy of Bruce Hugo and Leslie Gaul, Adam Block, NOAO, AURA, ASF; 17.3b:

Courtesy of Hubble Heritage Team, AURA / STScI / NASA; 17.3c: ©European Southern Observatory; 17.4a,b: ESA, NASA and E. Peng (Peking University, Beijing); 17.5(both): Courtesy of The Australian Astronomical Observatory; photograph by David Malin; 17.6(both): Courtesy of The Australian Astronomical Observatory; photograph by David Malin; 17.7(both) & 17.8: Courtesy Sloan Digital Sky Survey; 17.13(top, four & bottom, left): (bottom, right): Courtesy of Joshua Barnes, University of Hawaii; 17.14a: NASA, ESA, and The Hubble Heritage Team (STScI / AURA); 17.14b: Courtesy STScI and NOAO; 17.15a: R Jay GaBany-Collaboration: A. Sollima (IAC), A. Gil de Paz (U. Complutense Madrid) D. Martínez-Delgado (IAC, MPIA), J.J. Gallego-Laborda (Fosca Nit Obs.), T. Hallas (Hallas Obs.); 17.15b: ©NRAO/AUI/NSF; 17.16: Courtesy Sloan Digital Sky Survey; 17.17: Roff / Lowell Observatory; 17.18a,b: Courtesy of NASA and The Hubble Heritage Team; 17.18c: Richard Griffiths (JHU), STScI, NASA; 17.19: R. Jay Gabany (Blackbird Observatory)-collaboration; D. Martinez Delgado (IAC, MPIA), J. Penarrubia (U. Victoria) I. Trujillo (IAC) S. Majewski (U. Virginia), M. Pohlen (Cardiff); 17.22a: ©Martin Pugh; 17.22b: NASA, ESA and J.P. Kneib (Laboratoire d' Astrophysique de Marseille); 17.23: NASA and The Hubble Heritage Team (STScI/AURA); 17.24: AURA/ STScI/NASA; 17.25a-c: ©NRAO/AUI/NSF; 17.26: NASA; 17.28: Courtesy of John Bahcall, Institute for Advanced Study; 17.29(left): Courtesy of W. Jaffe, Leiden Observatory, and H. Ford. Johns Hopkins University, Space Telescope Science Institute and NASA; (right): ©NRAO/AUI/NSF; 17.35: Color image creation by Jean-Charles Cuillandre, The Canada-France-Hawaii Telescope Corp. Copyright © 2001 CFHT; 17.36: Courtesy of A. Fabian (IoA Cambridge) et al., NASA; 17.37: ©Stephen E. Schneider; 17.39: Numerical simulations performed at the National Center for Supercomputing Applications, NCSA, Urbana-Champaign, Illinois; by Andrew Kravtsov (The University of Chicago; and Anatoly Klypin, New Mexico State University; 17.43: Courtesy of NASA; 17.44: Courtesy of STScI; NASA / STScI; Magellan/ U. Arizona / D. Clowe et al.

Chapter 18

Opener: Springel et al. (2005)/Max-Planck-Institute for Astrophysics.; p. 477: NASA/ESA/STScI/W. Zheng (JHU), and the CLASH team; 18.1a: Two Micron All Sky Survey, a joint project of the University of MA and the Infrared Processing and Analysis Center/CA Institute of Technology, funded by NASA/National Science Foundation/T.H. Jarrett, J. Carpenter, & R. Hurt. Perha; 18.10(both): Arcminute Cosmology Bolometer Array Receiver; 18.17(right): based on Planck 2013 results. I. Overview: Fig. 19, ESA/Planck, and the Planck Collaboration; 18.18: ©CERN; 18.23b: Courtesy NASA/WMAP Science Team.

Essay 4

Figure E4.2: © David Wacey; E4.3a: © Paul Hoffman; b: Courtesy of Donald R. Lowe; E4.4(mushroom): © Gallo Images-Nigel Dennis/ Digital Vision/Getty Images RF; (butterfly): © Corbis RF; (giraffes): © DV/Getty RF; E4.6: Courtesy of Ken Eward/BioGrafx; E4.8: © Steven Brooke Studios; E4.9a,b: © Dudley Foster/Woods Hole Oceanographic Institution; E4.10: NASA/JPL-Caltech/Cornell/MSSS; E4.14: © Seth Shostak.

TEXT AND LINE ART CREDITS

Chapter 1

Page 33: Data from NASA's eclipse website: http://eclipse.gsfc.nasa.gov/. Partial solar eclipses and "penumbral" lunar eclipses, are not listed.

Chapter 2

Figure 2.3a: Aristotle (2010) *"On the Heavens and On Generation and Corruption"* Digireads.com Publishing; p. 47: William of Ockham; 2.18: Nicolaus Copernicus (1543) *"De revolutionibus orbium coelestium"*.

Chapter 3

Figure 3.4a & b: Courtesy of NASA.

Chapter 4

Table 4.4: Lodders, Data on relative number of atoms drawn from Lodders (2003) *The Astrophysical Journal,* vol. 591, pp. 1220–1247.

Chapter 11

Figure 11.5: Courtesy of E.L.G. Bowell, Lowell Observatory; 11.9: Courtesy of E.L.G. Bowell, Lowell Observatory.

Chapter 16

Page 409: Chaucer, G. (1894). *The house of fame.* Oxford: Clarendon Press; 16.23: This sequence was generated by P. Seiden (deceased) and H. Gerola, *Ap. J.* 233, 56, 1979. Reproduced by permission; 16.26a & b: Clemens, Dan P. Adapted from diagram by Dan P. Clemens, *Ap. J.* 295, 422.

Chapter 17

Figure 17.11b: Data courtesy of Robert P. Kirshner, CFA.

Chapter 18

Page 477: Genesis (1:1–3), King James Version; 18.9: Based on curve supplied by NASA/Goddard Spaceflight Center: COBE Science Working Group; p. 496: Frost, R. (1920) "Fire and Ice" in *Harper's Magazine,* Vol.142, Harper's Magazine Company.

Essay 4

Figure E4.5: Raven and Johnson, *Biology,* 2/e. 1989. Copyright © The McGraw-Hill Companies. All rights reserved; p. 509: Lucretius; Sir Martin Rees; Epicurus; p. 514: Carter, B. "Large Number Coincidences and the Anthropic Principle in Cosmology", p.291–298. in *Longair, M. S. (1974). Confrontation of Cosmological Theories with Observational Data.,* Softcover reprint of the original 1st ed. 1974 edition (September 30, 1974) Dordrecht: Springer Netherlands.

Appendix A

Table A.10: Listing based largely on the SEDS Messier compilation of Local Group data (http://messier.seds.org/more/local.html); A.11: Galaxies in this table were selected from the HyperLeda database (http://leda.univ-lyon1.fr) based on their total visual apparent magnitude.

Index

SPRING NIGHT SKY

WHEN TO USE THIS STAR MAP

Early March:	2 A.M.
Late March:	1 A.M.
Early April:	midnight
Late April:	11 P.M.
Early May:	10 P.M.
Late May:	dusk

This star chart is most accurate if used within an hour or so of the times listed and is plotted for observers located between 30° and 50° north latitude. All times are standard time; if daylight-saving time is in effect, add one hour.

To use this chart, hold it in front of you and rotate it so that the yellow label corresponding to the direction you are facing is positioned at the bottom, right-side up. The stars in the sky should match those depicted on the chart. The center of the chart is the zenith, the point in the sky directly overhead.

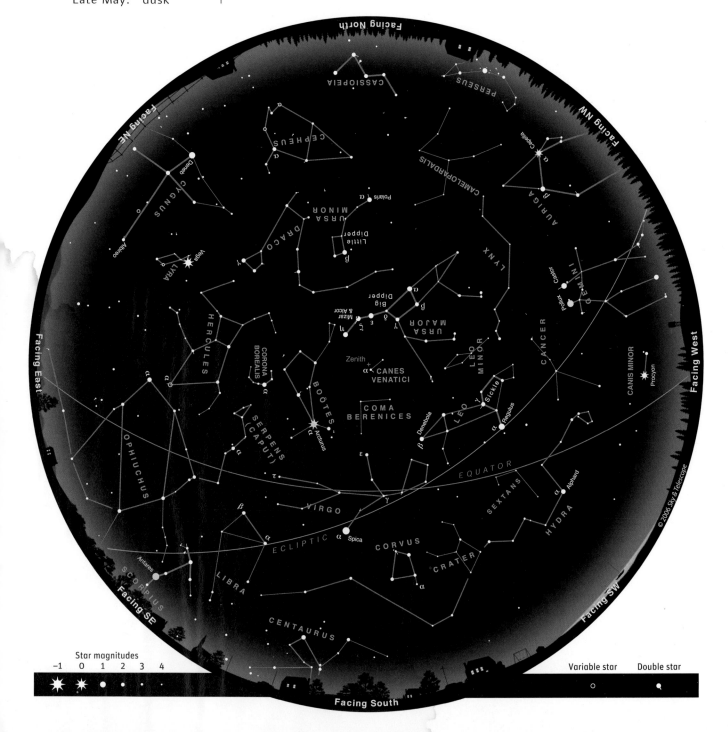

SUMMER NIGHT SKY

This star chart is most accurate if used within an hour or so of the times listed and is plotted for observers located between 30° and 50° north latitude. All times are standard time; if daylight-saving time is in effect, add one hour.

To use this chart, hold it in front of you and rotate it so that the yellow label corresponding to the direction you are facing is positioned at the bottom, right-side up. The stars in the sky should match those depicted on the chart. Ignore all the parts of the map above horizons you are not facing.

WHEN TO USE THIS STAR MAP

Early June: 1 A.M.
Late June: midnight
Early July: 11 P.M.
Late July: 10 P.M.
Early August: 9 P.M.
Late August: dusk

Star magnitudes
−1 0 1 2 3 4

Variable star   Double star

AUTUMN NIGHT SKY

WHEN TO USE THIS STAR MAP

Early September:	midnight
Late September:	11 P.M.
Early October:	10 P.M.
Late October:	9 P.M.
Early November:	8 P.M.
Late November:	7 P.M.

This star chart is most accurate if used within an hour or so of the times listed and is plotted for observers located between 30° and 50° north latitude. All times are standard time: if daylight-saving time is in effect, add one hour.

To use this chart, hold it in front of you and rotate it so that the yellow label corresponding to the direction you are facing is positioned at the bottom, right-side up. The stars in the sky should match those depicted on the chart. The farther up from the map's edge they appear, the higher they'll be shining in your sky.

WINTER NIGHT SKY

This star chart is most accurate if used within an hour or so of the times listed and is plotted for observers located between 30° and 50° north latitude. All times are standard time.

To use this chart, hold it in front of you and rotate it so that the yellow label corresponding to the direction you are facing is positioned at the bottom, right-side up. The stars in the sky should match those depicted on the chart. Remember that star patterns will look much larger in the sky than they do here on paper.

WHEN TO USE THIS STAR MAP

Early December:	midnight
Late December:	11 P.M.
Early January:	10 P.M.
Late January:	9 P.M.
Early February:	8 P.M.
Late February:	dusk

Star magnitudes
−1   0   1   2   3   4

Variable star   Double star

Facing South